Newton *and the* Origin *of* Civilization

Newton
and the
Origin *of* Civilization

Jed Z. Buchwald & Mordechai Feingold

PRINCETON UNIVERSITY PRESS
PRINCETON AND OXFORD

Published by Princeton University Press,
41 William Street, Princeton, New Jersey 08540

In the United Kingdom: Princeton University Press, 6 Oxford Street,
Woodstock, Oxfordshire OX20 1TW

press.princeton.edu

Jacket art: *Gli Astronomi,* inv. 269; Niccolò Tornioli.
Courtesy of the Galleria Spada, Rome.

Library of Congress Cataloging-in-Publication Data

Buchwald, Jed Z.
Newton and the origin of civilization / Jed Z. Buchwald and Mordechai Feingold.
p. cm.
Includes bibliographical references and index.
ISBN 978-0-691-15478-7 (hardcover : acid-free paper) 1. Newton, Isaac, Sir, 1642–1727. 2. Newton, Isaac, Sir, 1642–1727—Philosophy. 3. Newton, Isaac, Sir, 1642–1727—Public opinion. 4. Scientists—England—Biography. 5. Philosophers—England—Biography. 6. Newton, Isaac, Sir, 1642–1727. Chronology of ancient kingdoms amended. 7. Chronology, Historical—History—17th century. 8. Civilization, Ancient—Philosophy. 9. Europe—Intellectual life—17th century. 10. Public opinion—Europe—History—17th century. I. Feingold, Mordechai. II. Title.
QC16.N7B93 2012
530.092—dc23 2012024733

British Library Cataloging-in-Publication Data is available

This book has been composed in Minion Pro and Centaur

1 3 5 7 9 10 8 6 4 2

CONTENTS

ILLUSTRATIONS

TABLES

ACKNOWLEDGMENTS

In the course of writing this book we have incurred many debts. We are grateful to friends and colleagues who generously assisted our research, responded to queries, or commented on draft chapters: Roger Ariew, Henk Bos, Diana Kormos Buchwald, Andreas Corcoran, Ginevra Crosignani, Steffen Ducheyne, Anthony Grafton, Chantal Grell, Niccolò Guicciardini, Kristine Haugen, John Heilbron, Rob Iliffe, Tom Levinson, Rhodri Lewis, Mike Mahoney, Scott Mandelbrote, William Newman, John North, Adam Perkins, Mac Pigman, William Poole, Larry Principe, Jamil Ragep, Alan Shapiro, George Smith, Noel Sugimura, Noel Swerdlow, and Nico Wey-Gomez. We also express our gratitude for assistance received from various libraries: the Caltech Archive, the National Library of Israel, the Bodleian Library, the Cambridge University Library, the Library of New College, Oxford, and the Huntington Library. Special thanks to the staff of the interlibrary loan system at Caltech, who managed to get for us every article and book we required. Finally, we also express our gratitude to the Andrew A. Mellon Foundation, for generously supporting two research programs at Caltech, aimed at investigating the manifold issues raised by the use of "evidence" and "reason" during the long eighteenth century. That agenda helped sharpen our own understanding of the changing nature of evidence during the seventeenth century, in general, and the unique evidentiary basis of Newton's investigations, in particular.

Newton
and the
Origin *of* Civilization

Introduction

> Newton's writings on biblical subjects seem to me especially interesting because they provide deep insight into the characteristic intellectual features and working methods of this important man. The divine origin of the Bible is for Newton absolutely certain, a conviction that stands in curious contrast to the critical skepticism that characterizes his attitude toward the churches. From this confidence stems the firm conviction that the seemingly obscure parts of the Bible must contain important revelations, to illuminate which one need only decipher its symbolic language. Newton seeks this decipherment, or interpretation, by means of his sharp systematic thinking grounded on the careful use of all the sources at his disposal.
>
> —Draft of a letter from Albert Einstein to Abraham Yahuda. September 1940[1]

In 2006, archaeologists announced that the ancient Minoan kingdom on the island of Crete was a century older than had been thought. Radiocarbon dating of tree rings and seeds, coupled to statistics, placed the volcanic explosion of Thera, which likely ended the Minoan period, to between 1660 and 1613 BCE. This had disturbing consequences. It had been long held that the Minoan period overlapped the New Kingdom in Egypt, which began about 1550 BCE, and that contacts existed between the two civilizations. The revised dating made this impossible, since at the earlier time the Egyptians were ruled by Canaanite foreigners, the Hyksos. Nevertheless, the *New York Times* reported, "early indications suggest that proponents of the later chronology are not backing down. Their main line of defense is the Egyptian historical chronology, derived from its written records as well as pottery and iconography. They insist that a chronology tied to the Egyptian record could not be off by as much as 100 years." Evidence drawn from a source that knows neither culture nor history—the traces of radiocarbon—suddenly battled in 2006 with the remnant words and art of antiquity. Still, the proponents of text and relic held their ground, while an archaeologist argued that "the dates offered in the textbooks for these periods have always been interpretations and estimates with little evidence." The proper solution requires

[1] ALS, 1p. [AEA 39-602], Albert Einstein Archives, by permission of the Hebrew University and Princeton University Press.

"realigning the Aegean and Egyptian chronologies for the period 1700–1400." Equipped with statistical methods, scientific archaeology makes chronological claims with which ancient words and arts must contend.[2]

Disputes over chronology with overtones of a divide between the text and the laboratory or field, between the humanist and the scientist, have been raging since the sixteenth century. Though even in antiquity bits of astronomy had intrigued those interested in chronology, only after the Reformation did calculation begin to intersect fruitfully with philology. Anthony Grafton notes that in Rudolphine Prague, where Johannes Kepler lived and worked for a time, astronomy and chronology were "fused into a single pursuit not identifiable with any modern discipline." The barriers between humanistic scholarship and computational science had not then been fully erected, and considerable interaction took place between philology and astronomy. Nevertheless, signs of discord were already visible. The great humanistic scholar Joseph Scaliger, though not by any means the first to make use of astronomical "tidbits" to reset chronology's clock, corresponded with both Kepler and Tycho Brahe in his attempt to elicit satisfying results by means of an astronomical armamentarium. A century later, the relationship between the two disciplines turned decidedly frosty when Isaac Newton intruded calculations into the precincts of history and philology by virtually turning certain texts into numbers and tampering with others to fit his reckonings.[3]

Around 1700 (and concertedly so by 1704), the fifty-eight-year-old Newton—then Master of the Mint—began applying himself seriously to technical chronology, supplementing the historical studies he had been working on for the past decade and a half. The extensive notes he took ranged from excerpts of, and commentaries on, such classical sources as Herodotus, Clement of Alexandria, Diodorus Siculus, and Eusebius, to material extracted from Scripture, and elaborate astronomical and genealogical computations. The astronomical material was drawn for the most part from Denis Petau's *Uranologion*, which included a Latin translation of Hipparchus' critical *Commentary* on Aratus' third century BCE poem on the heavens. The *Commentary* provided Newton with a spectrum of remarks, many of which could, he thought, be transformed into numbers. The motivation for doing so derived from his concern with issues in antiquity that bore on the origin of civilization. Having first followed common tradition, in fixing the origin of kingdoms shortly after the Deluge, he became convinced that elaborate kingdoms and city life emerged only toward the beginning of the first millennium BCE. Gripped to a greater extent than his contemporaries by questions relating to the amount of time needed for the earth's population to recover from the devastation of the Flood, Newton convinced himself that neither Egyptian nor, certainly, Greek civilization could have existed much before the time of Solomon's reign.

As a historical chronologer, Newton seems hardly the person whose name would become an eponym for natural philosophy during the Enlightenment. Odd though it may seem for the modern observer, his pursuit of chronology is not the only seeming idiosyncrasy. As recent scholarship has demonstrated, Newton was thoroughly im-

[2] Kwan, 2006; Wilford, 2006. Archaeologist Sturt Manning of Cornell University.

[3] Grafton, 1991b, p. 186. See Grafton, 1993, esp. 3.4.

mersed in ancient prophecies, Church history, and alchemy. These investigations raise several questions: what links his interest in such matters to his investigations in optics, mechanics, and mathematics? Was Newton in his alchemical laboratory the same Newton who analyzed the passage of light through a prism and who measured the behavior of bodies falling through fluid media? What did the Newton who interpreted the *Book of Revelation* have to do with the man who wrote the *Principia Mathematica*? And how does the Newton who pored over ancient texts square with the author of the *Opticks*?

In several respects, Newton as natural philosopher differed little from numerous other early-modern observers, experimenters, and calculators who toiled incessantly in their workrooms and at their desks. Surviving manuscripts attest to the passion that drove them to formulate and to solve problems of nature and of mathematics. Galileo tangled intensely, and repeatedly, with questions of motion; the young Newton computed logarithms to dozens of places with evident joy; Boyle avidly probed the workings of air pumps, the combinations of "chymistry," and the complexities of colors. The sheer power of an irresistible need to observe, to experiment, to formulate, to grapple with a problem and crack it shines through their labors. And yet, whoever spends enough time with Isaac Newton realizes not only the pleasure he took in calculating and in experimenting (and in little else besides), but his conviction that the cosmos reveals the presence of an active deity. What unites the Newton who probed nature's secrets and delighted in the intricacy of computation with the Newton who hunted the secrets of prophecy and of divinely guided human destiny? Are these two different mind-sets in a single body? Is the Newton who pondered divine activity the "real" Newton, whose theological convictions informed the calculator and experimenter? Or is there another possibility? To us, this late seventeenth-century genius was not simply driven by *this* to do *that*; rather, his way of working reveals a mode of thought and practice which underpins both his efforts to unravel the workings of a deity in history and to grasp the innermost mysteries of mechanical nature.

Frank Manuel is the only one to date to have engaged seriously with Newton the chronologer, seeking always to forge a unity out of his apparently disparate life's work. Subsequent scholars have built on and extended Manuel's insights regarding the coherence of Newton's work and persona. Since the 1970s, several have placed particular emphasis on the character of Newton's theology and on his investigations into the Apocalypse.[4] Others have opened the way to serious consideration of his alchemical investigations.[5] In this context, Newton's procedures in the alchemical laboratory have been replicated in recent years, together with analyses that demonstrate how thoroughly his work there was grounded in a vision of matter composed of corpuscles, an understanding that has a long history among alchemists.[6]

Still, the nature of the connections between Newton the mathematician, the experimental philosopher, the corpuscular alchemist, the expositor of the Apocalypse,

[4] See, e.g., Iliffe, 1994, 1995; Snobelen, 1999, 2003a, 2003b, 2004; Mandelbrote, 2002; Delgado-Moreira, 2006b, 2006a.

[5] Dobbs, 2002; Figala, 2002a.

[6] Newman, 1998; Newman and Principe, 2001, 2002, 2003; Newman, 2006.

and the originator of a new system of chronology remains obscure. Manuel wondered what effect Newton's "scientific method" had on such matters, but at the time, the precise meaning of that "method"—if it ever existed—remained unclear. Since then, scholars have discovered that the Newton of the *Principia* did exploit a particular manner of thinking about and investigating nature, which has been termed the "Newtonian style." Analysis of that style has shown that Newton had a most specific way with data and conjecture.[7]

It was during Newton's earliest years at Cambridge that he developed the core of his method for generating and working with trustworthy knowledge—a core that carried forward directly into his chronology and indeed into his writings on the Apocalypse and the origins of civilization. We will not be asserting that a single, unchanging Newton can be followed through the nearly nine decades of his life. How could that even be? How could a man in his seventies and eighties be the same person in all respects as the youth who arrived at Cambridge in 1661? Changes certainly occurred over the years, as the introverted young man grew into the socially adept and ever-vigilant president of the Royal Society. Nevertheless, Newton's mature approaches to nature, history, and theology are rooted deep in his undergraduate career. Attitudes forged in those years evolved a way of trusting—and distrusting—evidence that produced a singular commonality among these apparently disparate realms. Critics of his chronology sensed this presence: hints of something inimical to, perhaps even dismissive of, the kinds of reasoning that philologists and theologians steeped in texts had long deployed. They were right to sense danger. Newton's ways with the past implicitly posed the questions: who has the right to command history? What sorts of evidence and reasoning should govern historical and theological understanding?

The old chronologer and the young natural philosopher deployed a complex conception of what is knowable, a conception that diverged radically from contemporary expectations of the knowledge that can be harvested from experiments—or from texts. We will see how Newton's scheme for human history is shot through and through with this structure, while his commitment to divine action in both nature and history cannot be separated from his understanding of how reliable knowledge in both domains can be gleaned or generated. Newton, we will argue, was not motivated by a conception of *divine action* to query *nature*; rather, a single conception of the probing character of *human knowledge* bound together a Newtonian triad of history, theology, and science.

In turning to chronology, Newton relied in the first instance on contemporary English work. Though the subject of technical chronology had not been extensively pursued in England during the seventeenth century, there was a rich native tradition in higher criticism, and a spectrum of ongoing controversies about the origins of language and of civilized life. These concerns had been greatly furthered by new reports of Chinese civilization, as well as by experiences of native cultures in the New World. Newton was certainly aware of such developments, but there is little evidence that he devoted much attention to them (except, in the case of China, when he briefly attempted to dismiss seeming countervailing evidence to his several claims). He re-

[7] Cohen, 1980; Smith, 2000a, 2002.

mained principally concerned with the chronologies of Egypt, Greece, and Mesopotamian empires. By the time Newton undertook extensive studies in the area, or in matters theological that were bound to ancient evidence, he had already developed his novel approach to what can be known through nature or through text.

We begin with the young Newton at Cambridge, where he grappled with new scientific ideas, subtly but distinctively transforming them. It was then that he developed a specific kind of skepticism concerning the reliability of the senses, one that powerfully governed his ways in the laboratory and in calculation. Convinced that the senses could not be relied upon to generate reliable knowledge, Newton developed a way to overcome this limitation by taking the extraordinary tack of increasing the number of measurements without discarding any. He alone forged a trustworthy resultant out of a discrepant set of numbers, each of which was inherently unreliable, by taking an average among them. Instead of discarding every measurement other than the one thought to be the very best, as his contemporaries usually did, Newton kept them all, thinking that a good number could be produced by combining a multitude of bad ones.

Newton's way to eradicate error passed strikingly, and seamlessly, into his theological and chronological works, shaping the particular forms of his reasoning. In his writings on the Apocalypse, Newton required multiple sources—each subject to various degrees of doubt—to be balanced against one another. Although that in itself was hardly a new procedure, Newton's skeptical attitude toward singular pieces of evidence resulted in a novel understanding of the significance of the remarks in the books of Daniel and Revelation. That understanding was directly connected to his conviction that the deity's activity followed an essentially law-like structure even in matters that engaged human affairs.

Newton's turn to chronology was stimulated in major ways by a concern that he shared with few contemporaries: namely the amount of time required to repopulate the earth after the Noachian Deluge. Difficult issues plagued discussions of these matters, and Newton grappled with them all. Increasingly convinced that regularities observable at his time must have prevailed among the survivors of the Deluge as well, Newton eventually developed a novel theory of the evolution of civilization, one that required chronology's clock to be radically reset. Human populations, he decided, expand according to certain rules that implicate specific stages in the development of civilization. Although ancient texts suggested to most interpreters a very different sequence and timing of developments, by the late 1690s, Newton had also decided that these sources could not be relied upon unless they had a particular pedigree—namely, unless they had come down through a trustworthy chain of transmission. The prime example of such a proper transmission for Newton was the Masoretic version of Scripture. Otherwise, he thought, texts—especially texts written in the form of poetry rather than prose—had to be treated with a great deal of skepticism, which licensed Newton's frequent, radical reinterpretations.

Newton's especial vehemence in this regard was likely furthered by his work in the London Mint, where he had for some time been in charge of prosecuting coin clippers and counterfeiters. His experience in taking testimony exacerbated his skepticism concerning the reliability of words, especially words that could not be turned into numbers or balanced against corroborating testimony. As he saw it, words from the past resembled the sorts of stories that he had heard from clippers and counterfeiters who could not be trusted. In tackling chronology, Newton accordingly rejected ancient remarks that struck him as "poetical fancies," thinking poetry to be, *sui generis*, a form of fictive storytelling. Nor could singular remarks even in prose be relied upon—unless they could be transmuted into numbers. That could be done by creatively manipulating Hipparchus' *Commentary*. From it Newton extracted statements concerning the parts of constellations passed through by the *colures*, the great circles that pass through the poles and, respectively, the equinoxes and solstices. After considerable effort turning text into numbers—work that required him to engage creatively (and, his critics claimed, arguably) with images of the constellations—Newton produced a set of discrepant numbers out of which error was drained by means of the average. Although he eliminated the computations from his published *Chronology*, Newton's extensive manuscripts, which were written and rewritten over more than two decades, have enabled us to reconstruct his reasoning.

The astronomical remarks that Newton used pertained, he believed, to the very first stellar sphere, which had been passed down to Eudoxus through the centuries. The result of his calculations gave him precisely what he sought: traditional chronology was too long by five centuries. In particular, the expedition of the Argonauts must have occurred ca. 939 BCE, and not in 1467, as one of his French critics maintained. Egypt fared similarly in Newton's hands. Here his argument relied principally on identifying the Egyptian pharaoh Sesostris with the Biblical Sesac—an identification that he found in the work of John Marsham—which enabled him to contract Egyptian history by six hundred years.

Newton's iconoclastic chronology generated antagonistic reactions in both England and France, several years prior to its posthumous publication in 1728. In France, a storm of controversy greeted his claims and methods, as they made their way there through various intricate paths that Newton himself may have had a hand in clearing. French reactions were especially pointed as the erudite members of the *Académie des Inscriptions et Belles Lettres*, who fought enough among themselves, perceived that Newton's *Chronology* attacked more than their elaborately constructed dating systems: it undermined the very foundations of their methods grounded for the most part in texts. During the seventeenth- and early eighteenth centuries, skepticism concerning the reliability of texts, and historical Pyrrhonism, had become rampant, and attempts were made to establish criteria for judging the reliability of evidence from the past.[8] Material relics were used increasingly to challenge or to buttress ancient texts. However, since Newton scarcely relied on inscriptions, medallions, or coins—his main periods of interest predated these forms of evidence—even those Academicians who had turned to these relics from the past to grant a competitive philosophical luster to their work felt threatened. In England, Newton's scheme for ancient history elicited powerful critiques, not least from the verbose but accomplished William Whiston, furious that Newton had not publicly avowed the Arianism that Whiston so vehemently preached. For almost a century, the immense reputation of the

[8] Though published more than six decades ago, Momigliano, 1950 remains a formidably readable and informative account of these developments, about which a great deal has since been written. See especially the several essays in Grafton, 1991a, 2001b.

great Newton forced historians, chronologers, and theologians to come to grips with the challenges that his *Chronology of Ancient Kingdoms Amended* posed. In the process, new foundations for the study of ancient history, archaeology, and Biblical exegesis were laid, rapidly eclipsing the venerable domain of chronology, which increasingly remained the preserve of orthodox theologians.

The Newton that is the subject of this book differs in striking ways from any scientist of the twenty-first century. But he differed as well from his contemporary natural philosophers, theologians, and chronologers. That difference is both our subject and our method, as we investigate its origin and then use it to produce a new understanding of Newton's worldview and its historical context.

Troubled Senses

In 1583, the Huguenot scholar Joseph Scaliger published his *De Emendatione Temporum*. There he examined the chronologies of Babylon, Egypt, and Persia, as well as those of Greece and Rome, arguing for an amended structure based in substantial part on antique reports that could be given astronomical significance, in particular eclipses, as well as on ancient calendars. Scaliger was hardly the only one to use astronomical evidence to date the past. Another Protestant, the theologian Heinrich Bünting, had employed eclipse reports with great technical proficiency in his *Chronologia* (1590).[1] Scaliger and Bünting were followed by others in the seventeenth century who also relied on eclipses to date the past, including the Protestant Sethus Calvisius and the Jesuit Denis Petau (Petavius). Grafton calls such men "systematic" chronologers—their technical expertise plied to produce treatises on ancient and modern calendars and epochs, and so markedly different from "humanistic" chronologers, editors of ancient texts, and antiquarians.[2] Calvisius, for example, remarked in 1605 that "historians frequently record eclipses, and they are often inserted into accounts of the deeds of kings and emperors, in such a way that they usually provide the most certain testimony both about the length of any given king's reign and the true course of events. Eclipses are of infallible certainty, and they can be dated and demonstrated by astronomical computation for any period."[3]

Eclipse reports from antiquity certainly raised issues of accurate reportage and textual corruption, but they seemed not to require an interpretative framework. Eclipses are after all singular events that, it seemed, could be connected to chronology by means of the technical tools of astronomy that had become available with the publication of Erasmus Rheinhold's *Prutenic Tables* in 1551, which made use of the basic numerical parameters that Nicholas Copernicus had deployed in his 1543 *De*

[1] See Grafton, 1991a, 1993 , and in particular Grafton, 2003, which notes that the fundamental methods of a technical chronology based on eclipses were set out in Paul Crusius' *Doctrina revolutionum solis* (1567). Grafton points out that while chronologers used eclipses solely for dating, many contemporaries continued to see in them astrological prognostications. Scaliger's originality lay in producing a "novel fusion" of astronomy with philology and "Oriental studies" (Grafton, 1991a, pp. 139–40).

[2] Though of course Scaliger required great humanistic training to pursue his technical work: cf. Grafton, 1991a, p. 120.

[3] Grafton, 2003, p. 228.

Revolutionibus Orbium Coelestium.[4] The use of astronomy in matters chronological produced occasional antagonism, but the assimilation of singular human to singular celestial event apparently did not raise philosophical problems, at least among the technical chronologers. Even when a celestial event was granted portentous significance, its astrological meaning derived from its place in an established system of knowledge and not from its signaling a natural novelty. This raises the question of the relationship between natural and historical knowledge in the late sixteenth- and seventeenth centuries, which, in turn, brings us to consider changes in the character of natural knowledge proper during this period—an extraordinarily large and complicated issue that has garnered considerable attention.

A great deal of historical work concerns the sense in which a specifically experimental form of knowledge emerged during the seventeenth century. It has been argued that until the last half of the century, experiment-based knowledge remained suspect. When it became respectable to glean information about nature from experiment, the argument continues, specific techniques had to be invented to make the process socially and (*ipso facto*) intellectually acceptable.[5] Three distinct but related strands wind through this argument. First and foremost, how could artificially produced experiences become accepted as foundational for proper knowledge when traditional scholastic categories sharply distinguished between the natural and the artefactual? Second, artefactual knowledge is not merely non-natural (in the sense of not being produced by unaided nature); it is also singular and specific, referring to the results of particular interventions. That too is thought to conflict with scholastic tradition, according to which proper knowledge of nature must be founded on experiences that are universal and common. Finally, how did experimentally derived knowledge, grounded in particulars, become conjoined with mathematical demonstration when the scholastics had long considered that mathematics applied only to those sciences whose objects partook in their essence of geometric qualities?

There have been various answers to these questions, but in respect to the first two, all presume that there was in fact a broadly enforced barrier between the experiment and proper knowledge. Yet even if such a division existed among some scholastics—and the claim remains controversial—it hardly follows that the boundary was effectively policed everywhere before the mid-seventeenth century; it certainly was not. Historians of alchemy have for example demonstrated that practitioners regularly and unproblematically produced and worked with laboratory-generated knowledge, and that one of the prime examples used to illustrate the novelty of the "matter-of-fact" as a privileged item, namely Robert Boyle, derived a good deal of his approach from the alchemist George Starkey.[6] Much of the difficulty in developing a full un-

[4] Copernicus incorrectly identified the Babylonian king Nabonassar with the biblical Salmanassar, thereby linking Ptolemy's dating scheme in his *Almagest* to the Bible. Rheinhold continued the identification: Grafton, 1991a, pp. 129–30.

[5] The most extensive, and one of the earliest, assertions along these lines is Schaffer and Shapin, 1985. Their emphasis on the specifically social production of truth from experimental reportage is further elaborated in Shapin, 1995; for a critique of the latter see Feingold, 1996. Dear, 1992 offers a different but related set of assertions. Also see Daston, 1991.

[6] On which see Newman, 1998; Newman and Principe, 2001, 2002, 2003; Principe, 1998.

derstanding of the emergence of widely pursued laboratory science may derive from concentrating on methodological prescriptions rather than on actual practice, which alone reveals what was being done; prescriptive pronouncements in contrast mostly uncover reactions to activity and not its generative sources.

Though experiment-based facts were not uncommon by the sixteenth century, and probably long before that, the question of how locally produced results could be incorporated into the foundations of knowledge systems inevitably arose—as they do to this day. Novelties always generate questions when they come into contact with an existing structure, however loosely built the system may be. An event of nature, whether generated in a laboratory or out of it, certainly occurs in time and place. It is also true that in mid- to late seventeenth-century England, "natural histories" were produced that recount specific events. However, many of these—such as Boyle's narratives concerning color of 1664—aimed to establish claims that, though limited in various ways, nevertheless transcended the place and time of their original production.

Consider the following example. At the beginning of the third part of his "history of colours," Boyle described an experiment performed on "October the 11. About ten in the Morning."[7] He continued by providing a perfect example of an experiment specified in time, place, and circumstance. But it was done in the service of a claim that holds *outside* locality; Boyle did it in the first place "because that, according to the conjectures I have above propos'd, one of the most general causes of the diversity of colours in opacous bodies, is, that some reflect the light mingl'd with more, others with less of shade ... I hold it not unfit to mention in the first place, the experiments that I thought to examine this conjecture."[8] Specifically local the experiments certainly were, but they were intended to be of a type that could be regenerated elsewhere.

In this respect, a historical event of the sort explored by chronologers is a very different kind of beast from knowledge about nature because it is inherently and inevitably singular: an event of history cannot be reproduced in other places and at other times unless it is taken to be exemplary of a type that transcends the specific event's locality. Caesar's crossing of the Rubicon in 49 BCE is unreproducible, but not simply because 49 BCE occurred only once; after all, Boyle's October 11 experiment also occurred just once. Rather, Caesar's incursion cannot be reproduced because the people involved and the technological, economic, political, environmental, psychological, and social circumstances irretrievably alter over time, whereas the character of the event is embedded in the complex specificities of its original occurrence. Boyle would in contrast assert that his October 11 experiment could be reproduced everywhere and by anyone—although perhaps with difficulty—precisely because he intended it to provide evidence for a locality-transcending claim. Natural events that occur without human intervention—which are of the sort that, to a pure-bred scholastic, are the only true products of nature—would constitute for Boyle an exception to the rule only in the sense that they might not be reproducible by human agency, though not because human actions might somehow step outside nature's course. Only divine interventions and spirits (in both of which Boyle firmly believed) could do that.

[7] Boyle, 1664, p. 186.

[8] Ibid., pp. 185–86.

The differences between historical and natural events have consequences for evaluations of novelty. In Boyle's world, an experiment designed to substantiate a claim that does not work as expected may provide evidence against the claim and suggest alternatives to it. Laboratory novelties have locally transcendent meaning just because they may, or may not, support claims or suggest new directions. Historical events cannot provide surprises or point out new paths, which is to say that they cannot constitute proper novelties, unless they too are linked into a wider scheme of knowledge that gives them general significance. To return to the river-crossing Caesar, if we knew enough about the circumstances to fit this and subsequent events into a class of military endeavors, then we could say that the event's significance transcends its locus and time of production—in which case the event would be an exemplar within a broader system of knowledge.

The technical chronologers of the early modern period did not produce knowledge systems in anything like this sense. They developed instead systems of concordances and sequences that located events of human history in time by means of their simultaneous occurrences with particular astronomical events, usually eclipses. Put differently, the likes of Scaliger aimed to use their collections of happenings to establish chronologies and not to uncover or demonstrate general theories about history. Neither Scaliger nor Bünting were embryonic Giambattista Vicos, who did produce just that, odd and fantastic though Vico's scheme appears to twenty-first-century eyes.[9]

It is precisely here that Isaac Newton, as a chronologer, differed programatically from his predecessors: he sought to use astronomical tools to mold singular events into a system for understanding ancient history, indeed for grasping the entire development of civilization—what's more, we shall see, a system that shared and exemplified the same evidentiary and argumentative structure deployed in his science. Consequently, Newton faced a new kind of problem—new, that is, in technical chronology. Historical remnants can be known of course only through the transmitted testimony of individuals whom no one living has ever met. Humanists had long developed methods for handling the inevitable results of corruption and fraud over the centuries. In Newton's chronological world, the chains of transmission ran back to a period before the very existence of written records in Greek, or at least to a period when Greek literacy was in its infancy, whereas the eclipses that the technical chronologers had used all derived from literate eras. More problematic still, the remnant testimonies that reached back to that pre-literate time did not concern precise astronomical events. They described instead characteristic, but comparatively inexact, features of the heavens, and they referred to them in words that required interpretation. For unlike eclipses, which become dates *via* the mathematics of astronomy, the features at Newton's disposal could not be easily transmuted into numbers. Most significant of all, even when Newton did effect his transformations, the numbers that emerged did not agree with one another.

Ultimately, Newton would produce from these discrepant numbers a determinate date that, if correct, would fundamentally alter all previous chronologies—and indeed would challenge contemporary understanding of the human past. That date did not emerge from just one among the numbers into which Newton had transformed

[9] See Grafton, 2001b.

ancient words, but from all of them together. This manufacture of harmony out of discord was not only new in historical chronology, but unprecedented in natural philosophy before Newton's own work in optics in the late 1660s and early 1670s. He alone had developed a method that not only permitted, but actually urged, the experimental production and subsequent amalgamation of discordant numbers—a method that directly informed Newton's historical chronology.

Newton's career spanned nearly seven decades, a period that witnessed profound changes in the many methods, techniques, conceptions, and practices that together constitute the late period of Early Modern mathematics and natural philosophy. He himself was responsible in several ways for a good number of these changes.[10] So protean a career as Newton's, which continued to generate new results through the 1690s, can hardly be expected to exhibit a single continuous strand of development. Nevertheless, by the early 1670s the young Newton had developed the contours of method and technique, as well as a number of specific conceptions in optics and mechanics, and (especially) mathematics, that the mature scholar adopted, adapted, reworked, and deployed in subsequent decades.[11] These several and disparate explorations were undertaken in the context of mid-seventeenth-century views concerning nature, mind, God, and the links among them. His own understanding of these matters evolved during his early years at Trinity College, Cambridge, which he entered in June 1661 at the age of eighteen and a half. Over the next half decade, the "solitary scholar"[12] encountered a considerable amount of seventeenth-century learning, and he systematically set down many of his thoughts and textual extracts in a notebook, part of which he titled *Questiones quaedam Philosophicae* (*Certain philosophical questions*). Here we find the young Newton first grappling with issues which had gripped so many for the previous half century, often involving questions concerning the proper ways to generate reliable knowledge.

Questions of this sort have long been associated with Newton, for he is often thought of as having refused to admit "hypotheses," Unlike Hooke or Huygens—or that great world-maker Descartes—Newton is said to have remained extraordinarily wary of conjecturing explanatory structures that were not strongly connected to the empirical world. His most celebrated remark about "hypotheses" appeared for the

[10] The works of I. Bernard Cohen, Alan E. Shapiro, Richard S. Westfall, and D. T. Whiteside have illuminated the evolving structure and content of Newton's mathematical, physical, and experimental endeavors; J. Bruce Brackenridge, Michael Nauenberg, and Curtis Wilson have probed the ways in which Newton built his mechanics, and George E. Smith has unveiled the complex structure of Newton's particular approach to the construction of what he considered to be reliable knowledge. For further references see Buchwald and Cohen, 2001 and Cohen and Smith, 2002.

[11] On the development of Newton's mathematics, see the introduction and, especially, the notes to Whiteside's magisterial edition of Newton's *Mathematical Papers*. Shapiro's introduction and notes to the *Optical Lectures* of 1670–72 insightfully display and explicate the early development, not only of Newton's specific optical conceptions and methods, but also his experimental methods and beliefs. Cohen's translation of the final edition of the *Principia*, accompanied by his and Smith's essays, joins Brackenridge's *Key to Newton's Dynamics* as an invaluable entrée into the complexities of Newton's mechanics.

[12] On Newton's early years in Cambridge as a "solitary scholar" see Westfall, 1980, pp. 66–104. We will diverge from Westfall below in considering just what Newton might have absorbed during these years.

first time in the *General Scholium* appended to the second edition of the *Principia* (1713), and remained unchanged in the third edition (1726):

> I have not as yet been able to deduce from phenomena the reason for these properties of gravity,[13] and I do not feign hypotheses. For whatever is not deduced from the phenomena must be called a hypothesis; and hypotheses, whether metaphysical or physical, or based on occult qualities, or mechanical, have no place in experimental philosophy. In this experimental philosophy, propositions are deduced from the phenomena and are made general by induction. The impenetrability, mobility, and impetus of bodies, and the laws of motion and the law of gravity have been found by this method. And it is enough that gravity really exists and acts according to the laws that we have set forth and is sufficient to explain all the motions of the heavenly bodies and of our sea.[14]

More has likely been written about this single passage than about anything else that Newton ever did. Yet only in the past few decades have scholars come to understand fully what Newton meant, and how his very public disdain for hypotheses connects with his experimental work and with his development of mathematical theory.

The roots of Newton's attitudes and methods in respect to experiments and hypotheses reach back to his years as a student at Cambridge; it was here that Newton evolved his own way to merge mathematical structure with experimental investigations—a way that, he would soon discover, was neither congenial nor fully comprehensible to many of his contemporaries. Here especially lie the origins of Newton's skepticism concerning human perception, as well as his concomitant attention to aspects of measurement that few at the time had probed. This attitude, together with the data-handling techniques to which it gave rise, had a profound impact upon Newton's manipulation of matters chronological and upon contemporary reactions to it.

The natural philosophy curriculum that the young Newton encountered at Cambridge in the early 1660s had for some time incorporated a great deal beyond the traditional structure of scholastic learning. The novice would begin with a comparatively short introduction to the elements of the Aristotelian system, but he would also soon be introduced to novel developments, such as the works of Pierre Gassendi. Shortly after arriving, Newton bought a bound book with blank leaves to record notes about his studies.[15] This "Trinity notebook" is dated June 1661 on the front flyleaf.[16]

[13] Namely, its proportion to "the quantity of *solid* [*sic*] matter" (and not to particles' surfaces), its infinite reach, and its decrease as the inverse-square.

[14] Newton, 1999, p. 943.

[15] The notebook (CUL MS Add. 3996) has been partly transcribed and thoroughly discussed in McGuire and Tamny, 1983, on which we draw.

[16] The editors date the notes on the traditional curriculum to 1662–63, based in part on their "youthful hand": McGuire and Tamny, 1983, p. 14. They also assert (see immediately below) that the Magirus and Stahl notes were done after the extracts from Eustachius, even though Magirus and Stahl respectively bracket Eustachius in the notebook. Nevertheless, Newton certainly did leave enough space to add subsequent material in at least one instance (see below, note 19).

Newton began by inserting passages in Greek, with marginal headers in Latin, from Aristotle's *Organon*, and Poryphry's *Isagoge*.[17] He continued with Johannes Magirus' *Physiologiae peripateticae*, a widely used compendium of scholastic material on causality, physics, and cosmology. The notes break off near the end of Magirus' fifth chapter, entitled "De meteoris apparentibus," and are followed by a very different set on astronomical matters, beginning with Galileo's estimate of apparent stellar diameters.[18] The notes continue with Aristotelian ethics, also drawn from the *Opera Omnia*, and then the ethics of Eustachius of St. Paul. Newton's most extensive notes were on Daniel Stahl's *Axiomata philosophica* (1645), which is concerned exclusively with core Aristotelian topics such as act and potency, the typology of causes, agents, and patients, and so on. While Magirus' text, as its title suggests, concerns natural specifics, Stahl's provides the fundamental terms of Aristotelian metaphysics. The traditional material ends with Vossius' rhetoric.[19]

For the most part, Newton's pages on the Magirus and Stahl texts are student notes with little specific commentary.[20] They do not necessarily reflect his developing views at this time, much less Newton's later understanding of related matters, but they do show that he had a reasonable grounding in scholastic materials. Westfall had good reasons to remark that this "was the first sophisticated system Newton met, and there is no reason to think that it failed to impress him initially as he emerged from his intellectual provinciality."[21] Nevertheless, there are signs even here that Newton was not engaging as deeply as, say, a somewhat older Galileo had, with the intricacies of peripatetic doctrine.[22] In his early notes Galileo had grappled at some length with the question of whether the heavens are formed of the same stuff as the (Aristotelian) elements; more to the point, he had delved as well into the complex arcana of Aristotelian forms in respect to heavenly bodies.[23] In so doing, Galileo followed the standard structure of a disputation, presenting the claim, then intricate arguments *pro* and *contra*. Newton in contrast expended few words, coming right to a bald statement, such as "A heaven ... has matter that is not fiery ... Its motion is simple and circular, natural by analogy ... in the sense that it has a natural aptitude to be moved in this

[17] Aristotle, 1654, containing the *Categories*, *On Interpretation*, and the *Prior* and *Posterior Analytics*. The *Isagoge* provides an introduction to the *Categories*. McGuire and Tamny, 1983, p. 15 remark that Newton's notes "merely reproduce essential information from the Aristotelian *Organon* as it is represented in a fine edition of Renaissance scholarship."

[18] These astronomical notes, which securely date to 1664/5, are transcribed and discussed in McGuire and Tamny, 1985.

[19] Eight notebook folios are devoted to the *Organon*, eleven to Magirus, five to astronomical issues which begin with Galileo on stellar diameters, three on Aristotle's ethics, three on Eustachius, twenty-nine on Stahl, and six on Vossius, for a combined total of sixty folios, excluding the five astronomical ones. The *Questiones*, which we will consider below, and which follow the Vossius notes, occupy forty-nine folios and, unlike the earlier notes, contain a good deal of commentary of Newton's own.

[20] See Wallace, 1988, p. 26 for examples, in both the original Latin and translated into English.

[21] Westfall, 1980, p. 85.

[22] See Wallace, 1977. These notes by Galileo on *De caelo* likely date from ca. 1590, when the twenty-six-year-old Galileo was already teaching at the University of Pisa (ibid., pp. v–vii), though it remains possible that some among them were composed earlier.

[23] Wallace, 1977, pp. 103–58; 1988, p. 29.

way by an intelligence, so the intelligence is similar to a nature." Indeed, "the very brevity of Newton's summary," remarks one historian, "conceals problems he has glossed over, and thus leads one to question the depth of his understanding of this problematic subject matter."[24]

Which is precisely the point. The only pages we have that concern Newton's studies of such matters do not show us a young man gripped by the power of a system to follow its many elaborate byways. Quite the contrary. Newton's absorption of scholastic material bears every sign of having been superficial, though in this he was likely little different from the majority of his contemporaries, given the much wider latitude of studies at both Cambridge and Oxford by his time. The young Galileo grappled intimately with Aristotelianism, but there is hardly any evidence to show that, nearly three-quarters of a century later, the young Newton did.[25] The difference is telling, because during the intervening decades alternative systems had taken shape, which, though different from one another, were nevertheless grounded on a common epistemology that diverged from scholastic understanding of human knowledge.

The philosophy that Newton encountered at Cambridge in the works of Stahl and Magirus, and probably in discussion as well, was certainly not identical with medieval scholasticism, but the late sixteenth and early seventeenth centuries had seen a revival that emphasized Thomistic Aristotelianism.[26] Renaissance humanists had long before criticized scholasticism for its concern with the formal structure of logic,[27] and varieties of Platonism, sometimes coupled to Hermeticism, too.[28] Though the new philosophical currents associated with the names of Galileo, Gassendi, Descartes, and Hobbes were widely discussed and taught at both Oxford and Cambridge quite early on, nevertheless students were expected to absorb the language and structure of Aristotelianism, which was after all a stable, systematic, and highly evolved system. Indeed, the very terminology and argumentative structure of the new philosophies of the seventeenth century were couched in ways that required knowledge of the scholastic alternatives. "I begin with the philosophy of the schools," wrote Ralph Bohun of

[24] Wallace, 1988, p. 29.

[25] In which case we cannot agree with Wallace that both the young Galileo and Newton "may be characterized as Aristotelians," even with the qualifier that the Aristotelianism in question was "progressive." Wallace has reasonably solid reasons for Galileo, given the form and substance of the notes, but the curt, to-the-point character of Newton's cannot sustain the claim where he is concerned.

[26] Nadler, 1998, pp. 516ff., who remarks that in "late scholasticism one still finds, above all, an unwavering commitment to the Aristotelian doctrine of substance."

[27] See Jardine, 1988a, who remarks (p. 188) that for humanists, logic "was a means to the analysis and understanding of canonical texts (literary and legal), and orations (ancient and modern; forensic and epideictic), and an aid to composition of discursive writings emulative of 'golden' Latin."

[28] On which see Copenhaver, 1988; Ingegno, 1988. For portraits of one polymath of the period see Grafton, 1999; Siraisi, 1997. Telesio in particular critiqued Aristotelianism, breaking especially "the Aristotelian relationship between the sublunary sphere and the heavens and between the heavens and the realm of the intelligible" (Ingengo, 1988, p. 252). Bruno rejected scholastic forms altogether and "saw matter as a real substantial principle—stable, persistent, capable of receiving all forms and therefore nobler and more durable than any limited form that must eventually disappear. Disregarding individual species and genus as illusions of merely logical diversity, he concluded that the forms indicated by these notional distinctions lacked substantiality" (Copenhaver, 1988, p. 293).

New College to John Evelyn, "which though I make it not my creed ... yet since Aristotle has so universally obtained in all the universitys of Christendome for so many ages [and] thus insensibly crept into all modern writers by the use of his terms, it's almost impossible, as things stand, to be either divine, physician, or lawyer without him ... how then can it be expected that we should understand the new philosophies without him, when the greatest part of their works consist only in confutation of his."[29]

Among the most fundamental of scholastic doctrines, and certainly one that could not be missed by even a cursory reader, was *hylomorphism*. For scholastics an entity consists of a material substance, or *hule*, which is conjoined to two kinds of *form*, or *morphe*. Forms are, very loosely, the characteristics of a given entity. Matter does not exist in the absence of form, and neither does form exist in the absence of conjoined substance.[30] There is, at the most fundamental level, an object's *substantial form*, which makes it a particular *this* rather than a *that*. A horse's substantial form differentiates it from the substantial form of a donkey. There is, in addition, *accidental form*, which does not alter the essence of an object but which may characterize a particular entity at some particular time. A piece of wood may be painted brown one day and blue the next, but it remains a piece of wood: color is not an essential part of its being a piece of wood. This way of thinking entailed a specific understanding of the process of cognition itself: of how knowledge of things comes about. In scholastic cognition, an object's accidents are brought to bear upon the sense organs through any of various means, including direct contact. However it occurs, the entity's form produces in the sense organ, proximately or directly, a change: a quality known as an "intentional species," particular to the form in question.

Hatfield concisely describes the next steps. The species is received by a sensory power in the percipient, which is thereby

> actualized to its characteristic sensory activity ... in the act of sensing, a kind of identity arises between the sensory power and the object sensed, which identity permits the power to be "directed toward" or "attentive of" the object, and so to cognise it ... The "common sense" discriminates among the objects of the special senses (e.g. it discriminates white from sweet). The species received by the external senses are retained as "phantasms"[31] in the internal senses. These phantasms are corporeal in nature; that is, they are states of the corporeal organs informed by the sensitive power of the soul ... Cognition of the natures or essences of bodies requires intellection. The immaterial intellect "illuminates" the phantasm and abstracts the essence or "common nature" of the represented thing. [And then] the "agent intellect", "together with the phantasm" produces an (immaterial) intelligible species

[29] Feingold, 1997, pp. 401–2.

[30] There are innumerable discussions of this; for a concise one see Wallace, 1978, p. 100.

[31] In Aristotle's original terminology: "As sight is the most highly developed sense, the name *phantasia* [imagination] has been formed from *phaos* [light] because it is not possible to see without light" (*De Anima* III.3, 429a).

> in the "patient" or "possible" intellect. Reception of the patient intellect completes the act of understanding.[32]

Perception itself, wrote Aristotle, "of the special objects of sense is never in error or admits the least possible amount of falsehood."[33] He meant that the forms that activate the correctly operating sense organ are always perceived as qualities that are, *qua* forms, the same as those of the perceived entity from which they arose. However, the entity whose form or forms these qualities originated from may be incorrectly *imagined* as such. So, for example, "while the perception that there is white before us cannot be false, the perception that what is white is this or that may be false."[34] Imagination is the faculty by means of which the activated qualities of sense organs produce knowledge of objects. Common to both beasts and humans, imaginations "remain in the organs of sense," i.e., they are not functions of the soul.

There were many intricate issues over which scholastics had long chewed concerning these matters, involving in particular questions concerning the manner in which an entity, not in direct contact with the sense organ, can engender its form. To that end, late scholastics carefully distinguished two ways in which forms may exist. They may exist, as it were, bound to an entity's natural material being. Or they may exist as forms connected to an entity's matter but without a corresponding natural binding, which allows the matter to carry the form without the corresponding quality characterizing the matter which bears it. These latter sorts of forms are "intentional," and they are (among some late scholastics) what affect the sense organs, which are therefore not in themselves characterized by the forms in question—an eye that yields the perception of red does not itself become red but does bear the intentional species of redness that the original (red) object impressed upon the air between itself and the eye, with the air carrying the form only intentionally and not naturally. Issues of this sort date back to the thirteenth century, but, intriguingly complex though they were, scholastics all held that qualities as such come to bear on the sense organs.[35]

Scholastic cognition accordingly involves a tight connection between perceiver and perceived, one in which the perceiver's sense organs are activated, in most cases *via* intermediaries, by the forms that endow the object's substance with those characteristics which the sense organ's own qualities adapt it to receive. When for example we stroke a soft, furry thing, our sense of touch receives its *softness* and *furriness*; our sense of sight receives the *redness* of a ripe apple, our sense of taste the *sweetness* of

[32] Hatfield, 1998, p. 956. Also see Smith, 2000b, pp. 316–19, which provides a careful and apposite example of the process as deployed in medieval optics. Nadler, 1998, p. 517, notes that in late scholasticism the "ultimate and only possible explanation of observable property *x* in a body *b* is the intrinsic presence in *b* of the real quality or form *x*-ness. Explanations of this sort were considered complete and satisfactory."

[33] *De Anima* III.3, 428b.

[34] Ibid.

[35] Although late scholastics had markedly mutated earlier doctrines, and though a considerable part of these earlier positions were themselves hardly present in Aristotle, nevertheless Aristotle's own views, albeit refracted through a scholastic prism, were read and reread, particularly in light of the Greek Aristotle that was first printed by the Aldine Press between 1495 and 1498. Thorough discussions of these matters occur in Simmons, 1994; Tachau, 1988. Also see Kessler and Park, 2000 and Park, 2000.

sugar. The perceived entity does not induce changes in the sense organ that merely correspond to the form in question; it produces instead an effect that is specific to the form itself and that can be produced in no other way if the sense organ is in proper working order and the body is not otherwise active.[36]

Magirus' text, over four hundred pages long, consists of six books, on four of which Newton took notes.[37] Stahl's contained twenty-two "titles" and ran to seven hundred pages. Both provide ample discussion of issues concerning matter and form, with Magirus investing the scheme in specific instances.[38] Moreover, Magirus's account of perception, based primarily on *De Anima*, was extensive, consisting of fifty-nine propositions followed by a commentary. It provides a good, concise account of the scholastic understanding of perception that we have just outlined. Newton, however, took no notes at all on Magirus' sixth book, which contains the account. Neither is there anything in Stahl on the subject. While Newton was accordingly introduced in some detail to Aristotelian physics and metaphysics, he may not have encountered—he certainly took no notes on—the complex issues concerning cognition that are treated in *De Anima*. Neither do Newton's notes show him to have been gripped in any substantial way by *Aristoteliana*, for his notes are mostly uncommented *précis* of the two texts. These are not the reflective remarks of someone who was interested in exploring the intricacies of scholasticism.[39] Newton may have drawn on elements in the standard vocabulary and conceptions of the period, using them as resources to forge his own notions,[40] but there is simply no evidence in the notes on Magirus and Stahl to suggest that Newton was doing anything beyond the normal note-taking of a Cambridge student.

Stahl and Magirus were hardly the only texts with which Newton grappled, for the *Questiones*, though (perhaps significantly) later than the Aristotelian material in

[36] An apposite example from *De Anima* illustrates the point. Referring to the sense of touch, which distinguishes the elemental properties of hot, cold, moist, dry, Aristotle noted that the organ of touch "is that part which is potentially such as its object is actually: for all sense-perception is a process of being so affected; so that that which makes something such as it itself actually is makes the other such because the other is potentially such" (*De Anima* II.11, 424a).

[37] Loosely, Magirus' first four books respectively concern motion and cause, celestial, terrestrial, and atmospheric affairs; the fifth concentrates on the mineral, animal, and vegetable worlds, while the sixth concerns the soul.

[38] Stahl's eight and ninth "titles" are, respectively, "circa doctrinam materiae," and "circa doctrinam formae"; subject and accident are treated in title 16.

[39] Wallace, 1988, asserts otherwise, remarking that Newton's "enduring concern was with the world of nature, and thus he was basically a natural philosopher. As such, his primary debt was to Aristotle, the rudiments of whose thought he quite correctly grasped" (p. 35). Newton may well have grasped these rudiments—after several hundred pages of Magirus and Stahl it would have been hard not to—but it would be misleading to "characterize" him as an Aristotelian at any time. To do so requires reducing Aristotelianism to a "concern with nature and causality," which does considerable injustice to the scheme's intricate complexities, and, in particular, ignores its grounding in hylomorphism.

[40] As, perhaps, in his subsequent articulation of space and time: see, e.g., McGuire, 1978; see also McGuire, 1968. Standard scholastic terminology, such as the intension and remission of qualities, often appeared in new philosophical discourse, for that was the available, and familiar, vocabulary of the period. The words and phrases were, however, shaken out of their Aristotelian context and twisted into novel forms, until over time the old vocabulary and syntax evaporated altogether.

the notebook, plunge deep into the worlds of the new philosophies. The *Questiones* are dated by the notebook's editors to, at the earliest, the summer of 1663 (or perhaps even spring 1664).[41] Newton continued to enter remarks until early to late 1665, treating topics ranging from motion and atoms to light, gravity, and comets. Throughout these pages Newton repeatedly turned, apparently for the first time, if we may judge by their absence from his Aristotelian notes, to connections between the senses, matter, and the "soul." Furthermore, in the texts that prompted these notes—and, we shall see, his engaged reflections—Newton encountered an often-corrosive anti-scholasticism.

Evidence from the *Questiones* shows that Newton was acquainted with, and in at least some cases carefully read, works by Boyle, Walter Charleton, Descartes, Kenelm Digby, Galileo, Joseph Glanville, Henry More, and Thomas Hobbes. He knew Boyle's *Spring of the Air* as well as his *Experimental History of Cold*, and—especially—Boyle's *On Colours*. He also knew Charleton's sarcastic and vigorously anti-Aristotelian presentation of Gassendi's Epicurean atomism, the *Physiologia*; Digby's *Two Treatises*, which mixed Aristotelian with Cartesian elements; Galileo's *Dialogue* (but not, most likely, his *Discourses*); Glanville's *Vanity of Dogmatizing*, which dissolved all secondary causes into the activity of the deity; Descartes' *Opera philosophica*; Hobbes' thoroughgoing materialism in his *De corpore*; and Henry More's critique of both Hobbes and Descartes, with its efforts to reinsert spirit into space.[42]

The very first pages of Newton's *Questiones*, headed "Of the first matter" followed immediately by "Of atoms," reflect his reading of Charleton and of More.[43] Newton grappled with several arguments raised in Charleton's thoroughgoing atomism. Though he did not take notes on other material in Charleton's text, even a quick perusal of the *Physiologia* would reveal Charleton's attitude toward Aristotelianism. One example drawn from many nicely illustrates the tone of his rhetoric. The "despot of the schools," as he at one point called Aristotle,[44] assimilated "sapours" (taste) to the unacceptable principles of hylomorphism, so that "the extreme slenderness of his

[41] McGuire and Tamny, 1983. Also see Westfall, 1980, pp. 83–95. McGuire and Tamny admit that their dating is conjectural, though there are several markers that provide *termini a quo*.

[42] McGuire and Tamny, 1983, pp. 20–25, note the sources, while their full transcription of the *Questiones* makes specific connections. A great deal in the *Questiones* reflects Newton's thoughts on reading Descartes, including both the *Dioptrica* and, it seems likely, *Meditation VI*. Newton had Descartes, 1656, which he extensively dog-eared (McGuire and Tamny, 1983, p. 23); this edition does not contain the *Geometrica*, but is otherwise extensive. Newton cites (p. 450) "Hobbs. Part 4th chap 1st," meaning the first chapter in part 4. Since this chapter is actually chap. 25 in the English *Elements of Philosophy* (Hobbes, 1655; 1656), it's clear that Newton did read Hobbes directly. He refers explicitly to, and read very carefully, More's critique (More, 1659). The evidence for Boyle, Charleton, Digby, Galileo, and Glanville is also either explicit or quite secure. Further, Newton possessed a 1664 edition of Diogenes Laertius' *Lives of the Philosophers*, in which he might have read Epicurus (ibid., p. 352). On the other hand Newton did not likely encounter Gassendi directly, though Charleton's *Physiologia* included translations of parts of Gassendi, 1649. Charleton's aim was to render Epicurean atomism acceptable by "demonstrating its theological purity" (Kargon, 1964, p. 186), something that had much engaged Gassendi, an ordained priest, considering that Epicurus had asserted the soul to be composed of fine particles and so to be as mortal as the body.

[43] McGuire and Tamny, 1983, pp. 336–43. See ibid., pp. 26–43 for analysis of what Newton grappled with from Charleton.

[44] Charleton, 1654, p. 183.

doctrine, touching the *essence* and *principles*, of sapours as well in general as particular; erected on that common imaginary base of immaterial qualities, hath given us occasion to suspect the solidity of his inference or conclusion; and left us cause to account that sentence, much more canonical, *that things most manifest to the sense, often prove most obscure to the understanding*."[45] Indeed, Aristotle's "endeavours afford so dim a light to our profounder inquisitions, as to leave us in the dark of insatisfaction." Though he deployed the common terminology of causes, including the formal, throughout his various discussions, Charleton aimed to cut quality away from its hylomorphic seat, and to replace it with the physical actions of atoms striking the organs of sense. In discussing vision, though Charleton disagreed with "the excellent *Monsieur Descartes* "concerning the nature of light proper, he fully adopted the account of vision itself. The "throne of the mind," Charleton remarked, judges the "nature of the object ... only by the variety of strokes given to the external organ [the eye], thence to the filaments of the nerve annexed thereto, thence to the presence chamber of the soul: we are informed of the particular qualities, and conditions of every sensible; the variety of these sensory motions being dependent on the variety of qualities in the object, and the variety of judgments dependent on the variety of motions communicate."[46] Digby, to whom Newton referred briefly in "of touching,"[47] and who was neither atomist nor Cartesian in respect to material substance, nevertheless also rejected hylomorphism, remarking that "the sensible qualities of bodies are not any positive real thing, consisting in an indivisible and distinct from the body itself but are merely the very body, as it affects our senses: to discover how they do, which must be our labour here."[48] He too, approved Descartes' "great and heroic attempts" to explicate "the nature of sense," agreeing that "by the great variety of knocks or motions that our brain feels (which rises from as great a variety of natures in the objects that cause them), we are enabled to judge of the nature and conditions of every thing we converse withal."[49]

[45] Ibid., p. 241.

[46] Ibid., p. 151.

[47] McGuire and Tamny, 1983, p. 393. "A man has been deprived of his feelings. Sir Kenelm Digby."

[48] Digby, 1669, p. 305. Newton's remark concerning the man deprived of feelings does not explicitly occur in Digby's section on touch.

[49] Ibid., p. 347. Digby parted company with Descartes in respect to what it is that conveys the motions set up in the sense organs to the brain, preferring the "more common way; and make the spirits to be the porters of all news to the brain: only adding thereto, that these news, which they carry hither, are material participations of the bodies, that work upon the outward organs of the senses, and, passing through them, mingle themselves with the spirits, and so go whether they carry them" (p. 384). Digby seems to have attributed hylomorphism not to Aristotle himself but to his "perverters" (the scholastics), asserting that "nothing is more evident, than that *Aristotle* meant by *qualities* no other thing, but that disposition of parts, which is proper to one body and not found in all" (p. 437), taking as an example "beauty," "which is nothing else but a composition of several parts and colours, in due proportion to one another." Digby was following the usual tack in such matters of attributing nonsense, not to the inerrant master but to his bastardizing followers.

Rejecting both Descartes' plenum, which locked substance and space together, as well as atomism, which separated substances by substance-free space, Digby seems to have given primacy to the notion of *density* itself: "to these two notions of rarity and density we must allow a great latitude, far from consisting in an indivisible state; for, since rarefaction makes a lesser body equal to a bigger; and all inequalitie

Hylomorphism accepts, indeed it embraces, an association between a form of an entity and a related formal property of the sensing organ. Mechanism, whether Cartesian or atomist, cannot do anything like this because qualities, properly speaking, exist in perception but not in the unperceived world. The Gassendian and Charletonian atoms that leave an entity and strike the senses engender the sequence of mechanical effects that lead ultimately to the perceiving mind's construction of the entity's qualities, but these qualities, as such, require perception for their very existence.[50] Mechanical philosophies of the seventeenth century, including renovated and theologically purified Epicureanism, accordingly break the link between a percept and an entity's property: even, e.g., shape must somehow be constructed by the soul or mind out of stimuli that in themselves are related to the entity's physical characteristics only through a chain of mechanical effects. This is the central problem with which mechanists had to grapple, with all of its ancillary difficulties, not least theological.

Scholastics certainly did think that the organs of sense might be active in ways that do not, or do not any longer, match the form of the originally stimulating entity. This can occur because "even when the external object of perception has departed, the impressions it has made [on the sense organs] persist, and are themselves objects of perception."[51] Aristotle provided an apposite example, akin to one to which, we shall see, the young Newton paid especially close attention: "if, after having looked at the

---

betwixt two bodies, has the conditions of a body; it follows, that the excess of one body over another consists in infinite parts, into which it might be divided; and, consequently, that what is rarefied passes as many degrees, as the inequality or excess hath parts. And the same law being in condensations; both dense and rare things muct be acknowledg'd capable of infinite variety and diversity of states, in regard of more or less in the same kind" (p. 33).

[50] See Asmis, 2005, p. 274 on Epicurean perception, who asserts that "apart from perception, there is no external red sphere; the coloured three-dimensional shape that we see exists only in perceptual interaction; outside us, there are fine networks of colourless atoms that are densely arranged in the so-called 'solid' source and very thinly distributed in the perceptual stream reaching from the source to the beholder." There is however some complexity to the Epicurean account, because Epicurus was loathe to remove qualities altogether from reality, and yet he does not seem to have developed, or at least deployed, the quintessential mechanical notion that qualities are entirely *sui generis*, their reality consisting in their construction either in the sense organs themselves or in the soul. See Sedley, 2005, pp. 380–82, who remarks that "where Democritus might well have concluded directly that colour is unreal, Epicurus declines to make that move, in line with his general refusal to privilege the atomic ontologically over the macroscopic."

The complexity may be connected to Epicurus' conviction that the "permanent" qualities of bodies, including shape, size, and weight, "are knowable by sensation of these very properties" (Diogenes Laertius, 1925, pp. 599: X, 569). Sensation directly apprehends these permanent qualities, whereas impermanent qualities, though not immaterial forms, nevertheless "are seen to be exactly as and what sensation itself makes them individually to be" (p. 601: X, 71). Admittedly obscure, Epicirus' remarks do seem to indicate that he did not distinguish between an object's permanent qualities and the sensations that are engendered by them. For permanent qualities the sensation matches the quality proper: atoms have shape, so *ipso facto* do objects formed from them, and sensation apprehends shape as it actually exists in nature. For Epicurus, it seems, as indeed for Aristotelians, there can be a true identity of kind between at least some properties in the world which exist independently of any sensing being and the sensations that they engender.

[51] Aristotle, *De Somniis*, II, 460b.

sun or some other brilliant object, we close the eyes, then, if we watch carefully, it appears in a right line with the direction of vision (whatever this maybe), at first in its own color; then it changes to crimson, next to purple, until it becomes black and disappears."[52] Further—the primary aim of Aristotle's account in *De Somniis*—"the stimulatory movements[53] based upon sense perceptions, whether the latter are derived from external objects or from causes within the body, present themselves not only when persons are awake, but also then, when this affection called sleep has come upon them, with even greater impressiveness."[54] Alternatively, there might be a complete mismatch between a stimulating entity and the perception induced in an organ of sense if the organ is not functioning properly—if, that is to say, the form activated in it does not correspond to the form of the activating subject, as when the hand mistakenly senses an intensely cold object for a hot one, having as it were been damaged by the intense cold.

The percepts that arise in the Aristotelian sense organ are, we noted previously, never in error: they are what they are and cannot be otherwise; the percept "green" is simply what it is, though the formal property of the eye that *is* the percept may not have been activated by a correspondingly "green" subject. Mechanical philosophers would certainly agree with the inerrancy of the senses in this meaning, and they would further agree that the same percept can be engendered by an entity's persisting effect on the organ as well as by its immediate action, or by causes internal to the organ itself. However, because they rejected the hylomorphist match between the forms of (properly working) sense and of the stimulating entity, mechanical philosophers also had a subtly different understanding of the errors to which the senses may give rise even when they are working correctly and contemporaneously with the engendering subject.

For the scholastic, a correctly operating sense organ possesses an activated form that necessarily matches the form of the external entity. The organ's form does not just come close to complementing the latter's property; the two are indissolubly united. Not only is the percept induced in a sense organ inerrant by its very nature, so must be the connection between the percept and the form of the stimulating subject, provided only that the organ is functioning properly. The reason for this lies deep within Aristotelian metaphysics, which does not break apart the perceived from the perceiver. If the sense organs are working well, then the percepts to which they give rise must directly match qualities that are the same ones that characterize the subject of perception. The imagination may err under certain circumstances as to the subject that produces the percepts, as when a circle seen edge-on with one eye may be perceived as a line, but this is the result of the senses not operating in conjunction with one another, for imagination is a result of their joint action.[55] The scholastic's eye, ear,

[52] Ibid., I, 459b.

[53] "Movement" can mean a physical displacement of the organ or a part of it, what scholastics termed local motion, as when sounding air moves the organs of the inner ear, or it might have the broader sense of change, as when a hand touching a hot object acquires in actuality the formal property of hotness that it previously had in potential.

[54] Ibid., III, 461a.

[55] For example: "when the fingers are crossed, the one object [placed between them] is felt [by the touch] as two; but yet we deny that it is two; for sight is more authoritative than touch. Yet, if touch stood alone, we should actually have pronounced the one object to be two" (ibid.).

smell, taste, or touch either operates properly, in which case it correctly captures the form of the stimulating subject, or else it does not. No *via media* crosses between the properly working and the faulty sense, and for just this reason scholastics did not have to face the problems of epistemology that arose with mechanism.

The sense organs of mechanical philosophy, unlike those of scholasticism, could not be said meaningfully to reproduce or to match the qualities of the external entity. Neither did mechanists locate percepts in the organs themselves, as Aristotle had. They certainly did envision links between the motions of the entity and the engendered motions of the sense organs, but the connections had to be purely mechanical and not ones of formal identity, because "motion" meant only one thing, namely the change of spatial relations between qualitatively unalterable parts. Error creeps ineluctably into the mechanically engendered perception because the sense organs are akin to *instruments* like the telescope or even the ruler and compass, differing from them in only one critical respect: unlike the devices of the natural philosopher, the sense organs produce motions that are transmitted by the "animal spirits" through the nerves and thence to some region of the brain.[56] And just as no craftsman can make an instrument to any desired level of perfection whatsoever, neither can the organs of sense achieve a perfect match between the motions produced in them and the originating movements in the external world. Vibrations engendered in the ear by a sounding entity cannot for example perfectly match the entity's, any more than one thing can force another to move in flawless mechanical harmony along with it. Perfect harmonious motions can occur only at the level of elementary interactions between the underlying mechanical bits and pieces that constitute the world, but not between the furniture of the perceived universe which is forged out of them. Only a Gassendian atom, or a microscopic Cartesian volume, can nicely equal another atom's or spatial bit's movement. The mechanist's perceived world must accordingly differ from its natural progenitor even when the organs of sense work as well as they

[56] To Aristotle's account of perception the physician Galen had added the assertion that material entities—the "animal spirits"—convey the forms induced in the sense organs to the seat of the soul. Aristotle had located the soul's seat in the heart, but Galen displaced it to the brain, which he divided into three distinct, and spatially sequential, physical parts, each of which has a specific function in respect to the soul, these being the construction of imagination (Aristotelian *phantasia*), reason, and memory. Robert Burton's *Anatomy of Melancholy* nicely encapsulates the Galenic structure as understood in mid-century: "The brain itself is divided into two parts, the fore and hinder part; the fore part is much bigger than the other, which is called the little brain in respect of it. This fore part hath many concavities distinguished by certain ventricles, which are the receptacles of the spirits, brought hither by the arteries of the heart, and are there refined to a more heavenly nature, to perform the actions of the soul. Of these ventricles there are three—right, left, and middle. The right and left answer to their site and beget animal spirits; if they be in any way hurt, sense and motion ceaseth. These ventricles, moreover, are held to be the seat of the common sense. The middle ventricle is a common concourse and cavity of them both, and hath two passages—the one to receive pituita, and the other extends itself to the fourth creek; in this they place imagination and cogitation, and so the three ventricles of the fore part of the brain are used. The fourth creek behind the head is common to the cerebral or little brain, and marrow of the back bone, the last and most solid of all the rest, which receives the animal spirits from the other ventricles, and conveys them to the marrow in the back, and is the place where they say the memory is seated" (Burton, 1652, section 1, member 2, subsection 4, "Dissimilar parts"). Note that the Galenic account locates the operations of the soul in the empty corridors of the brain—empty, that is, of brain matter, but filled with animal spirits.

possibly can. And so inevitable error stalks the sensed world of the mechanical philosophy, whereas harmony between perception and subject governs the universe of healthy scholastic sensation.

By the late sixteenth and early seventeenth centuries, a devotee of form could nevertheless dispense with the apparatus of formal transformation that undergirded fully fledged hylomorphism. The atomist Daniel Sennert's explorations of alchemical processes were grounded both in his conviction that forms do exist and that they do not transform. "Sennert himself," notes William Newman, "did not view substantial forms as an unfortunate but necessary consequence of scholastic hylomorphism—rather, they were active witnesses of God's power and beneficence in the world … Sennert's explanation of chymical reduction locked the forms safely within their material vehicles, the atoms, and allowed them to persist in the face of the technological assault stemming from those striking agents of qualitative change, the mineral acids … From Sennert's perspective, then, his atomism had saved the substantial form from the unworkable theorizing of an excessively metaphysical type of Aristotelianism that mired itself in speculation to the detriment of evidence."[57] Forms could be allowed, even insisted on, but only so long as they no longer served any functional role. "All qualities flow from form," Newman continues of Sennert, "and yet the opaque and shiny tint of silver disappeared when the metal dissolved into a clear solution in nitric acid, as he observed no less astutely."[58] Moreover, during the century even "the dominant scholastic position became somewhat more dualistic than hylomorphic, with matter being endowed with being," while "another trend—perhaps in the opposite direction—was the shifting of one of the principal functions of matter [individuation] to form."[59]

Conservative hylomorphs nevertheless did not evaporate, even as scholastic terminology was twisted and broken into the frameworks of mechanism. Existing words and syntax are rarely, if ever, abruptly dropped, for they constitute the substance of conventional speech; new ways of talking evolve within an older universe of expressions, with discursive slippage, overlap and inconsistencies only gradually sorting out. Many did not lock forms away in Sennert's fashion, and neither did they transmogrify matter and form into space-filling substance and shape. Consider for example the 1672 essay of the Jesuit, Antoine Rochon, in a "letter" to "a Cartesian among his friends." Rochon's prime concern was a common one among Catholic theologians throughout the seventeenth century, namely that the Cartesian world was difficult, and perhaps impossible, to reconcile with the demands of transubstantiation.[60] Ro-

[57] Newman, 2006, p. 152.

[58] Ibid., p. 153.

[59] Ariew and Greene, 1999, p. 83.

[60] That is, the doctrinal presumption that in the mystery of the Eucharist the bread takes on the substance of Christ's flesh, and the wine his blood, while the bread and wine accidents, viz. their apparent qualities, remain unaltered. There is a considerable literature about this difficulty, which does not arise in the same way in most Protestant circles because of the different conception of consubstantiation, according to which Christ's substance comes to cohabit the bread and the wine with their original substances. On the specifically Cartesian difficulties for Catholics see, e.g., Ariew, 1999; Clarke, 1989; Gaukroger, 1995, and Jolley, 1998. Descartes thought he had an appropriate answer within his scheme, though it required a

chon's theological issues with Cartesianism orbited about the issue of matter and form. He pointed out, with good reason, that Descartes' mechanisms were purely hypothetical, that "if I ask you [a Cartesian philosopher] what is that certain figure, that certain manner, that certain juice, and those certain parts, you have nothing to tell me that you didn't already know." And so, Rochon continued, driving home his point, "you see well, Monsieur, that up to that point we [you Cartesians and we 'ordinary,' i.e. scholastic, philosophers] are equal, and that your philosophy and ours are alike: the subsequent difference is that you stop there, without even wanting to recognize that there's anything else in nature; whereas we, we would think to stop at the first surface, if we didn't try to penetrate further to discover that beyond all external appearances there also lies within something that is the principle of all dispositions and effects, which we call *Form*: and so long as you don't come to it, we will always have the advantage, and the right to make the same reproach to you, that Aristotle did to previous philosophers who stopped like you at these first appearances and explained things in your fashion. *Here's the cause of your error*, he told them, *that you do not know Form at all*."[61] For Rochon, qualities continued to inhere essentially in nature, so that, for example, Aristotle, were he alive—Rochon apparently being a good substitute—would (rightly) tell Cartesians that "heat cannot be explained by the effusion of little igneous bodies," as they had it.[62]

Forms continued to live in England as well as France. John Sergeant was educated at Cambridge and became secretary to the Bishop of Durham. The Anglican bishop's unfortunate sermons had the remarkable effect of convincing Sergeant to turn Catholic priest in 1650, after which he wrote extensively in attempts to rescue scholasticism from the clutches of Descartes, and, especially, John Locke. At the advanced age of seventy-eight, he published a book aiming to show the "unreasonableness of atheists, deists, anti-Trinitarians, and other sectaries," which provided a vigorous scholastic draft. "The essence of body," Sergeant insisted, "or of such an ens as consists of matter and form, is chiefly taken from its proper act, or its form. For, since what distinguishes a thing from all others, does, consequently make it this, and no other; and matter being of itself perfectly indeterminate, cannot distinguish or determine a thing to be of this kind, much less to be this individuum; it follows, that the essence of all bodies, and consequently of body in common, is chiefly taken from the form or act."[63] Neither were Catholics, native-born or converts, the only English adherents to form in the last half of the century.

The society physician Gideon Harvey published in 1663 a wordy "new principles of philosophy" that was saturated with distinctions and definitions grounded in cause, matter, and form and that provided innumerable one-sentence "proofs" of such propositions as that "the form giveth a being, not a power of being to matter."[64]

perplexing kind of supernatural intervention given his rejection of real accidents (though here too he temporized, recurring again to humanly incomprehensible actions of the deity—Ariew, 1999, p. 150).

[61] Rochon, 1672, pp. 145–46.

[62] Ibid., p. 154.

[63] Sergeant, 1700, p. 37; also see his preceding monographs, Sergeant, 1697; 1698. Sergeant was particularly exercised by Lockean ideas.

[64] Harvey, 1663, p. 77.

Descartes did not impress him. He was "a great proficient in the mathematicks" who "laboured much to reduce all philosophical conclusions to demonstratons, depending from certain hypotheses; but wherein they excelled the ordinary, or peripatetick ones, either in truth, certainty, or evidence, I have hitherto not yet learned."[65]

For those who continued to think in terms of matter and form, certain kinds of instruments could pose a problem because they intercede between object and sense, possibly distorting the concord between them. Not all devices might have nefarious effects; indeed, before the seventeenth century none of the instruments used for terrestrial or astronomical distance and angle measurements would do so, and neither would scales or other such devices. These mathematical instruments do not intervene between object and sense because they do not intercept the object's form; they only guide the skilled measurer to capture as best can be done an object's quantitative aspects. Error in measurement was certainly thought to occur (how could it not?), but it had to do rather with an instrument's degree of perfection and the skill of its user than with the device's transformative action or with errors of the senses proper.[66] Telescopes and microscopes, on the other hand, do intervene between object and sense, and this could and certainly did raise difficulties of a new kind. Harold Brown notes that the "idea that our senses are subject to error is an old one, but a new element is added once we recognize that we are capable of constructing instruments which can correct these errors." Moreover, "once we have made this step, it seems natural to begin to think of these senses themselves as instruments that operate in ways that are very similar to the ways that our eyes function."[67] The telescope's neutral effect on an object's visible form nevertheless quickly achieved widespread acknowledgment as the early devices were improved, in particular by Galileo.[68]

When the sense organs come to be thought of as instruments, which is an inevitable consequence of every mechanical philosophy, they become subject to the same kinds of infirmities that devices exhibit. For the mechanist, even a properly functioning eye must inevitably be subject to error, just as even the best-crafted telescope must

[65] Ibid., p. 23.

[66] By the late sixteenth century, an English designer of a measuring instrument might attribute inadequacies in the plan of a device, though not in its execution, to ignorance of formal knowledge on the part of the artisan. Thomas Bedwell, for example, insisted that his instruments were much superior to ones in common use, precisely because they were designed on the basis of arithmetic and geometric knowledge (Johnston, 1991, pp. 325–26). The divide between artisanal and formal knowledge that Bedwell insisted on would over time shift attention from the quality of a craftsman's product as a whole—as a unitary outcome of the artisan's skill and knowledge—to questions concerning the adequacy of the artisan as the executor of a design over which he had little control and for which he could take little credit. The divide between the mathematicians and the instrument makers themselves was clearly not rigid. Indeed, it has been argued that the two groups "seemed to have developed a symbiotic (rather than adversarial) relationship" (Biagioli, 2006, p. 165). The relationship in question, though, concerns the connection between describer (or even designer) and maker, and here there was obvious room for contention even if the two depended on one another.

[67] Brown, 1985, p. 499.

[68] Helden, 1974, p. 52. Van Helden notes that "in a way the telescope was a blessing to conservative astronomers" because it permitted a split between religious and astronomical beliefs, since, e.g., lunar mapping could proceed in a craftlike manner without raising explicit cosmological issues.

distort the visual quality of an object, whether because of imperfections in its construction or because of the properties of light itself. Whereas for scholastics a properly functioning sense organ exhibits a form that correctly matches the engendering form of the object, for mechanists the senses can only move when activated by whatever material motions the stimulating object produces. They can as a result convey only a *relationship between two motions* and not an *affinity between two forms*. In which case the eye, as a device moved, like all devices, by motions, must be treated in the first instance as an instrument.

Descartes thoroughly understood this unavoidable mechanistic slippage from instrument to sense organ, and he probed it in considerable detail in his *Dioptrique*. Nature, Descartes remarked, has "used many means" to make the eye work in a way that mitigates as much as possible the unavoidable defects that all instruments must exhibit. Thus, "by filling the eye with very transparent liquids which are not tinted with any color, she has enabled the impulses coming from outside to pass to the back of the eye without being changed. And by the refractons caused by the surfaces of these liquids, she has made it possible that among these rays along which these impulses are conducted, those coming from a single point converge to a single point against the nerve," and so on.[69] However, error on this last score "cannot be totally avoided," though the function of the pupil is to reduce the fault by stopping down the eye's aperture to the extent that enough light remains with which to see. Descartes continues with ways to remedy the eye's "deficiencies through art," the goal of the immediately succeeding discourse being to "tell you more exactly in what way we must make these artificial organs in order to render them as perfect as they can be."[70]

"Artificial organs" can distort the effect that they are intended to induce in the sense organ, and so too can the sense organ itself misrepresent. Since, Descartes remarked, "the impressions which come from without pass to the common sense by way of the nerves, if the position of these nerves is constrained by some extraordinary cause, it can make us see objects in places other than where they are."[71] If for example one eye "is forced by the finger" away from its position in looking at an object, while the other is not, then "the parts of the brain where its nerves originate are not disposed in quite the same way," which will cause the object to "seem to be double."

Descartes famously insisted on the separation of mind from matter, so that the function of the sense organs is to produce appropriate motions in a part of the corporeal body (for Descartes, the pineal gland) out of which, put crudely and inadequately, the mind apprehends the external world. These motions in the body constitute "images" in the sense of correspondences, and then "it is only a question of knowing how they can enable the mind to perceive all the diverse qualities of the objects to which they refer."[72] The mind, he continued, "is caused to perceive as many different qualities

[69] Descartes, 1965, p. 115.

[70] Ibid., p. 127.

[71] Ibid., p. 108.

[72] Descartes, 1965, p. 90. Further, though the "picture" formed on the retina and "transmitted into our head, always retains some resemblance to the objects from which it proceeds ... we must not hold that it is by means of this resemblance that the picture causes us to perceive the objects, as if there were yet other eyes in our brain with which we could apprehend it; but rather, that it is the movements of which the picture

in these bodies, as there are varieties in the movements they cause in the brain." Accordingly, if the sense organ is disturbed so that the motion that it would normally send to the brain after stimulation by a particular motion is altered, then the mind will no longer be able to perceive the appropriate quality in the engendering object. And since the sense organs have become, as it were, organic instruments, even under the best of circumstances they will not be able to transmit through the nerves perfect correspondences to the stimulating motions. For Descartes, error in perception arises in the first instance whenever the mind does not have correctly corresponding brain motions with which to work. For scholastics, error arises when sickly or deformed sense organs fail to acquire the correct form.

The difference between the Cartesian and the late scholastic on this score may seem nugatory, since Descartes could hardly pinpoint brain motions, but it has an important consequence for experimental practice. Sensual error for scholastics was a deformation of properly acting nature, whereas for Descartes it signaled either unavoidable deficiencies in the construction of an instrument (namely, the eye), or else that the instrument was not working properly. Deficient designs can be improved by means of artificial devices, and mechanical imperfections can be ameliorated. The mind, as a distinct substance, remains unaffected in itself by defective or poorly working sense organs, though it receives from them flawed information. Were the mind to have the same corporeal existence as the sense organs and nerves proper, then it might be problematic to concentrate, as Descartes did in the *Dioptrique*, on ways to ameliorate defects in the body's instrumentalities.

Thomas Hobbes did consider the mind to be entirely corporeal. His views are complex and, where the mind itself is concerned, not altogether explicit. The fourth part of Hobbes' *De Corpore*, which Newton read, contains a chapter entitled "of sense and animal motion." There Hobbes set out his identification of sensation with motion. "Sense," he concluded, "in the sentient can be nothing else but motion in some of the internal parts of the sentient; and the parts so moved are parts of the organs of sense. For the parts of our body, by which we perceive any thing, are those we commonly call the organs of sense. And so we find what is the subject of our sense, namely, that in which are the phantasms; and partly also we have discovered the nature of sense, namely that it is some internal motion in the sentient."[73] This much differs little from Descartes (with whom Hobbes engaged in arguments and priority disputes),[74] but Hobbes went further, rejecting the existence of anything at all, whether soul or mind, that was not corporeal.[75] The motions generated in the sense organs by external con-

is composed which, acting immediately on our mind inasmuch as it is united to our body, are so established by nature as to make it have such perceptions" (ibid., p. 101).

[73] Molesworth, 1839–1845, vol. 1, p. 390. Newton read it either in the original Latin (Hobbes, 1655), or in the English translation published a year later (Hobbes, 1656).

[74] Tuck, 1988, p. 16 argues that Hobbes' early work "was developed very largely in response to the issues raised by Descartes in 1637," when Digby sent Hobbes Descartes' *Discours de la méthode*. Tuck differs from the older views set out in Brandt, 1928. On this point see Martinich, 1999, p. 169, who asserts that both Hobbes and Descartes "were vain, glory-seeking, self-absorbed, self-proclaimed geniuses. This statement is purely descriptive. Hobbes, like Descartes, was obsessed with being first."

[75] Hobbes' views on spirit are complicated by his explicit separation of philosophy from theology. There has been much discussion of this by historians of philosophy; among the most recent is Pacchi, 1988.

tact are conveyed by "certain spirits and membranes, which, proceeding from the pia mater, involve the brain and all the nerves; also the brain itself, and the arteries which are in the brain; and such other parts, as being stirred, the heart also, which is the fountain of all sense, is stirred together with them."[76] "Sense"—perception for Hobbes—arises when, the transmitted motion having reached the brain, the brain reacts as all matter does to a shove and pushes back; the pushback *is* the perception proper. Which led Hobbes to the question of why "everything that reacteth" doesn't perceive. To find an answer, he turned as always to motion: only bodies in which the motions that constitute the sensory "phantasms" persist can perceive, bodies which accordingly have Hobbesian memory. "Wherefore sense," he concluded, "as I here understand it, and which is commonly so called, hath necessarily some memory adhering to it, by which former and later phantasms may be compared together, and distinguished from one another."[77]

Hobbes' thoroughgoing mechanism rendered the deity material and measurement problematic. A corporeal mind must be doubly subject to errors of sense, because it may exhibit the same sorts of infirmities and troubles as the sense organs themselves. Whereas for Descartes the mind may misjudge as a result of sensory error, for Hobbes the mind itself may go wrong as well, in which case proper knowledge should certainly not rely overmuch on material processes that may easily lead to mistakes. Measurements made during experiments were not frequent during the seventeenth century since most experiments were designed rather to produce an interesting effect (say, the death of a mouse in an evacuated air pump) than to generate numbers. But any measurements that were made would necessarily share the epistemic status of the experiment itself. For Hobbes, experiments were extremely doubtful sources of knowledge, since they cannot, he thought, lead to the proper causes behind appearances.[78]

"Hobbes's God," Pacchi writes, "is a kind of transcendent warrant of Hobbes's conception of reality as a material world of moving bodies, causally connected in a necessary determined sequence of events" (p. 183). In his extensive reply to the accusation of atheism leveled against him by Bishop Bramhall, Hobbes wrote: "Either God, he [Bramhall] saith, is incorporeal or finite. He knows I deny both, and say he is corporeal and infinite: against which he offers no proof, but only, according to his custom of disputing, calls it the root of atheism; and interrogates me, what real thing is left in the world, if God be incorporeal, but body and accidents? I say there is nothing left but corporeal substance. For I have denied, as he knew, that there is any reality in accidents; and nevertheless maintain God's existence, and that he is a most pure, and most simple corporeal spirit" (Molesworth, 1839–1845, vol. 4, p. 306).

[76] Molesworth, 1839–1845, vol. 1, p. 392. According to Hobbes, the heart pumps 'round the transmitting corporeal spirits.

[77] Ibid., p. 393. Neither Descartes nor Hobbes produced their new philosophies *ex nihilo*. Both wrote to a great extent in the vocabulary and syntax common among late scholastics, and to a certain extent their views, though explicitly critical of hylomorphism and its reification of form, nevertheless contained powerful, if transformed, scholastic elements. For Hobbes see, in particular, Leijenhorst, 2002, who notes affinities between Hobbes' views in his *Short Tract* and the Italian naturalism of Telesio and Campanella. He continues, however, that the "subjectivity of sensible qualities distinguishes seventeenth-century mechanism from both scholastic and naturalist realism. It is the logical consequence of the reduction of the external world to matter and motion, and had profound metaphysical and epistemological repercussions" (p. 72).

[78] Hobbes' attitude toward experiment was connected to his overall philosophy, which embraced politics, religion, and ethics as well. It has been argued that he rejected experiment as a proper source of knowl-

"If indeed philosophy were (as it is) the science of causes," he wrote in attacking the protaganists of the air pump, "in what way did they have more philosophy, who discovered machines useful for experiments, not knowing the causes of experiments, than this man who, not knowing the causes, designed machines? For there is no difference, except that the one who does not know acknowledges that he does not know, and the others do not so acknowledge."[79] The mechanic may build a machine without knowing its principles (according to Hobbes); the experimenter then uses the machine to manufacture experiments without knowing the causes that underly them. Hobbes accordingly thought experiment to be of little import. The proper way to work is to produce causal accounts based on the conjecturing of hidden mechanisms that are consistent with the appearances of things.[80] And if experiment was no proper source of knowledge, then experiments combined with measurements were even worse, since measurement shared the dubiety of the experiment as well as the potential errors of sense and mind.[81]

---

edge because it involved a closed community of privileged practitioners, no one of whom could compel authoritative assent, and which policed the boundary between matters of nature and matters of social life: "It was fitting," write Schaffer and Shapin, "that philosophical places should have masters who determined right philosophy, just as it was right and necessary that the commonwealth should have a master. Indeed, Leviathan could legitimately act as a philosophical master" (Schaffer and Shapin, 1985, p. 338). The tension, they continue, between "democratic ideals and the exigencies of professional experience" required the experimental community to introduce a technique, that of "virtual witnessing," which artificially increased the community of those who could participate in experiment. This last claim, which is critical to Schaffer's and Shapin's overall argument concerning the connection between "the nature of knowledge" and "the nature of the polity" (ibid., p. 344), is challenged in Principe, 1998, pp. 106–11. Principe argues that the "probation by witness accounts" was hardly unique to Restoration society, and in particular had been deployed by alchemists, who also "turned to the accumulation and retelling of experiential evidence, most notably codified in the tradition of transmutation history."

[79] Schaffer and Shapin, 1985, p. 383, translation by Schaffer from Hobbes' *Dialogus physicus* of 1661.

[80] Jesseph, 1996, p. 101 remarks that for Hobbes the ideal would be that "through a series of definitions that reveal the causal origins of things, a perfect natural science could derive all the phenomena of nature from unquestionable first principles," but "the best we can hope for is to employ a process of analysis to lead from the phenomena to be explained to the hypothetical causes of such phenomena." The appropriate kind of phenomenon to be discussed in this way would be general and broad, say, the reflection or refraction of light, and not the kinds of limited experiences produced by specially constructed devices like the air pump.

[81] This is not to say that Hobbes denigrated or suspected the telescope, because he certainly did not. Galileo, Hobbes wrote in an unpublished manuscript of the 1640s, was "the greatest scientist not only of our own century but of all time" (Jones, 1976, p. 123), and so hardly likely to have produced a doubtful device. The manuscript is a critique of White, 1642. White was a Catholic priest, a friend of Digby, and a fellow resident in France at the time. His *De mundo* was, like Galileo's *Two Chief World Systems*, in dialogue form, and the 11th problem of its 1st dialogue states that (in Hobbes' words) "the telescope is at its maximum development" (Jones, 1976, p. 114). Hobbes proceeded to assert otherwise. He prefaced his account of White's errors with a précis of his own theory of the apparent distance of an object when viewed with one eye—something that cannot be measured and that is in fact instrumentally meaningless (see Malet, 2001 on this, though his discussion is nearly as opaque as Hobbes' own). White thought that the failure of anyone to contrive hyperbolic lenses, *per* Descartes' method, showed that no improvements at all were possible. Hobbes disagreed. Even if, he wrote, "the said method is not the true one, does this mean that

Descartes had separated spirit from matter by refusing it extension, while Hobbes materialized it, but the Cambridge Platonist and enthusiast for spirit apparitions Henry More feared both alternatives as tending too easily to atheism.[82] Certain that spirit is active in the world, More insisted that it, like matter, must occupy space. To do so, More distinguished penetrable substance from impenetrable, and divisible from indivisible. Matter is both impenetrable and "discerpible," More's term for divisible. Spirit, on the other hand, can occupy the same space as matter, and so must be penetrable, but unlike matter it cannot be divided into parts, and so is "indiscerpible." The activity of the sense organs directly implicated spirit.[83] Though More had absorbed the reduction of qualities to the effects of mechanical contact on sensory organs, he insisted that the soul pervades the body, including the sense organs themselves. "Sensation," he wrote, "is made by the arrival of motion from the object to the organ; where it is received in all the circumstances we perceive it in, and conveyed by virtue of the soul's presence there, by virtue of whose continuity to those in the common sensorium, the image or impress of every object is faithfully transmitted thither."[84] The "chief seat of the soul" is located in the "animal spirits" which occupy the fourth ventricle of the brain, and it is here that "imagination" is exercised, which essentially follows the traditional Galenic specification.[85] Memory poses a problem because More reluctantly admitted that it has something to do with the "pith of the brain" itself. But this tends too far toward Hobbes, or even Descartes; More instead admitted only that the brain may be impressed with "some small dots here and there standing for the recovering to memory a series of things that would fill, it may be, many sheets of paper to write them at large." These dots, as it were, stimulate memory in the soul proper, so that the brain itself has at most a series of mnemonic devices, or what he called "brachygraphie." Moreover, these "small dots," should they exist at all, are not made by mechanical brain actions, but by the "plastic faculty" of the soul,

---

no method exists for the task?" (p. 118). Hobbes had no quarrels with a device designed to improve a sense, anymore than Descartes had. The senses are, after all, both faulty and weak. New things can be discovered by means of artifical organs, but it's another matter altogether to contrive experiments, and *ipso facto* measurements, that provide knowledge of underlying cause. It's one thing to spy the moons of Jupiter through a telescope; it would be another to make claims about colors by measuring the displaced images of colored threads seen through a prism, as Newton would do in the 1660s. Jupiter's moons are objects, and they can be seen by a healthy eye through a correctly working telescope; the purpose of the telescope is not to distort the object's instrument-free appearance but to render it visible. A prism used to view a colored thread makes nothing new visible beyond the mere shift by refraction of the thread's position. Hobbes would likely have said that no useful knowledge about the underlying mechanical cause of color could be gained by attempting to measure the image's displacement.

[82] On More, see Gabbey, 1982, the introduction to Jacob, 1987 (1662), and Hall, 1990.

[83] Hall, 1990, p. 128 notes that in the seventeenth century "the protean term *spirit* signified a principle or substance that is active, elusive, impalpable, fugitive and mysterious." Like many of his contemporaries, More was a particular enthusiast for ghostly apparitions and a convinced believer in witchcraft, in both of which he spied spirits at work in the world. More added material to the posthumous reissue in 1681 of Joseph Glanville's 1666 *Saducismus Triumphatus*—which Hall translates literally as *Agnosticism Overcome*—in which spirits are proven through witchcraft and demonic actions (ibid., pp. 137–38).

[84] Jacob, 1987 (1662), pp. 140–41.

[85] See above, note 59.

a faculty whereby it can affect the body's organization, albeit without perception of doing so.[86]

■

McGuire and Tamny remarked on Newton's "extensive use" of Descartes' *Opera philosophica* in his notebook *Questiones*, and they emphasized Newton's interest in Hobbes and More on sensation. Newton particularly engaged with what McGuire and Tamny refer to as a "physicalist" program, according to which "many of the things we attribute to the soul and its judgments are really products of the human body and its motions."[87] Drawing especially on Hobbes' *De corpore*, and on More's critique of Hobbes in his *Immortality of the Soul*, Newton considered that the material organs whose behaviors eventuate in sense perception are acted on by moving matter, which produces corresponding motions in them. These motions are then transmitted along nerves to a bodily structure that is the physical seat of the "common sensorium." Unlike Hobbes, but like Descartes, Newton did not identify mind with the motions generated there, but considered instead that a distinct and immaterial entity, the "soul," derives information about objects from the motions in the sensorium. A primary question for him, note the editors of the *Questiones*, was "how to demarcate what belongs to" the soul from what pertains to the body.[88]

The first remark in the *Questiones* concerning the senses drives directly to the point. "The nature of things," Newton wrote, drawing directly on Hobbes' *De Corpore*, "is more securely and naturally deduced from their operations upon one another than upon our senses. And when by the former experiments we have found the nature of bodies, by the latter we may more clearly find the nature of our senses. But so long as we are ignorant of the nature of both soul and body we cannot clearly distinguish how far an act of sensation proceeds from the soul and how far from the body."[89] By the "senses" Newton meant not only the organs of sense proper, but also the motions that they eventually produce in the sensorium—itself a bodily organ—and out of which the soul constructs its knowledge of objects. Indeed, the sense organs themselves are among the very things whose "operations one upon another" must be examined prior to considering "the nature of our senses."[90] Which is to say that the motions in the sensorium that are generated there by the motions transmitted along the nerves from the sense organs must themselves be understood before it's possible to disentangle judgments of the soul that may derive from perturbations to the body from judgments that are based on its orderly working.

[86] More's "plastic faculty" is not utterly transparent in meaning, but it served a necessary function for him and his Cambridge friend and sometime rival, Ralph Cudworth. Cudworth, in his tedious and verbose refutation of every form of atheism, wrote that it is "something in the bodies of animals superior to mechanism ... which keeps the more fluid parts of them constantly in their same form and figure, so as not to be enormously altered in their growth by disproportionate nourishment ... [and] which restores flesh that was lost, consolidates dissolved continuities ..." etc. (Cudworth, 1678, p. 167).

[87] McGuire and Tamny, 1983, p. 218.

[88] Ibid.

[89] McGuire and Tamny, 1983, p. 377.

[90] See the remarks in ibid., p. 218.

Two aspects of perception particularly captured the young Newton's interest. Assuming, as he did, that the motions in a properly working system of sense organs, nerves, and sensorium track the motions of the perception-inducing objects outside the body, what occurs when the bodily system is out of order? And how can faulty-system percepts be distinguished from others, when both ultimately involve motions transmitted along the nerves by the "spirits"?[91] Perceptions can also be produced by the internal workings of the sensorium itself, by motions that are generated within it by the internal action of the mechanism that Newton, using common terminology, called imagination or fantasy. How can these be distinguished from motions generated by external action? How, in other words, does the judging soul separate motions that are properly produced by extra-body actions from motions that are either internally generated or that are deformed by a damaged or distorted transmitting mechanism? Further, once a motion has been produced in the sensorium by whatever device, how does memory work? How do the motions that allowed the soul once to judge produce their effects in memory, allowing the judgments to be recreated? And why do not memories lead to exact reproductions of the judgments engendered by the original motions? Newton did not explicitly set down each of these questions, but they are quite apparent throughout his several remarks, scattered across the *Questiones*. To find anwers Newton almost always concentrated on the exceptional, the abnormal.

Oddities of memory were the first among these several issues to engage Newton. He found several in More's *Immortality of the Soul*. "Messalina Corvinus forgot his own name. One by a blow with a stone, forgot all his learning. Another, by a fall from a horse, forgot his mother's name and kinfolk. A young student of Montpellier, by a wound, lost his memory." Newton compared these defects of wounded memory with the ways in which forgotten things can be recalled, as "by meeting with other things of like nature, as dreams never thought upon in the morning at the time of awaking are remembered by some actions of the like nature met with in the day time." Conversely, "forgetfulness" occurs "sometimes out of the want of thinking of things." Thinking about events past—"meditation," Newton called it, perhaps himself thinking of Descartes' *Meditations*—"remind a man of actions, and actions of meditations." But how does memory work, and—to Newton an inevitably related question—where does it work? Memory, he suggested, may perhaps be "done by characters in the brain," but that cannot be all there is to it since "the soul remembers too." What is it that the soul recalls? "She must remember," he avers, "those characters" produced by the originating motions in the bodily sensorium in order to regenerate a previous judgment.[92]

[91] The terminology and basic conceptual structure that Newton deployed are set out clearly in Descartes' *Dioptrique*, 4th Discourse, "On the senses in general," where we read that the tubes which structure the nerves are filled with "animal spirits, which are like a very subtle wind or air which, coming from the chambers or concavities in the brain flows away ... throughout the muscles" (Descartes, 1965, p. 88). These "spirits" were for Descartes decidedly material entities, whereas Digby demurred on this point. Newton did not read French, but the optics was translated into Latin as the *Dioptrice* and included in the 1656 edition of Descartes' works which he used: see Shapiro, 1990, p. 7, n.19.

[92] Ibid., pp. 393–95.

The nature of the soul was bound to the question of memory. If the soul were "nothing but modified matter" then memory would have to "consist in action," of motions in the material soul. But this cannot be—though Hobbes thought otherwise—because once the action ceased, so too would the "phantasm," that is, the perception associated with the motion, because we have "no principle within us to begin such a motion again." Insisting once again on the connection between memory and the soul, Newton, echoing More's critique of Hobbes on this point, wondered how the soul remembers "the signification of those characters" previously generated by motion in the sensorium. He had no immediate answer, and so he turned to a prior but related question, namely, how the soul judges—one might say constructs—an object once given its "image in the brain," which is to say the effect in the physical sensorium that is produced there by the motions which flow through the fine substance ("spirits," of which more below) in the nerve channels.

The images of objects are not only located in the physical sensorium; they are sent there from motions created by the object in the sensing organs proper. There is accordingly a mapping established between a physical effect in the sensorium—the image—and a corresponding stimulus, or motion, in the sense organ itself. The judging soul must somehow deduce from the image the original object's properties. Moreover, the soul must also be able to interpret images in such a way that the corresponding objects are placed outside the body itself and not within it, although the images are physically located in the bodily contained sensorium. Posed in this manner, the question of how the soul properly recreates the world took Newton in thoroughly novel directions, even though for the case of vision the issue dates at least to Kepler and was discussed at some length by Descartes in his *Dioptrice*. Newton clearly studied the latter quite carefully,[93] in particular Descartes' fourth through sixth discourses, which respectively consider "the senses in general," "the images that form on the back of the eye," and "vision."

The question, Descartes remarked, is to understand how the "images" (the physical brain states that are produced by the motions transmitted by the animal spirits which fill the nerves) can provide the means by which the soul senses "all the diverse qualities of the objects to which [the images] refer."[94] The sense organs nevertheless cannot inerrantly transmit the motions produced by objects in the external world because no instrument can be physically perfect. The eye provides an apposite example, for it cannot flawlessly reassemble at a point of the retina all of the rays which diverge from a luminous point in the visual field. Nor is this the only "defect" in the "picture" formed on the retina, for among others the "picture" has its parts "reversed," "in a position quite the opposite to that of the objects."[95] Moreover, the senses can be distorted by deliberate manipulation; if for example one eye is turned forcefully to the side, the other being left in its proper state, then the object "will seem to be double" because of the asymmetric effect on the nerves leading from each eye to the brain.[96] The "faults of the eye" with which Descartes was primarily concerned involve its fail-

[93] There is no known evidence that Newton ever read either of Kepler's optical works.

[94] Descartes, 1965, p. 90.

[95] Ibid., p. 97.

[96] Ibid., p. 108.

ure to produce a proper image on the retina, faults "which consist in our inability to change sufficiently the shape of the crystalline humor or the size of the pupil." Some of these faults can be remedied by "practice," Descartes thought, because they are caused by problems with the muscles that control the lens and pupil, and muscles can be trained. Descartes assigned these sorts of remedies "to medicine, whose purpose is to remedy the deficiencies of sight through the correction of natural organs." He was himself concerned only with remedies by means of "Optics, whose purpose is only to minister the same deficiencies through the application of other organs that are artificial." Descartes treated the eye, and by extension every sense organ, as a device whose output, as it were, could be rendered less imperfect only by constructing compensatory instruments. Like any faulty mechanism, the senses produced untrustworthy results, but these could be rendered dependable by means of artfully designed apparatus. Hooke, and indeed every other natural philosopher of the second half of the seventeenth century who worked with apparatus, had essentially the same opinion and were little interested in sensory "faults" precisely because they considered such things to be instrumental imperfections that required external means to remedy.

To judge from his notebook entries, the young Newton was unusually interested in precisely those problems that concern the reliability of the "natural organs" themselves. An entry labeled "Of sensation," based on More's *Immortality*, began with the remark that "the senses of diverse men are diversely affected by the same objects according to the diversity of their constitution."[97] This raised the question of just where in the body the "common sensorium" might be located. Newton listed ten possibilities, among them "the orifice of the stomach" as well as the heart and the brain. But the brain seemed particularly important because experiments on frogs indicate that the frog "loses both sense and motion" when its brain is pierced, but "it will leap and have sense, though its bowels be taken out." Moreover, the "least weight upon a man's brain when he is trepanned makes him wholly devoid of sensation and motion." This interest of the young Newton's in the sensory effects of the physical manipulation of the "common sensorium" evolved into the actual probing of his own perceptual system as the notebook progressed. And the problem of the senses was inevitably connected to the problem of measurement, because measurements are produced by a particular kind of object, namely instruments, which must in turn be perceived through the senses. Here we spy just where Newton was beginning to diverge from his contemporaries.

Boyle was not overly concerned with measurement, largely because he remained skeptical about the general applicability of any numerical relationship. Though Huygens thought all knowledge to be only probable, he aimed to probe or to produce material structures that exemplified mathematical relations. Hooke was obsessed with devices, and he thought to achieve as perfect a result as possible through their perfection. Boyle, Hooke, and Huygens shared skepticism about the trustworthiness of the senses. For these natural philosophers, the way to overcome the problem was to improve the observer's instruments and his experimental or observational skills, to improve, in other words, the externalities of the experimenter's world. Yet Newton's

[97] McGuire and Tamny, 1983, pp. 383–85.

Questiones show no concern at all with skilled measurement, or for that matter with instruments per se. Time and again in the notebook Newton turned to the senses, but not to the problem of how to replace these blurred and shifty windows through which the "common sensorium" generates knowledge of objects. He sought instead to understand just how the senses—in particular the visual apparatus—produced misleading knowledge of objects.

Newton's interest in vision might seem to be akin to Hooke's, for in the mid-1660s Hooke was interested in determining the limitations of human visual acuity, and he examined members of the Royal Society to find those limits. Yet the difference between them runs deeper than their common concern with vision. In measuring visual acuity Hooke aimed to demonstrate how utterly pointless it was to improve naked-eye telescopes beyond a certain point, because human vision could only go so far. Uninterested in the specifics of visual behavior, focused only on its limitations, Hooke sought to replace the naked eye with the micrometer-endowed telescope. For him, sensual limitation led inevitably—and (given that his career orbited about devices) happily—to the production of new and better instruments.

Newton had different concerns in mind, for he did not take the limitations of vision as givens. He literally probed vision by manipulating the eye in order to see just what the effect would be on the knowledge of objects generated by vision. In a section of the notebook entitled "imagination and fantasy and invention," dated by the editors to mid-1664, Newton began with fantasies generated by the "boiling blood of youth" or by the "dry" brain of old age and then turned to the sun's action upon vision. And here, it seems, Newton decided to experiment on himself. "When I had looked upon the Sun all light colored bodies appeared red," he recorded, "and dark colored bodies appeared blue."[98] He listed two more effects: white paper viewed directly now seemed red, but the sun, "through a very little hole," "looked green," and "after the motion of the spirits in my eye were almost decayed, so that I could see all things with their natural colors, I shut it, and could see no color or image till I heightened my fantasy of seeing the Sun."[99]

By late December, Newton was also making his first astronomical observations, for a comet had appeared early that month. Hooke and Wren observed it, as did others. Newton must have heard of the comet's appearance soon after it first became visible, because on Saturday, December 10 (20) "by a subtle observation" he "found the distance of a comet from the center of the Moon to be 9°48′. Its altitude 3°40′ or 4°. The moon's altitude 8°40′." He crossed out the next sentence, which had placed the comet in "Capricorn 26°20′ or else Aquarius 5°."[100] The lunar altitude gives the time of the observation as 5:20 PM, looking southwest, the sun having set that day at 3:49. The observation had to be "subtle" because the moon was in crescent phase at 9°, so that Newton had somehow to locate the center from the arc of its limb. The comet, how-

[98] The passage is curiously close in phrasing to a description in Artistotle's *De Somniis*, which reads: "if, after having looked at the sun or some other brilliant object, we close the eyes, then, if we watch carefully, it [the sun] appears in a right line with the direction of vision (whatever this may be), at first in its own color; then it changes to crimson, next to purple, until it becomes black and disappears" (I, 459b). Newton likely read this passage and decided to try for himself.

[99] McGuire and Tamny, 1983, p. 397: f109r.

[100] Ibid., p. 357.

Figure 1.1. The evening sky over Cambridge at 5:20 PM on Dec. 10(20), 1664. Image produced using TheSky, Astronomy Software Version 6, copyright 1984–2003, Software Bisque.

ever, was not what Newton saw, for it was in an entirely different part of the sky, over 120° away, and 61° below the horizon, in the constellation Crater.[101] Figure 1.1 depicts the configuration in the evening sky over Cambridge at that time.

The evening's crescent moon cast little light, while both Venus and Jupiter were still brightly visible above the horizon.[102] Twenty minutes earlier Mars had accompanied them, just before setting. Newton was a neophyte observer, and never before having seen a comet, he likely mistook Venus for it.[103] At about 5:03 PM, with Mars setting, Venus was 3°40′ in altitude, which matches Newton's claim. The moon, on the other hand, was about 10° above the horizon, not Newton's 8°40′. The notes that he took listed first the distance between the "comet" and the moon, followed immediately by the "comet" altitude, and only then by the moon's height. Since the entire business of astronomical observation was new for him, it may have taken Newton ten or fifteen

[101] The mistaken observation was noted in McGuire and Tamny, 1985, p. 350, and originally in McGuire and Tamny, 1983, p. 298, though in the latter with the incorrect remark that Capricorn and Aquarius were not visible—corrected in the later publication. The comet at that time had right ascension 11h5m56s and declination −26°0′15″.

[102] Venus at magnitude −3.9, Jupiter at −2.

[103] The respective distances of Venus and Jupiter from the lunar center were 9°58′ and 11°44′, compared to Newton's measurement of 9°48′. Their altitudes, again, respectively, are 1°45′ and 1°32′ (again at 5:20 PM), compared with Newton's 3°40′ or 4°. The planets were, however, close to the horizon, where altitude measurement becomes difficult. The distances are not affected by horizon problems, and so it seems highly likely that Newton mistook Venus for the comet. McGuire and Tamny had suggested M30, but it is first of all over 13° away at an altitude of 6°45′, and in addition very faint (magnitude 6.9), which would have required a telescope, whereas Newton makes no mention of having used one.

minutes to adequately perform the "comet" altitude and distance observations, and only then to measure the moon's height.

Newton soon recognized that he hadn't seen the comet; he found it a week later. His note on the observation is worth quoting in full because it may well be the very first exact one that he ever made. "On Saturday at 30^{min} past 4 of ye clock in ye morning Decemb^{r} 17th 1664," he wrote, "A Comet appeared Whose distance from Sirius was 30^{d}, 0′. from procion 38^{d},45′. There was little or noe difference twixt y^{e} time of its & Sirius his setting, it setting about 2′ after him. Soe y^{t} its Right ascension was about 126^{d}, 32′. & its declinacon soutwhard 31^{d}. The length of its tayle was about 34^{d} or 35^{d} & pointed $^{\text{below}}_{\text{toward}}$ procion or almost to y^{e} horizon North pole cutting y^{e} horizon at an angle of about 35^{d} or 40^{d} & y^{e} Ecliptick at 47^{d}."[104] Both Sirius and the comet were close to the horizon at the time of Newton's observation, their distance apart being about 28°36′. Sirius was slightly lower in the sky than the comet, and its position would not have been easy to determine.[105] Procyon, on the other hand, was far above at an altitude of about 24°, its distance from the comet being 38°37′, or only 8′ less than Newton's measurement. Apparently the newly fledged astronomer had learned within a week how to take accurate distances.

Indeed, he had learned considerably more. Newton's crossed-out note on the previous week's spurious observation ended with the sentence "The longitude of the moon" without any attempt to specify either the moon's coordinates or, therefore, the comet's. Either one would have been difficult—the moon's because determining precise lunar coordinates was difficult, the most accurate theory being that of Jeremiah Horrocks, which however was available only a decade later, in Horrocks' *Opera posthuma* of 1673 (Newton had it privately in 1672) . Even given lunar coordinates, Newton would have needed to observe the direction as well as the magnitude of its distance from the comet, in order to compute the comet's location. Direction measurements of that sort were out of the question. Instead, within a week Newton had not only improved his observational technique, he had also decided to measure distances to two, not one, stars. From these, given the known coordinates of the two reference objects, the comet's location could also be computed. The computation is not difficult, given reasonable knowledge of spherical trigonometry, which Newton had clearly developed by this time.[106] Further, he knew how to both work with right ascension, declination and longitude, latitude, and to transform between them. The comet's right ascension (in degrees) and declination at the time of observation were, respectively and to the nearest minute, 125°5′, −31°2′, where the difference between the comet's true right ascension and Newton's value reflects the comparatively large error in his measurement of its distance to Sirius.

We do not know with certainty the source from which Newton learned so quickly how to work spherical trigonometry, though he might well have developed the ability himself after a cursory glance at some of the possible sources. Newton owned at least five at his death, including Gunter's manual for using a sextant, quadrant, or cross-

[104] McGuire and Tamny, 1983, p. 356.

[105] Sirius was actually *below* the horizon by about 1°23′ and was visible only because of atmospheric refraction. It would set nine minutes later. The comet would have been about 45′ above the horizon.

[106] See below, note 109.

staff, Wing's "succintly handled ... Trigonometricall Part" of astronomy, and Ward's *Trigonometriae*.[107] The relations of spherical trigonometry that he found certainly contain all that Newton needed to compute the comet's location, but astronomical calculations require standard moves that are not usually specified explicitly in the available texts. Rather, they are exemplified through the solutions to typical problems. Observation provided Newton with two legs of a spherical triangle;[108] the third had to be computed from the coordinates of his reference stars. This would have been easy to do. The next step was not quite so simple, because the star-comet-star triangle cannot, by itself, yield the comet's coordinates. To find these requires inscribing other triangles on the stellar sphere in a manner available in various contemporary sources. Wing in particular provided one that shows what must be done in a situation like Newton's: namely, to inscribe three legs running from the pole of the ecliptic and respectively through each of the comet and two stars to the ecliptic itself. Relations among these then yield the comet's locus.[109] By reconstructing Newton's computation, and comparing as well his observations of the comet's positions with values computed using the comet's orbital parameters,[110] we find that his observational accuracy was between 10′ and 30′, and his computations about half a minute or so.[111] Figure 1.2 (top) maps the comet's actual position for most of Newton's observations; at this scale his results are virtually indistinguishable from the true loci.[112]

[107] Gunter, 1662; Norwood, 1645; Oughtred, 1657a; Ward, 1654; Wing, 1651. Streete, 1661, on which Newton did take extensive notes in a separate section of the *Questiones* notebook, assumes a very good familiarity with spherical trigonometry. Newton's copy of Wing's *Harmonicon* has marginalia, and we shall see that he was almost certainly using it at just this time. There is also good evidence that Newton was "deeply read" in Oughtred in 1665–1666, which provides a solid account of spherical trigonometry (Whiteside, 1967–1981, vol. 1 pp. 5–6, 466).

[108] That is, the comet's distances to each of the two stars.

[109] See appendix B.ii for the procedure using Newton's Dec. 27 observation. Wing's chapter 16 teaches how to compute the longitude and latitude of a star given its meridian altitude and its distance from a known star. The procedure is of course different from what Newton needed, in that he had the distance to a second known star and not the unknown's meridian altitude (which, given the latitude, at once gives the unknown star's declination). Nevertheless Wing's example provides an entry into this type of problem.

[110] The comet's orbital parameters were first calculated by Halley and published in 1705 (Halley, 1704–1705, p. 1886). He found the following: ascending node = ♊21.14.0; orbital inclination = 21.18.3; on Nov. 24 (OS-Dec 4) at hour 11:52 = .4944UT, perihelion = ♌10.41.25; distance to sun at perihelion = 1.025755 (in mean radii of the earth-sun distance). Results obtained by Lindelhof and updated to epoch 2000, yield for Dec. 4, 1664 (in terms of standard contemporary parameters): angle of perihelion = 310.6748; ascending node = 86.0525; i = 158.6965; q = 1.025755. The comet approached closest to the sun in its parabolic orbit on Dec. 29 at .1699 AU distance. Lindelhof's parameters were provided by Brian Marsden (Marsden, 1972) and updated to epoch 2000 by Gary Kronk (Kronk, 2006), whose results we have used.

[111] Newton's calculations had to involve five sequential uses of the spherical cosine formula for a leg, or an equivalent, combined with look-ups in a table for the logged cosines. There was much room here for error, and in any case multiple approximations due to the use of tables would accumulate unless Newton went so far as to interpolate, which is not likely.

[112] The annotated Bayer in figure 1.2 (bottom) was acquired as a gift in the 1930s by the California Institute of Technology. It had at one time belonged to the Radcliffe Observatory at Oxford. This comet generated considerable discussion at the time, primarily because Hevelius' location for it on Feb. 18 generated a controversy given Auzout's observations for Feb. 17 and 19 (the Parisian sky was obscured on the 19th),

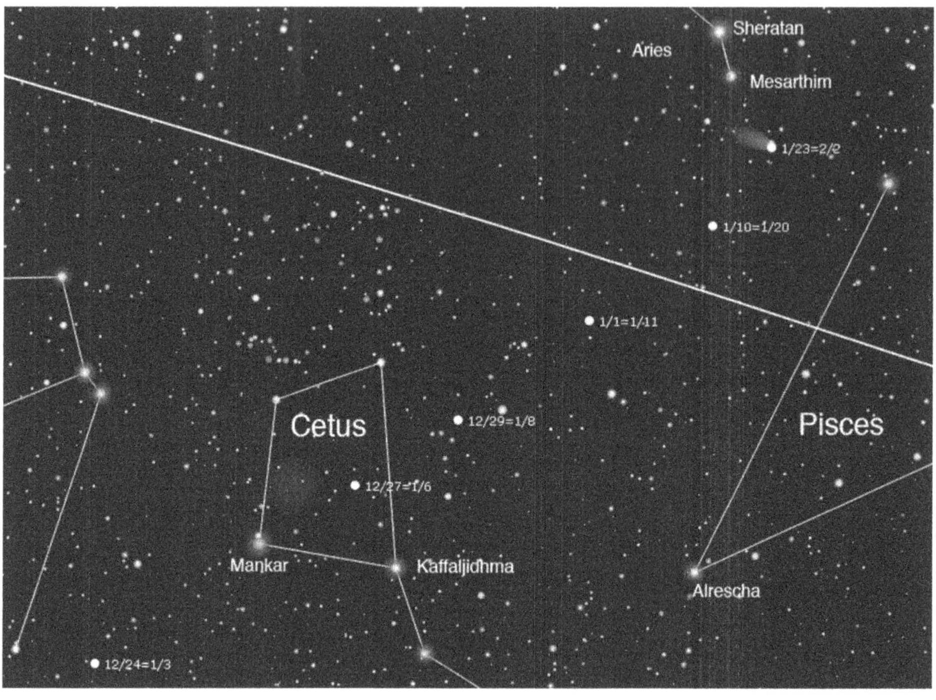

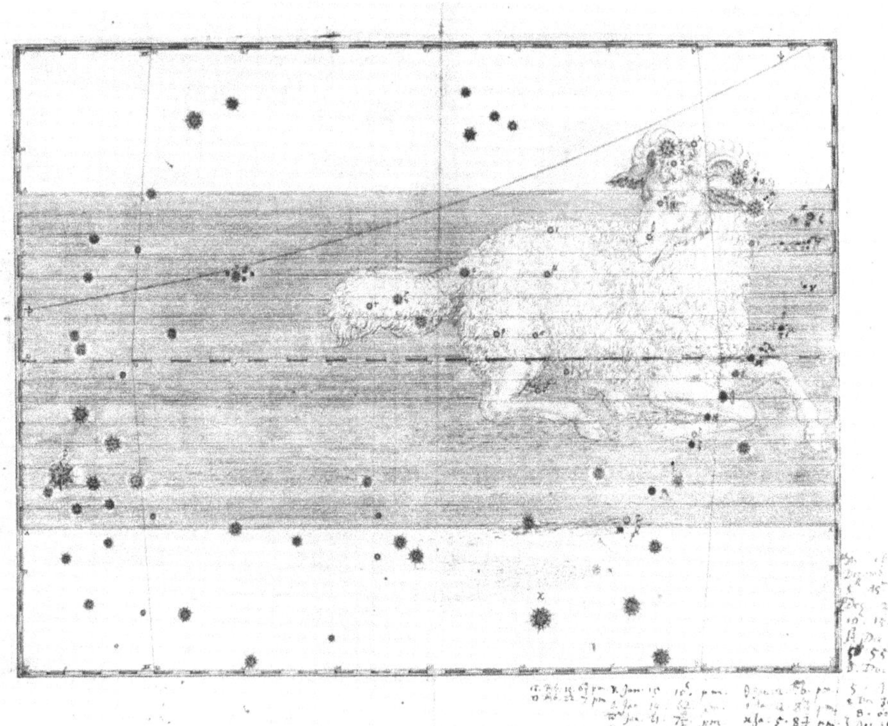

Figure 1.2. The comet of 1664/1665 at times of several of Newton's observations (top), and, accompanied by dates of observations, as drawn in by an unknown contemporary hand on a 1603 Bayer. Image produced using TheSky, Astronomy Software Version 6, copyright 1984–2003, Software Bisque.

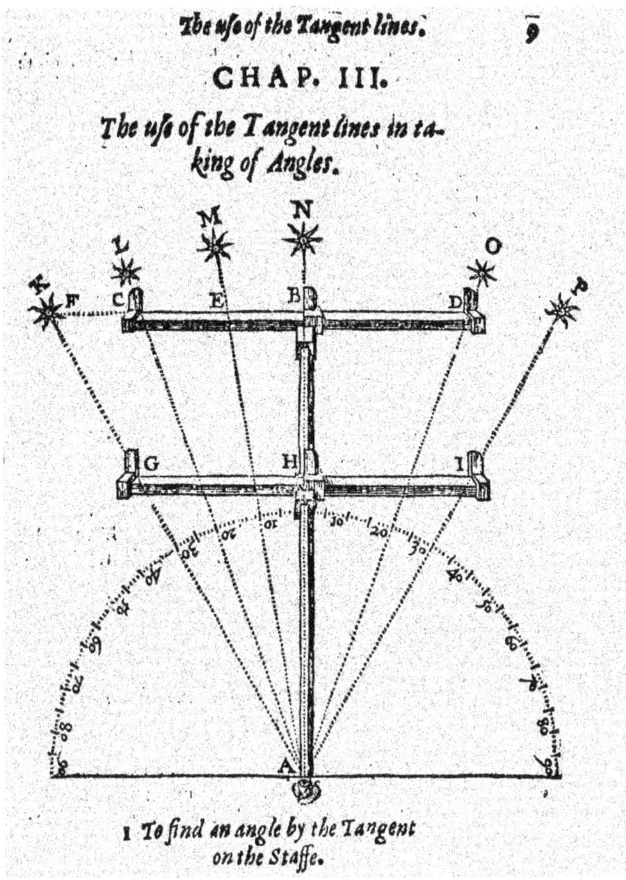
The use of the Tangent lines. 9

CHAP. III.

The use of the Tangent lines in taking of Angles.

1 To find an angle by the Tangent on the Staffe.

Figure 1.3. The cross-staff, as depicted in Gunter, 1662, p. 9, chap. 3: "The first book of the crosse-staffe."

By the end of 1664, then, Newton had begun to learn the difficulties of astronomical measurements. He had also acquired experience with sequences of computations based on them. Although we do not know what instrument Newton used for his observations, it's possible, and likely, that he deployed a cross-staff (figure 1.3). Gunter described the cross-staff's operation in detail and included an engraving of its use for angle measurement—indeed, specifically for angles between stars. The staff consisted of two pieces. One (AB) was in Gunter's description a yard in length; the other, a crosspiece (GI or CD), 26½ inches long, was attached to the first at its middle and could slide along it. Moveable sights were fastened at the ends of the crosspiece and a fixed one at its center. To find the distance between a pair of stars, K and P say, the end

which would have required the comet to deviate and then turn back, something deemed unlikely at the time, even given a widespread view that comets were subject to all sorts of perturbations. On the controversy see Shapin, 1995, pp. 266–91.

A was brought to the eye, and the crosspiece moved along AB until K and P were sighted through the outer-sights G and I or through G and the central sight H. The angle between the stars could then be read directly off a scale marked on AB. Alternatively, for small angles the crosspiece would be fixed in one of two marked positions along AB and the angle found by sliding an outer-sight along it.

Given Gunter's dimensions, the minimum angle that could be observed by sliding the crosspiece with an outer-sight fastened at its end would be about 20.2°. The minimum angle using both outer-sights would therefore be 40.4°. Gunter specified drawing a line along AB and marking it to the left with divisions running from 20° (at B) through 90° (at A), and to the right running from 40° through 180°. The divisions corresponded to the tangents of the measured angles, so that the smaller the angle the more exact the measurement.[113] For a crosspiece of half-length equal to Gunter's (13.25 inches), a one-minute change in an angle measurement of 20° would produce a shift along AB of about one millimeter, while a minute's change at 38° would produce a shift of about a quarter millimeter. Newton's angles did not exceed 38° (meaning that he would have to have used the central sight), so it's hardly surprising to find him giving results to the nearest minute—though this would have been his accuracy in reading the instrument's scale, not in setting the crosspiece to sight the comet and star. He also measured angles smaller than 6°, for which he would have had to use a sliding outer-sight and a scale marked on the crosspiece itself. Newton may well have fabricated the device for himself, since the inhabitants of Grantham and Woolsthorpe vividly recalled his dexterity in building models of all sorts, including windmills and carts.[114]

Within months, Newton had not only begun to think about issues concerning sense perception; he had also gained experience in taking measurements by sight. These early astronomical observations were probably the first ever undertaken by Newton with the aim of submitting them to computation. Moreover, the comet observations had provided him with experience in the limits of vision both in measurement and in the proper distinguishing of one object from another (as in his early mistaking of Venus for the comet). At about the same time he also began to probe the nature of visual perception itself: to manipulate his eye to find what perturbing it would do to sight. Hooke never did anything like this, and neither Boyle nor Huygens were interested in examining perception itself. Newton alone among his contemporaries decided to undertake what can only be described as intervening experiments in human visual perception. Motivated and guided to some extent by the common mechanical-philosophy conviction that objects produce nervous motions that are then judged by the mind or soul, Newton explored the ways in which the motions, once disturbed, effect sensations.

Hooke, and for that matter Boyle, Huygens, and many others reaching back to Bacon, also distrusted the senses, whose limitations they considered to be inherent and essentially invariant. A public controversy that erupted between Hevelius and Hooke in the early 1670s illustrates this particularly well and illuminates by contrast

[113] Since a change $\delta\alpha$ in the angle produces a shift of magnitude 13.25 $\delta\alpha/\sin^2(\alpha)$ along AB using an outer sight fixed at the end of the crosspiece together with the central sight.

[114] Westfall, 1980, pp. 60–63.

Newton's considerably different way of thinking about the confines of perception. Before continuing with Newton, we shall discuss the Hooke-Hevelius controversy, following this with remarks on Huygens' and Boyle's attitudes toward perception and measurement. In that way we will be able to understand with greater clarity how different Newton's emerging attitudes toward perception, and the very process of measurement itself, were from those of his contemporaries. These differences played a decisive role in Newton's eventual treatment of testimony, both present and historical. They are moreover strongly linked to his unique way of working back and forth between data and theory, one that forged the skeletal structure of Newton's attempt to rework ancient chronology.

Troubled Numbers

In 1673, Hevelius published the first part of his *Machinae coelestis*, which provided verbal descriptions and elaborate plates of the naked-eye devices that he had constructed for determining stellar coordinates. The very next year Hooke took umbrage at the publication and attacked it in print in a series of *Animadversions*. The *Animadversion*'s fulsome praise for Hevelius' skill weakly compensated its obvious disdain for the astronomer's methods. For Hooke was not only certain that naked-eye observations could not possibly match those performed with a telescope equipped with cross-hairs, he had through Oldenburg previously urged Hevelius to adopt the new apparatus.[1] Hevelius nevertheless published the *Machinae*, and Hooke was incensed. If Hevelius had only taken the advice, wrote Hooke, "which I long since communicated to him, I am of opinion he would have done himself and the learned World a much greater piece of service, by saving himself more than 1/10 of the charge and trouble, and by publishing a [stellar] Catalogue ten times more accurate."[2]

Hooke had emphasized the necessity of using a telescope with focal-plane sights for highly accurate observations in a 1670 Cutlerian Lecture, where he discussed the zenith device that he had constructed, and whose use to prove that the earth moves forms the immediate background to his swords-drawn attack on Hevelius' observational methods.[3] Despite Hooke's unqualified assertion that Hevelius could not possibly have achieved the accuracy that he claimed, when the young Halley visited Hevelius in 1679 he found otherwise. As he informed Flamsteed, "I assure you I was surpriz'd to see so near an agreement in [distances measured with Hevelius' six-foot sextant], and had I not seen, I could have scarce credited the relation of any. Verily I have seen the same distance repeated severall times without any fallacy agree to 10″, and on Wednesday last I myself tryed what I could doe"—with the convincing results mentioned above.[4]

[1] "Something to this purpose I communicated to *Hevelius* in the year 65. And hoped that I might have thereby somewhat assisted him in his great and laborious Work, first by easing the eye, and next by making it capable of distinguishing more exactly, I having hinted to him the way how to reform and obviate that inconvenience by Telescopical or Perspective Sights" (Hooke, 1674a, p. 5).

[2] The Hooke-Hevelius contretemps has been discussed many times: see, e.g., MacPike, 1937, pp. 82–87, Volkoff, et al., 1971 , Cook, 1997, pp. 93–100, and Bennett, 2003, pp. 85–100. The correspondence between Halley and Flamsteed pertaining to Halley's visit to Danzig is in MacPike, 1932, pp. 42–43.

[3] Hooke, 1674b, pp. 8–9.

[4] Forbes, vol. 1, p. 694.

Historians have at least twice examined these results. In 1971, Volkoff and colleagues computed from modern tables the error in the eight stellar distances that Hevelius had challenged Hooke to produce, and that Flamsteed eventually did. The standard deviation among Hevelius' errors is about half a minute, though in five instances the error lies below 14″. This compares to a standard deviation among Flamsteed's own distances of 18″, which is certainly not as profound an improvement as Hooke would have expected from a micrometer-equipped telescope.[5] Cook in 1998 found that the observations completed over three days during Halley's visit have a standard deviation of only 10″.[6]

These computations are persuasive enough, and a sample of three named stars chosen from Hevelius' list does in fact confirm the claim: his observations were accurate to better than half a minute overall, and often better than half of that.[7] This much would seem not to raise any significant issues beyond implying that Hooke was overly taken with the power of sighted telescopes and too convinced that the eye was necessarily a poor, defective instrument. However, Hooke was in fact correct, or at least his claims were well grounded. Modern studies have shown that uncorrected normal vision—20/20 as measured by the standard Snellen chart—corresponds to a human visual acuity of 1′ of arc, just as Hooke asserted. How then was it possible for Hevelius (and for that matter even Halley, using Hevelius' apparatus and working with him) to do so much better?

To unravel the mystery, and to grasp its significance for issues of perception and measurement around the time that Newton's own attitudes were taking shape, we begin with Hooke's nearly unprecedented experiments in human visual perception, to which he subjected the members of the Royal Society in their meetings during the years after the Great Fire had forced the Society mostly to meet at Henry Howard's Arundel House on the Strand. Inwood evocatively remarks that Hooke had each Thursday to "carry or cart his experimental equipment down Threadneedle Street and Cheapside, around the ruins of old St. Paul's and along the full length of Fleet Street."[8]

Before performing public examinations, Hooke must have experimented on the acuity of his own vision, or perhaps on that of an assistant, because his scheme to "prove the motion of the earth" depended on it. More than any other of Hooke's experiments, his work here illustrates the reciprocity, as he saw it, between the fixed limitations of human perception and the open-ended perfectibility of instruments that are designed, built, and worked by skillful hands.

The project was animated in principle by what Hooke believed to be continued resistance to the Copernican system. "Whether the Earth move or stand still hath been a Problem," he wrote, " that since *Copernicus* revived it, hath much exercised the Wits of our best modern Astronomers and Philosophers, amongst which notwithstanding there hath not been any one who hath found out a certain manifestation either of the one or the other Doctrine." Though the "knowing and judicious" had

[5] Volkoff, et al., 1971, pp. 46–47.

[6] Cook, 1997, p. 96.

[7] The stars are Spica in Virgo and Castor and Pollux in Gemini.

[8] Inwood, 2002, p. 95.

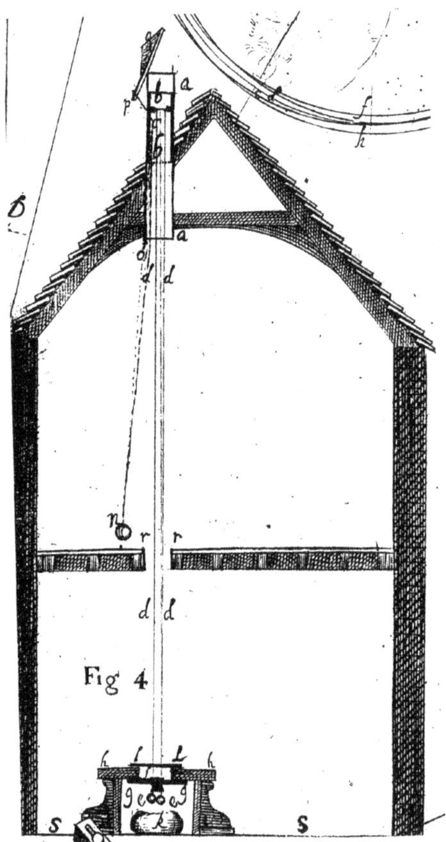

Figure 2.1. Hooke's parallactic telescope. From Hooke, 1674b, pp. 1–2.

"many plausible reasons" for adhering to Copernicus, many others "either out of ignorance or prejudice, have rejected it as a most extravagant opinion." There's little hope to change the opinions of those "who understand not the grounds and principles of Astronomy" since "a man shall as soon perswade them that the Sun doth not shine, as that it doth not move; and as easily move the Earth as make them believe that it do's so already."[9] Hooke aimed rather at the unpersuaded "Learned," and he intended to convince them by means of an "experimentum crucis" which would decide the matter once and for all. To that end he decided to measure stellar parallax, even though everyone agreed that it would be exceedingly small, and indeed beyond even the best observer's capabilities, however finely crafted his instruments—unless, that is, the observer was Hooke himself, and the instruments were built precisely to overcome what he took to be the unavoidable limitations of all earlier devices.

The measure of Hooke's commitment to the enterprise is nicely represented in his diagram (figure 2.1) which shows the holes that he cut in both the roof and the

[9] Hooke, 1674b, pp. 1–2.

second-story floor of his Gresham College residence to accommodate the very long telescope that he felt would be needed. The device incorporated focal-plane sights, because, Hooke was convinced, naked-eye pointing devices, however large, could never achieve the requisite accuracy. Tycho and Riccioli, he remarked, had both asserted that parallax does not exist. Nevertheless, he insisted, "there is no one that can either prove there is, or that there is not any Parallax of that Orb amongst the fixt Stars from the Suppellex [*sic*] of observations yet made either by*Ticho*, *Riccioli*, or any other Writer that I have yet met with from the beginning of writing to this day."[10] How did Hooke know this to be so?

Hooke canvassed past claims, noting first that both Kepler and Riccioli "hypothetically affirm" that any parallax must be less than a minute of arc. Tycho, though "a man of unquestionable truth in his assertions," thought he could construct instruments that would reach ten seconds of arc, while Riccioli and Francesco Grimaldi "affirm it possible to make observations by their way, with the naked edge [of a pointing sight] to the accurateness of five Seconds." Kepler on the other hand didn't think anything better than twelve seconds was possible. To have any hope of providing a convincing measure for parallax (or, for non-Copernicans, placing a satisfyingly small upper limit on it), Hooke had to reach below a minute of arc. Others thought that it could be done with naked-eye sights and sufficiently large quadrants, but Hooke was skeptical on two grounds.

His predecessors were mistaken, Hooke claimed, first "from divers inconveniences their wayes of observations were liable to."[11] Materials shrink and stretch so much in the cold of winter and the heat of summer that "a minute may be mistaken in taking an altitude of fifty Degrees." The heavy weight of the large quadrants and sextants that would be necessary will bend and warp them, and moreover even the usual method of marking the divisions is "liable to many inconveniences" that make it difficult, if not "altogether impossible," accurately to divide to a minute "for the whole Quadrant." Not, however, wishing to be accused of unduly denigrating skills of construction and design, Hooke turned to a second problem that no amount of skill, he thought, could possibly overcome: the inherent limitations of visual acuity.

There "was one inconvenience," he wrote, "which was worse then all the rest, which they [Tycho and Riccioli] seem not to have been sufficiently sensible of, from whence proceeded all their own mistakes, and their imposing upon others, and that was from their opinion that the sight of the naked eye was able to distinguish the parts of the object as minutely as the limb of the Quadrant (of what largeness soever) was capable of Divisions; whereas 'tis hardly possible for any unarmed eye well to distinguish any Angle much smaller than that of a minute: and where two objects are not farther distant then a minute, if they are bright objects, they coaless [*sic*]and appear one."[12] The only solution was "to assist my eyes with a very large and good Telescope, instead of the common sights, whereby I can with ease distinguish the parts of an object to Seconds."

[10] Ibid., p. 4.

[11] Ibid., p. 8.

[12] Ibid., pp. 8–9.

The force, novelty, and even probity of Hooke's attempt to measure parallax depended critically on his claim that vision was limited to near a minute of arc, a value which he must already have ascertained (not have guessed) because we know today that he was quite correct. Or, rather, we know that Hooke was correct in asserting that two objects placed a fixed distance apart cannot be distinguished if they are separated by less than a minute of arc. His claim that only a device of the sort that he had constructed could possibly ever measure parallax, and that therefore all earlier remarks were hopelessly flawed, depended on the inability of the eye to reliably gauge any separation below a minute. Otherwise, doubts that Hooke could not have easily dismissed would persist. But if the limitation was absolute, inherent in the very nature of human vision, then Hooke's device would prove triumphant, demonstrating not only Hooke's skill in construction and design, but his unparalleled understanding of inherent human limitations that only his perfectly designed instruments could possibly overcome.

With so much at stake Hooke could hardly hope to persuade skeptics, including some members of the Royal Society who perhaps thought him addicted to braggadocio. And so, to make his point unbreakable, Hooke decided to use the Fellows themselves as test subjects, thereby inventing what must certainly have been one of the earliest experiments to use living subjects to examine perception itself. Only Galileo preceded Hooke in describing a similar test, and for essentially the same reason: to argue that imperfections of the eye produce adventitious effects that only artificial devices (such as the telescope) can eliminate.[13] Issues surrounding errors produced by the senses, indeed whether properly operating sensory organs could traduce nature—and what effects a device like the telescope has in this respect—had for decades been central elements in the construction of mechanical philosophies. We will turn to these issues in the next chapter, for they captured the young Isaac Newton's attention in markedly novel ways.

The first public test of Hooke's claim took place at the usual Thursday meeting of the Society on January 15, 1674.[14] Although the meeting was "thin," the experiment, Hooke enthused, "succeeded beyond hope." According to Oldenburg's minutes, the test, "with a ruler divided into such parts, as being placed at a certain distance from the eye, appeared to subtend a minute of a degree; and being earnestly and curiously viewed by all the persons present, it appeared, that not any one present, being placed at the assigned distance, was able to distinguish those parts, which appeared of the bigness of a minute, but that they appeared confused." Hooke had been very clear indeed about his purpose, which was to show (Oldenburg wrote) "that we cannot by the naked eye make any astronomical or other observation to a greater exactness

[13] Galileo argued that the fringing radiation around stars which seems to amplify their size and which hides the Milky Way's myriads of stars in an undifferentiated glow is due to imperfections of the eye. In the *Dialogue* he described a procedure for measuring the unfringed (and so, he asserted, optically correct) apparent size of a star by means of a string which, just covering the star, also eliminates the fringing that the star's bright light against a dark background causes the eye to produce (Galileo, 1632 (1962), pp. 361–62): cf. Brown, 1985.

[14] The very day on which, Hooke recorded in his diary, he had "bought of Oldenburg, Hevelii *Machina Coelestis*, and paid him presently 18 sh." Robinson, 1968, p. 80.

than that of a minute," with the unavoidable consequence "he alledged, that whatever curiosity was used to make the divisions of an instrument more nice, was of no use, unless the eye were assisted by other helps from optic glasses."[15]

Hooke's test did not immediately convince everyone present at the "thin" meeting, or at least some doubts remained, which is hardly surprising considering the claimed consequence that there was no hope of improving traditional methods by more finely dividing a quadrant or sextant, or by making them larger. The old and well-trodden path to instrumental improvement ended, Hooke argued, in the swamp of imperfect vision. And so to ensure conviction Hooke returned to the test at the next meeting, and this time Oldenburg unambiguously stated that the experiment "proved what it was designed for."[16] Hooke, who that very day also "read part of Lecture about Hevelius," noted triumphantly in his diary that he "shewd Experiment about Sight. Convinced all."[17]

Hooke must have been very convincing indeed, and he no doubt emphasized his points in lecturing on Hevelius, because he succeeded as well in gaining support for "the making of a new kind of astronomical instrument of his own invention for the taking of hights, angles, and distances, of celestial bodies by one observation more exactly than was ever done, viz. to a second."[18] An instrument that would have the most useful advantage of being so excellent that it "might, upon occasion, be duly ordered by the astronomical observer, so as not to rely upon the credit or skill of the instrument-maker." The instrument would not be overly expensive compared to its less-exact predecessors (less than ten pounds, Hooke told the meeting), and he happily recorded that "Lord Brouncker ordered making a quadrant after my way."[19]

Before discussing the implications of Hooke's insistence on the advantage to be gained by avoiding the "credit or skill of the instrument maker," we need to consider the human test that he performed. Oldenburg's description provides its basic structure, which Hooke explained in greater detail in his attack on Hevelius' *Machinae Coelestis*.[20] Hevelius had not been the target of Hooke's "An Attempt to prove the motion of the earth," which had been delivered as a lecture four years earlier, because Hevelius had not as yet published anything about methods for stellar observations, while Hooke still hoped that Hevelius would take his advice to abandon the naked eye for the sighted telescope. Only when he found otherwise in the *Machinae Coelestis*—with its elaborate and finely detailed plates showing just the sort of instruments that Hooke was attempting to make obsolete—did he embark on the warpath. "Who is there," he wrote, "that by his bare eye can distinguish any of the Telescopical spots in the Moon, though some of them are above a minute in Diameter? As for instance, Who can see Mount Sinai, so call'd by Hevelius, which is a bright spot in a dark field, and consequently must appear near two minutes in Diameter to the naked eye? Or

[15] Birch, 1968, vol. 3, p. 120.
[16] Ibid., p. 121.
[17] Robinson, 1968, p. 82.
[18] Birch, 1968, vol. 3, p. 121.
[19] Robinson, 1968, p. 82.
[20] Hooke, 1674a, p. 8.

Figure 2.2. Hooke's test of visual acuity. From Hooke, 1674a, preceding p. 1.

who can see the Palus Mareotis or the Lacus niger, which are two dark spots in light fields, and each more than a minute in Diameter?"[21] To avoid any possibility of doubt, he carefully described his way of proving the eye's failure

Hooke took a "sheet of white paper" (figure 2.2) and drew on it a pair of parallel lines OO, PP separated by four or five inches. Between and perpendicular to these he drew a set of lines; he then blackened the region to the left of the first line (counting from the left in the figure), skipped the region between lines one and two, blackened the space between two and three, and so forth, thereby producing a series of alternating white and black stripes. Hooke did not specify in print the distances between the stripes, but he required that the paper be set "against a Wall open to the light, and if it may be so that the Sun may shine on it," after which the test subject,

> removing himself backwards for the space of 287⅓ feet, let him try whether he can distinguish it, and number the dark and light spaces, and if his eyes be so good that he can, then let him still go further backwards and backwards from the same, till he finds his eyes unable any longer to distinguish those Divisions, there let him make a stand, and measure the distance from his eye to the aforesaid Paper, and try by calculation under what Angle each of those black and white spaces appears to his eye, for by that means it will be manifest how small an Angle his eye is capable of distinguishing, and beyond which it cannot reach: Which being once known, he hath a Standard, by which he is able to limit the bigness and exactness of his Instruments, if he make use of common Sights, beyond which all magnitude and curiosity is not only useless, but of much detriment upon many accounts.[22]

The initial distance Hooke specified corresponds to one minute of arc if the stripes are set one inch apart, and so Hooke has specified in effect a test for 20/20 vision or better.

Hooke acknowledged that the rarest of individuals might have vision good enough to distinguish half a minute, but Hevelius had claimed even higher accuracy. Hooke was particularly intent on proving his point because Hevelius had dismissed Hooke's advice concerning the use of sighted telescopes, in words that Hooke knew to be directed at him: "You may perceive by comparing this slender Refutation with his Letter before [sent in 1665 in reply to Hooke's early advice], who he means by the

[21] Ibid.

[22] Ibid., pp. 8–9.

Assertor of Telescopical Sights."[23] According to Hooke, only if Hevelius possessed eagle eyes could he possibly achieve half a minute, and that only on the rarest of occasions. And yet, Hevelius regularly reached half a minute, and occasionally more than that. To understand what happened, and why it is significant for Hooke's distinctions between sensory apparatus and built instruments, we must return to his white-striped test paper.

Hooke thought that visual acuity might depend upon whether a bright object was seen on a dark background, or a dark object on a white background, which is why his test paper consisted of alternating white and black stripes—he apparently thought to combine both possibilities into a single pattern. However, he did not think to investigate whether his experimental configuration well matched naked-eye stellar observations. Hooke had determined perfectly well that human visual perception could not reliably distinguish two objects if the distance between them is much less than a minute. But did that correspond to what Hevelius (or others before him, such as Tycho) were doing?

It did not. In Hevelius' naked-eye observations, a bright spot (the star) moved across a dark field to pass close to the tip of a slightly less dark object—the pointer. Two characteristics at once distinguish this from Hooke's test: first, motion—not static separation—is involved, and second, Hevelius had to perceive the *offset* between two objects, not the distance between them. Modern experiments have shown that under these circumstances human visual perception becomes *hyperacute* and is capable of distinguishing extraordinarily small offsets, normally reaching five seconds of arc.[24] A nineteenth-century translator of sections of the *Pars Prior* remarked with some justice that by "constant practice ... Hevelius had acquired such wonderful skill and confidence in his eye and hand, that remonstrance and opposition to his method were alike unheeded by him."[25]

That was precisely the problem, as Hooke saw it, because skill "with eye and hand" were just what he aimed to render obsolete through improved instrumentation. Perhaps more than any other natural philosopher at the time, Hooke knew intimately the delicate and hard-won abilities that craftsmen needed to fabricate instruments,

[23] Ibid., p. 7.

[24] Geisler, 1984, p. 775, who remarks that "observers can reliably detect an 8–10-sec instantaneous displacement of a short line or bar, a 2–4-sec misalignment in a vernier-acuity task, a 6-sec difference in the separation of two lines, and a 2–4-sec difference in the separation of lines in a stereoacuity task." Further, experiments with the separation of point sources have shown that "human observers perform as well in this task as in the vernier tasks" (p. 778), that anything over a 5-second separation can be detected in the two-point case (a situation that corresponds reasonably well to Hevelius' observation, treating the tip of the pointer as itself a point).

Moreover, a study of Tycho's observations of the star α Ursa Minor (i.e., Polaris) indicates that he was capable of repeating these observations to an accuracy of 13″ or even 8″ for those which he himself considered to be the best of all (Maeyama, 2002). There seems in addition to have been a systematic error in the setting of the meridian plane of about a third of a degree, which translates into a coordinate error for α Ursa Minor of about half a minute. Tycho himself apparently claimed to have achieved an extraordinary accuracy of about 5″ in measuring the star's altitude (ibid., p. 112), which is not out of line with what he does seem to have generated.

[25] Prince, 1882, p. 72.

abilities that also were necessary for the operation of expensive artisanal objects. Hooke sought to avoid expense and to obviate the inevitable limitations of human perception. Highly skilled himself, and on familiar and friendly terms with the artisanal world, Hooke nevertheless sought profoundly to alter the relationship between instrument maker and instrument user by avoiding "trouble, clutter, and Apparatus."[26] In "animadverting" against Hevelius, Hooke showed "how to make the Plain of the Instrument, that it shall not be subject to bending or warping, and yet be so light as to be easily manageable.... How to make the Divisions on that Instrument, so as to distinguish certainly and exactly to Seconds, without any trouble, or wearying the sight ... how to make the Sights of that Instrument, so as to distinguish the parts of the Object to Seconds, if need be, even by those who cannot distinguish to Minutes with common Sights, certainly, and without fallacy or error ... How to make the Sights, so as to see two Objects, though never so far distant, with one glance of the eye."[27] Hevelius' instruments gave him a perfect target, for "though they be never so large, never so accurately divided, of never so choice and convenient materials, and never so tractable for use, and never so skillfully and industriously used," must fail because of the "limited power of distinguishing by the naked eye."[28]

Hooke concluded quite early in his career that human senses are untrustworthy. "As for the actions of our Senses," he noted in the preface to the *Micrographia* (1665), "we cannot but observe them to be in many particulars much outdone by those of other Creatures, and when at best, to be far short of the perfection they seem capable of: And these infirmities of the Senses arise from a double cause, either from the disproportion of the Object to the Organ ... or else from error in the Perception."[29] To remedy sensational infirmity Hooke urged "the adding of artificial Organs to the natural"—and not only telescopes and microscopes. For "as Glasses have highly promoted our seeing, so 'tis not improbable, but that there may be found many Mechanical Inventions to improve our other Senses, of hearing, smelling, tasting, touching." But instruments can only be perfected by recognizing that "the Arts of life have been too long imprison'd in the dark shops of Mechanicks themselves, & there hindered from growth, either by ignorance, or self-interest."[30]

Hooke's attitude toward "Mechanicks"—despite his constant frequenting of their workshops—was hardly unique. Huygens, too, had vexed relationships with artisans and, in his case, issues concerning the probability of instruments-based knowledge interacted with concerns over social probity and ownership of knowledge. In a letter to Pierre Perrault in 1673, for example, Huygens had this to say about the reliability of natural knowledge and how to produce it:

> In the matter of physics there are no certain demonstrations, and one can know causes only from effects by making suppositions founded on some

[26] Hooke, 1674a, p. 10, referring to Tycho and Hevelius' large instruments.

[27] Ibid.

[28] Ibid., p. 5.

[29] Hooke, 1665, Preface.

[30] Ironically, in view of his later troubles in extracting promised compensation, here Hooke praised Sir John Cutler, founder of the Cutlerian lectures, whose avowed aim was to have Hooke lecture on these "Arts," rather than on the manner in which improved Arts can produce new knowledge.

> known experiments or phenomena and then testing if other effects agree with these same suppositions ...
>
> Nevertheless, this lack of demonstration in physical things must not cause us to conclude that everything is equally uncertain, but it is necessary to take account of the degree of probability that is found according to the number of experiments that concur to confirm for us what we have supposed.[31]

Two decades later Huygens amplified his views in the preface to his *Cosmotheoros* (1698):

> All the same, the things more hidden which we have principally in mind here are not, we know, of such a kind that they could be perfectly brought to light by our efforts. Consequently we do not here advance anything with complete conviction (how could we?), we are content with conjectures concerning whose likelihood each is free to judge for himself. If someone were to say that we trouble in vain and uselessly in proposing conjectures about things which we ourselves admit cannot be comprehended with certainty, I would answer that the entire study of Physics, to the extent that it is concerned with finding the causes of phenomena would be subject to disapprobation for the same reason, the greatest glory being to have found likely theories; the quest itself, as much of principal subjects as of things hidden, constitutes its charm. But there are many degrees of likelihood of which some are nearer the truth than others; it is above all in the evaluation of these degrees that one must show good sense.[32]

For Huygens (as for all mechanical philosophers, including the young Newton) the world can only provide "effects," the hidden "causes" of which cannot be accessed directly. Huygens had in mind mechanical causes which, he believed, underlay all physical phenomena, but his notion of "cause" embraces also a wave front in optics, for it too can only be accessed through its effects. In fact, Huygens did not clearly distinguish between a wave front as "cause" of optical appearances and ethereal matter as "reason" for the front's very existence.[33] Nevertheless, he went beyond the common practice of the day when he insisted that causal "suppositions" must be "tested" against other "effects," which had not been used in suggesting the original conjecture.

According to Huygens' methodological remarks, after first setting down suppositional causes in connection with some given set of effects, one should proceed to generate novel effects from the causes and then test to see whether they occur. Huygens

[31] Translated in Shapiro, 1989, p. 226 from Huygens, 1888–1950, vol. 7, pp. 298, 300.

[32] Huygens, 1888–1950, vol. 21, p. 688.

[33] At the very beginning of the *Traité* Huygens remarked that he would "try to give, by means of received principles in contemporary philosophy, more clear and more probable reasons, first for these properties of directly-spread light; second for that which reflects in meeting other bodies ... Then I will examine the causes of the strange refraction of a certain crystal brought from Iceland" (Huygens, 1888–1950, vol. 19, pp. 459–61).

did just this in working with Iceland spar, for he conjectured the existence of a prolate wave front and then deduced several of its implications for the paths of rays within the crystal. Towards the end of the eighteenth century, the novelty-generating stage of such a process usually involved a quantitative comparison between the implications of the cause and the observed effects. And yet we find very little by way of numerical comparison in Huygens' published *Traité de la lumière*, or in his manuscript notes on light. Elsewhere, as we saw above, Huygens did measure in order to test a claim, but here, in the boldest and most novel conjecture of his career, he hardly did. For the most part Huygens did not produce numerical comparisons; he generated instead numerical parameters, and the weight of supporting his conjecture was more often born by the exhibition of new effects than by their calculated values.

Huygens had reasons for working this way. From his point of view, physical knowledge—whether remote causes, such as the hidden mechanisms of nature (e.g., the structure of the ether), or such proximate causes as their behavior (e.g., wave fronts, which are movements of the ether)—is at best only "probable."[34] It cannot have the demonstrative certainty of geometry, the foundations of which in "clear and distinct perceptions" seemed to be comparatively secure (if nevertheless still only "probable"[35]), because very little in natural philosophy is "clear and distinct." Some conjectures can be rendered more probable than others by virtue of their generating larger numbers "of experiments that concur to confirm for us what we have supposed." But this at best confers a higher degree of probability.

Indeed, more than any other natural philosopher or mathematician of the period, Huygens insisted on, and frequently made use of, the "probable." Late in life he wrote (in Latin) "On the probation of the probable," where he remarked that "nearly everything falls under this head, perhaps even the demonstrations of the mathematicians. It is not well to identify certitude with clear and distinct perception, for it is evident there are, so to say, various degrees of that clearness and distinctness. We are often deluded even in things which we think we certainly understand. Descartes is an example of this; it is so with his laws of communication of motion by collision of bodies."[36] Indeed, Huygens' main intellectual problem with Descartes, despite their shared belief in mechanism, lay in the latter's claims to certainty. "Mr Descartes," wrote Huygens in 1693, "found a way to take his conjectures and fictions for fact. And so those who read his *Principles of Philosophy* experience something rather like those who read pleasing novels which give the impression of being true tales."[37]

Huygens' probabilism[38] did not explicitly rank conjectures among one another in respect to their distances from effects, even though the bulk of his actual work concentrated rather on proximate than on remote causes—for example on the behavior

[34] It might seem peculiar to characterize both a thing (say the ether) and a property of the thing (say a wave front) as a "cause" of an observable effect, but Huygens scarcely distinguishes among the two *qua* causes. However, see below, note 39.

[35] Cf. the passage quoted immediately below.

[36] Translated in Elzinga, 1972, p. 37 from Huygens, 1888–1950, vol. 21, p. 541.

[37] Huygens, 1888–1950, p. 403.

[38] Which accords in several respects with contemporary French "mitigated skepticism," as noted in Shapiro, 1989, p. 226, who cites Pierre Gassendi, Marin Mersenne, and Le Mothe le Vayer as other exemplars of the tradition. On seventeenth-century skepticism, see Popkin, 2003 . See also Dear, 1998, pp. 161ff.

of wave fronts rather than on the details of moving ether parts. But as he saw it, both kinds of cause[39] shared the same epistemological status as conjectures that could be rendered probable, and the way to do that was, in principle, through experiment. As he intimated to Leibniz in 1691, the continued production of new results in geometry "should not impede works in physics, for which I think we know enough, and more geometry than is needed; but it is essential to reason with method about experiments, and to amass new ones, following approximately the project of Verulamius [Francis Bacon]."[40]

This way of thinking about the reliability of physical knowledge had at least two substantial results for Huygens' scientific work. Since he had no method for assessing probability beyond the production of as many new effects as possible, there was little incentive to probe any individual novelty in quantitative detail. Its demonstrated existence would be enough, as we can see from the paucity of numerical comparisons in his *Traité*. Yet, in what only seems to be a contradiction, Huygens' emphasis on the exhibition of fresh properties puts a premium on accurately characterizing the proximate cause, for it is the guarantor and generator of novelties. Here we spy an essential separation between experiments that generate parameters, which characterize the cause, and experiments which demonstrate new effects. To measure a parameter as best one can was, for Huygens, to come as close as possible to the cause itself—but no closer than probabilism allowed. Once the parameters had been generated, there was no need to probe just how well they agreed with their associated cause by going beyond the qualitative production of novelties. His way of doing science accordingly offered little room for a dynamical interplay between experiment and theory.

Epistemology did not float freely in an abstract realm, for it seems likely that Huygens' mature views were connected to his experiences with measurement and the construction of instruments. Throughout his life Huygens was closely involved in the design and fabrication of telescopes and clocks. At the age of twenty-five (1654) he experimented with lens grinding and polishing; the next year he built a microscope. In 1661, he worked on pendulum clocks (having discovered the isochrony of

[39] Though Huygens probably did not think of a wave front as a "cause" in the same sense as moving ether. The front was after all the behavior of the moving ether, a characterization of a particular form of motion. It could be dissociated from a particular mechanism, and Huygens was well aware of the fact. Indeed, Huygens in the *Traité* distinguished yet a third layer, intermediate as it were between front and moving ether. He argued that whatever the particular underlying mechanism might be—whether hollow canals as Descartes thought, or rapid motion of smaller parts, as Huygens maintained—the "springiness" of the matter in which the front propagates had a higher probability than any particular mechanism: "though we shall ignore the true cause of springiness we still see that there are many bodies which possess this property"; Huygens, 1690, vol. 19, p. 472; cf. Huygens, 1690, p. 559.

[40] Huygens, 1888–1950, vol. 10, p. 190. Huygens' remark about knowing enough geometry refers to Leibniz's work on the calculus. Huygens was certainly capable of using infinitesimal methods, and had indeed done so in the route in 1659 that led him to discover that the cycloid is the tautochrone, though by the end of that year he was able to remove the "infinite spaces" of this discovery path; his publication years later (1673) followed this final procedure: see Yoder, 1988, p. 62, who remarks that in discovery Huygens "use[d] indefinitely small particles of time and space, including the technique of summing infinitely many line segments to achieve an area." In the end, Huygens came to appreciate the new methods as mathematically productive in ways that his own simply could not match (ibid., p. 176).

a pendulum that traces a cycloidal path in 1659), and the next year he invented a new type of telescopic eyepiece. "Among the diverse parts of mathematics," Huygens wrote in a 1667 or 1668 précis for the astronomical charter of the Académie Royale des Sciences, "those which provide the most activities to our assembly are the ones that are not purely theoretical but that apply to matter and produce useful effects."[41]

Huygens was also thoroughly committed to the production of mathematical structures that would govern abstract systems freed from worldly imperfections. One specialist in the history of instruments argues that he was "not really interested in mechanisms as such: he was a mathematician first and last, and only when he saw a way to apply the rules of mathematics to a mechanical construction did he turn to practical machinery."[42] Another claims that Huygens' extensive practical activities with the telescope indicate that he was "fascinated by the instrument for its own sake."[43] Apparently the *relicta* of Huygens' life can yield different, even contradictory, stories. Yet beneath the contradictions lies a hidden formation whose probable outlines we can (in agreement with Huygens' moderate skepticism) sketch, and here we will find the links that bind together his epistemology, his mathematical structures, and his engagement with instruments.

Huygens may have designed clocks, but he knew perfectly well that no real pendulum can possibly follow a perfect cycloidal path because no geometrically perfect cycloid can exist. Neither do real colliding bodies obey in an absolute sense Huygens' laws of collision.[44] This much would have surprised no one by the middle of the seventeenth century, and neither would it have surprised the Greek astronomer Ptolemy to be told that instruments cannot be fabricated to a geometric perfection. With Huygens, however, the issues ran deep and had important consequences for his way of working. According to Joella Yoder, he always proceeded by reducing a physical problem "to a few choice parameters capable of mathematical expression and hence manipulation, and the results of the mathematical derivations are presumed to say something valid about the real world."[45] But just what does mathematics say about the real world of moving bodies, subject always to the inevitable influences of factors that lie outside the purview of Huygens' structures, such as air resistance or other forms of friction? "Is it," Yoder continues, the cycloid that is isochronous or the clock whose bob follows its path? The terminology can be loose because the pairing of mathematical entity and physical object is so precise," but only when care is taken to ensure that the physical matches "the mathematical as closely as possible. For example, Huygens' clocks tended to have relatively lightweight bobs and cords, thus approximating the ideal weightless situation ... In Huygens' world, an isochronous

[41] Huygens, 1888–1950, vol. 19, p. 264.

[42] Leopold, 1980, p. 221.

[43] Dijksterhuis, 2004, pp. 104–5.

[44] Huygens had begun to think critically about collisions as early as 1652, following his rejection of Descartes' rules. These rules do not *ex hypothesi* hold even in the abstract Cartesian plenum since they require the absence of contact between any but the two colliding bodies, which is for Descartes impossible. They are accordingly doubly idealized: first, because observable bodies are always constructed out of many different parts, and second, because even if a body were built out of a single part it would still always be in contact with many others, thereby violating the conditions for the rules to hold.

[45] Yoder, 1988, p. 167.

clock must have a regulator, such as a pendulum, that moves according to the mathematics of perfect oscillating motion. The curved [cycloidal] plates were dispensable when they were the result of experimentation, but they became essential once they were the product of mathematical analysis. Whatever objections might be advanced regarding the actual value of cycloidal plates, such as the negative effects of the added friction, are irrelevant to this view. Physical aberration should be met by additional derivations, not by the abandonment of the design proved by mathematics."[46]

Cycloid or clock? Mathematics or matter? The answer for Huygens would be both and yet neither, but to different degrees. We have already seen that for him not even mathematics could attain perfect certainty, though in the hierarchy of probable truths it ranks the highest. Instruments such as clocks, with their congeries of moving and interacting material parts, cannot perfectly match the requirements of geometry, and so their workings must fall much further from the probable truth than the abstractions to which they were designed. But Huygens always sought to increase probability, and to do so by means of instruments meant to build them with so great a skill that their deviations from mathematical perfection fell within the bounds of the probable. And those bounds were themselves determined by the practical limits of skill.

Consider for example Huygens' 1658 design for a pendulum clock. Michael Mahoney points out the design's special character.[47] The device that he invented links the pendulum to the driving mechanism through a small crutch. The crutch transmitted force to the pendulum in such a way that the weight of the pendulum's bob could be increased without simultaneously adding frictional drag to the pallet that regulated the motion of the crown wheel. Increased bob weight meant reduced sensitivity to the unavoidable variations that would occur in the driving mechanism and therefore better guaranteed continued isochrony. At the same time, the crutch itself could be adjusted to minimize the amount of force on the pendulum. The result was a form of "oneway sensitivity" that enabled the mathematical analysis of small pendular perturbations.[48] A material defect was ameliorated by means of a material structure, thereby permitting motions which could have mathematical significance to shine through. Huygens, one might say, used matter to fight its own imperfections in order to marry it as closely as skill permitted to geometry.

To think that matter must be fought, that it must be forcibly milled into as perfect a form as possible—a form that could only reach for but never achieve the ideal—had consequences for Huygens' relations with craftsmen. Since so much depended upon skilled working, Huygens was especially sensitive to craftsmen's abilities, particularly clockmakers, who had to produce complex mechanisms. He refused to consider that these artisans had anything beyond their skills and speediness in fabrication to contribute, and on occasion he complained about that.[49]

[46] Ibid.

[47] Mahoney, 1980, pp. 237–38; Mahoney, 2004.

[48] Mahoney, 2004. The crutch has a fork whose two tynes are alternately contacted by the oscillating pendulum, thereby bringing an arbor pallet into contact with the driving crown wheel. In between contacts the pendulum oscillates freely. We thank Mahoney (†2008) for clarifying the issue (private communication).

[49] Leopold, 1980, pp. 223–31.

Skepticism on the part especially of mathematically inclined mechanical philosophers toward craftsmanship is already apparent in Descartes' *Dioptrique*, and indeed before that in Descartes' difficult relationship with the artisan Ferrier in the late 1620s and early 1630s.[50] Descartes had designed an automatic machine for the grinding of plano-hyperbolic lenses which, if perfectly shaped, would focus parallel rays entering through the plane side to a single point, thereby avoiding the defect of spherical aberration. From Descartes' point of view, the signal virtue of his design was that it replaced the artisan altogether. Ferrier was necessary, but only for the purpose of constructing the machine that would replace him and his ilk. Descartes aimed thereby to displace the locus of craft from the frequently repeated manual grinding of lenses to the one-time production of a lens-grinding device. Craftsmanship would accordingly be restricted, though not eliminated altogether.

Ferrier disappointed Descartes, and indeed no one ever produced practical hyperbolic lenses, with the possible exception of Hevelius.[51] Hevelius was nevertheless the exception that proves the rule, because he was both fabricator and user—a mathematical adept endowed with the skill of the artisan. That combination was unlikely to be replicated elsewhere, and certainly not reliably. Huygens himself briefly essayed the production of hyperbolic lenses in 1654, but abandoned the quest. Instead, as he learned the techniques and complexities of lens-grinding from Gitschoven,[52] Huygens

[50] See Burnett, 2005, pp. 41–59 for an account of the Ferrier-Descartes contretemps, and in general for a pointed discussion of the complex of issues raised when Descartes, Huygens, and even Hooke aimed to replace the artisanal craft of lens-grinding with automatic machines that would perfectly incarnate their mathematically precise requirements. Although there is a *prima facie* case for connecting these issues to the natural philosophers' conviction that the world is itself a mechanism—and therefore best investigated by mechanical entities such as automatic machines—the issues may more directly concern the difficulties raised by mechanism's construal of the sense-organs as failure-prone machines. Burnett (ibid., pp. 107–21) notes for example that the Capuchin Cherubin d'Orléans (François Lasséré), who remained profoundly scholastic, produced—partially *in consequence* of his scholasticism—"the most exhaustive treatise on lens making in the seventeenth century"; he also produced the first binocular telescope. For in it d'Orléans did not remove the artisan, though he disdained "vulgar" craftsmen whose techniques were inevitably defective. Instead, he designed machines that would, he wrote, "direct the hand of the Artisan in the working of optical lenses" (ibid., pp. 108–9, quoting from D'Orléans, 1671, p. 380). These machines had "essential" attributes, namely their parts, as well as "accidental" ones—to wit, the artisans' hands and feet, and d'Orléans' scholastic distinction between the essential and the accidental kept the artisan in the system but in a way that attached him as "accident," that is, in effect as a skilled operator, to the machine. Descartes would have abolished artisanal skill altogether.

[51] Hevelius, 1673, chap. 23. See the informative discussion of the various attempts to produce automatic or nearly automatic machines for grinding hyperbolic lenses in Burnett, 2005, pp. 71–121. Among other signal points, Burnett remarks the frequent absence of the artisan's hand in depictions of these several machines, though traditional machine illustrations, in particular those of lathes, usually included images of craftsmen or hands or feet working the devices.

[52] Huygens and his brother began lens grinding themselves in 1654 following a disappointing experience the previous year in having a "master Paulus" in Arnhem construct a telescope for them. They obtained craft knowledge through correspondence with the Augsburg optical instrument maker Johann Wiesel, and also with the Louvain scholar Gerard van Gitschoven. In 1667–1668 Constantijn employed Cornelius Langedelf to polish glass and grind eyepieces, and in 1682, one Dirk van der Hoeven in The Hague, whom he called "the chimney sweeper." In Paris, Christiaan was unable to extract what seems to have

aimed to improve the process as much as possible by minimizing the artisan's role, at first by inventing a device to prevent the lens-blank from moving irregularly during grinding, and eventually by designing a nearly automatic machine.[53]

Turning from lenses to clocks, and considering the role played by motion in both, provides an even more pointed example of the vexed relationship between artisans and mathematically inclined natural philosophers because it raises issues of knowledge-making proper. The production of lenses does of course require motion—whether of the artisan or, in a Cartesian paradise, of perfect machines—but the lens itself operates without any internal mechanical movements. Clocks, however, have moveable parts, and here the artisan's skills are even more on display in the finished product than with lenses, for each moving piece must interact in just the right way with its siblings. This opened a door for clockmakers to intrude into the realm of knowledge-making itself, thereby generating a particularly vigorous reaction from Huygens.

He took umbrage, for example, at knowledge claims made by the clockmaker Van Ceulen. "On 17 December 1683," he wrote, "I took to Van Ceulen the clockmaker the model that I made of the movement of the cylindrical pendulum, so that he would thereby change the 2 clocks that I had him make for the East India Company. I asked my brother de Zeelhem to come with me because the said clockmaker imagined that he had himself found the same thing that I had after I had told him something vague about it. But having seen [my] model he admitted that what he had modeled didn't resemble it at all."[54] Van Ceulen's skill, and that of others like him, may have been necessary, but Huygens did not think highly of their activity. In another example, Newton's friend Fatio de Duillier had asked Huygens to look into lens-grinding lathes that he had heard about in Delft. Huygens described the lathes (and their cost) but remarked that he doubted that "you'll want them, since they don't have what's needed to make objective glasses, because I can't imagine that you want to spend the time amusing yourself making things with a lathe when it can be spent on much better things."[55] Despite Huygens' own acquisition of similar skills, he certainly did not hold the activity in high regard, and this in turn raised significant issues for him (and others like him) in respect to the probity of knowledge produced by, or attested to by, instruments.

Huygens' skeptical conviction that natural knowledge can only approach, but never reach, absolute truth put an inevitable premium on the perfection of the craftsman's handiness, and yet he thought that the craftsman would always be the weakest link. The objects produced by craft were, in a sense, obstacles that stood in the way of knowledge claims, because they were inevitably troubled by material imperfections

been new craft knowledge concerning grinding from Philippe-Claude Lebas, and no lenses were made by him during his years there. The most productive period of lens-making by Huygens and his brother took place from 1682 through 1687 when they produced at least 41 telescopic objectives. Local craftsmen prepared the glass or ground eyepieces; the brothers ground the objectives themselves. Helden and Gent, 1999.

[53] Burnett, 2005, pp. 96–103. Huygens' machine was described in the posthumous Huygens, 1703.

[54] Huygens, 1888–1950, vol. 18, p. 532; discussed in Leopold, 1980, p. 230 and in Mahoney, 2004, p. 304, n.346.

[55] Huygens, 1888–1950, vol. 10, p. 277; noted in Leopold, 1980, p. 224.

that could only be improved by the perfection of skill. This, in turn, required either that Huygens develop skills himself or that he rely on the efficacy of craftsmen—unless, best of all, craftsmanship could be replaced altogether by an automatically operating machine. Artisans were furthermore deeply concerned to establish proprietary interest in their products, which again raised issues for Huygens, who developed different notions of ownership. He certainly exhibited proprietary interests in devices that he claimed to have been the first to invent—which led to a number of controversies, not least with Robert Hooke over the spring-balance clock—for Huygens, it seems, the primary motivation was to secure ownership of the knowledge that the apparatus embodied. He regarded devices as material incarnations of the (geometrical) schemes on which they were based, while artisans had the inferior, though crucial, task of matching matter to geometry. "Having affirmed the equality and constancy of our automaton," Huygens wrote after concluding his description of the new pendulum clock, "we will stop our description, leaving to the industry of artisans things that could be added: instructed in the principle of our invention, they will without difficulty find the way to apply it to clocks of diverse types, even to those which had previously been constructed in the old way."[56]

The probability of knowledge depended in consequence on the skills of those who themselves had no purchase on knowledge-production. This produced an inevitable tension between the high claims of geometry, as Huygens saw it, and the messy world of artisan-produced mechanisms which carried over into his attitudes toward measurement itself. Measurements require instruments, and instruments have to be fabricated, either by the measurer himself or else by craftsmen. In either case the quality of a measurement depends upon the quality of the instrument and upon the ability of the measurer. Skill unavoidably enters, raising concerns for someone with Huygens' views of how measurements—which are inevitably troubled by the imperfect skills of the instrument-maker and the observer—can be used to increase the probability of knowledge-claims. By the late eighteenth century the opinion was growing that imperfections were to be expected but that their effects can be ameliorated by making the same measurement many times and then combining the results, even though no one among the measurements could be known *a priori* to be any better than any other. From Huygens' point of view, however, there would have been no point in combining admittedly defective measurements when the goal was to produce the best-crafted measurement. No measurement, he knew, could be perfect, but there would always be a hierarchy among measurements, which reflected the degree of practiced work that went into each one's production. Just as a well-crafted clock might nicely tick the even beats of a cycloidal path, so might an expertly done measurement nicely reflect the geometric structure that governs the phenomenon. Though sharing some of Descartes' disdain for artisans, Huygens, the maker of knowledge, shared with the artisanal manipulator-of-matter the conviction that only skill could produce a well-crafted product, whether it be a designed object or a measured number—unless skill could be incarnated in an automatic machine, thereby avoiding altogether the imperfections of sense and hand.

[56] Huygens, 1888–1950, vol. 17, pp. 70–72.

Another consequence of these attitudes concerns the purpose of experiment itself. Huygens produced few experiments, and those he did produce aimed at probing or buttressing specific knowledge-claims. His most extensive experimental work concerned optics. In the late 1670s in particular he worked hard on the properties of Iceland spar, as we noted above. Here there was no wide-ranging series of exploratory experiments, but rather a tightly controlled set of parameter-measurements that were explicitly bound to his geometrical theory. When Huygens measured the force of falling water, he took a single number for each fall and then looked among them all to see whether they seemed as a group nicely to fit the claim that the impact ran as the square of the velocity. When he wished to use the lengths of seconds-pendula at different latitudes to ascertain the earth's degree of oblateness (much smaller on Huygens' theory than on Newton's), he was particularly concerned with the trustworthiness of each one of the individual numbers that had been produced. Preparing for the 1686 expedition to the Cape of Good Hope, "Huygens not only wrote elaborate and precise instructions for the installation and handling of his clocks, but he also gave detailed instructions to [Thomas] Helder [who was responsible for the clocks] for measuring the length of the seconds-pendulum along the course of the voyage."[57] Guaranteeing correct and appropriate skill was no small matter when the probability of knowledge was involved.

Robert Boyle differed in many respects from both Hooke (who had begun his career as assistant to Boyle at Oxford) and Huygens, not least in his skepticism concerning the uses of mathematics. And yet he held similar attitudes toward perception and measurement. Boyle wrote extensively about his philosophy of experiment and engaged in substantial exchanges with, among others, Thomas Hobbes about the character of experimental knowledge, indeed about whether or not experiments could be said to produce trustworthy knowledge at all.[58] In 1661, Boyle produced a striking image in support of his claim that experiments can produce reliable knowledge, but that they must be chosen judiciously. "Experiments," he wrote, "ought to be estimated by their value, not their number; and that a single Experiment ... may as well deserve an entire Treatise, as a great many less considerable ones. As one of those large and orient Pearls that are fit to adorn a Monarchs Crown, may outvalue a very great number of those little (though true) pearls that are to be bought by the ounce in Goldsmiths and Apothecaries shops."[59] Boyle's striking image of the "orient Pearl" emphasizes the signal importance of the knowledge that can be gleaned from one supremely revealing experiment. Little droplets of knowledge may no doubt fall from lesser experiments, but their combined mass cannot outweigh the "Orient Pearl."[60]

[57] Schliesser and Smith, 1996.

[58] Boyle's conceptions and practice have been discussed many times in recent years, particularly in respect to putative connections between his social status as a gentleman and the probity of the knowledge that he produced, on which see especially Schaffer and Shapin, 1985. Sargent, 1995, provides a nuanced and careful understanding of Boyle's experimental philosophy, and in what follows we have adapted many of her conclusions. We have also relied on the astute discussions of Boyle's "chymstry" in Principe, 1998.

[59] Hunter and Davis, 1999–2000, pp. 112–13.

[60] The "Orient Pearl" as a trope was in frequent sixteenth- and seventeenth-century use, notably by Milton in *Paradise Lost*: his chapter "The Argument" begins "Now Morn her rosy steps in the eastern clime;

Boyle performed and described at length a great many experiments, ranging from productions in the "chymical" laboratory, through manipulations with the new air pump, to his "considerations touching colours." Each required the exercise of skill in order to ensure that the faculty of judgment could be properly exercised in drawing appropriate conclusions. For Boyle shared Francis Bacon's view that "to the immediate and proper perception of the sense ... I do not give much weight; but I contrive that the office of the sense shall be only to judge of the experiment and that the experiment itself shall judge of the thing."[61] Boyle himself insisted on what he termed the "imbecility of the visive faculty," and in unpublished papers probably dating from the 1680s he listed a four-step sequence that ended in judgment:[62]

1. The change made by the impression of the object on the exterior organ
2. the conveyance of that impression to some internal parts of the brain (or seat of common sense)
3. the reception of the impressions there, on which follows *in man*
4. the perception of the mind and the judgment etc. it makes on occasion of that perception

"Judgment" works on perceptions, which arise from impressions made by an object on the senses and then conveyed to the seat of "common sense." Because the senses are unreliable, and because reason itself is easily deceived by prior convictions and diversions,[63] the faculty of judgment requires well-honed skills, which can only be gained from repeated experience. Boyle studied and visited artisans and craftsmen, and he developed an appreciation of the ways in which their work required time and continual effort to achieve proficiency.[64] In fact, the very notion of a "mechanical philosophy" was in Boyle's hands intimately associated with the model provided by the operations of machines themselves. "Unlike substantial forms," notes William Newman, "whose precise nature even the scholastics viewed as escaping the senses and hence remaining forever 'unknowable', machines and their parts had properties that were capable of easy visualization and comprehension. These properties or 'affections' are 'mechanical' because in the traditional discipline of mechanics, descending

---

Advancing, sowed the earth with orient pearl." Shakespeare (Richard II, Act 4) wrote of "liquid drops of tears that you have shed shall come again, transformed to orient pearl."

[61] Sargent, 1995, p. 51.

[62] Boyle, 2006, vol. 18, transcribed in Sargent, 1995, pp. 146–47.

[63] Demonstrative reasoning is particularly prone to deception: "For though, in pure mathematics, he, that can demonstrate well, may be sure of the truth of a conclusion, without consulting experience about it; yet because demonstrations are wont to be built upon suppositions or postulates; and some things, though not in arithmetic or geometry, yet in physical matters, are wont to be taken for granted, about which men are liable to slip into mistakes; even when we doubt not of the ratiocination, we may doubt of the conclusion, because we may of the truth of some of the things it supposes" (Hall, 1966, p. 171).

[64] Boyle often refers to his interactions with tradesmen. One apposite example that indicates his close knowledge of their work and habits appears in his "Certain Physiological Essays" (1661): "There are some Mechanical Experiments wherein I have purposely omitted some manual Circumstances, because I was unwilling to prejudice some ingenious Trades-men, who make either a Livelyhood, or at least a gain, by the sale of the productions of such Experiments" (Hunter and Davis, 1999–2000, vol. 2, p. 31).

from the Hellenistic engineers, they are all that one needs to explain the operations of machines."[65]

Machines had to be built by skilled artisans in Boyle's day, and just as artisans had to work hard to achieve proficiency in construction, so experimenting natural philosophers had to develop expertise to produce judicious conclusions from experiment, for there was no other way to avoid the Scylla of sensual deception and the Charybdis of prejudicial reason. Once appropriate skill has been achieved, then a concurrence among many well-judged experiments might lead to a morally certain result. Though the conclusion could not be said to have demonstrative certainty in the logical or mathematical sense, nevertheless an agreement among many experiments gives it, Boyle averred, an equally firm seat in reasoned conviction. As Sargent put it: "The purpose of concurrence, as both an epistemological and a methodological strategy, is not the simple quantitative multiplication of witnesses. Rather, it is required as a way by which to evaluate the reliability of the information attested to by such witnesses. Because of the inherent fallibility of human senses and reasoning, such a qualitative check is absolutely necessary."[66]

Boyle's conviction that only skill can compensate the inherent deceptiveness of the senses affected his attitudes toward instruments, and toward the production of numbers by means of them. Yet even skill is no guarantor of success; the best tradesmen may sometimes fail to produce the same quality in their products. "If there be any Trade that obliges the Artificers to be assiduously conversant with the Materials they employ," Boyle wrote, "it is that of the Glass-men; and yet even to them, and in their most ordinary operations, there happen now and then little accidents, which though they know not well to what to ascribe, are yet capable of hindering them from doing sometimes what they have done a thousand times."[67] These unavoidable "little accidents" affect experiment and even observation. All the more is this true when numbers are involved, for numbers apply to a world of mathematical objects, and not with perfection to the world of matter. "I have observed," Boyle continued,

> even Mathematical Writers themselves to deliver such Observations as do not regularly hold true. For though it hath been look'd upon as their priviledge and glory to affirm nothing but what they can prove by no less than Demonstration; and though they use to be more attentive and exact than most other men in making any kind of Philosophical Observation; yet the certainty and accurateness which is attributed to what they deliver, must be restrain'd to what they teach concerning those purely-Mathematical Disciplines, Arithmetick and Geometry, where the affections of Quantity are abstractedly consider'd; but we must not expect from Mathematicians the same accurateness when they deliver Observations concerning such things wherein 'tis not only Quantity and Figure, but Matter, and its other Affections, that must be consider'd.[68]

[65] Newman, 2006, pp. 186–89.

[66] Sargent 1995, p. 212.

[67] Hunter and Davis, 1999–2000, vol. 2, p. 64.

[68] Ibid., pp. 74–75.

Since all "material objects," Boyle believed, must possess qualities that escape "Quantity and Figure," their nature may "deceive the Expectations grounded on what [the mathematician] delivers." Moreover, "so may the like happen by reason of the Imperfection of the Instruments which he must make use of in the sensible observations whereon the mixt Mathematicks (as Astronomy, Geography, Opticks, &C.) are in great part built." This can go quite far because "the Observations even of skilful Mathematicians may hold so little, or disagree so much, when they pretend to give us the determinate *measures* of things, that I remember of three very eminent modern Mathematicians, who have taken upon them by their Experiments to determine the proportion betwixt Air and Water, that one makes not the weight of Water to exceed above 150 times that of Air, the other reckons Water to be between 13 and 14 hundred times, and the third no less than 10000 times the heavier."[69]

Boyle's judicious attitude toward the results of even qualitative experiments verged on the skeptical when it concerned actual measurements. His 1662 *Defence of the Doctrine Touching the Spring and Weight of the Air* contains "a table of the condensation of the air" generated from experiments with a J-shaped tube whose last two columns contain respectively the pressures measured (in inches of mercury) and the pressures that should have been measured "according to the hypothesis that supposes the pressures and expansions to be in reciprocal proportion."[70] The differences are generally about half an inch (above 35 inches).[71] "Although we deny not," Boyle remarked, "but that in our table some particulars do not exactly answer to what our formerly mentioned hypothesis might perchance invite the reader to expect; yet the variations are not so considerable, but that they may probably enough be ascribed to some such want of exactness as in such nice experiments is scarce avoidable." Nevertheless, Boyle's purpose was not to demonstrate that the proportion holds everywhere and with minute exactness—he reserved judgment about that[72]—but rather to furnish evidence against the arguments of his "adversary," the British Jesuit Franciscus Linus, who countered Boyle's claims concerning the spring of the air.[73] Numbers had a noumenal character for Boyle, because material things cannot by their very nature be captured by perfect abstractions. Skilled labor can probe how far mathematics may usefully work, but once skill has reached its limits—imposed both by human limitations and by the material reality of instruments—so too have the uses of numbers. In this respect Boyle differed from Hooke and Huygens, both of whom were willing to take the best result of skilled investigation as sufficient to represent hidden

[69] Ibid., p. 76. Also see Sargent, 1995, p. 169.

[70] Boyle, 1662 : cf. Hall, 1966, pp. 340–42; Hunter and Davis, 1999–2000, vol. 3, pp. 59–60.

[71] More precisely, the mean difference is about 0.54 inches, with a standard deviation of about 0.53 inches.

[72] "I shall not venture to determine whether or no the intimated Theory will hold universally and precisely" (Hunter and Davis, 1999–2000, vol. 3, p. 60).

[73] "our Adversary may plainly see that the Spring of the Air, which he makes so light of, may not only be able to resist the weight of 29. Inches [of mercury], but in some cases of above an hundred Inches of Quicksilver and that without the assistance of his Funiculus" (ibid.). Linus' anti-mechanical "funiculus" was a construct designed to avoid the spring of the air. For a discussion of Boyle's response to Linus, see Schaffer and Shapin, 1985, pp. 156–69, esp. pp. 168–69, which emphasizes Boyle's concern to defend the weight-bearing power of the air's spring, and which also notes Boyle's use of "the mathematical regularities displayed in the resultant tables relating spring to compression" to refute Linus.

exactitude. Boyle remained doubtful about perfection, whether occult or overt; he was persuaded that matter remained too limber to be caught in a mathematician's web. Even so, Boyle shared with Hooke and Huygens the conviction that a properly produced experiment depends critically on the experimenter's skill, which alone can guarantee its worth. None of them would have welcomed a quarrelsome group of numbers into the house of experiment (though Boyle would simply have ignored what he took to be minor discrepancies among measurements, whereas Hooke and Huygens would have selected the best of the lot).

Indeed, to this way of thinking the public display of discrepant measurements would have appeared to be a confession of poor skill, for the early modern observer or experimenter seems to have thought that skill in observation could be improved nearly *ad infinitum* by hard work combined with ever more accurate instruments to compensate the inherent limitations of the senses. At a certain point the result would be good enough for all practical purposes, and then it was really a matter of on-the-spot local judgment, which among the numbers was chosen. Only skill, it seemed, could domesticate wild numbers. What might account for such an attitude?

The example of Huygens provides a suggestive clue, for he worked directly (if, often, agonistically) with artisans in the crafting of pendulum clocks and an intricate planetarium, and both he and his brother Constantijn had extensive experience in lens grinding. This collaborative—if often vexed—link between the observer or experimenter and the artisan may in part account for what appears to have been a common attitude to the handling of data, whether astronomical or experimental, until well into the eighteenth century. Moreover, the personal link between artisan and observer or experimenter became less common as measuring devices were increasingly shopped from instrument makers, while during the same period the use of averaging techniques gradually increased.

The artisan aims to produce a concrete result, an *object*, which has certain properties. The rings of an armillary sphere must be just circular enough, and their pivots well-placed and properly balanced, to exhibit in a satisfactory way the places of the stars and the ecliptic in relation to a horizon. If the rings are too far out of round, or if the pivots are rough and ill-placed, then the device will be useless. As he works with metals and woods the craftsman applies his experience and skill to hone, polish, carve, clip, slice, bend, and melt until the result feels right and proper. Along the way he may discard unacceptable pieces and remake them; he may find that one or another element has to be reworked. Even if an artisan did make several pieces, like the astronomer who makes multiple observations he will in the end choose just the ones that best fit together. And he certainly could not in any meaningful sense "average" the pieces precisely because they are material objects.

The early-modern astronomer, and many of his counterparts in the laboratory, knew intimately the ways and methods of craftsmen. Their own handling of data seems to have been, if not modeled on artisanal practice, then developed in a way that reflects what one might call an ethos of artisanal perfection. Just as the craftsman sought to produce as nearly perfect an object as possible, so did someone like Huygens look for that one best number. Such a number could not be the result of averaging, because that would imply ignorance of which measurement was best. This would be rather like a craftsman handing over three or four armillary spheres and telling the buyer to use all of them and then combine the results, which would be an economic absurdity and a confession of failure.

Like skilled craftsmen, Huygens, Hevelius, and Flamsteed provided singular products (numbers) without indicating how they selected them from among the several results that they had obtained. Hevelius' production of stellar coordinates provides a particularly striking example of this attitude. Although fire on September 26, 1679 destroyed Hevelius' observatory and most of his library, many of his papers were saved, apparently by his daughter. A manuscript in his hand, entitled *Catalogus Fixarum*, contains all the coordinates that were eventually printed in the *Prodromus*; furthermore, Hevelius published the measurements from which the later manuscript and catalogue were generated in the *Pars Posterior* of his *Machinae Coelestis*,[74] and four manuscript volumes containing the raw data from 1657 to 1679 (on which the 1679 publication was based) were also saved.[75] However, his computations are apparently lost, as it seems are any explicit remarks concerning how he decided on final coordinates, if indeed he ever wrote down these sorts of judgmental considerations (on which see below).

The *Prodromus* lists Hevelius' coordinates, as well (where they exist) as those from Ptolemy, Tycho, Ulugh Beg, and the Landgrave of Hesse—all adjusted to epoch 1660. Hevelius additionally transformed his ecliptic into equatorial coordinates. Although the *Prodromus* provides just one longitude and latitude for Hevelius' observations, the manuscript provides at least four, and often more, values for each star observed (figure 2.3) , which raises the question of how Hevelius produced the single number that was eventually printed.

We know with certainty what Hevelius did not do: he didn't combine in any way the several numbers in the *Catalogus*, because in every case the printed coordinate corresponds to a single number in the manuscript; someone (most likely Hevelius himself) put a red asterisk next to that number. How then did he decide which observations to choose, and do the ones listed in the manuscripts exhaust all of the observations that were made? Consider his instruments. The *Pars Prior* of his *Machinae Coelestis*, printed in 1673, described their design and use. All of the instruments were mounted on a platform that traversed the roofs of Hevelius' home and two neighboring houses that he also owned (figure 2.4).

The star-measuring devices were of two kinds: a six-foot brass sextant, and a five-foot brass quadrant. The sextant (figure 2.5 left and center), which covered a sixth of a circle, measured the distance between a pair of stars, and it therefore had two sighting lines. Its use required two people. First, the apparatus was adjusted so that its plane included the two stars. Then one of the two observers adjusted the position of the sector, using a screw device to keep a sight locked onto one of the stars, while the

[74] Hevelius, 1679.

[75] The *Catalogus* is in the library of Brigham Young University, while the four volumes of raw data are in the library of the Observatoire in Paris. For details see Volkoff, et al., 1971. A convenient source for comparing Hevelius' with other coordinates is Baily, 1845a. The data in the *Pars Posterior* are presented first by date and then, in summary at the end, by zodiacal sign and, within each sign, by constellation. A comparison with the volumes in the Observatoire shows that the data there were transferred directly to the *Pars Posterior.* The manuscript and the printed *Prodromus* follow custom in listing by constellation, with each of its listed stars located within a zodiacal sign.

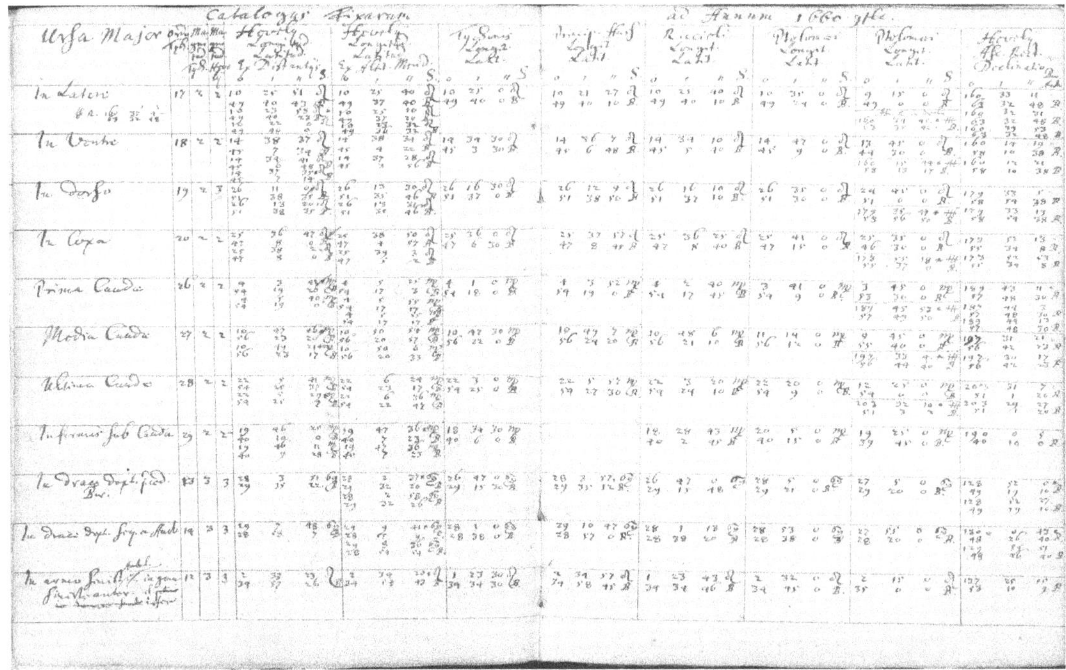

Figure 2.3. A page from Hevelius' manuscript for the *Catalogus Fixarum* containing multiple observations for *Ursa Major.* By permission of the L. Tom Perry Special Collections, Harold B. Lee Library, Brigham Young University.

other observer adjusted a moveable sighting rule over the curved scale to point to the second star.[76] To judge from the plates, Hevelius operated the sighting rule. He was often assisted in these highly technical procedures by his much-younger second wife, Elizabeth (*née* Koopman); she is depicted in figure 2.5, center.[77] Like Tycho's device, Hevelius' was mounted in a ball-and-socket arrangement to make motion in any direction possible. Cables, which can be seen in both figures, were run through pulleys to counter-weights to keep the device balanced. Observations had to be made rapidly.

The azimuthal quadrant, whose plane must remain as close to vertical as possible, measures the elevation of a star as it passes across the local meridian. The quadrant was mounted on an elaborate stand which was topped by a horizontal azimuth scale for setting the device in the meridian; it was especially important for the quadrant and scale to remain orthogonal to one another (figure 2.5, right). From the star's altitude, given the observer's latitude, the star's perpendicular distance from equator or ecliptic can be found. However, to find the position of the star parallel to the equator—its right ascension—or the distance in the same direction between two

[76] In order to transform sextant distances into stellar positions one needed two stars of known coordinates. The sextant measured the distances between each of these two and a third star, from which its coordinates could then be calculated. For a succinct explanation see Cook, 1998, pp. 439–40.

[77] The plates were done by Hevelius himself, who was a talented engraver.

Figure 2.4. Hevelius' rooftop observatory platform. Hevelius, J., 1673. Image by JZB.

stars requires a clock. Since the heavens rotate around the equator at the rate of one minute of arc every four seconds, to achieve a repeatable right-ascension measurement to a quarter minute of arc requires a clock that can beat seconds accurately for the time interval between meridian traversals; by the 1660s pendulum clocks could do so.[78] Hevelius had two with short-throws that were small enough to sit on a table.[79]

Both sextant and quadrant require two observers: in the case of the sextant both of them look at stars; in the case of the quadrant one of them observes a star while the other marks the time of its meridian passage. Since astronomers have long since abandoned both devices, as well as pendulum clocks, it would be hazardous to at-

[78] Christiaan Huygens fabricated the first pendulum clock in December 1656, though Galileo had designed one that was apparently never built. Despite Huygens' efforts with cycloidal cheeks, keeping the pendulum's arc short was the only practical way to achieve useful seconds accuracy. Tycho had used mechanical clocks, including a huge one, and by 1581 he noted that seconds could be measured. But sufficient accuracy for astronomical purposes could not be achieved, and Tycho reverted to a clepsydra for transit interval determinations: see Landes, 1983, p. 104, citing Zinner, 1979, pp. 21–22. Landes also notes that William of Hesse (whose astronomical observations were often listed alongside those of Tycho and Riccioli) had seconds-marking clocks made by Jost Bürgi which apparently achieved accuracies of one minute per day.

[79] According to the young Halley, writing to Flamsteed from Danzig on June 7, 1679: cited in Cook, 1997, p. 91 from Flamsteed 1995–2001, vol. 1, pp. 684–89; MacPike, 1932, pp. 42–43.

Figure 2.5 Hevelius' six-foot sextant (left); measuring stellar distances with Elizabeth (center); the azimuthal quadrant (right).

tempt to gauge their comparative accuracy, especially since no one today possesses the requisite skills to work the equipment properly. One might think that the quadrant observations would be more reliable, since the plane of the device was fixed in the meridian, and only the pointer (or alidade) was moved. Hevelius' *Prodromus* does not list which device produced the coordinates. However, the manuscript from which the *Prodromus* was certainly prepared makes it clear that many of the asterisked observations were done with the sextant.

The data of the *Pars Posterior* provide the distance between two stars measured by the sextant, or else by the quadrant—in the latter case from the time and altitude of the stars' meridian traversals. Hevelius had to transform these measurements into latitudes and longitudes, which was not a trivial task. Since Hevelius' notes were lost, it would be difficult to match by computation the date in the *Pars Posterior* with the coordinates in the manuscripts and in the *Prodromus*. Nevertheless, it's quite obvious that the number of listed coordinates in the manuscripts do not always match the number provided in the *Pars Posterior*. Evidently Hevelius had some way of choosing which distances to use in producing the penultimate set for the manuscripts. But there is something significantly common between the manuscripts and the *Pars Posterior*: the latter often prints an asterisk next to one of the stellar distances (see figure 2.6 for an example from *Ursa Major*). The same notation appears in the Observatoire manuscripts, which seem to have been marked at about the same time as the observation.[80] From this we see that Hevelius decided to privilege some observations over

[80] The *Pars Posterior* lists a total of six distances for the first of the two stars (located *In Ventre*); the MS lists five coordinates. For the second asterisked star (*Cauda media*), the *Pars Posterior* provides five different distances, whereas the MS lists four. We have not attempted to determine by computation whether the

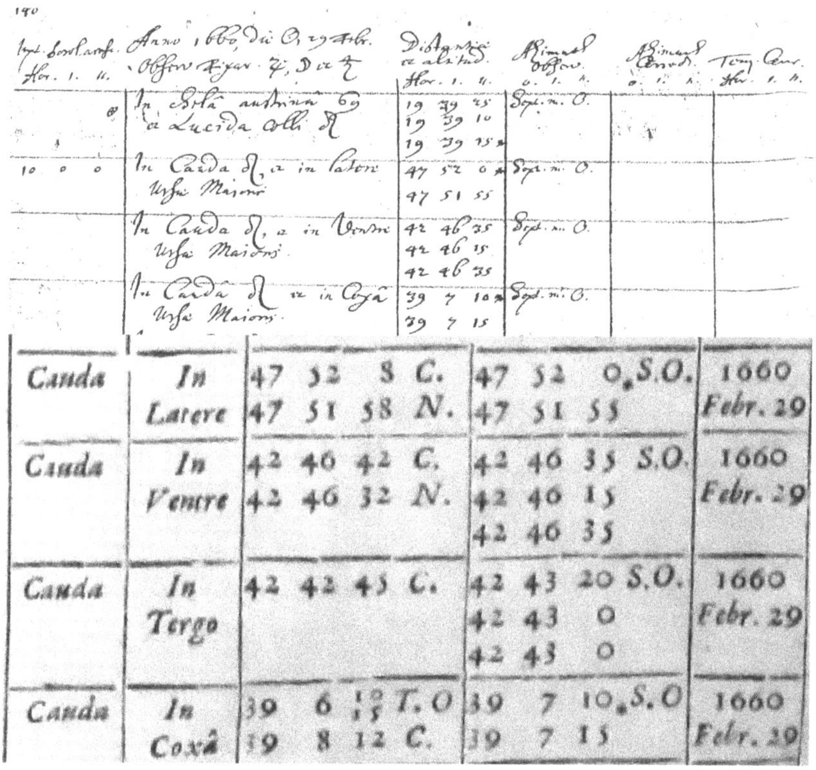

Cauda	In Latere	47 52 8 C. 47 51 58 N.	47 52 0*S.O. 47 51 55	1660 Febr. 29
Cauda	In Ventre	42 46 42 C. 42 46 32 N.	42 46 35 S.O. 42 46 15 42 46 35	1660 Febr. 29
Cauda	In Tergo	42 42 45 C.	42 43 20 S.O. 42 43 0 42 43 0	1660 Febr. 29
Cauda	In Coxâ	39 6 10 15 T.O 39 8 12 C.	39 7 10*S.O 39 7 15	1660 Febr. 29

Figure 2.6. (top) From Hevelius' manuscript showing asterisks next to distances to *Regulus* for two stars in the constellation *Ursa Major* located in the zodiacal sign *Leo*. By permission of the Bibliotèque de l'Observatoire de Paris. (bottom) The same listing, unaltered, in the printed *Pars Posterior.*

others, probably on the spot.[81] His decisions about which observations to mark were certainly based on a skilled assessment of the human and natural conditions at the time. Whatever his selection criteria may have been, he apparently never combined his observations to produce an unobserved number. Every distance listed in the *Pars Posterior*, and almost certainly every asterisked coordinate in the manuscript (and so in the *Prodromus*), was itself the direct result of a *singular* observation, and accordingly required an act of judgment on Hevelius' part rather than the rote application of a numerical algorithm.

The very fact that Hevelius published his raw observational data and only indicated with a mark which item was the best raises an intriguing question. Although the manuscript that Hevelius worked on for the *Catalogus Stellarum Fixarum* provides several values for a star's longitude and latitude (figure 2.3), each of which had

asterisked distances in the *Pars Posterior* lead to the marked and printed coordinates in the MS and the *Prodromus*, though it seems probable that they do.

[81] The Observatoire volumes consist of ruled pages bound together. Hevelius likely had the volumes ruled and bound and entered the data directly on observation, marking the best among them at once.

to be computed from the corresponding raw data, the posthumously printed version provides only single values—the very ones that Hevelius had marked with an asterisk in the manuscript. Since Hevelius had worked extensively to prepare the final version for printing, it seems that he had decided to publish only one computed set of coordinates for each star. Why then had he earlier printed the several raw observations? Why, after all, print bad observations?

We have come face to face with an instance of the early modern attitude toward data. Sometime toward the end of the eighteenth, or the beginning of the nineteenth, century, scientific reports began regularly to include tables of discrepant data—of numbers, that is, which did not agree among themselves. A particular experiment, or a series of observations, would yield a multiplicity of values for something that was believed to constitute a natural (and therefore uniquely valued) phenomenon, say a planetary position at a given moment, a magnetic declination, or an angle of refraction. By that time, as it is today, this was thought to be quite normal and indeed to be expected. It was not considered to indicate a failure of experimental or observational skill but rather to be an inevitable by-product of the very processes of instrument construction and of measurement. Opinions concerning the trustworthiness of astronomical (and other) numbers were quite different during the seventeenth and eighteenth centuries, even though—since antiquity—measurers and observers had been concerned to make their instruments as precise as possible.

In comparison with later periods, observers and experimenters were few and far between until the late eighteenth century, while instruments and measuring devices were often unique items produced by local craftsmen. Tycho, William of Hesse, Hevelius, and Flamsteed all had their instruments specially fabricated. One might think that disputes over numbers would consequently revolve in perpetuity with little possibility of resolution other than by social *fiat*. For the most part we find nothing of the kind. Astronomers for example did produce different stellar coordinates from the time of Tycho onward, but we rarely find any disputes over whether or not observer *x* worked his apparatus correctly in producing number *y*. There was a great deal of discussion concerning ways to improve the accuracy and precision of instruments, and these did take into account what the later observer thought to be the range of trust to be placed in previous results. Usually the later astronomer accepted the skilled judgment of his predecessor in such matters. Hevelius, for example, did not undertake his program in order to challenge Tycho's numbers; he sought to improve them. Certainly each succeeding observer thought his products to be more precise, and more accurate, than his predecessors', but not because he challenged the integrity of their numbers.[82]

Located far from the centers of early modern science, Hevelius was trained neither as a philosopher nor as a mathematician, though he had traveled widely and had strong contacts with members of the republic of letters. The son of a successful Danzig brewer, Hevelius himself had become an important and wealthy figure in this busy trading city, with sufficient resources to construct an elaborate observatory and printing shop. How could he, given his background and location, convince others that his observations should be accorded trust? One way to do so would be to provide

[82] See for example Flamsteed's respectful assessments of both Tycho and Hevelius in Chapman, 1982.

evidence that would enable others to judge the quality of his work by providing examples that were less than the best. The presence, one might say, of bad numbers could lend weight to the good ones among them.

If we look a bit earlier, we can find evidence of a related attitude toward good and bad measurements in a context that seems in other respects to be closer to the later understanding of what to do with deviant data. Though he did not combine discrepant results to produce a final number, Galileo did develop a sophisticated method for choosing among such things in his argument against the anti-Copernican (and anti-Tychonic) Scipione Chiaramonti's claim that observations proved Tycho's new star of 1572 to be located below the orbit of the moon.[83] To counter Chiaramonti, Galileo assumed that observational errors are inevitably present to one extent or another, that their deviations from the true value are scattered, but that the larger the deviation the less frequently it occurs. On this basis he rejected Chiaramonti's claim since it required a much larger value for the total correction that would be needed to place the star below the moon than the one necessary for the superlunary claim (by a factor of nearly eight to one). Though there is no hint here of any connection to the average, nevertheless, as Hald points out, Galileo had developed criteria that are similar to those later used by Boscovich in minimizing the sum of the absolute deviations. Yet the core point of Galileo's argument was not to show how to take account of all the available data, but rather how to exclude unacceptable results in order to choose the best ones.

Singular numbers—the golden results of measurement—are also strikingly evident in Huygens' work in optics. In the early 1670s, Huygens read Erasmus Bartholin's description in 1669 of the crystal Iceland spar's strange double refraction.[84] Bartholin was professor of mathematics and medicine at the University of Copenhagen, having studied mathematics at Leiden under Frans van Schooten. Though he published a number of articles, he mainly occupied himself with editing Tycho Brahe's observations. Bartholin and Huygens had met in Holland during the 1650s, but they had had no subsequent contact.

Jean Picard, the French astronomer, was in Copenhagen in 1671 to measure the distance between the new observatory in Paris and Brahe's Uraniborg on the island of Hven. Bartholin and his son-in-law, Ole Rømer, assisted him. Picard brought back to Paris both Rømer and several crystals of spar. On July 8, 1671, Fr. Ignace Pardies wrote Huygens about the crystal's strange optics, and Huygens soon began his own investigations. Iceland spar has the then-unique property of visibly doubling an image seen through it. One of the two images behaves in the usual way of an object seen through glass or water, but the other—the extraordinary ray, as it became known—does not: it rotates around when the crystal is turned and, as Bartholin showed in some detail, does not obey the ordinary law of refraction (figure 2.7).

[83] For an informative analysis of Galileo's discussion in the 1632 *Dialogue Concerning the Two Chief World Systems* see Hald, 2003, pp. 149–60. We thank Noel Swerdlow for further discussion.

[84] The original publication is Bartholini, 1669. It is reprinted in facsimile with a translation and an introduction that briefly discusses Bartholin's discovery in Archibald, et al., 1991. Henry Oldenburg at the Royal Society received a copy of Bartholin's book and summarized it in the *Philosophical Transactions* (Oldenburg, 1670/1).

Figure 2.7. Iceland spar's double refraction.

In 1672, Huygens began working with the crystal, in the process discovering at least one new phenomenon, namely, that light which has passed through one crystal is somehow changed in a way that makes it behave differently when it traverses a second crystal. Huygens does not seem to have made any significant measurements of his own at this time. Near this time he had been introduced (probably by Pardies) to the concept of light as a spreading surface, and to the ray as something perpendicular to the surface. A contemporary drawing from his notes indicates that he already perceived a difficulty posed by the extraordinary ray for such an account.[85]

Occupied for the next several years with publishing his account of isochronous pendulums (whose results were by then widely known), with sea-trials of his pendulum clocks, and with a priority dispute with Hooke over the spring-balance, Huygens did not return to Iceland spar until 1676, while in the Hague. By August 1677, he had

[85] Huygens, 1888–1950, vol. 19, p. 414. Namely, that the ray seemed to be oblique to the front: "How can the perpendicular ray become oblique by refraction [referring to the fact that a normally incident ray is turned through an angle in extraordinary refraction], because the rays won't be at right angles to the direction of their extension or emanation, contrary to what our hypothesis concerning light requires." See Dijksterhuis, 2004, pp. 140–52 and Shapiro, 1973.

developed in full the fundamental concepts of a wave-based optics, explaining the behavior of spar by replacing the expanding spherical light wave of ordinary refraction with an ellipsoid—in the process reducing the optical ray to a line drawn from the center of emission to a plane tangent to the expanding front, thereby replacing a problem with a necessity, for now the extraordinary ray had to be oblique to its front in uncut crystals of spar.

Huygens' manuscript notes include a detailed diagram, showing a section of his new ellipsoid together with numbers for its several radii and for an unspecified refraction. We can use these numbers to understand what Huygens did, or did not do, by way of optical measurements at the time.[86] And here we find a most interesting thing. Despite the fact that Huygens was developing a profound revision—both physical and mathematical—to the very heart of contemporary optics, he undertook no measurements to test his new theory. Huygens worked hard to transform his ellipsoid into a tool for calculating refractions, but he apparently did not at that time apply it to anything that he had himself measured. Instead, he assumed that the one specific refraction which Bartholin had looked into—namely the ray that is not deviated in extraordinary refraction—was indeed parallel to a crystal edge, as Bartholin claimed.

There are many numbers on Huygens' 1677 diagram, but the only measurement of a refraction angle that he made at the time was for a normally incident ray, and even then only because Bartholin had not specified its value. Huygens did make one other optical measurement, since he needed a value for the radius of the sphere that governs light in air, and that could only be found by working with the crystal. Yet even here he did not actually measure a refraction angle, but seems instead to have used observations of the visual heights of images in the crystal. In 1677, then, Huygens measured refractions only where Bartholin had not done so or where the demands of the new ellipsoid required him to do so.

Huygens did eventually undertake careful refraction measurements with spar in 1679, when Rømer queried his claims concerning the crystal. Even then, however, his notes do not provide lists of the several observations that by his own account he had made to measure the crucial parameter of the light sphere's radius in air. Moreover, when, late in the summer or early fall of 1679, Huygens refined even these measurements, he left no manuscript record of the process, and his 1690 *Traité de la Lumière* provides no information at all about the set of observational numbers that Huygens must have worked hard to produce, or of how he chose among them. His actual observations have vanished altogether from the record. Although Huygens was highly sensitive to difficulties of measurements and paid close attention to what might be concluded from them,[87] he was clearly not concerned to combine the several results of a given measurement done over and over again. He certainly did perform measurements several times, but he apparently kept at it only until he was confident that

[86] For a reconstruction of Huygens' measurements and computations, see Buchwald, 2007a.

[87] His concern appears particularly well in his notes and report on a sea voyage which carried his pendulum clocks, and from which he concluded that they worked well enough to determine longitude. Central to Huygens' argument was his assertion that the evidence from the voyage was consistent with his contention that gravity varied over the earth as a result solely of the dependence of centrifugal force on distance from the earth's axis. See Schliesser and Smith, 1996, for a discussion of what Huygens did conclude, and what he might have concluded.

the observation was being done as well as could be. And, in the end, he always provided a number that resulted from a single, well-done observation, and not one that had been fabricated out of a congeries of measurements.

Huygens' production of experimental numbers was hardly unique. By the third quarter of the seventeenth century, groups of numbers were appearing with increasing frequency in laboratory experimentation, where they inevitably raised issues of choice, though demonstrations of novelties were more common than the production of elaborate measurements.[88] Yet, though those who produced experimental results inevitably encountered discrepancies among their measurements, in the end almost no one attempted to generate a final result by putting together a host of discordant observations.[89] This was so whether the measurements were all of the same process done several times (e.g., a given star's position) or whether they were of different processes, all of which should according to some theoretical account produce one and the same number (e.g., water's index of refraction measured at different incidences). In neither case was it common to combine an entire group of measurements.

The goal seems accordingly to have been the production of the best single result; the others were simply thrown out as bad products. That number stood alone because the early modern measurer considered it to represent the best that his skillful work could produce. In what seems to have been a common understanding, the limitations imposed on skill by the unavoidable defects of the senses, particularly as the senses came to be thought of as themselves akin to instruments, placed a premium on achieving the best result possible with the artificial and organic devices available to the measurer. To this way of thinking, the limitations of the senses had no essential bearing on the result of measurement, beyond setting limits to the outer possibilities of skill. It was important to ascertain those limits, which were assumed to be essentially fixed. The only question was just what the limits were, which, as we have seen, provoked a heated controversy between Hooke and Hevelius. When Hooke lined up

[88] See Meli, 2004 for a discussion of numerical tables in Galileo and Mersenne. The Jesuit astronomer Riccioli, whom we will encounter again, printed a table "of experimental data on bodies of various materials and sizes falling from the Asinelli tower" which "showed that in no case among the twenty-one investigated did the spheres fall at exactly the same rate" (ibid., p. 185), for which he "tried to provide an Aristotelian explanation" of the discrepancies. It would be interesting to know the extent to which discrepancies in certain kinds of measurements before the subsidence of scholasticism were linked to expectations that mathematical structures can never fully match a world in Aristotelian flux. To someone who thinks this to be a normal situation, anything like an "average" could have little meaning, because the very point of the average is to eradicate as much as possible the difference between a measurement and a "true" value. If there is no such thing as a true value then the average, or any procedure designed to do the same thing, would usually be pointless. Riccioli did however use an arithmetic average in the traditional sense—i.e., a cut exactly midway ("media") between two numbers—when each of the two had been produced by trustworthy computers: see Riccioli, 1651, p. 139, where in the table of values for the solar year he calculated a "media" in four cases. This is not an average but a judicious refusal to choose between two reliable values.

[89] For details and further examples ranging from Tycho's stellar catalog to Antoine Parent's experiments with the pressure-volume relation for air see Buchwald, 2006. Tycho may have used a method for the sequential processing of observations, one that did not combine them en masse but which sought instead to give the greatest possible weight to more recent ones (and so to ones that were perhaps more skillfully performed).

the luminaries of the Royal Society to measure their visual acuity, he was looking for that fixed limit; he did not try to see whether different conditions—of the seen object or of the seeing eye—might affect the limits, might indeed render them sufficiently uncertain that there would be little point in attempting to capture them in a number which represented the limits of skill. The notion that the defects of the senses can be corralled by pushing skill to its limits suggested, in an apparent paradox, that sensual imperfection could be ignored, that it could be buried deep within the golden result of the best, the most skillful possible of measurements.

That practical conviction returns us to the young Newton, who did not share it. For him the connection between an object and its visual perception always had the potential to be deceptive because of the many ways in which physical perturbations can affect the motions that result in sensation. The eye's inherent limits were of comparatively minor interest by comparison, and indeed Newton did not discuss them at all when, around the time of his youthful comet observations, he began to probe the vagaries of vision and to investigate the nature of color.

Several folios past Newton's comet observations in the *Questiones* he began a section titled "of colors." The editors of the notebook date this section to mid-1664, and so about a half-year before the appearance of the comet.[90] The material in this part of the notebook describes his first investigations of colors, undertaken after reading Boyle's recent book, on which he took careful notes.[91] Newton had already stared at the sun to produce afterimages, and Boyle had extensively discussed the production of color sensation in the absence of a colored object—what we now call phosphenes. To provide evidence for the claim that light produces color "but [only] as it produces such a determinate kind of local motion in some part of the brain,"[92] Boyle described several arresting situations, including colors seen by plague victims. "An ingenious man, that was so bold, as without much scruple to visit those that were sick" of plague, Boyle wrote, noted that before they showed symptoms the victims "would tell him that the neighbouring objects, and particularly his cloths, appear'd to them beautifi'd with most glorious colours, like those of the rainbow." Moreover, "when a man receives a great stroak upon his eye, or a very great one upon some other part of his head, he is wont to see, as it were, flashes of lightning, and little vivid, but vanishing flames, though perhaps his eyes be shut." Newton took this much further and experimented on himself. He pressed his eye and described in careful detail the colors that he saw, including their shapes (figure 2.8). He even found that "by putting a brass plate between my eye and the bone nearer to the midst of the tunica retina than I could put my finger, I made a very vivid impression, but of an elliptical figure because the edge of the plate with which I pressed my eye was long and not round like my finger."[93]

[90] The section "of colors" (McGuire and Tamny, 1983, pp. 430–33) must postdate the publication of Boyle's *Touching Colours* early in 1664, and likely precedes Hooke's *Micrographia*, published the next year, since Newton doesn't mention it (Shapiro, 1984, p. 5). The comet section precedes "of colors" by five full folios, but the first mention of the comet, with Newton's spurious observation, occurs twenty-one folios before his longer series. He obviously jumped around in the notebook, so that the physical position of a section is not a reliable guide to its position in chronological sequence.

[91] See Shapiro, 1984, pp. 5–7.

[92] Hunter and Davis 1999–2000, vol. 4, p. 29.

[93] McGuire and Tamny, 1983, pp. 438–39.

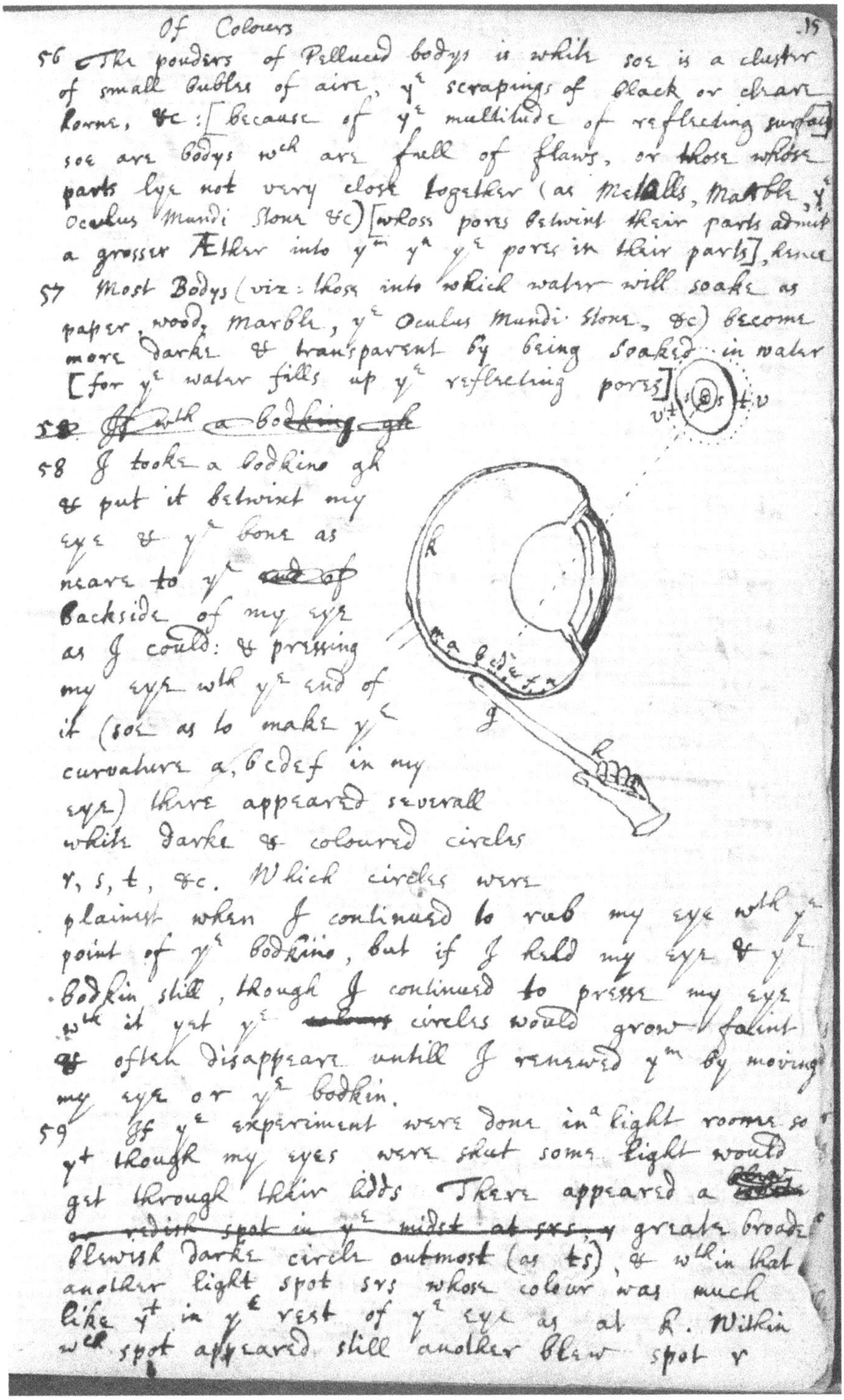

Of Colours .15

56 The powders of Pellucid bodys is white soe is a cluster of small bubles of aire, ye scrapings of black or charre horne, &c: [Because of ye multitude of reflecting surfaces] soe are Bodys wch are full of flaws, or those whose parts lye not very close together (as Metalls, Marble, ye Oculus Mundi Stone &c) [whoss pores betwixt their parts admit a grosser Æther into ym yn ye pores in their parts], hence

57 Most Bodys (viz: those into which water will soake as paper, wood, marble, ye Oculus Mundi Stone, &c) become more darke & transparent by being soaked in water [for ye water fills up ye reflecting pores]

~~58 If with a bodking gh~~

58 I tooke a bodkine gh & put it betwixt my eye & ye bone as neare to ye ~~end of~~ backside of my eye as I could: & pressing my eye wth ye end of it (soe as to make ye curvature a, bcdef in my eye) there appeared severall white darke & coloured circles r, s, t, &c. Which circles were plainest when I continued to rub my eye wth ye point of ye bodkine, but if I held my eye & ye bodkin still, though I continued to presse my eye wth it yet ye ~~colours~~ circles would grow faint & often disappeare untill I renewed ym by moving my eye or ye bodkin.

59 If ye experiment were done in a light roome so yt though my eyes were shut some light would get through their lids There appeared a ~~[illegible]~~ ~~a redish spot in ye midst at sxs~~ greate broade blewish darke circle outmost (as ts), & wthin that another light spot srs whose colour was much like yt in ye rest of ye eye as at k. Within wch spot appeared still another blew spot r

Figure 2.8. Newton pressed his eye some time ca. 1666. ULC MS 3975, f123v. With permission of University of Cambridge University Library.

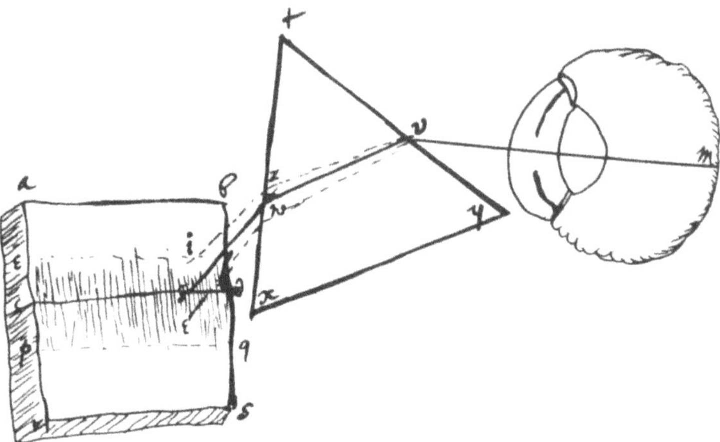

Figure 2.9. Object, prism, and eye. ULC MS 3996, f122r.

Newton's careful descriptions of self-produced phosphenes appeared in the same section of his notes that described his use of a prism to view the boundary between two differently tinted regions painted white, black, red, or blue. Newton had two shades each of white and black, and he paired a white or black region with a red or blue one or a black with a white. Three objects appear in the sketch that he drew of the experiment: the painted object itself, the prism, and the viewing eye (figure 2.9). Each region sends its light to the prism, which brings them together and sends them on to the eye. There, in the eye itself, a new color is made manifest. If, for example, the upper region is blue and the lower one black, then the boundary appears to be "green or red," though neither of these two colors shows when the regions are seen without the prism.[94] Moreover, the color depends upon whether the white or black is painted above or below the red or blue. White above and blue below produces red, while red above and white below produces blue. But black above and blue below produces more blue. White above and black below produces red, while black above and white below produces blue. Blue above and white below on the other hand produce an even bluer boundary.

Newton conceived that the prism has separated light in ways that empower the eye to perceive colors that are not seen absent the prism. The essential notion was, first, that the only difference between visibly black and white regions is the amount of light that each sends to the prism. Second, the light from a white region contains a mixture of rays which differ among one another in ways that permit the prism to effect a separation, refracting some more, others less. The optic nerve is set into motion when a ray strikes it, and a given ray sets up a specific movement. Newton's idea at this stage

[94] McGuire and Tamny could not produce Newton's green at the boundary between upper blue and lower black (1983, p. 247), but we have been able to do so. We were able to see a narrow band of green above and red below. Much depends on the color blue used.

was that the more varied the motions that are engendered in the nerve by the rays that strike it, "the more bodies appear white, black, or gray." Conversely, the more uniform the motion engendered "the more bodies seem to be colored red, yellow, blue, green, etc."

Newton's prism, then, alters the light entering it in such a way that the exiting beam consists of rays that individually make the colors of the spectrum, with the visibly blue light refracted the most and the visibly red the least. The border experiment, as he understood it, accordingly sends various mixes of rays to the eye from the prism. For example, if the upper region is black, and the lower white, then the upper one sends few rays of any sort to the eye, while the visibly blue rays that emerge from the prism will be bent in such a way that they appear to come from the upper, black region, with the result that the eye sees blue in the border region.[95]

Turn again to Newton's diagram (figure 2.9), and imagine that we remove the lines which represent the light rays. We are left with three objects: there are the painted block, the prism that examines it, and the eye that examines the prism. Both object and prism are artificial, made things; the eye is a natural being. Yet the prism is to the object as the eye to the prism, for the eye is instrument to the prism, as the prism is instrument to the object. There is, however, a signal difference between prism and eye, for the function of the prism is to alter the effects produced by the object, enabling the eye thereby to probe visually what it could not otherwise perceive at all. The eye itself becomes an active instrument in virtue of the transforming power of the prism. For if we now put Newton's light rays back into his diagram we see that the Newtonian eye assembles into new patterns the light from the object that the prism had disassembled.

The form of the relationship in Newton's experiment reflects the specific function of the prism in it. If we replace the prism with a telescope, for example, then the eye is no longer to the instrument as the instrument is to the object, because the function of the telescope is to improve the eye's ability to distinguish features of the object that the unaided eye could perceive were it able to come close enough. For Descartes, for example, the telescope was an "artificial organ"; the eye on the other hand was not, as it were, a "natural telescope."[96] More to the point in respect to Newton's early experimental practice, consider Descartes' 1637 account of the rainbow in his *Météores*, which was included in the 1656 *Opera* that was in Newton's library, and that he used

[95] Newton's diagram (figure 2.9) depicts the rays entering the eye from the prism along a single line, resulting in a perceived color at point *m* of the retina. If the upper region were white, and the lower blue, then, since the blue would be darker than the white, the less refracted and more intense light that makes red from the top would enter the eye along with the more-refracted but less intense blue from below, resulting in perceived red. He assimilated light to "globuli" that move with various speeds. No one speed produces a specific color; "blue, sky color, and purples," for example, "arise from the slow ones," whereas "from the swift ones red, yellow," and green from intermediates. Further, he concludes that colors "are made in bodies" by their altering the mixture of speeds in light that strikes them, "redness, yellowness, etc.... by stopping the slowly moved rays without hindering of the motion of the swifter rays." Consequently, as noted in Shapiro, 1984, p. 6, Newton holds at this point the notion common among mechanical philosophers that bodies make colors by modifying light.

[96] Descartes, 1965, p. 127; Adam and Tannery, 1897–1910, vol. 6, p. 165: this is the 8th Discourse of the *Dioptrique*.

in 1664. To understand the rainbow's colors, Descartes turned first to device-assisted observation—to a round-bottomed and water-filled glass flask, in effect a globe of water with a uniformly thick glass surround.[97]

Though comparatively short, the discussion is extraordinarily complex in its mixture of observation, geometry, and mechanical model. It produced a great deal of confusion at the time, and has done so ever since. We can be quite certain that Newton read it, as well as the *Dioptrice*, carefully, not least because the foundation of Descartes' account lies in his observations of boundary colors. The better to understand the production of colors by raindrops, Descartes decided to explore the colors produced by prisms. To do so he set up the configuration of figure 2.10.

Among the several complexities in Descartes' account, one in particular seems especially to have caught Newton's attention: the use of an aperture, placed by Descartes on the bottom of the prism, to restrict the beam of sunlight, which then shines on the screen (*PHGF*). In Descartes' narrative, the production of colors required the conjoint presence of both aperture and refraction; if either is absent then (in his configuration) colors do not visibly occur. Furthermore, the extent of coloration depends markedly on the size of the aperture; if it is small then colors appear throughout the image, but if it is too large then the colors recede to the borders of the beam.

Although complex and difficult to reduce to its several elements, anyone who carefully reads Descartes' account would perceive two central elements in it. The first, exemplified by his experiments with the prism, was the signal importance that Descartes placed upon the aperture as the generator of colors in refracted light. But the second would be Descartes' reconceptualization of the aperture as a function rather than as a physical item. Thinking over the difference between the wide and the narrow, Descartes apparently concluded that the important thing about apertures is not absolute—the delimitation of the lit from the unlit by a physical barrier—but relative. He began to conceive of apertures not as physical blockages of light but instead as regions wherein the interactions of neighboring rays change from point to point. Sometimes the interactions change slowly (wide apertures), sometimes more rapidly (narrower ones), and sometimes swiftly indeed (very narrow apertures). A physical object that cuts out light would now be an *instance* of an aperture and not its sole exemplification. Another instance would be a region where the density of rays changes quickly, for this too provides a situation in which whatever interactions among the rays take place must change from point to point. The drops of water that produce the rainbow have no physical borders, but Descartes' geometry demonstrated that they cluster rays of light, thereby producing a functional aperture.

Descartes' reconceptualization transformed the color-generating boundaries of the aperture into generic places where rays in one state abut rays in another state. He went no further with this beyond applying it to the rainbow. The young Newton, however, did. He chose the same two colors—red and blue—that appear at the boundaries in a Cartesian wide-aperture experiment and then painted them individually on a board abutting a region painted either white or black, with black coming in both

[97] Descartes' excursions with the rainbow have been discussed many times. See in particular Armogathe, 1987, 2000; Boyer, 1959; Garber, 1993; Gaukroger, 1995; Lee and Fraser, 2001; Sabra, 1981; Shea, 1991. The points made below are developed in Buchwald, 2007b.

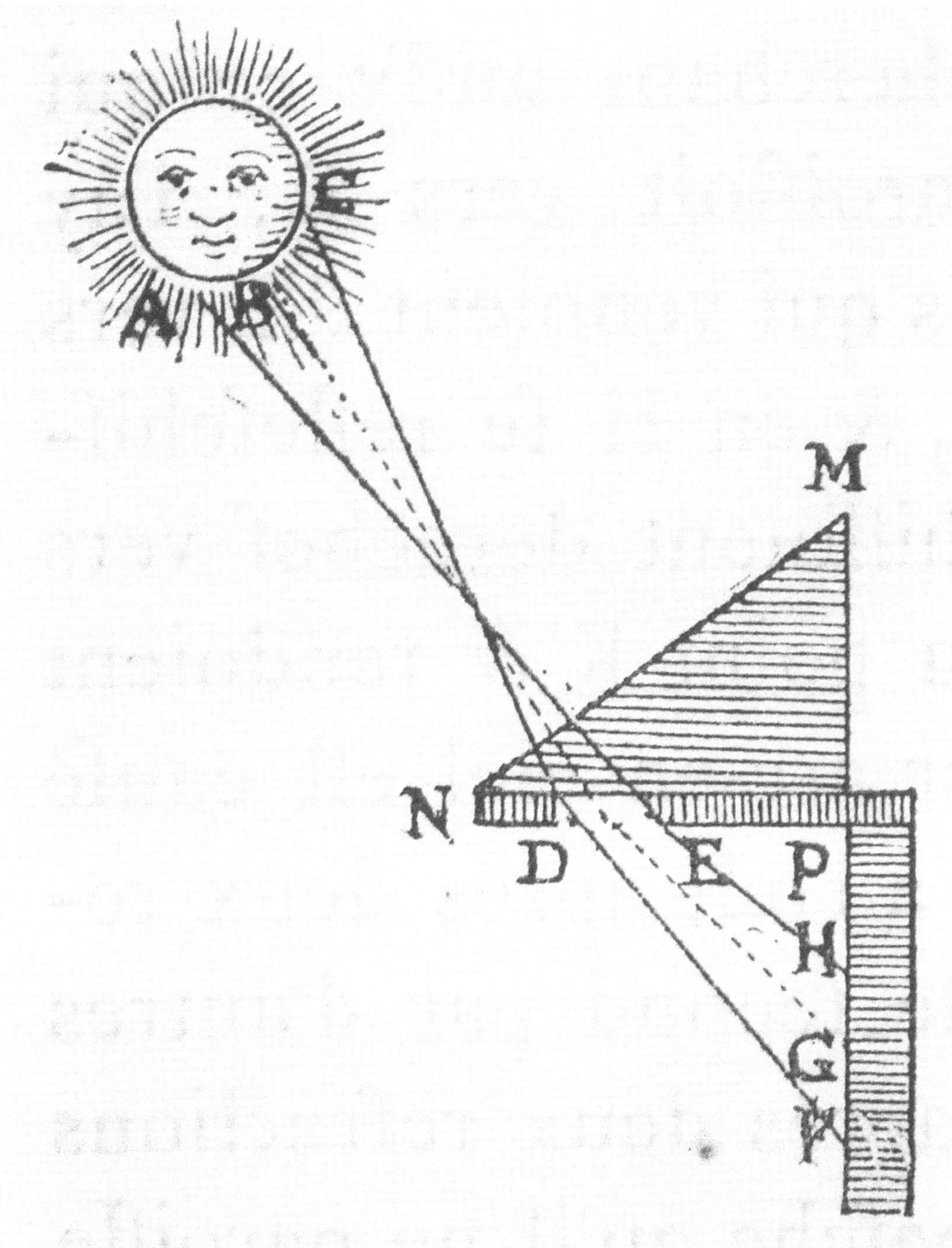

Figure 2.10. Descartes' color-generating prism, exit-aperture, and attached screen. Adam and Tannery, 1897–1910, vol. 336, p. 330.

"black" and "blacker." Looking at the boundary region through a prism, he then drew up a table with the colors that he saw.

There's a noticeable difference between Newton's way of experimenting and Descartes', even though at this early point in Newton's career he was looking to much the same phenomenon that had engaged Descartes. There is first of all the most obvious difference between them: in Descartes' understanding, "white" corresponds to a normal state for light, whereas colors correspond to deviations from the norm. Newton, however, understands color as the normal—indeed the only—physical state of light, whereas white or gray always represents a mixture of these states.[98] More significant for our purposes here, Descartes' prism experiment manipulated only one parameter—the width of his aperture—and then only to determine that the aperture must be sufficiently narrow in order for full tinting to become visible. Newton instead manipulated the object under observation—the bordering colors—as well as the prism itself since (to judge from his diagram) he did not fix the prism in place with respect to his painted objects. Where Descartes, as it were, engaged in a form of nearly static experimentation, Newton actively changed his configuration.

As Newton continued to work with colors in 1665–66, he transformed his system even further, eventually passing the sun's light through a small aperture in a shade-covered window, and then through a prism, which cast the solar image onto a wall 260 inches away. The prism was set close to the aperture ("one little round hole") and oriented so that the rays were "equally refracted" on either side, in which case the image on the wall should have been circular but was instead "oblong".[99] The experiments described in these notes form the foundation of what would grow over the next few years into Newton's fully fledged new theory; by the latter part of 1668 he was also likely constructing his reflecting telescope. The manuscripts provide little information on the details of his evolving experimental technique or of how he worked with numerical data, but they do show him to be constantly engaged in altering the experimental design to reveal aspects that he deemed important. Descartes never worked in that fashion. And there is very little evidence to show that anyone else did. Huygens certainly examined the properties of his crystals in various ways, for example, but he was seeking to establish parameters for a nearly fully developed geometrical theory. Newton by contrast was engaged in truly exploratory experimentation, changing first this and then that factor to probe the effects.[100]

[98] Although Descartes believed that white represents a normal state, he also thought that white would occur if every ray in a beam deviated from the norm, as long as the deviations were random—see Buchwald, 2007b for discussion. It's unlikely, though possible, that Newton could have picked this up by reading the *Dioptrica* and the *Meteora*, however, because Descartes elaborated the point in Latin only in correspondence with the Dutch Jesuit Jean Ciermans. Although the correspondence was printed in 1666–67 by Clerselier, he translated it into French and did not include the Latin until the edition of 1724.

[99] McGuire and Tamny, 1983, pp. 468–69: Tamny and McGuire date this second "Of colours" (CUL MS Add. 3975, fols. 1–22) to 1665–66. Shapiro, 1984, pp. 12–13 dates it "to sometime between the beginning of 1666 and 1669, but most probably closer to the former than to the latter."

[100] Several of his probes seem particularly designed to explore further the production of colors at boundaries in ways that Descartes' understanding of border effects could not easily accommodate.

During the years between 1666 and 1671, Newton developed a novel approach to experimentation, and, we shall see below, to the evaluation of quantitative data. The two novelties were actually different sides of the same coin. To understand them we must jump forward in time to the lectures that Newton gave after he became Lucasian professor in the fall of 1669. The first of his *Lectiones opticae* was delivered in January 1670; together with the second, it expounds the famous prism experiment that he had first undertaken in 1666, and here he provided revealing details.[101]

The letter that Newton wrote to Oldenburg on February 6, 1671/2, and which Oldenburg printed in his *Philosophical Transactions*, differs in a number of details from the previous year's Cambridge lectures.[102] Both the lectures and the published version nevertheless betray Newton's particular ways of working with data, but with a significant difference. Consider first the account in the lectures.[103] There Newton described two experiments. In the first, a prism is placed inside and next to a small hole in a window or wall through which the sun shone into an otherwise dark room, its prismatic image being cast onto the opposite wall. In the second, the prism was placed between a hole in an "opaque body" and the sun. Newton specified a 60° prism, but he gave no other dimensions or settings, remarking only that the image is "a very oblong figure.... specifically, one whose length ... is four times and more its breadth." Four was accordingly a minimum ratio among the images produced at different prism orientations for the particular distances that Newton was using. He continued by noting that if all the rays that strike at equal incidences had been refracted at equal angles—if, that is, the usual law of refraction held for all rays of the sun with the same value—then "in a certain position of the prism [the solar image] would appear to the senses completely circular."[104] Since this does not occur, he argued, "at equal incidences some rays undergo a greater refraction than others."

The phrase "would appear to the senses" suggests the inherent limitations on what the senses can perceive, further indicating that Newton had developed techniques to work with these limitations rather than to seek as perfect a result as developed skill might provide. There was, however, one skill that Newton had evolved to a high degree, and that he deployed, namely, geometrical demonstration. He accordingly argued that a position of the prism does exist such that the breadth of the image in a plane that contains a principal section of the prism[105] and that passes through the center of the image (call this the vertical length) is "sensibly equal" to its breadth in the orthogonal plane through the image's center (the horizontal length). More precisely, and significantly, what Newton claimed to show was that the two lengths do "slightly" differ from one another but that, in Newton's specified orientation of the prism, the horizontal length cannot be smaller than its vertical counterpart—whereas experiment shows the precise opposite to be the case. Newton did not prove the point

[101] Shapiro, 1984, pp. 46ff.

[102] There are two versions of the lectures: a shorter one, the *Lectiones opticae*, which Newton retained, and the longer *Optica*, which he deposited at the University Library in October 1674: on which see Shapiro, 1984, pp. 16–20.

[103] The accounts in the retained *Lectiones* and the deposited *Optica* are essentially the same.

[104] Shapiro, 1984, p. 53.

[105] A principal section is perpendicular to a base of the prism and to the opposite edge.

in detail, and so he argued that in any case the difference between the two lengths is so slight "that they may be considered as sensibly equal"—equal, that is, insofar as the ability of the eye and hand to discriminate between them is concerned.[106]

Skill in geometry has underpinned the unavoidable limitations of instrument and sense by demonstrating that an unobserved difference which must exist falls below the threshold of detectability. This sort of argument—the explicit use of geometrical reasoning to delimit the regime of measurement—had rarely been used before Newton, and certainly not in so thoroughgoing a manner.[107] Nor did he stop here, for he went on in this first lecture to suggest a way to ensure that the prism is properly set. He did not provide geometrical backing for the suggestion until the ninth lecture (delivered the following October).

That instruments are imperfect, or imperfectly placed, adjusted, and used—and that even skillful observers cannot achieve perfection—was obvious to any measurer or experimenter. The difference between this form of observational or experimental practice and Newton's is that he used geometry in two novel ways: to demonstrate what cannot sensibly be detected, and, as we will see now, how to use the sensibly undetectable as a resource for establishing an experimental claim.

In the optical lectures Newton carefully specified the position of the prism at which the horizontal breadth of the image should sensibly equal its vertical length if the law of refraction held with a constant value: set the prism between the sun and the hole in such a way that the ray from the sun's top which is refracted by the prism, and then passes through the hole to the bottom of the image, makes the same angle with the side of the prism at which it enters that the ray from the sun's bottom and then to the top of the image does with the prism face at which it exits. But how can this be ensured? How can the experiment be done so that the angles in question will be "equal" to one another? Moreover, can this be done in such a way that unavoidable inequalities will not "sensibly" affect what should be the apparent equality of the image's vertical and horizontal lengths? "Even if," Newton asserted, "the prism's position were other than I have described, as long as the rays do not undergo a particularly unequal refraction on each side, the shape of the image will nevertheless hardly be changed because of that."[108] How did he know?

In his lecture Newton did not deploy geometry to answer the question; instead he recurred directly to experimental manipulation to specify how the prism should be set. Turn it about its axis "with a slow motion," he instructed, and "you will see the colors that it makes shifted with a continuous motion from place to place, so that in fact they appear to move now forward and now backward." Between forward and backward "you will observe a mean" where the colors "now having moved forward and just about to move backward appear to stand still. When you see that, stop the prism and fix it in that position."

[106] The full demonstration was produced in 1896 by Joseph Larmor: see Shapiro, 1984, pp. 58–59, n.28.

[107] The recognition goes far back, though the practice of estimating errors was not likely common. The last Director of the Library at Alexandria, Theon, in a rare example, did consider the errors in solar coordinates that would be produced by an instrumental error of 6 minutes; Pedersen, 1974, p. 148, n.111 referring to Rome, 1937.

[108] Shapiro, 1984, pp. 59–61.

Stop where the colors "appear to stand still." That phrase enfolds a procedure both for setting the prism and for explaining why the position of the image in that orientation of the prism is insensitive to slight rotations of it.[109] The geometric underpinning of Newton's claim was given expressly and briefly in lecture 9, which discusses a method for measuring the ratio of sines for a specific entering ray, namely one that enters and leaves symmetrically. Newton merely specifies the condition, but he then explains the "advantage of this method of measuring refractions," namely that "a small departure from the required position is almost inconsequential, insofar as the deviation [of the ray from its initial direction] ... will not be perceptibly changed from this, as will be clear to anyone trying it." The reason for the insensitivity to prism orientation is provided by geometry: at symmetric passage the deviation is a minimum, "and in quantities generated by motion, when they are either a maximum or a minimum—that is, in the moment of regression—their motions are generally infinitely small."[110] The phrase "as will be clear to anyone trying it" amounts to a manipulative method for testing whether the prism has been set well enough: the appropriate position will be the one at which the image moves very little when the prism is wiggled.

Newton had transformed a consequence of his developed understanding of the relations between quantities generated by motion into a tool for ensuring the probity of an experimental or observational claim. No one before him had done anything quite like this. A good case in point concerns Descartes' and Thomas Harriot's (the latter unpublished) accounts of the rainbow, for neither of them had recognized or at least noticed the usefulness for experiment of the fact that a comparatively large change in one quantity, whose value determines a second, corresponds to a very small change in the second when the latter is near its minimum or maximum value—should it have one.[111] On the basis of this reasoning, Newton knew, he did not have to go to elaborate lengths accurately to set the prism's position since—the important and novel point—it made no "sensible" difference to the configuration of the image. Here we detect a new attitude to experiment, or, better put, an implicit method for the proper construction of experiments or observations that produce numbers: use whatever mathematical tools are available to construct situations in which the measurements respond comparatively weakly to perturbations in the configuration of the apparatus.

[109] See a discussion of Newton's claims, and the purported accuracy of his measurements, in the 1671/2 letter to Oldenburg. Laymon (1978) places particular emphasis on Newton's not having specified a method based on image position for setting the prism in the 1671/2 letter, whereas he did there explain the rotation method for testing the position; conversely, in the *Lectiones*, Laymon writes, Newton specified seeking the point where the image reverses, but leaves out the rotation test. However, as we will now see, Newton did include a remark that amounts to the rotation test in Lecture 9 of the *Lectiones*.

[110] Shapiro, 1984, p. 181.

[111] Although a simple method for determining the extrema of polynomials had been developed by Johann Hudde in the 1650s (Katz, 1998, pp. 473–74), there does not seem to have been explicit discussion of the property in question, though this is hardly surprising since it has little purely mathematical significance. That a property which changes with another will alter the least when the latter is at a maximum had been recognized by Kepler in his *Nova Stereometria* (Frisch, 1858–1871, vol. 4, p. 612): "near a maximum the decrements on both sides are in the beginning only imperceptible," translated in Struik, 1969, p. 222. We thank Jesper Lützen for the reference.

The contrast between Newton's discussion in the *Lectiones* and in the 1671/2 letter to Oldenburg reflects contemporary attitudes to measurement, which Newton must by then have understood. The *Lectiones*, recall, provides just one actual measurement—the prism angle—and that only approximate, namely "about 60°." The letter to Oldenburg provides what amounts to an experimental recipe: the wall on which the image was cast was 22 feet from the prism, which was placed close to the hole through which the sun shone; the hole was ¼ inch in diameter, the image 13¼ inches long by 2⅝ wide; the deviation of the rays which struck mid-image was 44°56′, the prism angle being 63°12′. Finally, "the refractions on both sides the prisme, that is, of the incident, and emergent rays, were as near, as I could make them, equal, and consequently about 54°4′."[112]

These numbers are too precise to represent what Newton could actually have measured.[113] Moreover, he knew perfectly well that his experiment established the failure of the sine law as applied to a common index for all the solar rays without requiring anything even remotely like this claimed accuracy. Further, even in the 1671/2 letter, Newton invoked the observational claim that wiggling the prism did not "sensibly" move the colors, showing "by that variation of incidence, the quantity of refraction was not sensibly varied"—note the use of the adverb "sensibly" twice in one sentence. Why then did Newton provide impossibly precise measures? The answer is not hard to seek in view of what we have previously seen concerning expected standards of the period: measurements were not to be nearly this or nearly that; they were expected to have specific values, which reflected the sureness of the experimenter's skill. Anything less would inevitably raise doubt, perhaps even doubt concerning the experimenter's trustworthiness. Newton seems to have recognized the danger by the late 1660s. And yet, knowing the actual impropriety of his claimed numbers, Newton shielded himself from potential criticism on this score by noting that even if he had set the prism somewhat off symmetric passage, nevertheless the image moved very little. In the printed letter this was presented by-the-by, as a confirmation that the prism had been correctly oriented. Instead, he had in fact used the image's positional insensitivity actually to *set* the angle. Had he made the procedure public, it would have seemed a testimony to inadequate skill rather than a novel method of experimental manipulation—a method that took advantage of a sensibly invariable position to achieve a good-enough setting for the apparatus. This was at the very least an unusual procedure, and may have been unprecedented.

The point and counterpoint that took place after the publication of Newton's letter are well-known, if still controversial, and we will not enter into the intimate details

[112] Cohen, 1958, p. 49; Newton, 1671–72, p. 3077.

[113] On which see Laymon, 1978, p. 243. To measure the refraction of water, Newton constructed a right-angle wooden prism through which he bored a passage at 45° to the sides of the right angle; the openings were covered with glass, and a third passage was bored through the remaining side which intersected the first. Water was poured through this last passage and the remaining opening plugged. In this case Newton did specify the deviation of the emergent ray to the minute in the *Lectiones* (Shapiro, 1984, p. 187). Here it was possible to measure more accurately since the angle of refraction was fixed at 45° by the configuration of the device, and the apparatus was by construction forced to minimum deviation.

except to remark two features that are pertinent here.[114] First, the unusual character of Newton's letter to Oldenburg: not only was it extremely brief and barely descriptive of experimental details beyond the provision of overly precise, but rhetorically necessary, numbers; Newton presented the account as a provocation, overthrowing "Snel's law of refraction—discovered only a generation earlier and now a new foundation of optics—rather than as an extension of it and thus needlessly unsettled his audience."[115] More important for our purposes, Newton's response to criticism by Hooke and Huygens of his claims concerning color nicely illustrates the novelty of his way of forging empirical claims on the basis of a profoundly skeptical attitude to sensual judgments.

Newton replied that there are "two sorts of colours. The one original and simple, the other compounded of these. The original or primary colours are red, yellow, green, blew, and a violet purple, together with orange, indigo, and an indefinite variety of intermediate gradations." The only way to determine whether a given color is compound or primary is to pass it through a device that will separate its components, e.g., a prism. Moreover, certain colors are the "same ... in specie with these primary ones," even though they are in fact compound, "for, a mixture of yellow and blew makes green; of red and yellow makes orange; of orange and yellowish green makes yellow." This is to say that colors which appear to the eye to be identical may only be sensibly so. To probe physical essence requires apparatus that is more able than the deceiving eye to unfold the true nature of a color. Most surprisingly of all, according to Newton even "whiteness ... [is] ever compounded, and to its composition are requisite all the aforesaid primary colours, mixed in a due proportion."

An "ingenious person in Paris"—Christiaan Huygens—wrote Oldenburg a letter concerning Newton's "theory of colours" in which he offered as "the most important Objection, which is made against him by way of *Quaere*... Whether there be more than two sorts of Colours."[116] Would not, Huygens continued, yellow and blue suffice "for all the rest"? Indeed, "it may possibly be, that *yellow* and *blew*might also be sufficeint" to produce white "which is worth while to try." And should it be so, then "it can no more be said [as Newton had] that all the Colors are necessary to compound

[114] Despite ongoing controversy over just what Newton claimed and what his readers thought he had claimed, Alan Shapiro has conclusively demonstrated that Newton's "experimentum crucis" was designed to show unequal refrangibility, and that his claims to have established a new theory, grounded in experiment, hinged primarily on this rather than on the separate—if clearly related—issue of color immutability, for which Newton required a more elaborate arrangement, albeit one that he did not explain until the *Opticks*. With the exception of a group of expatriate English Jesuits at Liège—who were regarded as generally incompetent, and who were largely responsible for Newton's exasperation with controversy by 1678—his assertions concerning refrangibility were widely accepted in Great Britain. For a thorough analysis see Shapiro, 1996, who writes, "The key to the adoption of [Newton's] theory was the initial adoption of the concept that sun light consisted of unequally refrangible rays. Once this was accepted, it was not a bold step to accept his more radical ideas on color, in particular that sunlight consists of rays of different color. Two groups, mathematical scientists and Scottish natural philosophers, emerge as the principal supporters of Newton's theory between its initial publication and that of the *Opticks* ... They came to support his theory without public debate or public testing of his experiments" (p. 62).

[115] Shapiro, 1996, p. 47.

[116] Cohen, 1958, pp. 136–37.

White." Note that Huygens did not object to the notion that white might be producible out of other colors, though it's not likely that he thought the two that he chose to be separately present within white.[117] He wanted, however, a mechanical hypothesis to explain colors, which he thought would be simpler were only two necessary to produce white, though he accepted the "accident (which is very considerable)" of the unequal refrangibility of colors.[118]

Newton did not take the remarks well. First, he asserted that "all Colors cannot practically be derived out of the *Yellow* and *Blew*," specifically "those which I defin'd to be Original," a point which he had in effect already made at some length in his reply to Hooke's (unpublished) report of his letter to Oldenburg.[119] If Huygens thought otherwise, then he, not Newton, should try the experiment. Moreover, Newton had not intended "to shew, wherein consists the Nature and Difference of colors." But even if Huygens could produce white "out of two *Uncompounded* colors," nevertheless "I will further tell him, why he can conclude nothing from that."[120] This unfortunate turn of phrase was particularly insulting, and the irked Huygens retorted: "seeing that he maintains his opinion with so much concern, I list not dispute." Nevertheless, "what means it, I pray," to assert that he (Huygens) could "conclude nothing from that," when Newton had expressly asserted "that to compose the White, all primitive colors are necessary"?[121]

At this point Newton recognized that he had to amplify and explain the distinctions that he had in mind. In explaining his apparently self-contradictory remark, Newton asserted that a white produced "were there any such" out of two uncompounded colors "would have different properties from the White, which I had respect to, when I described my Theory, that is, from the White of the Sun's immediate light, of the ordinary objects of our senses, and of all white *Phaenomena* that have hitherto faln under my observation." Pass the "white" produced by the composition of two uncompounded colors through a prism. The light that emerges from the prism will consist only of the original two, whereas the "white" of the sun, when passed through the prism, resolves into a full spectrum. Huygens' proposed white, should it be possible to produce it, might "to the naked eye … appear like other whites," but would nevertheless differ from these others unless the prism resolves it into the same components.[122]

[117] Since it was entirely possible to consider that yellow and blue in due proportion (or, in Huygens' words, appropriately "charged"), might each destroy the effect of the other, producing white as a neutral resultant, in which case there would be little sense in asserting the yellow and the blue to be present in white. Hooke objected to Newton's claims concerning white on similar grounds, for "we may say indeed and imagine, that the rest or streightness of the string is caused by the cessation of motions, or coalition of all vibrations; and that all the vibrations are dormant in it: but yet it seems more natural to me to imagine it the other way," Cohen, 1958, p. 111.

[118] Presumably Huygens had in mind that a given color, produced by the mechanical resultant of the two external mechanical correlates of yellow and blue, would in an appropriate mechanical model behave in a specific way on refraction.

[119] Cohen, 1958, pp. 143–44.

[120] Ibid., p. 145.

[121] Ibid., p. 147.

[122] Ibid., pp. 137–38.

The eye cannot be relied upon to judge the physical character even of the purest white, or indeed of any color whatsoever. Only a device that has the power to destroy the phenomenon can reveal what it truly is, and the senses have no such abilities. The eye, for example, operates less well than a telescope for seeing far, or than a microscope for seeing near, and so it may be considered to be a defective instrument when responding to its appropriate stimulus. But Newtonian senses are not merely defective, as Descartes, Huygens, or Hooke thought, for they cannot probe the aspects of things that lie beyond their ken.The eye cannot for example unpack the physical nature of color.

Newton accordingly came to think that the senses, and especially vision, inevitably and irremediably deceive. The senses can be artificially stimulated to produce perceptions that do not differ from those engendered by external objects. Or they can be overstimulated (e.g., vision by the sun). This was known to Newton's mechanically minded contemporaries, or indeed to scholastics. Aristotelians considered these sorts of effects to be illusions engendered by an improperly working sense. Mechanical philosophers, on the other hand, thought like Newton; whatever stimulated a sense in a given way would necessarily produce the same effect, with the inevitable consequence that even a properly working sense could deceive unless compensated by an appropriate device. Newton's early distrust of sensory information was accordingly exacerbated by experiments which convinced him that what seems a pure, unitary white to vision must be an occult congeries of colored lights. Only a device that did not improve, but that actually replaced, the ordinary operations of vision could detect the true physical nature of a colored light.

The senses are accordingly thrice problematic for Newton: they may be weak or defective even in their normal operations; they are unreliable in that they may—in uncontrollable ways—exhibit effects that are not a result of the stimulation of the external objects under investigation; and finally, they are inherently unable to probe the hidden structure of phenomena. Weak, unreliable, and inadequate senses can make the design, implementation, and interpretation of experiments, especially quantitative ones, a difficult and problematic affair. Pursued in the manner of Huygens or Hooke, namely by attempting to reach as good a number as possible through skillful work and (in the case of Hooke) elaborate apparatus, experiments could never wipe away the taint of sensory caprice.

It is probably not a coincidence that Newton's very first quantitative endeavor in the laboratory engendered a method to use capricious sense against itself. He had no way to orient his prism precisely to the minute, nor perhaps even to measure its angle as accurately as that.[123] Any reasonably simple attempt to do so would inevitably be compromised by shaky hands and imprecise vision. Worse still, the image that the prism produced was hardly regular or evenly illuminated, making any attempt to determine the location of its center inherently problematic. Newton's method of setting the prism by wiggling the image until it reversed direction took twofold advantage of sensory weakness. He must soon have recognized that the eye is much more

[123] For comparison, Huygens very carefully measured the interfacial angle of Iceland spar and found it to be 105°, whereas a more accurate value is 105°5′. See Buchwald, 2007a.

sensitive to motion than to static position. If the prism is oriented anywhere but near symmetric passage, then the image moves rapidly when it is wiggled.

Newton's method in these early experiments nicely avoided the problem of what to do when the physical conditions do not cooperate in ameliorating the unavoidable deficiencies of the senses. Other experiments that he performed, first in 1666 and then in 1671, labored under different problems. Here, fortunately, the record of actual data that Newton produced illustrates what he did when his measurements of the same phenomenon conflicted with one another. The experiments concerned "ye coloured circles twixt two contiguous glasses"—later termed "Newton's rings," though first described by Hooke in his *Micrographia*.

Pressing a spherical lens against a plate of glass, Newton had observed a sequence of colored rings in both reflected and transmitted light. In 1666 he had asserted, on the basis of comparatively crude measurements, that the distances between spherical lens and glass plate increase arithmetically from ring to ring.[124] In about 1671, he pursued more precise experiments; the remnant manuscript provides the final results of a series of measurements whose individual results were not specified.[125] The 1666 measurements were connected to an undeveloped physical model involving ether vibrations, which remained present in 1671, though not explicitly developed. There Newton set down six propositions, the first of which expressed the arithmetic relation. The propositions were followed by a series of experiments. In the first, he pressed the glasses hard together and then "measured ye diameters of ye darke circles in 100th parts of an inch (soe many as were conspicuous) beginning at the inmost or y^{t} nearest to ye darke spot in the middle." He then produced a first table with six columns: the first contained the sines of the angles between the plane surface and a line from the central dark spot to the eye; the second the ring's diameter (measured with a compass); the third its calculated square; the fourth the difference between succeeding squares; the fifth the difference between the average among all the differences and the square of the first diameter—what he termed the "meane difference"; and the sixth contained the difference between the "meane difference" and the square of the first ring's diameter (table 2.1).[126]

According to Newton's Proposition 1, the successive differences between the squares of the ring diameters for a given incidence should be constant and equal to the square of the very first diameter. In assessing whether or not this holds, Newton did something unusual, and even, as Alan Shapiro notes, "almost [we would say entirely] unprecedented in the 17th century"[127]: he *averaged* all of the differences, and he then calculated the difference between the average and the value for the first ring.[128]

[124] McGuire and Tamny, 1983, p. 476. To link ring diameter to the gap between the glass lens and plate, Newton used an approximation derived from an exact relation in Euclid, *Elements*, III, 36, according to which the square of the diameter is proportional to the size of the gap. See Shapiro, 2003, pp. 44–45.

[125] The MS (CUL MS Add. 3970, fols. 350–353^{v}) is transcribed in Westfall, 1965, pp. 191–96.

[126] The table is reproduced in both Shapiro, 2003, p. 48; and Westfall, 1965, p. 192; Shapiro additionally contains an image of the manuscript page (p. 49).

[127] Shapiro, 2003, p. 48.

[128] This last column was left out of the transcription by Westfall, probably because, as Shapiro notes (ibid., p. 63, n.12), it appears past the fold on the same (folded) sheet.

TABLE 2.1.
Newton's first table for ring diameters

	Circulorum diametri						*Quadrata diametrorum*						*Quadrat diff*					*Med. Diff.*	
	1	*2*	*3*	*4*	*5*	*6*	*1*	*2*	*3*	*4*	*5*	*6*	*1.2*	*2.3*	*3.4*	*4.5*	*5.6*		
1000	26⅓	37	45	51⅔	57⅓	62⅔	693	1369	2025	2635	3287	3927	676	656	610	652	640	647	46
667	32	44⅔	54⅔	62¾	70	*	1024	1995	2988	3938	4900	*	971	993	950	962	*	969	55
435	39½	55	67	77½	86⅔	*	1561	3025	4489	6006	7511	*	1464	1464	1517	1505	*	1488	73
320	45	62	77	89	99	*	2025	3844	5929	7921	9801	*	1819	2085	1992	1880	*	1944	81
206	57	80	98	*	*	*	3249	6400	9604	*	*	*	3151	3204	*	*	*	3177	72

Of this he remarked, "observe ye differences of ye squares ye fourth table [column] how nere they approach ye meane diffs in ye 5th collum wch proves ye first rule."

Since Newton did not keep records here of his raw observations, we do not know with certainty whether he repeated a measurement of a given ring's diameter several times and then averaged the results. That would have been unprecedented at this early date for more than two observations or measurements—though, as we shall see, there is evidence for Newton having done just that. What Newton did explicitly carry out in the manuscripts is even more striking, because he averaged the values for what on theoretical grounds should be the same number obtained from the measurements of different rings. Moreover, he explicitly compared the average to the number that it ought to be (the squared diameter of the very first ring).

In subsequent investigations he put a water drop between the lens and the plate in order to test his Proposition 4, according to which the magnitude of the gap at a given ring should vary reciprocally with the index of refraction of the medium that fills it.[129] Here the obtained results differed sufficiently from the usual value for water (4/3), that he performed the experiment several times, attributing the differences to the glasses having been pressed together to different degrees. From these and other results Shapiro has concluded that even at this early date Newton expected optical experiments to vary by no more than about 5 percent, and that he aimed for 2 percent or 3 percent.[130] When results deviated by more than that, Newton tried to pinpoint reasons for the discrepancy; finding one or more, he would alter the experiment accordingly. This could go quite far. Newton suspected that differences in the convexity of the glass lens between its two sides, as well as variations in curvature from point to point along a given side, might be responsible for some of the variations in what should be a constant number. He accordingly took three results—one each from either side of the lens, and the third from a different part of one of the sides—and averaged them to obtain the proper number "if the glasses had been regular & also wthout pressure for that noe doubt had some effect."[131]

Newton not only took averages among numbers that should, on theoretical grounds, be the same; he took averages of averages. None of this reached print. The 1671/2 letter to Oldenburg provided singular measurements for the prism experiments, while Newton's second paper on light and colors of 1675 listed single numbers for the experiments with rings. The public presentation of singular measurements should no longer be surprising: it was not only common at the time, but to do otherwise would have been tantamount to a public confession of inadequate skill. Newton certainly avoided hinting in print that his law of arithmetical progression was adduced by anything other than the most skillful and precise of measurements. "I measured the diameter of the first six rings at the most lucid parts of their orbits," he wrote, "and squaring them I found their squares to be in arithmetical progression of the odd numbers 1.3.5.7.9.11."[132] No hint there of any discrepancies among measurements.

[129] Newton's proposition reads "reciprocally as ye subtility of ye interjected medium" (Westfall, 1965, p. 191), with "subtility" varying reciprocally with the index.

[130] Shapiro, 2003, p. 51.

[131] Ibid., p. 52, and Westfall, 1965, p. 195.

[132] Cohen, 1958, p. 204.

Yet elsewhere in the same publication, he stated explicitly that in one case at least he had made several discrepant measurements, and that he had formed the "mean" among them. This exception is particularly significant because it reveals Newton's skeptical attitude toward the ability of the senses to provide reliable data. "Some years past," he wrote, having cast prismatic colors onto a wall in a darkened room, "I desired a friend to draw with pencil lines cross the image, or pillar of colours, where every one of the seven aforementioned colours[133] was most full and brisk, and also where he judged the truest confines of them to be, whilst I held the paper so, that the said image might fall within a certain compass marked on it."[134] Why had Newton resorted to an "assistant"? "And this I did," he continued, "partly because my own eyes are not very critical in distinguishing colours, partly because another, to whom I had not communicated my thoughts about this matter, could have nothing but his eyes to determine his fancy in making those marks." Poorly discriminating eyes and the prejudicial effect of prior knowledge together demanded an unbiased observer, equipped with a better color-perceiving sense. But no one's visual apparatus is for Newton truly reliable; it must always be errant, and his way to compensate error was via multiple replication: "This observation we repeated divers times," he reported, "both in the same and divers days, to see how the marks on several papers would agree." What did he then do with all of these observations, which were difficult and inaccurate, because "the just confines of the colours are hard to be assigned, because they pass into one another by insensible gradation"? Even so, he continued, "the *differences* of the observations were but little, especially towards the red end, and taking means between those differences, that were, the length of the image ... was divided in about the same proportion that a string is, between the end and the middle, to sound the tones in the eighth."

Newton's "mean"—the average—was the weapon with which he slew the inevitable dragons of sensual error. It was a most paradoxical weapon for the times, because it amounted to a method by which error seems to be reduced by committing it repeatedly. No such method appears elsewhere at the time, and it would certainly have seemed odd, to say the least, to most practitioners of the period. Newton himself was reluctant to use it in print when he presented actual numbers—except in this one place, where he exposed the method he had utilized when taking a group of measurements. Why advertise the method? The reason is not hard to find: Newton knew perfectly well just how difficult it was to pin down the regions of the spectrum filled by one of his seven colors; it was inevitable that anyone else attempting to do so would encounter the same problem and might well object to a bald claim that the lengths are such-and-such solely on the grounds that he "found" them to be so—as he asserted in respect to the arithmetic progression for his colored rings. His "assistant" with exceptionally good and unprejudiced eyes, though he indubitably existed, also amounted to a marvelous rhetorical stand-in for the reader himself, a presence in the laboratory with Newton that went beyond the detailed narrative accounts, such

[133] Newton had decided on a seven-color typology for the spectrum, namely red, orange, yellow, green, blue, "indigo, and deep violet" which he associated with a theoretical division based on a particular harmonic analogy, on which see Shapiro, 1994, pp. 111–13.

[134] Cohen, 1958, p. 192.

as those produced by Boyle, and which were designed to bring the reader virtually into the author's experimental space. For here Newton had an active participant *in situ*—more than a participant, in fact, but the very author of the observations on which Newton rested his claims. He had, as it were, created a physical displacement followed by a virtual transposition, having replaced himself with the assistant, and then the assistant with the reader.

But Newton went even further. He mentioned a series of discrepancies out of which he forged what he claimed to be a reliable result by means of a mathematical procedure: the taking of an average. We have no contemporary record of the reasoning by which he justified this unusual method, one that most natural philosophers of the period would surely have seen as an illogical method. Nevertheless, it may be that Newton thought of the procedure as a mathematical analog of his manipulative practice in experimentation. Consider again what he had done to set the prism near symmetric passage. Here Newton had not tried precisely to set the light's incoming angle by means of a carefully deployed protractor. He had instead rotated the prism back and forth, at first grossly, then more finely, until the sun's image seemed visually insensitive to slight perturbations. Each rotation, each twist of the prism, might be thought of as a deviation from the true position, until at last all of the twists converged on the best possible location. Perhaps Newton thought of the average in a similar manner, as a method in which the several measurements together participate in a way that converges to the true position. A mathematical concordance of serial measurements replaced a physical sequence of converging manipulations. That does not explain why Newton chose the *average* rather than some other way to combine measurements, a point to which we will shortly return.

The average was being used by others in Newton's day, though not for his purposes, and with a considerably different rationale. It is worth briefly considering these uses in order fully to appreciate Newton's singularity. By the third quarter of the seventeenth century, measuring population series was becoming common in England. Most noteworthy were the compilations and analyses of plague mortality records, published by John Graunt in 1662.[135] Graunt also compiled numbers of burials, christenings, and weddings in "a certain Parish in Hampshire" for a ninety-year period. In so doing, he noted that "According to the *Medium* of the said whole 90 years, there have been five *Christenings* for four *Burials*, although in some single *Years*, and *Decades*, there have been three to two, although sometimes (though more rarely) the *Burials* have exceeded the *Births*, as in the case of *Epidemical* Diseases."[136]

Although Graunt's "medium" is not a meaningful value for the average yearly ratio of christenings to burials over the entire ninety years,[137] he did believe he had ob-

[135] Graunt, 1662. Graunt was a good friend of Hooke (Inwood, 2002, p. 31).

[136] Graunt, 1662, p. 64.

[137] He obtained his 5 to 4 ratio by adding together the total number of christenings and then dividing by the total number of burials during the entire period. Graunt then noted that the comparable ratio for "some single Years, and Decades" is different. But he did not compute a ratio for each year, and so neither is his "medium" the average of the yearly ratios. A proper ratio for comparison with this one would take another nine decades of data collecting. Designating the number of christenings in year i of n years by c_i, and the number of births by b_i, Graunt had inappropriately assimilated $(1/n)\Sigma_i c_i/b_i$ (which is the average among the yearly ratios) to $\Sigma_i c_i/\Sigma_i b_i$, which is the ratio taken over the total number of years.

tained a "medium."[138] Demographic statistics of this sort were soon thought to reflect inherent, even divinely ordered regularities.[139] By the early 1690s, Halley was using improved statistics from Breslaw to compute the value of annuities. He calculated the "odds" of survival at any given age by dividing the number of a specific age group alive in a given year, by the number of people of that age group who had died in the previous year. Toward that end, Halley divided the total number of deaths at a given age over five years by five to obtain a yearly rate. However, unlike Graunt, he did not proceed to compute average ratios of births to deaths.[140]

Human statistics were not the only kinds that had been compiled during the latter part of the century; variable natural phenomena, such as rainfall, were measured and tabulated as well. In 1694, Richard Townley informed the Royal Society "on the quantity of rain falling monthly, for several years successively."[141] Townley had carefully collected rainfall, measuring it in a glass cylinder marked at the height of a pound weight. Given both the bottom area of the cylinder and the volume of a pound of rainwater (22.7368 cubic inches, according to Townley), the equivalent of a weight in inches of the cylinder could be found. Townley first summed the measures for a given month across the ten years from 1677 through 1686, and then summed across the years. Repeating this for the years from 1689 through 1693, Townley then formed the annual "means" over the ten- and five-year periods separately, obtaining 41.413" and 41.78" respectively. These, he wrote, "do strangely agree, and both considered do give for the mean by all the Fifteen years 41.516 Inches in height."

Townley had certainly found the yearly rainfall averages, though he wrote little about the meaning of the results, other than to remark that the decanal and half-decanal averages "strangely" agreed with one another. Here, it seems, we encounter proper averages, whereas Graunt's "mediums" were not. Of course, Townley was not concerned with effective ratios, but with the means of a particular measure, so that one perhaps ought not attribute a substantially different understanding of the "mean" to him than to Graunt. In fact, if we compare what the two of them were measuring, we detect a resemblance between them, which distinguishes both from measurements taken in the laboratory or observatory.

Both Graunt's birth and death records and Townley's rainfall inches were tabulations of processes that varied naturally from one time period to another. No one expected exactly the same number of children to be born, or precisely the same amount of rain to fall, in any given year. Nevertheless, there was a presumption that beneath the changing numbers lay regularity at some appropriate scale. Were it not for vagaries—strong and hot winds in one year, an epidemic in another—the rainfall or birth-death ratio would be nearly the same given the right time interval. When Graunt (however unsuccessfully) or Townley calculated a "mean," they were looking for the natural

[138] Hald, 2003, pp. 81–105, discusses Graunt's statistics and, referring to a comparable count of male and female births in London and Romsey, notes that Graunt computed only the ratio for the entire periods and not for individual decades, for which Hald provides values from Graunt's tables.

[139] By the first half of the eighteenth century, regularities in human statistics were explicitly associated with divine order: see Stigler, 2003, referring to William Derham and to Johann Peter Süssmilch. See Derham, 1713; Süssmilch, 1741.

[140] Halley, 1693.

[141] Townley, 1694.

invariant: for the true number that would always occur were it not for disruptions that differed from year to year in unpredictable ways. They both sought a number which, when multiplied by the number of years, would produce the same total result as the sum of the actual numbers. Christiaan Huygens' brother, Lodewijk, had precisely the same notion in 1669 when, in a letter to Christiaan, he mentioned Graunt in the course of describing his own method for building a table of life expectancies: "Here then is the method I've used. I first of all count the years which all these 100 people have together lived, which are altogether 1822 years … These 1822 years, divided equally among the 100 people, give for each 18 years and about 2 months, which is the age of each created or conceived person, the one yielding the other."[142] Like Graunt's and Townley's "means," Huygens' "age of each created or conceived person" erased natural variability and was justified by the assumption that natural order underlies apparent irregularity.

Graunt used the variations he found to reason about adventitious effects. "We shall make it probable," he asserted, "that in years of plague a quarter part more dies of that disease then [*sic*] are set down."[143] And Townley thought it significantly "strange" that yearly means calculated for decanal and half-decanal periods should so strongly agree—implying that he expected a greater variability in the natural amount of rainfall (reasonably so, given English weather) than he found. But he never doubted that there would be a natural value, or at least a value for the particular region he inhabited—"'twould be unjust, without further observations of like nature in other parts, that all England should be esteemed to abound as much in rain as these parts do."[144]

For Graunt and Townley, means made sense because they could be used to gauge nature's constancy, from whence to discuss the significance of deviations from it. Though Graunt was well aware of problems in collecting statistics, and Townley estimated the systematic error in his measurements, the deviations that they were interested in derived from nature, not from the observer. And the deviations were significant; from them Graunt could estimate the effects of epidemics, and Townley could wonder why differences in rainfall were not larger than they turned out to be. Human

[142] Huygens, 1888–1950, vol. 6, pp. 515–16; cf. Hald, 2003, pp. 106–7 and Véron and Rohrbasser, 2000. Lodewijk's calculation of life expectancy at a given age is based on the notion that, over a sufficiently large time span, the total number of years lived by a population of given size is fixed by natural law. So, to compute life expectancy at at an age such that the fraction of the population that has perished is x for a population of size n, all of whom were born in the same year and who altogether live a total of y years, Lodewijk first subtracted from n the total number of deaths nx that have occurred up to that age, and from y the total number yx of years that the deceased had lived. He then divided the remaining number of years by the remaining population to obtain a life expectancy at age x of $(y - yx)/(n - nx)$. Christiaan famously altered his brother's calculation by introducing chance, arguing that the aim of the computation is after all to see what bet should be placed on an individual's remaining time to live in order to compute insurance or annuities. Christiaan computed the chance that a population member will live to age x as the ratio $(n - nx)/(nx)$ of the number who survive to age x divided by the number who have died by that point. It would hardly be meaningful to place a "bet" on the number that a subsequent measurement would produce, not least because physical measurement has nothing *prima facie* to do with insurance or annuities. Moreover, precisely because Christiaan was concerned with computing numbers that had meaning in terms of bets, Lodewijk's average life expectancy at age x was of little import to him, for it did not reflect the likelihood of reaching x in the first place.

[143] Graunt, 1662, p. 13.

[144] Townley, 1694, p. 54.

error seemed entirely different from natural variability, because skill reigned over error, whereas factors essentially beyond human control governed rainfall and even the ratio of births to deaths.[145] Moreover, skillful work, unlike natural variability, could be expected to improve with time, as the craftsman gained experience.

Newton had no such expectation, or at least he thought that skill had very definite limits. He used the average to find a reliable number despite the inevitable uncertainty of sense. At the time, he seems to have been unique in doing so, though in the following decade we find another instance, albeit with a considerably different outcome from Newton's. In a 1682 letter to Flamsteed, William Molyneux, who measured the distances reached by mortars elevated to 60°, remarked that the "Reason I take the Mean between so many shotts is by Reason of the uncertainty of our shotts, for at the same elevation of 60 degrees with the same powder and ball in a Most calm day I made these eight differing shotts following. Feet 722 686 656 647 630 627 The mean between these is 651."[146] There was a caveat: Molyneux regretted having to use a mean at all, taking it rather as a symptom of inadequacy than as a useful technique, for he rejected the entire set of measurements: "you may perceive that between the greatest and least shot there is 142 feet, which in so little a Randon [*sic*] I count intolerable, and therefore I do not rely upon these experiments."

Molyneux's correspondent, Flamsteed, did occasionally use a mean, though his nineteenth-century biographer, Francis Baily, failed to recognize this.[147] For example, in a 1699 letter to Newton, in which he determined the latitude of the Royal Observatory, Flamsteed explicitly took the "mean value" among eight observations that had been carried out in the same manner from 1689 through 1696.[148] The result that he gave (51°28′45″) is indeed the average of all eight. But why form the average? Why not just choose one among the set? After all, three of the observations are in fact precisely the same (42″). From a modern standpoint the answer is clear: statistical error theory demonstrates that the average represents the best estimator of the true value. Neither Flamsteed nor anyone else before the middle of the eighteenth century had any such notion at all.

Yet Newton used averages early on; he used them frequently and, it seems, consistently. He may have been unusually skeptical of the reliability of the senses, and the perfectibility of skills and instruments, but that alone does not explain why he chose

[145] Townley was certainly well aware of measurement error, and he took great care to avoid it—not by making multiple measurements and combining them, which he surely thought would only make matters worse, but by means of a procedure which involved only two monthly measurements between which he would pour rainwater into a jar until full and then empty it, keeping track of the number of fills. This procedure was designed explicitly to avoid the errors inherent in multiple measurements: "Whereas should I write down the rain that falls between two observations, I might be subject to make as great a mistake in every one of them, and consequently be much more uncertain of the quantity of rain fallen in many of those added together" (Townley, 1694, p. 53).

[146] Flamsteed, 1995–2001, vol. 2, p. 53. Molyneux lists only six numbers in the letter, apparently having forgotten to give two of them.

[147] Baily, 1835, p. 376. Though one historian conjectures that "in cases noticed by Baily, Flamsteed just did not consider his work finished" (Sheynin, 1973, p. 110), and another noticed at least one use of the mean among three measurements intended to estimate the systematic error of his mural arc (Plackett, 1958, p. 133).

[148] Flamsteed, 1995–2001, vol. 2, p. 765.

the particular technique of the average to effect harmony among a group of discordant results. Why did Molyneux and Flamsteed, a decade or two later, do so as well—however reluctant Molyneux may have been about it, and however infrequently Flamsteed did so? Is there evidence as to what underpinned the average, decades before statistical notions became widespread?

In the autumn of 1684, Newton was working hard on the manuscript that evolved into the *Principia Mathematica*. The original tract, entitled *De motu* (On motion), contains a scholium immediately following a demonstration of Kepler's third (or harmonic) law, for a body orbiting in an ellipse about a fixed center which attracts it inversely as the square of the distance. The scholium provides a method for determining one of the two foci of an elliptical orbit given the other (viz. the sun), together with the length of the orbit's transverse axis as well as at least observations that yield the planet's location in relation to the sun. For each observation, draw a circle centered on the planet's position with a radius equal to the difference between transverse axis and the planet's distance from the sun. Do the same for the next observation. Then the orbit's other focus will lie at the intersection of the two circles. The problem is what to do with more than two observations when, as is almost certainly going to be the case, "all their intersections do not coincide." In that case, Newton cryptically remarked, "you will need to take a mean point for the focus." The "advantage of the technique," he continued, "is that a large number of observations, no matter how many, may be employed to elicit a single conclusion and speedily be compared with one another."[149]

Here we have something similar to, and yet suggestively different from, Newton's earlier forging of means between computed values. Each of the ring differences produced a single number, and the average was forged out of this set. Here, however, we do not have singular numbers; we have instead singular *positions*—a sequence of points scattered in space. Newton proposes to forge a "mean" among them all. How was this to be done? To what procedure does the construction of a mean position among scattered points correspond?

We would today think the answer to be obvious: establish an appropriate Cartesian coordinate system and assign the corresponding coordinates *xi, yi, zi,* to the planet's *ith* position. Then calculate an average value for each of the sets *x,y,z*, thereby producing a mean position. Newton certainly would have deployed a functionally equivalent procedure, but he probably thought about it in a way rather different from the simple calculation of an average. He may well have had a mechanical analogy in mind to justify the computation, since this is just what we find years later in Roger Cotes' posthumous work.

Cotes had become the first Plumian Professor of astronomy at Cambridge in 1707, at age twenty-five, on the recommendations of Newton, William Whiston—Newton's successor as Lucasian Professor since 1702—and Richard Bentley, Master of Trinity College. Flamsteed opposed the choice. Two years later Cotes undertook the editing of a second edition of Newton's *Principia*, which began an extensive correspondence between the two. Newton originally assumed that Cotes would not do much more than correct printing errors from the first edition. Yet, over four years, Cotes under-

[149] Whiteside, 1967–1981, vol. 6, pp. 51–53.

took to examine carefully Newton's arguments and demonstrations. The exchange between them was on the whole polite and productive until, near the end, Newton sent a set of corrections to the printer without first informing Cotes. Though Newton did not thank Cotes in his "Author's Preface"—despite having drawn up a draft which did so—he approved of Cotes' composition of a long (16 pages) and influential "Editor's Preface."[150]

Although an observatory was built for his use over the King's gate at Trinity College, Cotes was not a prolific observer, despite owning a fine sextant—produced by John Rowley—and though Newton himself paid for a good clock at Cotes' request.[151] Nevertheless, Cotes was quite concerned with instruments, having for example designed a screw-mechanism for a transit telescope. Most significant for our purposes here, he undertook the production of a group of formulae for the resulting errors in a spherical-trigonometric computation, given the error in the observed angles. To do so, he worked purely synthetically, by assuming the differences between certain lines to be small, or that a line is nearly orthogonal to a particular side. Because Cotes worked entirely in geometry, he introduced for the first time the radian measure of an angle in order to form a ratio between it and a line.[152]

Cotes' deductions, entitled the *Aestimatio Errorum*, were printed posthumously in the *Harmonia Mensurarum*, where they had a substantial circulation among French astronomers in the eighteenth century. Cotes' concern with the effect of a measurement error on a computed quantity reflects his engagement with the processes and instruments of observation proper, an engagement that dates at the latest to 1707 and that accordingly precedes his correspondence with Newton on the second edition of the *Principia*. Given this close interest in error, it's not surprising to find that Cotes also thought about ways to work with the discordant measurements that observation (or experiment) inevitably generates. Instead of adopting the usual attitude of the time, which would place a premium on achieving the single best measurement, Cotes explicitly set out a method for combining an unruly set which is worth quoting in full, because it is related to Newton's own approach, and moreover provides, if not a demonstration, then an implicit argument by analogy for using the average:

> Let p be the place of some object defined by observation, q,r,s the places of the same object from subsequent observations. Let there also be weights P,Q,R,S reciprocally proportional to the displacements arising from the errors in the single observations, and which are given by the limits of the given errors; and the weights P,Q,R,S are conceived as being placed at p,q,r,s, and their center of gravity Z is found: I say the point Z is the most probable ["maxime probabilem"] place of the object.[153]

150 See Gowing, 1983, pp. 2–19.

151 Gowing, 1983, pp. 80–90. The clock remains in the Master's Lodge at Trinity.

152 Ibid., pp. 91–107 provides a thorough account, including computations (p. 100) of the percentage errors in Cotes' expressions.

153 Translated in Gowing, 1983, p. 107.

According to Cotes, then, the "most probable" value of an object's position can be computed by treating each measurement as a distance at which a weight is hung, with the magnitude of the weight representing the reciprocal of the observer's estimate of the imprecision with which the measurement in question can be made. Then, the locus of the "most probable" value will be the system's center of gravity. Cotes' "places" could be distributed arbitrarily in space, and so to compute that center of gravity required a method equivalent in result to finding the center of gravity for the components of the places.

Although Cotes remarked that the resulting center was the "most probable" value for the true position, he did not provide an explanation of his statement. One might suppose that he was thinking of a procedure capable of choosing a place which minimizes a quantity that represents all of the errors taken together. We might proceed by assuming that the true location lies at some arbitrary point, and then form the differences between the coordinates of this point and those of the measured loci. Since we are concerned with the magnitude of the errors, and not with whether the measured coordinates are smaller or greater than the true ones, we might naturally square each of the errors and then add together the results for the three sets of components separately. Suppose next that Cotes' "most probable" meant the least possible value for each of these sums. In that case the resulting locus of the true value would be the center of gravity of Cotes' system of weights—but only if all of his weights had the same value, i.e., only if all of the measurements had the same estimated precision. Nevertheless, Cotes' procedure is in fact extraordinarily close to the result implied by a statistical theory of errors for the best estimate of the measurement locus. That result presumes a Gaussian distribution for the squared errors, and Cotes certainly could have had no such notion in mind. If, however, we look briefly at his other work, in particular his lectures on hydrostatics, then we may spy a hint as to what underlies the procedure—which may furthermore account for the use of the average by Newton, and eventually by others in the decades before the concepts and tools of statistical error theory had been developed.[154]

In or about 1708, Cotes gave a series of lectures on hydrostatics and pneumatics which, according to his successor as Plumian Professor, Robert Smith, who published them in 1738, were "often performed before large assemblies at the Observatory in Trinity College Cambridge; first by the Author in conjunction with Mr. Whiston, at that time Professor of the Mathematicks, then by the Author alone, and after his decease by myself."[155] The lectures nicely illustrated the principles of fluid statics and particularly emphasized the uses of a hydrostatic balance, including a procedure for weighing the air itself, since "the immediate evidence of sense has always something

[154] See ibid., p. 108. Gowing, and before him Whiteside (Whiteside, 1967–1981–, vol. 6, p. 51, n.62), noted that the procedure just outlined is just a least-squares minimization for the simple case of single-parameter measurements, which yields the average. If there are n position measurements, and the imprecision of the *nth* measurement is σ_n, then according to Cotes' procedure the *ith* component C_i of the "most probable" resulting locus will be $\Sigma_n(1/|\sigma_n|)L_i^n/\Sigma_n(1/|\sigma_n|)$, where L_i^n is the *ith* component of the locus of the *nth* measurement. If all of the σ_n are equal among one another, then this produces the same result as minimizing the sum of the squares of the errors, i.e., the average. Note that according to statistical theory based on the Gaussian error curve (namely, that the probability for measuring L_i^n is proportional to $e^{-(L_i^n-C_i)^2/2\sigma_n^2}/\sigma_n$), the best estimate for a weighted set of measurements is $\Sigma_n(1/\sigma_n^2)L_i^n/\Sigma_n(1/\sigma_n^2)$.

[155] Smith, 1738, Preface.

in it, which does more powerfully affect and convince us, than the united strength of the greatest number of inferences drawn out by a series of reasoning.[156]

The series was determinedly non-mathematical, aiming for the most part to draw "general truths and conclusions from a select number of simple experiments, first represented to our senses, and then explained to our understandings."[157] The sole exception occurred in the third lecture, concerned with how to calculate the force on an immersed body or set of bodies, the surfaces of which are inclined at various angles to the surface of the fluid. "The pressure upon any surface whatever," Cotes demonstrated, "however it be situated, is equal to the weight of a body of water whose magnitude is found by multiplying the surface proposed into the depth of its center of gravity under water. So the pressure upon any number of surfaces of different bodies, however differently situated, is equal to the weight of a body of water whose magnitude is found by multiplying the sum of all those surfaces into the depth of their common center of gravity under water."[158] Smith illustrated Cotes' procedure with an apposite diagram (figure 2.11).

Central to our purposes here is that the "center of gravity" replaces a set of actual surfaces and distances by a single surface and distance in order to calculate their joint effect. The replacement of a group of weights by a single one was as old as Archimedes, but it had taken on a new importance in the hands of Simon Stevin and, especially, Huygens, who had computed centers of oscillation for compound pendula. Although the difference between these replacements and that of finding a single observation to substitute for a discordant set is obvious, nevertheless the fact is that Cotes specifically linked the two procedures by introducing weights to represent the reciprocals of observational imprecision. We suggest that the justification for using the average accordingly rested upon an analogy between a balance among weights and a *balance among measurements*.

This is Cotes in the early eighteenth century, not Newton in the 1670s or 1680s. Cotes, however, was extraordinarily close to Newton's way of thinking and working, having spent years editing the *Principia*. We have already seen that the young Newton deployed the average early on in his optical work. Moreover, he used it again in the draft *De motu* of 1684 out of which the *Principia* emerged, where he asserted that a focus of a comet's orbit can be found by forming a "mean" among the intersection of a set of circles, each of which is determined by observation. In both instances Newton forged a single value for a theoretically inspired parameter from a discordant set. There is little further evidence in manuscript before 1700, or in the first edition of the *Principia*, concerning Newton's handling of discrepant data—although we shall presently consider the complex, and much discussed, issue of his views concerning the appropriate relation between data and theory. We shall argue that his construction of a new chronology exhibits essentially the same pattern that he developed for natural phenomena.

In February 1699/1700 Leibniz requested the Royal Society's help in amending Kepler's Rudolphine Tables in order to better perfect the Gregorian calendar for its adoption in Denmark and Germany's Protestant states—with the hope that Britain

[156] Ibid., p. 154.

[157] Ibid., Preface.

[158] Ibid., p. 31. By "pressure" Cotes here meant the force on the surface in question.

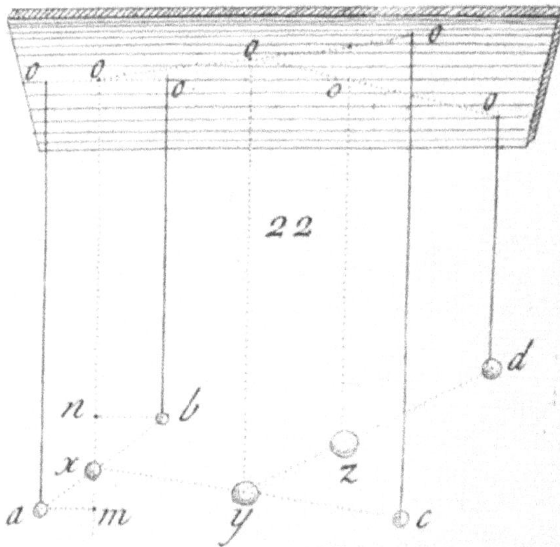

Figure 2.11. Smith's illustration of Cotes' procedure for replacing a series of immersed weights by their center-of-gravity. The weights *y,z,a,b* etc. are themselves the centers-of-gravity of the distributed surfaces.

would sign on as well. The Society's secretary since 1693, Hans Sloane, was ordered to write Newton about it, and in the event Newton produced a series of drafts along the way to his final response to the request.[159] For present purposes, the most significant considerations—which never passed out of his notes—involved Newton's engagement, for the first time, with ancient astronomical data, and his manipulation of it in altogether novel ways.

In order to provide appropriate corrections, Newton decided to examine ancient data concerning the autumnal and vernal equinoxes. Ptolemy's *Almagest* provides the times given by Hipparchus for a total of nine equinoxes ranging in date from −162 to −128. Although Newton did not as yet have a copy of the *Almagest*, nevertheless all nine times were available to him in convenient tabular form in Riccioli's *Almagestum novum*.[160] The question that Newton investigated was the quality of the Hipparchan data. Were these times consistent with one another and, if not, could they be corrected in some reliable manner?[161]

The original data as reported by Ptolemy specify times as sunrise, noon, midnight, and so on, which Newton translated into hours (figure 2.12A). It would have been

[159] Belenkiy and Echagüe, 2005 discuss the details of Newton's proposal, as well as the contents of the several drafts contained in Yahuda MS 24.

[160] Riccioli, 1651, p. 134. This is undoubtedly what Newton used, because his manuscript (reproduced in Belenkiy and Echagüe, 2005, p. 244) copies columns of Riccioli's table quite precisely.

[161] Although elsewhere in the drafts Newton wrote down several possible values for the solar year, in considering the Hipparchan data he settled in the end on Horrocks' 365d5h49m, as noted in Belenkiy and Echagüe, 2005, p. 227, who first uncovered the procedure discussed immediately below.

quickly obvious to Newton that the times were not in fact mutually consistent, given any acceptable value for the solar year, thereby implying defects in the original observations.[162] How then to correct the data? To judge from the procedure encoded in his table, Newton first decided that two of the equinoxes would be the most reliable: among the six autumnal, the one for −158 was observed at noon, while among the three vernal the one for −135 was observed at midnight. In the former case the sun was at zenith, in the latter at opposition to the zenith, and Newton estimated these to be more reliable than observations at sunset or sunrise. Assuming, next, that these two are indeed the most accurate, Newton used Horrocks' solar year to calculate when the equinoxes should have taken place in each of the remaining seven years (figure 2.12B). He then computed the differences between the original and the corrected times (figure 2.12C).

The next steps show Newton engaging in an altogether unprecedented manner with discordant data. Having calculated the seven deviations from the presumptively accurate observations, Newton then *averaged the deviations* separately for the autumnal and vernal equinoxes. He then treated each of the two deviations as the amount by which all of the observations should be respectively corrected—*including the two that he had chosen as the most accurate of all.* The corrected times were accordingly computed (figure 2.12D).

Newton's extraordinary procedure took his error-slaying method into new territory, for he had in effect used the original data to correct itself by choosing what he took to be the most reliable elements among them—but elements that were themselves taken to be mistaken by the very amount that he had used them to calculate in the first place. The situation here differed from earlier ones that he had encountered in the laboratory. Previously Newton needed to produce the best value for a parameter that had theoretical significance from among a set of measurements. The individual measurements were important only as elements in the averaging that produced the final result. Here, on the other hand, Newton required good values for each of the original equinoctial times, and so he devised a method that allowed them to, as it were, self-correct. A signal characteristic of the procedure involved separating the original data into distinct sets that he deemed to be unequally trustworthy. We will see in chapter 8 that he did just that when faced with ancient data—again provided, albeit at second hand, by Hipparchus—for establishing chronology.

Eight years before his efforts in calendrical computations, Newton discovered that the theory he had been developing of diffraction for the final book of what became the *Opticks* (1704) failed badly in view of experimental results obtained by passing light through the narrow, pointed region formed by crossed knife blades. These experiments countered the results with light past a hair, and impeded the completion of the *Opticks* for a decade, until Newton abandoned his quest for a workable account of the phenomenon.[163] Here we find amply displayed Newton's ways with data and mathematical theory, though there is no evidence in the surviving manuscripts that he needed to form averages (not least because the theory that he initially developed did

[162] For the comparatively short interval of 34 years, Newton naturally ignored both the precession and proper motion of the solar apsides. Newton's table sets noon as hour 0.

[163] See the incisive account in Shapiro, 2005, pp. 66–71.

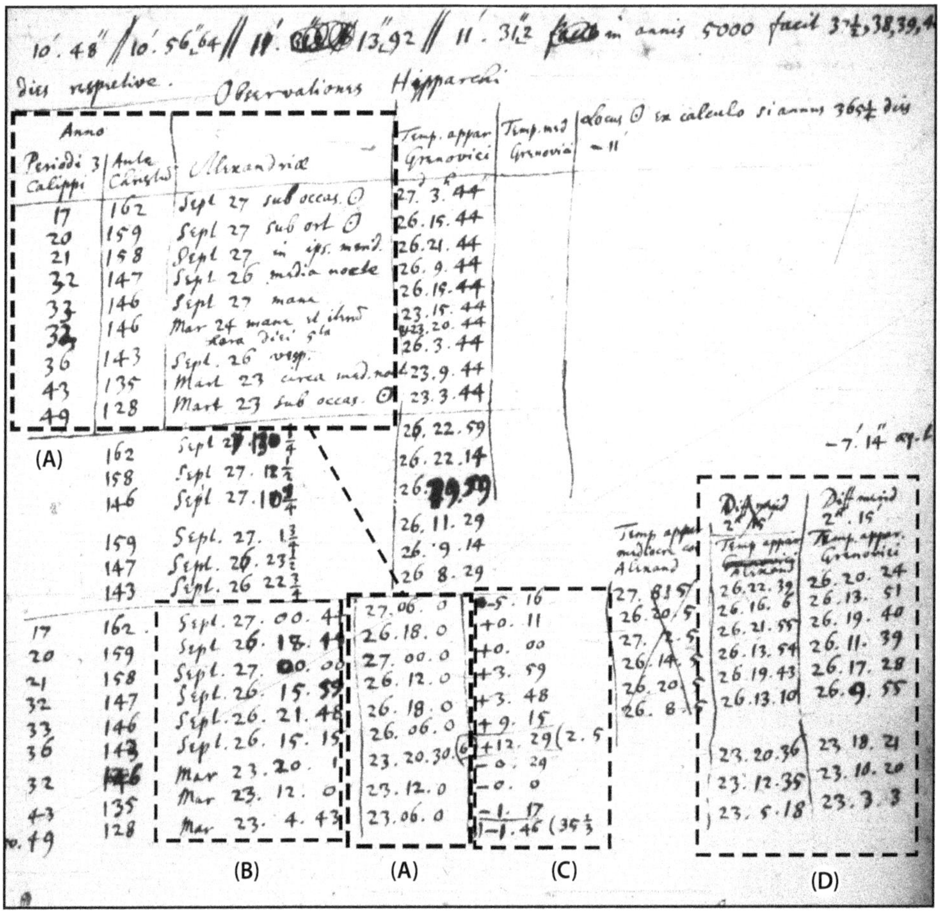

Figure 2.12. Newton's MS corrections of Hipparchus' equinoctial data. Yahuda MS 24. Courtesy of the National Library of Israel.

not lend itself to the computation of a relevant parameter by multiple measurements of different phenomena). In a manner that escapes full characterization, writes the editor of Newton's optical papers, he "plays off his measurements against his mathematical descriptions ... He did not change them at will to get perfect agreement, but controlled them by an awareness of his experimental error and by additional experiments."[164]

Ancient data posed a different problem from diffraction, the rings formed by a lens pressed against a glass plate, or from the motion of a body in a fluid. For with data from the past, especially astronomical, Newton had no direct control over the origi-

[164] Ibid., p. 70.

nal observations. However, as we shall see in the next two chapters, Newton developed an extremely skeptical attitude toward words from the past, one that required twisting them into shapes that fit a theory he had begun to formulate concerning the origins of civilization. Astronomical vestiges from antiquity were more trustworthy, but they too had to be manipulated in order to mesh with an evolving theory.

Since Newton famously rejected the formulation of "hypotheses" in favor of inductively established generalizations, we seem to have an overt conflict between his public statements and his private beliefs. Or do we? Careful research on Newton's *Principia*, first by Bernard Cohen and later by George Smith, have unveiled the intricacy with which Newton triangulated data and theory—or, more precisely, the manner in which he turned data into evidence for claims that, from his point of view, were not hypothetical in the same sense that Newton took Cartesian mechanisms to be. The point is worth developing because his efforts in chronology exhibit the same pattern.

According to Smith, Newton worked "through a series of idealizations, each of which is used to draw conclusions from phenomena, and which together comprise successive approximations, where residual discrepancies between theory and observation at each stage provide an essential basis for the next stage."[165] This approach, dubbed the "Newtonian style,"[166] appears in particularly sharp outline in Book II of the *Principia*, devoted primarily to the motion of bodies in resisting media. There Newton developed what Smith terms his "working hypothesis": "resistance forces can be represented within an idealized mathematical framework that superposes three distinct terms."[167] The third term, proportional to the square of the velocity, represents the supposed dragging effect of the fluid's inertia, while the second is directly proportional to velocity. The first term is a constant. Newton offered no expression for the proportionality constant in the second term. For the third he discussed what he termed a "continuous fluid," obtaining through a series of doubtful assumptions and steps its proportionality to the product of the fluid's density by the frontal area of the immersed body. The apparent lack of rigor here, as compared to the theorems produced for attracting forces in Book I, often led readers to conclude that the aim of Book II was to provide evidence for a dubious hypothetical structure—to conclude, in other words, that Newton was here engaged in a form of hypothetical reasoning not substantially different from Cartesian procedures. But this misconstrues Newton's primary goal.

Crucially, Newton's initial concern was to examine empirically his expression for the form of the constant in the dominating third term, a term that results from the fluid's inertia. "In other words," Smith argued, "Newton's continuous-fluid model is itself best regarded as a working hypothesis. The [experiments with bodies falling vertically] showed it too to be promising as an idealized first approximation to the physics underlying fluid resistance." If experiments did not sustain the form as a good approximation, Newton would abandon it together with its deductive source. But if

[165] Smith, 2000a, p. 251. See also Smith, 2002; 2005.

[166] Cohen, 1980.

[167] Smith, 2000a, p. 252.

the deviations of the form from experiment were sufficiently small, but not so small as to be within the range of what Newton took to be his experimental accuracy, then the approximate form would as a "working hypothesis . . . open an evidential pathway for reaching empirical conclusions about the microstructure of fluids from phenomena of motion under resistance."[168] This differs, not only from Cartesian model-making, but from existing experimental practice. It is not that Newton shunned hypotheses *simpliciter*; rather, he abjured hypothetical reasoning that could not be tied to experimental or observational structures, in a way that opened a path for correcting the original conjecture by successive approximation.

In this and the previous chapter we have seen that by the early 1680s, at the latest, Newton had stabilized a complex and, in contemporary context, unique approach to the production and use of experimental (or observational) knowledge. He would begin with a working hypothesis in the sense of a relationship that had consequences, which he could examine in the laboratory. Newton would then pursue experiments to see the extent to which the relation seemed to hold. To overcome the unreliability of the senses, he worked with the entire set of discordant data, producing from it a single number by means of the average. If the experimental conditions seemed to him to be sufficiently malleable, Newton would also, where necessary, manipulate the data into fitting form. Further experiments might require abandoning the working structure altogether and replacing it with another (as occurred with diffraction). If the initial structure was promising despite its deviations from experiment, then Newton did not abandon it. Instead, he modified subsidiary assumptions while leaving the basic structure intact. For example, he treated Kepler's laws in Book III as a first approximation based on the idealization of a fixed attracting point.

The production of single numbers out of a discordant set, the manipulation of data to turn it into workable evidence, the tenacious hold on a working hypothesis, as well as its modification in the light of further data—every one of these characteristics, we shall see, applies to Newton's attempt to rework ancient chronology. It, too, was based on a firmly held hypothesis: namely, that organized civilization evolved around the beginning of the first millennium BCE. It, too, manufactured single numbers out of discrepant observations to support the fundamental claim, while elaborating the initial structure in the light of other (in this case non-quantitative) data as well. And here, too, the entire structure rested on a deeply held conviction that the evidence of the senses must be treated skeptically, with the added, and critically important, factor that here the evidence derived from ancient testimony—often of a sort that Newton regarded with profound doubt. His iconoclastic edifice for antiquity radiated this novel skepticism, but it hardly evolved in a vacuum, for Newton was well aware of older, as well as contemporary, controversies. We accordingly turn next to the seventeenth-century English background within which Newton's refiguring of Biblical and classical antiquity took shape.

[168] Ibid., 282.

Erudition and Chronology in Seventeenth-Century England

Only once did Isaac Newton address his chronological studies in public. Angered by the unauthorized 1725 French translation of the abstract of his *Chronology of Ancient Kingdoms*, he lashed out at the Venetian abbé Antonio Conti—the "friend" who had "betrayed" Newton's confidence—and then proceeded to downplay the nature and extent of his interests: "When I lived at *Cambridge*, I us'd sometimes to refresh myself with History and Chronology for a While, when I was weary with other Studies."[1] As we shall see, for Newton chronology was anything but a diversion; indeed, the lion's share of Newton's investigations was carried out *after* his move to London in 1696. That Newton sought to portray his chronological studies as the dabbling of a weary mind partly reflects what he understood to be the relatively low status accorded chronology in the pyramid of learning. Thomas Pie articulated that outlook in 1597. Despite its centrality for the study of history, he concluded, chronological investigations ultimately amount to "λεπτολογια [quibbling] and lesser studies," or to "smaller and lower arguments."[2] This perception imbues the early-modern fixation with chronology—both in learned circles and in popular culture—a fixation informed by a consensus concerning chronology's subservience to theology and history. Hence, profundity in reconstructing ancient calendars or royal genealogies was neither encouraged nor pursued. Accordingly, technical expertise was explicitly excluded from the formal course of university studies. Newton, then, notwithstanding his avid interest in chronology, and the technical expertise he eventually attained, chose to depict himself of much the same cast of mind as his contemporaries. To further complicate Newton's relationship to chronology, his interest in the subject matured relatively late in life, by which time his insufficient grounding in humanist learning—owing to an unusually early devotion to the sciences—had become evident to friends and foes alike.

Late Renaissance perceptions about history and chronology, and their place in the encyclopedia of learning, are conveniently recapitulated in Jean Bodin's *Method for the Easy Comprehension of History* (1566). Underlying the French Jurist's program is a clear distinction between the mature author of histories and the inexperienced

[1] Newton, 1726, p. 320.

[2] Pie, 1597, sig. A4–A4^{v}.

(student) reader intended to enjoy the benefits of his *Method*. To ease the student into the arena of history—and to mitigate the danger of premature concentration—Bodin prescribes a gradual and accumulative approach in which chronology plays an important, if ancillary, role. An introduction in the form of a brief chronological chart—which, at this stage, need not be entirely accurate—is to be followed by a brief universal history. "Then from the general we move on little by little to the details, still in the order in which they are arranged in the tables of chronicles." Thus, in a paced and methodical manner, the student acquires knowledge of the succession of nations and empires, always according to the dictum the "general" before the "particular," the "brief" before the "detailed." Hence, Florus before Livy, Justin before Diodorus Siculus, the sense being that each coupling produces "an unparalleled advantage." Bodin further elaborates on the centrality of chronology to history in the penultimate chapter of his *Method*, wherein he presents his "System of Universal Time":

> Those who think they can understand histories without chronology are as much in error as those who wish to escape the windings of a labyrinth without a guide … the principles of time, the guide for all histories, like another Ariadne tracing the hidden steps with a thread, not only prevents us from wandering, but also often makes it possible for us to lead back erring historians to the right path.[3]

The contours of Bodin's advice—albeit stripped of his theorizing about time and early civilizations—were embraced by successive generations of English educators and authors; each recommended chronology as an indispensable, yet auxiliary, branch of history, both sacred and profane. The learned John Hales pointed out how ignorance of times and places would "breed confusion in your reading, and make you many times grossly to slip and mistake in your discourse"—a situation he believed could be remedied by a perusal of the *Theatrum historicum et chronologicum* by the Giessen professor of theology and Hebrew Christopher Helvig (Helvicus). The first Camden professor of History at Oxford, Degory Wheare, varied on much the same principle. A student should begin his historical studies with a compendium of chronology, so that he might gain a broad view of "the series of times and Ages, the succession of Empires, and the greatest changes which have happened amongst Mankind." The German historian Johann Sleidan's *Abridgement of the Four Chiefe Monarchies* was deemed particularly apt for such an overview; so, too, Petavius' recent *Rationarium temporum*, albeit written by a Jesuit, was considered excellent, furnishing as it did a complete narrative of sacred and profane history from the creation to 1632, "confirm'd with Chronological Proofs." For those wishing to delve deeper into chronology, Wheare recommended the works of Sethus Calvisius, Heinrich Bünting, Johann Funck, and Helvicus.[4]

In advocating chronology, authors waxed eloquent on the discipline's utility and virtues. The English geographer Richard Hakluyt coupled geography and chronology

[3] Bodin, 1969, pp. 20–27, 303.

[4] Hales, 1765, pp. 164, vol. 161; Wheare, 1685, pp. 38–41.

as "the Sunne and the Moone, the right eye and the left of all history." Expanding on what by now had become a conventional trope, Henry Isaacson in 1633 extolled chronology as the "chiefe light and Eye of *History* … the very *Load-star*, which directeth a man out of the Sea of *History*, into the wished for *Haven* of his Reading." Thomas Baker followed in 1700 with what was by then a cliché: "Chronology and Geography have been lookt upon as the two eyes of History, if these shine dim, our History must be yet more obscure; without these it lies in confusion, is only a heap of indigested matter, flat and insipid, and will neither profit nor delight in reading."[5] Many rhetorical flourishes were simply variations on—and embellishments of—Joseph Juste Scaliger's defiant tribute to the discipline he had so powerfully affected. As paraphrased by Edward Stillingfleet in 1662, chronology is "*the life and soul of History*, without which *History* is but a *confused lump*, a *meer Mola*, an *indigested piece of flesh, without life and form*." That cantankerous Scottish pedant, Alexander Ross, had paraded his rendering of the topos a decade earlier: "*History*, indeed, is the Body, but *Chronologie* the Soul of *Historical Knowledge*; for History without Chronologie, or a Relation of things past, without mentioning the Times in which they were Acted, is like a Lump or *Embryo* without *articulation*, or a Carcass without Life."[6]

While the frequency and casualness of such tropes might give the impression that English educators and authors spurred students to follow closely on Scaliger's heels, what was actually recommended was the acquisition of terms of art and familiarity with the general periodization of chronology. To be sure, some educators expected their students to acquire a more solid expertise in the technical aspects of chronology at some stage in their career. For most commentators, however, the grounding in interpretative chronology sufficed. And it was this rudimentary chronology that was integrated into the curriculum, with many Oxbridge tutors distributing their own manuscript digests, either as preparatory to a more detailed introductory printed text, or even as a substitute. John Crowther of Magdalen Hall, Oxford, for example, compiled in 1631 several such introductory manuals for the benefit of Ralph Verney—including manuals on astronomy and chronology. So, too, John Gregory undoubtedly composed his "De Aeris & Epochis" for the benefit of his students. Six decades later John Byrom's tutor at Trinity College, Cambridge, also furnished his charge with a manuscript chronology manual.[7]

Tutors' letters of advice shed additional light on prevailing practices. Daniel Waterland, master of Magdalene College, Cambridge, assigned the morning hours of two months in the student's third year to the perusal of the elementary section on chronology in Edward Wells' *The Young Gentleman's Astronomy, Chronology, and Dialling*, and of William Beveridge's more comprehensive *Institutionum chronologicarum libri II*—in the belief that "Chronology is a necessary part of learning, and ought to be well understood." Those wishing to pursue the subject further were directed to Aegidius Strauch's *Breviarium chronologicarum*. Small wonder, then, that many students followed the example of John Williams, the future Archbishop of York, who: "moiled a while in Chronology … yet he stuck not long in the Briars of Chronologers, from

[5] Hakluyt, 1599–1600, p. sig. *4; Isaacson, 1633: "To the Reader"; Baker, 1700, p. 124.

[6] Stillingfleet, 1662, pp. 89–90, preface.

[7] Verney, 1892, p. 123, vol. 121; Gregory, 1683, p. 172; Wordsworth, 1877 (1969), p. 25.

which a Man can never pluck out his Feet, but referr'd his own Collections to such Computations of Years as himself like best, by which he knew well enough in what Rank to find his own Notions.[8]

Newton's surviving papers suggest a similar route. Freshly arrived at Cambridge, he acquired two textbooks designed to introduce him to history and chronology: Johannes Sleidan's *The key of history. Or, A most methodicall abridgement of the foure chiefe monarches, Babylon, Persia, Greece, and Rome*, and Edward Halle's *The Union of the Two Noble and Illustre Famelies of Lancastre [and] Yorke*. This course of studies was cut short, however, when by his third year Newton had turned his attention almost exclusively to the mathematical sciences. (Indeed, Newton appears to have discarded both books: many years later, immersing himself in chronology, he acquired the 1686 Cambridge edition of Sleidan.) A glimpse into Newton's agreement with the existing mind-set regulating chronology instruction may be gleaned from his acquiescence in common educational habits. His scheme of university education indicates that he expected the college tutor to explain the "principles of Geography & Chronology in order to understand History"; the mathematics lecturer was enjoined to examine the principles of these two domains and, if the tutor were to be deficient, to offer the requisite instruction as well.[9]

The expectation on Newton's part that the duties of the mathematics lecturer should include the examination of chronological matters suggests a more general practice. John Bainbridge covered chronology in his capacity as Savilian Professor of Astronomy at Oxford,[10] but neither Isaac Barrow nor Newton appear to have lectured on the subject. Nevertheless, chronology was undoubtedly integrated into public lectures and exercises by other means. Archbishop James Ussher, for example, recalled that ca. 1614 "having occasion in a publick Lecture in the Colledge, to speak of the beginning of *Daniel's* Seventies: [he] laboured to prove that *Artaxerxes Longimanus* began his reign toward the end of the third year of the 77th *Olympiad*." For his part, Henry Dodwell, Camden Professor of History at Oxford, devoted certain of his lectures on the *Historia Augusta* to chronology.[11] Nor was chronology absent from college and university disputations. In the mid-1630s, Thomas Crosfield noted in his diary that MA candidates at Queen's College, Oxford, disputed in the Hall on such topics as optics, anatomy, chronology, and astronomy. Concurrent with the opening of the Sheldonian Theatre at Oxford, the heads of house decreed in May 1669 that the grand new building should serve as the forum where all learning, including "mathematical, critical, and chronological problems ... may be propounded and discussed ... by all persons of whatever degree." As late as 1803, questions in history, chronology, and geography were still an integral part of the Cambridge Greek tripos.[12]

[8] Waterland, 1843, pp. 410–11, vol. 414; Hacket, 1693, p. 13.

[9] Trinity College, Cambridge, MS R.4.48c; Harrison, 1978, p. 239, vol. 237; Newton, 1978, p. 370.

[10] A story is told of Bainbridge posting an announcement of his forthcoming lecture "de Polis & Axis," to which a student added a droll: "Doctor *Bambridge*, came from *Cambridge*,/ To read *De Polis & Axis*./ Let him go back again, like a Dunce as he came,/ And learn a new *Syntaxis*." Pope, 1697, pp. 10–11.

[11] Parr, 1687, p. 43 (of the letters); Dodwell, 1692.

[12] Queen's College Ms. 390, fol. 37^{v}; Ward, 1845–1851, vol. 1, pp. 73–74; Wordsworth, 1877 (1969), p. 117.

The actual chronological pursuits of students who became enamored of the topic—beyond what might be recommended or considered proper—are more difficult to ascertain. Archbishop Ussher's biographer remarks that while an undergraduate at Trinity College, Dublin, the fifteen-year-old youth "made such a proficiency in Chronology, that he had drawn up in *Latin* an exact Chronicle of the Bible, as far as the book of Kings, not much differing from his late *Annals*, excepting the enlargements in some more accurate Observations and Synchronisms of Heathen stories." Equally precocious was Christopher Wren who, while still an undergraduate, contributed to an Oxford edition of Helvicus' *Chronology* a brief method "to find any particular year required upon giving the cycles"—testimony to his proficiency in the intricacies of calendrical matters. A generation earlier, John Bainbridge had established his own reputation during the BA Act, when he disputed a question concerning the Julian and Gregorian calendars.[13] Isaac Barrow's route, in contrast, proceeded in the opposite direction. Having read Scaliger's remarks on Eusebius as an undergraduate, Barrow "perceived the dependence of Chronology on Astronomy, which put him on the study of *Ptolomy's Almagest*, and finding that Book and all Astronomy to depend on Geometry, he applied himself to *Euclide's Elements*, not satisfied till he had laid firm foundations; and so he made his first entry into the Mathematicks." As far as we know, Barrow never returned to chronology.[14]

The circumscribed role of chronology within English learned culture arose as much from the subordinate position of the discipline with respect to sacred and civil history as from the composition of the English scholarly community—predominantly divines. This identity crucially determined the course and focus of studies, as the students' extensive learning, philological or otherwise, was invariably employed within the confines of their future theological pursuits. For a theologian, even more than for a historian, the reconciliation of the "seeming Contradictions" in Scripture necessitated familiarity with chronology, for otherwise the infallibility of Scripture could not be maintained.[15] Such an instrumentalist approach tended to discourage interest in chronology per se, and to temper aspirations to enter a field considered to have reached its limits with the likes of Joseph Scaliger and Dionysius Petavius. Why bother with recondite and not altogether safe subject matter when student and scholar alike might content themselves with deferring to such authorities—or, better yet, to subsequent authors who distilled the relevant information for easier consumption? John Gregory expressed this very sentiment when commenting on the considerable improvements that Helvicus and Calvisius had made to Scaliger's technical chronology, rendering it thereby "everie waies absolute," and at such state of perfection "as needs not to bee added unto." Certain defects undoubtedly crept into their tables, Gregory admitted, but the differences were "but verie small and insensible, that it cannot bee much amended though never so much care should bee taken, and that by tampering it may bee made much wors"—as Petavius' "learned, infinite and equally unprofitable pain" had famously demonstrated. Given that degree of perfection, Gregory was of

[13] Parr, 1686 , p. 4; Wren, 1651 , prolegomena; Ward, [1740] 1967, p. 107. Pope, 1697, p. 11.

[14] Barrow, 1683–1687, vol. 1, sig. B.

[15] Barlow, 1693, pp. 28–29.

the opinion that a certain stamp of authority be conferred on chronology, so that it "might go currant in general Opinion, without further clipping or defacing upon whatsoever specious and pretending reformations."[16] Two decades later, the translator of Petavius' *Universal History* insinuated much the same argument when he extolled the Jesuit's superb distilling of "the innumerable testimonies of all the Ancient and Modern Authours" as well as his furnishing innumerable celestial phenomena, both ancient and medieval—"which are most infallible demonstrations of the truth of those passages and relations herein delivered." The translator concluded that "there was never yet an Historicall Book so Compleat as this, Extant in any Age."[17]

If the seeming recent perfection of chronology was not impediment enough to fresh and rigorous pursuit, the incessant quarrels chronology had engendered also tended to put off potential devotees. Joseph Scaliger himself—who allegedly "admired no man's works but his own"—may be blamed for much of the animus that pervaded the discipline. The vehemence with which Scaliger defended his work, and derided the work of pretty much every one else, caused consternation, even among his Protestant supporters. The witty early seventeenth-century epigrammatist John Owen carped with evident delight on Scaliger's temper and his reformation of time:

Renowned *Scaliger*, in the worlds Eye,
Was the Refiner of Chronologie:
The shrivel'd Face of Time is washt. The Man
That will correct the Manners; finde who can.[18]

Scaliger's reputation in seventeenth-century England had been further marred by his skewering of Thomas Lydiat, the most ingenious English chronologer of the first half of the seventeenth century. John Worthington, the editor of Joseph Mede's works, articulated a widespread belief that Lydiat's writings had "so manifestly worsted" Scaliger that the "great *Goliah* of Literature" felt compelled to "forsake his Weapon, and to betake himself unmanly to his Tongue." The other titan of chronology, Petavius, was only slightly less pugnacious—albeit his animus had been directed almost exclusively at Scaliger—insisting that "not one thing" in the *Emendatione temporum* was "not liable to just reproof" except for the grand edifice of the Julian period. Petavius' English translator regretted the latter's tendency to be "somewhat too liberal in his lashing of *Scaliger*," but mitigated his criticism with a reflection: Scaliger, though "a man of great knowledge," succumbed to "too much doting upon his own parts."[19] Hardly surprising, then, that in seventeenth-century parlance "when two men be irreconcileable, they are … compared to *Chronologers*."[20]

[16] Gregory, 1683, p. 172.

[17] Petavius, 1659, "To the Reader."

[18] "O Tempora, O Mores! / Scaliger annosi correxit tempora mundi: / Quis iam, qui mores corrigat, alter erit?" Pecke, 1659, p. 4.

[19] Mede 1672, p. XLIX; Petavius, 1659, "To the Reader."

[20] Isaacson, 1633, "To the Reader"; Contemporaries vied with each other to furnish variations on the maxim: "Chronologers agree like Clocks. Scarce two of one mind throughout," Fisher, 1643, sig. A2; "*Clocks* may sooner be agreed then *Chronologers*," Heylin, 1652 , pp. 2–3. Cf. Iacobus Curio's version—which may well

Archbishop Ussher, citing St. Basil, held that the perplexities of chronologers were by-products of their "love of contention."[21] Sir Thomas Browne detected a direct proportionality between the obscurity of past events and the determination of chronologers to force them into a coherent, tidy framework. As he saw it, the striking variance among the historical reckonings of ancient nations made it equally impossible that they be all "in the right" or that any one of them be proved to be so. Consequently, Browne embraced Petavius' pious disclaimer that to arrive at an exact chronology "without inspiration it is impossible, and beyond the Arithmetick of any but God himself." Hence, asked Browne, "what satisfaction may be obtained from those violent disputes, and eager enquirers in what day of the month the world began?" Moreover, since the ultimate aim of chronology is to point at "the Advent or Passion of Christ," the precise timing of His coming is irrelevant compared with the "that in the fulness of time he came."[22]

John Donne shared Browne's skepticism. He viewed with equal weariness those who considered it possible to render a coherent chronology from Creation to Incarnation, and those who had taken refuge in claiming "that prophane history cannot clear, but Scripture can." A poetic sense of Donne's ambiguous attitude may be found in the good-natured mocking epigram he affixed to his copy of Scaliger's *De Emendatione Temporum*: "Set to improve the time, Joseph, you will hardly succeed where laws, rewards, punishments and priesthood have failed. You must be satisfied if you don't make the times worse." Far more mordant was Donne's consignment of Christoph Clavius to Hell, in no small part on account of "the great paines" the Jesuit had taken "in the *Gregorian Calendar*, by which both the peace of the Church & Civill businesses have beene egregiously troubled."[23]

Browne's and Donne's skepticism regarding the relevance of chronology—technical or otherwise—to theology paled in comparison with Biblical literalists who rejected any part of universal chronology that appeared to call into doubt Scriptural chronology. Consider the case of Hugh Broughton, an erudite Hebraist who upheld an extreme and uncompromising stance regarding the immaculateness of Scripture, down to the last comma. In marked contrast to Scaliger, who had been convinced that no perfect universal chronology was possible without a thorough synchronization of the Bible with a wide range of pagan sources, Broughton maintained that Scripture needed neither emendation nor amendment from any external source: "The whole Booke of God," he thundered, "hath so great an harmony, that every part of it may be knowen to breath from one Spirite." He set down his "proofs" in *Concent of Scripture*, which purported to offer a precise chronology from Adam to Christ, based solely on Holy Writ. To Broughton's mind, his "success" demonstrated the absolute truth and

---

have inspired subsequent flourishes: "You will find it easier to make the wolf agree with the lamb than make all chronologers agree about the age of the world." *Chronologiearum rerum lib. II*, Basle 1557, p. 8, cited in Grafton, 1985, p. 102.

[21] Ussher, 1658, sig. A5. Cf. Leigh, 1656, p. 34: "some say the holy Ghost did obscure some things in Chronology to sharpen mens wits."

[22] Browne, 1964, vol. 2, pp. 404, 409–10.

[23] Donne, 1952, p. 18; Sparrow, 1958, p. 115; Keynes, 1958, p. 108: "Emendare cupis Joseph qui tempora, Leges/ praemia, Supplicium, Religiosa cohors/ Quod iam conantur frustra, Conabere frusta;/ Si per te non sunt deteriora sat est"; Donne, 1969, pp. 17–19.

coherence of the Biblical narrative, rendering it infinitely superior to any pagan text—so much so, in fact, that the Bible could serve to correct the lacunae and inanities of pagan sources.[24] In contrast, the absurdities of secular reckonings, the Olympiads in particular, were such that "Jupiter was never so frightened with Phaetons running of the Sunne out of his course, least all shoulde be brought by flames into the olde Chaos, as the course of the holy story is disturbed by seeking helpe at these forlorne Olympike recordes."[25] Broughton's extreme stance commended him to Scaliger's scorn—he dubbed the Englishman "furiosus et maledicus" (an abusive madman)—in addition to being taken to task in the public lectures of Edward Lively at Cambridge and of John Rainolds at Oxford.[26] Broughton derived much of his scholarly inspiration—and perhaps some fanatical zeal—from Matthieu Béroalde's *Chronicum* (1575), a chapter of which Broughton had translated into English to bolster his campaign against Rainolds and Lively. A French Huguenot who settled late in life as professor of Hebrew in Geneva, Béroalde (Beroaldus) sought primarily to establish the chronology of Daniel's Seventy Weeks and its Christological import. His guiding principles were reduced by his biographer into two seemingly irrefutable syllogisms. First, wisdom is the most precious of all goods; but God is the author of all wisdom; Therefore, we must seek it in God's word. Second, profane historians are inconsistent among themselves; but the Holy Scripture conforms with itself; therefore one cannot go astray when following it.[27] Thus, convinced that "secular history is full of lies"—and that ancient chronology "was an Augean stable of error which needed a Hercules to clean it"[28]—Béroalde proceeded to expurgate antiquity of all kings not named in Scripture. In what mattered to him most, the history of the Persian monarchy, he forbore to follow either Herodotus or Ctesias, "because we have certaine testimonies out of the Scriptures, of these kings who reigin [*sic*] *Persia* vnto *Xerxes*. So, after *Cyrus* we doo not name *Cambyses*, nor after him *Magus*, nor after him *Darius Histaspis*: which names, because they are never found in Scripture, by us are pretermitted [omitted]. For we must wholly rest in the authoritie of holie Scripture, which handleth the historie of these times, & neither to *Herodotus*, nor anie other whosoever hee bee, that saith he bringeth a true Historie from the *Persian* Records, as *Ctesias* is bolde to professe of himselfe."[29] The Persian Empire, he insisted, lasted for only 130 years. Small wonder that Joseph Scaliger expressed nothing but contempt for Béroalde, the actual target of the *Emendatione temporum*.[30] Scaliger naturally viewed subsequent rigid Scripturalists as of the same ilk—David Pareus, for example, whom he branded

[24] It is necessary to make secular history conform to Scripture, John More argued, "and not cleane contrarie, to wring the account in holy Scripture unto the account of prophane writers, as many chronographers have done hitherto," More, 1593, sig. A6^{v}.

[25] Broughton, 1588, "dedication"; Broughton, 1592 : Sig. G3^{v}. Citing Broughton approvingly, William Nisbet varied on Broughton in 1655: "it is said, that Circe did not so much bewitch Ulisses companions as the Olympiade hath done many learned men." Nisbet, 1655, p. 88.

[26] Scaliger, 1695, p. 68; Lively, 1597.

[27] Béroalde, 1575, "Preface"; Haag and Haag, 1846–1859, vol. 3, p. 9.

[28] Béroalde, 1575 , p. 150 (= Béroalde, 1590 , p. v^{v}; Béroalde, 1575, p. 208, cited in Grafton, 1993 , p. 268).

[29] Béroalde, 1575 , p. 166, translation taken from Béroalde, 1590 , p. 14.

[30] Grafton, 1993, pp. 233, 308–11, 343–44, 356.

"Hierophantam Beroaldinum and worse."[31] For our purposes here, however, the "Beroaldistae" are relevant not only because of their influence on English Calvinists, but because Newton himself occasionally resorted to their line of reasoning, albeit not on the contested duration of the Persian Empire.

In the *Chronology*, Newton explicitly outlines the history of the Persian monarchy according to Greek and Latin authors: "the *Jews* knew nothing more of the *Babylonian* and *Medo-Persian* Empires than what they have out of the sacred books of the old Testament; and therefore own no more Kings, nor years of Kings, than they can find in those books." In fact, he believed, those Jews who have considered Herod—and later Bar Kochba—as the Messiah foretold in Daniel's prophecy, "seem to have shortned the Reign of the Kingdom of *Persia*," thereby rendering it "very imperfect." Hence the need to resort to profane histories and to Ptolemy's *Handy Tables* "for stating the times of this Empire." In turn, such "secular" history furnishes not only "a better ground for understanding the history of the *Jews* set down in the books of *Ezra* and *Nehemiah*" but means for "adjusting it" as well.[32]

Newton found it easy to synchronize Daniel's prophecy with the correct reckoning of Persian history. To begin with, his Masoretic time line, following Ussher's, fixed Creation at 4004 BCE, whereas Béroalde, Broughton, and John Lightfoot determined Creation to have occurred on 3928 BCE. More importantly, in contrast to their dating of the commencement of the Seventy Weeks (=490 years) to the issuing of Cyrus' decree permitting the exiled Jews to return to Judea, Newton, in the *Observations upon the Prophecies*, explicated Daniel 9:24 to signify an interval of 490 years "from the time that the dispersed *Jews* should be re-incorporated into a people" to the Resurrection.[33] And since the Jews regained their polity only upon Ezra's return to Jerusalem, the prophecy necessarily commenced in the seventh year of Artaxerxes Longimanus, which Newton dated 456 BCE, in conformity with his rather unconventional setting of Christ's crucifixion at 34 AD.

Though Newton felt no need to follow the Beroaldistae in contracting Persian history in order to make it conform to Daniel's prophecy, he resorted to their tactics when handling the antiquity of Assyrian and Egyptian empires. More significantly, Newton was not averse to embracing Béroalde's (and Broughton's) habitual rejection of historical evidence on the grounds that Scripture had failed to mention it—though he did so with greater subtlety. As contemptuous of the "fabulous Ctesias" as Béroalde was, Newton rejected the long series of Assyrian kings marshaled by the historian from Cnidus for no better reason than that such a list comprised names that "are not *Assyrian*, nor have any affinity with the *Assyrian* names in Scripture." Moreover, he faulted Ctesias for failing to mention "the true Empire of the *Assyrians* described in Scripture." To top it off, Newton availed himself of an ambiguous verse in Nehemiah,[34] in order to interpret the events therein as having occurred "since the time of the

[31] Scaliger, 1695, p. 620. See Grafton, 1993, pp. 611–13.

[32] Newton, 1728b, pp. 356, 358.

[33] Newton, 1733, pp. 130–31.

[34] Nehemiah 9:32: *"let not all the trouble seem little before thee that hath come upon us, on our Kings, on our Princes, and on our Priests, and on our Prophets, and on our fathers, and on all thy people, since the time of the Kings of* Assyria, *unto this day."*

Kingdom of *Assyria* or since the rise of that Empire." In other words, "the *Assyrian* Empire arose when the Kings of *Assyria* began to afflict the inhabitants of *Palestine*; which was in the days of *Pul*."[35]

Béroalde and Broughton may have been extreme in their attitude toward non-Scriptural sources, but their faith in the absolute truth and coherence of the Bible resonated broadly, as did their tendency to discard or devalue pagan sources whenever they appeared incongruous with Scripture. This propensity resulted in an emphasis on the conjectural basis of all ancient chronologies—with the exception of Biblical chronology, of course—and the concomitant tendency to interpret pagan sources in the light of Scripture: in other words, to pick and choose seemingly at will—a practice characteristic in places of Newton as well. The courtier and author Lodowick Lloyd provides another example. He considered profane writers to have been "wrapped in errors by missing the true warrant of time"; in contrast he, intent on accurately tallying ancient chronology, had boldly waded into "those strong streames that have caried the best learned to a labirinth of errors, who in seeking consent of time by uncerteine computations of the Gentiles, have missed the square and perfect frame of the Prophets, the streight and perfect line from Adam unto Christ."[36] Such manifest uncertainty of secular ancient chronologies prompted William Nisbet to question the very purpose behind attempting to determine the precise length of the Persian and Greek monarchies. It suffices to know what "may be gathered from Scripture, either in expresse termes, or by consequent" concerning "the maine Period"—the chronology of Daniel's Weeks.[37]

Nisbet's comment underlies the tenacious hold of Daniel's chronology on the minds of contemporaries. Fairly typical is Andrew Willet's explication of the extent to which the reckoning of the Olympiads can offer "a certaine direction for the vnderstanding of Daniels weekes." His elucidation harkens back to the literalism of Béroalde while attempting to mitigate the latter's anti-intellectualist stance. In regard to "Greeke affaires," Willet pronounces, "that computation may safely be received." It must not, however, determine the commencement or termination dates for the reigns of Persian kings. Indeed, not even astronomy can serve to determine such dates. Willet had in mind to establish of the duration of the Persian Empire at 230 years on the basis of the computed time that had elapsed between the recorded eclipse observed during the seventh year of Cambyses' reign, and the eclipse observed on the eve of the battle of Gaugamela. To Willet's mind, astronomical evidence was well and good as long as it conformed to traditional sacred chronology: "But although the Sunne be most sure in his course, there is a Sunne which is more true and stedfast, … the Sunne-light of the truth which shineth in the Scriptures, and in this prophesie of Daniel, which calculateth but 490 yeares from the word going forth, which was the commandement

[35] King's College, Cambridge, Keynes MS 146, fol. 64; New College, Oxford, MS 361.1, fol. 95^{v}; Newton, 1728b , pp. 5, 265–67.

[36] Lloid, 1590, "Dedication." John Owen echoed similar sentiments in a 1680 sermon where he urged his auditors to "Take heed of *computations*": "How wofully and wretchedly have we been mistaken by this! We know the time is determined,—its beginning and ending is known to God; and we must live by faith till the accomplishment." Owen, 1850–1853, vol. 9, p. 510.

[37] Nisbet, 1655, p. 182.

of Cyrus, unto the Messiah."[38] In fact, Willet added, Varro's famous threefold division of the ages of the world—the unknown (prior to the Flood); the mythical (from the Flood to the first Olympiad); and the historical—ought to be inverted: "Rather, the contrarie is evident: that the historie of time from the beginning of the world, untill somewhat after the Olympiads, is most certaine out of the propheticall writings: but there is no certaintie of the times following, because the propheticall writings were ceased."[39]

Even more moderate scholars opted at times to ignore the complexities of chronology or eschew precision. Case in point, the treatment of Egyptian chronology. Samuel Purchas, for example, avowed his unwillingness to take it upon himself "to bee umpire and decider of those many altercations amongst Chronologers," opting instead to simply follow Scaliger, "whose verie name is able to shield [him] from contempt, if not to yeeld [him] commendation." As he explained, he aimed primarily to treat the history of religions, and in such an endeavor to "precisely to determine in what yeare of the world every King began his reigne, and to dispute the same with all opponents, would bee somewhat tedious to the Reader." Similar reasoning informed John Greaves' attempt to get a handle on Egyptian antiquity. Had he attempted to examine all the arguments concerning the duration of Egyptian history, Greaves wrote, "we shall finde our selves intangled in a Labyrinth, and a Maze of Times, out of which we cannot, without much perplexitie, unwinde our selves." Furthermore, cognizance of how few of the names found in Manetho can be found in Herodotus and other Greek historians and "that which is of greater consequence, how difficult it is to reconcile these Names, and Times, to the *Aegyptian* Kings recorded in the *Scriptures*, we shall finde our selves beset, and as it were invironed on every side, with great and inextricable doubts." Greaves recommended, therefore, fastening on "some common, and received *Epocha*, in which either all, or most agree"—unsurprisingly, he chose the Israelites' migration out of Egypt—and that epoch should serve as "guide in matters of so great antiquitie."[40]

For many who shared this cast of mind, the discovery of Ptolemy's *Handy Tables*—a Latin version of which Scaliger published in his *Thesaurus temporum* (1606)—proved reassuring. As even John Preston, the puritan master of Emmanuel College, Cambridge, explained in the 1620s, the exact chronology of Scriptures "and the agreement of them with the Heathen Histories"—a harmony that to him proved "that there is a *God* that made the world"—perfectly demonstrated the verity of the Bible. Hence his excitement with Ptolemy's *Canon* (quite unknown to Scaliger he believed), "which doth exactly agree with the Scripture," setting down the precise dates of Nebuchadnezzar and Cyrus, thereby demonstrating a most fortunate agreement with Daniel and Jeremiah (as well as with dating the Fall of Jerusalem): "the greatest testimony that the Scripture can have from Heathen men."[41] Preston's parading of a pagan

[38] Newton naturally concluded that these eclipses had settled the dating of these kings beyond dispute. Newton, 1733, p. 141.

[39] Willet, 1610, pp. 298–300.

[40] Purchas, 1613, pp. 59–60; Greaves, 1646, p. 17.

[41] Preston, 1634, p. 55.

chronological text as authenticating Scripture turns our attention to how Englishmen reacted to the revolution in the technical aspects of chronology.

Ptolemy's *Canon* appealed especially to a budding community of technical chronologers that came into being in the first half of the seventeenth century. Inspired by Joseph Scaliger, they shared the conviction of Sethus Calvisius that the Canon's provision of an orderly succession of Assyrian, Babylonian, Egyptian, Median, and Persian reigns made it "more precious than all gold." Indeed, he continued, had it "become known in the past, chronology would be in a far better state, and chronologers would not have split up so much into divergent sects." John Bainbridge shared these sentiments, preparing his own edition of "that golden List of *Reigns*" in 1620.[42] Such confidence prompted James Ussher to cite in the introduction to his *Annals* the German Calvinist David Pareus' dismissive view of technical chronology in his *Oratio Chronologica* (1606)—"those who promise to give us an exact Astronomical Table of Time, from the first point of the Creation unto Christ, seem to me more worthy of encouragement than praise, in that they attempt a thing above human capacity"—only to refute it.[43] Ussher conceded that such knowledge was difficult to attain, but not impossible for those skilled in Scripture and in astronomy. Ussher's confidence rested on the important contributions that several English mathematicians—including Thomas Lydiat, John Bainbridge, and Henry Briggs—had made to his chronological system, a system that sought to break the traditional mold of Scriptural chronology based on internal logic.[44] Henry Isaacson also advocated this new approach for he, too, believed that what bedeviled chronologers was not just the imperfect record of antiquity, but more fatally, their seeking "to perfect *Chronology* by *History*, whereas it is more properly and aptly rectified by *Astronomie*."[45]

The allure of astronomy and mathematics appeared to promise the resolution of disputes through the provision of incontrovertible knowledge. The giant steps taken by Scaliger and Petavius emboldened many a mathematician to publicly pronounce the indispensability of their profession to chronological studies, and thereby raise its dignity. In this regard, Martinus Hortensius' 1634 inaugural oration at the *Athenaeum Illustre* in Amsterdam is instructive. The "reliability of historical knowledge is uncertain and suspect," the professor proclaimed, "unless from astronomy it receives its strength and foundation." Two decades later, the young Christopher Wren echoed similar sentiments in his inaugural oration upon appointment as Gresham Professor of Astronomy:

> A day would not suffice me to run over the entire realm of the sciences, showing you the great sway of astronomy that obtains everywhere. Indeed theology, that queen of the sciences, admits her debt to astronomy, her most trusty handmaiden, for it has arranged sacred history in chronological order.

[42] Grafton, 1993, p. 727; Bainbridge, 1620, p. ¶3^{v}.

[43] For Scaliger's critique of Pareus—whom he dubbed "a certain prophet"—see Grafton, 1993, pp. 611–13, 663.

[44] Ussher, 1658, sig. A4. Those attempting to compute the age of the world, noted William Greenhill, "are rather to be pityed then praised." Allen, 1659, sig. a^{v}.

[45] Isaacson, 1633, "To the Reader."

> And since it is a part of this science, so it greatly relies on astronomy's observations (I mean its commentaries on the deeds of the ancients), which, with no injury to faith, ratifies phenomena such as eclipses, the more important conjunctions, and the like. In the absence of which, history both sacred and profane would be an impenetrable labyrinth, having no entrance or exit. But, coming to the rescue, chronology provides a kind of picture: it has drawn the times as if on a geographical map, by the help of which it is granted us to revert to the bygone millennia, to associate, if I may put it thus, with men of remote centuries, and hunt after unknown things, no less than nowadays it is permitted us to associate with the men of distant climes by the help of hydrographic charts.[46]

By the turn of the seventeenth century, a more assertive tone with respect to the bearing of astronomy on chronology can be detected, as in John Arbuthnot's youthful *Essay on the Study of Mathematics*. The soon-to-be celebrated wit and member of Newton's inner circle invoked traditional stock-phrases—for example, chronology and geography "are indispensable preparations for History: a relation of matter of fact being a very lifeless insipid thing without the circumstances of time and place"—before switching gears and insisting on the necessity of a certain mastery of the scientific principles of both domains "if a Man [is to] be thoroughly skill'd in History." Only knowledge of "the Technical or Doctrinal part of Chronology" allows one "to unravel the confusion of Historians," Arbuthnot insisted, marshalling as proof the recent successes of Edmond Halley and Henry Dodwell in using astronomical data to resolve thorny historical problems. Such successes further emboldened Arbuthnot to claim that theologians, too, were in need of astronomical chronology.[47]

Growing confidence in the ability of the mathematical sciences not only to transform chronology but to be the very arbiter of the discipline naturally appealed to the early proponents of "modern learning" in England. Already in the 1630s, George Hakewill embraced chronology as one of the examples illustrative of the superiority of the "moderns." Present-day chronologers and geographers, he boasted, "have so farre exceeded the *Ancients*," that in comparison the latter seem to have seen nothing. In respect to chronology, he asked rhetorically, "how dim sighted are the *Ancients* in the *computation* of *times*, how miserably doe they wander up and downe in the darke, and knock their heads each against other, and how excellently have latter Writers, and specially *Joseph Scaliger* in that most elaborate worke of his *de emendatione Temporum*, cleared those mists, and chased away that darknes."[48] Six decades later, William Wotton's dogged championing of modern scientific learning was also extended to erudition. The industry of the moderns had exposed the numerous "Mistakes in the Assertions of the Ancients about Matters not done in their own Times, but several Ages before they were born." Wotton marveled at the clarity with which "the *Old Chronology* and *Geography* [had] been stated by Modern Criticks and Philologers; and the Mistakes and Carelesness of many Writers detected, who were esteemed Authentick

[46] Imhausen and Remmert, 2001, p. 123; Ward, [1740] 1967, Appendix p. 31.

[47] Arbuthnot, 1701, pp. 22–24.

[48] Hakewill, 1635, p. 278.

even in the Times wherein they lived." The moderns' elucidation of antiquity went beyond the domain of pagan history to incorporate Scripture as well, so much so that none of the ancients "ever unravelled the Chronology of the Old Testament" as Archbishop Usher and Sir John Marsham did.[49]

Remarks like these, combined with the assertion that Scripture and prophecy should be subjected for accuracy to the exact sciences, engendered heated reaction. Traditional arguments regarding the uncertainty of non-Biblical ancient chronologies—which were used to dismiss pagan time-reckonings and to insist on the exclusive accuracy of Scripture chronology—gave way to renewed insistence on the subordination of chronology to the internal logic of Scripture. The critique may not have been as rabid as John Harvey's 1588 denunciation—in the charged apocalyptical atmosphere of the Armada year—of all "*chronological* computations, *Theologicall* constructions, *Cabalisticall* traditions, and *Mathematicall* speculations" as "ambiguous, uncertaine, fallible, erroneous, deceitfull."[50] But critical it certainly was. During the Puritan Revolution, radical sectarians disparaged chronology, and secular learning more generally, as irrelevant and even downright pernicious. As the Baptist minister John Horne put it: the mathematical sciences may afford "Some lesser usefulness and curiosities"—for example the size of the Ark and the measure of Solomon's Temple—while history and chronology might help in elucidating certain passages in the Book of Daniel. And yet, he concluded, "there is so great incertainty in them too as to what is Heathen, that they rather trouble then help."[51]

The Non-juror Thomas Baker concurred. He dismissed as fabulous and uncertain all pagan accounts of time before proceeding to attack technical chronology. As far as he was concerned, the manner in which Petavius sought to expose Scaliger's numerous "Errors and Hallucinations" clearly demonstrates that the true chronology remained "in the same obscurity" as previously; only now it had been arranged "in better order" under the umbrella of Scaliger's Julian period. That both Scaliger and Petavius committed numerous errors, "tho' both of them pretend[ed] to demonstrate, and in many of their Calculations proceed[ed] with Mathematical assurance," attests to the deficiency of technical chronology. Further to expose the liabilities of technical chronology, Baker pointed out that whatever progress had been made in the intervening decades concerned the historical part of chronology.[52] John Milner, a fellow non-Juror, went further, taking it upon himself to defend James Ussher—the paragon of piety and erudition—against the calumnies of recent pretenders to chronology. The "greatest part of chronology," Milner assured his readers, "is plain uncertain." He excepted only what Scripture "plainly expressed"; what could be deduced from Scripture "by a clear and undeniable consequence"; and whatever pagan chronology had been "consonant to the Scripture Chronology." Such uncertainty he believed to be the cause of the incessant controversies among chronologers, with pernicious effects on religion.[53]

[49] Wotton, 1694, pp. 311, 315, 325.

[50] Harvey, 1588, p. 34.

[51] Horne, 1654, p. 160.

[52] Baker, 1700, pp. 121–33.

[53] Had chronologers been cognizant of such endemic and irresolvable uncertainty, for example, "*Scaliger* would not have fallen so foul upon our Mr *Lydiat*, endeavouring to expose him, and triumphing over him with the greatest scorn and contempt." Nor would the world have "been burthen'd with many tedious and

Milner's attempt to stem the tide of novel and disputatious chronological studies threatening orthodoxy reflected a broader concern: the seemingly insidious consequences of unrestrained erudition and the dubious motivations of those engaged in it. The High Churchman John Edwards articulated this concern when defending Scripture during the 1690s. Whereas Milner singled out Archbishop Ussher as the epitome of orthodoxy and scholarship, Edwards invoked the equally pious Cambridge Hebraist John Lightfoot, contrasting the latter's judicious treatment of Scripture chronology with the treatment by Thomas Hobbes who, he asserted, aimed at nothing less than to deploy seeming contradictions in Biblical chronology in order to undermine Holy Writ as a whole.[54] Edwards chose wisely, for Lightfoot, the greatest Hebraist of the seventeenth century, never wavered in his belief that the only way to settle such matters as the age of the world and the birth of Christ was "to take the plain and clear account and reckoning of the Scripture, which hath taken a peculiar care to give an exact and most certain Chronicle to this time, and not to rely upon the computation of *Olympiades*, Consuls, or any other humane calculation; which, it cannot be doubted, must, of necessity, leave the deepest student of them in doubting and uncertainty."[55]

As previously noted, the composition of the English scholarly community—predominantly divines—proved crucial in determining the course and focus of studies. In publications, they appear invariably indifferent to "mere" philological and technical issues that did not touch on some aspect of sacred chronology (with the exception of those few whose publications are explicitly pedagogical). This dearth of contributions to classical philology and technical chronology, however, should not be equated with a general absence of appropriate talent. A close study of the state of English scholarship during the seventeenth century demonstrates that what England lacked was not competent scholars; rather an environment that might have enabled them to embark upon secular studies without impediment.[56]

It might have been otherwise, for in the half-century that followed the publication of *De emendatione temporum* several Englishmen drew inspiration from the book, notably Edward Lively, Thomas Lydiat, John Gregory, and John Bainbridge. Selden's case is particularly telling. His early works—*De Diis* Syriis and *The Historie of Tithes*—were inspired by, and greatly expanded on, Scaliger's work. Selden did little to conceal either his admiration for the great man or his ambition to be recognized as partner in the same scholarly enterprise that Scaliger had elevated to such prominence. During

voluminous writings." So, too, many talented men who had "spent a great part of their life in the study of Chronology" might have been "imploy'd in more useful Studies, [where] they might have been very serviceable both to the time in which they liv'd, and also to future generations." Milner, 1694, pp. 32–34.

[54] Controverting the chronology of a pagan writer, Edwards continued, would have been an admirable enterprise. "But in the Sacred Writings there is nothing that looks like such Defect in Synchronism: both Time and Place are truly assigned, though sometimes by reason of the things before mentioned we cannot presently discover the Truth of it, and make it appear how it is." Edwards, 1693, p. 410.

[55] Lightfoot, 1684, p. 383. George Bright, Lightfoot's editor, marshalled the Hebraist's position against Richard Simon: "Wee'l take leave to argue from the Chronology of the Scripture, especially where all Copies and Translations agree, notwithstanding the assertions and conjectures of the late famous Critick, *That no* exact *Chronology (what! for no time?) can be stated upon the Authority of these Books*; till he lays surer foundation for his Opinion, and more particularly explains it." sig. b^{v}.

[56] Feingold, 1995, pp. 234–61; Feingold, 2002.

the 1610s, however, Selden had largely steered clear of technical chronology for, despite his interest, he recognized his limitations. Thus, in his learned notes to Michael Drayton's *Poly-olbion*, Selden not only corrected several mistakes of "that great Dictator of knowledge *Joseph Scaliger*," but announced his commitment to a historical methodology based on proper chronology. His procedure, Selden proclaimed, intended to weigh the credit of the reporter of events against the testimony of "more perswading authority"—utilizing, above all, synchronisms, "the best Touch-stone in this kind of Triall." A notable example of his methodology can be found in his disentangling the Gaelic chieftain Brennus, who had sacked Rome ca. 390 BCE, from his namesake who died while attempting to pillage Delphos over a century later. Only a "confusion of time and actions" permitted the conflation of the two, Selden noted. Nor would "*British* chronologie endure our *Brennus* to be eyther of them." Three years later, in *De Diis Syris* Selden plunged even deeper. In the course of a curious (and misguided) digression, he sought to demonstrate that the Egyptian worship of Osiris and the feast for the golden calf established by King Jeroboam (Kings 12: 28–32) were one and the same, as well as to claim that both festivals had been assigned identical dates. Noteworthy, however, is Selden's feigned characterization of his excursus as an instance of "vain curiosity."[57]

The arrival in England of the Parian marble as part of the Arundel collection of Greek inscriptions in early 1627 offered Selden an opportunity to produce a significant new chronological work. The Marmor Parium included a dated list of events stretching from the (mythical) foundation of the Athenian kingdom by Cecrops in 1582 BCE to the archonship of Diognetus, over thirteen centuries later. Selden sought to supplement such an invaluable source of Greek ancient history with material deriving from an equally priceless new document: a copy of the Samaritan Pentateuch which Archbishop Ussher had acquired in 1625. In early September 1627, Selden requested from Ussher transcripts of Genesis 5 and 11—chronicling, respectively, the lives of the antediluvian patriarchs and Noah's progeny—indicating his seeming intention to embed his edition of the inscription within a novel universal chronology that would augment Scaliger's chronology. Unfortunately, Selden's letter took more than two months to reach Ussher, and by the time he received the desired information in December, Selden had abandoned his ambitious plan. Nevertheless, the *Marmora Arundellian* (1628) included, in addition to the transcription and the rendering of the material into an accessible "Canon Chronicus," an extensive commentary on the text and related matter. Contemporaries immediately recognized the significance of the book. John Bainbridge, for example, expressed hope "for some light in the Persian Chronology … which of all others I most desire to be illustrated, being so necessary to the connexion of Sacred and Profane History."[58] Newton, however, despite owning a copy of Humphrey Prideaux's edition of the work, studiously turned his back on it after perusing it during the 1680s, and for obvious reasons. The entire tenor of the *Chronology* ran counter to the evidence he found in Selden's work: Cecrops ascended to the throne nearly six centuries before Solomon; Deucalion's Flood took

[57] Drayton, 1961, vol. 4, pp. xi*, viii*, 155; Selden, 1617, pp. 52–53; Toomer, 2010, pp. 112, 226–27.

[58] Ussher, 1847–1864, vol. 15, pp. 380–87, 394; Selden, 1628 , pp. 92–119 and passim; Toomer, 2010 , pp. 360–87.

place in 1529 BCE; Ceres introduced crops to Greece in 1409 BCE; Troy fell in 1209 BCE, and so on.[59]

Selden never returned to chronology in a systematic way, and the shift in his scholarly interests reflects the more general waning of interest in technical chronology in England. The example of Thomas Lydiat, Selden's elder contemporary, is instructive in this regard. Lydiat's attempted corrective to Scaliger, in the form of *Tractatus de variis annorum formis* (1605), was instantly rebutted by Scaliger: Lydiat had plagiarized from him what had been of value, and the rest amounted to inanities. Lydiat spent the next thirty years smarting from the insult and seeking to vindicate his claims. Lydiat's exact contemporary, John Bainbridge, was a better astronomer, as well as a more judicious chronologer, whose labors eventually extended and corrected aspects of Scaliger's system.[60] His success in demolishing much of Scaliger's notions regarding the Egyptian calendar—thereby undermining the basis of the Julian calendar—earned him a fulsome tribute (by John Greaves) on his tombstone at Merton College as one "who corrected Scaliger with more success than Scaliger himself emended chronology."[61]

Averse until the end to publication, Bainbridge died in 1643; his *Canicularia* was published by John Greaves in 1648. Lydiat and John Gregory died three years later. Their passing put an end to the abortive flowering of technical chronology in England until Newton. Not that the number of narrative chronologers declined; just the opposite. They often substituted self-confidence for what they lacked in talent. Thomas Pie acknowledged in 1597 that many scholars had attempted to compute the chronology from Creation to Christ "with great paines and no lesse praise, whose candle my selfe am scarse worthy to hold." Since none of these computations persuaded him, however, he suggested none should blame him for "wading further in the broken yce to finde out the trueth, or for divulging my opinion and reasons in this naked shadow and short delineation." Likewise, the clergyman John Swan readily admitted some sixty years later that Scaliger, Petavius, Calvisius, and others had "outstript" his efforts "by the lustre of their own Names." Still, he protested, "though [he] came after them in time, yet a Dwarfe set upon a Gyants shoulders, may see further then he that bears him." As for the far more talented William Beveridge a generation later, he boldly advertised his success in simplifying and methodizing the noble discipline that Scaliger, Petavius, and others had, in his opinion, so badly damaged. They had bedeviled chronology with controversies and conjectures and failed to provide a clear method, thereby making it impossible to master the subject without the utmost difficulty and expense of time. In contrast, his own far plainer system consisted of only what was absolutely necessary, "without any perplexed disputes or intricate questions"—and all in 173 pages![62]

[59] Harrison, 1978, p. 187; Yahuda MS 17.2 fols. 1, 24.

[60] Lydiat, 1605, 1609; Bainbridge, 1620; 1648.

[61] "qui Scaligerum felicius correxit, quam Scaliger emendavit Tempora" Wood, 1786, p. 20; Grafton, 1990, pp. 294–95; Grafton, 1993, pp. 207–8.

[62] Pie, 1597, sig. A4; Swan, 1653: "The Epistle Dedicatory"; Beveridge, 1669: "ad lectorem." Henry Oldenburg's review of the book succinctly conveyed Beveridge's aim: Beveridge found Chronology to be "obscur'd with many intricate Questions, fill'd with many knotty Controversies, stuffed with a multitude of uncertain Comments, and deliver'd in such an odd and dark method, that the knowledge thereof was

We have, as it were, come full circle. Anthony Grafton wrote vividly of how, before Scaliger, chronology "was generally pedagogical in intent and elementary in execution," the chronological texts aiming primarily to furnish neophytes with "help in reading the Bible and the ancient historians." Nor were its practitioners motivated to pursue chronology for its own sake. They sought instead to discover God's hand in history "by revealing a divine order in events." Scaliger used technical chronology to champion an alternative route to the study of history: "not to find moral order in the past, but simply to reconstruct the past."[63] As we saw, notwithstanding the efforts of several talented Oxbridge men, his goal was certainly not realized among the generality of Englishmen in the century that separated the publication of Scaliger's *De emendatione temporum* and the inauguration of Newton's immersion in chronological studies.

Several factors contributed to the protracted decline of chronology, but almost all implicate the rapid demise of the ideal of erudition that had informed the education of gentleman and scholar alike. John Locke best articulated such a shift. Seizing upon latent prejudice against erudition, Locke constructed a powerful, and effectively anti-intellectual, educational program that appealed to the nobility by recommending broad competence over expertise. When Locke argued that "the business of education is not ... to make them perfect in any one of the sciences, but so to open and dispose their minds as may best make them capable of any, when they shall apply themselves to it," he shrewdly appropriated the language of the humanists in their ideal of education only to strip it of the scholarly substance with which they had invested it. Locke certainly sounded much like his predecessors when extolling history as "the great Mistress of Prudence and Civil Knowledge," which "ought to be the proper Study of a Gentleman, or Man of Business in the World." Or when heaping platitudes on Geography and Chronology without which history would be "only a jumble of Matters of Fact, confusedly heaped together without Order or Instruction." Yet Locke quickly clarified his intention to recommend only "the general part" of chronology in order to afford "a view of the whole current of time, and the several considerable *Epochs* that are made use of in History." The rest he discarded as superfluous, even cacophonous.[64] Hence, to his mind, a judicious selection from Strauchius' elementary *Breviarium Chronologicum*—with an occasional glance at the tables in Helvicus' *Chronology*—would suffice.[65]

Locke's advice reverberated in numerous treatises, both within and beyond England. The Parisian Professor of eloquence, and member of the Academy of Inscriptions, Charles Rollin, included a section on chronology as a matter of course in his

---

not to be obtain'd without much difficulty, and loss of Time. Whereupon he declareth his Design to be, To deliver the same from Controversy, and only to treat of what concerns meerly the Distinction of Time, omitting matters of less concernment, and yet nothing of what may be requisite to *Chronology*." *Philosophical Transactions*, Vol. 4 (1669), p. 956.

[63] Grafton, 1975, pp. 161, 169–70.

[64] "When I speak of *chronology* as a Science he should be perfect in, I do not mean the little Controversies, that are in it. These are endless, and most of them of so little Importance to a Gentleman, as not to deserve to be inquir'd into ... And therefore all that learned Noise and Dust of the Chronologist is wholly to be avoided."

[65] Axtell, 1968, pp. 292–93.

influential *Traité des Etudes*, but strictly limited its scope:[66] Equally blunt was the English educator John Clarke. As far as he was concerned, profane history hardly existed before Cyrus, and those wishing to inquire into it might peruse Marsham, Perizonius, and Newton.[67] By the early eighteenth century, chronology and geography stood to history in much the same relationship that Greek and Latin stood to literature; the prescription of "little Latin and less Greek" for the latter found its parallel in Richard Steele's injunction: "a short scheme of Chronology" and "a little view of Geography" was all that was required for the study of history.[68]

The rapid marginalization of erudition in society adversely affected technical chronology, which had been transformed into a formidable and recondite discipline by Scaliger and Petavius—as were all domains of expert learning, including mathematics and philosophy, natural or otherwise. Indeed, by 1738 one educational theorist accepted that settling ancient chronology with any exactness would require much greater application "than the young Student of Fortune [would] be willing to submit to," and more time "than the Clergyman [had] to spare from the necessary Business of his Function."[69] Not that interpretative chronology ceased to fire the imagination and aspiration of contemporaries. Throughout the eighteenth century numerous divines and pious laymen tried their hands at harmonizing conflicting parts of Scripture and synchronizing them with pagan chronologies—with the occasional *eureka*. (John Jackson claimed in 1752 that he had reconciled *all* ancient chronologies.)[70] Nevertheless, the subject attracted only a handful of technically inclined scholars. Among them, Newton alone embarked on a thorough-going attempt to reduce all ancient pagan chronologies to the temporal framework offered by Scripture. In doing so, he mobilized his exceptional mathematical and astronomical knowledge as well as exacting rules governing reasoning and evidence. The outcome was a theory that set the origin of civilization in the eleventh century BCE.

[66] Rollin, 1734, pp. 116, vol. 113: "When I recommend the study of chronology, I am far from desiring to engage youth in the examination of such difficult and knotty questions as this matter is capable of admitting, and of which the discussion properly belongs only to the learned. It is sufficient, if they have a clear and distinct idea, not of the precise year of every particular fact, for that would be endless and extremely troublesome, but in general of the age wherein the most considerable events fell out."

[67] He cautioned, however, that "there is so much confusion and uncertainty in the broken remains of Antiquity" concerning that period that he did not "think it worth any Body's while to trouble his Head much about them." Clarke, 1731, pp. 171–72.

[68] Bond, 1965, pp. 395–96, vol. 392.

[69] Boswell, 1738, pp. 78, vol. 71.

[70] Jackson, 1752, vol. 1, p. xi.

Isaac Newton on Prophecies and Idolatry

On February 9, 1674/5, Newton left Cambridge for London, remaining there some five weeks. It was probably his first visit to the capital since 1668 and he made the most of it. On February 18 he was formally admitted as a Fellow of the Royal Society—to which he had been elected three years earlier—and he attended at least two of its weekly meetings as well as conferred privately with several distinguished members, including Robert Boyle and Robert Hooke. Newton also may have taken the opportunity to inspect a major new tourist attraction that had just come to London: the spectacular model of Solomon's Temple that Rabbi Jacob Judah Leon (Templo) had shipped from Amsterdam shortly before he died there. Members of the Royal Society, such as Christopher Wren and Henry Oldenburg, had been involved in recommending the model to the King and in advertising the display more generally, in response to a request by Constantijn Huygens (père) in October 1674. Judging by an advertisement posted in the *City Mercury* of February 10–17, 1675, the replica was unveiled shortly after Newton's arrival in London: "The Model of the Temple of *Solomon* and the Tabernacle of *Moses* with all their Instruments, according to the Holy Bible. Made (after several years study) by a great Rabbi *Leon* of the *Hebrew* nation, may be seen every day in the Week, except Saturday and Sunday, near the Synagogue, against the Tree in great Dukes-Place in *London*. The said Model hath been already seen by his Majesty and several persons of quality."[1]

If Newton did visit the exhibit, the impact of the spectacular model on his thinking about the Jerusalem Temple nevertheless lay in the future. At the time, Newton was more interested in procuring a Royal dispensation from the statutory requirement of Trinity College that senior fellows take holy orders within seven years of election or forfeit the fellowship. As Newton had been elected senior fellow on March 16, 1668, the matter of ordination had become rather urgent. Newton had in fact sought a less dramatic (and less costly) solution to the problem two years earlier, when in February 1673 he attempted to exchange his fellowship for the Trinity law fellowship, recently vacated by the death of Robert Crane. The College statutes exempted the incumbents of lay fellowships (law and medicine) from ordination and, understandably, those reluctant to enter the Church coveted them. As it happened, Robert Uvedale succeeded

[1] Offenberg, 2004; Bunyan, 1989, p. xxx. A contemporary chronicler wrote ca. 1680 that the model "has been, and still is, common to be seene in London." Shane, 1977, p. 128.

to the law fellowship on account of his College seniority—and because it was obvious that Newton had no intention to dedicate himself to the study of law.

While Barrow's sense of propriety and concern for College affairs prompted him to deny his protégé and friend an improper request, the Master was fully cognizant—from his own experience, as we shall see—of the incongruity between scientific and theological vocations. Barrow himself had drawn up the statues of the Lucasian professorship a decade earlier, loosely modeling them after those of the Savilian professorships at Oxford. Since he himself had been ordained before the founding of the professorship, Barrow neglected to include in the statutes an exemption from the requirement to take holy orders granted by the Savilian specifications. Consequently, faced with Newton's plight, Barrow resolved to amend the Lucasian statutes to include just such a dispensation, not only for Newton, but for all subsequent professors. To that end, Newton arrived in London armed with a draft patent that Barrow and he had composed in order to speedily obtain the required royal sanction. The draft found favor in Court and was referred on March 2 to Attorney General Sir Francis North—elder brother of John North, fellow of Trinity College and Barrow's close friend—who quickly pronounced favorably on the matter. The draft was adopted almost verbatim and the official letter patent was signed April 27, 1675. The relevant section reads: the professor "shall not take up holy orders unless he himself desire to, and that he not, by failing to take up holy orders, be considered to give up his fellowship, nor that he be forced to do so. Rather, he should enjoy this immunity as long as he hold his office, just as the fellow professing medicine and either civil or canon law enjoys this immunity, whatever college statute, custom or interpretation notwithstanding."[2]

Newton's eagerness to obtain a dispensation from ordination is significant, for it has prompted scholars to impute religious motivation to the action. It has been argued that by 1673, at the latest, Newton had become a heretic and that his concentrated effort to evade ordination was spurred by his newly found anti-Trinitarianism.[3] No specific evidence exists to substantiate this claim. As we shall see, Newton's initial forays into heresy commenced only in the late 1670s, gaining intensity in subsequent decades. His course of action with respect to the ordination was motivated, not by religious objection per se, but by a strong sense of propriety shared widely by contemporaries who found themselves deep into secular pursuits and painfully aware of the resulting incompatibility with a calling to the ministry. John Locke faced this dilemma in the mid-1660s. As he explained to a friend, "should I put my self into order[s] and yet by the meanesse of my abilitys prove unworthy such expectation, (for you doe not thinke that Divines are now made as formerly by inspiration and on a suddain, nor learning caused by laying on of hands) I unavoidably loose all my former studys and put my self into a calling that will not leave me ... and from whence there is noe desending without tumbleing." Isaac Barrow articulated similar sentiments

[2] Feingold, 1990, p. 88; Newton, 1959–1977 , vol. 3, p. 146 (misdated); Stewart, 2003, p.473; Clark, 1904, p. 171.

[3] Westfall, 1980, 1982; Snobelen, 1999; Iliffe, 2007, p. 72. Dating Newton's heresy so early necessitated an otherwise arbitrary dating of the manuscripts as well.

to account for his abandoning mathematics along with his resigning the Lucasian chair in favor of Isaac Newton. As his friend and first biographer, Abraham Hill, put it: Barrow "was afraid, as a clergyman, of spending too much time upon Mathematics; for … he had vowed at his ordination to serve God in the Gospel of his Son, and he could not make a bible out of his Euclid, or a pulpit out of his mathematical chair."[4]

Though Newton was dispensed from the requirement to take holy orders, the fellowship itself had not actually been rendered lay. Consequently, almost immediately he found himself compelled to conform to University statutes which mandated that Art masters of four years standing—"except those who by the statutes of their colleges are destined to medicine or civil law"—be required to take part in the theology exercises. Uvedale was excused on June 11, 1673 from disputing precisely because he now held a lay fellowship.[5] Newton, however, was hounded by Joseph Beaumont, recently elected Regius Professor of Divinity, who sought to enforce statutory requirements as part of his attempt to impose a conservative vision of religious orthodoxy on the University. Judging by his holding out until 1677, Newton appears to have resisted this requirement, too. It is also possible that Barrow tried to assist him. Certainly, in the summer of 1676, Barrow, by then vice-chancellor of the University, delivered a public oration in which Beaumont was singled out for a series of backhanded compliments, possibly denoting disapproval of his rigid and outmoded high-churchmanship. Tongue in cheek, the witty orator portrayed the Regius professor as

> so eminent for the power of his mind and the keenness of his genius; for the solidity of his judgment; for the shining beauty, limpid richness, and muscular force of his eloquence; for the breadth and copiousness of his learning; and finally for his expertise and diligence—that in himself he could bless, distinguish, and illumine any University. I shall say no more lest I offend his own modesty or appear to be pouring light upon the sun. But this Hercules of a Theologian requires no praises, especially among you, to whom his gifts are and will be in no way unknown; if, that is, you are mindful of past events, or you pay due attention to present ones.[6]

On February 8, 1677, Newton disputed on two questions, one of which, pertaining to moral theology, required him "to consider the orthodox case against Socinianism," which rejected the divinity of Christ and therefore the Trinity.[7] Preparing for the disputation offered Newton his first taste of serious scholarly engagement with theological issues. Whether the exposure to Socinian sources planted the seeds for his future anti-Trinitarianism is doubtful; the evidence suggests that his immersion in heresy belongs to the 1680s. Regardless, in the late 1670s Newton was still deeply engrossed in chemical researches. In addition, his attention turned in a different theological direction—to prophecies. The impetus for this new focus is unclear. Per-

[4] Locke, 1976–1989, vol. 1, pp. 303–4; Feingold, 1990, pp. 80–81.

[5] Heywood, 1840, p. 14 (second pagination); Edleston, 1850, p. xlix.

[6] Barrow, 1859, vol. 9, p. 219; For Beaumont, see Gascoigne, 1989, pp. 31–32.

[7] Mandelbrote, 2006, pp. 282–83.

haps the near-paranoia that swept England following the "exposure" of the Popish Plot in the fall of 1678—prompting various scholars, including Henry More, to embark on new investigations into prophecies—turned Newton's attention to such studies as well. Or perhaps the catalyst was Newton's six-month sojourn in Lincolnshire during the second half of 1679. He had left Cambridge in mid-May 1679 to nurse his ailing mother, and remained home following her death on June 4 to settle affairs of estate. Devoid of his laboratory and books, and emotionally vulnerable, Newton may well have turned to an intense study of Scripture and the prophecies.

Whatever the catalyst, once embarked on such studies in the late 1670s, Newton's absorption proved swift and complete. He probably began by carefully perusing Joseph Mede's *Clavis apocalyptica*—almost certainly in the English translation, which he may have located among the theological books belonging to his long-deceased stepfather[8]—but he read few, if any, other commentaries on the prophecies.[9] Newton surely described his own course of study when prescribing, at the outset of an early treatise on Revelation, the prerequisite unmediated absorption with Scripture: "search the scriptures thy self & that by frequent reading & constant meditation upon what thou readest, & earnest prayer to God to enlighten thine understanding if thou desirest to find the truth." This nearly exclusive reliance on Scripture informed Newton's researches for the rest of his life. He certainly failed to cite any contemporary commentaries on the prophecies—so much so that readers of the posthumous *Observations* took note. William Whiston, for one, lashed out: Newton "seems to have digged long in the deepest Mines of Scripture and Antiquity for his precious Ore himself; and very rarely to have condescended to make use of the Thoughts or Discoveries of others on these Occasions."[10]

Newton's relentless mining of Scripture was propelled by a firm conviction that the prophecies, suitably construed, had verisimilitude. That conviction manifested itself early and remained unshaken long after he had become a skillful Biblical critic with a firm grasp of how Biblical texts were composed and transmitted. Though he accepted that Moses was not the sole author of the Pentateuch, and that the historical books of the Old Testament were written by different hands and at different times—subsequently redacted by Samuel or Ezra—Newton remained convinced that the prophets from Isaiah "down to the days of the second Temple" had collected their own writings "into one body." As for the Book of Daniel, its first six ("historical") chapters—he asserted—were written by other authors, but the final six, comprising the prophecies, were "written at several times by *Daniel* himself" and the whole collected into a single book by Ezra. Nor was Newton troubled by the fact that by the time of Judas Maccabeus "ye sacred books were much destroyed & dispersed," so that a major effort was required to reconstitute the text of the Old Testament contributing to a certain corruption of the original text. As far as Newton was concerned, Daniel's

[8] Newton probably acquired the 1672 edition of Mede's complete works later.

[9] He certainly did not cite secondary sources.

[10] Yahuda MS 1.1 fol. 2; Whiston, 1734, p. 70. In the same year Arthur Young, who "thought it not improper to warn" unlearned readers of the perils involved in accepting Newton's "conjectures" regarding the origin and progress of idolatry in the Church of Rome, further complained of Newton's unwillingness to take notice "of any one's having ever consider'd the same Things before him, for new and important Discoverys." Young, 1734; vol. 2, p. 265.

prophecies were spared such vicissitudes. Newton never articulated this conviction explicitly, but his position becomes evident from his repeated pronouncements on their centrality: "The predictions of things to come relate to the state of the Church in all ages: and amongst the old Prophets, *Daniel* is most distinct in order of time, and easiest to be understood: and therefore in those things which relate to the last times, he must be made the key to the rest." Newton reiterated his position more emphatically when asserting that "to reject [Daniel's] Prophecies, is to reject the Christian religion. For this religion is founded upon his Prophecy concerning the *Messiah*."[11]

Similar confidence informed Newton's acceptance of the Book of Revelation as genuine, despite his increasing suspicions that various texts of the New Testament had been tampered with in order to furnish Scriptural basis for the new Trinitarian doctrine. To alleviate the problem, Newton went to great lengths to argue that Revelation was written before the destruction of the Jerusalem Temple; hence the many Hebraisms it contained—in contrast to the later New Testament which shunned such forms of speech—and hence the allusions to Revelation found in St. Peter's epistles as well as in the Epistle to the Hebrews, suggesting the absence of later interpolations or corruptions. Moreover, for Newton the genuineness of Revelation was attested to by the fact that "all true *Christians* in that early age received this Prophecy: for in all ages, as many as believed the thousand years, received the *Apocalypse* as the foundation of their opinion." In an early eighteenth-century manuscript, Newton went so far as to claim that "God was so concerned that John get the text right that he sent Jesus, the messenger of God, to watch over John as he wrote down the prophecies."[12] The ultimate proof of the genuineness of Scripture prophecies, of course, rested in their fulfillment, as Newton stated repeatedly.[13]

Within a relatively short time Newton was convinced that he had become adept at prophetic understanding. The opening sentence of Newton's earliest treatise on prophecies conveys unbounded confidence in his achievement joined by a mission to teach others. "Having searched after knowleg [*sic*] in the prophetiq[ue] scriptures," he wrote, "I have thought my self bound to communicate it for the benefit of others." Initially, Newton stated that he had "obteined" such knowledge "by the grace of God," but almost immediately thought better of it and deleted the boast. Nevertheless, the manuscript everywhere displays confidence in his success in understanding prophecy. To begin with, Newton numbered himself among those "few scattered persons which God hath chosen" to unravel "the prophesies concerning ye last times"—the chosen few who, "without being led by interest, education, or humane authorities, can set themselves sincerely & earnestly to search after truth." In contrast, most commentators, with the notable exception of Mede, failed, in Newton's words, properly to methodize the Apocalypse and have "immediately fallen upon giving interpretations." These interpretations, in fact, "have been so botched & framed wthout any due proportion,"

[11] Newton, 1733, pp. 10–11, 15, 24, Yahuda MS 10b fols. 11–12; Yahuda MS 8.2 fols. 5–5^{v}.

[12] Newton, 1733, pp. 238–39, 248; Popkin, 1990, p. 109.

[13] God "gave this [Revelation] and the Prophecies of the Old Testament, not to gratify men's curiosities by enabling them to foreknow things, but that after they were fulfilled they might be interpreted by the event, and his own Providence, not the Interpreters, be then manifested thereby to the world. For the event of things predicted many ages before, will then be a convincing argument that the world is governed by providence." Newton, 1733, pp. 251–52.

that Newton came to doubt whether their authors "so much as beleive their own interpretations." Even Mede erred occasionally in the *Clavis*—and "had that been perfect, the rest would have fallen in naturally"—the implication being that Newton made the necessary perfecting corrections.[14]

A striking paragraph in the "Rules for methodising the Apocalyps" sets into sharp relief Newton's self-assurance as well as his discernment of the degree of certitude he had attained:

> if any man shall contend that my Construction of ye Apocalyps is uncertain, upon pretence that it may be possible to find out other ways, he is not to be regarded unless he shall show wherein what I have done may be mended. If ye ways wch he contends for be less natural or grounded upon weaker reasons, that very thing is demonstration enough that they are fals, & that he seeks not truth but ye interest of a party. And if ye way wch I have followed be according to ye nature & genius of ye Prophesy there needs no other demonstration to convince it. For as of an Engin made by an excellent Artificer a man readily beleives y^{t} ye parts are right set together when he sees them joyn truly with one another notwithstanding that they may be strained into another posture; & as a man acquiesces in ye meaning of an Author how intricate so ever when he sees ye words construed or set in order according to ye laws of Grammar, notwithstanding y^{t} there may be a possibility of forcing ye words to some other harsher construction: so a man ought wth equal reason to acquiesce in the construction of these Prophesies when he sees their parts set in order according to their suitableness & the characters imprinted in them for that purpose

True, Newton admits, "an Artificer may make an Engin capable of being wth equal congruity set together more ways then one, & that a sentence may be ambiguous: but this Objection can have no place in the Apocalyps, becaus God who knew how to frame it without ambiguity intended it for a rule of faith."[15]

Nevertheless, even in his earliest writings, where Newton seemed inclined to include explicit expressions of confidence in the accuracy of his own interpretations, he refrained from claiming that his elucidation was analogous to mathematical demonstration: "I could wish they would consider how contrary it is to God's purpose y^{t} ye truth of his religion should be as obvious & perspicuous to all men as a mathematical demonstration. Tis enough that it is able to move ye assent of those wch he hath chosen; & for ye rest who are so incredulous, it is just that they should be permitted to dy in their sins." Some two decades later Newton is said to have refused to see Richard

[14] Yahuda MS 1.1 fols. 1, 8.

[15] Yahuda MS 1.1 fols. 14–15. A few pages later Newton added: "I cannot but on this occasion reprove the blindness of a sort of men who although they have neither better nor other grounds for their faith then ye Scribes & Pharisees had for their Traditions, yet are so pervers as to call upon other men for such a demonstration of ye certainty of faith in ye scriptures that a meer naturall man, how wicked soever, who will but read it, may judg of it & perceive ye strength of it wth as much perspicuity & certainty as he can a demonstration in Euclide." Ibid., fol. 18. We thank Steffen Ducheyne for reminding us of this passage.

Bentley for an entire year because the latter dared to inquire whether Newton "could *demonstrate*" that a prophetic day denoted "a *year* in their completion." According to William Whiston, Newton was greatly offended by Bentley's challenge because he interpreted the request "as invidiously alluding to his being a *mathematician*; which science was not concerned in this matter."[16] Though Newton's interpretations were, he insisted, not mathematical, they nevertheless were "natural" and "grounded" on strong reasons.

A telling example of the public perception of Newton's assuredness can be gleaned from Henry More's account of their exchange over prophecies. It appears that no sooner had Newton returned to Cambridge from Lincolnshire than he shared his new notions regarding the Apocalypse with More. The latter found these to be "peculiar conceits" and attempted to convert Newton to his own ideas by handing over to him a manuscript of his *Apocalypsis Apocalypseos*. Subsequent discussion led More to believe that Newton not only approved his "Exposition as coherent and perspicuous throughout from the beginning to the end," but that Newton became "in a manner transported" by reading More's treatise. More undoubtedly mistook Newton's enthusiasm, for he quickly discovered that the Lucasian professor had "recoyle[d] into a former conceit he had entertained" over such issues as the commencement of the seven Vials and the prophecy of the seven churches. The "truth" of his own interpretation of the latter prophecy, More boasted, had been highly commended by an unnamed Cambridge mathematician as having been demonstrated "with Mathematicall evidence."[17] Newton's and More's divergent conceptions of certitude, as well as Newton's manner of handling different sorts of evidence in his scientific investigations and in his textual exegeses, will be discussed at greater length below. First, we consider the extent of his borrowing from Mede and the role that millenarian calculations played in Newton's prophetical writings.

Newton was uncharacteristically candid in acknowledging a great debt to Mede: "It was the judiciously learned & conscientious Mr Mede who first made way into these interpretations, & him I have for the most part followed. For what I found true in him it was not lawful for me to recede from, & I rather wonder that he erred so little then that he erred in some things." Elsewhere he affirmed more pithily that "Mr Mede layed the foundation & I have built upon it."[18] Certainly Mede furnished Newton with a powerful conceptual structure and the linguistic tools necessary to decipher the visions of Daniel and Revelation. His schema permitted Mede (and later Newton) to substitute the sequential arrangement of visions found in Scripture with a synchrony between images and symbols—irrespective of their order in the sacred texts—which, in turn, helped determine the true order of the prophecies and affirmed the general coherence of the Biblical texts. Small wonder that Mede considered the synchronisms he discovered to be law-like. As his biographer relates, having initially held a preterist view in considering the thousand years' prophecy to have begun during the reign of Constantine, Mede was "forced" into a millenarian position "by the unresistible Law of *Synchronisms*" that he had discovered.[19]

[16] Yahuda MS. 1.1 fol. 19; Whiston, 1753, p. 94.

[17] Nicolson and Hutton, 1992, pp. 478–79.

[18] Yahuda MS. 1.1 fols. 8, 15.

[19] Mede, 1672, p. x.

Mede himself referred to his discovery as "the Law of *Synchronistical* necessity." Hence his insistence that the interpretation of prophecies be subjected to the dictates of a comprehensive structure of Scripture prophecies:

> If the *Order*, *Method* and *Connexion* of the *Visions* be framed and grounded upon supposed Interpretation; then must all Proofs out of that Book needs be founded upon begged principles and humane conjectures: But on the contrary, if the *Order* be first fixed and setled out of the indubitate Characters of the letter of the Text, and afterward the *Interpretation* guided, framed and directed by the *Order*; then will the variety of Expositions be drawn into a very narrow compass, and Proofs taken from this Book be evident and infallible, and able to convince the Gain-sayers.[20]

Newton embraced these prescriptions without reserve. Heeding Mede's dictate, he labored first to establish his own prophetical structure by immersing himself in the study of Scripture and by availing himself of Mede's citations from the Targum (Chaldaic paraphrase) and Achmet's *Oneirocritica* (on the interpretation of dreams) in order to access the meaning of near Eastern symbolisms and further demonstrate the "consent" of his interpretations with Scripture. Having acquired the requisite structure, Newton proceeded to apply it to history. Here, too, Mede's work inspired him. In particular, Newton was drawn to the license provided by Mede to read Scriptural prophecy virtually as a secular text—an inspired narrative of political and ecclesiastical history. The Book of Revelation, Mede stipulated, should be considered

> according to the letters only; as if it were a certain bare history of things done, and not a prophecy involved with mystical allegories, and types: yet to be furnished by the holy Spirit, with such signes and characters through the whole narration, that thence the right course, order and Synchronismes, of all the visions, according to all things done in their time, may be found out, composed, and demonstrated, and that without the supposition, or help of any interpretation granted.[21]

Mede's editors enthused about the significance that he attached to the application of prophecy to past events, and his considerable success in doing so. William Twisse, for example, noted the "great skill in History" that was required for the proper "accommodation of the matter of fact in this prophecie thus interpreted by him, unto its proper time." Twisse also recalled that those familiar with Mede's work "would give him the bell for this, as herein out-stripping all others." John Worthington concurred. No interpreter before Mede, he believed, had "with equal clearness accommodated those *Prophecies* and *Visions* to their proper *Events*."[22]

Newton was determined to surpass Mede, and his manuscripts attest to the maturation of this resolve over nearly half a century. Whereas the structure for interpreting

[20] Ibid., pp. 583, 581.

[21] Yahuda MS 1.1, passim; Mede 1650, p. 27.

[22] Mede, 1650, sig. A4; Mede, 1672 , sig. *4^{v}.

Biblical prophecies that he had perfected by the early 1680s remained largely unchanged in subsequent reiterations, Newton's command of the historical record broadened and deepened considerably. For the elucidation of prophecies mastery was essential, Newton echoed Mede, for prophecies are nothing but "histories of things to come." Accordingly, the interpreter must apportion "the most notable parts of Prophesy to the most notable parts of history, & ye breaches made in a continued series of Prophesy to ye changes made in history."[23]

Shaping the visions and symbols of Biblical prophecies into a coherent structure, however, proved relatively straightforward compared with the need to impose that structure on past events. As would be the case with chronology—to which Newton also arrived with a schema in hand—the chosen framework included unalterable constraints that left little room for maneuver. Just as the fixed (and small) number of years that the Masoretic version of the Old Testament allocated to the duration of world history since the Deluge constrained Newton in his accounting for the rise of civilizations, so the visions of Daniel set a fixed point of historical origination for all prophecies—and also furnished a small set of numbers, interpreted as years, to denote the duration of each. These numbers included, in addition to the Seventy Weeks of Daniel, the following: 1260 years for the scattering of the holy people; a 1290 year duration from whatever date is determined for the setting up of the abomination of the Temple; 1335 years heralding the Day of Judgment; and 2300 years leading to the cleansing of the Sanctuary. Equally important, Daniel prophesized the rise and demise of four monarchies—invariably interpreted as the Assyrian, Persian, Greek, and Roman monarchies—on the ruins of which, Revelation added, the Kingdom of God would mount.

Not surprisingly, the conviction that scriptural prophecy had historical meaning for both past and future made the matter of their interpretation of great concern to contemporaries, especially in times of crisis. Newton had accordingly entered a vibrant and contentious field, inhabited by interpreters who were as confident as he in their abilities to decipher the visions of Daniel and Revelation, and to chart the interval stretching from Daniel to the end of time. Unlike Newton, however, most interpreters were keen to determine "the structure of the future by locating the present through systematic interpretation of the past."[24] Newton was instead a "passive" millenarian, one who made little effort to locate either the present or the future on the prophetic continuum of time. As we shall see, the few oblique references he made about dating the future indicate that Newton conceived a less cataclysmic millennium than did his contemporaries, one that was located much further in the future than was commonly held.

For most contemporaries the temptation to forecast the coming of the Millennium, or to incorporate current events into their calculations, was simply too hard to resist. *Any* date chosen to denote the commencement of a prophecy necessarily entailed a termination date as well. To give but one example, when Johann Heinrich Alsted settled on AD 69—the year in which Jerusalem fell—as the start of the "abomination of desolation" (Daniel 12:11), he felt justified to add to that date the 1290 and

[23] Yahuda MS 1.1 fol. 16.

[24] Almond, 1993, p. 193.

1335 prophetic years mentioned in the same prophecy, to conjure up AD 2694 as the Day of Judgment, which implied that the Millennium would commence in 1694. Virtually all commentators followed similar practices. Even Joseph Mede, who usually avoided predictions—admonishing interpreters of Revelation to heed "the known shipwrecks of those who have been too venturous"—found in the news regarding the military successes of Gustavus Adolphus a providential omen: "Is not this he, whom the Lord of Hosts hath destinated to execute the work of the Phial? So I hope, and heartily pray." In private, the exciting new chronological information provided by the decipherment of the Arundelian marble also prompted Mede to speculate, this time on the verity of Elisha's prophecy, that the world would come to an end after 6000 years. In such a schema, he conjectured, the Millennium could commence in 1736, 1260 years after the Fall of Rome.[25]

Attempts to prophesize naturally found critics. Joseph Hall was typical. In 1650 he considered Alsted's choice of AD 69 as a too "weak and sandy foundation" upon which "to raise so high a structure." Noting the proliferation of dates purported to denote the start of the Millennium, Hall wondered how anyone could determine which of "such diversities of probable judgments"—based, as they were, on "misconstructive conjecture"—was correct? As far as he was concerned, rather than engage in vain speculations, it would be far better "to be ever in a perpetual posture of readiness for that awful and glorious coming." From the opposite side of the confessional divide, John Owen reached a similar conclusion: "Take heed of *computations*," he admonished his auditors in 1680. "How wofully and wretchedly have we been mistaken by this! We know the time is determined,—its beginning and ending is known to God; and we must live by faith till the accomplishment." As for the witty Robert South, he believed that the more Revelation is studied "the less it is understood, as generally finding a man cracked, or making him so."[26]

While Newton proceeded to identify key historical dates as denoting specific prophecies, he studiously avoided adjoining completions to them for he was wary of playing the prophet. A case in point is his focusing in an early manuscript on AD 381 as the year in which "wthout all controversy" the "strange religion" introduced by the Council of Constantinople began to spread, "ushering in the worship of the "beast & his image." This year, he concluded, "in all respects was ye first of ye general Apostacy & consequently ye beginning of this seale." The date selected attests to Newton's budding anti-Trinitarianism—for that Council confirmed the Nicene Creed and embraced the Trinitarian doctrine—but Newton was careful not to suggest that the required reckoning of 1260 years commenced that year. It also may have occurred to him that the corresponding completion would have fallen on the eve of his birthday, in 1641![27] Elsewhere, Newton had ascribed the "end of darkness" to AD 607, when emperor Phocas conferred on Pope Boniface IV "the universall Bishopric"—a "uni-

[25] Mede, 1650, p. 117; Mede, 1672 , pp. 734, 603. Ironically, the habitually "rash" Bishop Lloyd, alongside Archbishop Sancroft, though they very "much approved" Mede's "way of interpretation," nevertheless chastised him for "resolving too hastily on the King of Sweden … success in Germany." Evelyn, 1955, vol. 4, p. 636.

[26] Hall, 1863, vol. 8, pp. 301, 310, 350; Owen, 1850–1853, vol. 9, p. 510; South, 1844, p. 377.

[27] Yahuda MS 1.4 fol. 50.

versal authority" that enabled the Pope to introduce idolatry into all churches. "And because the Church of Rome began now to reign over the ten kings & enticed them to this idolatrous religion & thereby became rich & potent, she is hence forward in this prophesy compared to a woman arrayed in purple & scarlet & decked with gemms."[28] True to form, Newton again desisted from suggesting that the rise of apostasy commenced that year, which would have fixed the termination of the prophesy on AD 1867. In another manuscript Newton selected three dates—AD 609, 788, and 841—as possibly denoting the commencement of the 1290 years of "abomination," whereas in the *Observations* he speculated that the 2300 years leading to the cleansing of the Sanctuary "may perhaps be reckoned either from the destruction of the Temple by the *Romans* in the reign of *Vespasian*, or from the pollution of the Sanctuary by the worship of *Jupiter Olympius*, or from the desolation of *Judea* made in the end of the *Jewish* war … or from some other period which time will discover."[29]

Only in the final two decades of his life did Newton begin committing to paper a few possible completion dates for key Scriptural prophecies. Significantly, he made sure to adjoin to virtually all of these speculations a disclaimer: he intended this exercise to serve as an admonition to those rushing to announce the coming of the millennium by showing the futility of such calculations. Newton made one of his clearest statements on the matter when speculating that if the period of 1260 years had begun in 800 AD, it would end in 2060 AD, though "it may end later." "I mention this period," Newton explained, "not to assert it, but only to shew that there is little reason to expect it earlier, & thereby to put a stop to the rash conjectures of Interpreters who are frequently assigning the time of the end, & thereby bringing the sacred Prophecies into discredit as often as their conjectures do not come to pass." Newton further observed that "Christ comes as a thief in the night, & it is not for us to know the times & seasons wch God hath put into his own breast."[30]

Newton jotted down a more elaborate (and famous) calculation on an envelope:

1. The 2300 prophetick days did not commence before the rise of the little horn of the He Goat.
2. Those day[s] did not commence after the destruction of Jerusalem & ye Temple by the Romans A. C. 70.

[28] Yahuda MS 1.6, fol. 47; MS 1.7 fol. 29; Keynes MS. 5 fol. 119. In the *Observations* Newton argued that in "the reign of the *Greek* Emperor *Justinian*, and again in the reign of *Phocas*, the Bishop of *Rome* obtained some dominion over the *Greek* Churches, but of no long continuance." Newton, 1733, p. 90.

[29] Keynes MS 5 fol. 138^v; Newton, 1733, 122. A partial exception may be found in Yahuda MS 1.7 fol. 65 where Newton—by assigning equivalent ratios to duration of regimes during the sounding of each trumpet—concludes that the Turkish Empire "may be expected to loose Constantinople its imperial seat in the year 1844, & perhaps the Emperors may then seat themselves for a time in Iudæa," in accordance with Daniel's prophecy. But this is done to conclude on analogies, as is clear when he goes on to say: "I should now proceed to the seventh Trumpet, but since that concerns the time to come, it will be better to return to the things that are contemporary to the Seales & Trumpets already explained."

[30] Yahuda MS 7.3g fols. 13–13^v. Westfall and the editors of the Newton project date the manuscript to be post 1700.

3. The time times & half a time did not commence before the year 800 in wch the Popes supremacy commenced
4. They did not commence after the reigne of Gregory the 7th. 1084
5. The 1290 days did not commence before the year 842.
6. They did not commence after the reign of Pope Greg. 7th. 1084
7. The diffe[re]nce between the 1290 & 1335 days are a part of the seven weeks.

Therefore the 2300 years do not end before ye year 2132 nor after 2370.

The time times & half time do n[o]t end before 2060 nor after [2344]

The 1290 days do not begin [this should read: end] before 2090 nor after*1374* [*sic* for 2374].[31]

On rare occasions Newton permitted himself to entertain different calculations. His treatment of the number of the Beast, 666, is particularly worthy of note, for Newton made an original contribution to the topic. He accepted the traditional identification, originated by Irenaeus, according to which the numerical value of the letters comprising the Greek name for Rome, Λατεινος, total 666— 30 1 300 5 10 50 70 200. The interpretation appears in the *Observations* as well: the Beast's "mark is †††, and his name ΛΑΤΕΙΝΟΣ, and the number of his name 666."[32] On several occasions Newton offered more detailed discussion. Since the mark of the Beast "is opposed to the name of God & to the number of his servants," he reasoned, it follows that there must exist an exact correlation between the number of their respective servants. And since God's servants numbered 144,000—a figure derived from "multiplying 12000 by its divisor 12, the number of the Tribes of Israel"—the number of the Beast (666) "is to be counted by multiplying some number by a divisor of that number. And there is but one way of counting it in this manner, wch is by multiplying 222 by its divisor 3 the number of the Tribes of them that worship the Beast." In another manuscript, where Newton reiterated his claim that there could be no other way to compute the number of the Beast, he explicitly rejected the interpretation put forth in 1642 by Francis Potter, according to which 25—the (approximate) square root of 666—is a better choice for Bestial interpretation than 12, the square root of 144. Potter found support for his conjecture by enumerating a host of characteristics of the Roman Church that were associated with the number: the 25 gates of Rome; the 25 articles of the Roman Creed, and so forth. Potter's invention proved all the rage, with Mede himself "exceedingly taken even to admiration, professing it to be the greatest mystery that hath been discovered since the beginning of the world."[33]

[31] Yahuda MS 7.30, fol. 8.

[32] Yahuda MS 7.2j fols. 28, 89^{v}; Newton, 1733, pp. 284, 320.

[33] Keynes MS 5 fol. 29; Yahuda MS 7.2j fol. 23; Mede, 1650, sig. A5; Mede, 1672 , p. 877. Henry More did so, too: the number of the Beast is "to be numbered no otherwise than by the extraction of the square root which is 25, which is marvelously applicable to the City and Church of *Rome*, as 12 the root of 144 to the *New Jerusalem*." More, 1680, pp. 134–35, 137, 342–44.

Newton, in contrast, rejected the extraction of roots as fanciful. He not only considered his own interpretation far more natural mathematically, but produced a list of proofs to demonstrate the ill-omened connotations of the number 3—as a direct response to Potter's claim that the number 25 had been "a fatall and unfortunate number":

> Now 222 consists of three binary numbers & in the Pythagorean philosophy the binary is a cursed number because it divides from unity.... And as the divisor 12 is ye number of the 12 Tribes of Israel so the divisor 3 is the number of parts into w^{ch} [the great city is divided at its fall. Tis also the number of Woes inflicted on the worshippers of the Beast & the number of their Gods for they worship the Dragon, the Beast & the Image of the Beast, & the number of descriptions of the seven plagues for they are described by ye seven Trumpets seven Thunders & seven Vials of wrath. & the number of parts into w^{ch}] the earth & trees the sea & ships, the rivers, the sun moon & stars & day & night of the great city are divided, & into wch also the great city is divided at its fall, & the same number is also a fit symbol of the Beast & his Image & the Dragon that old serpent.[34]

Other half-hearted efforts to synchronize numbers included estimations of the possible significance of ratios between the duration of certain prophecies. For example, in one place Newton described how Daniel "hath described by steps the chief actions & revolutions within the compass of the Greek Empire in all ages from the reign of Alexander the great to the resurrection of the dead. And in doing this he hath distinguished the times into three or four cardinal points." Describing the several abominations that were imposed on Jews and Christians during these periods, Newton noted that the interval between the third and fourth periods of abomination was 1010 years—the sacking of Jerusalem by Antiochus Epiphanus in 168 BCE to the restoration of the veneration of images by Empress Theodora in AD 842—equivalent to the ratio between the 1290 and 2300 prophetic days. Evidently Newton found little traction with this observation, for he let the matter drop after remarking that "these numbers admit of several interpretations & it must be left to time to discover the truth."[35] Similarly, he noted elsewhere that the Prophesy of the first Trumpet lasted 12 years, the second 21 years, the third 49 years, the fourth about 160 years, and the fifth about 620 years—leading to the sounding of the sixth Trumpet in 1258. "According to this progression," Newton concluded, "the sixt may last seven or eight hundred years or above"—a speculation made with the cautionary caveat that "it is not for us to foreknow." Realizing perhaps the inadequacy of his calculation of exponential growth for the final interval, Newton revised somewhat his estimate for this interval, and set it to be between 800 and 1,000 years—still far below what the series would have required.[36]

[34] Yahuda MS 7.3i fol. 12^{v}.

[35] On another occasion he credited Daniel with dividing the time "into five cardinal periods." Yahuda MS 7.3c fols. 20–21; MS 7.3n fol. 2.

[36] Yahuda MS 7.2j fol. 19; MS 7.3h fol. 2.

These forays into numerology were few and far between. But even they exhibit the structural mind-set that always informed Newton's prophetical researches. While it is clear that his confidence in the millennium was sincere and unwavering, Newton did not engage in end-time calculations or speculations—except as an exegete. Quite possibly his growing wariness of rash prophesiers derived from the penchant of friends, such as Fatio de Duillier and William Whiston, to indulge in such speculations.[37] Newton addressed these and other would-be seers in his *Observations*: "The folly of Interpreters has been, to foretel times and things by this Prophecy [of the Apocalypse], as if God designed to make them Prophets. By this rashness they have not only exposed themselves, but brought the Prophecy also into contempt. The design of God was much otherwise. He gave this and the Prophecies of the Old Testament, not to gratify men's curiosities by enabling them to foreknow things, but that after they were fulfilled they might be interpreted by the event, and his own Providence, not the Interpreters, be then manifested thereby to the world." But, Newton continued, "the time is not yet come for understanding them perfectly, because the main revolution predicted in them is not yet come to pass."[38] In his manuscripts, he occasionally drew attention to the fact that the present exhibits none of the preconditions necessary for the imminent fulfillment of prophecies. Before the seventh trumpet is to sound, he pointed out in one manuscript, there is to be a universal preaching of the Gospel, but this has not yet been fulfilled: "there has been nothing done in ye world like it, & therefore it is to come." Likewise, he remarked some years later, God's indignation and dispersion of the Jews "hath already lasted since the days of Hadrian 1590 years, & is not yet at an end."[39]

Incisive work by several recent scholars into the substance of Newton's analysis of prophecies makes it unnecessary to cover such ground here.[40] We accordingly turn now to probe his perception of the extent to which prophetical studies were capable of demonstrativeness and his manner of handling evidence. Frank Manuel's pioneering

[37] As early as February 1692, Newton was alarmed upon hearing from Fatio of his "odd thoughts" regarding several books of Scripture, "wch yet [he thought he] could make out beyond all manner of exceptions": "I am glad you have taken ye prophesies into consideration," Newton responded, "& I believe there is much in what you say about them, but I fear you indulge too much in fansy in some things." Newton, 1959–1977 , vol. 3, pp. 242, 245.

[38] Newton, 1733, pp. 251–52. Whiston retorted indignantly in his critique of the *Observations*: "that Sir *Isaac Newton* should here call it the *Folly* of Interpreters to endeavour to *foretel* Events to come by this or any the like Prophecies, as if they did thereby affect to appear to be themselves *Prophets*, is wholly groundless, and throughly confuted by Sir *Isaac Newton's* own Procedure.... If upon Examination, any one thinks he hath discovered the ten Kingdoms of the *Roman* Empire, foretold both by *Daniel* and *John*, and when the last of them arose and compleated that Number; whose Duration is known from them both to be 1260 Years; he cannot easily avoid thinking he hath discovered when that Period will end. Nor can Sir *Isaac Newton* well avoid making the like Discovery of his Opinion for the End of the 2300 Years in *Daniel*; when, as we have seen, *p.* 303. *prius* he supposes them to be solar Years or very nearly so, and dates them either at *A.D.* 70. or 132. or at the latest 136. The Consequence being thence too obvious to be concealed, that they will end, either at *A. D.* 2370. or 2432. or at the latest 2436. Yet do not I imagin that Sir *Isaac Newton* did at all hereby pretend to be himself a *Prophet*." Whiston, 1734, pp. 329–30.

[39] Yahuda MS 1.4 fol. 2; MS. 7.2a fols. 12^{v}–13.

[40] See, for example, Iliffe, 1994; Mandelbrote, 2007; Hutton, 1994; Snobelen, 2003b; Delgado-Moreira, 2006a.

work on Newton's theological writings, as well as on his chronology, drew attention to the manner in which Newton "applied what might be called scientific criteria to the interpretation of the books of prophecy, particularly the law of parsimony":

> [Newton] showed not only that every notable political and religious occurrence conformed exactly to some vision in prophecy, but that his set of equivalents had totally exhausted the possible meanings of each of the objects and images appearing in any prophetic verse. There was nothing left over, no random words still unexplained, no images that were superfluous. The system was enclosed, complete, and flawless. Newton saw his 'methodising of prophecy' as an ideal scientific structure, exhibiting the greatest possible simplicity and harmony.[41]

Richard Westfall likewise asserted that Newton had aimed above all for "certainty" in his theological work, a certainty similar to the demonstrative character, as Westfall saw it, of his science.[42]

Precisely in what ways Newton thought his natural laws to be "certain" depended critically upon their evidentiary basis. Consider his *Observations upon the Prophecies*. This mature (and somewhat sanitized) synthesis of his earlier work abounds with historical discussion of the kingdoms and empires which, he thought, corresponded to Daniel's arcane dreams.[43] Newton's account of the origin of civilization, the unpublished *Original of Monarchies*, is similarly filled with data, as we shall see below. Both efforts are replete with statements that give the texts a seemingly demonstrative character. The *Prophecies* deploy a series of rules for interpretation, while the *Monarchies* lay out a group of social laws. In both cases textual data were synchronized with rules or laws.

Seeking certainty and conferring on the texts a law-like appearance does not imply that Newton considered either the *Prophecies* or, now, the *Monarchies* to be mathematically demonstrative. Despite appearances, these works did not have the same epistemological status for Newton as, for example, the deduction of motion from a configuration of bodies and forces. There is, argues Delgado-Moreira, a "mismatch between the mathematical cloak that dresses up Newton's prophetic interpretations, and the level of certainty that can be attributed" to them.[44] Nevertheless, Newton's works on prophecy and kingdoms did lay out what appear to be general assertions from which others seem to follow, rather like the demonstrations in his *Principia*. He was hardly unique in doing so. John Napier promoted his own exposition of Revelation as a work treated "in form of proposition, as neer the analytick or demonstrative manner as the phrase and nature of holy Scripture will permit," while the explanation and interpretation will be "proved, confirmed and demonstrated, by evident proof and coherence of Scripture agreeable with the event of histories." The summary table

41 Manuel, 1974, pp. 97–98.
42 Westfall, 1980, p. 326.
43 The printer's copy of the *Observations* is CUL MS Add. 3989.
44 Delgado-Moreira, 2006b, p. 14.

for the conclusions Napier intended to prove in the first treatise promised that the propositions which appear in the table in their "naturall order" would be placed in the work itself, "Demonstratively, to the effect every Proposition may be proved by the former Propositions."[45]

As for Joseph Mede, the very title of his book undertook to furnish a "key" to Revelation: "Searched and demonstrated out of the Naturall and proper Characters of the Visions." In the body of the work, Mede often embraced the language of mathematicians, occasionally even utilizing a verbal form of QED. Thus, a paragraph in the sixth synchronism (regarding the Inner Court that St. John was told to measure) concludes: "the very point I was to demonstrate." So did another, treating the temporal equivalence of the 1260 days and the forty-two months, "which was the very point I was to demonstrate." Significantly, when challenged in 1624 on his interpretation of the vision of the Inner Court, Mede admitted that the force of the objection weakened his argument so "that it appears not to be of so sufficient strength as may force assent." Nevertheless, he regarded the objection as insufficient to cause him to abandon his position, "unless the Arguments shewn for that part do appear of more force and probability than himself grounded upon." Anticipating Newton, Mede went on to clarify: "Probability stands in place of a Demonstration, till a greater Probability can be brought to shoulder it out." Or again, his position "follows not by Apodictical necessity, but it may perswade morally as a probability."[46]

If, now, the rules or laws in Newton's works on prophecy and the origins of civilization were not intended to be *demonstrative* in the sense of geometry or logic, then what functions did they serve? There was something unique about Newton's efforts, something that did involve rules or law-like statements but in ways that were nonetheless considerably different from anything familiar to contemporary scholars and natural philosophers. Newton obsessed over his prophetic interpretations, mixing and matching different strands and bits of evidence over time. We have already seen that few if any natural philosophers of Newton's time, or before, systematically deployed experiments in efforts to tie down or, especially, to modify their law-like claims. Newton, in contrast, did precisely that, and his way of working with evidence perplexed many readers for whom the purpose of an experiment was either to measure a parameter whose meaning was given by prior reasoning, or to demonstrate that something that should occur, did so. Newton's quite different methods apply equally well "to his method of prophecy, suggesting that, to paraphrase Harper and Smith, his scheme could only be corroborated or proved by historical events, which would have the role of experiments so to speak, while his prophetic scheme would become entrenched by virtue of allowing further reading and understanding of history."[47]

Consider Newton's work on the motion of bodies in resisting media. There he set down a fundamental proposition that he did not question or modify: in this case that the forces acting in such circumstances can be separated into additively distinguished components. But what these components were, and what their relative importance

[45] Napier, 1593, Sig. A2–A2^{v}, A4^{v}.

[46] Mede, 1650, pp. 1, 12, 15; Mede, 1672, p. 586.

[47] Delgado-Moreira, 2006a, p. 244, referring to Harper and Smith, 1995. See also Smith, 1999, 2000a, 2002, 2005 .

might be, remained open to investigation. To that end, Newton had developed a particular way to work with data by using what Smith termed "residual discrepancies" between theory and measurement, thereby to evolve ever more exact structures by amplifying the available data, all the while retaining his fundamental assumption concerning additively separate forces.[48] Although nothing precisely like this series of successive modifications can be found in his account of prophecy, nevertheless there, too, we find a tight interconnection between an assertion and a spectrum of supporting evidence for it. Moreover, the presumption that prophetic remarks had necessary, if obscure, meaning was unalterably fundamental, as was the assumption that an appropriate interpretative scheme had to be developed for them. The question was what that scheme should be, and how it could be based on, or bound to, an evidentiary base.

Prophecy obviously offered particular difficulties in respect to evidence, in part because some of the prophetic claims refer to events that had not as yet taken place. The bulk of his work on the subject, however, concerned events that had occurred, and here the evidentiary basis consisted of whatever past events could reasonably be brought to bear. To do so, Newton wended his way through ancient texts in an effort to extract data for a consistent interpretative system. We have seen that he followed Mede in working to a set of rules designed to effect consistency among visions and symbols, and that he developed an elaborate series of correspondences which he carried across a broad evidentiary base. Newton tended to generalize, effecting interpretations that avoided particularities. For example, instead of connecting the sun in a prophecy to an individual, Newton linked it to "the whole species and race of kings"; the moon was similarly a generic stand-in for "the body of the common people," and so forth. He would then particularize by bringing evidence to bear. In Daniel's vision of an "image composed of iron and clay," the image's feet were first interpreted as metaphors for kingdoms, after which Newton turned to specific detail, in this case linking the corresponding kingdom to the Dacian region conquered by the Romans.[49]

This building of a scheme that first rendered the singular plural and hence *general* (e.g., a body of people for the decidedly singular moon) followed by the provision of elaborate, *specific* detail characterizes Newton's efforts at prophetic interpretation. Moreover, it tallies well with his ways of working with experiment. Recall how he worked with prisms. To draw a general conclusion from his experiment, Newton specified that the prism should be set in such a way that the light passed through it symmetrically. Only after setting this *general* requirement did Newton give the *specific* refraction involved. Huygens or even Hooke would instead have specified the refraction angle from the first.[50]

Here then we see how Newton's way with generalization and data fed nicely into an area that is otherwise quite distinct from his experimental work. However, Newton did not think his work on prophecy to be demonstrative in the way that, he always

[48] Smith, 2000a, pp. 250–51.

[49] Newton, 1733, chap. V.

[50] Though it is in any case unlikely that either Hooke or Huygens would have thought to set the prism in such a way, since Newton's reason for doing so was both to explore the situation in which the solar image should remain circular, and to minimize error.

claimed, his assertions about light's unequal refrangibility were. After all, the world of Newton's laboratory was limited in both space and time, and it could be controlled—forced, that is, to an appropriately malleable condition. On its face, the world of prophecy was quite different. Its fundamental data—the visions—were given once and for all. They could not be altered or controlled in any way; they could only be interpreted. It is as though at some time long ago, in some now unfamiliar place, an experiment had been produced with unknown equipment, an experiment that generated data the meaning of which had become obscure. One way to understand this otherwise murky intelligence from the past would be to calibrate it against other, known data, through a series of suitably general interpretative conjectures. Which is what Newton did: he *calibrated prophecy* against what he took to be reliable historical data by means of a series of conjectured decipherments.

Some prophetic remarks applied to events that had taken place, meaning that the calibrating data were at hand. Other remarks, pertaining to events that had not yet occurred and whose nature remained obscure, Newton largely ignored. As a result, Newton's effort to calibrate prophecy was not closed in anything like the way his prism experiments had been. The latter, he had argued, entailed the general scheme of unequal refrangibility. Prophetic visions, though manipulated into a coherent structure, could not be treated similarly because they remained either incomplete (having not as yet been fulfilled) or else insufficiently specified. Past events could at best be made compatible with an incompletely specified scheme. In unavoidable consequence, any prophecies that applied to the future had to be intrinsically problematic in a way that the prediction of an optical experiment was not. The development of civilization itself raised issues that were different from, but related to, those posed by prophecy. The goal in prophetic interpretation was to devise a system that could match prophetic symbol to historical event. In the case of civilization's evolution, the problem as Newton saw it was to seat events within a general scheme. The interpretative system for prophecy acted to colligate symbols with events, as it were, whereas in the case of civilization's development the point was to fit events directly into a scheme that had to be conjectured in the light solely of the events themselves, there being no corresponding symbolic structure with which to work.

■

The Origins of Idolatry

By the mid-1680s the scope of Newton's theological studies had expanded to include an investigation into the origins of idolatry. The motivation for such a probe may well have been to substantiate an *idée fixe* that Newton developed while composing the *Principia*: namely, that the ancients were in possession of the "true frame of nature," a cosmology he considered to be at the core of the original (true) religion.[51] Newton intended to make public his views on ancient science in his introduction to the

[51] This represented an extension (and expansion) of Newton's earlier attempt to uncover the roots of analysis in the writings of the ancients, as part of his campaign against Cartesian mathematics. See Guicciardini, 2009, pp. 79–107.

aborted version of Book III of the *Principia*, "On the System of the World." The heliocentric system, he wrote, was "the ancient opinion of not a few in the earliest ages of philosophy," most notably embraced by Anaximander, Philolaus, Aristarchus, the older Plato, and the Pythagoreans. He continued:

> The *Egyptians* were early observers of the heavens. And from them, probably, this philosophy was spread abroad among other nations. For from them it was, and the nations about them, that the *Greeks*, a people more addicted to the study of philology than of Nature, derived their first, as well as soundest, notions of philosophy. And in the Vestal ceremonies we may yet trace the ancient spirit of the *Egyptians*. For it was their way to deliver their mysteries, that is, their philosophy of things above the vulgar way of thinking, under the veil of religious rites and hieroglyphick symbols.[52]

The publication of Leibniz's *Tentamen de motuum coelestium causis* in the February 1689 issue of the *Acta eruditorum* fortified Newton's resolve to establish the ancient origins of the true frame of nature. Leibniz not only pretended, disingenuously, to have formulated his planetary theory before reading the *Principia*; he opened the *Tentamen* with a thinly veiled jibe at the burgeoning cult of Newton. After claiming that the ancients failed to "understand the splendour of nature"—and that even the proto-heliocentric system of the Pythagoreans had been "proffered tentatively rather than correctly determined"—Leibniz appropriated the tribute that Claudian had conferred on Archimedes in order to bestow it on Kepler: the "first mortal man to make public 'the laws of the heavens, the order of nature, and the precepts of the Gods.'" Contemporary readers would have noted immediately this deliberate retort to Halley's encomium to the *Principia*, which improvised on Lucretius in order to deify Newton. Interestingly, in the *Tentamen* Leibniz mentioned Newton but once, when claiming that he had gathered from the review of the *Principia* in the *Acta eruditorum* that Newton, too, appeared to understand the inverse square law. Surely, the intended implication was that it fell to him, Leibniz, to complete the revolution that Kepler had initiated. Indeed, in his correspondence, Leibniz rendered the insinuation explicit when he boasted to both Huygens and Bernoulli of an independent and prior discovery of the law, simply by means of a priori reasoning! As for the ancients, Leibniz pronounced, they had "not yet understood the splendor of nature," and even the Pythagoreans' heliocentric hypothesis had been "proffered tentatively rather than correctly determined."[53]

Leibniz's aggressive canvassing of his planetary theory—which included a direct provocation of Newton himself in a letter of March 1693, in which Leibniz contended that universal gravitation was "caused or regulated by the motion of a fluid medium, on the analogy of gravity and magnetism as we know it here"[54]—included an intensi-

[52] Newton, 1728c, pp. 1–2.

[53] Meli, 1993, pp. 126–27, 138. Meli demonstrated that Leibniz had deliberately reverse-engineered the *Principia*.

[54] Newton, 1959–1977, vol. 3 p. 258; Huygens chastised Leibniz for claiming for himself results that properly belonged to Newton.

fication of his rejection of atomism and the void. In a series of publications and private letters, throughout the 1690s and beyond, Leibniz boasted of his recovery from his youthful infatuation with these doctrines and his mature recognition that atomism, at best, satisfies the imagination.[55] Such challenges undoubtedly contributed to Newton's resolve to revise the *Principia*, complete with a defense of the antiquity and orthodoxy of atomism.

From the start, Newton sought to trace the origins of this "true" philosophy as far back in time as possible.[56] In 1694, he intimated to David Gregory his conviction that the Egyptians had been committed to heliocentrism, as "their religion and hieroglyphics and images of the Gods" made clear to him.[57] The same conviction informed the first proposition of Newton's unpublished "Theologiae Gentilis Origines Philosophicae," which delineated the philosophical character of Gentile theology—comprised primarily of astronomy and the physics of the world system. As "proof," Newton cited Clement of Alexandria's description of Egyptian religious ceremonies, wherein the science of the heavens seemed to feature prominently.[58] Elsewhere, Newton insinuated the Phoenician origins of atomism:

> That all matter consists of atoms was a very ancient opinion. This was the teaching of the multitude of philosophers who preceded Aristotle, namely Epicurus, Democritus, Ecphantus, Empedocles, Zenocrates, Heraclides, Asclepiades, Diodorus, Metrodorus of Chios, Pythagoras, and previous to these Moschus the Phoenician whom Strabo declares older than the Trojan war. For I think that same opinion obtained in that mystic philosophy which flowed down to the Greeks from Egypt and Phoenicia, since atoms are sometimes found to be designated by the mystics as monads.

Unlike many of his contemporaries, however, Newton rejected the temptation to identify Moschus with Moses.[59]

Newton clearly recognized the slender evidentiary basis for crediting the Egyptians, Phoenicians, or Chaldeans with profound understanding of the heliocentric system or "modern" natural philosophy. His few public statements on the matter were either cautious—"The *Egyptians* were early observers of the heavens"; the Chaldeans were "the most learned astronomers of their time"—or insinuative. Hence his conclusion to the third edition of the *Opticks*: "And no doubt, if the Worship of false Gods had not blinded the Heathen, their moral Philosophy would have gone farther than

[55] Leibniz, 1989, pp. 130, 133, 136, 142, 139, 162, 165, 171. For Leibniz's anti-atomism, see Wilson, 2008, pp. 158–68. Fatio de Duillier may have drawn Newton's attention to Leibniz's contestation of atomism in the early 1690s.

[56] In contrast to many contemporaries, Newton refused to speculate on the antediluvian origins of astronomy, which Josephus had attributed to Seth. See note 121 below.

[57] Newton, 1959–1977, vol. 3 p. 384. The very names that Thoth gave the planets, Newton averred, prove "that he was a believer in the Copernican system." Ibid., p. 338.

[58] Yahuda MS 16.2 fol. 1–2; Yahuda MS 17.2 fols. 18^{v}–21. The editors of the Newton project date MSS 16–17 to the late 1680s and early 1690s. Parts of MS 17 appear to be prior in composition.

[59] CUL MS Add. 3965.6 fol. 270, cited in McGuire and Rattansi, 1966, p. 115. Newton still accepted the traditional dating of the Fall of Troy, 1184 BCE.

to the four Cardinal Virtues; and instead of teaching the Transmigration of Souls, and to worship the Sun and Moon, and dead Heroes, they would have taught us to worship our true Author and Benefactor, as their Ancestors did under the Government of *Noah* and his Sons before they corrupted themselves."[60] Here Newton availed himself of received opinion regarding the transfer of learning from Egypt and Mesopotamia to Greece as proof of the Eastern origination of scientific knowledge. His manuscripts attest to the pains he took to convince himself of the manner in which the "mystic philosophy … flowed down to the Greeks from Egypt and Phoenicia."[61] Yet Newton's disciples, much like their master, cited primarily the Greeks, who were credited with elevating to new heights whatever they learned. This programmatic choice is evident in Nicolas Fatio de Duiller's February 1692 letter to Christiaan Huygens, informing the Dutch physicist that, according to Newton "the ancients" (i.e., the Greeks), were in possession of "all the demonstrations" underlying universal gravitation—albeit they made "a great mystery of their knowledge."[62]

Newton continued to scour ancient texts for material to include in his projected revision of the *Principia*, informing David Gregory in 1694 that he intended to "spread himself in exhibiting the agreement of this philosophy with that of the Ancients and principally with that of Thales," as well as to demonstrate his conviction that the "philosophy of Epicurus and Lucretius is true and old, but was wrongly interpreted by the ancients as atheism."[63] Natural philosophy, in contrast, receded into the background of Newton's investigation of the genesis and progress of idolatry. In the "Original of Religions" (ca. 1691), as we shall see below, Newton would insist on the interconnectedness of primitive religion and true astronomy. During the second half of the 1680s, however, he focused primarily on correlating Egyptian and Greek myth with the Book of Genesis, to which end the names and qualities of stars and elements served to uncover the "earliest memories" of postdiluvian men. Newton's investigation evinces his unqualified embrace of the theory promulgated by Euhemerus of Messina in his *Sacred History* (ca. 300 BCE), according to which all ancient deities were notable heroes who were posthumously deified in recognition of their deeds. Euhemerus' rationalization of myth was a by-product of his observation of the rapid process in which Alexander the Great and his immediate successors had been deified. For his own part, Newton found in Galileo's naming the moons of Jupiter after the Medici a more recent instance of deification.[64] As Manuel noted, Euhemerism had been "so profoundly ingrained in Newton's thinking that he described the tendency to transfigure heroes as a natural proclivity of mankind, operative even among the Israelites"—evidence for which he found in the concealment of Moses' burial place "for fear of idolizing him."[65]

[60] Newton, 1728c, pp. 2–3; Newton, 1721, pp. 381–82.

[61] McGuire and Rattansi, 1966. For more extensive notes regarding the transmission of Egyptian astronomy to Greece, see CUL MS Add. 3965 fols. 328–89, 654–56. Early on, Newton credited Orpheus as the conduit for such transmission: Yahuda MS 17.2 fols. 6^{v}, 15, 18; MS 16.2 fol. 2^{v}.

[62] Huygens remained sceptical, despite the fact that Fatio presented Newton's opinion as tentative and subject to corrections. Newton, 1959–1977, vol. 3 pp. 193, 196.

[63] Newton, 1959–1977 , vol. 3 p. 338. See McGuire and Rattansi, 1966; Casini, 1984; Schüller, 2001.

[64] Yahuda MS 16 fols. 11, 15, 46.

[65] Manuel, 1963, p. 106; Keynes MS 2 fol. 5^{v}.

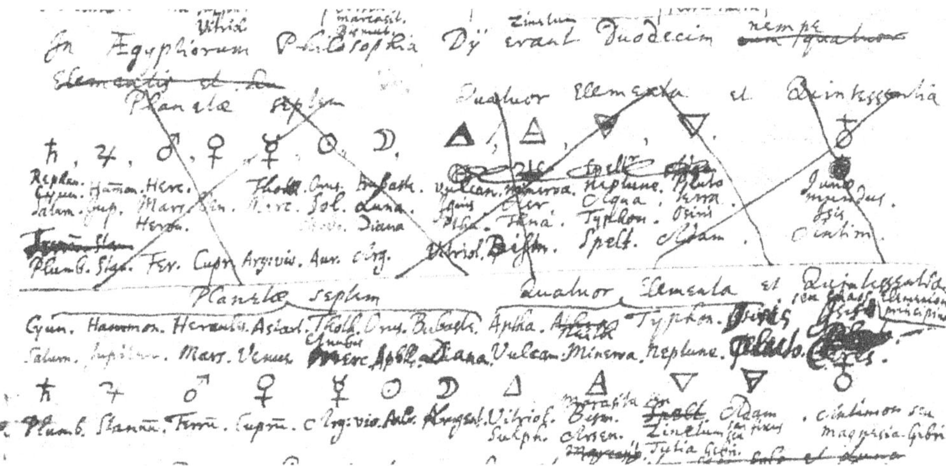

Figure 4.1. Newton's genealogy for Egyptian deities. This item is reproduced by permission of the Huntington Library, San Marino, California. Babson Newton Collection, Burndy, 420.

Convinced that ancient myths recorded the earliest history of mankind, Newton proceeded to correlate the first four generations after the Deluge to Hesiod's four ages of man, as narrated in the *Works and Days*. According to Hesiod's theogony, both gods and mortals sprang from a single source, and soon thereafter the immortal Olympians "made a golden race of mortal men who lived in the time of Cronos when he was reigning in heaven." After their death, these men became known as "pure spirits," and were considered kindly dwellers on the earth, "delivering from harm, and guardians of mortal men." Then followed the Silver Age, the inhabitants of which were cast away for failing to honor properly the Olympians—yet they, too, were "called blessed spirits of the underworld by men." The Bronze Age that followed comprised a "terrible and strong" race of men, whose infatuation with violence wreaked their self-destruction. The Fourth Age belonged to the "demi-gods": "Grim war and dread battle destroyed a part of them, some in the land of Cadmus at seven-gated Thebe when they fought for the flocks of Oedipus," and some in Troy.[66]

Newton applied Hesiod's structure to the postdiluvian era. "The history of the first ages," he believed, had been "couched in that of ye 12 Gods." And since all ancient authorities agreed that those deities were first worshipped in Egypt, clearly they were to be located in the history of Ham's posterity. Accordingly, Newton constructed the following Pantheon of deities and their corresponding heavenly bodies or elements: Noah (Saturn); Ham (Jupiter) and his four sons (Chus (Mars), Mizraim (earth), Canaan (fire), and Phut (water)); Canaan's sister/wife Astarte (Venus); Mizraim's consort Isis (quintessence); her three children (Thoth (Mercury), Orus (sun), and Bubaste (moon)); and the goddess Neith (air). Figure 4.1 represents Newton's genealogy of Egyptian deities:[67]

[66] Hesiod, 1914, pp. 11–15. Versions of the myth could be found in Ovid as well.

[67] Burndy MS 420; Yahuda MS 17.2 fols. 14^v, 20–1; Yahuda MS 41 fols. 26^v–27^v.

The centrality of myth to alchemical imagery, and the location of figure 4.1 at the beginning of an important alchemical manuscript known as "Praxis," has previously suggested the dependency of Newton's forays into ancient religions and chronology on alchemy.[68] For evidence scholars have pointed, in particular, to Newton's close study of Michael Maier's works, where the mythical foundations of celestial and elemental symbolism were explicated in great detail. Thus, it is noted, as early as 1669 Newton extracted material from Maier's *Symbola Aureae Mensae* (1617), including the myth retelling Typhon's murder and dismemberment of his brother Osiris, and the successful effort of Typhon's mother/sister/wife Isis to retrieve the body parts and resurrect him.[69] Yet, not only are the diagrams shown in figure 4.1 unrelated to the rest of the "Praxis," but Newton's acknowledgment of this and related myths as constituting significant symbols of alchemical processes was only indirectly related to his treatment of myth in his religious and chronological studies. It should be remembered that Maier and other alchemists drew upon well-known ancient historical and literary sources for their imagery. Diodorus Siculus, in particular, informed Maier's *Symbola*, even more his influential *Atalanta fugiens* (1618), which Newton read while working on ancient religions.[70] Maier's explanatory discourse to emblem XLIV of the *Atalanta fugiens*—Typhon's slaying of Osiris—for example, brings into sharp relief his divergence from Newton. Whereas Maier insisted on the necessity of comprehending the myth strictly within an alchemical context—thus explicitly rejecting the possibility that Osiris could have been either an Egyptian king or a deity—Newton's Euhemerist reading of the original sources of the myth extracted from the myth precisely that interpretation.[71]

We can reconstruct Newton's schema for the early history of mankind on the basis of his extensive notes on ancient history and myth (Yahuda MSS 16–17), and the more polished (and condensed) summary in the "Original of Religions" (Yahuda MS 41).[72] According to Newton, in the aftermath of the Deluge, Noah sojourned with his family in Mesopotamia, where they remained united throughout the building of the Tower of Babel 101 years later.[73] Following the confusion of tongues, Noah moved to Shinar, where he proceeded to divide the world among his three sons—as his family "grew too great to live quietly together." Keeping by his side Shem, the eldest, Noah dispatched Ham to Egypt and Japhet to Asia Minor. When it fell to Ham to apportion his dominion among his four sons, he followed his father's example: Chus, the eldest, remained with his father on the Arabian side of the Nile; Mizraim was dispatched to

[68] Dobbs, 2002; Figala, 2002b.

[69] Keynes MS 29 fols. 1–2.

[70] Keynes MS 32 fols. 25–48. See also Keynes MS 30 fols. 8–9, 49, 65, 78, 87.

[71] Maier, 1618, p. 186. Maier referred his readers to his *Arcana Arcanissima* (1614) for a fuller explication of his alchemical interpretation of Egyptian mythology. It, too, relied on Diodorus Siculus. See De Jong, 1969, pp. 273–78.

[72] Based on the content, handwriting, and watermarks, the editors of the Newton Papers Project dated Yahuda MSS 16–17 to the period ca. 1684–1692, and Yahuda MS 41 to the early 1690s.

[73] The Tower, Newton conjectured, had been raised "so high that in seeking for food for themselves & their cattle they might see the same from all parts of the great plane of the land of Shinar & know whether to resort least they should be scattered from their main body." Yahuda MS 25.2f fol. 22; New College MS 361. 3 fol 110.

Thebes, whence to expand southward; Phut received the western side of the Nile and was expected to spread out into Northern Africa; Canaan received lower Egypt, with a view to expand eastward toward Syria.[74]

Up to this point, Newton followed the conventional interpretation of Noah's "reign" and subsequent division of the world among his sons—respectively, the golden and silver ages—which had been formulated most recently (and thoroughly) in Samuel Bochart's *Geographia sacra*. In marked contrast, Newton's narrative of the third and fourth ages—the focus of his entire investigation—took a decidedly idiosyncratic turn, as little in either Scripture or Jewish and Christian traditions licensed Newton's creative attempt to graft the Titanomachia and related pagan myths onto Genesis 9–11. His conviction, therefore, that the mythical Four Ages recall the affairs of a single branch of Noah's family over four generations, caused Newton to labor hard to locate strife among Noah's grandsons and great-grandsons during the Bronze and Iron ages. The key figure in the story turns out to be Chus—also known as Hercules, Mars, Belus, Moloch, Bacchus, and Pan—whom Newton described as the "greatest warrior of his age & ye father of ye most warlike nation of all those wch descended from Ham." Chus distinguished himself first during the war of gods against the giants (titans), a war that Newton interpreted as an internecine feud between Ham's sons. The "villain" in that war was Phut (also known as Typhon, Python, Briareus, and Neptune), who treacherously murdered his brother Mizraim (Osiris, Apis, Serapis, Menes, Pluto, and Hyperion) and usurped his kingdom. Phut reigned over Egypt until Chus, who came to assist his brother's family, defeated him, and drove him and his allies out of Egypt.

An even broader internecine conflict followed on the heels of the Titanomachia, when Chus, "not content wth his own inheritance," turned against Shem's family—as well as against his own. For after he "warred wth a club in Egypt for the Gods against ye Giants," Chus embraced the new technology of iron weapons in order to wage war upon the Egyptians and Canaanites. He "conquered them in ye regions about Chaldea" and after expelling their inhabitants, Chus became the first King of Chaldea, peopling the country with the Arabian progeny of his son Havilah. He "placed his sons round about the Persian gulf, their seats extending as far as Carmania in ye borders of India," before advancing north to assist his son Nimrod—also not contented with his dominion in Shinar—in the conquest of Assyria. Then, while Nimrod busied himself with the building of Nineveh, Chus embarked on his triumphant Indian expedition.[75] The outcome of the expedition, however, proved tragic. Chus (Dionysus) had previously had a liaison with Semiramis (Venus), yet while he warred in India, she "enjoyed the company" of Adonis (Nimrod-Ninus). Learning of the betrayal upon his return, the envious Chus slew his own son.[76]

The determination to establish the existence of a vast empire, stretching from Egypt to India, consumed Newton. He persisted in the face of occasional inconsistencies and confused synchronizations, as his tidy genealogy of deities relied entirely on the sequence and contours of Chus' conquests. Hence, according to Newton, every

[74] Yahuda MS 41 fol. 9; MS 16.2 fol. 19.

[75] Yahuda MS 41 fols. 10–14; New College MS. 361.1b fols. 94v–95; Keynes MS 146 fols. 2–3.

[76] Yahuda MS 16.2 fols. 21–21v, 70v–71; MS 16.1 fol 3–3v.

nation that deified its kings, conferred the name Jupiter on "him whom they had most in honour," while their fathers were known as Saturn. Curiously, in developing his system, Newton took his cue from an Annian forgery—Xenophon's *De Equivocis temporum*:

> The most ancient members of the families of the noble kings who founded cities are called Saturns. Their first-born children are Jupiters and Junos. And their strongest grandsons are Herculeses. The fathers of the Saturns are the Heavens, their wives are Rheas, and the wives of the Heavens are Vestas. Therefore, there are just as many heavens, Vestas, Rheas, Junos, and Herculeses as there are Saturns. Also, the same man who is Hercules to some peoples is Jupiter to others. For Ninus, who was Hercules for the Chaldeans, was Jupiter to the Assyrians.[77]

Along these lines, then, in Egypt, the birthplace of idolatry, Ham became known as Jupiter Hammon, Noah as Saturn, and Chus as Hercules. The Chaldeans ordered their gods "one age lower than the Egyptians," and conferred the names of Saturn, Jupiter, and Hercules on Ham, Chus, and Nimrod, respectively. And lest "they should omit Noah they called him Caelus or Uranus." The Assyrians, in turn, who placed their deities one age lower still, "gave the names of Caelus, Saturn, Iupiter & Hercules to Ham Chus Nimrod & ye son of Nimrod & lest they should omit Noah called him Hypsuranius & Eliun that is the most High. For they made Hypsuranius to be ye father assigned to Uranus to have a father Hypsuranius & in ye genealogy of their Gods no nation ascended higher." Newton was clearly delighted with his creation: "this difference between ye theologies of the three nations is notably confirmed by ye light it gives to many difficulties & seeming contrarieties in history." Thus, the seemingly conflicting accounts regarding the identity of Venus' parents and the circumstances of her birth could be synchronized by recognizing that the several myths recounted versions of the Egyptian, Chaldean, and Assyrian theologies. Likewise, concluded Newton, "Saturn Jupiter & Hercules were father son & grandson in the line wch decends directly from Noah to Ninus, & y^t^ In ye Assyrian theology they are Chus Ninus & ye son of Ninus: in the Chaldean Ham, Chus & Ninus & in ye Egyptian Noach, Ham & Chus: the latter theology placing them all one age higher then the former. So The Egyptian Hercules is therefore Chus the great Jupiter Belus of the ancients."

Having suitably Euhemerized the contours of ancient theogonies, Newton turned to analyze the nature of the true religion that had been corrupted through the deification of great ancestors. He never doubted that true religion preceded error or idolatry, thus drawing on a long tradition of commentators. Early in the third century, Tertullian had already codified the axiom of antiquity as the yardstick of truth: "that principle of *time*, which rules that the authority lies with that which shall be found to

[77] Yahuda MS 41 fol. 13; MS 16.2 fols. 42, 49^v^, 67, 69; Berosus, 1612, fols. 109–109^v^. On MS 41 fol. 14^v^ Newton scribbled "Xenophon teaches." Newton utilized other Annian forgeries—for example, when he cited Myrsilus of Lesbos on the practice of the Prytanea among the ancient Tyrrhenians. Yahuda MS 41 fol. 1; Berosus, 1612 , fols. 154, 156, for which see Grafton, 1997, p. 131.

be more ancient; and assumes as an elemental truth, that corruption [of doctrine] belongs to the side which shall be convicted of comparative lateness in its origin. For, inasmuch as error is falsification of truth, it must needs be that truth therefore precede error. A thing must exist prior to its suffering any casualty; and an object must precede all rivalry to itself."[78]

Tertullian's "principle" reverberated throughout the early modern period. Like Newton later, Mathew Hale scarcely doubted that religion was "as ancient as Humanity it self." Inasmuch "as Truth is certainly more ancient than Errour," he explained, "we have reason to think that even before the ancientest Form of Idolatrous Worship in the World, even that of the Heavenly and Elementary Bodies, there was a True Worship of the true GOD." In fact, both Hale and Newton drew on John Tillotson's observation that the "idolatries of the heathen ... came in upon the world by insensible degrees, and did not oppose the corruptions of men, but grew out of them; and being united to the vicious temper and disposition of mankind, they easily gained upon their ignorance and superstition, by custom and example. They were just such a corruption of natural religion, in such times of darkness and ignorance, and by such insensible steps, as there hath been since, of the Christian religion in some parts of the world, which we all know."[79] Humphrey Prideaux concurred. When rebutting in the late 1690s the burgeoning deist challenge, he controverted the notion that all false religions were the product of impostures —as exemplified by Islam. Just the opposite was true, he insisted. They all originated "by Corruptions insensibly growing on from that *Religion* which was first true." Man received the original religion from God, a religion that had been "imprinted on his very Nature" with Creation. All "*Idolatry, Polytheism*, and other *false Worships*, which after arose in the *Heathen World*, were all by such corrupt Deviations therefrom, as the Superstitions of Men, the unfaithful Way of transmitting *Divine Revelations*"; decay through imagination after Noah, choosing angels or deceased men as deities."[80]

Newton imagined the primitive religion, prior to its corruption, to be that of the Prytanea: "The religion most ancient and most generally received by the nations in the first ages," he pronounced at the outset of "The Original of Religions" (ca. 1691) "was that of the Prytaneum or Vestal Temples. This was spread over all nations until the first memory of things." Thucydides and other Greek and Roman sources provided him with information on the proliferation of Prytanea in every town and small kingdom, only to be abolished and replaced by a central Prytaneum after they unified under a single ruler—as happened in Greece following Theseus' foundation of the Athenian Kingdom. These sources also informed Newton that the Prytanea were used for civil as well as religious purposes, and that the respective magistrates, not the priests, performed the sacrifices. Working his way back, Newton proceeded to mine other sources for information regarding the centrality of Prytanea to earlier nations. The sources cited in Eusebius' *Evangelicae Praeparationis* confirmed Newton in his belief that the ancient Egyptians dedicated a perpetual fire and offering of herbage to their deities; Leviticus 26.30—"And I will destroy your high places, and cut down

[78] Tertullian, 1868, p. 184.
[79] Hale, 1677, p. 168; Tillotson, 1820, vol. 4, p. 369.
[80] Prideaux, 1698, pp. 143–46 (second pagination).

your images"[81]—served as a demonstration that "Prytanea were used in ye Cities of Canaan & Syria before the days of Moses." Likewise, the altars that Abraham and Jacob built, respectively, in Beersheba and Bethel, undoubtedly indicated that for the patriarchs an altar stood in the same relation to "what a temple was afterwards to a City." Or, as Newton generalized later, any mention in Scripture of altars with perpetual fire should be understood to be of "ye same nature wth the Prytanea of other nations."[82]

Newton continued to systematically mine sources for evidence concerning the ubiquity of Prytanea throughout the inhabited ancient world, from Egypt to Scandinavia, from China to the British Isles. He regarded the universality of the Prytanea as the best proof of its antiquity: "There are many instances of other nations receiving other religions after this," he pointed out, but to the best of his knowledge no other nation received the prytanea "after any other" religion, and no other religion that "sprang up later [had] become so general as this." To Newton's mind, it would have been impossible to explain the universality of such a religion "had it not been propagated wth mankind in ye beginning." After all, he reasoned, it overspread the world rapidly and peacefully, while the fortunes of both Christianity and Islam offered ample testimony of the near impossibility of imposing a worldwide religion by force.[83]

Having established to his own satisfaction the antiquity of the religion of the Prytanea, Newton laid down several propositions regarding its nature. First, this religion had been Noah's religion, and from him it "spread into all nations at ye first peopling of the earth." Second, "ye sacrificing clean birds & beasts by a consecrated fire in a consecrated place was ye true religion till ye nations corrupted it. For it was ye religion of Noah, & tis not to be doubted but that ye religion wch Noah propagated down to his posterity was the true religion." Third, Moses taught the Jews nothing else but the "religion of Noah purged from the corruptions of ye nations."[84] With these propositions in place, Newton turned to expound his idea regarding the heliocentric foundation of ancient religion. On the strength of some vague comments made by St. Paul and Josephus,[85] Newton concluded that just as Moses designed the Jewish Tabernacle to serve as "a symbol of ye heavens ... so were ye Prytanæa amongst ye nations."

[81] The reference is to the worshipping of Baal.

[82] Yahuda MS 41 fols. 1–2^{v}; Eusebius, 1903, vol. 3, part 1, pp. 32, 102. On Newton and the Prytanea, see Iliffe, 1995; Delgado-Moreira, 2006b.

[83] Yahuda MS 41 fol. 3.

[84] Yahuda MS 41 fols. 4–5. To substantiate the last point, Newton cites John Spencer, who had "shewn y^{t} Moses retained all ye religion of ye Egyptians concerning ye worship of ye true God; & rejected only what belonged to ye worship of their fals Gods." If so, he continued, "then its certain that ye old religion of the Egyptians was ye true religion tho corrupted before the age of Moses by the mixture of the worship of fals Gods wth that of ye true one: & by consequence ye religion of ye Iews was no other then that of Noah propagated down in Egypt till ye age of Moses. (Ibid., fols. 3–3^{v}, 5).

[85] Newton does not provide a reference, but he seems to have Hebrews 8.5 in mind: "Who serve unto the example and shadow of heavenly things, as Moses was admonished of God when he was about to make the tabernacle; for, See, saith he, *that* though make all things according to the pattern shewed to thee in the mount." The "proportion of the measures of the tabernacle proved to be an imitation of the system of the world; for that third part which was within the four pillars, to which the priests were not admitted, is, as it were, a heaven, peculiar *to* God." Josephus, 1841 , p. 88.

Indeed, since the ancients reckoned the "whole heavens ... to be ye true & real Temple of God," they surely endeavored to frame it in such a way as to represent the universe "in the fittest manner," thereby rendering the "Prytanæum to deserve ye name of his Temple." This observation, Newton exclaimed, constituted a "point of religion then wch nothing can be more rational." Newton turned to Macrobius' commentary on Scipio's Dream as corroborating the rational structure of such a religion:

> [Cicero's] designation of the universe as the *temple* of God was appropriate, too, and was for the edification of those who think that there is no other god except the sky itself and the celestial bodies we are able to see. In order to show, therefore, that the omnipotence of the Supreme God can hardly ever be comprehended and never witnessed, he called whatever is visible to our eyes the temple of that God who is apprehended only in the mind, so that those who worship these visible objects as temples might still owe the greatest reverence to the Creator, and that whoever is inducted into the privileges of this temple might know that he has to live in the manner of a priest.

What Newton failed to mention was that Macrobius commented on a passage in *De re publica* in which Cicero describes the earth as a motionless sphere, ever fixed at the center of the universe. Not that this mattered to Newton. He considered the corruption of true astronomy by later Greeks and Romans, as we shall see below, to be an inevitable consequence of the corruption of religion. What mattered to him was that even the Roman geocentric-based religion still retained the centrality of temple fire: "From this comparison the fire in the middle of ye Prytaneum was taken for a symbol of ye center of ye world, & thence ye generality of ye Latines took Vesta for ye earth."[86]

Since the true religion had been designed in such a way that its essence would be "proposed to mankind" through the "frame of ye ancient Temples," continued Newton, the priesthood had been vested with the preservation and transmission of both true religion and astronomy. Hence the scientific proficiency observed among the ancient priests of all nations, which explained to him why the Greeks "went to ye Priests" when sojourning to Egypt to learn astronomy and philosophy there. Hence, too, the reason for the preservation of whatever the Greeks understood about the "true knowledge of nature" by their own priests. Clearly, concluded Newton, "the first religion was the most rational of all others till the nations corrupted it. For there is no way (wthout revelation) to come to ye knowledge of a Deity but by the frame of Nature."[87]

How, then, did the true and rational religion become corrupted? Incrementally, argued Newton. Initially, the Prytanea's very representation of the heavens "as ye real temple of the Deity" led men "by degrees to pay a veneration to these sensible objects," and then "to worship them as the visible seats of divinity." Before long the elements,

[86] Yahuda MS 41 fols. 6–7; Macrobius, 1952, p. 142; Cicero, 1999, p. 98. See also Schüller, 2001, p. 243.

[87] Yahuda MS 41 fol. 7. Newton lifted from Marsham, without acknowledgment, testimonies by Clement of Alexandria and Seneca regarding Greek mysteries: Marsham, 1676, p. 267.

also "parts of that universe wch is ye temple of God," morphed into objects of veneration as well. Newton never doubted that idolatry commenced with the adoration of the heavenly bodies. Scriptures and other ancient accounts confirmed this for him, as did reason; mankind had been more prone to worship the heavenly bodies than dead men or statues, and "therefore [the former] spread further." Specifically, intent on honoring their ancestors, the postdiluvians imposed the names of these ancestors upon stars, elements, cities, and rivers, representing those names hieroglyphically. Then "they feigned" that the souls or spirits of these ancestors had been "translated into ye stars" and animated them. To render credibility to these hypotheses, "they feigned that the Stars by vertue of these souls were endued wth ye qualities of ye men & according to those qualities governed the world." Soon, the new gods began to be worshipped in the Prytanaea alongside the true God. Conscious of their blasphemy, they attempted to "colour over this worship" by adding God's true name—Iah, ἰαω, Iehova, ἰαὼ ἐυω ἰουὰ, Iuba—to the new deities. Thence derived Jupiter from ἰου-πατὴρ, Bacchus from ἰουὰ Χοὺς, and Janus from Iah-Noah.[88]

Newton located the origins of idolatry firmly during Noah's lifetime. Abraham, he noted, left Chaldea 357 years after the Flood—a mere seven years after Noah's death—by which time the worshipping of dead men, imported from Egypt, had been well entrenched in Mesopotamia. Newton grounded his dating on Joshua 24:2 and on a common Jewish tradition that took Terah, Abraham's father, not only to be an idolater, but a manufacturer of idols to boot.[89] Equally important, Newton's conclusion that the first men to be worshipped were the "first fathers" of the nations mandated that they had lived during "ye ages of ye Gods," for later kingdoms "were too small to make them considerable." Nevertheless, the rulers of those smaller kingdoms actively promoted the veneration of their predecessors, seeking thereby to enhance their authority and, by virtue of their relation to the gods, their dignity—a process that ultimately resulted in their own deification as well, and accounts for the proliferation of gods.[90]

This burgeoning process of idolatry also helped Newton to account for the parallel corruption of natural philosophy. The very institutions of ancient religion, he wrote, had been designed to ensure that "the God of Nature should be worshiped in a temple which imitates nature, in a temple which is, as it were, a reflection of God. Everyone agrees that a Sanctum with a fire in the middle is an emblem of the system of the world."[91] And just as idolatry had originated in Egypt, so too the perversion of the true system of the world first appeared there. Newton believed that the practice of true philosophy was integral to the domain of "sacred philosophy," which had been conveyed emblematically to adepts.[92] As he summarized in a late manuscript: "So then twas one designe of ye first institution of ye true religion to propose to mankind

[88] Yahuda MS 41 fols. 9–9^{v}.

[89] "And Joshua said unto all the people, Thus saith the LORD God of Israel, Your fathers dwelt on the other side of the flood in old time, [even] Terah, the father of Abraham, and the father of Nachor: and they served other gods." Newton cited the reference under the heading "Idolatry" in his commonplace book (Keynes MS 2 fol. 5).

[90] Yahuda MS 41 fols. 9^{v} –12.

[91] Yahuda MS 17.3 fols. 9, 12; Westfall, 1980 , p. 354.

[92] Yahuda MS 16 fol. 1.

by ye frame of ye ancient Temples, the study of the frame of the world as the true Temple of ye great God they worshipped. And thence it was y^t^ ye Priests anciently were above other men well skilled in ye knowledge of ye true frame of Nature & accounted it a great part of their Theology."[93]

Given that the knowledge of the true system of the world had been vested in the priesthood, Newton placed its corruption squarely in their hands. Having initiated the veneration of stars, the priests inaugurated astrology and the cult of transmigration of dead souls. Other forms of magic proliferated, too, and these practices quickly migrated eastward. Newton carried this belief in double corruption into several nuanced statements he inserted in the "General Scholium," which he added to the second edition of the *Principia* (1713), and to the final query of the *Opticks*, first published in 1717. In the latter, as we noted above, he averred that "no doubt, if the Worship of false Gods had not blinded the Heathen, their moral Philosophy would have gone farther than to the four Cardinal Virtues; and instead of teaching the Transmigration of Souls, and to worship the Sun and Moon, and dead Heroes, they would have taught us to worship our true Author and Benefactor, as their Ancestors did under the Government of *Noah* and his Sons before they corrupted themselves."[94]

This précis of Newton's conceptualization of the early history of mankind fails to do justice either to the scope of his investigation or to the intense labor he invested in erecting the edifice, intermittently, over a period of seven or eight years. True, as was the case with his alchemical studies, Newton relied heavily on the labors of far more erudite scholars, but the agenda and the manner of execution were distinctively his own. The surviving manuscripts, amounting to more than 130,000 words, furnish precious insights into his methodology and working habits. In general, Newton's researches into the history of early humanity employed the same techniques he used in his study of prophecies. With a relatively clear conception in mind, he began by extracting supporting material from a variety of sources, turning them into a more coherent narrative in successive iterations. Particularly remarkable is the quiet confidence that pervades his writings. Just as in his study of prophecies, Newton appears to have been morally certain that he had succeeded in unveiling the mystery and obfuscation that shrouded ancient myths, a feat that eluded other scholars who failed to apply his rigorous criteria of examination and evidence to Biblical and historical sources. Manuel's characterization of Newton's exacting standards in the domain of prophecies applies equally to his other studies:

> Newton applied what might be called scientific criteria to the interpretation of the books of prophecy, particularly the law of parsimony. He showed not only that every notable political and religious occurrence conformed exactly to some vision in prophecy, but that his set of equivalents had totally exhausted the possible meanings of each of the objects and images appearing in any prophetic verse. There was nothing left over, no random words still unexplained, no images that were superfluous. The system was enclosed, complete, and flawless. Newton saw his "methodising of prophecy" as an

[93] Yahuda MS 41 fol. 7.

[94] Newton, 1721, pp. 381–82.

ideal scientific structure, exhibiting the greatest possible simplicity and harmony.[95]

The fixation with simplicity and harmony is everywhere evident in Newton's synchronization of Myth and Scripture, beginning with his distinctive Euhemerist practices. As noted previously, like Newton, most contemporaries were convinced that mythical heroes and deities were corruptions of Biblical figures. Unlike Newton, however, they did not seek an ironclad coherence within and between various theogonies. Nor did they insist, as Newton did, on rendering all myths to be time- and place-specific. Rather, they might be better characterized as "promiscuous" Euhemerists, who cared greatly about establishing the equivalence between pagan deities and Biblical figures, but cared less about the precise identities of the Biblical figures in question. This scripture-based Euhemerism became part of time-honored Christian apologetics, which insisted on the derivation of the most valuable aspects of pagan culture from the Jews—a topic to be discussed in greater detail below. First, it is necessary to compare Newton's Euhemerist ventures to those upon whom he drew.

Most commentators, for example, agreed that Saturn had been modeled on Noah, as both were considered to be the common parents of mankind who divided the world among their three sons—Shem (Pluto), Japheth (Neptune), and Ham (Jupiter). Bochart, in fact, proffered no less than fourteen parallelisms which, he claimed, established the correlation beyond reasonable doubt. Multiple parallelisms also existed to establish the similitude between Moses and Bacchus: They were both born in Egypt, both rescued from water, and in due course both fled to Arabia. The dog who served as Bacchus' devoted companion undoubtedly recalled Moses' faithful follower Caleb (from כלב (kelev), the Hebrew for dog), while Bacchus' nickname "bicornis" (double-horned) derived from the mistranslation of Exodus 34.29: "And it came to pass, when Moses came down from mount Sinai … the skin of his face shone." (קרן (karan) in Hebrew, meaning shone or radiated, is a denominative verb from the root קרן (keren), "horn").[96] Regardless, the identity of Bacchus lent itself to different interpretations as well. He may well have been Nimrod, based on the etymology of his name: בר-כוש (Bar-Chus), the son of Chus. He may have been Noah, as both were reputed to have been the first makers of wine. The Noah-wine correlation could equally be applied to Janus, whose name "most probably derived from יין [Yain, wine] because of *Noah*'s planting of vine."[97] And so on.

Such divergence of opinion regarding the precise identification of Biblical and mythical figures—though not about the necessary correlations between them—explains why not everyone accepted as definitive the most common coupling: Noah and Saturn. Sir Walter Raleigh preferred the opinion of those who claimed that Cain must have been the ancient Jupiter—by virtue of his building Enoch, the first city in the world—which thus made Adam the first Saturn. Theophilus Gale concurred. He freely acknowledged that scholars had variously identified Saturn as Noah, Abraham,

[95] Manuel, 1963, pp. 97–98.

[96] Vossius, 1668, pp. 115–20; Huet, 1690, p. 108.

[97] Bochart, 1681, p. 13; Gale, 1672, part 2 p. 31; Patrick, 1695, p. 189; Stillingfleet, 1662, p. 594. For Newton's linking Saturn, Janus, and Noah (via Bochart), see Yahuda MS 16.2 fols. 12, 16.

and Adam. Yet this was hardly cause for concern. "We may indeed take in each of these relations," he explained. "For its wel known, that these poor blind Heathens were wont to attribute *Traditions* and *Relations*, originally different, to one and the same person, according as their inclinations led them. Hence they framed more than one hundred *Jupiters*, by applying different stories to one and the same *name*, and *person*."[98] Nevertheless, the similitude between Adam and Saturn seemed to Gale most convincing. Adam is called "son of God" and Saturn called "son of Heaven"; Adam had been formed out of the dust of the Earth, while Saturn's mother was called Tellus, Earth; both Adam and Saturn taught mankind husbandry; and just as God cast Adam out of Paradise, so Saturn had been "expelled from his Dominion." Elsewhere Gale borrowed from Vossius another shared characteristic of Adam: "From *Adams* hiding himself from the face of God came the name *Saturne*: For סתר *Satar signifies, latere, to hide*: whence also *Saturne* was stiled *Latius*."[99] As if to bolster his case, Gale recorded a conversation he had with Bochart in the early 1660s, in the course of which Gale cited a passage from Plato's *Politicus* myth, in which the philosopher described the reign of Saturn during the Golden Age in terms that, to Gale's mind, "plainly related to Adam and *Eves* state in *Paradise*." Bochart, he avowed, "seemed to grant [him], that this storie of *Saturn* might be referred to *Adam*; though (as he said) 'twas not proper for him to mention it; because he referred *Saturne*, rather to *Noah*."[100]

As these examples illustrate, a considerable amount of scholarly arbitrariness, based on taste or agenda, informed the choices of early modern Euhemerists. Case in point, the controversy over the extent to which the Phoenicians fashioned their god Kronos on Abraham. In the *Evangelicae Preparationis*, Eusebius cites Philo Byblos' account of the Phoenician theogony—based on Sanchoniathon's Phoenician History—including a passage relating that when plague ravaged his dominion, "Kronos offers his only begotten son as a whole burnt-offering to his father Uranus, and circumcises himself, compelling his allies also to do the same." Understandably, early modern scholars delighted at this seeming proof for the Phoenician veneration of Abraham—who not only prepared to sacrifice his only son but, according to Scripture, had been the first to introduce circumcision (Genesis 17:10–27)—especially when Philo noted that Kronos' other names included Israel, and that his allies were named "Eloim" (God).[101] However, by 1640, Hugo Grotius cautioned against such a facile reading of Eusebius:

> The most high God is named by the Assyrians and other nations from that particular star of the seven ... by which mankind are governed, which is

[98] Raleigh, 1829, vol. 2 pp. 165–67; Gale, 1672, part 2, p. 5.

[99] For Newton's appropriation of this etymology, also via Bochart, see Yahuda MS 17.2 fols. 13–13^{v}; Yahuda MS 16.2 fol. 14.

[100] Gale, 1672, part 2, pp. 6–7; part 3, p. 58. Though Vossius furnished evidence for identifying Saturn as, variously, Adam, Noah, and Abraham (Vossius, 1668, pp. 72–5), he ultimately opted to identify Saturn, the son of Caelus and Tellus, as Adam. Quite rightly, commented Sir Edward Sherburne, "for who besides him was the Son of Heaven and Earth?" Sherburne, 1675, p. 7.

[101] Eusebius, 1903, vol. 3 part 1 pp. 42–43. For Newton's references to Eusebius see Yahuda MS 16.2 fols. 36, 50; Yahuda MS 41 fol. 24.

> moved in the highest orb, and with the greatest force: or certainly the Syriac word, איל Il, which signifies *God*, was therefore translated Κρόνος *Kronos*, by the Greek interpreters, because he was called איל Il by the Syrians. Philo Biblius, the interpreter of Sanchuniathon, hath these words: "Ilus, who is called Saturn." He is quoted by Eusebius: in whom it immediately follows from the same Philo, "that Kronos was the same the Phoenicians call Israel;" but the mistake was in the transcriber, who put Ἰσραήλ *Israel*, for ῖλ Il, which many times amongst the Greek Christians is the contraction of Ἰσραήλ; whereas ῖλ is, as we have observed, what the Syrians call איל *Il*, and the Hebrews אל *El*.[102]

Grotius' reasoning proved inconvenient for most contemporaries, who chose either to ignore Grotius or controvert him. Edward Stillingfleet endorsed the opinion that "the Phoenician antiquities seem to have preserved the memory of Abraham's sacrificing his son Isaac." Cognizant of Grotius' objections, Stillingfleet nevertheless believed that "the circumstances of the story make the ordinary reading not *improbable*." Gale followed suit, without even bothering to mention Grotius, as did John Lightfoot.[103] For his part, Sir John Marsham opted to follow Grotius, primarily because he sought to locate the origins of circumcision earlier than Abraham, and hence it suited him to correlate Kronos with Noah instead.[104]

Even among early-modern willful Euhemerists, Pierre Daniel Huet proved to be exceptionally arbitrary, judging by the tenacity with which he sought to establish in his *Demonstratio evangelica* (1679) that most pagan gods were fabulous representations of Moses, while pagan goddesses were modeled on his wife Zipporah. Small wonder that Leibniz found it easy to criticize much of the Euhemerist enterprise when faced with the extravagant labors of the truly erudite bishop of Avranches. The attempt to show "that pagan idolatry was copied from the history of Moses," he wrote to Johann Friedrich, Duke of Brunswick, amounted to a "*jeu d'esprit*." The history of the mythic gods, he continued, "is so confused that one can make of it anything one wishes.... It seems advantageous to our religion to maintain that the erudition and theology of the pagans came from the Jews. Some of the ancients and Vossius in our time have said fine things on the subject. But one should not flatter oneself in these matters at the expense of truth."[105]

Huet's reductionism may have been extreme, but his unbounded confidence in his ability to demonstrate correlations hardly differed from that of his contemporaries. Consider Samuel Bochart. Determined to defend the antiquity and supremacy of Scripture by unveiling the pagans' extensive borrowings from it, he pronounced that no one had "sufficiently demonstrated" the correlation of Moses' history of Noah

[102] Grotius, 1644 vol. 1, p. 154; Grotius, 1829, p. 44.

[103] Stillingfleet, 1662, p. 595; Gale, 1672, part 1, p. 22; part 2, pp. 5, 7–8; Lightfoot, 1822–1825, vol. 8, p. 433. See also Bochart, 1681, p. 790; Vossius, 1668, p. 74.

[104] Marsham, 1676, pp. 73, 78–79. Half a century later Shuckford attempted to refute Marsham's opinion with some vehemence: Shuckford, 1728–1730, vol. 1 pp. 326–27; vol. 2 pp. 25–26.

[105] Cited in Manuel, 1992, p. 171.

with that of ancient fables as fully as he did. Equally confident, Theophilus Gale proclaimed that he furnished "very probable, if not *scientific, notices*," concerning the "chiefest *Names*, *Attributes*, and *Offices*" in the pagans' fabulous genealogies, notices that "were assumed in *imitation* of, and *derivation* from, some *Hebrew Names*, *Attributes*, *Persons*, and *Traditions*."[106]

The same confidence permeates Newton's writings. Nevertheless, even as he helped himself to the erudition of his contemporaries—primarily Bochart, Vossius, Marsham, Huet and, probably, Stillingfleet—his approach differed from theirs in significant ways. First, whereas for most Euhemerists the establishment of individual correlations was the ultimate goal, Newton sought to establish specific temporal and familial correlations between several ancient theogonies. Second, Newton eschewed unrestrained multiplication of correspondences between Myth and Scripture. To argue, for example, that Hercules was Nimrod, and Joshua, and Samson, seemed to him to defeat the purpose of deciphering myth, especially in view of his conviction that before the invention of writing, myth perpetuated the memory of the original great heroes, not their less illustrious successors. Consequently, every choice and identification Newton settled on was designed to fit into an intricate genealogical and historical structure of early nations, thereby establishing, as it were, a consistent overall scheme anchored to evidentiary specifics. The single-mindedness with which Newton approached his sources makes clear that his citations from them invariably indicate approval—in contrast to earlier fellow travelers, who often furnished multiple mythical correspondences to Biblical figures, regardless of their own preferences, in order to augment the primacy and superiority of Scripture over myth. Third, Newton's interpretative strategies of myth positioned his correlations squarely within the realm of military and political affairs, squarely in line with his understanding of prophetic symbols to denote conflict. Thus, for example, given that both prophets and mythologists interpret Heaven to mean throne and dominion, and thunderbolt to mean war, Jupiter's striking Typhon with a thunderbolt surely means, reasons Newton, Phut's revolt against Ham and his attempt to conquer Egypt. Analogously, the satyrs that were said to accompany Bacchus on his Indian campaign—on account of their representation in myth as part man and part goat—refer to "the Arabians which led wild lives in the feilds & deserts like Goats wandring up & down in tents after their flocks & living by hunting & rapine & being Goats in lust."[107]

Equally significant, whereas Scriptural Euhemerists thought their enterprise to consist solely in the locating of Biblical events, or the attributes of Biblical figures, in pagan myth, Newton reversed the process. Where Scriptural Euhemerists began with Scripture and went looking for myths, Newton instead began with events and persons that inhabited myth and sought Scriptural parallelisms—which oftentimes necessitated the embellishment of Scripture in pursuit of symmetry. His version of the pantheon of twelve Egyptian deities, for example, included three female deities for whom there is no Scriptural correlation. If the identification of two of them, the consorts of Canaan and Mizraim, could be framed as a legitimate inference, the further

[106] Bochart, 1681, p. 1; Gale, 1672, part 2, p. 97.

[107] Yahuda MS 16.2 fol. 62; Yahuda MS 41 fol. 14.

claim that Ham's sons married their sisters—in conformity with what Newton took to be common Egyptian regal practice[108]—verged on creative license. More audacious still, Newton's determined exposition of Semiramis' liaisons with Canaan, Chus, and Nimrod—never mind the suggestion that the envious Chus engineered the slaying of his son—were inspired fiction, altogether invented by Newton and fed by his compulsion to harmonize his version of the history of Noah's progeny with myth. Likewise, the Biblical allusion to Ham's rebellion offered Newton fertile ground to graft onto the early history of mankind elements of the Saturn-Jupiter myths. These included accounts detailing Jupiter's dethroning and expulsion of his father, which Newton furiously and obsessively located in the myth of several nations: He identifies Noah as the Saturn who had been driven out of his kingdom by his son Jupiter (Ham), and who, according to legend, took refuge in Italy;[109] he relied on Sanchoniathon to relate how Caelus (Ham) "was vanquished & banished by his son Saturn or Chus (w^{ch} … happened in ye end of the brazen age just before Chus & his sons went into Babylonia)"; and he narrated how, a generation later, having established Ninus in Babylon, Belus retired to Arabia, most probably because his son drove him out.[110]

In addition to his extensive synchronization of individuals and events, Newton occasionally tried his hand at etymological exercises, implicitly conferring greater credibility on the connections he devised and perhaps demonstrating that he, too, could contribute to recondite erudition. A few examples are instructive. Part of Newton's effort to demonstrate that the ancients sought to conceal their idolatrous veneration—by affixing the name of the true God to the names of their ancestors—involved his correlation of various members of Ham's family. Consider Newton's manipulation of the names of Chus. From his reading of Dionysius Periegetes, Clement of Alexandria, and Natale Conti's *Mythologiae* Newton had learned that during the festivals in honor of Bacchus the frenzied worshippers repeatedly exclaimed "Ἐυὰ-Χους" — from whence, Newton concluded, came Evechous, the Chaldean name of their first king. More elaborately, Newton detailed the metamorphosis of Chus into Bacchus: "In ye festival of Bacchus they used to exclaim Ἐυὰ & ἐυὰ Bacchus & Io Bacchus & from this exclamation Bacchus was called Deus Euoius & Ἰobacchus & so from ye exclamation Ἐυὰ Χους & Ἰο Χους came Bacchus & Ἴακχος. And because these exclamations were used only in the feasts of Chus wch were celebrated wth great revellings therefore Bacchus became his name as he is ye God of wine."[111] Similar linguistic manipulations yielded the correlation of names in Egyptian and Phoenician mythology with the names of Chus' brothers. In line with common opinion, Newton understood Mizraim to be not the true name of Ham's son, but rather the plural designation of the

108 "The Egyptians also made a Law, they say, contrary to the general custom of mankind, permitting men to marry their sisters, this being due to the success attained by Isis in this respect; for she had married her brother Osiris." Diodorus, 1939, vol. 1 p. 85.

109 Ham's rebellion is expressed in Genesis by his glee at the sight of his father's nakedness (Genesis 9:22).

110 Yahuda MS 16.2 fols. 14, 24, 70^{v}. In recounting the casting of Saturn into Tartarus, Newton elaborated: "Jupiter is said to have started a war against his father either because his release from the paternal yoke and his departure for Egypt were attended with some kind of violent rupture, or because the war of Nimrod (who was also called Jupiter by the Assyrians) was confused in the end with that of Jupiter Hammon." Yahuda MS 16.2 fols. 17–18.

111 Yahuda MS 41 fol. 13^{v}; Conti, 2006, vol. 1 p. 407; Clement of Alexandria, 1919, p. 31.

Egyptian people. The name of the Egyptian Jupiter "in ye singular number" was Masor or Sor or Siris—whence "the star Syrius & the name Siris of ye river Nile wch star & river were dedicated to him." Masor also connected Egyptian and Phoenician mythology for, Newton believed, the name undoubtedly stood for Misor—the son of Ammon (or Ham)—whose brother Χνα (Chna) evidently designated Canaan. Furthermore, the Hebrew form of the Phoenician name (כנע) also denoted the curse of eternal servitude imposed on כנען (Canaan) to pay his brethren (Genesis 9.25).[112]

Newton also proved creative in other Hebrew etymological constructions. Sometimes he modified readily available suggestions, as he did with Bochart's explication of Phut's name. According to the French Huguenot, the Hebrew letters ט and צ were interchangeable and hence the equivalency of פוט (Put/Phut) to פוצ (putz) or to נפוץ (scattered). But whereas Bochart applied the "scattering" to the nomadic condition of the Libyan people—Phut's descendants—Newton found in the etymology evidence for the scattering of Phut's African army by Hercules (Chus). He further construed נפט to be equivalent to נפצ, thus connoting Nephtys, the wife of Python/Phut.[113] Equally creative was Newton's attempt to prove the identicalness of Nimrod and Ninus. The parallelism, of course, had existed since antiquity. But Newton added his own spin. He reasoned that "if Nimrod be נן רודה Nin-rod, that is Ninus dominus, the names agree." Several years earlier he had attempted a more detailed explication: "And the names fit excellently well if the word Nimrod was said either corruptly or ironically for the original word נן רודה Nin-rode, that is Ninos-Lord. Certainly the founder of the kingdom of Assyria, whom Moses calls Nimrod, is called Ninus by all the pagans with a single voice. ... Certainly Assyria, whose capital Nineveh took its name from Ninus, is what the Hebrews called the land of Nimrod, Micah 5:6." In all likelihood, Newton improvised on two entries he had found in his copy of Buxtorf's *Lexicon Hebraicum et* Chaldaicum, which translated רדה (rada) as "dominari, Dominium, Dominatum, exercere," and רוד (rod) as "Dominatus est."[114]

Newton never doubted his ability to establish veritable and internally coherent pagan theogonies that, at the same time, harmonized with Scripture. Indeed, he could scarcely resist conferring a degree of quasi-mathematical certitude on his results. For instance, his correlation of Ham and Mizraim as the Ammon and Osiris of the Egyptians (basing himself on Diodorus Siculus), with Sanchoniathon's affirmation that Chna was Isiris's brother—which established the former to be Canaan—elicited the exclamations "Quod erat probare" and "Quod erat indicandum."[115] On other occasions he couched his attempted synchronizations in the language of demonstrativeness. "I shall now prove with the following arguments," he promised, that Typho is the same as Antaeus. Elsewhere he concluded that he'd "proved" that Belus, Ninus' father, was

[112] Yahuda MS 41 fols. 9^{v}, 16; MS 17.2 fols. 9, 10^{v}; MS 16.2 fols. 12, 13^{v}, 30, 56, 76.

[113] Yahuda MS 16.2 fols. 8, 15^{v}, 40; MS 41 fol 20. Bochart, 1681, p. 335.

[114] Yahuda MS 41 fol. 14; MS 16.2 fols. 68 70, 70^{v} Buxtorf, 1689, pp. 716, 720.

[115] Yahuda MS 17.2 fol. 2. In a later version Newton presented the case without reference to the authorities he sought to harmonize: "The word Chanaan, exactly like the word Hamon, is expanded by a grammatical termination. The root word is Chana, humble ... and, in contracted form, Chna. Therefore Osiris is the brother of Chanaan, and therefore son of Cham, and is Misraim, father of the Egyptians. This is what had to be demonstrated." Yahuda MS 16.2 fol. 31.

Hercules, and that Antaeus was Phut.[116] This leap from conjecture to "proof" positions Newton as worthy successor to his guide in such matters, Samuel Bochart, who, Newton insisted on several occasions, had "abundantly proved" his synchronizations.[117]

Noteworthy of Newton's early forays into the study of early religions and idolatry was his lack of concern to establish a time line for deep antiquity beyond his focus on the first four generations after the Deluge. From occasional remarks, however, we surmise that at least until the early 1690s, he adhered to traditional accounts of ancient history. He found the early dating of the foundation of the Assyrian Kingdom unproblematic, and he certainly endorsed the contemporaneousness of Ninus, Ham's grandson, and Semiramis. As for Greek chronology, he dated the commencement of Cecrops' reign over Attica "about 60 years before ye Israelites came out of Egypt," by which time cities had proliferated there. Cities flourished in Italy, too, he believed, long before the founding of Rome—according to the accepted date of 753 BCE—while Troy had fallen four centuries earlier than Rome's foundation.[118] Newton's acceptance of traditional chronology can also be inferred from his uncritical use of Manetho and Sanchoniathon, sources that were destined to become problematic for him. The latter, especially, elicited some uncharacteristic encomia from the otherwise sparing Newton: "vir imprimis eruditis"; "Scriptor longe antiquissimus et curiosus originum historicarum investigator" (by far the earliest Writer and a careful investigator of historical origins).[119]

Even as he obsessed over the early history of Ham's family, Newton began to develop strong misgivings concerning the foundations upon which his narrative rested. To begin with, Newton postulated that the divinized heroes of antiquity had been members of Noah's family, and that they had been deified while their memories "were fresh" in the minds of contemporaries. "For it is not likely that while they were in memory men would omitt them & worship only others who were neither so ancient nor had so large territories nor founded so many cities nor were ye the common fathers of so many nations & tribes."[120] True, early on Newton thought that ancient mythology furnished sufficient clues to suggest that certain common traditions had survived in the records of Eastern Nations, just as it "had passed down in a clearer form" from Noah, to Abraham, and then to Moses. He even argued that the Egyptians, Chaldeans, and Phoenicians were in possession of writing long before the Greeks which, he thought, might enable the sifting of their traditions from Greek fictions. But Newton did not date the origin of writing, nor did he make use of Josephus' account of the invention of writing and astronomy by the virtuous sons of Seth—knowledge that survived the Flood thanks to the preservation of one of the pillars upon which such knowledge had been inscribed.[121] Rather, he appears to have be-

116 Yahuda MS 16.2 fols. 33, 67; MS 17.2 fol. 17.

117 Yahuda MS 16.2 fol. 12; MS 17.2 fols. 4^{v}, 10, 13^{v}, 16.

118 Yahuda 16.2 fol 21; Yahuda MS 41 fol. 1. Elsewhere he cited, presumably approvingly, Clement of Alexandria's determination that the Exodus occurred during the reigns of Amosis in Egypt and of Inachus in Argos. Yahuda MS 17.2 fol. 1.

119 Yahuda MS 16.2 fols. 31, 32, 74; Yahuda MS 13.3 fol. 1.

120 Yahuda MS 41 fol. 12.

121 Yahuda MS 16.2 fols. 49^{v}, 52; Josephus, 1841, p. 32.

lieved that ancient memories were preserved in emblematic form and, as he pointed out elsewhere, in the absence of distinct written records memories fade and become liable to be replaced by the deeds of new fathers of cities or nations. How to explain, then, the exceptionally long perpetuation of the memories of Noah's progeny?

More problematic still, the tacit acceptance of the historical narratives furnished by Diodorus, Ctesias, Sanchoniathon, and others, eventually raised in Newton's mind the problem of the scale of ancient populations. How large, for example, were the armies involved in the epic battle between gods and giants in Egypt? How many warriors did Chus command in his Indian expedition? And what exactly is meant by the statement that Belus embroiled the entire world with complex wars? These were hardly trivial issues. For example, he found in Diodorus, citing Ctesias, an estimation of the size of the army that Ninus had raised against the Bactarians: "one million seven hundred thousand foot-soldiers, two hundred and ten thousand cavalry, and slightly less than ten thousand six hundred scythe-bearing chariots." Ninus' successor, Semiramis, led nearly twice as many men on her Indian campaign, including "three million foot-soldiers, two hundred thousand cavalry, and one hundred thousand chariots." Diodorus' reassurances that such immense armies should "not seem at all impossible to any who consider the great extent of Asia and the vast number of the peoples who inhabit it," failed to render the figures more credible.[122] In contrast to the pagan authors upon whom he relied, Newton felt constrained by the brevity of Masoretic chronology, which allotted a mere century and a half to the period between the Deluge and Ninus (Nimrod's) death—an exceedingly brief interval for Noah and his three sons to people Asia by natural means.

The conundrum appears to have struck Newton after he completed, or so he believed, the grafting of the fabulous Four Ages of Men onto the history of Noah's family, on the basis of a purely hermeneutic synchronization of Biblical and mythical texts. Only when he turned to consider the correlations between his tidy idealized schema and the known history of the postdiluvian world, did he stumble upon some unwelcome consequences and inherent contradictions. Most exasperating, how to make sense of the number of people when Newton appears not to consider the ancient world to have been overly populated? Surely Newton indicated as much with a gloss on the pervasiveness of incestuous marriages among members of Ham's family: During the age of the gods, "brothers & sisters *for want of further choice* became husbands & wives," a condition that "agree[s] best to ye times next ye flood."[123] Sensitivity to the population factor and to the plausibility that empires existed shortly after the Deluge proved seminal, we believe, to Newton's decision to radically restructure his early history of mankind. But before turning to his conceptualization of the origins of monarchies, we must examine early modern conceptions regarding population, for these impinged crucially on contemporary views—including Newton's—regarding chronology, and the veracity of Scripture more generally.

[122] Diodorus, 1939, vol. 1, pp. 363, 405.

[123] Yahuda MS 41 fol. 12 (emphasis added).

Aberrant Numbers

The Propagation of Mankind before and after the Deluge

The Bible was universally perceived as the only authoritative account of "prehistory," its divine authorship compelling unqualified assent. The first chapters of Genesis accordingly entailed that mankind had expanded quite rapidly from a common ancestor—both before and, especially, after the Deluge. After all, had not God blessed Adam and Eve and commanded them to "Be fruitful, and multiply, [and] replenish the earth" (Gen. 1:28)? Had not the same blessing been repeated twice, in the same chapter, to Noah and his sons (Gen. 9:1, 7)? And were not benedictions made subsequently to the Hebrews? More specifically, had not Sarah been promised that her seed would proliferate exceedingly, and "not be numbered for multitude"? That Abraham's progeny would multiply "as the stars of the heaven, and as the sand which is upon the sea shore"? And was not the same promise reiterated to Isaac and Jacob?[1] The conviction that these benedictions should be understood literally and accepted almost as an article of faith was shared by virtually all commentators. John Shute, first viscount Barrington, encapsulated the prevailing conviction: "I take the only meaning, that these words, 'Be fruitful, and multiply, and replenish the earth,' when spoken to Noah and his sons, can well be supposed to have, so as to operate, is, that as the earth should be more fruitful than it had been; so should his sons and descendants be also more fruitful than their antediluvian ancestors. And perhaps, in order thereto, the sorrows that were multiplied on Eve ... might be lessened and abated after the flood."[2]

In view of the assumption, which amounted to a religious conviction, that the earth was populated rapidly, both before and after the Deluge, the computation of details hardly seemed necessary. Nevertheless, Scripture did provide particulars for anyone eager to "demonstrate" the reality of a truly exponential growth. Consider the evidence pertaining to Jacob's progeny. The Exodus narrative related that from the small company of seventy that went down to Egypt, there emerged 215 years later a nation of more than 600,000 arms-bearing men, in addition to women, children, and Levites—conservatively estimated by early modern commentators at more than 2 million people—all in the face of brutality and oppression. Additional figures confirmed the prodigious increase of the Israelites. The census ordered by King David

[1] Genesis 16:10, 17:2, 22:17, 26:4, 35:11.

[2] Barrington, 1828, vol. 2, p. 468.

toward the end of his reign revealed that the number of those who "drew the sword" stood at 1.3 million—800,000 for Israel and 500,000 for Judea (2 Samuel 24:9). The corresponding account in 1 Chronicles 21:5 gave an even higher figure: 1.1 million warriors for Israel and 470,000 for Judea—a total of 1.57 million. By the time Jehoshaphat ascended the throne a century later, the small kingdom of Judea alone boasted an army of 1,160,000 strong (2 Chronicles 17:14–19).[3]

The sole source for the early history of mankind, the genealogies set forth in Scripture, were perceived as essential to Christian faith. In the words of Martin Luther, Scripture enumerated those "in whom it pleased God the holy line should be continued," from Adam to Christ—a conviction universally shared by contemporary commentators.[4] God's benediction "Be fruitful, and multiply" embodied the commandment to procreate abundantly—and thus also became the source of Luther's excoriation of the Catholic Church for its willful veneration of chastity.[5]

That God willed abundant procreation sufficed for virtually all exegetes to claim the actuality of the earth's replenishment. The manner in which the stocking and restocking occurred—or, for that matter, the precise numbers generated—became trivial compared to the edifying lessons to be derived from contemplating the fulfillment of the commandment. According to John Calvin's seminal commentary on Genesis, what mattered most in the enumeration of genealogies was their apportioning the duration between the Creation and the Flood, and between the Deluge and the covenant with Abraham. The patriarchs mentioned by name were primarily those who had preserved pure worship, yet they were but a fraction of the "greate or rather huge multitude of men" that had then existed. Calvin believed it futile to elaborate on a topic fit only for "scoffing persons and common deriders" who, in their refusal to accept God's creation of the world, chose to conceive the "souden replenishing and inlarging of the world" as incredible, worse yet, to ridicule it as fabulous. The restoration of the human race, he thundered, was an act purposely designed to display "the unspeakable power of God." What else, he asked, did God intend "than to declare, that the posteritie of three men was so increased by the wonderfull power of God, and not naturally, nor after the common order, that it replenished and overspread the whole earth?"[6]

[3] Andrew Willet insisted on the literal interpretation of *Exodus* 12:37, for "thus the Lord made good his word unto *Jacob: I will there make of thee a great nation*" (Gen. 46.3). He went on to argue that since only two of those who exited Egypt entered Canaan, a prodigious number of people must have been born during their years of wandering. Willet, 1633, vol. 2 pp. 138–39. Likewise, William Perkins posited that those claiming that 600,000 armed men could not possibly "be propagated out of the familie of Jacob" within two centuries should be reminded "that they might, because then polygamie was in use, and then they did beget children from twentie yeares of age untill they were fourscore and more, and then was the multiplication of seede promised." Perkins, 1609, pp. 766–67.

[4] Luther, 1958, vol. 1, pp. 119, 427.

[5] Ibid., pp. 89–90, 163.

[6] Calvin, 1578, pp. 238–39. Mathew Poole emphasized the significance of Biblical genealogies, too: "This Chapter [Genesis 10], though it may seem to some unprofitable as consisting almost wholly of Genealogies, yet hath indeed great and manifold uses: 1. To shew the true original of the several Nations, about which all other Authors write idley, fabulously and falsly, and thereby to manifest the Providence of God in the Government of the World and Church, and the Truth and Authority of the Holy Scriptures. 2. To

Whereas Calvin viewed the propagation of mankind as having occurred largely within the institution of monogamous marriage, some of his English disciples were not nearly as reticent. Andrew Willet believed that God's blessing of Noah mandated for its fulfillment not only "lawfull marriage," but polygamy and the "obscene fecundity" of "unlawfull copulations." The very fact that the benediction was rehearsed twice, he insisted, "sheweth the certainty of it," a certainty that gained external corroboration from the size of the immense armies that the Assyrian king Ninus commanded within three centuries of the Deluge. As in the case of Calvin, the moral lesson that Willet drew from the Genesis narrative was that the re-peopling of the earth by the progeny of Noah's "three men" was little short of miraculous, the very purpose of which was to display God's "wonderfull providence." The miracle involved, however, did not imply propagation by means other than the normal course of nature. As clarified by Alexander Ross, the "supernaturall" element of mankind's extraordinary increase was the "shortnesse of time" in which it was accomplished, and the "multitude that were begotten." Yet, in respect "of the worke it selfe," the process was perfectly "naturall."[7]

Neither Calvin nor Willet nor Ross cared to speculate on the actual number of mankind either before or after the Flood—beyond vague statements regarding its prodigiousness—reticence that was characteristic of most exegetes on Genesis. They were equally vague about the length of time necessary for generating these multitudes. Exegetes wavered only on whether the building of the Tower of Babel, and the ensuing dispersion of mankind, had actually occurred at the time of Peleg's birth—101 years after the Flood. They recognized that Peleg's name signified the division of people,[8] and so must have occurred no later than his birth, but lingering doubts regarding the likelihood of so rapid a propagation explain why Calvin and many of his followers opted ultimately to avail themselves of Jewish commentators who placed the dispersion toward the middle or end of Peleg's life.

While commentators on Genesis, especially the more zealous Protestants, saw little reason to engage in vain speculations regarding the number of mankind—or to synchronize sacred and secular histories beyond what Scripture itself required—a handful of chronologers and academic theologians began treading this territory at the turn of the seventeenth century. A pioneer of this new approach was Jean Du Temps (Johannes Temporarius), a French Protestant jurist with a Biblicist frame of mind, who surely merited better than to be dismissed by the great Scaliger as "that shitty Temporarius, a completely stupid and incompetent man."[9] In his *Chronologi-*

discover and distinguish from all other Nations, that People or Nation in which Gods Church was to be, and from which Christ was to come. 3. That *Noahs* Prophecy concerning his three Sons might be better understood, and the accomplishment of it made evident. 4. To explain divers prophetical predictions and other passages of Scripture, as will appear in the sequel." Poole, 1683, sig. F2.

[7] Willet, 1633, vol. 1, pp. 86, 91; Ross, 1626, p. 75 (second pagination). Commentators repeatedly emphasized God's special providence in matters of population. Samuel Purchas viewed the rapid multiplication of the Jews in Egypt as a feat accomplished "by naturall meanes though by singular providence." Purchas, 1905, vol. 1, pp. 164–65. Thomas Burnet, too, was certain that the longevity of the patriarchs had been "providentially design'd for the quicker multiplication and propagation of mankind." Burnet, 1684, p. 23.

[8] In Hebrew פלג: (palag) denotes "split" or "divided."

[9] Grafton, 1993, p. 498.

carum demonstrationum, Du Temps imagined that no sooner had the Flood waters receded than Noah's three sons began producing twins annually—a male and a female—and they, in turn, began procreating at the same rate upon reaching their twentieth year. Hence, he computed that upon Peleg's birth the earth numbered 1,554,420—518,140 descendants for each son. Temporarius also estimated that Nimrod commanded a group of 21,289, a twenty-fourth part of Cham's progeny. Sixty years later, according to Du Temp's computation, Cham's posterity alone numbered no less than 172,724,085, with Nimrod commanding 7,196,836 of them. In his *Chronologia sacra*, Archbishop James Ussher found nothing faulty in Temporarius' reasoning or calculations, other than to question the necessity of his assumption regarding the annual birth of twins. Even half the number of people enumerated by the Frenchman at the time of Peleg's birth, Ussher believed, would have sufficed to build the Tower of Babel.[10]

Among Catholics, a similar trend became visible. Benedictus Pererius, the celebrated Jesuit commentator of Genesis, deemed it unnecessary to provide figures. No one could doubt, he wrote in 1589, that the human race multiplied rapidly and prodigiously before and after the Flood. The Biblical account of Jacob's posterity bore out an extraordinary fecundity, even when men's lives were considerably shorter and their bodies more feeble, while the size of Ninus' massive army 250 years after the Flood corroborated the extraordinary increase of mankind.[11] Three decades later a fellow Jesuit, Denis Petau, whom we will encounter again, pontificated more forcefully on the issue, proposing a population estimate that did not require the assumption that women began procreating at twelve or thirteen, or that they begot twins or triplets—however reasonable that assumption might be. Petau suggested instead that the progeny of two of Noah's sons consisted solely of males, while the third son furnished an ample number of females for them. He further postulated that women began procreating at seventeen, and in the subsequent eight years gave birth annually to one child. Under such conditions, Petau envisaged an eight-fold increase of the male population every twenty-three years: 32,768 upon Peleg's birth; over a million when Belus (= Nimrod) seized power 153 years after the Flood; to a total 623,612,358,728 males within 285 years of the Deluge— shortly before Abraham was born (figure 5.1).[12]

Athanasius Kircher's figures were even more extravagant, commensurate with the dictate of his various learned projects. For example, in order to substantiate his claim that the Egyptian dynasties originated within two or three centuries after Creation, the Jesuit was obliged to demonstrate that the globe was sufficiently populated by then to account for the formation of large kingdoms. He did so by invoking Petavius's population figure for the globe's population—1,247,224,717,455—as a low benchmark

[10] Temporarius, 1596 p. 130; Ussher, 1847–1864, vol. 11, p. 533. A century later, Arthur Bedford estimated the number of mankind on the eve of dispersion at no more than 100,000. He further calculated that if the Tower of Babel measured 120 million square feet, it would have required 2 billion bricks and forty years to build. Bedford, 1730, pp. 190, 193, 205.

[11] Pererius, 1601, vol. 1, p. 376.

[12] Petau, 1703, vol. 1, pp. 18–19. Petau committed a calculation error upon reaching the 215th year. His table reads 1,065,741,824 when it should have been 1,073,741,824. The error is compounded, and the correct total number of males should have been 80 billion higher: 705,601,770,057. The total number must be doubled to account for women.

DIAGRAMMA
propagationis hominum.

Anni à Diluvio	Numerus filiorum
VIII	8
XXXI	64
LIV	512
LXXVII	4096
C	32768
CXXIII	262144
CXLVI	2097152
CLXIX	16777216
CXCII	134217728
CCXV	1065741824
CCXXXVIII	8525934592
CCLXII	68207476736
CCLXXXV	545659813888
Summa omnium collecta	
	623612358728
Duplicata	1247224717456

Figure 5.1. Petau's table for sons born after the Flood.

for inferring the multiplication of mankind in the more wanton and more robust antediluvian world. A quarter of a century later, Kircher's estimates became more excessive still, undoubtedly because he determined by then that the Tower of Babel was built within a century and a half of the Flood. To conjure up a prodigious population by such an early date, Kircher abandoned Petau's figures and embraced Temporarius' presuppositions that each of Noah's sons begot twins annually, and that the offspring began procreating at the same rate upon reaching the age of thirty. According to his calculations, within thirty years of the Flood the world numbered 360; 21,600 after 60 years; and 1,944,000 after ninety years. By the time the Tower was built a generation later, mankind had reached the staggering mass of 23,328,000,000—clearly an error for 233,280,000, reflecting a characteristic sloppiness on Kircher's part, in addition to over-eagerness to come up with a convincing figure.[13]

To enlightened minds, notably Voltaire's, calculations of this sort seemed preposterous, and he often targeted Petau's estimates for ridicule. In his satire on the physiocrats, *L'homme aux quarante écus*, the character of a doctor of the Sorbonne is

[13] Kircher, 1652–1654, vol. 1, pp. 69–70; Kircher, 1679, p. 9.

made to cite Petau to the effect that a single son of Noah had 623,612,358,000 descendants within 285 years of the Flood—only to allow the doctor's interlocutor to dismiss the figure as an example of absurd calculations and fictitious systems conjured up by pedants. More biting are quips made in *La Philosophie de l'Histoire*—where Voltaire lashed out at the veracity of the likes of Petau, who created children "with the stroke of a pen," and who are indistinguishable from those relating that Deucalion and Pyrrha "peopled Greece by throwing stones"—and in the *Philosophical Dictionary*, where the "good priest" Petau was ridiculed for evidently knowing very little about how children were begot and reared.[14]

Yet most early modern commentators, including Voltaire's own contemporaries, found nothing fanciful in such estimates. Well into the eighteenth century and beyond, the devout would have endorsed as a matter of course Michael Drayton's musings over antediluvian children who, "at the instant of their birth," were already "halfe men," and who begot children so early that a single man "ten thousand in his time might see, / That from his loynes deriv'd their Pedegree."[15] Drayton's contemporary, Sir Walter Raleigh, was no less forceful on the matter in his *History of the World*. Raleigh was particularly concerned to establish the populousness of the postdiluvian world, but in the process he reasoned from analogy to an earlier age: given that the postdiluvians became so numerous within three centuries of the Deluge, he asked, how exceedingly plentiful must have been their more robust and long-lived predecessors?[16] For Raleigh, this reasoning warranted the conclusion that the globe was fully peopled before the Flood—thereby rebutting those who doubted the universality of the Deluge—and that the Egyptian dynasties had been established within three centuries of Creation. Nevertheless, Raleigh shunned precise figures. Nor did he attempt to divine the number of postdiluvian people in his more elaborate treatment of the origin of Assyrian and Babylonian monarchies. Like Pererius before him, Raleigh accepted the historical narratives of Diodorus Siculus and Ctesias as credible resources for the external substantiation of the Biblical injunction to procreate.

To further facilitate the synchronization of sacred and secular history, Raleigh offered a creative interpretation of the origin and dispersion of the postdiluvians, informed by the need to account for the rapid emergence of prodigious populations. Contrary to received opinion, Raleigh argued, Mount Ararat must refer to Paropanisus (Hindu Kush), one of the peaks of the Caucasus range south of ancient Bactria. Had the dispersion of mankind originated in Mesopotamia, after Nimrod's arrival at Shinar from neighboring Armenia, he reasoned, whole nations must have been begot "without the helpe of time"—for it would have been impossible to account "by any multiplication natural" for India's huge population "two reigns" later, a multitude that enabled king Staurobates to raise a vaster army than the two million troops that Semiramis brought along with her from Babylon. One ought to presume a reverse migration, Raleigh suggested, as well as amend Biblical chronology. If one accepts Berosus' claim that Nimrod usurped power 131 years after the Flood, and follows the

[14] Voltaire, 1968– , vol. 66, p. 401; vol. 69, pp. 172–73; Voltaire, 1901, vol. 12, p. 253.

[15] Drayton, 1961, vol. 3, p. 327.

[16] Raleigh, 1829, vol. 2, pp. 301, 303.

Byzantine historian Glycas in allowing forty years for the building of the Tower of Babel, then sufficient time is allowed for the peopling of Mesopotamia, while the Asian population continued to grow apace.[17]

Raleigh's decision to shun figures is somewhat surprising given that his confident generalizations regarding the number of mankind before and after the Deluge were based on precise calculations prepared for him by his friend Thomas Harriot: England's preeminent mathematician during the first half of the seventeenth century. In a surviving manuscript Harriot set out to prove that "the issue from one man & one woman in 240 yeares may be more than can inhabit the whole earth." That conclusion was based on the assumption that both parents and their progeny remained alive throughout the period, and that every new couple began procreating upon reaching twenty, begetting males and females in alternate years. Under such conditions, Harriot reckoned, a single pair would spawn 10,068,606,874 descendants. Were such a rate to continue for another century, Harriot also calculated, their number would have exceeded the maximum number of people that can stand on the surface of the earth—42,490,552,800,000.[18]

In the mid-1640s, it fell to Sir Thomas Browne to confute those who perceived the Deluge "needlesse to bee universal," and who considered it to have been confined to the region where the Ark was constructed. Browne viewed such an opinion to be "not only injurious to the text, humane history, and common reason, but also derogatory unto that great worke of God, the universall inundation."[19] To obviate this pernicious error, he set out to demonstrate that longevity was conducive to "populosity." Considering that mankind is so numerous at present when men's lives extend to three- or four-score years, Browne wrote, the number of generations would have been "proportionably multiplied" had their lives extended upwards of 800 years. Under such circumstances, the "first Original" would have become "a stranger unto his proper progeny," yet would continue to live alongside them. Indeed, "if half the men were now alive, which lived in the last Century, the earth would scarce contain their number." To illustrate "how powerfully the length of lives conduced unto populosity," Browne proposed a computation that "no reasonable spirit may contradict": Allow two or three centuries for the provision of a sufficient number of "women fit for marriage"; assume that they conceived their first child at sixty, and that in the course of the following forty years they brought forth twenty children. If such a cycle were to prevail for 700 years, Browne figured, the globe's population would reach 1,347,368,420 people—several centuries before the Deluge![20]

Browne carried his presuppositions regarding the exponential growth of people into the postdiluvian period as well. Arguing "from probabilities, and several testimonies of Scripture and humane Authors"—though without offering figures—Browne concluded that Mankind had "mightily increased" already at Peleg's birth. He inferred the existence of these multitudes by reflecting on the vast numbers that would have been required to build a city and a Tower "whose top should reach unto the

[17] Ibid., pp. 220–25.

[18] Sokol, 1974, p. 210.

[19] Browne, 1964, vol. 2, p. 555. The passage was excised from subsequent editions of the work.

[20] Ibid., pp. 427–31.

heavens," in view of the myriads that were employed in building "inferiour structures" such as Solomon's Temple or the pyramids. Likewise, thc huge armies that Ninus raised against the Bactrians, and his wife Semiramis against the Indians, suggested to him that within several centuries after the Deluge, the world was just as populous as it had been on the eve of the Flood.[21]

Noteworthy is that Browne, and Raleigh before him, considered the evidence from population as a powerful weapon against those who denied the universality of the Deluge prior to the publication in 1655 of Isaac De La Peyrère incendiary *Prae-Adamitae* and *Systema theologicum*, which rendered such "evidence" crucial as a counterweight. La Peyrère made time, or rather its absence, central to his claim that the Bible narrated only the history of the Jews, and that mankind had existed long before Adam was created. He accepted as true the assertions of great antiquity that ancient nations had claimed for themselves, and boldly contended that Biblical chronology was simply too short to account for the vast age of Chaldean, Egyptian, Chinese, and American people—not to mention the inability of such a chronology to accommodate the millennia necessary for the acquisition of the astronomical and astrological expertise that those nations possessed. Even the Deluge, La Peyrère charged, "was peculiar to the Jews, not universal in all Nations," and its effect was limited to the "Holy Land"—effectively, the area between Mesopotamia and Egypt. As for the propagation of mankind, La Peyrère dismissed it as improbable that Noah's posterity was capable of peopling the entire globe.[22]

From postulating a vast duration for the world to conferring an eternity upon it was but a small step, and La Peyrère certainly skirted that possibility. Understandably, the reaction that greeted his twin books was swift and resounding. As early as 1643, Hugo Grotius—who was privy to La Peyrère's ideas in manuscript—sought to refute the latter's conjectures regarding the antiquity and distinctive origination of early Americans by bestowing a Scandinavian descent on them. To claim otherwise, Grotius protested, was tantamount to considering native Americans as "the Off-spring of no Nation; which is as much as to believe, with Aristotle, that they were from eternity; or born of the earth, as is reported of the Spartans ... or that there were some men before Adam, as one in France [La Peyrère] lately dream'd."[23]

Grotius' indignation was characteristic of the numerous refutations that flowed from the European presses in the following three decades. For our purpose here, however, it is Isaac Vossius's audacious attempt to defuse La Peyrère's challenge that proved crucial for the fracturing of the nearly canonic stature of the Masoretic version of the Pentateuch as well as for its contribution to the grudging willingness to confer a longer span on Biblical chronology. Vossius has often been portrayed as a libertine, who may well have been a furtive adherent of La Peyrère's ideas. More likely, however, is that Vossius was cocksure of his superior erudition and of his ability to defend Scripture through ingenuity—while at the same time synchronizing it with Chinese history.

[21] Ibid., pp. 433–37.

[22] Peyrère, 1655, pp. 244, 249 and passim.

[23] Grotius, 1643, pp. 13–14, translated in Peyrère, 1655, p. 278.

Sinophilia had overtaken Vossius along with the resolve to rebut La Peyrère, both of which were fired during the sojourn in Amsterdam of the Jesuit Martino Martini, who there shepherded through the press his *Atlas Sinensis* (1655) and *Sinicae Historiae* (1659). The exotic tales that Martini related enthralled Vossius to such a degree that he often "wished he had been born a Chinese rather than an European." The orientalist Eusebius Renaudot attributed to Vossius' penchant for the "Marvellous," his reckless eagerness to lay down "as a certain Fact, that the *Chinese* History was much older than the Books of *Moses*," notwithstanding the protestations of his Jesuit informant to the contrary. Nor did Vossius "trouble his Head about the Consequences that might attend his Assertion," or "perceive what a Handle it might be made by Free-Thinkers and Libertins."[24] Vossius' unqualified acceptance of the antiquity and brilliance of Chinese history, coupled with his realization that the refutation of La Peyrère's ideas demanded more than simple dismissal, certainly prompted his postulation of a "partial Deluge": a universal flood appeared to him incommensurate with Chinese annals, and he doubted that the globe contained sufficient water to generate a worldwide inundation. Even more audaciously, Vossius maintained that the Masoretic version of the Old Testament had been corrupted by the Jews, and proceeded to advocate its substitution by the Septuagint version which, Vossius announced, was divinely inspired!

The militant espousal of Septuagint chronology in the *Dissertatio de vera ætate mundi* (1659) and *De septuaginta interpretibus eorumque translatione* (1661) was prompted by Vossius' determination to make Chinese history fit Scriptural time line without recourse to preadamite theory. To further justify his choice, Vossius pointed out that the Masoretic chronology was simply too short to credibly account for the multiplication of mankind and colonization of the globe—unless one supposed that men "sprouted like Gourds, and could beget Children at twelve Years of Age or sooner." Only a duration of more than a millennium between the Deluge and the birth of Abraham—as the Septuagint chronology offered—could adequately explain the emergence of ancient kingdoms, including the Chinese empire that, according to his computation, was founded 531 years after the Flood.[25]

Vossius disavowed any aspiration for innovation; he aimed only to restore "an ancient chronology that has been abbreviated rashly," a chronology that ensured the conformity of ancient histories to Scripture, thereby confirming it. Contemporaries failed to be reassured. Georg Horn chided him that pagan authorities ought not to serve as ultimate arbiters when it comes to Scripture: "It has always seemed contrary to Christian piety that many prefer to draw from the labyrinth of their own brain or from the stagnant waters of the pagans instead of from the crystalline fount of Moses and the Prophets."[26] More troubling to others was Vossius' relentless assault on the Masoretic text. To the Oxford orientalist Edward Pocock, this was nothing short of a deliberate attempt to raise doubts concerning the very possibility that "we have any such Thing, as a true Bible at all." His friend Narcisus Marsh concurred, adding that

[24] Bayer, 1730, preface p. 30; Renaudot, 1733, pp. 258–60. See also Weststeijn, 2007.

[25] In the course of the debate that followed, Vossius pronounced that Noah's three sons could have produced a progeny no larger than 350 people within a century of the Deluge. cited in Rossi, 1984, p. 150.

[26] Rossi, 1984, pp. 145–46, 151; Horne, 1668, dedication.

Vossius posed an even greater danger by virtue of his being the celebrated son of a far greater scholar: "whatsoever Dr. *Vossius* says, because his Name is *Vossius*, *ipse dixit*, is enough to make it be believed; which seems to me the more insufferable, because [contemporaries] cannot, or else will not make any Distinction between *Gerard* and *Isaac Vossius*.... If they would do but thus much, I believe, *ipse dixit*, would quickly stand for nothing, and that *Isaac* would not long pride himself with the Plumes, wherewith *Gerard*'s Fame has adorned him."[27]

Vossius courted controversy in other domains involving population as well. In 1685 he published a treatise on the size of ancient Rome wherein he not only set the city's population during its imperial heyday at 14 million, apparently making it twenty times larger than present-day London and Paris combined. Similarly, Vossius reckoned the population of Nanking at its height to be nearly 20 million—assuming twelve inhabitants for each of its 1.6 million houses—which meant, if Nanking's suburbs were added, that it was more numerous than the entire European population at his time, set by him at 27 million (excluding Russia).[28] Nor was Vossius unique in his high estimates; Lipsius conjectured the Roman population at 4 million, while James Howell set the population of the Chinese capital at 10 million (assuming ten people to each of its alleged one million houses).[29] Still, Vossius was more extreme than most and, judging by contemporary accounts, he evidently regaled London audiences with his estimates.[30] Small wonder that Charles II grew merry at Vossius' credulity. He was "the strangest man in the world," the King quipped, for "there is nothing which he refused to believe, except the Bible."[31]

Lord Macaulay hardly exaggerated, therefore, when commenting that in the absence of accurate censuses "men were left to conjecture for themselves; and, as they generally conjectured without examining facts, and under the influence of strong passions and prejudices, their guesses were often ludicrously absurd."[32] Certainly, this uncritical and impressionistic vogue gave rise to what became the first attempts to quantify population. Leading the way was John Graunt, whom we encountered above, and who reminisced that it was upon hearing an alderman "of eminent Reputation" pontificate in 1660 that London numbered "two Millions of People more" than it did prior to the 1625 plague, that prompted him to undertake a quantitative analysis of official records in order to ascertain the true number of inhabitants in the Capital.[33]

A careful analysis of birth records and tables of mortality, and the application of elementary (if vexed) averaging techniques, enabled Graunt to establish a 12/13 male

[27] Pocock, 1740, vol. 1, pp. 74–75.

[28] Vossius, 1685, pp. 56–68. Vossius also set the world population at 500 million, 60% of whom lived in Asia. For his part, Riccioli estimated Rome's population during Augustus's time to be 9.4 million: Riccioli, 1672, p. 678.

[29] Lipsius, 1630, pp. 113–20; Howell, 1657, p. 383.

[30] St. Evremond, 1714, vol. 3, p. ciii: "At the salon of the duchess of Mazarin/ Of China, Vossius a grave Treatise brings,/ Divine the Nation, and her Thousand Kings:/ Old Rome, as in his spacious Chart design'd,/ You Forty Times as big as Paris find."

[31] Addison, 1989, p. 100.

[32] Macaulay, 1866, vol. 1, p. 221.

[33] Graunt, 1662, p. 59. James Howell estimated the population at one and a half million. Howell, 1657, p. 403.

to female ratio for London—and 15/16 elsewhere in England—as well as the 5/4 ratio between births and deaths. Graunt also pioneered the life table, which permitted him to estimate the relative size of various age groups in the population, including the calculation that "fighting men" (age 16 to 56), comprised 34 percent of the population. On the basis of the information he gathered, Graunt concluded that the population of England in 1662 was just under 6.5 million, and that of greater London stood at 460,000. London proper, he reckoned, numbered 384,000 persons, including 24,000 "pair of Breeders"—one-eighth of the population, which indicated that the city would double its population in sixty-four years. Graunt was not prone to speculate beyond his figures. However, he reflected in passing that had the London rate of growth held constant throughout the 5610 years that had elapsed since Creation—he accepted Scaliger's reckoning of 3949 BCE for the event—the progeny of Adam and Eve would have exceeded the number of people that "are now in it." Such a conclusion convinced him that "the World is not above 100 thousand years older, as some vainly imagine, nor above what the *Scripture* makes it."[34]

Graunt's reluctance to speculate on the long durée of universal population was not shared by other early modern demographers. His friend and collaborator, William Petty, was drawn to such cogitations after a Dublin theologian had requested his assistance in 1680 to refute skeptics who argued against the reality of the Resurrection with the claim that "the whole Globe of the Earth could not furnish Matter enough for all the Bodies that must Rise at the last Day, much less would the surface of the Earth furnish footing for so vast a Number." Petty obliged by calculating that some 20 billion people had died since Creation, and that 20 percent of Ireland's surface would suffice to furnish graves for all those bodies—reckoning 4 million graves per one square mile—while two mountains in the island "were as weighty as all the Bodies that had ever" been born."[35]

Petty was adamant in insisting that his calculations of population growth since Biblical times offered "a brave argument against Scripture Scoffers and Prae-Adamites." Building on Graunt's *Observations*, Petty gauged the population of England in 1682 at 7,400,000, and of London alone at around 670,000. He also posited that London doubled its size every forty years, while the country doubled every 360 years. Turning to tabulate the increase of mankind since the Flood, Petty hypothesized that variations in the number of births and deaths over time engendered fluctuations in the rate of multiplication—as much as between 10 and 1200 years. This, Petty believed, rather tended to corroborate "*Scriptures* and all other good *Histories* concerning the *Number* of the People in Ancient Time," while, concurrently, establishing a rationale for the current size of the globe's population—320 million. Petty admitted that he had taken "discretionary liberty" in his presuppositions, yet he reiterated that his calculation not only "solves all the Phenomena of Scripture upon ground Experimentally true," but is unavoidable, given that no single rate of doubling could bear out the number of Israelites upon the Exodus out of Egypt or during David's time, without necessitating too large multitudes for subsequent generations. For example, a fixed doubling every 150 years would have set the number of people during David's time at

[34] Graunt, 1662, pp. 62–63 and passim.

[35] Petty, 1683, pp. 22–24, 44–47; Petty, 1967, p. 92.

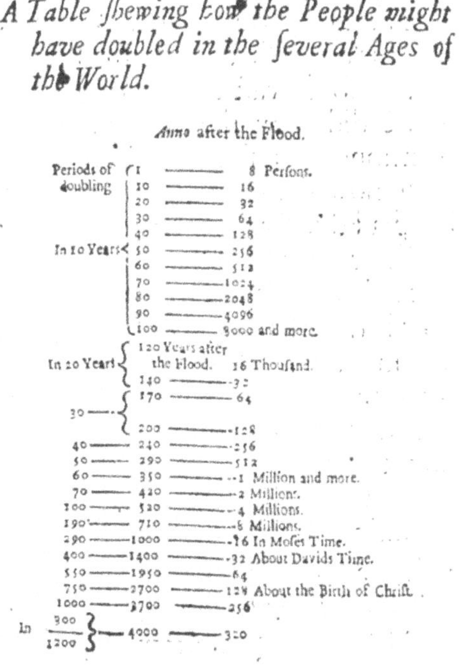

A Table ſhewing how the People might have doubled in the ſeveral Ages of the World.

Periods of doubling	*Anno* after the Flood.	
In 10 Years	1	8 Perſons.
	10	16
	20	32
	30	64
	40	128
	50	256
	60	512
	70	1024
	80	2048
	90	4096
	100	8000 and more.
In 20 Years	120 Years after the Flood.	16 Thouſand.
	140	32
30	170	64
	200	128
40	240	256
50	290	512
60	350	1 Million and more.
70	420	2 Millions.
100	520	4 Millions.
190	710	8 Millions.
290	1000	16 In Moſes Time.
400	1400	32 About Davids Time.
550	1950	64
750	2700	128 About the Birth of Chriſt.
1000	3700	256
In 300/1200	4000	320

Figure 5.2. Petty's doubling table.

a mere 8,000. Conversely, doubling every 100 years would have "over-peopled the World" as early as AD 1000. Hence Petty's "Progressive doubling," which increased Noah's progeny to 8,000 within a century of the Flood; to over a million 250 years later; and to 16 million in the age of Moses—by which time the population doubled every 290 years (figure 5.2).[36]

Petty's exasperation over his failure to come up with a credible rate of population growth was also keenly felt by fellow demographer Gregory King. In his *Natural and Political Observations and Conclusions upon the State and Condition of England* (1696), King dealt only cursorily with ancient population. He suggested that the number of mankind ca.1700 BCE—six centuries after the Flood—was between one and two million. In addition, assuming that the English population in 1260 CE amounted to 2.75 million, he speculated that were the country originally settled by a small colony, the event would have occurred six or seven hundred years after the Flood; if the colony numbered several hundred people, the settlement would have taken place two centuries later.[37]

King's manuscript commonplace book, however, attests to the obsessiveness with which he sought to correlate ancient and contemporary population figures, only to abandon the effort in desperation. Assuming that the world was created in 3935 BCE

[36] Petty, 1899, vol. 2, pp. 465–69; Petty, 1967, pp. 92, 115.

[37] King, 1804 [1696], p. 41 (second pagination).

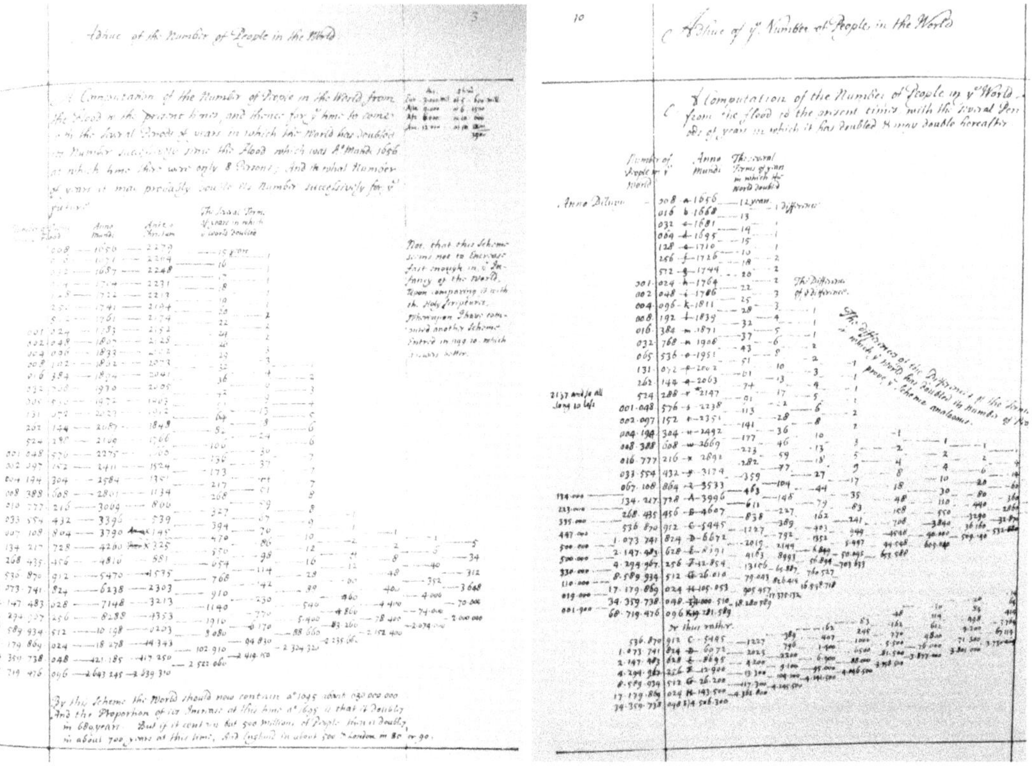

Figure 5.3. King's first correlations.

(Andrew Willet's figure), and that the world's population in AD 1695 amounted to 630 million, King attempted to configure a rate of growth by availing himself of Petty's method of progressively diminishing rates of doubling. His first table produced 512 inhabitants 105 years after the Flood (AM 1761 = 2174 BCE); 65,536 people 211 years later (AM 1972 = 1963 BCE); over half a million by AM 2169 (1766 BCE), by which time the rate of doubling occurred every 106 years; and nearly 8.4 million people AM 2801 (1134 BCE) when the population doubled every 268 years (figure 5.3). King obviously realized the difficulty of accounting for the number of people immediately after the Deluge, and jotted down in a side note that "this Scheme seems not to Encrease fast enough in y[e] Infancy of the world, upon comparing it with the Holy Scripture." So he drafted another scheme "which answers better"; it lowered the rate of doubling for the first hundred years, but the difference proved negligible.[38]

Abandoning these attempts, King attempted instead to configure meaningful ratios to account for the globe's population before and after the Deluge. He conjured up one sliding ratio for the antediluvian period—generating 67 million people on the eve of the Flood—and inferred from it that "the Terms of Doubling the Number of People" ranged between 10 and 338 years before the Flood; between 15 and 410 years

[38] King, 1973, p. 3 (of the manuscript section).

Number of Doublings	Persons	Anno mundi	Post Diluvium	Term of years for Doubling
0	070	2298	642	13
1	140	2311	655	13
2	280	2324	668	13
3	560	2337	681	14
4	1·020	2351	695	14
5	2·040	2365	709	14
6	4·080	2379	723	15
7	8·160	2394	738	15
8	16·320	2409	753	15
9	32·640	2424	768	16
10	65·280	2440	784	16
11	130·560	2456	800	17
12	261·120	2473	817	18
13	522·240	2491	835	20
14	1·044·480	2511	855	2
15	1·200·000	2513	857	

Figure 5.4. King's second attempt.

from the Flood to Christ's time; and between 410 and 680 years in the following two millennia (figure 5.4).

If constructing universal tables of population was not exasperating enough, the necessity to accommodate the number of the Israelites proved almost insurmountable. Consider the Exodus. When Jacob and his family of seventy went down to Egypt AM 2298 (1637 BCE), the world population stood at 1.4 million. Upon the Exodus 215 years later (AM 2513=1422 BCE), the world's population scarcely doubled to 3 million, but the Israelites swelled to 1.2 million![39] King was frustrated by "so considerable and extraordinary an Increase," for according to his calculations even during the more fertile antediluvian age, an increase from 70 to 1.2 million people would have taken 460 years—and should not have occurred after the Deluge in less than 600 years. Yet veneration of Scripture mandated that such an increase be rendered within the realm of possibility, so King produced a table that illustrated how the Israelites might have multiplied every fifteen years on an average: "Suppose every couple produces 8 children or 4 Couple in a generation or 30 years. Then 70 Persons or

[39] King interpreted 600,000 "men" to include women, too, and only added an equal number for children under sixteen.

nber of doblings	Persons	A° M.	Post Dil.	Term of years for Doubling	
0	070	2298	642		
1	140	2308	652	10	0
2	280	2318	662	10	0
3	560	2328	672	10	0
4	1.020	2338	682	10	0
5	2.040	2348	692	10	1
6	4.080	2359	703	11	0
7	8.160	2370	714	11	0
8	16.320	2381	725	11	0
9	32.640	2392	736	11	1
10	65.280	2404	748	12	1
11	130.560	2417	761	13	2
12	261.120	2432	776	15	6
13	522.240	2453	797	21	24
14	1.044.480	2498	842	45	64
15	2.088.960	2.607	951	109	134
16	4.176.920	2850	1194	243	241
17	8.353.840	3334	1678	484	391
18	16.607.680	4209	2553	875	

Figure 5.5. King's lowered rate of doubling.

35 Couple will every 30 years produce 140 Couple, By which means 70 persons will Encrease in 215 years in this proportion."

Finally, King needed to lower sharply the Israelites' rate of doubling in order to account for David's census 434 years later (AM 2987 = 948 BCE). Reversing himself on his earlier decision to double the number of "fighting men" in order to derive the number of Israelites upon leaving Egypt, he now more than trebled the number of men who bore arms in David's army in order to produce the more plausible figure of 5–6 million people, at a time when the world swelled to 13.7 million inhabitants. Obviously, such a figure necessitated further tampering with earlier and later rates of multiplication, so King constructed a table in which the rate of doubling was lowered for the first 130 years in Egypt—between 10 and 20—and then gradually decreased to about every 500 years at David's time. At this point King (no doubt wisely) decided to abandon his calculations (figure 5.5).[40]

The demographers' efforts to establish the study of population on a more rigorous (and quantifiable) footing played only a tangential role in the growing number of

[40] King, 1973, pp. 4–10. A decade and a half earlier, Edmond Halley estimated the number of "Fencible Men" to exceed a quarter of Breslau's population, a ratio he thought to "pass for a Rule for all other places." Halley, 1693, p. 601.

English scholarly responses to the combined challenges posed by La Peyrère and Vossius. As a rule, post-Restoration authors retained the Biblicist mentality that characterized their predecessors, for they, too, sought to defend the integrity of the Genesis narrative with as few modifications as possible. Thus, already in 1657, Brian Walton, the editor of the London Polyglot Bible, advocated the superiority of the Septuagint chronology partly for its ability to better resolve problems of population.[41] More influential was Edward Stillingfleet, who propounded in his *Origines sacrae* (1662) that just as God had "punished the world by destroying mankind, so after the flood he did in a particular manner bless Noah and his sons," and his benediction must have had "an extraordinary effect." The future Bishop of Worcester, and John Locke's nemesis, set for himself the task of confuting the "Atheists of our age," especially those alleging, on the one hand, the "irreconcileableness of the account of times in Scripture, with that of the learned and ancient Heathen nations" and, on the other hand, "the inconsistency of the belief of the Scriptures with the principles of reason." Stillingfleet considered the peopling of the earth to be an important article of faith, on account of the ramifications of the issue for "the truth of the Scriptures, and the universal effects of the fall of man ... For as it is hard to conceive how the effects of mans fall should extend to all mankinde, unless all mankind were propagated from Adam; so it is unconceivable how the account of things given in Scripture should be true, if there were persons existent in the world long before Adam was." But whereas upholding the common ancestry of mankind was central for faith, there was no "urgent necessity from Scripture to assert, that the Deluge did spread itself over all the surface of the earth." Stillingfleet despaired of ever seeing it proved that the entire globe was peopled before the flood; how could it be fully populated in 1656 years, he wondered—or even over 2000 years if the Septuagint chronology were accepted—"when in so much longer a space of time since the flood to this day, the earth is capable of receiving far more inhabitants, then now it hath"? Surely, then, it sufficed to accept that the Flood was universal in effect, albeit not in scope, for it would have been needless to extend it "beyond the occasion of it, which was the corruption of mankind"—namely beyond Mesopotamia and neighboring lands.[42]

When turning to consider the postdiluvian world, Stillingfleet remained consistent in his understanding of a slow and gradual growth of population. Hence, he found it nigh impossible to accept the chronologers' claims that the Assyrian monarchy was established within a century of the deluge. Nor was his discomfort much alleviated by Petavius' dating of the event half a century later. It was inconceivable, Stillingfleet stated bluntly, that Noah's three sons could have produced the multitudes necessary to account for such an empire and for the deeds of Ninus and Semiramis. True, God's benediction to Noah ensured "more than ordinary multiplication of the world" after the Flood. Yet, he remained skeptical of the figures that Petavius and Ussher had marshaled, and considered the much longer chronology of the Septuagint best suited to remove difficulties: "there will be sufficient space given for the propagation of mankind, the building the Tower of Babel, the dispersion of Nations, the founding the Assyrian Empire, the plantation of Egypt, China, and other places."[43]

[41] Walton, 1654–1657, vol. 6, fol. 68^{v}.

[42] Stillingfleet, 1662, vol. 1, sig. b3; vol. 2, pp. 534, 539–40.

[43] Ibid., pp. 554–58.

Stillingfleet's tacit endorsement of parts of Vossius' position remained a minority opinion among English scholars, and it soon generated a host of rebuttals by laymen and clerics alike, all of whom strove to defend the verity of the Masoretic text and the universality of the Deluge—often through the deployment of population arguments. Chief Justice Mathew Hale pronounced confidently in *The Primitive Origination of Mankind* (1677) that he had no need for the chronology of the Septuagint in order to account for the rapid propagation of mankind. Considering the longevity and coexistence of the postdiluvians, "we may without the help of a miraculous fertility find that in 104 years Mankind descended from *Noah* and his three Sons and their Wives, might arise to a stupendious multitude by that Arithmetical Progression [*sic*] that would be found in their Generations." Hale found it unnecessary to compute their number as Temporarius and Petavius had already done so, and he contented himself by citing Kircher, who colorfully imagined such a vast multitude to exist within 200 years of the Deluge "that if they were cast into a square *Battalia*, allowing to every person but one square foot of ground, the side of that Square would be 372 Astronomical miles, or 25 Heavenly degrees." Nevertheless, taking his cue from Graunt, Hale now tried his hand at calculations of a different sort. He postulated the average life span at sixty, and the begetting of two children by each individual, male or female, by thirty. Then, allowing each of these children to produce two offspring before the death of their father, he concluded that within a single generation "they become increased in a quadruple proportion, and all coexisting." And while the parents die, within another generation "by a Geometrical Proportion their Increase is multiplied proportionable to the Excess of their number above Two." Naturally, he added, if procreation started earlier and continued later the numbers would increase exponentially. Having satisfied himself with the force of his reasoning, Hale proceeded to generalize along the same lines about an era in which men's lives—and coexistence—was considerably longer. Petavius and Temporarius, he wrote, furnished "a plain Demonstration" that within 215 years of the Deluge, Noah's progeny "might without a Miracle increase to prodigious and incredible multitudes"—the former explicating "with very clear evidence" how a single son of Noah had spawned 1,219,133,512 people. Hence, Hale concluded, Hebrew chronology is more than adequate to account, without recourse to miracles, for the beginning of the Assyrian monarchy 153 years after the Flood, and the subsequent battles of Ninus.[44]

Nor did Hale regard his reasoning for a near instantaneous proliferation of people to be at odds with Graunt's gradualist model of population growth. Indeed, Hale deemed the content of the latter's *Observations* to be based on such solid evidence and strict analysis as to offer far "greater Demonstration of the Gradual Increase of Mankind upon the face of the Earth, than a hundred notional Arguments." Yet just as the empty postdiluvian earth was propitious for the absorption of a vast number of people in conformity with God's benediction, so His "infinite Wisdom" enabled "extraordinary Occurrences and Correctives"—such as wars, plagues, famines, floods, and conflagrations—to serve as means for reducing "the Numbers of Men to an Equability." Clearly, since Hale considered the gradual increase of population to be "the most sensible Evidence of Fact against the Eternal Succession of Mankind," it

[44] Hale, 1677, pp. 203–5.

was incumbent on him to demonstrate that the "correctives" he had enumerated were at play throughout all eternity.[45]

During the 1690s Richard Cumberland, Bishop of Peterborough, attempted his own rebuttal of those denying the possibility of a rapid peopling after the Deluge, and in turn substituted the Septuagint for Masoretic chronology. In an unpublished essay—"Concerning the possibility of a sufficient Increase of men from the three sons of Noah, to a number large enough to found all the nations mention'd in the eldest credible histories; and that in the times assign'd to their foundation, agreeably with the Hebrew accounts"—Cumberland set himself the task of computing the number of mankind upon the death of Peleg, the first to die after the Flood (figure 5.6). He proceeded from "reasonable suppositions," namely that the average life span of the first three generations of postdiluvians was 400 years, and that their constitution was exceptional. Hence, one may easily allow each of Noah's three sons to begin procreating immediately after the Flood and annually produce a child over the course of twenty years—evenly divided between males and females as Graunt's tables demonstrated. Upon the youngest attaining twenty, the cycle resumes at the same rate, and recurs every forty years. Under such conditions, he calculated, upon Peleg's death 340 years after the Flood, the world contained 3,333,333,330 couples. Cumberland thought he was quite moderate in his calculations: he allowed a hiatus of twenty years between cycles and he refrained from considering the births of twins. Furthermore, his table included only the decedents of Shem, Ham, and Japheth, without taking into account the possibility that their parents continued procreation, which could have multiplied mankind many times over! Not surprisingly, he was convinced that he managed to account for sufficient numbers of people to establish kingdoms, without "violently lay[ing] hands on all possible methods of multiplying men, but left out very many ways whereby we might have increas'd our numbers." His editor and son-in-law, Squire Payne, concurred. Cumberland's computation, Payne was certain, demonstrated not only what was possible "in theory," but what could be deduced from the "fact" recorded by Scripture regarding the proliferation of the Israelites at a later age, "when both the strength and lives of men were very much impair'd." If the Israelites managed to propagate so prodigiously in 215 years, and under "great oppression," how much greater must have been the increase of population after the Flood in 340 years, so that "there could not be want of a sufficient number of men to begin the first monarchies."[46]

Whereas Stillingfleet, Hale, and Cumberland wrote in the venerable tradition of Christian apologetics, Thomas Burnet and William Whiston—authors who solicited Newton's comments prior to publication—broke new ground in their handling of the reality of a universal Deluge and the matter of ancient populations. Burnet charged that those who sought "to avoid an objection from reason"—Vossius and Stillingfleet—ended up by denying a "matter of fact" by opting for a regional inundation in order to avoid the necessity of explaining whence came the vast body of water that produced a universal Deluge. Equally erroneous was their assumption that the antediluvian population was sparse and confined to Judea and adjacent regions. Like his predecessors,

[45] Ibid., pp. 206–45.

[46] Cumberland, 1724, pp. 145–54, x–xi.

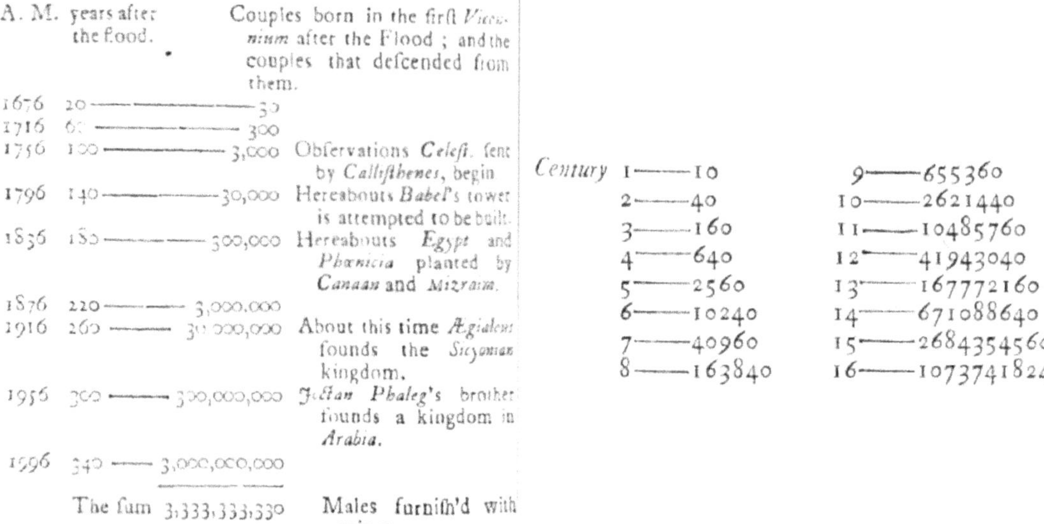

A. M.	years after the flood.	Couples born in the first *Vicennium* after the Flood; and the couples that descended from them.	
1676	20	30	
1716	60	300	
1756	100	3,000	Observations *Celest.* sent by *Callisthenes*, begin
1796	140	30,000	Hereabouts *Babel's* tower is attempted to be built.
1836	180	300,000	Hereabouts *Egypt* and *Phœnicia* planted by *Canaan* and *Mizraim*.
1876	220	3,000,000	
1916	260	30,000,000	About this time *Ægialeus* founds the *Sicyonian* kingdom.
1956	300	300,000,000	*Joctan Phaleg's* brother founds a kingdom in *Arabia*.
1996	340	3,000,000,000	
	The sum	3,333,333,330	Males furnish'd with wives.

Century			
1	10	9	655360
2	40	10	2621440
3	160	11	10485760
4	640	12	41943040
5	2560	13	167772160
6	10240	14	671088640
7	40960	15	2684354560
8	163840	16	1073741824

Figure 5.6. Cumberland's numbers (left) and Burnet's centuries.

Burnet took his cue from analogy. Given that, after the Flood, Noah's progeny rapidly filled Asia, Europe, and Africa, it stood to reason that during the infancy of the world, when the longevity and fruitfulness of mankind were greater—which "seems to have been providentially design'd for the quicker multiplication and propagation of mankind"—they would do as well. Indeed, he thought, it is more likely that the earth was overstocked than thinly populated. Burnet proposed an "easie supposition" to substantiate his estimate: Had Adam and Eve given birth to but "ten pair of Breeders" within the first century after Creation, and these multiplied at "the same decuple proportion [as] the first pair did," there would arise in fifteen hundred years "a greater number than the Earth was capable of." Hence he lowered his estimate to a "quadruple proportion" for mankind's multiplication, which produced over 10 billion people (couples) on the eve of the Deluge. Burnet realized, of course, that the product was "too excessive high" compared to the population at his time—"betwixt three and four hundred millions"—yet he remained confident that his was a fairly reasonable figure, and that it sufficed to render groundless any "conceit" that the Deluge was confined to the Holy Land and environs.[47]

William Whiston entered the Deluge fray a decade later. A staunch adherent at that time of "Hebrew Verity" as codified by Archbishop Ussher, Whiston berated both Samaritan and Septuagint chronologies for their multiplying years "without reason," and "contrary to the Truth, and to the Sacred Writings together." The youthful Whiston was more confident than other contemporary cosmogonists of his ability to pin down the precise day on which the Deluge began. Thus, the "seventeenth day of the second month" of Noah's six hundredth year, could be "nearly determin'd by the place

[47] Burnet, 1684, pp. 22–24.

of the Perihelion, and exactly by the Astronomical Tables of the Conjunction of the Sun and the Moon" to have occurred on the day separated by that interval from the Autumnal equinox in 2349 BCE. From such a felicitous congruence between the Biblical narrative and astronomy there

> arises a very surprizing and unexpected Confirmation of the Verity of the Scripture History. Here is a great and signal instance of the wonderful Providence of God indeed, and of his care for the Credit and Establishment of the Holy Books; that he has left us means sufficient, after above Four thousand Years, of examining and ascertaining the Veracity of the most Ancient of its Writers, and in one of the most scrupled and exceptionable Points of his Narration, that of the Universal Deluge; and that from unexceptionable Principles, the Astronomical Tables of the Coelestial Motions.

Proof of this sort not only invalidated Samaritan and Septuagint chronologies, but confuted "all the pretended immense numbers of Years, which the Annals of some Nations recount."[48]

The aspiration to ground the reality of the Biblical Deluge on quantitative foundations also informed Whiston's determination to prove that the antediluvians were "more numerous than the present Earth either actually does, or perhaps is capable to contain and supply." His evidence amounted to an extrapolation from the number of Israelites that exited Egypt onto the populousness of the world on the eve of the Deluge. If Jacob's posterity after 266 years exceeded 600,000 men bearing arms, Whiston reasoned, then the Israelites' male population must have totaled about 1,764,700—based on Graunt's observation that the 16–56 age group comprises 34 percent of males. Likewise, as Graunt had also demonstrated that the number of females is one-fifteenth lower than that of males, 1,635,300 Israelite women must have exited Egypt, thus making a nation of 3,343,000 people, once the 43,000 Levites were also tallied.[49] The total implied a doubling of the Israelites every fourteen years. Whiston realized that such a growth rate, if applied to the antediluvian era, would have generated too prodigious a population to be believed. So he conjured up a more "modest" computation that could nonetheless establish "the probability, if not certainty," of mankind's origination from a common ancestry, and the full population of the earth long before the Deluge. His solution was in the form of "fair and modest *postulata*," according to which Adam's posterity increased as fast as Jacob's in the first 266 years after Creation, but thereafter it doubled only every 280 years—the rate that Graunt had established for the doubling of England's population. The renovated calculation, however, produced too low a figure for Whiston's purpose, since he believed that it failed to take into account the antediluvians' longevity and coexistence, and their "more numerous posterity." Hence, he multiplied by twenty the total of 100,290,000 that he derived—

[48] Whiston, 1696, pp. 123, 151–2.

[49] Whiston was sloppy in his calculations. The number of women should have been 1,646,353, giving a grand total of 3,454,053 people.

to factor in coexistence—and that figure by forty to compensate for fecundity, setting the number of mankind on the eve of the Deluge at an astronomical 82,232,000,000.[50]

Whiston was certain that his computation made obvious "how vastly numerous, [and] according to the regular method of humane Propagation, the Offspring of a single person may certainly be; and this on a Calculation from undoubted matter of fact, not from a meer possible *Hypothesis*, (according to which numbers prodigiously greater would still arise)." Moreover, given that Hebrew chronology easily accounted for a fully inhabited world 1656 years after Creation—229 times more numerous than the globe's population at Whiston's own time—Whiston found no need to resort to the longer chronology of the Septuagint in order to produce the large number of people that "any Authentick Histories of those Ancient Times do require us to suppose."[51]

While his account was in press, Whiston's attention was drawn to William Petty's *Multiplication of Mankind*, and he hastily composed a postscript in which he accepted Petty's figure of 360 years as the mean measure of population doubling, since it permitted him to produce another scriptural "demonstration." If the Israelites doubled themselves every 360 years—postulating that the length of human life was set at the present rate following the Exodus—then the number of men bearing arms should have risen after 473 years from 603,000 to 1,576,666, a figure "wonderfully near" to the 1,588,000 generated by the census that King David ordered.[52]

Concerned as he was with the Flood, for the most part Whiston ignored in 1696 the ramifications of his calculations for the postdiluvian population, except to reiterate his position regarding the sufficiency of the Hebrew chronology to account for a rapid increase of mankind in the first centuries after the Deluge, given that the average life span until Abraham, 427 years later, was three hundred years. The verity of a "numerous Stock of Inhabitants" was central to his argument, he wrote, "or else no Account were to be given of the present numbers of Men upon the Face of the Earth; whereby the Verity of this Proposition, the Veracity of Moses therein, the great importance thereof, and the necessity of the present Solution, and of that Theory on which it is built, are mightily confirm'd." Six years later he elaborated. He now assumed that world population at the time he wrote stood at four billion, and that the rate of doubling had been a constant 400 years ever since "Human Life was fix'd in the days of *David*." Naturally, the rate of doubling was far greater in the thirteen centuries that separated the Deluge from David—when men's lives had been up to seven times longer—and Whiston devised a diagram to illustrate how an average doubling every sixty years until David's time, and every 400 years since, would produce the prerequisite figure of four billion people (figure 5.7).[53]

Whiston maintained that his scheme accounted perfectly for the earth's population in his day, at the same time furnishing "as many [people] in every Age preceding as any Authentic Accounts of Ancient Times do require." Yet, strict Biblicist that he was,

[50] Whiston, 1696, pp. 174–79. Whiston was sloppy again. Instead of computing a five-fold doubling, he multiplied the number of Israelites by thirty. The correct figures should have been 106,976,000 and 85,580,800,000, respectively.

[51] Ibid., pp. 284–85.

[52] Ibid., pp. 383–87.

[53] Ibid., pp. 284–85, 338; Whiston, 1702, pp. 65–68.

Number of Mankind.	Years after the Flood.	Years of doubling.	Series.
16	4	4	(1)
32	9	5	(2)
64	15	6	(3)
128	23	8	(4)
256	35	12	(5)
512	50	15	(6)
1024	70	20	(7)
2048	95	25	(8)
4096	125	30	(9)
8192	160	35	(10)
16384	200	40	(11)
32768	245	45	(12)
65536	295	50	(13)
131072	350	55	(14)
262144	410	60	(15)
524288	475	65	(16)
1'048576	545	70	(17)
2'097152	620	75	(18)
4'194304	700	80	(19)
8'388608	800	100	(20)
16'777216	1000	200	(21)
33'554432	1300	300	(22)
67'108864	1700	400	(23)
134'217728	2100	400	(24)
268'435456	2500	400	(25)
536'870912	2900	400	(26)
1073'741824	3300	400	(27)
2147'483648	3700	400	(28)
4294'967296	4100	400	(29)

Number of Mankind.	Years of the World.	Years of Doubling.	Series.
4	2	2	1
8	6	4	2
16	12	6	3
32	20	8	4
64	30	10	5
128	42	12	6
256	56	14	7
512	72	16	8
1024	90	18	9
2048	110	20	10
4096	132	22	11
8192	156	24	12
16,384	182	26	13
32,768	210	28	14
65,536	240	30	15
131,072	272	32	16
262,144	306	34	17
524,288	342	36	18
1,048,576	380	38	19
2,097,152	420	40	20
4,194,304	462	42	21
8,388,608	506	44	22
16,777,216	552	46	23
33,554,432	600	48	24
67,108,864	650	50	25
134,217,728	702	52	26
268,435,456	756	54	27
536,870,912	812	56	28
1,073,741,824	870	58	29
2,147,483,648	930	60	30
4,294,967,296	992	62	31
8,589,934,592	1056	64	32
17,179,869,184	1122	66	33
34,359,738,368	1190	68	34
68,719,476,736	1260	70	35
137,438,953,472	1332	72	36
274,877,906,944	1406	74	37
549,755,813,888	1482	76	38

Figure 5.7. Whiston's first (left) and second series.

Whiston focused almost exclusively on Scriptural genealogies and prophecies, rarely bothering to incorporate profane history into his narrative—except to identify Chinese emperors as the patriarchs of Genesis. One wonders, for example, whether it occurred to him that if the dispersion of mankind occurred 101 years after the Deluge, there would have been little reason for the measly 2,100 men, women, and children that, on his reckoning, then inhabited the world to disperse, not to mention the impossibility that such a company could have embarked on the building the Tower of Babel. Likewise, a population of 300,000 at the time of Abraham's birth was nowhere near enough to account for Ninus's mighty kingdom.

By the late 1710s, Whiston had lost confidence in Hebrew chronology. The change of heart was partly owing to the publication in 1715 of Edmond Halley's ground-breaking table of comets, and especially to Halley's suggestion that the 1680/1 comet returned every 575 years. Whiston quickly convinced himself that this very massive comet was the one that God had used to induce the Deluge and, accordingly, he focused on Josephus' and the Samaritan's "truer Chronology" in order to fix the event 577 years earlier, at 2926 BCE.[54] Whiston may also have been motivated to abandon Hebrew chronology as a result of his falling out with Newton, including on matters of chronology.[55] In any case, the new dating necessitated correction of antediluvian population figures, since the Samaritan chronology fixed the Deluge at 1556 years after Creation—a century sooner than the Hebrew chronology. Concluding again that since King David's time the world population doubled every 400 years, Whiston now proposed that, since the antediluvians' life span was ten times longer, it stood to reason that the rate of their doubling was ten times shorter—every forty years on an average, on a sliding scale between two years to seventy-six—according to which he concluded that "the number of the Antediluvians before the Deluge would easily amount to above 500,000 millions." Actually, it should have been nearly 1.1 trillion people, for Whiston conveniently omitted to add the final period of doubling necessary to reach the deluge at AM 1556.[56]

Curiously, the book that drew Whiston's attention to Petty's calculations in 1696 was William Nicholls's *A Conference with a Theist*, which offers a revealing articulation of Nicholls's awareness of the problems that population posed to those eager to brandish it against scoffers of religion—real or imaginary—not to mention the relative ease with which these complexities could be ignored. Nicholls was not concerned with the annals of ancient empires. He simply sought ammunition to refute arguments regarding the eternity of the world—an issue central to his book—and thus Petty's seeming "demonstration" of a slow and gradual increase of people appealed to him as particularly useful. Accordingly, Nicholls's persona in the dialogue contends that, as late as the days of Homer and Hesiod mankind "led a sort of a Pastoral Life," often roving from one territory to another, which attests to how thinly the world had been populated. Indeed, had the world been four or five thousand years older than what the Mosaic account allowed, it would have swarmed with people by now. The interlocutor in the dialogue rebuts with a common "atheist" trope: recurrent catastrophes had reduced populations to an equilibrium from all eternity. To which Nicholls responds that no evidence exists to substantiate periodic devastations; Petty's calculations demonstrate that, notwithstanding plagues and other devastations, "Mankind doubles it self once in 360 years."

The introduction of such "fine Arithmetical Argument" amuses the interlocutor. The devout, he was certain, would not find such reasoning pleasing, while the "infidels" would draw courage from it: "For settle the Increase of Mankind how you will, make the Period of doubling as large or as narrow as you please, you will find your

54 Whiston, 1725, pp. 143–44, 188, 22–23. Halley's "A Synopsis of the Astronomy of Comets" was first published as an appendix to David Gregory's *The Elements of Astronomy Physical and Geometric*. Gregory, 1715, vol. 2, pp. 881–905.

55 By then Whiston, perhaps in reaction to Newton's fidelity to it, was convinced that Jewish chronology was a deliberate corruption.

56 Whiston, 1725, pp. 247–50.

Mosaical Account will stand miserably loose, upon that Bottom." In response to his interlocutor's questioning, Nicholls conjectures that the population of the world in 1696 amounted to 960 million—three times larger than Petty's estimate. Very well, his opponent responds, but if doubling occurs every 360 years, then AM 1404, shortly before Exodus, the world's population stood at 3.75 million, with the Israelites comprising half; and 720 years earlier there needed to be over 900,000 people—twenty-nine years after the Deluge! Nicholls responds by invoking Petty's method of progressive doubling, but his opponent is hardly convinced: "This seems to be all Banter," he cries.

> You make your Periods as you please, and Mankind must either double or treble as you have a Mind to it, to serve your Hypotheses. Sir, I believe the Generation, like the Age and Stature of Mankind, is governed by a steddy unalterable Law, and it is not to be turn'd about to go either fast or slow like a *Dukes-Place* Clock. I find all of you when you have but a new Hypothesis to advance, will take Nature as well as your Bibles by the Nose, and lead them which way you please, to serve a turn.[57]

Nicholls refuses to budge: This is not "an empty Hypothesis, but a necessary Truth to confute the Calumnies of Unbelievers against the Mosaical Books; which is not only consonant to the tenour of those writings themselves, but to experience and good reasoning." God's special benediction to Noah must compel belief that the "blessing took effect," and reason dictates that when the world was empty there was urgent need to fill it, a need that later abated. Moreover, he avers, the recent invention of all arts and sciences, and especially the late invention of writing, without which mankind could certainly not exist from all eternity, further substantiates the short duration of the world. Otherwise one must assume "an eternal Race of the most stupid Blockheads imaginable." Not surprisingly, Nicholls summarily dismisses the claims regarding the great age of Chinese and Egyptian empires on the grounds that "the pure assertions of Nations as to their Antiquity, without good History to support them, have always been very little regarded."[58]

Drawing evidence from population to combat notions concerning the eternity of the world became rampant by the second half of the seventeenth century, not only in response to the erudite arguments by the likes of La Peyrère and Vossius, but also in response to a possible implication of Cartesian mechanism, which insinuated the eternity of matter. "If the World had been eternal," John Wilkins asked in the early 1670s, "How comes it to pass, that it is not every-where inhabited and cultivated?" Extrapolating from the fact that much of the globe was still unknown and that the rate of multiplication in the new world was slow, he concluded: "the beginning of the world was much about the time mentioned by Moses for the Creation of it; whereas had it been Eternal, it must long ere this have been over-stocked."[59] Two decades later

[57] Nichols, 1696, pp. 78–79.
[58] Ibid., pp. 80–95.
[59] Wilkins, 1675, pp. 72–73.

Richard Bentley devoted his third Boyle lecture to refuting the likelihood that an infinite number of generations had come and gone, or that the world was eternal. The former idea he considered to be rife with contradictions and paradoxes, the latter he damned as outright atheistic and "repugnant also to matter of fact." Jewish and Roman history as well as contemporary bills of mortality, Bentley argued, attest to the "gradual increase" of mankind notwithstanding periodic retardation owing to wars, plagues, and other calamities. Conversely, even if one grants that billions of people are alive today, they must nonetheless descend from a single couple—an irrefutable proof against the eternity of the world, he believed, be it millions of years old or 6000.[60]

By the time Newton embarked on his early history of mankind, the matter of population had become central to contemporary debates over the historicity of the Biblical narrative of remote antiquity. Owing to the inextricability of sacred and profane history, contemporaries had grown increasingly savvy about the inherent contradictions involved in all population estimates, be they impressionistic or grounded in rigorous quantification. Indeed, therein lay the dilemma. Maintaining fecundity and prodigious growth of mankind after Creation—necessary to explain from whence there came Cain's contemporaries and to populate the world by the time of the Flood—helped to confute those seeking to deny the universality of the Deluge. However, reasoning on that basis necessitated a corollary assumption: postdiluvian patriarchs must have proliferated at virtually the same prodigious rate as their forefathers. Conversely, inferences drawn by the new breed of demographers concerning the gradual increase of population could help confute those arguments in favor of the eternity of the world, but they also offered fodder to those eager to deny the universality of the Deluge. Along the same lines, the "gradual increase" argument—assuming the demographers' estimates to be correct—could not account for large kingdoms shortly after the Deluge. Espousal of the Septuagint chronology somewhat alleviated the dilemma, but only at the cost of accepting the proposition that a translation was superior to the original—a choice that was anathema to the devout.

One example must serve to illustrate the inherent contradictions that plagued these calculations. In his Commentary on Genesis 4:14, Matthew Poole sought to rebut the "conceit" of those who thought that there existed no more men on earth than mentioned in Scripture. Rather, he believed that upon "rational grounds and suppositions" one could credibly conclude the opposite, as God's blessing of Adam and Eve ensured that their progeny "might have two, three, four or more [children] at a time . . . which was then expedient for the replenishing of the World." Consequently, by AM 130, when Cain slew Abel, "how vast and numerous an off-spring might have come from *Adam*, none can be ignorant that can and shall make rational computation." In contrast, when he came to comment on Genesis 7:19, Poole silently reversed himself, as he wished to maintain the local nature of the Deluge. Poole considered it unnecessary to assume that the entire globe had been covered with water, for why inundate vast uninhabited regions? Following Stillingfleet and Vossius, he considered it quite unreasonable to suppose such an increase of mankind before the deluge, one

[60] Bentley, [1836] 1971, vol. 3, pp. 62–65.

that would have required the habitation of the entire globe—far beyond Syria and Mesopotamia.[61]

This conundrum underlies Newton's own interpretation of ancient history. His *Chronology* forcefully championed a Masoretic framework that allowed for sufficient time to synchronize sacred and profane history. Put differently, whereas proponents of Septuagint chronology sought to expand the boundaries of time in order to accommodate the great antiquity of ancient kingdoms, Newton contracted the historical record of those kingdoms in order to render more credible traditional reckoning. The choice was forced upon him. Newton's commitment to the Masoretic version of Scripture and to the lateness of civilization prevented him from entertaining the wholesale substitution of Septuagint chronology for Hebrew chronology. Nevertheless, Newton's quantitative and analytical frame of mind grasped both the contradictions inherent in traditional attempts to reckon historical time from Scripture, and the corresponding difficulties invariably raised by the demographers' quest for mathematical precision and solid reasoning. He found himself, therefore, constrained to follow the relatively limited number of commentators who determined the population during the first few centuries after the Deluge to be quite small—albeit none were as committed as Newton to synchronizing Biblical narrative with secular annals of ancient nations. This, we shall see, led the generalizing Newton to frame a theory concerning the development of civilization itself.

One predecessor in population matters whom Newton likely read was the eccentric Aylett Sammes, who argued in *Britannia antiqua* (1676) that the original inhabitants of England were the Phoenicians. Sammes differed from earlier historians of Britain who ascribed an immense antiquity to the Island, substantiating his dissenting opinion with the argument that Scripture makes no mention of an "extraordinary encrease of Mankind" after the Flood. The sons of Noah, Sammes insisted, begot only those offspring mentioned in the Bible. With a similar aim in mind, Sammes interpreted certain passages in Genesis to mean that Canaan, Armenia, and Egypt were thinly peopled as late as the times of Abraham and Jacob. Indeed, so meager was the number of Egyptians that "the single Progeny of Jacob" exceeded within two centuries the population at large—and yet accounted for, at most, two-thirds of the present population of London. As for Ninus' large army 250 years after the Deluge, Sammes found it a fable. How else might one explain that a century afterward the Assyrian empire had "grown so low, as with its Confederates, not to be able to resist Abraham and his Three hundred men?" He preferred to follow the historian Justin, who made Ninus rule later than Tanaus, King of Getes, who himself reigned about the time of the Argonauts and, according to Ussher, 1081 years after the Flood (1267 BCE).[62]

Newton owned a copy of Sammes' book and would reason in an analogous manner to him in the *Chronology*. Even more influential was the work of Joseph Mede, whose views on prophecy and the Apocalypse indelibly marked Newton's own thoughts on those topics. Newton also found in his copy of Mede's *Works* support for the argument

[61] Poole, 1683, vol. 1, sig. Dv, E1; Poole, 1669–1675, vol. 1, pp. 62–63, 99. Noteworthy is that Poole took care to endorse the localism of the Deluge only in the Latin version of his commentary.

[62] Sammes, 1676, pp. 7–8.

concerning the meagerness of the postdiluvian population. Mede composed two brief discourses on Genesis 10:5—"By these were the isles of the Gentiles divided in their land"—wherein he suggested that the reluctance of Noah's posterity to leave the place of "paradise" (Shinar) prompted God to sever the common bond of language that had united them as "the best means to force mankind into a *Plurality of Societies*." Mede was as certain that the dispersion of mankind occurred at Peleg's birth as he was certain that within a century of the Deluge "there were no *Families* so much encreased that they could be called *Tribes* or *Nations*." At most, he estimated, their number then amounted to 7000 men, in addition to women and children, "so that it is not like they took the whole world before them." Mede also established several "rules" that could help determine the location of the territories into which Noah's progeny dispersed. In order to conform to the language of Scripture, these needed to be "divided by Sea from *Egypt* and *Palestine*," though contiguous to the habitation of the Jews. The dispersion was also confined, in view of its small number, to a relatively limited area.[63]

Mede acknowledged that "the order of this Dispersion and Manner of this Confusion" was difficult, and that such difficulty had discouraged serious inquiry. However, he insisted, this was not "needless" knowledge, for without it "the Events and Complements of the Prophetical Blessings of the Patriarchs and the particular Predictions of the after-Prophets can never be understood." He remained troubled by the perceived early history of mankind, partly because he recognized that chronologers were reluctant to date the dispersion a century after the Flood, "as seeming too small a time for eight persons to multiply unto such a number as may be presumed to have been at the building of the Tower of *Babel*, and at their dispersion thence." It would have been helpful, he thought, if some portion of the Septuagint or Samaritan chronology was allowed to supplement the Hebrew one—and not only because such an extension of time would render large numbers of people more credible but, more importantly, because it would verify his reckoning of the Apocalypse![64]

Mede arrived at this conclusion in 1628, after he had an opportunity to reflect on the chronology of the Samaritan version of the Pentateuch that Archbishop Ussher had procured from Syria. Mede's excitement at his discovery is evident in his letter to Ussher, wherein he announced that if the excess of 311 years of the Samaritan chronology is added to the Hebrew one, then the fulfillment of the so-called Elijah's prophecy, according to which the world was to endure for 6000 years, would happen in 1736—he availed himself of Scaliger's date for the Creation—which is the very year in which "the 1260 years of the Beast's reign will expire." Such an admirable synchronization prompted Mede to contemplate whether the discrepancy between the Hebrew and Samaritan chronologies was "not ordered by a special disposition of Providence, to frustrate our Curiosity in searching the time of the day of Judgment." He proceeded to incorporate his thoughts on the matter into a brief essay, in which he inquired "Whether the computation of the years of the world before the Promise made to *Abraham* were or could be certainly known or not." If only it were possible to add 350 or so years to Hebrew chronology, he concluded, the "Tradition of the *Seventh Thousand year to be the Day of Judgment and of the glorious Reign of Christ*,

[63] Mede, 1672, pp. 271–76.

[64] Ibid., p. 896.

will . . . have good probability of Truth: Otherwise, I cannot see how possibly it can be admitted."[65]

Newton himself never wavered on the verity of Masoretic chronology—even in order to facilitate his computations of the Apocalypse—but he tacitly did come to share Mede's position on the gradual increase of population in the aftermath of the Deluge. Building on this position, Newton posited a theory of primitive mankind as a people barbarous in nature, whose refinement proceeded at an even slower rate than their numerical growth. Hence, he titled a projected first chapter of the *Chronology* "Of the times before the Assyrian Empire" and began it with a grim view of postdiluvian conditions in Egypt and Mesopotamia: "The first men . . . lived in caves of the earth & woods & planes well watered by rivers for feeding their heards & flocks." By degrees, "they cut down the woods & learnt to build houses & towns of brick in the planes & to live in society under laws & government." Elsewhere Newton furnished a more detailed chronology of the process: "The first kingdom of great extent was ye Assyrian & this grew great not long before ye captivity of ye ten tribes. Which . . . affords a good argument y^t ye world was peopled not long before ye age of Abraham." True, Scripture taught "that Nimrod was potent in ye earth in ye first age & reigned first in the land of Shinar & then in Assyria." But the world at the time "was not yet peopled [and] we are not to conceive y^t he founded a standing kingdom" comparable to those that had arisen later. Nimrod, Newton asserted, aimed "rather to people & plant wth his posterity" than reign over "a standing kingdom & therefore he . . . is said to found divers cities that is to plant families in several places wch at length grew into cities."[66]

This chronological and cultural framework permitted Newton to preserve the broad outlines of his early history of mankind. Noah's partitioning of the earth among his three sons was followed by the division and subdivision of those lands among succeeding descendants of Shem, Ham, and Japheth. Chus could still be made to encroach on Shem's territories, and plant "his sons round about the Persian gulf from ye furthest part of Arabia felix to ye furthest part of Carmania"—all according to what Bochart "has shewn" in his *Geographia sacra*." The custom of dividing territories among all sons—a cornerstone of Newton's "Original of Monarchies," to be discussed in the next chapter—explains why in the process of inheritance none was made lord over his siblings. Hence also "ye cause why the world in ye times of Abraham & for a good while after became divided into almost as many kingdoms as cities: & why we have so little history of those early times. For whilst kingdoms were so small their wars & other actions were scarce considerable enough to deserve lasting memory."[67]

In other manuscripts Newton developed further the correlation between the very slow rise of kingdoms and the scarcity of people. The fertile plains of the Tigris and Euphrates "seem to have been planted" with cities, but they were "thinly peopled before the days of Abraham." Less fertile areas became inhabited only later. The Holy

[65] Ussher, 1847–1864, vol. 15, pp. 407–8; see "Answer concerning a Discourse inferring from the Septenary Types of the Old Testament and other Arguments, That the World should last 7000 years, and the Seventh Thousand be that happy and blessed Chiliad," Mede, 1672, pp. 892–96.

[66] New College MS 361.2 fols. 238, 186.

[67] Ibid., fols. 186–86^v.

Land, for example, had been so "thinly peopled with villages in Abraham's days that he & Lot fed their heards & flocks wherever they pleased." Nor were Egyptian towns built earlier than those in the Holy Land, or lower Egypt "peopled much sooner." Newton submitted as evidence for the smallness of city-nations—or, more precisely, extended family units—the ease with which kings mentioned in Scripture routed each other, and how Abraham defeated the confederate armies of four kingdoms with a mere 318 men.[68]

As will become apparent in the next chapter, Newton evolved a forceful theory about the evolution of ancient kingdoms, one that presupposed a considerable population before the foundation of cities, let alone kingdoms. Newton repeatedly characterized the inhabitants of the territories upon which the great ancient monarchies were to rule as relatively small groups of nomads who lacked all organized lifestyle and structured government for centuries after the Deluge. After the dispersion, and shortly before Abraham was born, mankind began "overspreading" the earth "with villages, towns and cities," and these grew "into Kingdoms, first smaller and then greater, until the rise of the Monarchies." It was a gradual process, analogous to, and conditioned upon, the growth of population, and Newton frequently pointed out how long this primordial state of affairs persisted.

The Jewish patriarchs were a case in point. They wandered in Canaan with their tents,

> and fed their flocks wherever they pleased, the fields of *Phoenicia* not being yet fully appropriated, for want of people. The countries first inhabited by mankind, were in those days so thinly peopled, that four Kings from the coasts of *Shinar* and *Elam* invaded and spoiled the *Rephaims*, and the inhabitants of the countries of *Moab*, *Ammon*, *Edom*, and the Kingdoms of *Sodom*, *Gomorrah*, *Admah* and *Zeboim*; and yet were pursued and beaten by *Abraham* with an armed force of only 318 men, the whole force which *Abraham* and the princes with him could raise.

For its part, "*Egypt* was so thinly peopled before the birth of *Moses*, that *Pharaoh* said of the *Israelites*; *behold the people of the children of Israel are more and mightier than we*: and to prevent their multiplying and growing too strong, he caused their male children to be drowned."[69]

The same condition prevailed everywhere in the Middle East. Libyan antiquity was not "much older" than the antiquity of Europe, its inhabitants remaining nomads until Sesac's father forced on them city dwelling. Nor had the marauding Medes walled towns before their revolt against the Assyrians 267 years after the death of Solomon. In Canaan, cities were established only shortly before Joshua conquered the land, and these resembled European cities in that each "had its own king ... before they conquered one another." Multiplicity of small city-kingdoms abounded in the upper parts of Egypt until Misphragmuthosis' reign during Eli's time, while in

[68] Yahuda MS 25.2f fols. 22^{v}–23, 36–37^{v}.

[69] Newton, 1728b, pp. 185–86.

Syria, according to Scripture, "about 72 years after the death of *Solomon*, *Benhadad* King of *Syria* had two and thirty Kings in his army against *Ahab*."[70]

As a rule, however, Newton found it safer to describe the slow, and much later, peopling of Europe. His "Short Chronicle" opens with an account of how, toward the end of the twelfth century BCE, the expulsion of the Shepherds from Egypt drove certain of their captains to colonize Greece. "Before those days," Newton wrote, "*Greece* and all *Europe* was peopled by wandring *Cimmerians*, and *Scythians* from the backside of the *Euxine Sea*, who lived a rambling wild sort of life, like the *Tartars* in the northern parts of *Asia*." Newton reiterated this conclusion following his lengthy discussion of the origins of the several Greek kingdoms: "All these footsteps there are of the first peopling of *Europe*, and its islands, by sea; before those days it seems to have been thinly peopled from the northern coast of the *Euxine sea* by *Scythians* descended from *Japhet*, who wandered without houses, and sheltered themselves from rain and wild beasts in thickets and caves of the earth."[71]

Whatever else might be said of Newton's novel scheme, he strove for consistency as he deployed diverse bits and pieces of textual evidence, all in the end leading up to what we shall see was his conclusion that the first large-scale kingdoms arose only during the reigns of David and Solomon. Egypt: during the Israelites' sojourn, "the cities of Egypt . . . must not have been much further asunder than our villages." Mesopotamia: "almost all Mesopotamia continued in villages except Babylon until the Greeks assembled them into cities."[72] Babylon: "Nimrod founded a kingdom at Babylon & perhaps extended it into Assyria; but this kingdom was but of small extent if compared with the Empires which rose up afterwards."[73] Greece: "continued divided into small governments till after the days of Solomon." Italy: primitive until "above 300 years after the death of Solomon & other places of Europe more westward & northward were still more rude & barbarous."[74]

Clearly, then, a population-driven argument pervades the *Chronology*, albeit implicitly. For Newton never did provide elaborate arguments concerning the differential rates of population growth that captivated so many of his contemporaries, preferring instead to develop ones for the ages of kingdoms on the basis of what he thought to be reliable evidence concerning regnal lengths, as we shall see below. Other aspects of the argument—the recent invention of letters, and of the arts and sciences more generally—confirmed Newton in his opinion: "the Original of letters, agriculture, navigation, music, arts and sciences, metals, smiths and carpenters, towns and houses," he pronounced, was not older in *Europe* than the days of *Eli, Samuel* and *David*; and before those days the earth was so thinly peopled, and so overgrown with woods, that mankind could not be much older than is represented in Scripture."[75] We shall address these issues below. Here, we may conclude with the observation that Newton's

[70] Ibid., pp. 181, 185.

[71] Newton, 1728b, pp. 10, 184. For earlier versions, see New College MS 361.3 fols. 163, 199^{v}, 200^{v}, 241, 244.

[72] Keynes MS 146, pp. 22, 26^{v}.

[73] Newton, 1728b, p. 138. The biblical Nimrod, the "mighty hunter" of Genesis 10, was the founder of Babylon, among other Mesopotamian places, and the great-grandson of Noah. Newton reduced him to the overlord of a small kingdom.

[74] Keynes MS 146, pp. 19, 20.

[75] Newton, 1728b, p. 190.

subtle and effective deployment of population-based evidence resonated well with readers, though not necessarily for the same reasons. It could appeal to those who strove to preserve the authority of Masoretic chronology as much as to those eager to find a cogent synchronization of sacred and profane history. It might also appeal to those troubled by the inability to correlate large postdiluvian populations with demographic laws. Understandably, owing to the scarcity of solid and agreed upon "evidence" to account for prehistoric times, Newton's ingenuity could offer, at least for a season, a compelling theory that had a great deal to recommend itself—not least to the authority of the great man himself.

Newtonian History

Newton penned the "Original of Monarchies" in early summer 1702, following the death of William III, judging by his calculating in the manuscript that 635½ years had elapsed since the coronation of William I in 1066, and twenty-eight monarchs had ruled over England.[1] At the heart of the new project was the question of whether large empires had existed prior to the four monarchies mentioned by Daniel, namely Babylon, Persia, Greece, and Rome. If not, and Newton believed this to be the case, how to account for the seemingly well-documented counter examples of ancient empires, namely Egypt and Assyria? For Newton's understanding of rates of population growth combined with his reading of the Masoretic text of Scripture to entail a highly modified chronology. At first, he reasoned, the unceasing division and subdivision of territories among Noah's progeny engendered a growing number of local lords in the world—"till the earth was planted with innumerable scattered families not subject to any other Lords then their own common fathers"—with subsequent, natural population growth in each of the localities. Eventually, conflicts over dominion arose among these small entities, and then, to thwart aggression, tribes or neighborhoods found it necessary to "consult together for their common safety & chuse out wise & valiant men to lead them against their enemies & fortify places with walls within which should be many houses for the people to resort unto … in time of danger." This set in motion a chain of events that culminated in kingdoms: "These fortified places became the first cities & the fathers of families became ye elders of the city composing a Council with ye same legislative & judicial power over the whole body of all their families wch every father had before over his own apart, & the captain of their forces being the most honourable & potent amongst the Elders became their King."[2]

In the early 1710s, Newton refined his scheme, and began as well to manipulate evidence concerning the ages of the Egyptian and Assyrian empires to effect consistency with it. The developed theory that Newton produced again began with the immediate aftermath of the Flood. Population growth during that early, postdiluvian era led to the rise of towns inhabited by multiple families. Coexistence of many families,

[1] The coronation of William I took place in December 1066, suggesting that Newton made his calculation in May or June 1702. At this stage, he did not consider Oliver Cromwell as a ruler. His final list of English kings attests to his continual struggle with the issue. He jotted down Cromwell's name, only to remove it from his final calculation.

[2] Keynes MS 146 fols. 1–2.

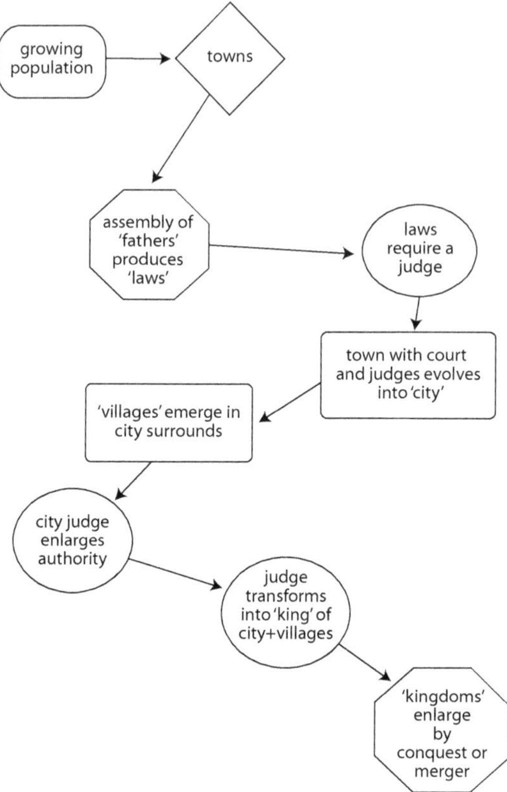

Figure 6.1. Newton's theory for the origin of civilization.

in turn, required governance, and so the "fathers of families" united to produce a common set of laws, eventually choosing a "judge" to adjudicate among the families on the basis of such agreed-upon laws. These governed towns later morphed into larger units, "cities," which were administered by councils of "elders." Villages arose in the vicinities of the cities, and the authority of the judges extended to them, thereby producing the first conurbation that could be termed a "kingdom." Over time these cities expanded further still, either by joining together under a common "captain," or through conquest, to form an empire.[3] Newton's scheme can be succinctly diagrammed (figure 6.1).

Once Newton had formulated the essential structure of his theory for the evolution of civilization, he treated it in much the same way that he had comparable issues in mechanics and optics. In those situations, as we have seen, Newton developed a basic structure, one that he held to firmly. That structure was in turn compatible with a spectrum of possible values for the variables that appeared in it, or even for the presence of specific terms. Consider again the example of fluid resistance. The assumption that resistance forces formed a set of additively separate terms remained

[3] New College MS 361.1 fol. 93^{v}.

basic and, barring extraordinary evidence to the contrary, essentially immutable. But the values of the distinct terms had to be obtained both from theory and from experiment, which Newton had undertaken in the 1680s. The situation in respect to his theory for the evolution of civilization was similar in structure, albeit different where its data set was concerned, since Newton could not, as it were, experiment in the usual sense with the past. He could, however, manipulate items from Scripture and, especially, from remnant Greek and other texts to fit the overall scheme, and that is precisely what he undertook to do in an extraordinarily elaborate effort to effect consistency among the several texts that he used under the controlling power of his evolutionary theory, which was itself grounded on population issues.

It is important to understand the evidentiary differences that distinguish Newton's scheme for civilization from his earlier work on prophecy as well as from his laboratory work. Whereas the interpretative scheme for prophecy could never be complete, or at least not until the Apocalypse, the data on which Newton drew for the origin of civilization were textually complete and therefore formed a closed system. But precisely because the system was closed (barring the discovery of new texts—and Newton had only textual sources in mind), it differed as well from his work on mechanics or on optics, which was open at subsidiary levels to the results of novel experiments. Nevertheless, even if Newton could not generate new data about the past, he could toil within the evidentiary base that he had to support a series of secondary assertions (secondary, that is, to the presumptive correctness of the basic system) that together revamped traditional chronology in major ways.

To see how Newton worked, we must again delve a bit into the forbiddingly elaborate details of his ways with particular events from the past. We begin once again with Newton's fundamental source, the one upon which he placed the greatest reliance, namely the Masoretic version of Scripture. As we saw in chapter 5, Newton discounted the reality of the kingdoms allegedly formed by Noah's descendants. Shem, Ham, Japheth, and their children were indeed bequeathed huge swaths of land, but the smallness of the population and the relentless division of land among all sons meant that whatever "kingdoms" Chus and Nimrod had established "became as small in those regions as in other parts of the world." Indeed, the earliest kingdoms "were single cities & those perhaps not so big nor so well peopled as o^{r} Villages." The reversal of this centrifugal process did not begin "till after the days of Abraham," and not before Moses' time did the first major city-kingdoms emerge—the continued growth of which ultimately gave rise to the "four monarchies."[4]

A source that may have left its mark on Newton's first forays into the "Original of Monarchies" was Robert Filmer's *Patriarcha* (1680)—either directly or through John Locke's powerful rebuttal of it in *Two Treatises of Government* (1690)—though neither work appears in the (incomplete) catalogue of Newton's library. The propositional format of Filmer's chapter titles would have appealed to Newton. Filmer's first chapter in particular would have resonated with him, entitled as it was "That the first Kings were Fathers of Families," in which the political theorist maintained the continuance of patriarchal power "even until the Egyptian bondage." According to Filmer, early nations comprised, not a multitude of "petty lords under some greater

[4] Keynes MS 146 fol. 3.

king," but numerous independent small kingdoms, commensurate with the minuteness of the population at the time: "These heaps of kings in each nation are an argument that their territories were but small, and strongly confirms our assertion that erection of kingdoms came at first only by distinction of families."[5] Nevertheless, whatever parallelism existed between the *Patriarcha* and the "Original of Monarchies," Newton rejected Filmer's monarchist political philosophy, which informed his scheme; Newton sought instead to argue for governments by representative councils prior to the amalgamation of petty states into kingdoms.[6]

In order to fit the requirements of his own theory, Newton turned to a close scrutiny of the Assyrian empire, and to the Egyptian monarchy, the latter of which had not been part of Daniel's prophecy. The rise of Assyria, he insisted, had to follow the contours of all cities; therefore, the process could not be considered complete much before the fall of the Kingdom of Israel. In postulating the lateness of the Assyrian monarchy, Newton not only considered the "laws" of population and Biblical accounts regarding the consolidation of power in the Near East to be on his side—"there is nothing in all antiquity wch can make this monarchy much exceed the bounds of an ordinary kingdom before the reign of Pul"—but the very language of Scripture, he believed, made it clear that the advent of the Assyrian menace with which the Israelites were threatened during Isaiah's time (8th century BCE) had been a thing "fresh in memory."[7]

Crucial for Newton's case was to prove that Nineveh, the Assyrian capital that Ninus (Nimrod) built as early as the second century after the Deluge, had not been a truly large metropolis—a prerequisite for the existence of an empire. True, Newton admitted, Nineveh "had been long a populous city," and "perhaps it might have been one of the greatest cities in the east." Nevertheless, he insisted, "it grew not up before the reign of Pul to that extent of Dominion wch was called the kingdom of Assyria & accounted one of the great monarchies." Besides, prior to Pul's reign (coincident with the time of Menahem, the 16th King of Israel) "there [was] not one word of this Monarchy in all the scriptures." Before the days of Jonah, the Bible never referred to the rulers of the city as "kings of Assyria; only as "kings of Nineveh," testimony, he believed, of its status as a city-kingdom rather than an empire. What's more, there must have existed a "greater number of kingdoms wch went to make up the body" of the Assyrian monarchy—and for which no records survived—since "almost all Mesopotamia continued in villages except Babylon untill the Greeks assembled them into cities." Newton found corroboration of his theory that a late-in-coming Assyrian monarchy had quickly usurped numerous small city-states, in the fact that following the collapse of the short-lived empire, those cities promptly reverted to their former state, "recovered their ancient liberty, and lived for a while under their proper laws, being every where divided into … little polities."[8]

In subsequent iterations of his attempt to invalidate the antiquity of the Assyrian Empire, Newton explained away the explicit description by the prophet Jonah of

[5] Filmer, 1991, pp. 8–9.

[6] Manuel, 1963, pp. 123, 286–87.

[7] Keynes MS 146 fols. 3^{v}–6.

[8] Keynes MS 146 fols. 5^{v}–6^{v}.

Nineveh's grandeur: "an exceeding great city of three days' journey" (Jonah 3:3)—a description widely interpreted to denote an area measuring sixty miles in circumference.[9] At some point Newton was tempted into using numbers. The greatness of the City, about 230 years before its fall, he equivocated, "was rather in extent of ground then multitude of people." For proof, he singled out another verse in Jonah (4:11): "that great city" was said to have been inhabited by "six score thousand persons that cannot discern between their right hand and their left hand." In accordance with current opinion, Newton interpreted the figure to refer to children under two and—perhaps aware of Gregory King's estimate that children under five comprised 15 percent of the population—he applied a 1:6 ratio in order to set Nineveh's total population at 720,000. Newton contrasted this seemingly imposing figure with the far more spectacular population of Thebes during the reign of Ramesses II: 700,000 men of war, so it was said, which, according to the 1:4 ratio that Halley had shown to exist between men of war and total population, implied a total population of nearly three million. Accordingly, Newton concluded, "we may reccon that the Kingdom of Assyria was now [ca. 840 BCE] grown a considerable one but not yet arrived to its greatness." Ultimately, Newton thought better than to use such a comparison in the *Chronology*. Instead, he ever-so-slightly distorted Scripture. Nineveh became a "a city of large extent, but full of pastures for cattle, so that it contained but about 120000 persons"—a manipulation that did not escape the attention of some critics, as we shall see.[10]

In establishing the origins (and short duration) of the Assyrian monarchy, Newton relied heavily on Sir John Marsham's *Chronicus Canon*. Marsham had made a number of useful points for Newton in respect to Assyria. Neither Scripture nor the monuments of the Egyptians, he noted, made any mention of the Assyrians prior to the collapse of the Egyptian empire in Asia—namely, following the death of Sesostris. Marsham therefore identified Pul as the founder of the Assyrian empire, an identification embraced by Newton.[11] Marsham's erudition appeared sufficient to Newton to dismiss Ctesias' claims regarding the antiquity of the Assyrian empire and to privilege the reckoning of Herodotus, who allotted 520 years to it. Herodotus, and Marsham more forcefully, also disputed the early dating of Semiramis' reign, with the former stating she reigned five generations before Nitocris, the supposed mother of Nabonides, the last Babylonian king.[12] The concurrence of the erudite Henry Dodwell may have also helped fortify Newton in his opinion. Dodwell's *A Discourse Concerning Sanchoniathon's Phoenician History* presented a devastating critique of the authenticity and credibility of Sanchoniathon's influential work. Given his own embrace of the Phoenician historian during the 1680s, Newton would have considered himself chastised by Dodwell's sneering at critics who had found Sanchoniathon to be "the *Faithfulest*, and *Antienest*, and consequently the most *useful Heathen Author*

[9] Newton himself, still subscribing to traditional reckoning, cited this verse as proof that Nineveh "grew immensely" after Nimrod had moved the capital of his huge empire to Nineveh. Yahuda MS 16.2 fol. 71^{v}.

[10] Yahuda MS 25.2f fol. 28; Newton, 1728b, p. 270; Halley, 1693.

[11] We know Pul as Tiglathpileser III, who reigned in the third quarter of the 8th century BCE.

[12] Marsham, 1676, pp. 503–23; Herodotus, 1975, vol. 1, pp. 127, 229. During the late 1680s Richard Cumberland composed an essay refuting Marsham's dating of Assyrian chronology, but did not publish it. Cumberland, 1724, pp. 231–61.

that was extant within the Memory of Learned Ages," or serviceable "in clearing several *Historical* and *Philosophical* passages of the *Old Testament*." Nevertheless, Newton would have found in Dodwell's treatise a powerful endorsement of Herodotus' late dating of the origins of the Assyrian empire, as well as of Semiramis' reign. And though we lack hard evidence for Newton's familiarity with Dodwell's critique, once he left behind his project on the origin of religions he refrained from citing Sanchoniathon again.[13]

The "original of kingdoms," Newton had asserted in 1702, had been much the same elsewhere in the East. The history of Greece and Italy demonstrated that the partition of patrimonies continued uninterrupted "untill there was no room for division"; only then did the small territories begin the process of unification "into greater polities." As late as the Trojan War, the Greek invading force comprised twenty-nine distinct, small nations, "each under the command of its Prince, & some (being not yet grown into single kingdoms) under the command of more Princes then one." Not surprisingly, Newton viewed the process of amalgamation to continue for nearly another millennium, until the successors of King Perdiccas of Macedon finally conquered "all Greece & ye east [and] erected the third Monarchy." Italy, too, had followed a similar process. The torturous manner through which Rome ascended to prominence, Newton wrote, furnished an "unquestionable record" of "how difficult it was to unite the divided cities into one polity."[14]

Having demonstrated how the cases of Greece and Rome conformed to the pattern he had established for the growth of empires, Newton lamented the poor state of these nations' records: "the Europeans had no Chronology ancienter then the Persian Monarchy. And whatever Chronology we have now of ancienter times has been framed since by reasoning & conjecture." His reasoning and conjecture was based on the fact that the Greeks, lacking records, "estimated times past" strictly by genealogies, which they computed without anchoring them to any established Era. As for the Latins, the Gauls burned their records six decades before the death of Alexander the Great, while the earliest Roman historian flourished only a century after Alexander's death. It followed, therefore, that a more exact reckoning of the ancient history of Greece and Rome necessitated the rectification of their chronologies, especially since "all nations before they began to keep exact accompts of time have been prone to raise their antiquities & make the lives of their first fathers longer then they really were." Such a "humour has been promoted by the ancient contention between several nations about their antiquity," Newton explained, affecting not only Egyptian, Chaldean, and Assyrian analysts, but even the seventy-two sages who allegedly translated the Old Testament into Greek had succumbed to such a convention and "added to the ages of the Patriarchs"[15]—an argument that also served to strengthen Newton in his allegiance to Masoretic chronology.

[13] Dodwell, 1691, pp. 1–2, 4–5 and passim. The treatise was first published in 1681, as an appendix to Dodwell's *Two Letters*. The *Discourse* received a second printing in 1691. In later years Thomas Hearne noted on several occasions Dodwell's continued efforts to prove that Semiramis could not have been Ninus' wife. Hearne, 1885–1921, vol. 1, p. 247; vol. 2, p. 233.

[14] Keynes MS 146 fols. 7–11, 14.

[15] Keynes MS 146 fol. 14.

Newton proposed to rectify chronology, in part by introducing what he thought to be a more powerful tool than words by themselves, for unless they derived from the Masoretic Scripture or could otherwise be traced reliably from version to version, these were ever subject to exaggeration and change over the millennia. But words that could be construed numerically, Newton was convinced, had a better chance of providing reliable data than their quantitatively barren siblings. To do that he introduced novel forms of evidence. The later of the two "monarchies" manuscripts contains several astronomically based assertions. Although the earlier version lacks claims of that sort, both manuscripts contain calculations based on regnal lengths. In the first, from mid-1702, Newton at one point tabulated the years from the time of Aesculapius to "the middle of the reign of Artaxerxes Longimanus ... when Hippocrates flourished."[16] He found the interval to be 481 years. That was based on there being seventeen generational "intervals" by, as he put it "the father's side," but eighteen "by the mother's."[17] He took "25 or 30 years to a generation," and a cut halfway between these (27.5) indeed produces his 481. That, he concluded, meant the interval from the mid-point of the period when Artaxerxes and Hippocrates "flourished" produced a date about fifty-five years after Solomon's death. Since Newton always took Solomon's death to have occurred in 980 BCE, the date for Aesculapius had to be about 925 BCE (and for Artaxerxes and Hippocrates, therefore, 444 BCE). This was near the time for the Argonautic expedition (and so long before the Greek city-states had achieved their later status as highly organized civilizations) because Hercules, "whose sons were at the Trojan war," flourished at the time of Aesclapius, and both Hercules and Aesculapius were themselves Argonauts.[18]

We'll examine below Newton's considerations of regnal lengths, but it is worth pointing out here that in 1714 or thereabouts he altered his earlier computation. In the later manuscript Newton obtained 508, not 481, for the time interval, because he there took "28 or, at the most, 30 years to a generation," again assuming male primogeniture. Taking a cut, i.e., twenty-nine years to a generation, produced 507.5, which Newton rounded to 508. This would seem to change the date for Aesculapius (and so for the Argonautic expedition) to about 900 BCE from 925. But Newton had also changed his mind about how to do the calculation. Thirteen years before, he had not specified a date for the mid-reign of Artaxerxes, though he must have taken it to be about 444 BCE, as we just saw. What he now did was to take the date for Artaxerxes and Hippocrates to be "the beginning of the Peloponesian war, at which time Hippocrates began to flourish." Moving back 508 years placed the Argonauts at forty-two years after Solomon, or 938 BCE, instead of the earlier 925. Why the two changes, first in the date for Artaxerxes and Hippocrates, and second in the length of a generational interval?

[16] This would be Artaxerxes I, whose reign as king of Persia is today assigned to 465–424 BCE. Hippocrates of Cos dates from ca. 460–370 BCE.

[17] A male count would begin with the son of a king; a female count would begin with the mother of the son and so would be one interval greater.

[18] Keynes MS 146 fol. 17. Newton did not here mention the Argonauts, but remarked that the destruction of Troy followed by a generation, or "about 88 years after the death of Solomon," putting the destruction at 892 BCE. On fol. 19 he placed the Argonauts "35 or 40 years older than the taking of Troy," so around 927–932 BCE.

Consistency with the new astronomical arguments that, we shall see, he had by then extensively developed, required the change. Here we need note only that his astronomical calculation in the later of the two *Monarchies* placed the Argonautic expedition about forty-five years after Solomon's death, whereas his earlier computation, based on regnal lengths, had it at fifty-five years. Ever concerned to match his several calculations as closely as possible, Newton could have altered the generational interval to accommodate the astronomically required date, since (leaving Artaxerxes and Hippocrates at 444) an interval of twenty-eight years would work fine. However, Newton was leery of using a single number for the generational interval; he clearly preferred to take a cut between a lower and an upper bound. Since he seems to have felt that thirty years remained the best upper limit, to produce twenty-eight without rounding up he would have to have used twenty-six for his lower limit. That would no doubt have worked, but in the meantime Newton had apparently also worried about the date for Artaxerxes and Hippocrates. Better to take it at a well-defined moment, he seems to have reasoned (the start of the Peloponnesian war), than an abstracted midpoint for their "flourishing." That date was set at 431 BCE.[19] Given this, the generational interval had to be readjusted to about twenty-nine years, entailing a new lower limit of twenty-eight, given the fixed upper one of thirty.[20]

Here we find Newton working in very much the same way that he did in the laboratory when generating measures that had theoretical significance. Consider what he did when revising for the second edition of the *Principia* the account of fluid resistance to a body's motion that he had produced for the first edition. New experiments, as well as other considerations, required changes to the respective contributions of the several terms for resistance. Newton now concluded on the basis of these new experiments that almost all of the resistance was due to fluid inertia.[21] Smith aptly characterized this way of working. Newton, he argued, would produce a theory that was supposed exact under appropriate circumstances, after which he would map phenomena to idealizations generated under the theory's constraints and implications. He continued thereafter by "focusing on the discrepancies between these idealizations and the real world, under the demand that every such discrepancy be physically significant, as judged from the point of view of the theory."[22]

The analog here of an exact Newtonian theory is his fundamental scheme for the evolution of civilization, which entailed a radical change in chronology. That, in turn, required the generation of a subsidiary series of generalizations concerning the sequence that led from isolated villages to fully fledged kingdoms. These subsidiary ele-

[19] See, e.g., Strauchius, 1704, p. 296. Strauch set the date of the war's onset at Julian 4283, hence 431 BCE, which is what Newton always used (Newton, 1728b, p. 41).

[20] Strictly speaking, Newton needed 45 years after Solomon, not the 42 that this new period produced. However, this was merely a matter of rounding: to obtain 45 with an initial date of 431 BCE and 17.5 generational intervals requires 28.8 years per generation. Newton obviously wanted his upper and lower estimates to be whole numbers—anything more precise would clearly be absurd—and so he had used 28 for the lower limit, producing a mean of 29, which is only 0.2 years off an astronomically required generational interval.

[21] A claim that turned out to be utterly wrong when d'Alembert years later demonstrated that an inviscid fluid exerts no forces whatsoever on a body moving through it.

[22] Smith, 2005, p. 134.

ments were highly idealized, and Newton went to considerable efforts throughout his analyses to expand the evidence for them.[23] Discrepancies arose here, and some of Newton's text was aimed at handling them. One example concerned a problem with the traditional date for the Argonauts (and so for the Trojan War) that arose when, as we shall see, astronomy was brought into the analysis. That was especially significant, because the late emergence of populous, highly developed kingdoms required that the Greeks could not until after the unification of Egypt have been sufficiently organized to carry out a large-scale, trans-Mediterranean campaign, nor could Troy then have been a true city. Consequently even comparatively minor dating inconsistencies could raise questions about the fundamental claim. Newton accordingly reworked his subsidiary social theory for the length of a generational interval, and he altered as well the point within a historical period (viz. the Artaxerxes and Hippocrates date of "flourishing") to which the interval was to be applied.

When in 1702 Newton sought to establish a radical new dating for the fall of Troy—about eighty or ninety years after the death of King Solomon—he was not yet driven by any desire to use the new chronology as proof for the primacy of Jewish civilization per se. That aspect of his system would follow several years later. Rather, Newton's new dating effected consistency with Marsham's identification of Sesostris as Sesac who, according to Scripture, invaded Judaea in the fifth year of Rehoboam's reign. To that end, undisputable dates drawn from Jewish history, like Solomon's death, served as benchmarks against which Newton's new dating system could be measured. Establishing the era of Troy was intended to be the cornerstone of Newton's new system. Subsequent Greek and Roman chronology could then be synchronized to it. The new dating enabled Newton not only to rectify these chronologies but, more importantly, to set Egyptian dating upon new foundations—since in Newton's mind Greek and Egyptian chronologies were interdependent: the foreigners who became the founders of the various Greek nations had been Egyptians, or Egyptian émigrés. To ascertain the time of their arrival in Greece would ascertain as well the identity of their family members and the sequence of events that brought about their departure from Egypt.

The "great antiquity of the kingdom of Egypt," Newton acknowledged, "makes it difficult to give an Acc[t] of its Original" yet, "some footsteps there are thereof in History." And this is precisely what the revised version of "The Original of Monarchies" (ca. 1714) sought to do: to delineate and expand upon such "footsteps."[24] Newton's confidence in being up to the task was grounded as much in what he considered to be Scriptural evidence as in his ability to extract from the adorned myths of the Greeks' crucial historical facts about the Eastern nations. He remarked already in the 1680s

[23] Compare, for example, the 1702 account of pre-kingdom Italy (Keynes MS 146 fol. 93[v]) with its later expansion (New College MS 361.1 fol. 11). The earlier account is limited to a single paragraph with about a hundred words that begins, "So also Italy consisted of many small dominions & those arose out of smaller" but cites no source for the brief data provided (concerning the "dominion of the Latines" at the time of Romulus). The latter begins "The rise of kingdoms in Italy was like that in other places." Though also a single paragraph, the latter account runs to over a thousand words and provides much more detailed evidence for the claim, drawing on Strabo and Dionysius Halicarnassus. The evidence provided by the latter, Newton wrote, constituted an "unquestionable record."

[24] New College MS 361.1 fols. 91, 97.

that "as the antiquity of the use of letters was greater among Chaldeans, Egyptians and Phoenicians than among the Greeks, the memory of things which those peoples retained was much more ancient. So that if it were possible to distinguish their scattered traditions from the fictions of the Greeks, and compare them with each other, I would hope that the history of the times that immediately followed the flood, could be written up with some degree of truth."[25]

Newton never wavered from his postulation that the great kingdom of Egypt, just like its Assyrian counterpart, came into being gradually, centuries after the fragmentation of the vast territory bequeathed to Mizraim (Menes). To support that claim, Newton seized on Scriptural accounts of the Israelites in Egypt. Case in point: the Egyptian kingdom under which the Israelites lived in servitude. It "comprehended the territories of Pithom & On [Heliopolis] & the regions between them & the Nile & the land of Zoan [Tanis] on the other side the Pelusiotic mouth of the river." Nor could it scarcely have been larger, Newton reasoned, for the Israelites were scattered throughout the land "in two days time ... to gather stubble in stead of straw" (Exod. 5.12, 14). As to the smallness of the kingdom's population, Newton cited Pharaoh's fear that the Israelites would grow "more & mightier then his people ... [w]hich is an argument that Egypt then consisted of several small kingdoms of wch this was but one." Equally to the point, the manner in which the Egyptians stored their harvests during Joseph's time reinforced Newton's image of the Egyptian cities as clustering "not much further asunder then our villages & by consequence as numerous & small as the ancient cities of Syria & δήμοι of ye Medes & Greeks. Which is an argument that the first constitution of Egypt was like that of other nations. For these cities like the δήμοι of Greece united under Common Councils & thereby grew into kingdoms."[26]

While Newton believed that Lower Egypt comprised multiple kingdoms during Moses' day, he ascribed to the territories south of the Nile Delta a swifter process of amalgamation. He cited approvingly Marsham's determination of the division of Egypt into four kingdoms—Thebes, This, Memphis, and Heliopolis—following Menes' death, before concluding: "the kingdom of Thebes swallowed up those of This & Memphis about the days of the Patriarch Jacob & thereby became the greatest kingdom in the world then known to ye Greeks."[27] Marsham, however, held that Egypt remained divided into four kingdoms until the Shepherds conquered the country—with the exception of Thebes—a generation after Jacob's death. Newton asserted a much later date for the conquest. In fact, he had convinced himself that the Shepherds (Hyksos) were none other than the Canaanites whom Joshua had expelled. Fleeing westward, the Canaanites conquered Lower Egypt, where they continued to rule until their expulsion in the days of Eli and Samuel.[28]

[25] Yahuda MS 16.2 fol. 52.

[26] New College MS 361.1 fols. 97, 130, 134; Keynes MS 146 fol. 22. Elsewhere Newton also commented that 370 years after the Flood, "the majesty" of the Egyptian Pharaoh "was not then grown so great but that he was inclinable to marry the supposed sister of Abraham the father of a lord of nothing more then his own family." MS 361.3 fol. 203^{v}.

[27] New College MS 361.3 fol. 93^{v}. In subsequent reiterations Newton would revise such an early date for the Theban kingdom — based on Homer's extolling the city "for riches & greatness."

[28] New College MS 361.1 fols. 130, 137, 140; MS 361.3 fol. 63; Newton, 1728b, pp. 9–11, 199, 201.

That interpretation proved significant in determining the chronology of the first unified Egyptian empire which, Newton believed, Scripture itself warranted. Whereas Samuel, he noted, "by one single victory over the Philistims restored liberty & a long peace to Israel," by the time that Saul ascended the throne a generation later, "the Israelites were again in bondage & the Philistims wth a very numerous army made a new vehement & lasting war upon Saul & David & could not be subdued till they had been beaten in many battels." Surely, Newton reasoned, events in Egypt occasioned a reverse migration by the Canaanites, which accounts for the sudden swelling of the Philistines' potency:[29]

Newton's radical interpretation of Near Eastern history went well beyond his identifying the Canaanites as the Shepherd kings or his exceedingly late dating of their reign. Newton also differed from most commentators in rejecting the widely held belief that the Canaanites expelled by Joshua were the Phoenicians who subsequently transplanted themselves throughout the Mediterranean. Samuel Bochart's massive *Geographia sacra* forcefully and influentially promulgated this theory. Bochart conjectured that the Canaanites, ashamed of their name because of the curse that Noah had pronounced against their father—as well as fearful of Joshua's relentless zeal in following God's injunction to annihilate the Canaanites—promptly changed their name to Phoenicians, and under that appellation colonized the Mediterranean, introducing Eastern learning wherever they went.[30]

Newton would have none of this. While acknowledging that some Canaanites settled in North Africa, he rejected the notion that they reached Europe at that time or that they might be identified as the harbingers of Greek culture. That role he reserved for the Edomites. Nevertheless, during the late 1690s Newton was still trying to reconcile his theory with elements of Bochart's interpretation. Case in point is his manipulation of an important piece of evidence that Bochart and his followers used to substantiate the equivalence of the Canaanites and Phoenicians—the account given by Procopius in his *History of the Wars* of the two pillars he claimed to have seen in the Numidian town of Tigisis (present-day Ain el Bordj in Algeria), upon which was engraved an inscription: "We are they who fled from before the face of Joshua, the robber, the son of Nun." Newton opened a projected chapter on "the Empire of Egypt" with a typical manipulation of the passage. When Joshua drove out the Canaanites, he wrote, they fled into Egypt and North Africa, where they erected two pillars with the inscription: "We are Canaanites & flee from the face of Joshua the robber"—substituting "Canaanites" for Procopius' "they"—Newton capped that with a rhetorical question: "Was not the kingdom of the sheppherds in Egypt erected by these Canaanites?"[31] Though confident of his identification, Newton initially offered a linguistic explanation for the widespread conflation of the Canaanites with the Phoenicians: "Phenicia is called the land of Canaan because peopled by Canaan." Or, more

[29] When the Philistines "found themselves too weak for the Hebrews & the shepherds being besieged in Abaris found themselves too weak for the Egyptians, the Philistims & Shepherds agreed to assist one another & the Philistims thereupon received the shepherds into their territories & joyntly with them made war upon & subdued the Hebrews till Saul revolted & he & David by a tedious & difficult war recovered the liberty of the Hebrews." New College MS 361.1 fols. 98^{v}, 142–43.

[30] Bochart, 1681, p. 341 and passim; Gale, 1672, part 1, p. 20.

[31] New College MS 361.3 fol. 63. See also fol. 246^{v}.

expansively: "we now give the name of Phœnicia only to the country in the North of Palestine wch was subject to the kingdom of Tyre yet anciently the Land of Canaan was also called Phœnicia & the Cananites Phœnicians."[32] Along similar lines Newton sought to resolve the confounding of the Phoenicians with the Egyptians. The inhabitants of both nations, ran the argument, "were in those days pretty much intermixed, many Phœnician shepherds being left in Egypt & many Phœnicians flying thither from David & many Egyptians being afterwards left in Phœnicia by Sesostris."[33]

Newton accepted that a certain number of Shepherds who had been driven out of Egypt in Eli's day—notably Cecrops (mythology's Athenian king, who instituted worship of the gods), Lelex (king of Laconia, husband to a nymph), and Inachus (king of Argos, son of Oceanus and Tethys, grandson of Gaia and Uranus)—proceeded to establish colonies in Greece. However, he claimed, their contribution to Greek culture was very limited. These Shepherds "brought in no manual arts unless perhaps of building houses of Clay or brick & cloathing men with the skins of Beasts." It was left to the Phoenicians who came with Cadmus during David's time to introduce letters and arts. Newton did not find this at all surprising. As he commented curtly in the *Chronology*, the Shepherds simply "minded not Arts and Sciences."[34] Regardless, they could not have arrived in Greece earlier "because Cadmus brought in letters & it is not likely that any thing done in Europe could be remembred before above three generations before the use of Letters." Newton's dispensing with the possibility that the Shepherds (Canaanites) could have arrived to Greece in Joshua's time explains his more assertive appropriation of Procopius in the *Chronology*. In addition to once again manipulating the content of the Tigisis inscription, Newton further embellished Procopius' account, since the Byzantine historian had mentioned the pillars while discussing the early history of the Phoenicians, stating explicitly that the inscription was written in the Phoenician tongue. "By the language and extreme poverty of the *Moors*"—ran Newton's creative gloss—"described also by *Procopius*, and by their being unacquainted with merchandise and sea-affairs, you may know that they were *Canaanites* originally, and peopled *Afric* before the *Tyrian* merchants came thither."[35]

Further to substantiate his version of Egyptian chronology, Newton utilized once more the comparative framework that permitted him to argue for an evolutionary process common to all nations, evidence from one thus applicable to others. Egyptian cities, for example, "were at first free absolute & independent like the first cities of other nations but in time of common danger united in common Councils for their safety like ye cities of Greece & set up captains over their armies wch in time became absolute kings for the common councils of the Greek cities were set up in imitation

[32] New College MS 361.3 fols. 186, 171^{v}, 210. Newton cites Bochart as his authority on these matters. Newton thought so already in the 1680s: "The Canaanites, who in the wars of Joshua migrated from the whole land of Chanaan into Africa, were called 'Punic or Phoenicians,' and the Septuagint Translators use Phoenicia for Canaan, and therefore Phoenicia and the land of Canaan are words with the same meaning. Hence the first Phoenician, or Father of the Phoenicians, is the same as Chanaan, Father of the Chanaanites, here called Chna." MS 16.2 fol 74.

[33] New College MS 361.3 fols. 136^{v}, 211.

[34] New College MS 361.3 fols. 200^{v}, 51, 197^{v}; Newton, 1728b, p. 82.

[35] New College MS 361.3 fol. 31^{v}; Newton, 1728b, pp. 200–201; Procopius, 1916, vol. 2, pp. 287–89.

of those set up before in Ægypt & the remains of such councils continued in several parts of Ægypt till the days of Herodotus." He further conjectured that these councils "seem to have laid the foundation of the Nomi or Provinces into wch Ægypt was divided & by consequence to have equalled them in number." And if the division of Egypt into Nomi occurred during Sesostris' day, as Newton came to believe in the early 1700s, it stood to reason that the Egyptians and Phoenician colonizers brought along with them institutions and traditions that had been prevalent in Egypt two or three generations before Sesostris.[36]

Newton's revisionist history sought to establish that Egypt had not become an empire before the 10th century BCE—indeed, that demographic and sociocultural conditions for more than a millennium after the Deluge made it impossible for any empire to evolve. "Cities grow great & populous in proportion to their dominion," Newton pronounced. When "ye Thebes by expelling the shepherds became lords of all Egypt, their city would grow greater then before, but not arrive to its greatness till they conquered the nations round about & built it accordingly."[37] And yet, while Sesostris' Egypt merited the appellation "empire," Newton did not consider that empire to be on a par with the four great monarchies of Daniel (Babylon, Persia, Greece, and Rome), owing to its short duration—grounds similar to those Newton had used to exclude Nimrod's kingdom from the list.

During the 1680s, when he still adhered to traditional chronology, Newton had credited Nimrod with the establishment of the largest and longest-lasting empire of the ancient world, ranging over not only Mesopotamia, but larger, and more populous, regions in the Mediterranean.[38] As the *Chronology* gestated, his opinion altered. "We must allow that Nimrod founded a kingdom at Babylon & perhaps extended it into Assyria," he acknowledged. Nevertheless, that kingdom "was but of small extent if compared with the Empires wch rose up afterwards. And if it had been greater yet it was but of short continuance"—owing to contemporary customs of inheritance. Even more important, Nimrod's kingdom was fundamentally different from those great monarchies that emerged later, for Nimrod "conquered not to reign over nations, but to plant his posterity in a good soile, according to ye example of his father Chus & the laws of those ages."[39] Analogously, Sesostris "invaded & spoiled Judea, Syria, Asia minor & all ye kingdoms round about, there being no kingdoms in those days great enough to oppose him. And this was the first great Conqueror we read of in ye world. But whilst he dissolved not ye kingdoms, he conquered but only spoiled them & made them tributary, he founded not a durable & well united Empire like the four wch followed, & therefore his kingdom is not recconed among them."[40]

Noteworthy is Newton's brief flirtation with the attempt to claim that there existed no early Chinese empire either, on grounds similar to those he used to reject an Egyp-

[36] Yahuda MS 41 fol. 28; Keynes MS 146 fols. 22–23.

[37] New College 361.2 fol 128.

[38] Yahuda MS 16.2 fol. 71^{v}.

[39] New College MS 361.1 fol. 128; Yahuda MS 25.2e fols. 60–61; Yahuda MS 25.1f/b fol. 1; Newton, 1728b, p. 268.

[40] Yahuda MS 16.2 fol. 23^{v} (later hand than the rest of the manuscript).

tian monarchy. According to Newton, the chronological tables that Philipp Couplet appended to his *Confucius Sinarum philosophus* (1687) clearly revealed China's division "into many kingdoms & that these kingdoms continued warring with one another & conquering one another till they became reduced into seven kingdoms." Around 230 BCE Xi Hoan-ti, one of the seven kings, conquered the others and founded the Chinese empire. However, since the new emperor ordered all books, except medical or juridical books, to be burned, "there are no histories extant in China but what were written about seventy years after this conflagration." Had any books "escaped the flames," Newton added on reflection, they might have related histories of a particular kingdom, not of China at large. Equally significant, not only did the Chinese way of writing become perfected only after the birth of Confucius (ca. 551 BCE), but the "historical books now most in repute amongst the Chinese are void of Chronology." On the basis of these facts, Newton dismissed the "story" that Hoan-ti founded a monarchy in China as early as 2697 BCE as nothing more than "a fable invented to make that Monarchy look ancient"; as for the length of reigns attributed to early Chinese emperors, they were simply "too long for the course of nature."[41]

In all likelihood, Newton's attempt to fit ancient Chinese history into his chronological scheme derived in part from Giovanni Domenico Cassini's "Doubts concerning the Chinese Chronology," which the French astronomer had contributed to Simon de La Loubère's *Du royaume de Siam* (1691, English translation 1693). In this brief essay, Cassini forcefully challenged the astronomical competence of the Chinese—attested to by their total reliance on the Jesuits in rectifying their calendar—as well as the astronomical basis of Chinese chronology. Specifically, Cassini noted that the great conjunction of the five planets, purported to have been observed by the fifth emperor—who allegedly ruled between 2513 BCE and 2435 BCE—actually occurred half a millennium later; so, too, the purported observation of the winter solstice in 2341 BCE (during the reign of the seventh emperor) also needed to be postdated by 500 years. For his part, La Loubère pronounced that the list of Chinese kings, stretching back beyond the Deluge, could not be accepted as true: "it may be presumed that in the succession of Kings, which they give us, they have put those who have reigned at the same time in diverse Provinces of *China*, when it was divided into several little Feudatary States under the same Lord."[42]

Newton did not own a copy of *The Kingdom of Siam*, but he would have found the arguments of both Cassini and La Loubère conveniently summarized in William Whiston's *A Short View of the Chronology* (1702), which the author had given him. Whiston availed himself of the two Frenchmen in order to demonstrate that "Chinese Chronology, when rightly understood, is exactly agreeable" to Masoretic chronology. Accordingly, he posited that all of the Chinese annals were false. He then summarized Cassini's arguments, before pronouncing that the "duration of the reigns and Lives" attributed to early Chinese monarchs could "by no means agree to the standard of Human Life in those Ages." Hence, Whiston deemed it "very probable" that

[41] New College MS 361.1 fols. 80–80^{v}; Yahuda MS 25.2e fol. 68^{v}. Newton's copy of Couplet's book shows signs of dog-earing. Harrison, 1978, p. 124.

[42] La Loubère, 1693, pp. 254–58, 252.

the lists of Chinese rulers should be taken to describe collateral, not successive, reigns—analogous to Marsham's reckoning of Egyptian dynasties.[43]

It is difficult to gauge the extent to which the force of Daniel's prophetic vision informed Newton's belief that "before the reign of Pul & the beginning of the Olympiads there were no great empires in the world."[44] Certainly, in the last two decades of his life, Newton contemplated incorporating sacred chronology into his universal history. Yet, as Newton prepared the final draft of the *Chronology*, he carefully expunged prophetical references, rendering his book virtually secular in character. Projected chapters on the "Chronology of the Gospel," on "The Chronology of the Books of Ezra & Nehemiah," and "Of the sacred History of the Persian Empire," never made their way to the final version—though Newton did incorporate some of the material, in more innocuous forms, into other chapters.[45] More telling are his excisions for publication of prophetic passages that had found their way into historical contexts. When discussing Jewish reckoning of time in the *Chronology*, for example, Newton noted that from the mid-point of Solomon's reign to the "vulgar *Aera* of *Christ*," there had elapsed 1000 years which, for him, suggested a connection of Jewish chronology to "that of later times." In his preparatory manuscripts he added: "from the middle of Solomons reign to the birth of Abraham were another 1000 years." Newton probably considered it wiser to remove the second sentence as possibly raising the famous prophecy of Elijah that postulated that the world would endure for 6000 years.[46] Elsewhere, when commenting on the Babylonian addiction to astrology and divination, Newton cited testimonies by Herodotus and Strabo on the lewdness of Babylonian women. The citations survive in the *Chronology*, but not the gloss that Newton had originally composed: "this leudness voluptuosness sorcery & idolatry is alluded unto the description of the great whore of Babylon (Apoc. XVII, XVIII) & so are the rivers of Babylonia & the great deserts between Judea & Babylon."[47]

Newton's strategy is set in even sharper relief in his projected conclusion to the chapter on the Persian monarchy:

> Of all things wch happened in the time of the Persian Empire the most memorable was the invasion of Greece by Xerxes, with an army of two or three millions of people. And this is thus described by Daniel. "There shall stand up yet three kings in Persia & the fourth shall be far richer then they all: & by his strength through his riches he shall stir up all against the realm of Greece." And as Daniel has noted the principal action of the Persian Empire, so he has noted the principal actions & changes in the Empires of the Greeks & Latins, & has done it with more understanding then any Greek or Roman writer ever did, & therefore in the short account & general ideas wch I intend to give of those Empires his descriptions of them deserves to be regarded.

43 Whiston, 1702, pp. 61–62. The theologian Robert Jenkin also derived these details from Whiston's book. Jenkin, 1708, vol. 1 pp. 322–24.

44 New College MS 361.1b fol. 129.

45 Yahuda MS 25.1 passim.

46 New College MS 361.3 fol. 107.

47 Newton, 1728b, pp. 328–29; Yahuda MS 26.1b fols. 8–9; CUL MS Add. 3987 fol. 97.

Little wonder that John Conduitt jotted below this cancelled passage the following remark: "This is a proof that he intended his prophecies as a sequel to the Chronology."[48]

In retrospect, Conduitt may have been correct to perceive, as Manuel suggested, that the *Chronology*, "when joined and connected with Newton's history of empires and churches since Daniel, forms one complete, universal history of mankind, both sacred and profane, since the Creation."[49] Such a grand conception—if ever it existed—may not have been foremost in Newton's mind as he waded his way through the annals of ancient empires. Nevertheless, as his views evolved, and as his conception of shorter Egyptian and Greek chronologies took shape, Newton came to recognize the relevance of the theory concerning the primacy of Hebraic civilization, which the early Church Fathers had promulgated in imitation of earlier Jewish polemics against Hellenistic culture.

Justin Martyr and his disciple Tatian inaugurated this apologetic tradition, which sought to sustain the superiority of the religious and moral teachings of Christianity over pagan superstitions and depravity while, at the same time, underscoring the venerable Hebraic tradition that had inspired the philosophical and literary productions of the Greeks. In his "Hortatory Address to the Greeks," Justin noted the general agreement over the great antiquity of Moses, "our first prophet and lawgiver," before presenting choice passages from Plato—testimony, as he saw it, of the philosopher's embrace of "the doctrine of Moses and the other prophets regarding the only God" that he had absorbed during his sojourn in Egypt. The influence of Jewish sources extended to the "prince" of Greek poets as well, noted Justin, for Homer himself had "received similar enlightenment in Egypt." The Assyrian Tatian followed suit, albeit more briefly. Do not condemn the "Barbarians," he exhorted the readers of his "Address to the Greeks." For "which of your institutions has not been derived from the Barbarians?" Tatian focused primarily on exposing the errors and vices of the pagans. Intermittently, however, he chided the Greeks for claiming for themselves, "without reason," the invention of all arts and sciences. He also regarded the antiquity of Moses as conclusive proof for the primacy of the Judeo-Christian tradition.[50]

Though familiar with both addresses, Newton was undoubtedly influenced more by the weightier, and more systematic, polemics of Clement of Alexandria and Eusebius of Caesarea. The former devoted much of the *Stromata* to illustrating the ancientness of Jewish learning and religion, and to driving home the charge that the Greeks had pilfered shamelessly from the Jews everything of value in their culture. Greek philosophers, he thundered, were "thieves and robbers" who "received fragments of the truth" from Hebrew prophets and, without acknowledgment, "claimed these as their own teachings, disguising some points, treating others sophistically," only occasionally discovering something new. Clement delighted in citing the Greeks' own acknowledgments, Plato's especially, of borrowings from the "barbarians." Like Justin before him, Clement availed himself of the famous passage in the *Timaeus*, where Plato recorded the humbling of Solon by an old Egyptian priest, whom Solon had

[48] CUL MS Add. 3987 fol. 123. The first half of the paragraph can also be found in New College MS 361.3 fol. 53^{v}.

[49] Manuel, 1974, p. 92.

[50] Justin Martyr, 1867, pp. 285–328; Tatian, 1867, pp. 5–45.

attempted to impress with accounts of Greek myth and chronology: "O Solon, Solon," retorted the priest, "you Hellenes are but children, and there is never an old man who is a Hellene.... in mind you are all young; there is no old opinion handed down among you by ancient tradition; nor any science which is hoary with age."[51] Clement also considered as crushing the testimony of the Greco-Jewish historian of the second century BCE, Eupolemus, who, in the (now lost) *On the Kings of Judea* pronounced: "Moses was the first wise man, and the first that imparted grammar to the Jews, that the Phoenicians received it from the Jews, and the Greeks from the Phoenicians."[52]

Two centuries later, Eusebius' *Evangelicae Praeparationis* traversed much the same ground. He commented repeatedly on the derivation of Greek philosophy and theology from the "barbarians" (Egyptians, Babylonians, and Phoenicians) who, in turn, had received such learning—directly or indirectly—from Jewish sources. Typical is Eusebius' determination that "the doctrines of the Hebrews have been plagiarised" by Greek philosophers; the Greeks "are not only proved to have stolen the other branches of learning from Egyptians and Chaldees and the rest of the barbarous nations, but even to the present day are detected in robbing one another of the honours gained in their own writings." Nor did Eusebius ever tire of insinuating the indebtedness of those Greek philosophers who traveled to Egypt or Mesopotamia to the Jews they there encountered. Thus, he noted, Solon's sojourn to Egypt occurred just as the "Hebrews were again dwelling" there. Pythagoras, too, "became a disciple of the Egyptian prophets, at that time when some of the Hebrews appear to have made their settlement in Egypt, and some in Babylon." Elsewhere, Eusebius claimed that the Syrians who reputedly invented letters were none other than the Hebrews, domiciled in that part of Phoenicia that later became known as Judea—whence other nations derived their letters. Most importantly for him, the Greeks derived their very "knowledge of the God of the universe" from the Hebrews. Evidence for these borrowings was, for Eusebius, a matter of doctrinal comparison: "if the Greeks should be found to hold the same doctrines with the prophets and theologians of the Hebrews, you may no longer be in doubt who were likely to have borrowed from the others."[53]

These impassioned apologetics enjoyed great traction among early modern scholars, not least because of the authority that the writings of the early Fathers conferred on latter-day defenders of Christianity. In addition, the early Fathers had drawn on a variety of sources that had since disappeared. For example, John Selden's influential *De Jure Naturali et Gentium*—a work Newton studied carefully—attempted to bolster the Fathers' exaltation of the antiquity of the Hebrews and their important contributions to gentile learning. Pre-Christian philosophers, Selden argued, greedily consulted the writings of the Hebrews, who enjoyed "unparalleled reputation for wisdom" in antiquity. Indeed, he added, when the Greeks spoke of barbarians or Assyrians, they often meant Jews. In this context, Selden found significant Iamblichus' account of Pythagoras' association, while in Sidon, "with the prophets who were the successors of Mochus the natural scientist."[54] By this, Selden concluded, Iamblichus meant

[51] Plato, 1875, vol. 3, p. 606. See also vol. 2, pp. 233, 250.

[52] Clement of Alexandria, 1869, vol. 1 pp. 397, 409, 421, 451, 455–58; vol. 2, pp. 1, 274–301, 319–23.

[53] Eusebius, 1903, vol. 3, pp. 491, 502–3, 505, 516.

[54] Iamblichus was a neo-Platonist philosopher of the late third and early fourth centuries CE.

the Jewish prophets, successors of Moses. Equally significant, Selden thought, was the related testimony of Numenius (fl. second century CE) who, not only nicknamed Plato Μωσῆς ἀττικίζων (the Attic Moses), but viewed him as the philosopher who "stole *everything he taught about God and the world from the books of Moses*." Selden attributed the absence of explicit "traces" of Hebraic ideas in the writings of Greek philosophers to the complex process by which their ideas were assimilated:

> Nor is it any wonder that not very many explicit traces of Hebrew knowledge are found in the Writings of the Greek philosophers—indeed, virtually none occurs there that sufficiently retains the true nature of its Hebrew origin. For at that time Greek teaching was intermingled with the sects of Barbarian Philosophy (of which the chief was the Jews'), and Greek philosophy itself was shredded into parts, like Pentheus by the Bacchantes, and pulled apart into factions of opinion. As a consequence, both what was drawn from the fountains themselves and what was derived from them (so to speak) by the channels of streams flowed into an Ocean of such multifarious teachings that they for the most part took on its nature, keeping very little of their own, as well as to the splintering of Greek philosophy to numerous warring sects, which engendered "an ocean of indistinguishable teachings which mostly took on the same insipid character, instead of retaining any of their own."[55]

Selden's forceful endorsement of the Jewish origins of pagan philosophy in general, and of the synchronization of Moschus with Moses in particular, was widely shared by English scholars. Archbishop Ussher cited Numenius' opinion in his *History of the World*—under the date of Plato's death—while William Cave determined that the Church Father had demonstrated how all Greek poets and philosophers "had drunk deep of the *Fountains* of the Prophets, and had forced their best Doctrines and Opinions from thence, though subtilly altering and disguising them, to make them look more like their own."[56] English philosophers followed suit. Ralph Cudworth relied on Selden's authority when identifying Moschus as Moses in order to claim with "good Historical probability" that the atomist philosophy was older, and more pious, than either Democritus or Leucippus. Robert Boyle concurred, as did John Wallis who—amidst a sermon aimed at disclosing the Jewish origins of the doctrine of the Trinity from "the footsteps thereof yet extant in Heathen Writers"—accepted as uncontroversial the opinion that those heathen writers also borrowed much of their learning "from the Jews though much Disguised, and sometimes Ridiculed by them." Plato, in particular, Wallis perceived, had pilfered so much "of his Philosophy, History, and Theology from *the Jewish* learning, as that he hath obtained the title Μωσῆς Ἀττικίζων, *Moses disguised in Greek dress*." And why did the Hebraic contribution go unacknowledged? Because the Greeks found the name of *Jews* "odious" and, consequently, they invariably referred to them as Barbarians, Syrians, Phoenicians, or Egyptians.[57]

[55] Selden, 1665, pp. 10–28.

[56] Ussher, 1658, p. 198; Cave, 1676, pp. 23–24.

[57] Cudworth, 1678, p. 12; Wallis, 1691, pp. 99–100.

The mind-set that produced this near unanimity among early-modern scholars over the derivativeness of Greek learning manifested itself in the penchant to view the literary productions of the Greeks (and the Latins) as regurgitations of Biblical narratives. That view informed the identification of mythical figures as corrupted Biblical figures—a practice Newton followed as well. Some scholars carried the tradition to extraordinary lengths. Justin's expounding on Homer's indebtedness to Scripture, for example, inspired English scholars to vie in attempts to elaborate. Sir Walter Raleigh believed that Justin's authority sufficed to substantiate a blanket assertion: "Homer had read over all the books of Moses, as by places stolen thence almost word for word may appear." Half a century later James Duport—Regius Professor of Greek at Cambridge and Isaac Barrow's tutor—launched a project intended to corroborate Justin's opinion. Toward that end, his *Homeri gnomologia* (1660) sought to demonstrate just how Homer had "transposed not a little of the sacred scriptures into his poetry." In documenting what he considered to be a remarkable correspondence between Scripture and Homer's sententious expressions of the *Iliad* and the *Odyssey*, Duport followed in the footsteps of Zachary Bogan who, two years earlier, had found extensive parallelisms with the Old Testament in both Homer and Hesiod. Bogan's labors, in turn, were paralleled by those of his exact contemporary at Oxford, Edmund Dickinson, whose *Delphi Phoenicizantes* minutely documented how all Greek myths pertaining to Apollo's oracle at Delphi drew on the book of Joshua.[58]

Newton steered clear of these sorts of excesses. Just as he had been previously disinclined to enumerate correlations between Biblical figures and ancient deities beyond what his schema of the postdiluvian Four Ages absolutely demanded, so he proved sparing when he took on cultural diffusion. In fact, before embarking on the "Original of Kingdoms," Newton had shown little curiosity in the genesis of ancient learning. He assumed that astronomy flourished already during Noah's time—judging by Newton's insistence on the centrality of heliocentric cosmology to original religion—but the issue remained peripheral. Nor did he devote much attention to the antiquity of writing, even as he grounded his interpretation of the first Four Ages after the Deluge on what he considered to be the antiquity of recorded history. Toward that end, he cited approvingly Josephus' claim that the Egyptians, Chaldeans, and Phoenicians "preserved the memorials of the most ancient and most lasting traditions of mankind" as part of their determination to take "special care to have nothing omitted of what was [remarkably] done among them." Ignoring the polemical agenda informing Josephus' claim, Newton accepted Josephus' statement as demonstrative of the great antiquity of writing. Undoubtedly, Newton wrote, there existed a "certain common tradition ... which had been passed down in a clearer form" from Noah to Abraham, and then to Moses, and that tradition "was preserved for a long time among the Gentiles too." If, therefore, "it were possible to separate the scattered traditions of those peoples from the fictions of the Greeks and to compare them with each other, I would hope that the history of the times which immediately followed the flood could be written with some element of truth."[59] With this mind-set, Newton also accepted as

[58] Duport, 1660, sig. C4; Bogan, 1658; Dickinson, 1655. Raleigh, 1829, vol. 2, p. 179; Hobbes, 2008, vol. 1, pp. xlix–xl.

[59] Josephus, 1841, p. 785; Yahuda MS 16.2 fols. 49^{v}, 51–52.

unproblematic the tradition that credited Thoth (Mercury)—the second son of Osiris (Mizraim, son of Ham, Euhemerized)—with the invention of writing as well as of other sciences. But, again, he did so only in passing; both topics were tangential to his preoccupation with the history of the postdiluvian ages.

Nonetheless, once Newton became convinced of the need for a shorter chronology, he felt bound to address the antiquity of learning more seriously; his system now mandated the late onset of civilization. If humankind had propagated slowly, and if the process of territorial amalgamation had been violent and protracted, it followed that no propitious conditions for the cultivation of learning could have existed in the first centuries after the Flood. Newton's line of thought paralleled Edward Stillingfleet's reflection on the original condition of mankind. There had been "no *likelyhood* at all of any great *improvement* in *knowledge*," argued the future bishop of Worcester. Quite the contrary. Whatever knowledge the descendants of Noah initially possessed, it must have vanished in the course of their scramble for subsistence, and the subsequent struggle for dominion. Only once mankind had settled into commonwealths could there emerge the degree of leisure necessary for the arts and sciences to flourish—"without which all certain *histories* of their own former *state* must *vanish* and *dwindle* into some *fabulous stories*."[60]

Newton never doubted that rudeness and barbarism were the natural condition of the early centuries after the Flood. Most vividly drawn, perhaps, is his description of the state of Greece prior to the arrival of the Egyptian and Phoenician colonizers: "houses towns cities & kingdoms, altars temples & religious statues were first erected in Greece by forreigners & arts & sciences introduced while the rude & ignorant natives continued Canibals & lived in dens & caves of the earth like wild beasts & came not out thence but to mix with the forreigners & live in with them in houses & in towns but by degrees, as they could be induced to leave of their savage customes & become civilized." Before Cadmus arrived, he amplified elsewhere, "Greece & all Europe must have been in a very barbarous & uncivilized condition; even more barbarous & rude then the Americans were when we first discovered them."[61] Prometheus, an Egyptian contemporary of Osiris, who introduced letters and arts to Mount Caucasus, also taught the natives how "to build houses in the room of caves & dens in which they lived before." Ammon, for his part, followed his conquest of Libya ca. 1034 BCE with lifting "that people from a wandering savage life to a civil one."[62]

For Newton it stood to reason that similar conditions prevailed in the birthplace of humanity. A chapter entitled "of the times before the Assyrian Empire," for instance, opened with a commentary to the effect that the "first men after the Flood lived in caves of the earth & woods," and "by degrees they cut down the woods & learnt to build houses." Though generally vague about when exactly early people grew to transcend their primitive state, the frequency with which Newton harped on the sparse habitation of much of the area stretching from Mesopotamia to Egypt—well into the time of the Israelites' captivity in Egypt—attests to his estimation of an extended period of barbarism. Just as the Cretan Dactyls lived in the labyrinths of Mount Ida, he

[60] Stillingfleet, 1662, pp. 17–18.

[61] New College MS 361.3 fols. 163, 230. See also fols. 224, 244.

[62] New College MS 361.3 fol. 241; Newton, 1728b, p. 14.

argued, so the early Egyptians inhabited mountainous "syringes or subterranean vaults," a way of life common to Mesopotamia and North Africa as well. Even so, Newton found it impossible to determine when the Egyptian "quitted their way of living in Caves" in favor of houses and towns. Perhaps, he suggested, such a move occurred after they had invented the sowing of corn.[63]

Significantly, whereas boorishness as well as primitive living conditions persisted in Europe for nearly a millennium after the Deluge, Newton noted, the Egyptians' custom of digging their habitation into mountains resulted in their discovering metals, followed by their learning of tool forging, masonry, painting, and navigation. "It seems therefore," he concluded, "that most arts & sciences began in Egypt and were propagated thence to the Arabians and from them by the red Sea to the Chaldeans." Notwithstanding all this, Newton refused to credit the Egyptians with the discovery of writing. Before the expulsion of the Shepherds, Newton pronounced, "the Egyptians had no Letters, and their hieroglyphicks are not understood."[64] Though this was calculated to justify his decision to ignore Egyptian dynasties prior to the Shepherds' takeover of the country—a strategy dictated by his burgeoning new chronology—Newton struggled to marshal a supporting argument: "Some think that the letters wch Cadmus brought from Phenicia were originally from Egypt, wch is improbable because Cecrops Erichthonius & Lelex came from Egypt before without letters."[65] As for hieroglyphics, Newton appeared to share Stillingfleet's dismissive view of their historical value: "It was the solemn custome of the *Aegyptians* to *wrap* up all the little *knowledge* they had under such *mystical representations*, which were unavoidably *clogg'd* with two inconveniences very *unsuitable* to the *propagation* of *knowledge*, which were *obscurity* and *ambiguity*: for it not only *cost* them a great deal of time to *gather* up such *symbolical* things which might *represent* their *conceptions*; but when they had pitched upon them, they were *liable* to a great *variety* of *interpretations*." As Newton put it, the Egyptians might boast of a 9000-year-old antiquity, but "before the use of letters they could not write down the names of their kings. They could only represent them by Cyphers & write down their histories of the men represented by those cyphers."[66] Similar views of hieroglyphs were entirely common until the middle of the eighteenth century, and indeed even later.

By 1700, then, even as Egypt remained central to Newton's new chronology, the radical shortening of the duration (and number) of Egyptian dynasties precluded, for various reasons, his acceptance of the early invention of writing. To begin with, despite his abridging Egyptian history, Newton remained committed to the mythological structure that had informed his earlier schema—and this structure necessitated the ascription to the family of Sesostris a crucial role in the civilizing process of humanity. Second, Newton based his revised chronology on the assumption that written chronicles could not have existed before the time of Moses—lest Ctesias and Sanchoniathon be credited with reliance on genuine records—and "there could be no memory of things done above three or four ages before the first use of letters." Thus,

[63] New College MS 361.2 fol. 238; Yahuda MS 25.2f fols. 35–36.

[64] Yahuda MS 25 fols. 35^{v}, 37^{v}.

[65] MS 361.1 fol. 153; MS 361.3 fol. 251^{v}. Newton dated Cecrops' migration to ca. 1100 BCE.

[66] Stillingfleet, 1662, pp. 18–19; New College MS 361.3 fol. 136a.

for example, he rebutted Ctesias' inflated reckoning of the antiquity of the Assyrian Empire by contending that "before the use of letters the names & actions of men could scarce be remembred above 80 or 100 years after their death." As for the great Bacchus, who conquered much of the inhabited world, he must have "flourished in the times next before the Argonautic expedition. Had he been much older his actions would not have been remembred for want of ye use of letters."[67] What remains unstated here is the conviction that the long-drawn-out growth of nations ensured that no great deeds worth remembering could have occurred before the twelfth century BCE.

In the early years of the eighteenth century, Newton still hedged about ascribing to the Hebrews the invention of arts and letters. Despite his reliance on Josephus, Newton refused to avail himself of the latter's crediting the descendants of Seth with the discovery of writing and astronomy—discoveries they had allegedly inscribed on stone and brick pillars, one of which, Josephus claimed, still survived in Syria. Neither did Newton ascribe these discoveries to Abraham, notwithstanding his belief in a "common tradition" that passed from Noah to Abraham and Moses.[68] Instead, Newton consistently proposed an evolutionary process to the growth of learning, one that followed the contours of the gradual rise of great kingdoms. Accordingly, since Egypt gave birth to the first great empire as well as to the twelve major deities, it surely had been the cradle of civilization as well. Specifically, Newton pinpointed the expulsion of the Shepherds and the consolidation of Theban rule over Egypt as the great turning point in the history of the country as well as of knowledge.

The Shepherds, Newton wrote curtly in the *Chronology*, "minded not Arts and Sciences." In contrast, Ammon, his son, and grandson, were addicted to learning. In focusing on Ammon, Newton followed Plato's famous account in the *Phaedrus*, wherein Theuth (Mercury) is credited with the invention of both writing and the mathematical sciences. According to Plato's narrative, Theuth presented his inventions to King Thamus of Thebes (whose god was Ammon), and requested that the Egyptians "might be allowed to have the benefit of them." Newton cited the passage repeatedly during the 1680s, while working on the structure of postdiluvian Egyptian theogony. There, he identified Thamus with Ham (Jupiter Hammon), the father of Sesostris (Mizraim), while making Theuth (Thoth) the illegitimate son of Sesostris: the product of Sesostris' rape of Maia, daughter of his brother Phut.[69] A decade later, with a new chronology in mind, Newton rendered Ammon as the successor of Amosis—the first king of unified Egypt—who ascended to the throne during the reign of King David. Sesostris (Sesac) was Ammon's son, while Thoth served as Sesostris' secretary.

[67] Newton, 1728b, p. 166; New College MS 361.3 fols. 6, 101, 208, 256. Stillingfleet articulated the idea succinctly: Nothing preserved the memory of ancient times "but the most uncertain report of *fame*, which alters and disguiseth things according to the *humours*, and *inclinations*, and *judgments* of those whose hands it passeth through." Stillingfleet, 1662, p. 578.

[68] Josephus, 1841, p. 32. According to Josephus, Abraham communicated arithmetic and astronomy to the Egyptians. Ibid., p. 39. Newton's position was consistent with his earlier refusal to accept the identification of Moschus the Phoenician.

[69] Newton, 1728b, p. 82; Plato, 1875, vol. 2 p. 153; Yahuda MS 17.2 fols. 14^{v}, 15^{v}, 21^{v}; MS 16.2 fols. 11, 14, 15, 26, 31, 46, 55, 68^{v}.

Newton credited Ammon and his descendants with important contributions to the progress of astronomy, a pursuit they turned to as an outgrowth of their maritime interests: "When the Egyptians applied themselves to Navigation, that they might have the sea coasts by wch men had hitherto sailed & guide themselves in the middle of the seas by the Sun Moon & starrs, their kings & Princes & chiefly their Admirals applyed themselves to the observation of the heavens & study of Astronomy." Ammon, in particular, "was much addicted to Astronomy & from him the study descended to his children & grandchildren Hyperion Sasyches or Sesac Atlas, Typhon Mercury, Memnon, Prometheus." The admirals in question were Atlas—"eminent for his skill in this science"—and Antaeus who, in addition to serving as tutor to Bacchus (Sesostris), also "observed the course of the Moon wch was the hardest part of Astronomy." In general, however, Newton proved short on specifics. Little is heard of astronomy, he noted on one occasion, "before the Egyptians under Ammon & Sesac applyed themselves to observe the heliacal risings & settings of the stars & reduce the year to a certain length for the sake of Navigation." He also credited Ammon and his children with the division of "the Zodiak into 12 signes & 360 degrees & forming the Constellations & twelve Asterisms of ye Zodiak & naming them & the sun Moon & Planets after the kings & great men of Egypt." Notwithstanding Ammon's contributions, however, his adherence to a luni-solar year of 360 days meant that a new year needed to be invented "for keeping an exact account of time before Astronomy could be brought to any competent degree of perfection."[70]

The "first attempt" at such perfection Newton claimed to have detected during Osiris' reign, a claim that he failed to substantiate: Osiris, after all, also adhered to the antiquated luni-solar year of 360 days, which Newton took as evidence that astronomy still remained "in its infancy." Whatever progress astronomy made during Sesostris' reign was owing to Atlas' discovery of the "Globe wth its circles & constellations in memory of wch it was ever after painted upon his shoulders." Newton dated the discovery to a short time before the death of Solomon (979/980 BCE), in imitation of which Orpheus, "soon after the Argonautic expedition," formed the asterisms on the Greek globe.[71] A more significant development occurred during the reign of Memnon (Amenophis, Menes), the first king to rule over Egypt after the gods. In 892 BCE, five additional days were added to the calendar, thereby enabling the formation of a new year that commenced "on the third of April which was then the first day after the vernal Equinox according to the suns mean motion."[72]

Newton must have recognized the flimsy evidentiary basis for his stringing together a few vague allusions from Homer, Ovid, and Diodorus Siculus to evoke a coherent narrative on the state of ancient astronomy.[73] Consider his attributing specific astronomical accomplishments to Antaeus and Atlas. Little evidence exists for

[70] New College MS 361.3 fols. 25^{v}, 66^{v}, 99, 129^{v}.

[71] New College MS 361.3 fol. 46.

[72] New College MS 361.3 fols. 156^{v}, 61^{v}. Later in the manuscript Newton revised his date for the institution of a 365-day Egyptian year. He previously had it at 892 BCE, but subsequent consideration that, as Newton thought, "it is not likely that the æquation of the suns mean motion should be known in the infancy of Astronomy" required a recalculation of the time to "ninety eight years after the death of Solomon," i.e., to 881BCE (fols. 184^{v}–185).

[73] See, for example, Diodorus, 1939, vol. 2, pp. 279, 431.

any of this, beyond the fabulous allotment to Atlas of proficiency in astronomy, a proficiency that he allegedly imparted to Hercules in gratitude for the latter's rescue of the daughters of Atlas. So, too, Newton vacillated on the very identity of Antaeus and Atlas. Recall that in the "Original of Religions" Newton determined that the two names denoted the same person—Phut (or Neptune), the evil brother of Sesostris. When working on his history of kingdoms, in contrast, Newton was caught between his desire to retain the structure of his previous political framework and his effort to attribute specific learned discoveries to his protagonists. Thus, on several occasions he considered Anthaeus and Atlas to have been two different individuals. At one point he identified Atlas as Antaeus' son and Sesostris' nephew. On another occasion, he claimed that Atlas and Sesostris were brothers.[74] Intermittently, he returned to his original idea that the two names designated one and the same individual, borrowing an etymological construction he had frequently used beforehand. As he explained in the *Chronology*, where he announced his support of this solution: "even the name *Atlas* in the oblique cases seems to have been compounded of the name *Antaeus*, and some other word, perhaps the word *Atal*, cursed, put before it."[75] Newton was well aware how arbitrary all of this was. Following another attempt to argue that Antaeus' age, dominion, and actions "agree wth Neptune, he wrote, "But as its difficult to state these things exactly so its of small consequence."[76]

Newton moreover considered the precise level of Egyptian astronomical knowledge also to be a matter "of small consequence." What mattered to him most were not exaggerated claims regarding the profundity of Egyptian astronomy, but to establish the derivativeness of Greek learning—astronomy especially—from Egypt. He made his intention explicit on several occasions. Since Endymion had been an astronomer—as Newton argued when making a case for the lateness of the Aeolian kingdom—and since "the native Greeks in those days were ignorant of all arts & sciences we may reccon that his grandfather Æolus came with his family from Egypt in the days Eli." Similarly, when commenting on the Greeks' invention of Octaeris, and the subsequent introduction of Meton's nineteen-year cycle, Newton observed that "in doing these things the Greeks received light from the Egyptians."[77] Note that Newton could hardly argue for more than a modest progress of astronomy in Egypt in the age of Sesostris and Memnon, given that he regarded Greek astronomy on the eve of the Argonautic expedition to be in its infancy, too.

Already in the late 1690s, Newton entertained the possibility that the Egyptians were themselves indebted to others for their knowledge of arts and letters. As he admitted on one occasion, no records survived to explain how and why the Egyptian court had become so "addicted" to the study of astronomy. Nevertheless, "if room

[74] New College MS 361.3 fols. 25, 54^{v}, 156^{v}, 238^{v}, 248, 252^{v}; MS 361.1 fols. 124, 126; Newton, 1728b, pp. 23, 99, 165, 229–33.

[75] Newton, 1728b, p. 234; Yahuda MS 16.2 fols. 15^{v}, 38^{v}; MS 17.1 fols. 11^{v}–12; MS 41 fol. 20; 361.3 fol. 53, 54^{v}.

[76] New College MS 361.3 fol. 72.

[77] New College MS 361.3 fols. 5, 129^{v}. Meton discovered that nineteen tropical years is also very nearly a multiple of synodic months, thereby leading him to adjust the lunar calendar by introducing an intercalary month seven times during the period in order to keep a usual year of twelve lunar months in synchrony with the tropical year.

may be allowed for conjecture I suspect that the Merchants trading upon the red sea were the first that found out letters numbers & Arithmetic & observed the stars, these things being usefull in their trafique, & that when the Court of Edom fled from David into Egypt they carried these things with them to ye Court of Pharaoh, & on that account were there enterteined wth extraordinary favour." Analogously, having pronounced that nothing had been heard of astronomy before the day of Ammon and Sesac, Newton felt it necessary to temper the statement by adding: "unless the Constellations mentioned by Job were a little older." He followed the qualification with a digression on Job's habitation among the Edomites in the land of Seir, from whom he "might have [had] his Asterisms & the Egyptians some of theirs," inasmuch as the Edomites had navigated the Red Sea since the days of Esau. Elsewhere, while noting how Ammon had been "much addicted to astronomy & from him the study descended to his children & grandchildren," Newton conjectured that the arrival of the Red Sea merchants triggered his passion—albeit astronomy attained a greater perfection only during the reign of Osiris.[78]

These Edomites, according to Newton, came to be known as Phoenicians, under which name they civilized the Mediterranean world. Newton's identification rested partly on the authority of scholars such as Vossius and Bochart, and partly on a creative reading of the ancient sources. Vossius, for example, equated φοῖνιξ with ερυθρος (red) in order to suggest that the Phoenicians formerly resided by the Red Sea. That Sea, Vossius also pointed out, was named after Esau's nickname, Edom (red): hence the derivation ἐρυθραῖον, Erythraean Sea, and thus the transplanted Tyrians from the Red Sea were called by the ancients φοίνικες, Phoenicians.[79] Without citing Vossius, Newton pronounced Edom, Erythra, and Phoenicia to be "names of the same signification," a determination that supported his narrative regarding the Edomites: having been "conquered and dispersed by *David*," they "fled to the *Mediterranean*, translating the word *Erythraea* into that of *Phoenicia*, give the name of *Phoenicians* to themselves, and that of *Phoenicia* to all the sea-coasts of *Palestine* from *Azoth* to *Zidon*."[80] Newton further invoked Herodotus and Strabo to support his interpretation. Specifically, he alluded to a comment by Herodotus regarding the Phoenicians, who "had formerly dwelt on the shores of the Erythreaen Sea, having migrated to the Mediterranean and settled in the parts which they now inhabit." Contrary to Newton's insinuation, however, Herodotus had made it perfectly clear elsewhere in his *History* that by the Eritrean Sea he meant not the Red Sea but the Persian Gulf—as is evident in his description of the fall of the Euphrates and Tigris into that body of water.[81] Likewise, when citing Strabo's relating how "some report that the Phœnicians and Sidonians were colonies of the inhabitants of the Ocean & that they were called Phenicians [Punici] because the sea is red," Newton neglected to add one important fact: Strabo considered such reports to be unfounded.

Onto this novel interpretation Newton grafted an Abrahamic connection: "Arabia Petraea & Nabatea," he wrote in 1714, "as well as Phoenicia have been peopled by the

[78] New College MS 361.3 fols. 99, 149^v, 154, 156^v.

[79] Bochart, 1681, pp. 340–42; Gale, 1672, pp. 18–21; Vossius, 1668, p. 131.

[80] New College MS 361.1 fol. 145; Newton, 1728b, pp. 12, 96–97, 101, 104, 108–9, 207–12.

[81] Herodotus, 1975, vol. 1, pp. 3, 225, 237; Strabo, 1917, vol. 1, p. 157.

seed of Abraham besides the nations sprung from Keturah whom Abraham sent eastward."[82] In other words, learning emerged among the progeny of Ishmael and Midian, Abraham's sons—as well as among the descendants of his grandson Esau—for they were merchants since Jacob's time, and the exigencies of trade prompted them to invent writing and the mathematical sciences. Whence Newton concluded that by 1497 BCE the Midianites and the Edomites "on the borders of the red sea had letters & Astronomy."[83] Newton scrupulously avoided ascribing to Abraham himself the knowledge of writing, much less its inventions. To reiterate: a claim that letters were in use less than four centuries after the Deluge risked conferring greater credit on the annals of Manetho and Sanchoniathon, while jeopardizing Newton's intricate theory regarding the slow rise of civilization. Conversely, the invention of letters could not be dated too late in view of the venerable opinion of Saint Stephen: "And Moses was learned in all the wisdom of the Egyptians, and was mighty in words and in deeds" (Acts 7:22). Like most of his contemporaries, Newton considered Saint Stephen's statement to be definitive, and on several occasions he ascribed an early knowledge of writing to at least some Egyptians. While the Thebans and the Ethiopians used only hieroglyphics, he contended, "letters seem invented long before in the lower Egypt . . . for the Egyptians were learned before the days of Moses, & he being skilled in all their learning wrote the law in letters upon tables of stone & in books."[84] Hence the significance of 1497 BCE as the date by which the Red Sea merchants had possessed writing: it predated the Exodus (according to Ussher's chronology) by six years.

At times, Newton's insinuation of an Abrahamic genealogy to Egyptian learning became explicit. The practice of "writing was in use among the Israelites in the days of Moses & perhaps some ages before," he wrote apropos of his dismissing the exaggerated claims for the antiquity of the Assyrians and Chaldeans. This "consideration alone [the greater antiquity of Hebraic writing] is sufficient to give a reputation to the sacred history so much above the histories of Manetho, Berosus, Ctesias, Herodotus Megasthenes Diodorus Sanchoniathon or any other of the heathen historians."[85] In the *Chronology*, Newton articulated his views curtly and bluntly: the art of sailing "by the shoar with oars, in round vessels of burden" was first invented on the shallow waters of the Red Sea "by the posterity of *Abraham*"; the luni-solar year had been "derived from *Noah* to all his posterity"; "there is no instance of letters for writing down sounds, being in use before the days of *David*, in any other nation besides the posterity of *Abraham*." Not surprisingly, readers of the *Chronology* understood Newton to confer on the Jews alone the dignity of the first civilized nation. As Manuel aptly noted: "if Newton could not successfully trace the inception of *all* knowledge

[82] New College MS 361.3 fol 50. In the *Chronology* he even claimed that the "ancient *Brachmans*" were also the descendants of Keturah (Abraham's wife after Sarah's death), and that they were "instructed by their father in the worship of One God without images," after which Abraham sent them "into the east," and from their descendants the Persians derived part of their religion. Newton, 1728b, pp. 350–51.

[83] New College MS 361.3 fols. 121, 162; MS 361,1B fol. 154.

[84] New College MS 361.1 fol. 116; MS 361.3 fol. 6.

[85] New College MS 361.3 fol. 188. Fleetingly, Newton contemplated making use of Eupolemus' attribution to Moses of the invention of letters, judging by his transcribing the famous passage from Hugo Grotius' *De veritate religionis Christianae*: MS 361.3 fol 192.

to the children of Israel, he might at least restrict the inventions to the progeny of Abraham."[86]

Newton's malleable way with the historical record should be apparent by now. He approached the interpretation of ancient historical sources with the same degree of assuredness that he exhibited when deciphering the symbolisms of Biblical prophecies. Remarking on the extent to which previous scholars had been confounded in their attempts to synchronize Persian "prophane history" with the narrative of Scripture, he declared: "I shall not stand to recite other men's opinions, but propose as shortly as I can what I take to be ye truth."[87] Newton thereby reserved to himself the right to determine not just which historical sources were credible, but which parts of even these credible sources were nevertheless untrustworthy and to be discarded. Newton might appear to be following the common practices of contemporary scholars, yet his idiosyncratic, and exceptionally focused, chronology mandates a closer look at the manner in which he handled textual evidence as he grafted the new chronology onto his scheme.

[86] Newton, 1728b, pp. 208–10; Manuel, 1963, p. 118.
[87] Yahuda MS 25.1d fol. 3.

Text and Testimony

Three sources particularly exercised Newton as he proceeded to drastically abbreviate ancient history: the *Persika* of Ctesias of Cnidus, the *Aegyptiaca* of Manetho, and the Marmor Parium—key sources for the history of, respectively, Assyria, Egypt, and Greece.[1] Before the late 1690s, Newton rarely cited these sources; he certainly did not consider them problematic. Yet as his revised chronology took shape, he engaged critically with them, undermining as much as possible their credibility, and explaining to himself—as well as to others—his reasoning.

Ctesias composed the *Persica* in an effort to refute Herodotus' account of the Persian wars, and the polemical stance that colored Ctesias' historical narrative made it easier for Newton to point out seemingly egregious errors which, he insisted, marred the *Persika* to the point of rendering it worthless. He faulted Ctesias for dating the rise of the Assyrian empire within sixty or seventy years of the Flood; for ascribing to the Assyrian empire a duration of 1360 years, when Herodotus had already come closer to the truth by holding it to endure for only 500 years—although even this figure Newton considered exaggerated; for offering up names of Assyrian kings that bore "no affinity wth the names of the Assyrians mentioned in scripture"; for making Semiramis as old as the first Belus, despite Herodotus' correct placing of her five generations before Nitocris, the mother of Labynetus (Nabonidus), the last Babylonian King; for claiming that Ninus and Semiramis founded, respectively, Nineveh and Babylon, "whereas either Nimrod or Assur founded those & other cities without giving his own name to any of them"; for predating the destruction of Nineveh by three centuries and erroneously ascribing to the empire of the Medes a duration of 300 years, when "it lasted but 72." Most outlandish in Newton's mind was Ctesias' failure to mention the "true Empire of the Assyrians described in scripture, … tho much nearer to his own times." Together, such omissions exposed Ctesias' obvious ignorance of Assyrian antiquities, not to mention his unrestrained eagerness to stretch the epoch of the Assyrian empire "out of vain glory" and a desire "to please his reader." No wonder such a contemptuous view of Ctesias' history elicited repeated drubbings of the *Persika* as "fabulous" and allowed Newton to ignore it.[2]

Newton found the *Aegyptiaca* to be equally flawed. The names Manetho assigned to the Egyptian kings were often both confused and out of order so "that its difficult

[1] Interestingly, Newton felt no need to rebut Sanchoniathon.

[2] New College MS 361.1 fols. 95^{v}, 127–28; Keynes MS 146 fols. 6, 15; Newton, 1728b, 5, 265–67.

to find above two or three together in due order of time." Indeed, "it was frequent wth Manetho to repeat the same kings several times" as well as to multiply dynasties; to corrupt many names of kings; and to confuse kings with siblings or secretaries. Crucially, Manetho feigned his dynasties to be successive when, in fact, they were collateral. This sort of chaos naturally suggested the proliferation of numerous kingdoms in Egypt before they were all "swallowed up by the kingdom of Thebes" prior to Sesac's ascension to the throne. What Newton found most egregious in the *Aegyptiaca*, however, was Manetho's failure to mention Ramesses, under whom the Hebrews were kept in servitude. Not only did Manetho show himself ignorant "of the peregrination & servitude of Israel in Egypt," he mistook the Shepherds who conquered Egypt to be the Israelites. The failure to mention the kingdom of Ramesses convinced Newton that most of the kings enumerated by Manetho postdated Moses: "For he being a Priest of Heliopolis wch was under the dominion of Ramesses, it may be presumed that he would be most diligent & particular in the antiquities of his own country & therefore would not have omitted the kings of Ramesses & servitude of Israel had his records reacht so high."[3]

Even as Newton pounced on Manetho's ignorance and errors, he grasped the grounds that led him astray. As Newton argued in one of his manuscripts, following his conquest of Egypt in 343 BCE, Artaxerxes III carried away the Egyptian archives, thereby making it impossible for Manetho, writing some eighty years later, to consult any official records. More damaging still, alleged Newton, Manetho's treatise had "since been altered" by the Christian historian Julius Africanus, tampering that further diminished the *Aegyptiaca*'s reliability. Undermining Manethos' credibility allowed Newton to correspondingly magnify the credibility of Herodotus, who visited Egypt a century before Artaxerxes' conquest and whose history purportedly drew upon "the records themselves kept in the Temples of Egypt & recited to him by the Priests."[4] Significantly, in the *Chronology* Newton reversed himself. He now ascribed the removal of the archives to the Persian Cambyses II—who conquered Egypt in 525 BCE—in the aftermath of which "the Priests were daily feigning new Kings, to make their Gods and nation look ancient." Consequently, Herodotus' account, too, could no longer be considered based on original records. Nevertheless, determined Newton in his characteristic manner, Manetho, Eratosthenes, and Diodorus wrote at a time when "the Priests of *Egypt* had corrupted their Antiquities much more than they had done in the days of *Herodotus*."[5]

Newton developed a more respectful, albeit equally determined, critique of the Marmor Parium. Composed in the mid-third century BCE, the inscription included dated events from the time of Cecrops to the Archonship of Diognetus at Athens—ca. 1582–264 BCE. He considered the failure of the inscription to mention the Olympiads—or "any other standing *Aera*"—despite its composition sixty years after the death of Alexander the Great, as evidence that the scribe reckoned "backwards from the time then present." Hence, his dates could not have been very accurate. To redress the problem, Newton proposed a drastic solution: as elsewhere he asserted

[3] New College MS 361.1 fols. 98, 133–37; MS 361.3 fols. 164^{v}, 248, 254.

[4] New College MS 361.3 fols. 62^{v}, 239, 254.

[5] Newton, 1728b, pp. 244, 263–64.

that the Greeks set a king's reign equal to a generation, and three generations to a century. In reality, he claimed (we shall see how he worked reigns and generations in the next chapter), this is incorrect in that seven reigns correspond to four generations. Moreover, since the Marble records events that preceded the beginning of the Persian Empire, correcting the calculation starting from "the Conquest of *Media* by *Cyrus, An.* 4, Olymp. 60" backwards compresses the temporal extent in the ratio of four to seven. This method of shortening, Newton commented after furnishing several instances, "may be used alone, where other arguments are wanting; but where they are not wanting, the best arguments are to be preferred."[6] Newton also deployed his favorite theory: "The Greeks corrupted their Chronology before the Marbles were made, so as to add to the antiquity of all things done before the war of the Persians against them." In the published *Chronology* he muted somewhat this claim, at least for two Greeks. The greater credibility of Hesiod and Homer, he suggested, derived from their having written their works "before the *Greeks* and *Egyptians* corrupted their Antiquities."[7]

To these extended critiques of three major sources for the history of the ancient world, Newton added briefer appraisals of other surviving ancient fragments. One example among many will suffice. Julius Africanus relied on the lost genealogies composed during the second half of the sixth century BCE by the Argos mythographer Acusilaus in order to claim that Ogyges—the mythical king of Greece, during whose reign the eponymous "Ogygian Deluge" occurred—and Phoroneus (son of Inachus and king of Argos), flourished 1020 years before the first Olympiad. Newton contested the credibility of the source for no other reason than that "*Acusilaus* was an *Argive*," who "feigned these things in honour of his country." In fact, Newton added, "To call things *Ogygian* has been a phrase among the ancient *Greeks*, to signify that they are as old as the first memory of things"! Consequently, while *Inachus* may well have been a contemporary of Ogyges, "*Acusilaus* and his followers made them seven hundred years older than the truth; and Chronologers, to make out this reckoning, have lengthened the races of the Kings of *Argos* and *Sicyon*, and changed several contemporary Princes of *Argos* into successive Kings, and inserted many feigned Kings into the race of the Kings of *Sicyon*."[8]

It hardly needs saying that the authority of Scripture trumped all others. True, in general Newton preferred to marshal historical evidence and reasoned arguments in support of his positions. But the word of Scripture always loomed in the background, evident to contemporaries who knew their Bible pretty much by heart. Newton certainly cited Scripture, but usually as a historical resource, often together with pagan works. Occasionally, however, he invoked Scripture as the ultimate arbiter. Newton felt justified to reject Ctesias' "feigned" list of Assyrian monarchs, precisely because the names furnished by Ctesias bore no "affinity wth ye Assyrian names in scripture." Nor was it possible for Nineveh to have been the seat of an empire before the reign of Pul, since "there is not one word of it in y^{e} sacred history" about it. And why could

[6] Ibid., pp. 47, 120–21.

[7] Yahuda MS 26.2 fol. 16^{v}; Yahuda MS 25.1a fol. 10; CUL MS Add. 3987 fol. 18; Newton, 1728b , p. 102.

[8] Newton, 1728b, pp. 172, 45.

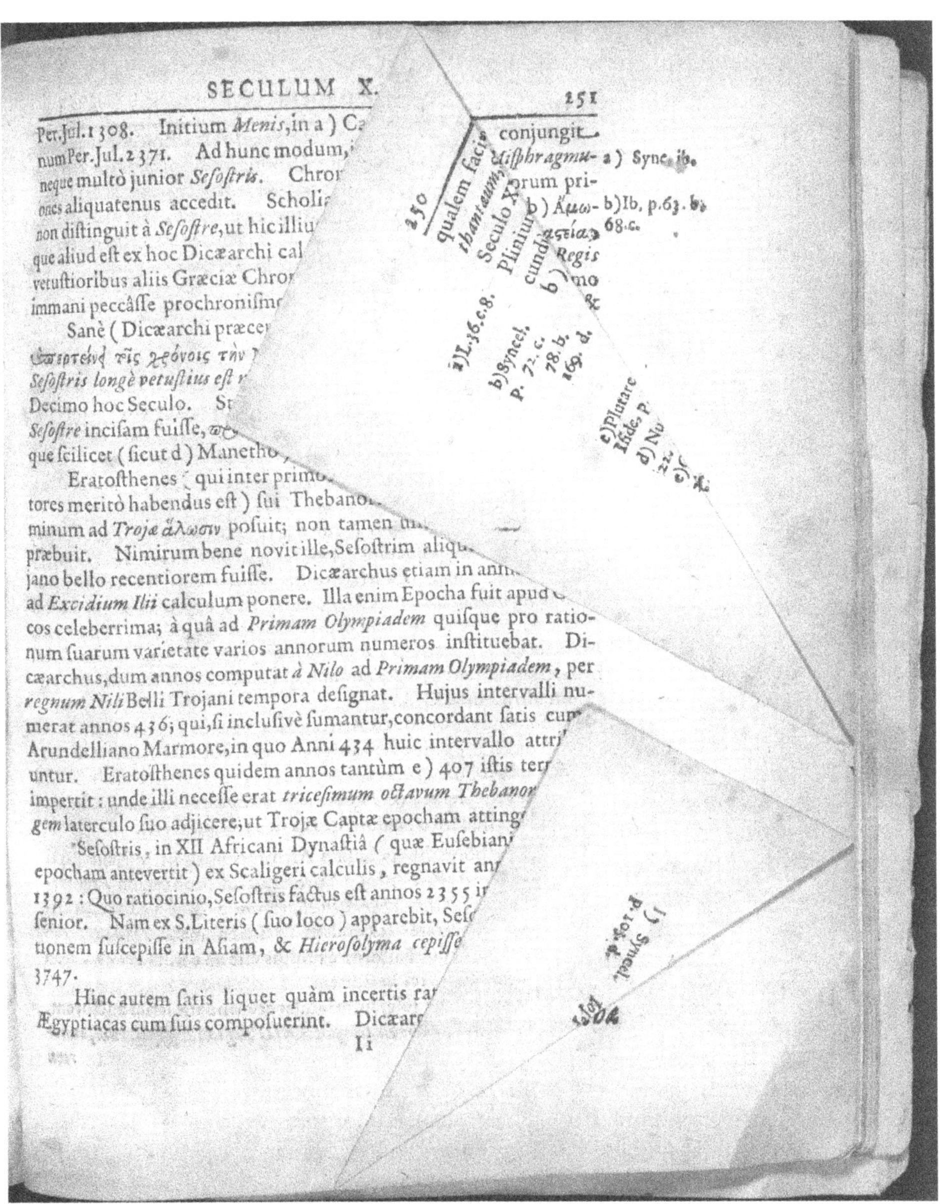

SECULUM X. 251

Per.Jul.1308. Initium *Menis*, in a) C…
num Per.Jul.2371. Ad hunc modum, …
neque multò junior *Sesostris*. Chron…
ones aliquatenus accedit. Scholi…
non distinguit à *Sesostre*, ut hic illiu…
que aliud est ex hoc Dicæarchi cal…
vetustioribus aliis Græciæ Chron…
immani peccâsse prochronism…
Sanè (Dicæarchi præce…
ὑπερτείνῃ τοῖς χρόνοις τὴν …
Sesostris longè vetustius est …
Decimo hoc Seculo. S…
Sesostre incisam fuisse, …
que scilicet (sicut d) Manetho …
Eratosthenes (qui inter primo…
tores meritò habendus est) sui Thebano…
minum ad *Trojæ* ἅλωσιν posuit; non tamen …
præbuit. Nimirum bene novit ille, Sesostrim aliqu…
jano bello recentiorem fuisse. Dicæarchus etiam in ann…
ad *Excidium Ilii* calculum ponere. Illa enim Epocha fuit apud G…
cos celeberrima; à quâ ad *Primam Olympiadem* quisque pro ratio-
num suarum varietate varios annorum numeros instituebat. Dicæarchus, dum annos computat *à Nilo* ad *Primam Olympiadem*, per
regnum Nili Belli Trojani tempora designat. Hujus intervalli nu-
merat annos 436; qui, si inclusivè sumantur, concordant satis cum
Arundelliano Marmore, in quo Anni 434 huic intervallo attri-
buntur. Eratosthenes quidem annos tantùm e) 407 istis ter…
impertit: unde illi necesse erat *tricesimum octavum Thebanor*…
gem laterculo suo adjicere, ut Trojæ Captæ epocham atting…
Sesostris, in XII Africani Dynastiâ (quæ Eusebian…
epocham antevertit) ex Scaligeri calculis, regnavit an…
1392: Quo ratiocinio, Sesostris factus est annos 2355 i…
senior. Nam ex S.Literis (suo loco) apparebit, Ses…
tionem suscepisse in Asiam, & *Hierosolyma cepisse*…
3747.
Hinc autem satis liquet quàm incertis ra…
Ægyptiacas cum suis composuerint. Dicæar…
Ii

Figure 7.1. Newton's dog-eared page on Sesostris from Marsham's *Canon*. By permission of the Linda Hall Library of Science, Engineering & Technology.

Sesostris be no other than Sesac? Simply because Scripture admits no other great conqueror between Abraham and Solomon![9]

On the rarest of occasions, Newton invoked the authority of a near contemporary in order to bolster his case. In his prophetical studies, he reserved that distinction for Joseph Mede. When he turned to the history of ancient kingdoms and to chronology he relied heavily, if not altogether uncritically, on Sir John Marsham—and for good reason. Of all the early modern chronologers, Marsham's mind-set most resembled Newton's own. Broad-ranging scholars, both Newton and Marsham sought to forge a novel system of ancient chronology grounded on the Masoretic text. Both, again, privileged Herodotus' account of Assyrian and Babylonian history over the chronicles of Ctesias, while radically compressing Egyptian history. Marsham's seeming success in doing so undoubtedly emboldened Newton in his own efforts to abbreviate the antiquity of early empires. Yet Newton's debt to Marsham is greater still. To Marsham he owes the centerpiece of his revised chronology: the identification of the Egyptian king Sesostris with the Biblical pharaoh Sesac (Shishak). In his defense of the antiquity of the Jews, Josephus, who downplayed Sesostris as a great conqueror, notably argued that Herodotus had mistakenly applied the deeds of Sesac to Sesostris.[10] But Marsham made the identity of the two central to his *Canon chronicus* (1672), and Newton's heavily dog-eared copy of the work clearly attests to the influence that Marsham exerted on both the structure and the details of the *Chronology of Ancient Kingdoms Amended* (figure 7.1).[11]

Marsham, and later Newton, struggled to fit Manetho's lists of Egyptian dynasties into their respective systems. Joseph Scaliger, who published the lists in his *Thesaurus temporum* (1606), deemed them to comprise a most "venerable monument." In fact, he considered Manetho to be far more authoritative on Egyptian affairs than Herodotus, notwithstanding Scaliger's great admiration for the latter. While the pious Scaliger remained perplexed that Manetho's dynasties stretched back some 5355 years prior to the defeat of Darius III by Alexander the Great—nearly 1700 years before the (Masoretic) date of Creation—he never wavered in his confidence in the authenticity of the lists. As he explained on one occasion, though the *Chaldaica* of Berossus and the *Aegyptiaca* of Manetho "could properly be criticized on account of the prodigious extent of the antiquity and long periods involved, yet both are not only to be kept but to be esteemed highly, for the sake of reverence for antiquity and because in the middle period the true events are contiguous with the fabulous ones." Scaliger's promotion of this seemingly impious opinion caused consternation among Protestant scholars, which in 1641 Gerard Vossius attempted to alleviate by suggesting that certain Egyptian dynasties were collateral, not successive.[12]

In his determination to defend Egyptian records from charges of falsehood, Marsham not only embraced Vossius' solution to the Manetho problem, but availed him-

[9] New College MS 361.1 fols. 18, 114, 127; MS 361.2 fol. 18; MS 361.3 fols. 136a, 208, 214; Keynes MS 126a; Newton, 1728b, pp. 5, 69–70, 220, 265, 267.

[10] Josephus, 1841, p. 233.

[11] Newton owned the 1676 Leipzig edition of the work. His copy is located at the Linda Hall Library in Kansas City, Missouri.

[12] Grafton, 1993, pp. 683, 708, 711–16; Vossius, 1668, p. 108.

self of another catalogue published by Scaliger: Eratosthenes' list of thirty-eight kings, from Menes to Amuthartaeus, who purportedly reigned altogether for 1076 years, beginning shortly after the Deluge. This "most venerable monument of antiquity," Marsham wrote, is "extraordinarily valuable for establishing the eras of the Egyptians."[13] Acting on his conviction, he utilized the list of Eratosthenes as a calibrator to which the dynasties of Manetho needed to conform. Marsham thereby managed to contract Egyptian history from Menes to Sesostris into a compass of 1400 years, a feat that elicited a boast from the chronologer that "if he could fit all the dynasties within the shorter period allowed by the Hebrew Bible, then he could obviously fit them into the Septuagint chronology."[14]

Newton accepted the broad outlines of Marsham's system and, judging by his well-worn copy of the *Canon*, helped himself to many of the particulars as well. For example, Newton highlighted passages in which Marsham noted that since both Apappus, twentieth king of Thebes, and Phiops, twentieth king of Memphis, reigned for nearly a hundred years—and were succeeded by kings who reigned for only one year, after whom reigned Nitocris, according to both canons—they must be one and the same. Newton incorporated the observation into his preparatory manuscripts for the *Chronology*, just as he helped himself to Marsham's excursus on the Egyptian calendar, or to his discussion of the rule of Nitocris over Thebes and Ethiopia. Newton may well have also borrowed from Marsham two additional arguments: the charge that Julius Africanus had tampered with the text of Manetho; and the argument that since Scripture makes no mention of an Assyrian empire before Pul, no such empire could have existed.[15] This considerable indebtedness accounts for the fulsome praise that Newton bestowed on "our great Chronologer, *Sir John Marsham*," both in his manuscripts and in print. On one occasion Newton came mighty close to claiming parity for the authority of Scripture and the authority of the erudite Englishman: "Scripture admits no other conqueror between Abraham and Solomon but Sesac. Our great Chronologer Sr John Marsham is also of opinion that Sesostris was Sesac."[16]

His indebtedness notwithstanding, Newton also rejected a great deal of the *Canon*. He did not accept the identification of Menes with Ham, nor did he subscribe to Marsham's determination that the Shepherds conquered much of Egypt in 1642 BCE—a little over seven hundred years after the Deluge—and reigned over the land for 511 years before they were expelled by Amenophis in 1131 BCE, thereby initiating the unification of Egypt under Amenophis' successor, Orus.[17] Newton, recall, had con-

[13] Marsham, 1676, pp. 3, 467.

[14] Marsham, 1676, p. 12, cited in Grafton, 1975, p. 178.

[15] Marsham, 1676, pp. 4–6, 87–91, 112, 141, 244–47, 503–12; New College MS 361.1 fols. 136, 158; MS 361.3 fols. 165, 239.

[16] New College MS 361.2 fol. 18; Newton, 1728b, p. 70. See p. 199 for Newton's reliance on Marsham's authority in rejecting the antiquity of the Assyrian monarchy. Such an assumption, he wrote, based solely on Ctesias, "has been justly called in question by Sr John Marsham." Yahuda MS 25.1d fol. 1^v^.

[17] Josephus cited Manetho to the effect that the Shepherds ruled over Egypt for 511 years (Manetho, 1964, pp. 85–87). Newton, who assigned far shorter duration to their reign, nevertheless attempted to resolve the discrepancy by suggesting that Manetho lumped together the Shepherds and the Israelites who invaded Egypt under King Asa into a continuous reign: "If the Jews, whom Manetho takes to be the shepherds, entered Egypt a year or two after the battel of Mareseh, & Amenophis after three or four years more

vinced himself that the Shepherds were the Canaanites whom Joshua had expelled ca. 1440 BCE, and who reigned in Egypt until the days of Saul. And since Newton also believed that the colonization of Greece coincided with the expulsion of the Shepherds, he necessarily rejected Marsham's dating of the foundation of the Kingdom of Argive to precede Jacob's birth, and the dating of the arrival of Cecrops in Greece before the birth of Moses. Finally, Newton never subscribed to two of the most contentious claims made by Marsham: that Jewish laws and customs derived from the Egyptians, and that Daniel's Seventy Weeks ended during the reign of Antiochus Epiphanes (i.e., in the second quarter of the second century BCE).[18]

Selective appropriation, here of Marsham, is characteristic of Newton. He picked and chose at will, regardless of the source in question. Even in the despised Ctesias who, he once stated, "made all things too ancient & taken too great a liberty in feigning names & stories to please his reader," there could be found "something of truth ... in the bottom of some of his stories as there uses to be in Romances."[19] For Newton assayed historical evidence in light of a schema he had conceptualized in advance, much as the operations of nature were utilized to test and to extend the domains of his scientific theories. That strategy enabled him to transplant the entire edifice of the Four Ages of Man—which had originally been designed to recount the history of Noah's progeny—onto the history of the world in the aftermath of the expulsion of the Shepherds, twelve centuries later. Newton's unwavering position on key methodological issues facilitated his wholesale grafting. First, he remained convinced that ancient myth captured, in a veiled form, the political reality of the first significant period in human history. Second, he insisted on interpreting the myth of the Four Ages as the history of a single family over the course of four generations, the exploits of which influenced the entire world. Third, in recognition of their great deeds, members of the family soon after death were deified; and since ancient sources were unanimous in ascribing the birthplace of the gods to Egypt, it followed that the historical events informing the myth also originated in Egypt. Fourth, the memory of these events recur in the myths of other ancient nations owing to Egyptian conquest and colonization—with idolatry spreading along the contours of conquest.

Newton predicated his Euhemerist interpretation of ancient texts on the assumption that the invention of writing postdated these events. The fables preserved by Diodorus Siculus, Herodotus, Lucian, Cicero, and numerous poets—the very sources Newton had used to construct his original system concerning the origins of idolatry—served again to revise chronology as Newton worked ancient texts in ways that fit his theory of civil evolution. A heap of manuscripts, totaling well over 500,000 words, bespeak the enormous time and effort he devoted to the new enterprise, and the extent to which he sought to build a synchronous, consistent account. Tellingly, the survival of a few pages from his original system among the chronology manuscripts

went up into Thebais & Ethiopia & reigned there 13 years before he returned back against Osarsiphus, this last expulsion of the shepherds will be about the 33th year of Asa. Count backwards 511 years & the first reign of the shepherds in Egypt will begin about ten years after the conquest & division of the land of Canaan by Joshuah. Which agrees well with what we said above of those shepherds being Canaanites driven out by Joshua." Yahuda MS 25.1f fol. 10. See also New College MS 361.1 fols. 122–23.

[18] For Marsham's discussion of Daniel's prophecy, see Marsham, 1676, pp. 610–19.

[19] New College MS 361.1 fol. 127.

attests to his wish to follow the original structure.[20] The similarity between the two structures is striking. The original rulers of each of the four ages—Noah, Ham, Chus, and Nimrod—morphed into the family line of Amosis, Ammon, Sesostris, and Orus, with the first three of each genealogy corresponding to, respectively, Saturn, Jupiter, and Mars (or Hercules). The Titanomachia and empire-building of the Iron and Bronze Ages, originally associated with Mizraim, Phut, and Chus, were now iterated to involve Sesostris, Japetus, and Hercules—only that Newton transposed the sequence of events, setting the formation of empire before the War of Gods and Giants. As for the twelve great deities, they corresponded perfectly to the new heroes: Ammon and his wife Rhea (Uranus and Titaea); Osiris and his wife/sister Isis; their children Orus and Bubaste; Thoth, Osiris' Secretary; Hercules and Pan, their generals; and Japetus (Neptune, Typhon). And so on.

The adjusted list of the major Egyptian deities signals Newton's revisiting his position on the origination of false religion. He now had to steer carefully between the need to argue, as the new theogony mandated, that the progress of idolatry followed the deification of the twelve gods—notwithstanding the explicit Scriptural references to much earlier practices of idolatry. As always, Newton concocted a clever solution. Idolatry, he acknowledged, had always involved the posthumous worshipping of the founders of cities and kingdoms and, as such, the practice did manifest itself early. In Mesopotamia it started "a little before the days of Abraham, most probably by the worship of Nimrod the first Monarch mentioned in history," after which it was transported to Canaan and Egypt. In Egypt, idolatry must have existed before Moses' time, since the second Commandment both described and prohibited it. Nonetheless, this early form of false worship proved transitory; the deities of the numerous towns and kingdoms of the postdiluvian centuries changed continually in the aftermath of conquest, so that the memory of the old gods faded.[21]

With this evolution in mind, Newton explained that when he spoke of "founders of cities," he meant not their first inhabitants, but those who "first raised y^{m} above other cities [and] made y^{m} seats of kingdoms & built them accordingly." The "worshipping of such founders of kingdoms gave the first beginning to Idolatry in Egypt, Chaldea, Syria, and the neighbouring nations from whom it spread into Europe & other places." Hence, the Egyptians "propagated" the worship of the twelve gods "into their conquests; and made them more universal than the false Gods of any other nation had been before," thereby meriting the designation "Dei magni majorum gentium." In this way, Newton could adjust the temporal ordering of the great idolatrous temples, since the building of imposing edifices always followed the rise of an empire. Thus, from Lucian's *The Syrian Goddess*, he gathered that Sesostris had built the Egyptian temples still then surviving, and that they were older than Assyrian temples. To Newton this proved not only "that the Monarchy of Assyria rose up after the Monarchy of Egypt"—in conformity with Scripture—but that the building of these structures did not mark "the first original of Idolatry, but only the erecting of much

[20] New College MS 361.3 fols. 77, 236, 236^{v}.

[21] When the Shepherds conquered Egypt they proscribed earlier forms of Egyptian worship, "and spread that of their own Kings" instead. Their idolatry, in turn, gave way to the idolatry of the kings of unified Egypt after the expulsion of the Shepherds. New College MS 361.2 fol. 238; Newton, 1728b, pp. 23–24, 161.

more sumptuous Temples than formerly to the founders of new Kingdoms: for Temples at first were very small."[22]

The name of Sesostris never appears in *The Syrian Goddess*; Lucian commented only on the cults of Osiris and Dionysus. To Newton's mind, however, these deities were cognomens for Sesostris.[23] More crucially, Newton's formulation attests to the centrality of temples to his mature evidentiary scheme. The Chronology includes some numerous references to temples, specifically designed to introduce a civilization-based argument to bolster Newton's re-dating of ancient history. As was the case with his reasoning from astronomy, Newton omitted from the Chronology much of the details informing his decisions. Hence it is necessary to examine his fusion of three distinct lines of argument into a single claim: The flowering of capitals of kingdoms, the prerequisites for the maturation of civilization; the acquisition of advanced skills in architecture and the arts; and an advanced process of deification of great heroes that warranted the construction of temples.

His starting point was the history of capital cities. In the process of time, to recap Newton's scheme, villages morphed into cities, and these combined to form a kingdom. The chief city became the capital, and its Prytaneum—now transformed into a temple—engrossed all judiciary and ceremonial functions of the new kingdom, rendering the Prytanea of other cities obsolete. A key piece of evidence that Newton deployed was Pliny's curt ascription to the Egyptians the invention of "royal cities." Newton had already utilized the term in the 1680s—for example, when describing the flight of the terrified gods from their Royal city ("urbe regia") following the advent of Phut's invading army[24]—but only in the early years of the eighteenth century did he come fully to appreciate the significance of the concept to dating the establishment of great kingdoms. Pliny's verdict, he reflected on one occasion, "is as much as to say" that Thebes had been "the first City which became the seat of a Monarchy."[25] Newton struggled hard and long to fix a date for this presumptive event. Recall his general agreement with Marsham's contention that Egypt had been divided into four kingdoms immediately after the death of Menes (Mizraim). As late as 1714, Newton argued in the second version of the "Original of Monarchies," the kingdom of Thebes "swallowed up those of This and Memphis about the days of the Patriarch Jacob & thereby became the greatest kingdom in the world then known to the Greeks." As proof of such an early dating for a royal city, Newton invoked Homer's famous celebration of the grandeur and wealth of Thebes, taking care to emphasize that the poet's failure to mention Nineveh or Babylon attests to their comparative insignificance at the time.[26]

The invocation of Homer represents another non sequitur. Homer's tribute to Thebes appears in the context of the attempt of several Greek commanders to prevent Achilles' withdrawal from the Trojan War—following Agamemnon's seizure of the beautiful Briseis. When Ulysses recounts the amends Agamemnon proposed, Achil-

[22] New College MS 361.2 fol. 110^{v}; Newton, 1728b, pp. 161–62, 220.

[23] Lucian, 1913, pp. 47, 56–47.

[24] Yahuda MS 16.2 fol. 62.

[25] Pliny 1949–1971, vol. 2, pp. 640–41; New College MS 361.1 fol. 97^{v}.

[26] New College MS 361.1 fol 93^{v}. See also Keynes MS 146 fol. 26; Yahuda MS 25.1f fol. 2.

les responds that not even all the riches of the famous Thebes—that city "of an hundred gates wherefrom sally forth through each two hundred warriors with horses and cars"—would appease him.[27] Surely, nothing in Homer's poetic license suggests that no other considerable cities then existed. Yet Newton persisted in interpreting the passage as proof of the primacy (and singularity) of Thebes, changing only the time line.[28] He now advanced an ingenious new account for the city's rise. Before the advent of the monarchy, Egypt had been divided into three major kingdoms: Lower Egypt, Thebes, and Coptos. The latter ultimately conquered Thebes, and it fell to the kings of Coptos, *Misphragmuthosis*, and *Amosis*, to expel the Shepherds. But only under the next king, Ammon, did Thebes become the Egyptian "Royal City," Newton insisted, as demonstrated by the dedication of the city to him as "No-Ammon & Ammon-No that is the city of Ammon, or as the Greeks render the word, Diospolis, the city of Jupiter." But if Thebes "grew great" under Ammon, it was his son Sesostris who "built it more sumptuously." The invocation of Homer to support all of this was then easy, as the poet flourished, according to Newton's reckoning, ca. 865 BCE—a century and a half after Sesostris had ascended to the Egyptian throne.[29]

Newton generalized from this sequence to form another rule: "Kings upon founding or much enlarging their kingdoms usually build their royal cities more large & sumptuous." The building of Jerusalem and its Temple by David and Solomon; the Egyptian cities and temples erected by Sesostris; Babylon under Nebuchadnezzar; and the Rome of Augustus all served to confirm his rule. Conversely, from the fact that King Hiram magnified the city of Tyre, Newton inferred "a new dominion of the Tyrians"—thereby accounting for Josephus' inconvenient fixing of the building of the city 240 years before Solomon's Temple; the earlier date, Newton explained, refers only to a "little town."[30] Noteworthy, too, is Newton's estimation that several nations reached the same stage of enlargement roughly at the same time. The age of Ammon and Sesostris, he remarked, had been "memorable for the building of new royal cities," which included not only Thebes, Jerusalem, and Tyre, but Athens and Damascus as well.[31]

It may seem strange that Newton considered it possible to argue that the ancient Theban nation and the barely formed kingdom of Israel had reached a similar stage of monarchical and cultural development simultaneously. That claim, however, is central to Newton's chronology. According to Manetho, cited by Josephus, Jerusalem was built by the Shepherds, following their expulsion from Egypt. Both Manetho and Josephus, of course, identified those Shepherds with the Israelites whom Moses had delivered from bondage, while Josephus dated the expulsion of the Shepherds to nearly 400 years before Danaus settled in Argos. Committed as he was to the idea that the Shepherds were the Canaanites whom Joshua had expelled centuries earlier, Newton nevertheless embraced Manetho's claim that the building of Jerusalem was

[27] Homer, 1928, pp. 409–11. Along similar lines, Newton argued elsewhere that since neither Homer nor Hesiod mentioned the great Pyramids that Moeris built, they were unknown to then and, consequently, of later date. Newton, 1728b, p. 32.

[28] New College MS 361.1 fol. 93^{v}; Keynes MS 146 fol. 26.

[29] New College MS 361.3 fols. 7^{v}, 19^{v}, 51^{v}, 52^{v}, 63^{v}, 136a, 157, 203, 239, 247; Newton, 1728b, pp. 6, 161, 242.

[30] New College MS 361.3 fol. 165; MS 361.1 fol. 144; Josephus, 1841, p. 218.

[31] New College MS 361.3 fol. 19^{v}.

contemporaneous with the expulsion, for it enabled him to substantiate his late dating of the Egyptian monarchy.[32]

With this in mind, Newton proceeded to postulate a mandatory correlation between empire and civilization. As cities grew great "in proportion to their dominion," he wrote, so Thebes, following the conquest of its adjacent nations, grew in size and grandeur. What enabled such magnificence was the accumulation of wealth and a subjugated labor force. In developing this theory, Newton relied on Herodotus and Diodorus, who commented on the "great multitude of captives" that Sesostris brought back from his expedition, and who built the great Egyptian monuments. Both historians also remarked on the vast tributes the conquered nations paid Sesostris, which accounted for the splendor of Thebes and of the monuments Sesostris built throughout Egypt.[33]

Sometime in the 1710s, Newton enunciated most clearly his intertwining of grandeur and empire. Just as Babylon and Rome "were adorned in the height of their emperors," he reasoned, "so doubtless it was in the reign of Sesostris & his successors that Egypt was adorned wth those magnificent works of Temples, Obelisks, Pyramids, Labyrinths & ye like"—works made possible by the "spoils and Tributes of the nations." Indeed, he continued,

> Herodotus refers ye Pyramids to kings who reigned a little after Sesostris & wth good reason, since ye kings who preceded had not wealth sufficient for such great works. And tho he places Moeris almost 900 years before his own age, yet the vast Lake wch Moeris built of 3600 stadia in compass & 50 paces depth where deepest, wth two Pyramids in ye midst of 50 paces height above ye water & upon each a Colossus on a throne representing him & his wife, are works wch agree far better wth this magnificent age then wth that difficult one when ye Shepherds reign over ye lower Egypt. And therefore I had rather attribute it to that Marrus whom Diodorus makes ye author of the Labyrinth.[34]

We might add that Diodorus also attributed the building of the famous Temple of Vulcan in Memphis to Sesostris, a fact Newton accepted when composing this manuscript. In general, however, Newton argued that it was Menes who commenced building the Temple sometime between the Fall of Troy in 894 BCE and his own death seven years later.[35]

The linking of royal cities and grandeur licensed his ignoring Egyptian history before the expulsion of the Shepherds. Newton repeatedly cited a pronouncement by Herodotus to the effect that though the Egyptian priests had recounted to him the names of 330 kings who reigned before Sesostris, he did not mention them as none—

[32] Newton, 1728b , pp. 16–67; Manetho, 1964, p. 89.

[33] Herodotus, 1975, vol. 1, p. 397; Diodorus, 1939, vol. 1, pp. 195–97; New College MS 361.2 fol. 128; MS 361.3 fol. 63; See also Yahuda MS MS 25.1g fol. 8: "As Babylon & Rome were adorned in the height of their empire so was Thebes & all Egypt in the reign of Sesostris & his successors, the captives, spoils & tribute of ye nations being employed in building Palaces, Temples, Obelisks, Pyramids & other works."

[34] New College MS 361.2 fol. 12.

[35] New College MS 361.3 fols. 6^{v}, 51^{v}, 136A^{v}; MS 361.1 fol. 124; Newton, 1728b, pp. 238–39.

with two exceptions—had done anything notable. Newton wholeheartedly approved: Herodotus "justly passes over in a few words all the ages of Egypt before Sesostris as obscure & containing nothing memorable & begins his history of the kings of Egypt with this king." Hence he, too, resolved not to "meddle" with them. In the *Chronology* he articulated more forcefully his decision to ignore rulers who had been kings in name only: "If with *Herodotus* we omit the names of those Kings who did nothing memorable, and consider only those whose actions are recorded, and who left splendid monuments of their having Reigned over *Egypt*, such as were Temples, Statues, Pyramids, Obelisks, and Palaces dedicated or ascribed to them, these Kings reduced into good order will give us all or almost all the Kings of *Egypt*, from the days of the expulsion of the Shepherds and founding of the Monarchy, downwards to the conquest of *Egypt* by *Cambyses*."[36]

The building of great monuments, in turn, demanded the attainment of sufficient scientific and artistic skills and these, too, developed in the aftermath of the expulsion of the Shepherds. Just as astronomy was hardly known before Ammon and Sesac turned their attention to these sciences—Newton stated in the *Chronology*—so nothing was heard of "good Architecture" before Solomon embarked on building the Jerusalem Temple.[37] Nor could other elaborate structures have been built earlier. Case in point: the great Temple of Vulcan in Memphis. "It is not likely," Newton determined, "that so magnificent a structure for archit[ect]ure & bigness could be founded" before the erection of the Egyptian empire. Likewise, the famous Labyrinth that Daedalus built for Minos in Crete could not have been built before the end of Solomon's reign, and the same held true for the pyramids.[38] Indeed, Newton argued, since the regulation of Lower Egypt into Nomes by Sesostris corresponded to the construction of the first temples, such regulation was "older then ye Pyramids whose very accurate & magnificent structures shew that architecture was then grown to a good degree of perfection."[39]

Newton derived the idea of the contemporaneity of the division of Egypt into nomes and the origination of temples from *Concerning the Jews* by Artapanus of Alexandria, fragments of which had been preserved by Eusebius.[40] As (erroneously) summarized by Newton, Palmanothes "was ye first that built a Temple in ye lower Egypt," in his capital, Heliopolis. His daughter educated Moses who, in due course, "found out the arts & philosophy of the Egyptians & divided the kingdom into 36 Nomi." At the time, Newton rejected the attribution to Moses, claiming for Joseph the honor of regulating the kingdom into nomes.[41] In another early manuscript, he ascribed the regulation to an Egyptian king who reigned "soon after ye days of Osiris

[36] Herodotus, 1975, vol. 1 p. 389; New College MS 361.1 fol. 158. See also MS 361.3 fols. 7^{v}, 18, 30, 99^{v}, 136A; Newton, 1728b, pp. 244–45.

[37] New College MS 361.3 fol. 99; Newton, 1728b, p. 213.

[38] New College MS 361.3 fols. 28^{v}, 34^{v}, 63^{v}, 67^{v}, 229, 247.

[39] New College MS 361.3 fols. 97, 98^{v}.

[40] Artapanus was a Jewish historian of the second century BCE; Newton mistook him for a Christian.

[41] Yahuda MS 41 fols. 25–25^{v}. Artapanus did not credit Palmanothes with being the first to build temples. Eusebius text reads: "first he built Kessa, and founded the temple therein, and then built the temple in Heliopolis." Eusebius, 1903, vol. 3, part 1, p. 462.

[Mizraim] & Isis."[42] However, once he identified Sesostris as Sesac, Newton stopped citing Artapanus. He still accepted the ancient regulation of Egypt into nomes, each representing a kingdom, with a capital city and a temple. Yet he now believed that after Sesostris unified Egypt, he "made a new regulation of the Nomes & built their Temples more sumptuously."[43] The published *Chronology* dispensed with all qualifications:

> *Sesac* using the advice of his Secretary *Thoth*, distributes *Egypt* into xxxvi *Nomes*, and in every *Nome* erects a Temple, and appoints the several Gods, Festivals and Religions of the several *Nomes*. The Temples were the sepulchres of his great men, where they were to be buried and worshipped after death, each in his own Temple, with ceremonies and festivals appointed by him; while He and his Queen, by the names of *Osiris* and *Isis*, were to be worshipped in all *Egypt*. These were the Temples seen and described by *Lucian* eleven hundred years after, to be of one and the same age: and this was the original of the several *Nomes* of *Egypt*, and of the several Gods and several Religions of those *Nomes*.[44]

Clearly, then, well into the early years of the eighteenth century, Newton found it unproblematic to acknowledge the existence of early temples. When describing the aggrandizement of Tyre by Hiram, for example, he considered it natural to assume that the new temples replaced the old temples demolished by the King. Similarly, he noted that Shepherds subverted the temples of the Egyptians for the worshipping of their own deities.[45] Nevertheless, once he formulated a symbiotic correlation of temples and empire, Newton strove hard to deny the existence of early temples. Having found "no mention of sumptuous Temples before the days of *Solomon*," Newton concluded that they simply did not exist. Surely, the fact that Moses "commanded" the destruction of "the Altars, Images, High Places, and Groves of the *Canaanites*, but made no mention of their Temples," attested to their absence in those days.[46] To this reasoning from the absence of evidence, Newton added linguistic agility. He acknowledged that the kingdoms of Lower Egypt had worshipped their dead rulers before the time of Moses—to which "the second commandment is opposed"—and that the Shepherds suppressed such worship by introducing the veneration of their own gods. But he purposely avoided mentioning that these practices were evidently carried out in temples. Only when citing Lucian's description of the ancient Egyptian temples—which Newton attributed to Sesostris—did he half-heartedly concede the existence of earlier temples. Nonetheless, he immediately added, "Temples at first

[42] New College MS 361.3 fols. 97, 202.

[43] Keynes MS 146 fol. 26; New College MS 361.3 fol. 203^{v}. At the time Newton dated the regulation into nomes and foundation of temples to 967 BCE, two years before Sesostris' death. MS 361.3 fol. 123.

[44] Newton, 1728b, p. 22.

[45] New College MS 361.3 fols. 211^{v}, 137.

[46] Newton, 1728b, p. 221. Newton relied on such instances as Exodus 23:24, 34:13; Numbers 33:52; Deuteronomy 7:5, 12:3.

were very small"[47]—a diminutive representation that in no way, or so he believed, affected his correlating the building of "sumptuous" temples with the rise of the first Egyptian monarchy.[48]

Newton found it easier to divine the dating of Egyptian temples from the dating of Greek temples, for he postulated that the latter imitated Egyptian practices. "Altars might begin to be erected in *Europe* a little before the days of *Cadmus*," he wrote in the *Chronology*, but temples were introduced only during the reign of Solomon. In support of this Newton pointed to Aeacus, King of Aegina, who "is by some reputed" to have been among the first to build a temple in Greece, and who flourished ca. 988 BCE. Elsewhere Newton claimed that the temples at Eleusis and Argos were "the first instances that [he met] with in *Greece* of Deifying the dead, with Temples, and Sacred Rites, and Sacrifices, and Initiations, and a succession of Priests to perform them." Newton dated the former to ca. 1007 BCE, and the latter somewhat earlier—"about the same time with the Temple of *Solomon*." As for Phemonoe, she became the first priestess of Apollo in Delphi, "about the middle of the Reign of *Solomon*."[49]

Attempts like these notwithstanding, Newton never addressed one crucial point. If the early Greek temples were coeval with the *building* of Solomon's Temple (1015 BCE), in what way could their veneration of dead heroes have derived from prior Egyptian practices—given that Sesostris only started to adorn Thebes after his ascension to the throne in 1002 BCE, and given that he only divided Egypt into nomes (and embarked on his massive building of temples), after returning from his expedition in 965 BCE?

Another aspect of Newton's elucidation of Egyptian history involved the interpretation of hieroglyphs. Newton used what he (like others in his time, in conformity with the form of interpretation, though hardly the specific meanings, outlined in Horapollo) took to be the arcana of Egyptian symbolism encoded in hieroglyphs to substantiate his history:

> In those days the Egyptians using to write in Hieroglyphick affected to represent all things by symbols as by putting a trident in the hand of an Admiral to represent the three squadrons of his fleet, a rod writhen about wth two serpents for the symbol of an Embassador reconciling two nations, a man wth a syth for a king of a country, a man wth rams horns for a king a country abounding wth sheep, a man riding upon an eagle wth a thunderbolt in his hand for a great warrior soaring high in dominion, the world for a kingdom, a giant for a man great in power a man wth many heads & hands for a king with his kingdom or a Captain wth his army, water for people, a flood for an invasion, a new world after a flood for a new kingdom after an invasion,

[47] Ovid's *Fasti* furnished him with evidence: When Rome was new, and its people poor, "Jupiter had hardly room to stand upright in his cramped shrine, and in his right hand was a thunderbolt of clay. They decked with leaves the Capitol, which now they deck with gems." Ovid, 1951, p. 17.

[48] Newton, 1728b, pp. 161, 220–21.

[49] Newton, 1728b, pp. 17, 134, 139, 143, 160. A few years earlier Newton was less certain: "the Temple of Iuno Argiva was built about the latter end of Davids reign, or the beginning of Solomons" Yahuda MS 25.1a fols. 20^{v}, 21 28.

> a golden apples for scarce & precious & valuable fruit, a Dragon keeping the gardens of the Hesperides for an army keeping a country abounding wth fruit-bearing trees, a horn of the sea for a river, the horn of Amalthea for a river with fertile meadows on both sides given by Ammon to his Queen for her maintenance.

From these and similar "symbols truly interpreted," concluded Newton, "the history of the mystical ages may receive some light." In the *Chronology* he furnished some concrete examples of his decipherment. The representation of Ammon with the horns of a ram, for example, "signify the King who conquered *Libya*, a country abounding with sheep," while the representation of his father Amosis as a scythe denoted a king who had "conquered the lower *Egypt*, a country abounding with corn." Analogously, Jupiter's depiction upon an eagle meant "to signify the sublimity of his dominion"; his depiction with a thunderbolt intended to "represent him a warrior." A Trident in Neptune's hand epitomized him as the "commander of a fleet composed of three squadrons," while the symbol of Aegeon—a giant with 50 heads and hundred hands—represented Neptune "with his men in a ship of fifty oars." What Newton chose not to publish were certain Scriptural analogues for his interpretations of symbols, such as instances in Scripture where "waters" are used "for peoples & multitudes & nations."[50]

Understandably, as was the case with his original system for the development of idolatry, Newton's quest for comprehensiveness, and his desire to incorporate as many fables as possible, introduced some inconsistencies. The identity of Hercules is illuminating. The new chronology required that Hercules be Sesostris—the great hero of the Iron Age, after whom the Pillars of Hercules were named—and Newton generally followed this identification. But mythology also celebrated a Hercules who defeated Japetus, the murderer of Sesostris.[51] Newton never resolved the conflict, juggling awkwardly with the existence of two Hercules. He also suggested, occasionally, that the latter Hercules, whom Sesostris had appointed governor of Ethiopia upon setting out on his great expedition (and who came to the rescue of Orus after the death of Sesostris), was also the Hercules who subsequently invaded Egypt and killed Orus. Newton identified him as Zerah the Ethiopian who, having conquered Egypt, proceeded to invade Judaea, only to be roundly defeated by King Asa.[52]

Newton certainly took liberties when using literary sources but, ultimately, he felt compelled to ground his historical narrative on what he considered to be solid and varied evidence. Before 1700, however, he did not, with one exception (involving the lengths of reigns and generations), attempt to turn the evidence of texts into something of a different kind. After 1700 a new emphasis becomes increasingly evident in the way he dealt with words from antiquity. It can be seen in Newton's first attempt to summarize his "success" in revising Greek chronology. He had determined the Trojan War to be eighty years older than the return of the Heraclides into the Peloponnesus,

[50] New College MS 361.3 fol. 27; Newton, 1728b, pp. 226–27.

[51] In the earlier scheme this posed no problem, for Chus is both the conqueror and the Hercules who avenged the murder of his uncle Mizraim.

[52] New College MS 361.1 fol. 122; MS 361.3 fols. 27, 63, 127^{v}, 187, 243; Newton, 1728b , pp. 24, 99–100, 235–36. Newton also considered Amenophis (Menes) to be Zerah's son.

and the wars of Sesostris in Greece two generations older still. In so doing, he made "Chronology suit with the course of nature & with sacred history & with it self without the many repugnances complained of by Plutarch."[53]

What is striking here is Newton's introduction of the "course of nature," which attests to a growing conviction on his part of the limitations of textual evidence per se.[54] Above all, it foreshadows Newton's increasingly elaborate efforts to secure what he thought to be a more certain basis for his investigation. Why the new emphasis and growing skepticism? As to the novelty, Newton may have been prompted to seek a very different kind of evidence by the progress he made during the 1690s in detecting—it might be better to say in creating—natural regularities in the varied domains of demography, the formation of kingdoms, the lengths of reigns, and the progress of idolatry. Newton came to consider the "course of nature" to govern human affairs in nearly as unbreakable a manner as a "law of nature" does in areas such as mechanics and optics. And, like laws of the latter sort, these too could be forged in a cauldron of evidence.

Many examples, both in the manuscripts and in the *Chronology*, attest to the centrality of Newton's conviction in the regular governance of human affairs by the "course of nature," as we will see most vividly in the next chapter when we examine his considerations of regnal lengths.[55] He had used the phrase in two of the Queries added at the end of the second English edition of the *Opticks* (1717). Query 30 asserted that the interconversion of gross bodies with light (which followed from Newton's belief that light consists of particles) "is very conformable to the Course of Nature, which seems delighted with Transmutations." Even more significant is its appearance in Query 31: "Have not the small Particles of Bodies certain Powers, Virtues or Forces, by which they act at a distance, not only upon the Rays of Light for reflecting, refracting, and inflecting them, but also upon one another for producing a great part of the Phænomena of Nature? For it's well known, that Bodies act one upon another by the Attractions of Gravity, Magnetism and Electricity; and these Instances shew the Tenor and Course of Nature, and make it not improbable but that there may be more attractive Powers than these. For Nature is very consonant and conformable to her self."[56]

[53] Yahuda MS 361.3 fol. 208. The remark anticipates the claim for success he made in the *Chronology*, except that Newton had not as yet claimed to make his dating scheme cohere with Herodotus and with astronomy.

[54] Newton probably followed Thomas Burnet, who made extensive use of the concept in *The Sacred Theory of the Earth*.

[55] For example: the ascription of a mean reign of between 34 and 42 years to the kings of Messene,"is much too long for the course of nature." The length of time between the return of the Heraclides and the first Messenian War "ought to be corrected & reduced to such a length as suits best wth the course of nature." A reckoning of eighteen kings reigning from the death of Sesostris and the reign of Amasis—over a period of 390 years—"answers to the course of nature, especially if another king or two be inserted." Reducing the reign of king Afrasiab "to such a length as exceeds not the course of nature" would enable the supposition that the Scythians' first invasion of Persia, "gave occasion to y^{t} revolt of the Medes & other nations from the Assyrians." New College MS 361.3 fols. 2, 30, 49^{v}.

[56] Newton, 1721, pp. 349, 350–51.

Newton's use of the concept "course of nature" in the *Opticks* was designed to make reasonable the extension of a relationship that he had solid grounds for applying in one area of material interactions (gravitational and chemical), to others that had apparently similar observable properties (i.e., electricity and magnetism to gravity and the interconvertibility of light and matter to chemistry). These were suggestive connections, not proven by evidentiary argument. Yet Newton considered them to be highly plausible, on the grounds that "nature" operates in similar ways in similar circumstances. Of course, precisely what counted as similar circumstances was controversial, which explains why Newton placed the connections in "queries" and not in the body of the *Opticks*. The case for applying the concept to human affairs was both weaker and stronger than doing so for material nature—weaker because Newton did not possess solid exemplars in one area of human affairs for a nearly fixed regularity of the right sort that could be applied analogically to another area. In each case he had to rely directly on specific evidence for a particular claim. However, the application of the concept was stronger because it did not require analogy. Regnal lengths and generations were to be considered on their own grounds rather than in analogy to something else. And for that, we shall see in the next chapter, Newton thought there was persuasive evidence. The critical assumption, which Newton shared with quite a few contemporaries ca. 1700, was that at least some human activities obeyed for the most part—a critical modifier—regular laws.

By itself, the recurrence to arguments of this sort does not explain why Newton also developed a more critical attitude toward ancient texts proper, with the exception of Masoretic Scripture. Newton grew astonishingly free in treating words from the past. He implicitly justified his occasionally extraordinary manipulations (usually unacknowledged) by framing many statements as "poetical fictions," words that could not be trusted. There were exceptions for poetically phrased accounts that were thought to depict specific events, such as elements of the Homeric tales. Even there Euhemerist interpretation, at least, would often be necessary. But, Newton thought, tales deriving from periods before the "invention" of prose were inherently untrustworthy. They were useful primarily for divining how myths encoded Euhemerist transformations, but they could not be relied upon in the way that Herodotus (for the most part) or, of course, Scripture could be. Both of the latter benefited from a continuing tradition as texts, and, equally important, both were in the form of prose (barring only such *excurses* as the Song of Solomon, which did not in any case convey events). And the nicely pedigreed Masoretic text of Scripture had the benefit as well of recounting events in which divinity played a direct role.

The very first sentence of the printed *Chronology* suggests that Newton had certainly developed an intense skepticism with regard to poetic phrasing absent its extensive reinterpretation: "The *Greek* Antiquities are full of Poetical Fictions, because the *Greeks* wrote nothing in Prose, before the Conquest of *Asia* by *Cyrus* the *Persian*."[57]

[57] Elsewhere in the *Chronology*: "Some of the *Greeks* called the times before the Reign of *Ogyges*, Unknown, because they had No History of them; those between his flood and the beginning of the Olympiads, Fabulous, because their History was much mixed with Poetical Fables: and those after the beginning of the Olympiads, Historical, because their History was free from such Fables." And again: "*Plutarch* tells us that the Philosophers anciently delivered their Opinions in Verse, as *Orpheus*, *Hesiod*, *Parmenides*, *Xenophanes*,

But what exactly did he mean by "poetical"? Was it just the way a statement was phrased, its tone and meter, or did he also object to the use of metaphor, metonymy, or synecdoche? Or did Newton have something broader in mind than poetic phrasing per se, something that he had begun to think about during the 1690s and early 1700s? Did his assignation of a late date to the invention of writing become associated with suspicion of any form of testimony that had primarily an oral pedigree, with "poetical fiction" representing the vestiges of orality? Might there have been a specific event in the late 1690s that triggered these associations, perhaps an encounter with certain individuals who exhibited exaggerated self-promotion and spoke in a distrustful manner?

Newton left Cambridge for London once and for all to become Warden of the Mint in April 1696. There he took charge of the massive recoinage that was designed to solve the problem of currency debasement as a result of coin-clipping and counterfeiting. Though he himself had different ideas about how to solve the problem of debasement from the ones ultimately adopted—which largely reflected the views of his friend John Locke—Newton performed his duties with enthusiasm, organizational skill and, surprisingly, an increasingly acute understanding of how to trump clippers and counterfeiters. Eventually he assembled a veritable rogue's gallery of informants, usually criminals themselves, meeting them in taverns and street corners to extract information by doling out threats and promises of leniency in equal measure. Early on, however, Newton nearly met his match in the champion criminal of them all.

William Chaloner could, it seems, talk his way out of nearly any scrape, fobbing off his own crimes on gullible associates, and inveigling others into activities from which he profited and they paid the price—the Tyburn gallows. It was Chaloner's misfortune to cross Newton early in his tenure as Warden.[58] Not long before Newton's arrival in London, Chaloner had talked his way out of his latest stopover in Newgate prison by accusing workers at the Mint of corruption and counterfeiting. His aim was not merely release from Newgate, at which he was nicely experienced, but infiltration of the Mint itself so that he could more profitably ply his trade. Chaloner's accusations went nowhere as further inquiries revealed a morass of conflicting testimonies, so he reverted to his old habits. Newton interviewed him in late summer or early fall, whereupon Chaloner offered himself to the new Warden as a useful helpmate in fixing the scandalous situation at the Mint. Unenthusiastic about the offer, Newton still did not properly appreciate Chaloner's cleverness. Likely he didn't know much about the man, given the period's unsurprising lack of communication among the several different parties engaged in criminal-catching.[59]

Empedocles, *Thales*; but afterwards left off the use of Verses; and that *Aristarchus*, *Timocharis*, *Aristillus*, *Hipparchus*, did not make Astronomy the more contemptible by describing it in Prose; after *Eudoxus*, *Hesiod*, and *Thales* had wrote of it in Verse. *Solon* wrote in Verse, and all the Seven Wise Men were addicted to Poetry, as *Anaximenes* affirmed. 'Till those days the *Greeks* wrote only in Verse, and while they did so there could be no Chronology, nor any other History, than such as was mixed with poetical fancies." Newton, 1728b, pp. 44–46.

[58] Newton at the Mint, and his encounters with Chaloner et al., are outlined in Craig, 1946, 1963 . Levenson, 2009 presents a vivid picture of these events. We thank Levenson for providing us with copies of depositions taken at Newton's direction.

[59] Levenson, 2009, p. 156.

Early in 1697, Chaloner once again hurled allegations of improprieties at the Mint. Testifying before a House of Commons committee, he gave a compelling story of Mint faults and offered his own solutions to them. A master coiner himself, he knew well how to tell a convincing tale while always taking care to develop plausible deniability on his own account. So convincing was he that the committee ordered the Warden to allow Chaloner to demonstrate his proposals at the Mint itself—which would nicely fit Chaloner's aim of direct access to the cornucopia proper. Newton immediately took notice. Angered by this attempt to "vilify" the Mint, where he had labored hard for nearly a year, and undoubtedly offended by the order to consider a dubious claim to expertise, Newton refused to obey. Instead, he demanded that Chaloner show him what he had in mind. Chaloner refused, so Newton replicated the claimed effect himself. By then, the now implacable Warden decided to destroy the master criminal.

It took time. At first Chaloner's brilliant duplicity thwarted Newton, but eventually he had him before the bar. The Warden had gathered witnesses, had induced (often questionable) testimony, had gone to extraordinary lengths to catch his man. Charged before a grand jury with three indictments in March 1699, Chaloner was unable this last time to counter the whirlwind of accusations against him—though many were in fact doubtful or exaggerated. Unable to examine beforehand Newton's witnesses (and perhaps talk them into different stories), Chaloner had only moments to reply, and for once he failed. He argued (correctly in fact) that the charges were incorrectly brought and tried, since the alleged crimes were London-based whereas the juries were for Middlesex County. To no avail. The smaller-fry, whom Newton had let go, or whose punishments he mitigated on the extraction from them of incriminating testimony, eventually trapped the too-clever, and too-talkative, coiner in a tight web woven of his own and others' deceits and crimes. On March 4, Chaloner was sentenced to hang. He climbed the gallows eighteen days later.

Chaloner had been particularly fond of boasting to fellow denizens of Newgate during his several sojourns there that he could easily "fun" Parliament and King. By which he meant that the government could be duped by someone so clever as he. Newton, in drawing up an answer early in 1698 to Chaloner's accusations against the Mint, was particularly incensed by his "boasting secretly that he would fun the Parliamt as he had done the King & Bank before."[60] To "fun" was to lie. To lie was to give false testimony. And false testimony posed a problem. Newton had personally deposed many witnesses, had heard their conflicting accounts, their obvious self-exculpations, their flowery and elaborate, if usually ill-phrased, stories aimed to elicit clemency by throwing a collaborator to the wolves.

The problems Newton faced in sorting and weighing evidence hardly ended with the Chaloner case. The mélange of lies and half-truths continued for years afterwards. Consider for example the testimonies of a Katherine Linton and a Thomas Trott taken on July 1, 1704 and that of a Henry Fitzgerald taken the following October.[61] Linton had rented a "garret" and a lodging room to a silversmith by the name of Richard Jordan some three years previously. Jordan had died five weeks earlier, in mid-May.

[60] Newton, 1959–1977, vol. 4, p. 261.

[61] Mint MS 17, depositions 488–90.

Linton had "frequently" seen the "gentleman now shown to her"—Trott, whose name she claimed not to know—visiting Jordan, and she had further seen them work together "in the gold or silversmith's trade." After Jordan's death, she continued, Trott visited the garret three or four times, where he built a "fire of charcoal" and where she heard him at work "making a noise striking with a hammer." Trott told her that he intended to take over Jordan's room and garret (and she had never asked his name?). There was another man whose name she did know, one Fitzgerald, who "frequently" visited Jordan when Trott was there. She also "identified" the "two pieces of lead" Newton showed her (these were used in counterfeiting) as ones she had seen in the garret.

The interrogation continued the same day with Trott, alias Carter, alias Williams. Trott-Carter-Williams had been arrested by warrant for "high treason"—the price for counterfeiting—at his lodgings in the Lamb's Inn "on the backside of St. Clement's." Found on the floor in his room were "21 counterfeit French pistoles [currency]," but Trott was at a loss as to how they came to be there. A trunk also found in his room, Trott went on, belonged to one "smith," who had gone to the country without leaving the key. Having some linen of his own in Smith's trunk, Trott decided to break it open, whereupon he discovered therein twelve shillings of "his own." But how the trunk also came to contain dies for forging French pistoles—well, that was a mystery to the innocent Trott. As was the container of mercury additionally found in the trunk (and used to replace silver in counterfeiting). He was admittedly acquainted with the deceased Mr. Jordan, but he had only visited him for the purpose of making some silver "ferrils" for fishing tackle. He had even visited after Jordan's death—no doubt to retrieve the aforesaid fishing tackle—where he had seen the leads and other objects placed by Newton before him, but he had no idea what they were used for. He knew Mr. Fitzgerald as well and had seen him with the still-absent Smith in Jordan's rooms while Jordan was yet alive. And Trott of course had no idea whatsoever as to how the "plate engravure used for counterfeiting bank bills came into the aforesaid trunk and who did engrave the same."

It seems to have taken some time for Newton to find Mr. Fitzgerald, though by October 7 he was in the Warden's hands and under interrogation at the Mint's own Tower prison. Henry Fitzgerald knew Trott well enough. About a "quarter of a year ago," and so about the beginning of the previous July (which was just before Trott had been apprehended), the two had been together—for reasons unspecified—in the town of Wandsworth. Known to Fitzgerald as Carter, Trott creatively warned him that the local bailiff, one "North or Norton" who was just then walking by, had it in for Fitzgerald. The bailiff, Trott expostulated, was "a great rogue, and would raise the mob against him." Why so, Fitzgerald asked? Because, answered Trott, "there was money bid for him." Meaning that money could be had for a thief-catcher who caught up with Fitzgerald. Trott no doubt knew enough about Fitzgerald to scare him, particularly with a bailiff in sight (and how convenient for Trott would be Fitzgerald's disappearance given Linton's having seen them together at Jordan's). Fitzgerald likely headed off at a smart clip; he was followed by the bailiff "who took him up." Trott had disappeared in the meanwhile, and it seems likely that the bailiff went after Fitzgerald because he saw him in Trott's company. "Carter," Fitzgerald claimed the bailiff told him, "was a great rogue and a coiner, and had stood in the pillory." Find the bailiff,

Fitzgerald implied, and you may find out more about "Carter." As for himself, why, Fitzgerald knew "nothing of the outlawry." Indeed, he was he thought "not then in England when Carter was outlawed."

With the possible exception of Linton, the landlady (who was likely interested in whatever schemes were being cooked in her leased rooms), all of these interrogants no doubt told half-truths or outright lies. Trott-Carter-Williams invented the mysterious, and conveniently absent "in the country," Mr. Smith, whose financial shenanigans were a complete mystery to the guiltless Mr. Trott. Fitzgerald played the innocent bystander in the whole affair, though what he was doing meeting with Trott in the first place, why he had been seen in Jordan's rooms with him, and where he had gone (if he had gone) out of England, were, to say the least, cause for suspicion. Newton had to sort through first Linton's, then Trott's depositions—depositions whose precise content was governed by the Warden's questioning. Smith was an obvious fiction, but Fitzgerald had been seen by Linton and so had to be found and questioned. Having caught up with him months later, Newton had more than enough to hold Trott to account, who had after all been found red-handed in possession of a trunk overflowing with counterfeiting devices and their products.

Newton's zeal in chasing down all the evidence that he could, in generating as much as possible by way of corroboration out of a farrago of self-serving testimonies, illustrates the extent to which he refused, with good reason, to trust what he heard from the mouths of his interrogants. He had to sort through conflicting claims; above all he had to balance one against another, eliciting evidence in order to winnow the wheat from the chaff. This kind of sorting was not altogether different from the testimony of experiment. There too some testimonies—some numbers—were likely more trustworthy than others, but there was not always a good way to choose among them. Instead, Newton for some time (and uniquely) had had recourse to a technique that amounted to a judicious weighing of all the evidence by means of taking the average among his numerical deponents. To do so required a reasonably large spectrum of numbers. So, too, depositions from multiple sources could as it were produce something like a weighted judgment. There was nothing new about that, in the sense that testimony had long and normally been taken from several witnesses or involved parties. New, however, were the extensive, even elaborate efforts that Newton made to construct an intricate complex of interlocking testimonies. Where one might veer off one way in an effort at self-exculpation, the other might tend in a countervailing direction. Deposition followed deposition, each unreliable but each containing some element that intersected with the others.

Newton's experiences with the vagaries of conflicting testimonies likely exacerbated what may already have been a strong tendency on his part to look skeptically at certain kinds of words from the past. "Poetical" words were particularly suspicious because, he thought, they encoded crumbs of Euhemerist fact in ambiguous ways. Chronologers had worsened this already difficult situation by "doub[ling] the persons of men." For example, "when the Poets had changed *Io* the daughter of *Inachus* into the Egyptian," they "made her husband *Osiris* or *Bacchus* and his mistress *Ariadne* as old as *Io*, and so feigned that there were two *Ariadnes*, one the mistress of *Bacchus*, and the other the mistress of *Theseus*, and two *Minos*'s their fathers, and a younger *Io* the daughter of *Jasus*, writing *Jasus* corruptly for *Inachus*. And so they have made two *Pandions*, and two *Erechtheus*'s, giving the name of *Erechthonius* to the first; *Homer* calls the first, *Erechtheus*: and by such corruptions they have exceedingly perplexed Ancient History."[62]

Even prose testimony raised doubts. If a continuous textual tradition existed, if a given item had a reliable literary pedigree, then it could be included in a catalog of, as it were, antique depositions from which probable truth could be extracted by comparison with other such items.[63] The most reliable words of that sort, Newton thought, were the Hebrew Scriptures proper. Not only for their ancient, sacred character, which could not be questioned, but because there existed a continuous textual tradition among the Jews. Moreover, it was possible to compare a latter-day version of the Pentateuch as used by contemporary Jews, with the version produced by the Seventy in Alexandria two millennia ago. In discussing the structure of Solomon's temple, for instance, Newton remarked that he had taken his "description of the Temple ... principally from *Ezekiel*'s Vision thereof." For confirmation: since "the ancient *Hebrew* copy followed by the Seventy, [differs] in some readings from the copy followed by the editors of the present *Hebrew*, I will here subjoin that part of the Vision which relates to the *Outward Court*, as I have deduced it from the present *Hebrew*, and the version of the Seventy compared together."[64]

Newton's "chronological table" was grounded on this comparative methodology, setting off one text against another to extract nuggets of truth—hardly a new approach, since it had long before been developed by Renaissance Humanists. Newton nevertheless brought to bear a uniquely skeptical attitude, one that not merely utilized, but that actively *required* multiple sources that could be cross-linked to provide a stable base of data. And all had to be consistent with "natural" requirements, including not only astronomical factors but ones grounded on human behavior and biological necessities. "I have drawn up the following Chronological Table," he remarked, "so as to make Chronology suit with the Course of Nature, with Astronomy, with Sacred History, with *Herodotus* the Father of History, and with itself; without the many repugnancies complained of by *Plutarch*. I do not pretend to be exact to a year: there may be Errors of five or ten years, and sometimes twenty, and not much above."[65]

Newton's skepticism with respect to words was hardly unusual among early modern natural philosophers. Bacon remarked in the *Novum Organum* that the "faulty meaning of words cast their rays, or stamp their impression, on the mind itself; they do not only make discourse tedious, but they impair judgment and understanding." Locke famously claimed in his *Essay* that "the very nature of words makes it almost unavoidable for many of them to be doubtful and uncertain in their significations." And in the eighteenth century George Berkeley's *Principles of Human Knowledge* required that one "draw the curtain of words, to behold the fairest tree of knowledge,"

[62] Newton, 1728b, pp. 4–5.

[63] Manuel remarked that "Newton respected only two types of literary testimony: accounts by historians contemporary with the events they described—Herodotus and Thucydides for example—facts about which they were presumed to have direct and immediate knowledge, and summaries by historians who had access to records and monuments of earlier times from which they copied their information" (Manuel, 1963, p. 51).

[64] Newton, 1728b, p. 343.

[65] Ibid., p. 8.

whose fruit is excellent, and within reach of our hand." Indeed, precisely that attitude is often taken as a hallmark of the period, one that had a decided impact on the rhetorical characteristics of philosophical discourse. Neither was Newton's concern with textual pedigrees at all novel. But his nearly obsessive cross-comparisons, the long paragraphs and pages that nicely thwart even the most intrepid reader's efforts to work through Newton's text—Manuel called the *Chronology* a "dry as dust narrative of events"—betray his particular ways with words from the past, ways that were grounded in his experiences both in the laboratory and in the taking of oral testimony.[66] In his view the best kinds of ancient words were ones that could be turned into numbers, for then Newton thought he could deploy the techniques that he had developed for handling laboratory data.

Above all, however, it was Newton's turning to astronomy in the late 1690s that generated a more skeptical attitude to words in general, and a determination to subject the uncertain domain of words to a far more rigorous scrutiny: "The surest arguments for determining times past are those taken from Astronomy," he pronounced. And having "brought Chronology nearer to the truth" by demonstrating how the "genealogies & reigns of kings" must serve as the foundation of ancient Greek history, he proceeded "to confirm o^{r} recconing by an argument" drawn from astronomy, as such arguments "are accounted the surest."[67] Stimulating such a renewed interest in astronomy was the debate over calendar reform.

On September 23, 1699, the Diet of Regensberg resolved to reform the German calendar and make it conform to the dating of the Gregorian calendar, with the proviso that Easter should be determined astronomically, based on the advice of "evangelical mathematicians." Three months later the Diet requested Leibniz to approach the Royal Society for advice on how best to carry out the necessary calculations.[68] On February 14, 1699/1700, ten days after Newton's appointment as Master of the Mint, Hans Sloane read aloud at a meeting of the Royal Society an account concerning the introduction of the new calendar into Protestant Germany. A week later, after reading Leibniz's letter that had just arrived, the Secretary mentioned hearing that Newton "had made a very good Calculation of the Year, and that the Settling that affair might be helpt by that." Sloane was asked to wait on Newton and inquire about the calculation (on which see figure 2.12 in chapter 2).[69]

English savants followed German proposals for some time, and many favored reform. John Locke, for one, expressed astonishment at England's failure to adopt the Gregorian calendar: "how inexcusable it is, that in soe learned an age as this, and in a country wherein astronomie is caried to an higher pitch than ever it was in the world, an error of this kinde should be sufferd to goe on." To remedy the defect, he had already proposed in 1698 to omit intercalary days between 1700 and 1744, so that the calendar in England might conform with the rest of Europe without the inconvenience

[66] Manuel, 1963, p. 50.

[67] New College MS 361.3 fols. 166, 168.

[68] Poole, 1998, pp. 42–43.

[69] Royal Society, Journal Book IX (1699–1702), pp. 194–95. The reference may have been to a conversation between Newton and Bishop Lloyd regarding the ancient year, a discussion mentioned during a further exchange between the two on the topic in 1713.

of eliminating eleven days in a single year. A summary of Locke's views appeared in the May 1699 issue of the *Nouvelle de la république des lettres*,[70] and it prompted Archbishop Thomas Tenison to solicit John Wallis' opinion. The Presbyterian professor of geometry at Oxford vehemently objected to the idea, not simply because he deemed it needless but, especially, because he believed it would prove pernicious to religion. The English Church had long renounced the "Pope's supremacy," he sermonized, and one should not embrace a calendar imposed by a Pope. Indeed, Wallis detected in the proposal "a latent Popish interest, which (under other specious pretences) sets it on foot; in order to obtain (in practise) a kind of tacit submission to the Pope's Supremacy, or owning his Authority. And though they be so wise as to say nothing of it at present (for the Bait is designed to Hide the hook till the Fish be caught,) they will please themselves to have gain'd *de facto*, what in words we disclaim." Wallis proceeded to launch a campaign intended to prevent England from following the German example. He privately lobbied William Lloyd, Bishop of Worcester, and Chief Justice John Holt, and then, without seeking permission, published his correspondence with Tenison and Lloyd in the *Philosophical Transactions*.[71]

Newton feigned agnosticism on the issue. His memorandum consisted of a brief history of Egyptian and Roman calendars, drawing on his own researches into ancient time-reckonings,[72] followed by his declining to "press neither opinion but whenever the latter shall be resolved on I beleive the best way may be to receive the new stile without the Gregorian Kalendar by an Act of Parliament"—toward which eventuality he made several practical recommendations. Newton also prepared brief "elements of the Motions of the Sun and the moon from the vernal Equinox," to assist in the astronomical part of the reform, though he opted not to communicate additional calculations he made.[73] Newton's report was read at a meeting of the Royal Society on April 25, 1700. Vice President Sir Robert Southwell moved that it be sent to Leibniz, and that both John Flamsteed and Wallis be asked for an opinion.[74] A week later Sloane communicated Leibniz' letter, and Newton's calculations, to Flamsteed. The latter's resentment at being contacted after Newton, on a matter that undoubtedly lay within the purview of his office, shines through Flamsteed's annotations of the letters he received: "This letter was imparted to me by Dr Sloane May 2 (1700) but the schedule of Mr Newtons was sent away, without expecting my answer." Flamsteed also corrected mistakes that Newton committed, while gleefully noting that Newton relied on observations carried out at the Greenwich Observatory.[75] As we shall see in the next chapter, this perceived affront, coming atop a growing acrimoniousness in their relations, forced Newton to resort to an inspired stratagem in order to extract from Flamsteed the astronomical information that his chronology required. But first we must turn to Newton's unusual ways with words from the past as he pursued the calculation of antiquity.

[70] Locke, 1976–1989, vol. 6, pp. 745–47; Milton, 2006.

[71] Wallis, 1699 , quote at p. 345; Beeley and Probst, 2005, pp. 453–57; *Philosophical Transactions*, vol. 21 (1699), pp. 343–54. For Wallis' opposition, see Poole, 1998, pp. 85–101.

[72] New College MS 361.3 fols. 60–60^{v}, 61^{v}, 70, 114, 149^{v}–150, 184^{v}.

[73] Yahuda MS 24; Newton, 1959–1977, vol. 4, pp. 328–30.

[74] Royal Society, Journal Book IX (1699–1702), p. 206.

[75] Flamsteed, 1995–2001, vol. 2, pp. 808, 822–23.

Interpreting Words

More than three decades separate Newton's explorations of astronomical chronology from the youthful engagement with problems of perception and measurement that we examined in chapters 1 and 2. By the time of his first computations in the area, shortly before the publication of the *Opticks*, Newton's understanding of measurement had been refined through years of experimental and computational experience, not the least of which occurred as he worked on the motions of bodies in fluids during the 1680s. The final composition of the *Opticks* brought him back to issues raised by measurement which, joined with his work on the calendar reform, form the immediate background to his earliest work with astronomical remnants in ancient texts. The previous decade had given Newton considerable familiarity with words from the past, and he had slowly developed a highly skeptical attitude toward ancient remarks that did not have a continuous textual ancestry, or that reflected what he considered to be unreliable "poetic fancies." Words with a proper ancestral pedigree that could be rendered into observations, and from observations into numbers, might avoid problems associated with "fancies." Thus, aiming to produce a compelling argument grounded in computation for his new chronology, Newton faced a treacherous triple problem: he had first to argue that the words with which he worked were originally produced near the time of the Trojan War; then he had to transform these words into astronomical data; finally, he had to deploy a technique for working with what he rapidly learned was a set of extremely discrepant observations. He labored over these problems until his death. In this chapter we will follow Newton as he transformed words and calculated, postponing to the next chapter the philological problems posed by the most fundamental assumption made by Newton, namely that the words reflected observations of the heavens by the Argonauts themselves.

The principal text with which Newton struggled was a *Commentary* by Hipparchus to a poem that was well-known in Greco-Roman antiquity, the *Phaenomena* by Aratus—A physician native of Soli, who was probably born about a century and a half before Hipparchus. Measuring some 1000 lines in hexameter, the *Phaenomena* describes the constellations, how to mark time using them and the planets, and how to interpret astronomical phenomena and animal behaviors that signal the weather. "Stylistically," notes a recent reviewer, "Aratus is notable for his clever manipulation of Homer, both in the usage of obscure Homeric words and in his deliberate reminiscences of Homeric passages. Despite its apparently practical subject, this is a poem for an extremely erudite audience."[1] According to the latest editor of the *Phaenomena*,[2] Aratus relied on the didactic poetry of Hesiod, which is particularly significant for our purposes since Newton also found astronomical data in Hesiod's *Works and Days*, as we will see.

The *Phaenomena* was perhaps the major source of astronomical knowledge known to the literate Greek and Roman public. More important for Newton, this sole surviving text by Hipparchus critiques Aratus and provides information concerning a lost treatise by Eudoxus of Knidos (408–355 BCE) on the constellations, on which the *Phaenomena* (according to Hipparchus[3]) was based. Written sometime between 140 and 130 BCE, Hipparchus' *Commentary* was available to Newton in a Latin translation by the chronologer Denis Petau, first published in his 1630 *Uranologion*.[4] Newton had a great deal of potential data to choose from in Aratus' poem and in Hipparchus' *Commentary*, since they combined provide by recent count 173 separate items.[5] The *Phaenomena* itself yields 123, of which 31 were, according to Hipparchus, drawn from Eudoxus, to which Hipparchus added fifty Eudoxan items that do not appear in Aratus. Hipparchus charged Eudoxus with inaccuracy, but in so doing Hipparchus assumed that his predecessor was using the zodiacal *signs* and not the asterisms with the same names. If instead the remarks are associated with asterisms, as Newton would do, then it is possible to read them as specifying observations originating before Eudoxus. Newton eventually decided that the producer of the first sign system had been the legendary Chiron the Centaur, one of the Argonauts, after which (according to Diogenes Laertius) Musaeus fabricated the first stellar globe:[6]

> Now *Chiron* delineated the σχήματα ὀλύμπου the *Asterisms*, as the ancient Author of *Gigantomachia*, cited by *Clemens Alexandrinus*, informs us: for *Chiron* was a practical Astronomer, as may be there understood also of his daughter *Hippo*: and *Musaeus*, the son of *Eumolpus* and master of *Orpheus*, and one of

[1] Hordern, 1999.

[2] Aratus, 1997.

[3] Not everyone agreed that Aratus had relied on Eudoxus. Hipparchus used parallel quotations to prove the point (Aratus, 1997, p. 16), which is why Eudoxus' remarks survived.

[4] Petavius, 1630. The *Chronology* does not specify the edition that Newton used. However, all the marginal references are included in the 1703 edition of *De Doctrina Temporum*, which Newton owned. And since the *Opticks* appeared in 1704, it seems likely that Newton did not work hard on chronology until shortly before or after its publication. Evidence from the surviving manuscripts, in fact, yields 1701 as the earliest date for astronomical calculations in chronology, and 1706 as the most likely candidate for his first Aratus-based ones. The 1701 computation concerned Hesiod's date and was based on the acronychal rise of Arcturus.

[5] Schaefer, 2004, pp. 169–70.

[6] In his manuscripts Newton had originally chosen Palamedes rather than Chiron as the originator of the asterisms. Palamedes was reputedly the son of the Argonaut Nauplius. See Manuel, 1963, pp. 81–87 on Palamedes and Newton. Of Newton's use of the Argonaut story, Manuel remarked "That the charming and psychologically intricate story of Jason and his Argonauts should in the hands of the precisionist Newton become a fact-book for genealogy and chronology is one of the less fortunate metamorphoses to which ancient texts were subjected by the pragmatizers of myth" (p. 81).

> the *Argonauts*, made a Sphere, and is reputed the first among the *Greeks* who made one: and the Sphere it self shews that it was delineated in the time of the *Argonautic* expedition; for that expedition is delineated in the *Asterisms*, together with several other ancienter Histories of the *Greeks*, and without any thing later.[7]

Many items in the combined list depend on the latitude at which supposed primordial observations were done since they specify which constellations rise and set together. On the assumption that the remarks do refer to some set of pre-Eudoxan observations, it is perhaps possible to use them to pinpoint the location at which the putative observations were made. That, however, would have involved Newton in computations which would have exceeded even his outsized tolerance for huge arrays of numbers.[8] The ones that are independent of latitude concern the positions of the various constellations in respect to the colures, the tropics, the Arctic and Antarctic circles, the equator, or the summer solstice. None of the observations in Aratus' own list concern the colures, which is hardly surprising since these—unlike the tropics and the equator—are comparatively difficult abstractions that are ill suited to a didactic poem.[9]

Newton used only the Eudoxan items that involve the colures.[10] He ignored the polar circles, and though he asserted that the tropical observations agreed with his claimed dating, he never produced a trial calculation using them, no doubt for the very good reason that the tropics do not permit as simple and direct a computation. In the end Newton mentioned altogether twenty distinct observations, but he calculated with only ten of these and wrote that four others fit his results; he ignored the

[7] Newton (1728b, pp. 83–84) relied on the *Stromata* of Clement of Alexandria, who did make Musaeus the teacher of Orpheus (chap. 21), though Musaeus' very existence as distinct from Orpheus has been questioned (Guthrie, 1993, p. 191). Diogenes Laertius had Eumolpus father the sphere-maker Musaeus, about whom he remarked, "These authors ["some who say"] forget that the achievements which they attribute to the barbarians belong to the Greeks, with whom not merely philosophy but the human race itself began. For instance, Musaeus is claimed by Athens, Linus by Thebes. It is said that the former, the son of Eumolpus, was the first to compose a genealogy of the gods and to construct a sphere, and that he maintained that all things proceed from unity and are resolved again into unity" (Diogenes Laertius, 1925, Prologue to Book I).Chiron appears as well in the Argonautica (Rhodius, 1960), which Newton used extensively, but (we shall see in chapter 11) problematically. Newton first noted the identification of Chiron as an astronomer in the 1680s (Yahuda MS 17.2 fol. 1^{v}).

[8] Machine computation makes it possible, in principle, to try different latitudes in order to see which best fits the entire array of data (see Schaefer, 2004).

[9] To grasp what the colures are, and why they are significant, requires a reasonably evolved understanding of the celestial sphere—the colures do intersect the equinoxes and the solstices, which have seasonal meaning, but they are otherwise defined as great circles which include the celestial poles. By contrast, the tropics intersect the path of the sun among the stars at the solstices, and the equator has meaning in the most elementary concept of the celestial sphere.

[10] Tables 2, 3, and 4 in Schaefer, 2004, pp. 164–70 list all of the items in Aratus, as well (table 4) as the non-Aratus ones in Hipparchus. Schaefer does not directly quote either Hipparchus or Aratus here, but extracts solely the relevant astronomical information. Aratus, 1997 provides both the Greek text and a (prose) translation into English, as well as many useful notes.

remaining six. We will soon see just why he dropped these six, and whether the four he mentioned do in fact fit his date, but first let's examine how the computation was done.

Newton's method itself (though not the conclusions he drew from it) did not depend on his prior argument that the original of star globes was produced for the Argonautic expedition and then accurately reported by Eudoxus centuries later. And because Newton did not use any latitude-dependent information, neither did his computations depend on where the observations had been made. The originators of the ancient texts accordingly seem to fade from view wherever Newton plunges into the heart of one of his deepest and longest-lasting passions, the production of numbers. Nevertheless, even the generation of numbers is hardly a neutral enterprise. Decisions must be made about how to do a calculation, about the trustworthiness of the numbers that must be fed into it, and most especially about how to turn antique words into data. For Hipparchus' remarks do not lead directly to data; they must be interpreted, and we shall see that Newton's interpretations were not the only ones possible. Though he never discussed the point in print,[11] Newton's manuscripts betray the several difficult choices that he had to make.

Newton's aim was to use the Eudoxan material to determine the date at which the observations upon which they were based had been made. It should be noted at once that his enterprise produces an improbably early result, even if we assume that the Eudoxan remarks do refer to original, or at least very early, observations by archaic Greeks.[12] The constellations are asterisms which lie along the plane of the ecliptic, and the earliest known reference to them appears in the Babylonian astronomical compendium entitled the MUL.APIN, which dates to ca. 750 BCE. By that time eighteen constellations were well-established, including all of those referred to in the Eudoxan material.[13] Because of precession (see immediately below) the constellations move around the ecliptic, but the Babylonians always used a system that was fixed with reference to the stars, never having discovered the phenomenon themselves.[14] Nevertheless, by the 5th century BCE they had introduced a system which divided the zodiac into twelve segments, each a fixed 30° in length, forming thereby the group of zodiacal signs in addition to the constellations, which of course were of different lengths.[15] Hipparchus is, so far as we know, the first to have recognized the existence of precession,[16] and as a result the zodiacal signs were eventually set with respect to

[11] The only exception concerns the appropriate stars to choose for calculating the colure through the constellation Aries, which we will examine below.

[12] Though an argument based, like Newton's, on the Aratus (Schaefer, 2004), but making use of the entire set of remarks, argues otherwise. See below, note 32.

[13] Rochberg, 2004, p. 127.

[14] See below, note 16.

[15] Rochberg, 2004, p. 130.

[16] See Neugebauer, 1975, vol. 1, pp. 292–96 for a discussion of Hipparchus on precession. He argues that the value usually attributed to Hipparchus, namely 1° per century, was a lower limit, and that consideration of Ptolemy on Hipparchus' lunar theory leads to a difference between the sidereal and the tropical year which implies a precession of 1° every 77 years. He further suggests that Hipparchus may himself have invented the theory of trepidation traditionally assigned to Theon of Alexandria, father of Hypatia, who discussed it in his *Small Commentary to the Handy Tables* [of Ptolemy] (the relevant portion is translated

the intersection of ecliptic with equator, detaching them altogether from their originating constellations.[17]

The fundamental phenomenon upon which Newton relied, and that had been used by other chronologers before him (notably, but problematically, by Scaliger)[18], was the very phenomenon discovered by Hipparchus—the *precession of the equinoxes*. Precession results from the fact that the earth's axis does not remain parallel to itself as the planet revolves about the sun; it executes a very slow conical motion about the earth's center. At the end of the seventeenth century, the period for this precession was known to be about 25,748 years (as compared to the 36,000 years given by the Alexandrian astronomer Ptolemy in the 2nd century CE); for his chronology computations Newton used the same period as the Jesuit astronomer Giovanni Battista Riccioli, namely 25,920 years, or one degree every 72 years.

Precession affects chronology in the following way. The plane of the earth's orbit cuts a great circle on the apparent sphere of the stars called the ecliptic, along which lie the zodiacal constellations. Since the sun appears to move along the ecliptic, during the course of the year it travels bit by bit through the zodiac. Twice a year the sun lies at the intersection of the ecliptic with the earth's equator, and at these equinoctial points the hours of day and night are equal. The two points that lie on the ecliptic at 90° to the equinoxes are the solstices, and here (in the Northern Hemisphere) the hours of daylight are longest (at the summer solstice) or shortest (at the winter solstice). Precession rotates the stars around an axis perpendicular to the center of the ecliptic, and as a result the position of the sun at the equinoxes and at the solstices with respect to the zodiacal constellations changes over time. For example, in about −3000 the summer solstice was in the constellation Leo, whereas by 1000 it had moved to Cancer (see figure 8.1).

Because the colures include, by definition, either the solstices or the equinoxes, as well as the poles, the constellations that pass through them will also change with time because of precession. If we know that at a certain otherwise unspecified date a colure included this or that part of a constellation, then we can (in principle) use the rate of precession to discover when this would have taken place. That was Newton's aim, and in using precession for purposes of dating he was not the first. Nevertheless, Newton was the first to use colures, and unlike everyone before him he pursued the goal with a uniquely single-minded attention to the details of calculation, and with, we shall see, a most particular understanding of how to turn words into numbers.

---

in Neugebauer, 1975, pp. 632, vol. 632). The suggestion that Babylonians may have discovered precession is controverted in Neugebauer, 1950 and Brack-Bernsen and Hunger, 1999. Ragep, 1993, pp. 398, vol. 392, arguing on the basis of Arabic texts, remarks that a now-lost ancient source may have linked trepidation to "Chaldeans" or "Babylonians:"; for another opinion see Neugebauer, 1975, pp. 297–98, vol. 291. On trepidation see appendix E.iii, note 27.

[17] The beginning of the signs was fixed by Ptolemy in the *Almagest*, who set 0° *Aries* at the vernal equinox (Ptolemy, 1998, p. 90 and Rochberg, 2004, pp. 131–33). Babylonian procedures also began the signs with Aries, but fixed the equinoxes and solstices sidereally, rather than tropically. Over time the cardinal points would therefore have moved away from their sidereal positions, but the Babylonians either ignored or never recognized the displacement.

[18] Grafton, 1993.

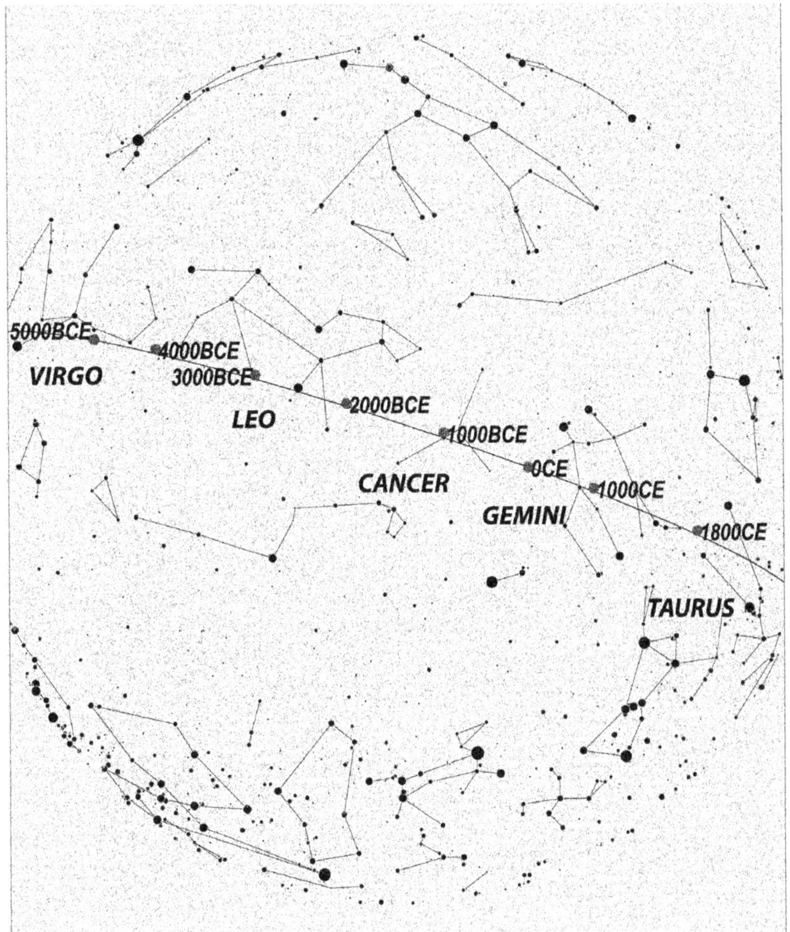

Figure 8.1. Precession of the summer solstice. Image produced using TheSky, Astronomy Software Version 6, copyright 1984–2003, Software Bisque.

Hipparchus' reports posed a problem of data that was different in kind from anything that Newton had dealt with before. In his optical experiments, as well as in the ones which he had performed on the resistance of fluids to moving bodies, Newton had himself generated the phenomena that he was investigating. Precisely because he controlled the data, Newton felt that he knew just how reliable it was likely to be. The apparatus (which he had often constructed), the way in which it was deployed, and any changes that were rung on it were all under Newton's sway. He had used astronomical data before he had begun to work hard on chronology (in particular on lunar motion), and he knew that data of the sort produced by astronomers differed in one important respect from experimental numbers.

Unlike Newton's optical or resistance data, the numbers generated by astronomers concerned objects that could only be observed. Although telescopes certainly do affect the light that passes through them, their design and behavior are subject to control, whereas the objects at which they peer are not. In his own experiments Newton had produced the actual phenomena by (for example) dispersing light through prisms, and he had also controlled the measuring devices that he had used to probe the effects. Both the prisms that produced his spectra and the apparatus with which he measured them had been within Newton's power. Astronomical data were quite different, for here the original objects remained pristinely intact, themselves free from the vagaries of human intercession. Just because astronomical objects are not within anyone's power to change, the resulting data depend entirely on judgments about observations and instruments, but not on procedures to generate the phenomena proper. This accordingly mitigates a factor that might raise questions since, when experiments are done by others, doubts concerning the production of the phenomena as well as their measurement can easily arise. Given an astronomical observer as reliable and as diligent as Flamsteed, who had worked very hard indeed to obtain the best possible instruments (as illustrated in figure 8.2), few issues concerning the manner in which the numbers had been produced were likely to crop up, and none did. Flamsteed's numbers were minted trustworthy.

Newton would have much preferred astronomical data of a similar sort from the past, but Hipparchus had not obliged. The *Commentary* is written in prose, not in arithmetic. Moreover, the observational devices and methods that presumably long predated even Eudoxus were unknown. In Newton's recounting, Hipparchus provided twenty significant remarks. Hipparchus, Newton wrote,

> tells us that *Eudoxus* drew the *Colure* of the Solstices through the middle of the *great Bear*, and the middle of *Cancer*, and the neck of *Hydrus*, and the Star between the Poop and Mast of *Argo*, and the Tayl of the *South Fish*, and through the middle of *Capricorn*, and of *Sagitta*, and through the neck and right wing of the *Swan*, and the left hand of *Cepheus*; and that he drew the Equinoctial *Colure*, through the left hand of *Arctophylax*, and along the middle of his Body, and cross the middle of *Chelae*,[19] and through the right hand and fore-knee of the *Centaur*, and through the flexure of *Eridanus* and head of *Cetus*, and the back of *Aries* across, and through the head and right hand of *Perseus*.[20]

Table 8.1 lists Newton's wording for the remarks that he chose to use together with their original form in Petavius' Latin rendering of Hipparchus' *Commentary*. Notice that Newton has omitted two remarks and, except in one instance, has ignored the qualifiers "longitudinem" (along) and "latitudinem" (across).

Descriptive words, not numbers, faced Newton here. Moreover, the remarks ascribed to Eudoxus by Hipparchus do not mention stars at all, but only positions

[19] *Chelae* refers to the claws of the Scorpion which were separated from it to form the separate constellation *Libra*, the remainder constituting *Scorpio*. At least as early as the time of Hipparchus in the second century BCE both *Chelae* and *Libra* are both used, and Ptolemy continued the practice.

[20] Newton, 1728b, p. 83.

Figure 8.2. Flamsteed. Thornhill ceiling. Photo by JZB.

within constellations. To turn prose into coordinates, Newton (or anyone else) had to make choices and had to work in ways that were quite unusual in astronomical context. For the constellations are not pinpoint objects in the heavens; they are large collections of stars, asterisms. Since, however, the colures are lines that pass among the constellations, transforming Hipparchus' words from Eudoxus into data required specifying a precise position within each constellation. To do so, Newton, and others

TABLE 8.1.
Newton's colure remarks and the originals in Petavius' Latin translation of Hipparchus' *Commentary*

Newton	*Petavius' Latin*[a]
solstitial: middle of Great Bear and of Cancer	Medium Ursae maioris, [et] Cancri
solstitial: neck of Hydrus	Hydri collum
solstitial: star between poop and mast of Argo	de Navi quod puppem inter [et] malum intercedit
solstitial: tail of South Fish	Caudam australis Piscis
solstitial: middle of Capricorn and Sagitta	medium Capricorni pariter ac Sagittae
solstitial: neck and right wing of Swan	Cygnum collum, alamque dextram
solstitial: left hand of Cepheus	sinistram Cephei manum
solstitial: *not included*	flexum serpentis
solstitial: *not included*	minoris Ursae caudam
equinoctial: left hand Arctophylax	sinistram Arctophylaxis manum
equinoctial: mid-body Arctophylax	medium ipse [Arctophylaxis] transire secundum longitudinem
equinoctial: mid-Chelae	Chelarum medium secundum latitudinem
equinoctial: right-hand and foreknee of Centaur	Centauri praeterea dextram manum, [et] anteriora genua
equinoctial: flexure of Eridanus	flexum fluminis
equinoctial: head of Cetus	Ceti caput
equinoctial: back of Aries across	Arietis terga secundum latitudinem
equinoctial: head and right hand of Perseus	Persei caput cumdextra manu

[a] *Petau, 1630, pp. 207–8.*

after him, eventually decided that the remarks refer to actual stars, and not just to unoccupied positions within a constellation overlaid on an asterism. That might very well be correct, but it was an assumption (and one that Newton did not initially hold firmly to, since his first computations worked rather with the "center" of Aries[21] than with a particular star within it). But even if the remarks are taken to refer to stars, appropriate choices among the several that might fit a particular statement must still be made. How was that to be done? What criteria should be used to govern star choice? Indeed, what stars were available to fit each constellation?

The only star catalog remaining from Greco-Roman antiquity had been compiled by Ptolemy and was included in his *Almagest*.[22] That catalog, it had long been

[21] The center being defined as the midpoint between *Prima Arietis* and *Ultima Caudae* in the asterism—see below.

[22] On Ptolemy's star catalog see Grasshoff, 1990.

thought, was closely connected to Hipparchus' own, which was otherwise lost.[23] There Ptolemy gave tropical coordinates, specifying a star's longitude—its distance along the ecliptic from the spring equinox—by its position within one of the twelve fixed signs of the zodiac through which the zodiacal constellations (after which the signs were named) move with precession over the centuries. The catalog had been thoroughly updated by Tycho at the turn of the sixteenth century, and most notably thereafter by Hevelius for epoch 1660, who also added eleven new northern constellations.[24] Manuals were also published during the seventeenth century that contained star catalogs, as well as information on computational procedures useful for navigation and in some cases for astrology.[25] Newton initially used Hevelius' coor-

[23] Ptolemy, 1998, pp. 341ff.

[24] Hevelius, 1690. Only seven of Hevelius' constellations were accepted after their publication in 1690: Canes Venatici, Lacerta, Le Minor, Lyn, Scutum, Sextans, Vulpecula; the rejected four are Triangulum Minor, Mons Maenalus, Cerberus, and Musca (a northern analog of the Southern Fly, whose stars are now incporated in Aries, but which was depicted on both the Senex globe and Newton's monument in Westminster). The first major star catalog since antiquity had been produced by Ulugh Begh and coworkers at his observatory in Samarkand between 1408 and 1437. Their catalog provided coordinates for 1000 stars to an accuracy of about a quarter of a degree. Wilhelm IV, Landgrave of Hesse, cataloged 400 stars. At Uraniborg, Tycho and his assistants cataloged 777 stars between 1588 and 1591 to an accuracy of between one and three minutes, except for nine reference stars where he achieved an accuracy of about half a minute. Riccioli's 1651 *Almagestum* (which Newton possessed and which, we shall see, he used to learn certain computational procedures) included many earlier star coordinates as well as a number that he had himself produced. At Danzig Hevelius cataloged 1564 stars to considerably better than a minute's accuracy. Neither Riccioli nor Hevelius used telescopes for measuring stellar distances (though Hevelius was an expert telescopic observer, having produced the first detailed lunar map; Hevelius, 1647). However, telescopes alone add nothing to stellar observations; they need to be supplemented with a device that permits a star accurately to be located. Cross-hairs placed at the focal point of the objective and eyepiece micrometers provided the necessary additions, though the earliest ones were not perhaps much better than naked-eye sights. See Bedini, 2005; King, 1955 (1979), pp. 93–100. Olmsted remarks that "the definitive application of telescopes to astronomical instruments had its origin in Paris in 1666 or 1667 in obscure efforts to improve 'open' sights. It reached its first or preliminary stage in the autumn of 1667 with the use of two convex lenses to form a 'tubeless' telescope. The second and final stage was not attained until the late summer or autumn of 1668 when Picard first utilized a completed astronomical telescope, including cross-hairs and a tube, in place of sights" (Olmsted, 1949, pp. 224–25). See the discussion of the Hooke-Hevelius controversy in chapter 2.

[25] Two examples that might have crossed Newton's path were Hartgill's tables, updated at mid-century by Timothy and John Gadbury (Hartgill, 1656) and Henry Philippes' *Sea-mans Kalender* (Philippes, 1676). The Gadbury-updated Hartgill provided longitudes and latitudes for 365 stars ca. 1650, as well as their magnitudes, astrological significance, constellation descriptors, and—of notable interest considering Newton's chronology—"their number in their proper constellation after Aratus" (on Hartgill, an astrological country parson, see Morgan, 1968). We have no evidence, however, that Newton ever did see the publication. The *Kalender* provided right ascensions and declinations for mid-century for 65 "principal stars" as well as a method for correcting coordinates for precession every century. Thomas Streete's *Astronomia Carolina* provided longitudes and latitudes, as well as descriptors, for 128 stars listed in order of their longitude; Newton did use Streete in 1664/5 (McGuire and Tamny, 1985). He also heavily annotated his copy of Vincent Wing's 1669 *Astronomia Britannica*, and in 1664/5 he certainly used Wing's *Harmonicon* of 1651 (on which see McGuire and Tamny, 1983, p. 300); both contain Tycho's stellar catalog

Figure 8.3. Tycho, Kepler, and Newton. Thornhill ceiling. Photo by JZB.

dinates in his chronological computations, although apparently he did not own a copy of the 1690 *Prodromus*. At some point during the 1690s, Newton received from Flamsteed a large set of coordinates[26]—in the early 1680s he had already obtained several of the Astrononomer Royal's epoch 1680 coordinates for use in cometary computations, though these would not have been the ones that he needed for the purpose of chronology.[27] Figure 8.3, painted with figure 8.2 at Greenwich in 1727 by James Thornhill, places Newton together with Flamsteed, Kepler and Tycho.

(epoch 1601) with the stars ordered as usual by constellation. Hartgill, Gadbury, and Wing were active astrologers.

26 These were almost certainly CUL MS Add. 3969, which contain, in Flamsteed's hand, a set of stellar coordinates for the year 1686, together with lunar observations which he had made in 1677. *Lucida Pleiadum*, for example, has 1686 longitude ♉25°37′13″, *lat* 4°0′16′, compared with Flamsteed's final (1689 epoch) coordinates of ♉25°40′8″, *lat* 4°0′37″. Three years of precession from 1686 would produce a longitude of ♉25°39′43″, so there's about a 25″ difference between Flamsteed's 1686 and 1689 observations. Hevelius' coordinates, precessed to 1689, are ♉25°40′1″, *lat* 4°1′18″. *Lucida* figured in Newton's use of an observation due to Thales (see appendix E.ii).

27 See, e.g., MS CUL MS Add. 3965.Sec14.590: "The right ascention of ye star in ye heel of Perseus calculated from Mr Flamsteed's longitude & latitude thereof. . . ."

Numbers could, however, only be found once the objects they attached to were known, for which Newton needed images of the constellations to find such places as "the flexure of *Eridanus*." And that posed a problem, because no images were available from antiquity, with the exceptions of the ones carved onto the globe acquired by Cardinal Alessandro Farnese (on which see below). Johann Bayer's *Uranometria*, originally printed in 1603 and available to Newton in a 1655 edition, drew maps of the stars among figures of the constellations, and Newton relied on Bayer to link stars to Hipparchus' Eudoxan statements.

Once stars had been selected from the engravings and then their coordinates obtained from Hevelius or Flamsteed, Newton could proceed directly to calculation, though here he needed two different methods. Stars on the solstitial colures were simple to work with, because their latitudes were irrelevant for the computation. Stars on the equinoctial colures were more complex, because their latitudes had to be taken directly into account (for details see appendix D). Since Newton's calculations depended upon his star choices, which in turn derived from the Bayer atlas, we turn first to the origins of Bayer's images, for they too have histories.

During the late sixteenth century and throughout the seventeenth century, images of all kinds flooded the world of print and manuscript. To be sure, there were many image-makers, such as the empirical antiquarian Nicolas Claude Fabri de Peiresc, who nevertheless aimed at avalanches of descriptive words.[28] But from Vesalius' anatomical drawings through Galileo's depictions of sunspots to Hooke's startling graphic of a fly's eye, the proliferation of pictures conveyed arcane experiences and promulgated their authors' claims. And though many drawings depicted monsters and impossible natural objects,[29] image-makers and description-producers thought to depict or to describe objects of nature that did, or at least that might, exist.

What kind of object is the image of a constellation? Its associated star group certainly exists and can be drawn on a globe. But the images themselves, the depictions of mythological creatures and of people, are human products. Though anyone can see the stars glitter in the night sky, no one ever spied Perseus dangling Medusa's head in the heavens. Perseus and his Gorgon live only in the imagination and in the sculptures and drawings that artisans and artists had produced over the millennia. They are purely historical objects, born in antiquity and carried through to the engraving plates of Bayer's *Uranometria*. To draw Perseus, or any such figure, in relation to the stars, the *Uranometria*'s artist had to rely on previous depictions—themselves based on even earlier drawings and descriptions—and on his own imagination, which was conditioned by Baroque canons of representation. Problems inevitably arose whenever drawings were used to generate data, given the uncertain historical evolution of images from verbal descriptions.

[28] Peter Miller writes of the indefatigable Peiresc: "There are drawings in Peiresc's own hand interspersed throughout [his] archive, and also two surviving volumes of drawings executed by artists at his behest. These contain bowls, vases and sarcophagi, among other things … And yet, in Peiresc's archive it is the word—and the verbal description—that dominates" (Miller, 2005, p. 361). In general, *ekphrasis* "the verbal description of the visual … links the various continents of Peiresc's world of learning." There is considerable recent literature on the production and use of images in the early modern period, such as Svetlana Alpers' on Dutch art and nature (Alpers, 1983).

[29] Daston and Park, 1998.

Figure 8.4. The Farnese Globe, from Bianchini, 1752 (top), and Roman mosaic, North Africa, ca. 150–200 CE. Scorpio on left, Libra on right (Musée du Bardo, Tunis. Photo by Niccolò Guicciardini).

Depictions of the constellations date at least from late antiquity,[30] although only three globes marked with them are known to exist. The earliest one currently extant is made of brass, 110mm in diameter, now in the Zentral Museum in Mainz. Based on "stylistic comparisons" this globe probably dates to ca. 150–220 CE and comes perhaps from Roman Egypt. A small one made of silver and of uncertain date, with crudely placed depictions, is currently owned by a Parisian antiquities dealer.[31] The third is the Farnese, a second-century CE Roman copy of what may have been a second-century BCE Greek original; also in the second-century CE we have a Roman mosaic from North Africa that depicts both Libra and Scorpio (figure 8.4).[32]

[30] Muris and Saarman, 1961.

[31] Hélène de Cuvigny believes that this globe "is the work of a goldsmith who was not knowledgeable in astronomy and was probably simply cop[ied] from an existing sphere, likely to have been a repaired celestial globe" (Cuvigny, 2005).

[32] Bradley Schaefer revived the claim that the constellations on the Farnese globe are the only remaining relics of what may be Hipparchus' star coordinates (Schaefer, 2005). Schaefer reached his result by mapping known stars to loci where the constellations depicted on the globe intersect colures or tropics (which

Medieval manuscripts nicely illustrate the close dependence of constellation images on contemporary canons of representation. Figure 8.5 (top) presents two pages from a twelfth-century manuscript with illustrations of Cancer, Leo, Libra, and Scorpio; its images are evocatively mediaeval in style and do not depict stars. An earlier manuscript, the famous and colorful ninth-century Leiden *Aratea*, also provides constellation images, and here stellar positions are clearly indicated (figure 8.5, center). The figure for Gemini, e.g., seems to show most of the stars in Ptolemy's catalog, though the vagaries of depictions and the inherent looseness in description prevent precise identification.[33] Figure 8.5 (bottom) represents the zodiac as envisioned in fifteenth-century Venice. Each drawing strongly reflects local artistic conventions in their respective times.

Bayer's 1603 *Uranometria* was the first and for a time the only elaborate stellar atlas. In 1627, the Augsburg lawyer Julius Schiller decided to fully Christianize the constellations, replacing for example the Gemini twins with an image of James, the apostle, thereby disconnecting the constellation from past depictions (figure 8.6).[34] Schiller's figures never became common (though he did work with Bayer to improve star positions). Reprinted in 1655, Bayer's atlas was not the only useful rendition of the constellations that was available by the late seventeenth and early eighteenth centuries, whether on a globe or in the form of an atlas. For example, Hevelius' (epoch 1660) stellar coordinates, which were printed at Danzig in the astronomer's 1690 *Prodromus*, were accompanied by a separate set of fifty-nine woodcuts of the constellations which had been printed three years earlier.[35]

---

are also carved on it). Apparently using the same technique that he deployed for the Aratus data (Schaefer, 2004), he arrived by statistical methods at a date of −125 +/− 50, which corresponds reasonably well to the date of −129 for Hipparchus' catalog. The first to suggest Hipparchus as at least an inspiration was Georg Thiele (Thiele, 1898). Valerio, 1987 argues instead that the "globe represents the state of knowledge at the time of the Almagest" and not Hipparchus. Schaeffer's attribution has been persuasively challenged by Dennis Duke in Duke, 2005a, and, intemperately, by Dennis Rawlins in Rawlins, 2005, primarily by pointing out that there are many more discrepancies between the globe and Hipparchus' remarks in the *Commentary* than Schaefer notes. See Duke, 2005b for a further critique, wherein he points out that Schaefer's statistical method will not detect a systematic defect, or what Duke refers to as an "intrinsic error in the underlying coordinate system," which would affect all stars equally and so evade a test which chooses a date based on minimum chi-square statistics. See Tabarroni, 1971 for a brief look at Cassini's early attempt to date the Farnese, of which nothing remains but a notice that he had done so. A further challenge, on different grounds, to Schaefer's interpretation appears in Dekker, 2008.

[33] Katzenstein and Savage-Smith, 1988 note that the Carolingian MS is likely "a faithful copy of a manuscript probably made in the mid-fourth or fifth century." The text is the Latin translation by Claudius Caesar Germanicus of the first-century AD. Eastwood, 1983, dates one diagram to March 28, 579 using planetary configurations on fol. 93^v, which, however, derives from a different tradition than the *Aratea* illustrations.

[34] Schiller, 1627: see Targosz, 1988, p. 160 for brief remarks.

[35] Hevelius, 1687 (1690). See Warner, 1979 for a catalog and discussion of star maps; on Hevelius' *Firmamentum* see ibid., pp. 115–26. Elaborate star charts included Pardies, 1674, and there were in addition planispheres, such as those in Cellarius, 1661; Sherburne, 1675 or, later, La Hire, 1705. There was in addition Lubieniecki, 1666–1668. Though dedicated to comets, Lubieniecki's *Theatrum* provided star maps drawn from various sources against which to display comet positions. The one for the comet of 1664, which the young Newton observed (see above, chapter 1, and below, appendix B.ii), was drawn directly from a Blaue globe.

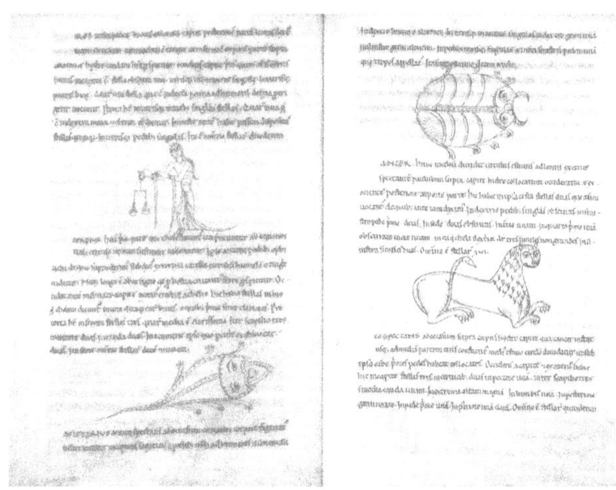

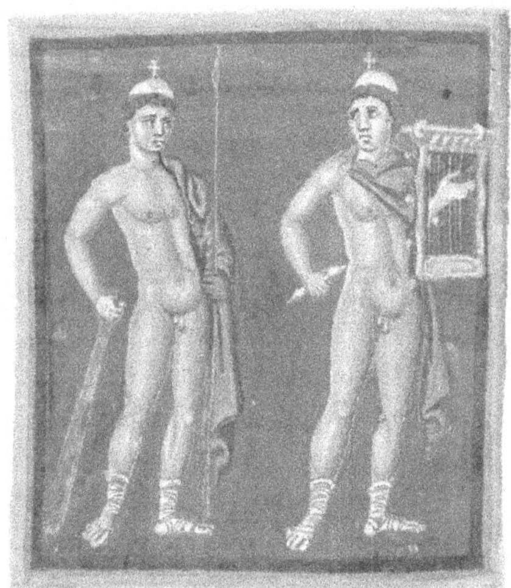

Figure 8.5. Top: Twelfth-century MS with Cancer and Leo to the right, and Libra and Scorpio to the left (Voss. MS Lat. Q92, folio 101 recto & verso). Center: ninth-century *Aratea* with Gemini (Voss. Lat. Q79, Fols. 16v). Both by permission of the Bibliotheek of the Rijksuniversiteit Leiden. Bottom: Zodiac from the Piazza San Marco, Venice. Photo by JZB.

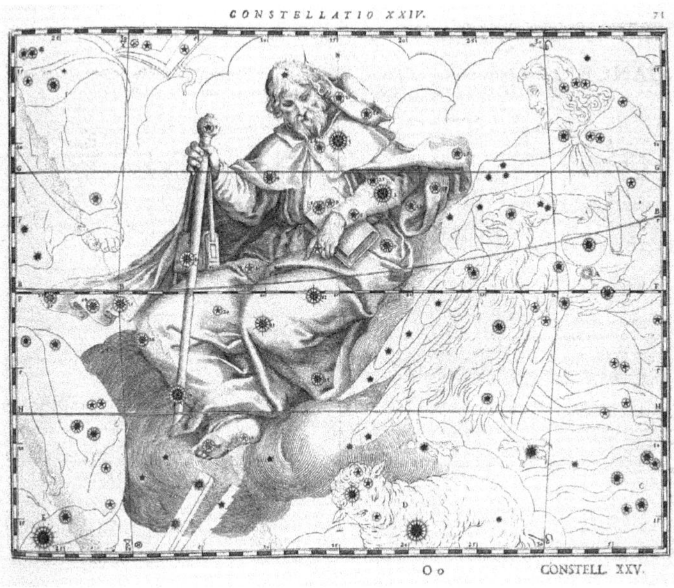

Figure 8.6. Schiller's 1627 depiction of *Gemini* as Saint James.

Bayer's *Uranometria* introduced Greek and Roman letters as star designators.[36] He relied to some extent on the thirty-nine woodcuts done by Jacob de Gheyn, which had been printed in Hugo Grotius' (Huig de Groot) edition of Aratus.[37] Grotius himself relied directly on the illustrations in the Leiden *Aratea* (figure 8.5, center), thereby introducing for the first time in print an "ancient style" that would later become quite common, albeit with many variations.[38] Hevelius used neither Bayer's star designators nor his illustrations. He relied for his drawings on (either) a celestial globe pro-

[36] The ordering first divides the stars into six groups by magnitude; the lettering begins with the stars in the group with the highest magnitude and then proceeds within that group in alphabetic order, loosely following either the graphic's periphery or longitude. The lettering continues sequentially with the next magnitude group: see Swerdlow, 1986, p. 192. Also see Argelander, 1842 for a careful discussion of Bayer's atlas. Swerdlow (1986) confirms Argelander overall, and further shows that a copy of Scheoner, 1561 in the John Carter Brown Library at Brown University bears the written name "Joannes Baÿer" as well as marginal lists of longitudes and latitudes for nearly all the stars in Bayer. Swerdlow concludes that this copy was owned by Bayer, and that the marginal coordinates may reflect Bayer's having had temporary access ca. 1597 to Tycho's catalogue before its publication in the 1602 *Progymnasmata*, and that he entered the values into his copy of Scheoner.

[37] Grotius, 1600. The images in the *Uranometria* were drawn by Alexander Mair, and those in Schiller's atlas by the Baroque painter Johan Mathias Krager (Targosz, 1988, pp. 159–60; Thieme and Vollmer, 1965).

[38] Dekker, 2004, pp. 54–63. The most elaborate verbal descriptions of the constellations appeared in the Latin *Poeticon Astronomicum* attributed in Newton's day to the first-century BCE Caius Julius Hyginus (but due to a subsequent author of the same name), which had been printed by Ratdolt in Venice in 1482 together with woodcuts in contemporary Venetian dress. Albrecht Dürer had drawn celestial maps in 1515, and Gerard Mercator had produced a celestial globe in 1551. There were others in the sixteenth century.

Figure 8.7. *Bootes.* Left: from a 1640 Blaeu globe now at the library of the University of Utrecht (photo by JZB). Center: Bayer atlas. Right: Hevelius atlas. The Blaeu globe and Hevelius atlas are depicted in globe view, whereas the Bayer atlas is in sky view.

duced by the Dutch cartographer Willem Blaeu, originally ca. 1600 (and based on Tycho's coordinates), or else on a 1648 wall map by Willem's son, Joan). Figure 8.7 shows depictions of *Bootes* by Blaeu, Bayer, and Hevelius.[39]

The Blaeu globes and map do not incorporate or come separately with a list of star coordinates, whereas Hevelius' maps were paralleled by his *Prodromus*. Neither did Bayer include coordinates, but on the reverse side of each map he noted the Ptolemaic star number that corresponds to the Bayer letter, as well as the star's magnitude and locus within the constellation. Bayer's atlas also depicted hundreds of stars that Ptolemy had not catalogued, most of which come from Tycho, Pieter Dirkszoon Keyser, and (he claimed) his own observations.[40]

The first star atlas that could yield reasonable coordinates, by Giovanni Gallucci, had been published in 1588.[41] It, and subsequent atlases, were based primarily on Islamic, and not on classical, images. Warner notes that "astronomers of the Renaissance had no easy access to correct classical constellation figures. The Farnese Atlas, the only extant globe from antiquity [but see above], was unearthed in the early sixteenth century, but its importance was not widely recognized until around 1700. There were no ancient copies of the *Almagest*, and medieval European ones were usually devoid of figures. The early European non-Ptolemaic astronomical manuscripts, which were often illustrated with constellation figures, were lost during the Renaissance, and not re-discovered until the nineteenth century ... With classical forms so rare, Renaissance artists worked largely through Islamic sources."[42] Warner identified no fewer than eight distinct traditions between 1500 and 1800 and notes that "cartographers presented vigorous figures, some appropriately dressed for a northern

[39] The actual figures on Hevelius' maps were drawn by the Danzig painter Andreas Stech, but the stars were entered by Hevelius himself. Stech had also drawn the illustrations for the *Machinae Coelestis pars prior*. The engraving for the *Firmamentum* was done by Charles de la Haye, a French emigré. See Targosz, 1988, p. 156 for details.

[40] The points made here are based on Warner, 1979, pp. xiii,18, 26–28, 93, 116.

[41] Warner, 1979, pp. x–xiii, 91.

[42] Ibid., pp. xii–xiii.

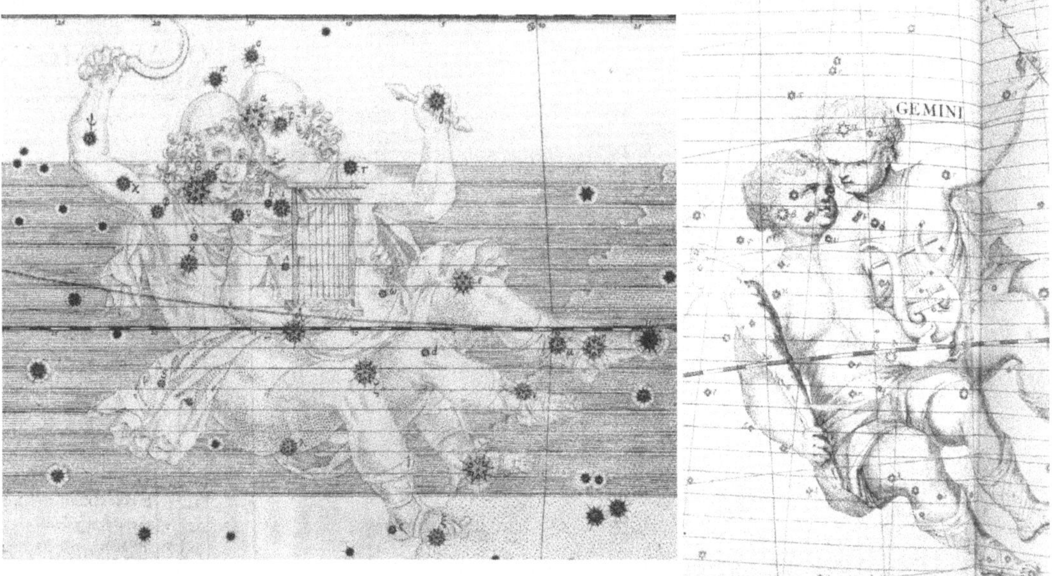

Figure 8.8. Bayer's (left) and Flamsteed's (right) *Gemini.*

European winter." Compare Bayer's drawings for *Gemini* with Flamsteed's figure 8.8), done about a century and a quarter apart.

These are quite different images. Flamsteed's twins look directly at one another; Bayer's stare out in different directions. Flamsteed's twins are nearly uncovered in an increasingly typical eighteenth-century representation of classical dress; Bayer shows billowing robes. Flamsteed's twins are uncapped and shorn; Bayer's are capped with flowing locks. One of Bayer's twins holds a sickle in a raised hand; Flamsteed's unraised arm does not. And so on. Compare also a sketch of Cassiopeia done at Uraniborg with the similar image published a few years later in the *Uranometria* (figure 8.9), where we can see how differently constellations might be depicted even when drawn just a decade or two apart.

Differences in representation could affect where stars were placed within a figure, as we shall see. Ptolemy's catalogue did, however, provide verbal descriptions of a star's location in relation to a figure, as well as its tropical coordinates. In *Pegasus*, for example, the first star listed has coordinates ♓ 17⅚, +26 (north of the ecliptic), has less than magnitude 2, and is described as "The star on the navel, which is [applied in] common to the head of Andromeda."[43]

In addition to questions concerning the placing of stars (particularly ones that are not mentioned in Ptolemy's catalogue),[44] the plates in Bayer's *Uranometria* occasion-

[43] Ptolemy, 1998, p. 358.

[44] Since most of the stars that Ptolemy lists are reasonably well-separated from one another, it's possible to map them to their modern equivalents. Some do remain ambiguous even so. For example, Ptolemy's star 14 in *Auriga*, the last in his list for that constellation, is described as "The small star over the left foot"

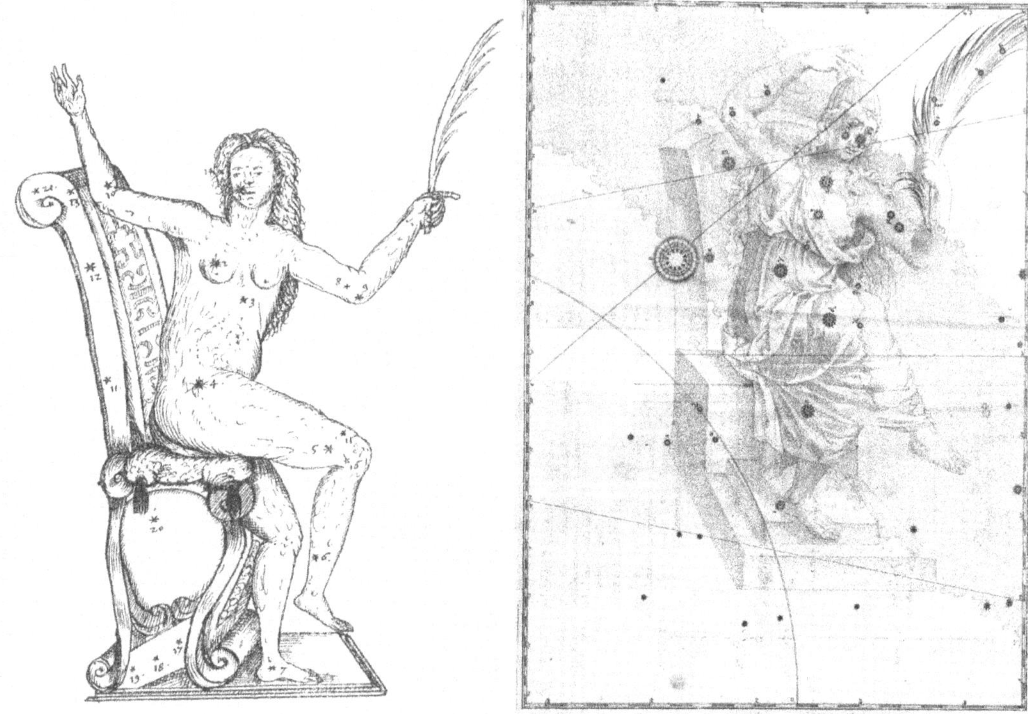

Figure 8.9. Uraniborg 1587 (left) and *Uranometria* 1603 Cassiopeia. Brahe, 1913–1929, vol. 211, p. 235.

ally suffered from an ambiguity, as Hevelius and Flamsteed both knew, and as Newton himself recognized. Compare for example the Blaeu globe's depiction of the constellation *Bootes* with Bayer's to see an example of the problem (figure 8.7). The depiction of constellations on maps and globes has a long and rather vexed history. Hipparchus had laid out the fundamental rule: the constellation should be so oriented as to reflect its proper disposition from the observer's position, looking up in "sky view" from the earth's center, or looking inward in "globe view" from a point beyond the heavens. A figure that appears in frontal view as seen from the earth should accordingly be drawn *on a globe* with its back to the viewer. As Dekker points out, Arabic cartographers drew their constellation maps from the front. This would be consistent with Hipparchus' rule provided that the depictions are correctly oriented as seen from the earth. Arabic cartographers, however, reversed their depictions left-to-right, producing mirror images of how they should appear in sky view.

(Ptolemy, 1998, p. 354). Because variant coordinates appear in the manuscript sources, its identity as the modern 14 *Aur* is, Toomer notes, "very uncertain" (ibid.). In Bayer's rendition of *Auriga*—seen from below, or in Newton's words "rightly delineated"—this could be either one of two stars. For a discussion of Ptolemy's catalogue see Grasshoff, 1990 and the early, but still invaluable, Peters and Knobel, 1915, which lists coordinate variants from extant manuscripts of Ptolemy's catalogue.

In the Latin *Aratea*, the twins of the Gemini constellation are reversed left to right as though they were meant to be drawn on a globe (and therefore to be seen from outside) but without depicting the images with their backs to the viewer (figure 8.5). The rule was usually correctly followed during the Renaissance, but Bayer, in Dekker's words, "made a complete mess of the principle" in the *Uranometria* because he was not consistent. Although his stars are always plotted in sky view, some of the constellations are nevertheless drawn from the back, in globe view, including two (*Perseus* and *Argo*)[45] that Newton used.[46]

Flamsteed attributed the difficulty with Bayer's images to problems in transforming words into images, and it's worth quoting his remarks in full to gain a flavor of his concern for philological as well as astronomical detail:[47]

> Tycho Brahe died in 1601; and two years after his death Bayer published his *Uranometria*, in which he gives maps of all the constellations. His figures are mediocre; but the stars are fairly well drawn in, according to their positions in Tycho's catalogue, and many other small stars are added which are not found in Tycho's catalogue, but which are inserted only on the basis of his own seeing them. It seems, indeed, when we compare them with the fixed stars which he adopted from Tycho's catalogue, that he used the same nomenclature, but since he drew all his human shapes (besides[48] Bootes, Andromeda and Virgo) with their back parts turned towards us, those stars which previous astronomers had placed on the right shoulders, ribs, hands, legs or feet, appear reversed in his figures ... The most likely reason for the error made by Bayer, is that when he met the phrases "the back" of "in his midriff" in the catalogue of Ptolemy, as often happens, and thus consulted a Greek lexicon to find their meaning, he always found νωτοσ given as "back" and μεταφρενον "back part" or "space between the shoulders". He concluded, therefore, that "interscapilium" [space between the shoulders] meant the space on the *back* between the shoulder blades, or the gap on the back between the lumbar region and the shoulder blades, so whenever either of these words occurs in any description of a constellation (except of Virgo or Andromeda) he represents the figure with its back towards us ... he thus reduced the most truthful observations to falsity and absurdity ... If Bayer had only drawn a map of Sagittarius, or any other of the human shapes, so that the stars, placed in hands, sides, feet and so forth, stood at the same parts in his figures then he would have perceived that every single one of them ought to turn its face to us. From this, he could have learned that in Ptolemy's Greek, the word νωτοσ means the "ribcage of the body"' or the ribs, and μεταφρενον means the space between the shoulders, or between the lumbar

[45] The prow faces right in Bayer's drawing, whereas it faces left in sky view.

[46] Dekker, 1992, p. 93; 2004, p. 56.

[47] Flamsteed himself noted that he was not the first to point out the problem, since it had been remarked by William Schickard, a friend of Kepler, who also produced the first mechanical calculator based on Napier's bones in 1623.

[48] Meaning "except for."

> region and the shoulder blades, not only at the back (tergum or dorsum) but also at the front of the body, being the space between the belly and the top part of the chest; and so there would be no difference between his figures and the descriptions of the ancients. Indeed, I am convinced that not only Ptolemy, but also Homer himself used these words in a broader sense than is generally admitted by the dictionaries. Nevertheless, in most maps of the fixed stars engraved after Bayer's, the figures are based upon those of Bayer and hence are subject to the same errors.[49]

Charging the long-dead Bayer with ignorance of Greek etymology, Flamsteed insisted on reinterpreting even Homer in the interests of matching verbal descriptions of the constellations to their graphical representations. Newton signaled his knowledge of the problem by inserting in his published text the qualifier "rightly delineated," which meant that the Bayer figure had to be reversed, with the star chosen accordingly.

Consider Bayer's drawing of the constellation *Perseus*. Hipparchus had specified that the equinoctial colure passes through *Perseus*' "head and right hand." The constellation as drawn by Bayer (figure 8.10) has this as the hand that holds the head of Medusa, and the figure is depicted in globe view. Newton accordingly shifted to the other hand, which holds the sword.[50] But a decision must still be made. Which star should be chosen? Only Ptolemy's catalog among ancient remains enabled the placing of stars in a constellation, yet many of the stars depicted here by Bayer do not occur in Ptolemy's list for *Perseus*, and where overlap does occur the correspondences are often very loose.[51] For example, though Bayer draws three stars on *Perseus*' sword, Ptolemy doesn't even mention the sword itself. In fact, Ptolemy catalogs only a "nebulous mass" on the "right hand," as well as one star each on the "right elbow" and the "right shoulder." The very dim star (k)—the only one that Bayer locates on the elbow—doesn't appear in Ptolemy's catalog at all, whereas the star that Ptolemy does place on the elbow is η, which Bayer draws on the forearm.[52] It was clearly not a simple, mechanical matter to choose a star in Bayer's drawings that corresponds to the verbal descriptions provided by Hipparchus.

[49] Chapman, 1982. Hevelius was well aware of Bayer's inconsistency, and he deliberately decided to maintain the representation as seen from outside rather than from the earth: see Targosz, 1988, p. 157, citing Hevelius, 1687 (1690), p. 5.

[50] In Newton's words, "In the right hand of *Perseus*, rightly delineated, is a star of the fourth magnitude called η by *Bayer*." Newton, 1728b, p. 89.

[51] Bayer depicts 39 stars in the constellation whereas Ptolemy lists only 26.

[52] See Ptolemy, 1998, pp. 352–53, which also gives the corresponding modern designator. This is usually the same as Bayer's, so that modern star catalogs can be used to find coordinates. Bayer's star k, e.g., has epoch 2000 right ascension 3h5′32.4″ and declination 56°42′21″, which in 1601 would put it at longitude 55°55′28″ and (at any time) latitude 37°29′29″. Bayer denotes relative magnitude by the graphic size of the star; η is fourth magnitude, whereas star β in Medusa's head is second magnitude. Star k would therefore be quite dim and likely invisible without a telescope or extremely careful observation. Nevertheless, Bayer extended Perseus' elbow to include it.

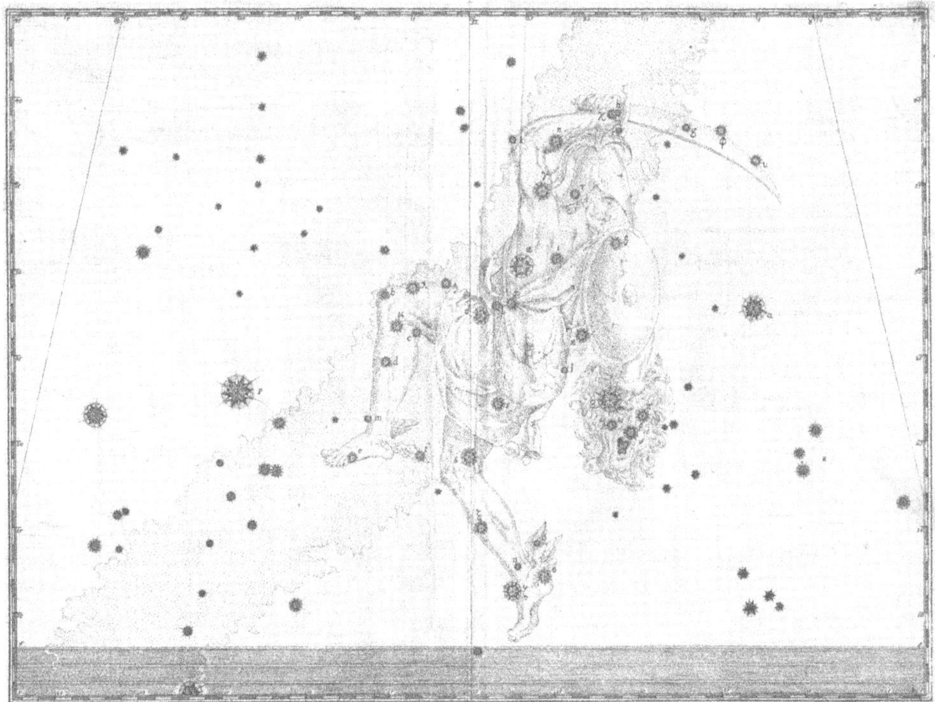

Figure 8.10. Bayer's star map and image for *Perseus.*

Flamsteed might have been even more severe with Bayer had he known that the maps of the *Uranometria* led him (Flamsteed) to place several stars in the wrong constellations. His nineteenth-century biographer, who also compiled the 1845 British Association star catalog, remarked perceptively of the consequence that "our modern map-makers . . . have suffered the evil not only to continue, but to increase to such a degree by subsequent innovations, that the celestial maps have at length become a system of derangement and confusion. For, a practice seems to have been adopted that whenever a modern astronomer has, in his catalogue, inadvertently introduced a star which he has designated by an erroneous constellation, the map-maker, or globe-maker (probably through ignorance) immediately extends the circuit of the constellation so as to embrace the star within its limits; although in so doing he causes the most inconvenient and absurd distortion of the boundary lines, and, in some cases, actually includes stars that ought not to have been disturbed; which consequently renders the map, or globe, a mass of confusion and intricacy, and totally unfit for accurate reference. An inspection of most of the modern celestial maps or globes will fully confirm this remark."[53] Bayer's, Hevelius', and Flamsteed's own maps exhibited these sorts of difficulties, all of which derive from the question

[53] Baily, 1845b, p. 54.

of which constellation a star should be assigned to, or, put differently, what stars should be included in a constellation.

To better appreciate the difficulties involved, and how Newton sought to resolve them, it is necessary first to analyze in some detail his relations with Flamsteed as he delved ever deeper into the intricacies of technical chronology. In 1694, Flamsteed began furnishing Newton with his observations of the moon, vital for the construction of a robust lunar theory. Before long, however, the relations between the two soured. The then Lucasian Professor of mathematics became exasperated over what he perceived as the obdurate refusal by Flamsteed to furnish him with additional lunar observations; for his part, Flamsteed resented what he viewed as lack of appreciation and reciprocity from Newton. In November 1698, Flamsteed committed another provocation. At the request of John Wallis, he composed an account of his observations of the parallax of the Pole star, which Wallis wished to publish in the third volume of his *Opera mathematica*. Inter alia, Flamsteed inserted a paragraph in which he attributed the slow progress of his observations to the assistance he had provided to others, including furnishing Newton with "150 places of the Moon deduced from [Flamsteed's] observations previously made, also with the places matching their times calculated from [his] tables ... as appropriate for the amendment of the Horroccian lunar theory." Upon being informed of this, Newton reacted strongly. He took umbrage not only at Flamsteed's furtive attempt to publicize his exchange with Newton, without first seeking permission, but at Flamsteed's reductive account, which insinuated Newton's efforts as a mere amplification of Jeremiah Horrox's lunar theory. On January 6, 1699, Newton demanded that Flamsteed remove all references to him—stating that he loathed "to be publickly brought upon the stage about what perhaps will never be fitted for the publick and thereby the world *put into an expectation of what perhaps they are never like to have*." Flamsteed attempted to mollify Newton by promising to remove the reference to Horrox's theory, and by explaining how he desired "to silence some busy people that are allwayes askeing, *why I did not print?*" Newton, however, remained adamant and Flamsteed instructed Wallis to remove the passage.[54]

Though no correspondence between the two exists for the next eighteenth months, Flamsteed visited Newton on several occasions. One such visit occurred on May 3, 1700, the day after Sloane approached Flamsteed for his opinion on the calendar reform. Judging by Flamsteed's account, the meeting was acrimonious, in no small part owing to his own belligerence. In fact, while waiting for Newton to receive him, Flamsteed decided to "cure" Newton's "suspitious" natural temper—and penchant "to be possest with calumnies especially such as are imprest with Raillery"—by inserting a piece of paper into a copy of a Bible he found lying on the table, on which he jotted a distich from Thomas Brown's "An Epitome of a Poem truly call'd *A Satire against Wit*": "A Bantring spirit has our men Possest / And wisdom is become a standing Jest." For greater effect, he further referred Newton to the first ten verses of Jeremiah 9, verses he found particularly fitting to describe Newton's evil minions.[55]

[54] Flamsteed, 1995–2001, vol. 2, pp. 735, 738–39, 742 (italics by Flamsteed).

[55] Ibid., p. 818. For example: "They be all adulterers, an assembly of treacherous men. (3) And they bend their tongues, like their bow, for lies; but they are not valiant for the truth upon the earth. For they proceed from evil to evil, and they know not me saith the Lord."

During this and other visits he paid Newton in spring 1700, discussion often revolved around Flamsteed's star catalogue, as the Astronomer Royal had informed Newton as early as January 2, 1699 of his "layeing in a stock of observations to rectifie the places of the fixed stars." By then, however, Flamsteed appeared to be rather secretive about details. Certainly, when informing one of his assistants of the addition of several stars to his catalogue, all calculated to the beginning of 1690, he quickly added a caveat: But "this I desire you to keep to your self, for my servants scarc know it and I purposly conceal it that if any one has a mind to steal it they may not know to what time they are done."[56]

Given his growing interest in ancient calendars, Newton may well have availed himself of the visits to consult Flamsteed about ancient astronomy. However be the case, by summer 1700, Flamsteed had embarked on a thorough examination of Ptolemy's star catalogue. As he informed the orientalist Thomas Smith—whom he had approached for philological assistance on March 21, 1701—for several months he had been engaged in translating "Ptolemys Catalogue of the fixed stars from the Greek in order to restore the antient figures." Nearly a year later, Flamsteed recounted to Abraham Sharp at greater length how he found all existing translations of Ptolemy to be "exceeding faulty by reason that neither the Arabs nor Copernicus were Critics in the Greek." He had also discovered "many errors committed in the different Copys of Ptolemys Numbers." Confident of his success in restoring the true text, Flamsteed bid farewell to philology: "I am fully tired with the Work of a Critick tis the Worst I was ever upon. and I am glad I am got through it for tis enough for me to prevent or correct my own faults." As noted above, Flamsteed also discovered major flaws in Bayer's and Hevelius' "altogether ungeometrical" celestial maps, which prompted him to commission a new set of maps.[57]

Newton was elected President of the Royal Society on November 30, 1703. The *Opticks* appeared on February 16, 1704, thus enabling Newton to seriously embark on technical chronology. With such a project in mind, we believe, Newton visited the Observatory on April 11, where he was shown Flamsteed's "books of observations," together with his "Catalogue with Tychoes and Hevelius's as also the Charts of the fixed stars." Newton "seemed pleased and offered to recommend them privately" to Prince George of Denmark, Queen Anne's consort. Flamsteed mistook Newton's benevolence to be a plot "to get the honor of all my paines to himself." But Newton truly wished to effect publication of Flamsteed's observations, not least because he sought to gain open access to them, and not be obliged to ask for them—thereby becoming beholden to Flamsteed as well as pressured to reveal the nature of his project. Flamsteed proved suspicious and obstinate. As he intimated to a friend, "haveing known [Newton] formerly and his sole regard to his own interests I was carefull to give him no encouragement to expect I should give him any thing gratis." So Newton pursued a different course. On November 30, 1704, he orchestrated the election of Prince George as Fellow of the Royal Society. Eleven days later the Prince charged Newton and other members of the Society "to inspect Mr Flamsteeds Papers and consider what is fitt for the press." Wasting little time, Newton summoned Flamsteed to dinner

[56] Ibid., vol. 2, pp. 738, 689.

[57] Ibid., vol. 2, p. 892.

on December 19 to discuss publication, and ten days later he spent five hours in the Observatory going through the papers of the Astronomer Royal, who had by then become quite excited by the project. On January 2, 1705, Newton finally received what he needed, Flamsteed's charts, not then his coordinates: "Yesterday I sent you according to your desire and my Promise," Flamsteed wrote,

> my Old Charts of Orion, Ophiuchus, Aquarius, and Pisces, these have all the circles of Longitude and Parallells of Latitude inscrib'd in them as you will find them in the New one of Orion, which I think is Compleately finish'd that by it you may judge of the Rest which I have design'd of the same bigness ... With these I have sent you my Greek Ptolemy and my Latin Version. Concerning which I must informe you that because I differ From the Common Translations I have thought it necessary to keep as close to the Greek as I handsomely could, and therefore you must carry the *Quae est* that begins the head or second Line of Every Constellation to the beginning of every Line Following it, And further that if you think it not advisable to Print Ptolemy's Greek text, it will be best to Put those Words, in whose interpretation I differ From others in the Margin, For if they be inserted with a Parenthesis, in the Text, they will make the Line too Long. I have also sent you a Bayer, that by compareing his descriptions and Figures with the Greek and Mine, you may see how (by takeing the Meaning of Greek Words From Lexicons) he makes all those Stars to lye on the left Sides, legs, Arms, etc which the Originalls and all the Printed Catalogues (except his Own) Put on the Right, and Vice Versa; and moreover Often thrusts such Greek Words into his Text as are No Where to be found in Ptolemy.[58]

Flamsteed elaborated on the topic in letters to close friends. He apprised James Pound on November 15, 1704 of his translating Ptolemy afresh, as well as of his opinion regarding the pictorial representation of the constellations. "Which causes me to think," he concluded, that "Ptolemy did not designe his Figures For the Globe but his Astrolabe, which they very well fitt. Bayers Mapps which our Globe-Makers Copy are all false; Hevelius's are onely fitt for Globes, and No body can learn the Constellations by them."[59] Nor did Flamsteed keep his opinion secret. David Gregory jotted down on January 21, 1704/5 extensive, and critical, remarks of Flamsteed's views, which the Astronomer Royal expressed during a visit that he and Newton paid to the Observatory a week earlier. Both then and in the following April, during another visit by Newton to the Observatory, Flamsteed displayed the new maps of the constellations that the draughtsman Paul van Somer had prepared for him in late 1703 and early 1704. Small wonder, therefore, that Flamsteed was foremost on Newton's mind as he delved deeply into the astronomical part of chronology.[60]

[58] Ibid., vol. 3, pp. 114, 116–17, 123.

[59] Ibid., vol. 3, p. 102.

[60] RS MS 247 fols. 78–78A^v; Baily, 1835, pp. 65–66. Flamsteed continued to broadcast his opinion widely. When in 1710 Zacharias von Uffenbach visited the Greenwich Observatory, he heard Flamsteed boast:

As for Flamsteed, his euphoria soon waned, owing to his own uncompromising views regarding the content and layout of the work, which ensured that little progress was made during the following five years. Increasingly exasperated by Flamsteed's obstinacy, Newton resorted to more drastic measures. First, he humiliated the Astronomer Royal by orchestrating his expulsion from the Royal Society in January 1709. Then, in December 1710, he secured for the Royal Society the role of Visitors of the Observatory. Almost immediately Halley assumed complete editorial control and the "spurious" *Historia coelestis* (Flamsteed's coordinates) was published in 1712. The details of this unsavory history need not detain us here. What is relevant is that Newton refused to return the manuscripts he had received from Flamsteed for a full decade, as long as he strove to perfect the astronomical part of his chronology.

We know that Flamsteed wanted his star catalog to include the positions determined by Ptolemy, Ulug Beg, Tycho, the Landgrave of Hesse, and (especially) Hevelius, as well as the "small catalogs of the Arabs," all of these to be placed in a third volume, as well as new translations of Ptolemy's descriptors. The manuscript that he eventually handed to Newton, and which was the outcome of fractious discussions, did contain Flamsteed's translations of descriptors, but it did not include previous astronomers' coordinates. We know as well that Newton had begun colure computations no later than 1706, when he used not Flamsteed's but Hevelius' coordinates. No copy of Hevelius' catalog is listed in the register of Newton's library, though he might have borrowed or seen it, or he might have obtained Hevelius' coordinates from Flamsteed himself. We do know, however, that by 1683 at the latest, Newton knew of Henry Philippes' *Kalender*, with its provision of right ascensions and declinations for 65 stars at epoch 1660, and he had long before utilized Wing's *Harmonicon* of 1651, which contains the full list of Tychonic coordinates (epoch 1601) together with stellar descriptors in English. And after 1669, Newton had Vincent Wing's *Astronomia Britannica*, which reprints Tycho's catalog, this time with Latin descriptors.

At some point before Newton had easy access to either Hevelius' or Flamsteed's stellar coordinates he prepared a table of longitudes, latitudes, right ascensions, and declinations of "ye fixed Starrs for ye yeare 1671. Of ye three first Magnitudes," though he did not also specify the source of the coordinates.[61] There is little doubt that the list was extracted by Newton from Wing:[62] the latitudes are all subsets of the ones printed by him (i.e., Tycho's), while the longitudes are all Tycho's plus exactly one degree. Newton apparently added a degree to the Tychonic longitudes and then computed

"he had entirely disproved Bayer and his constellation or 'Asterismos,' and he lamented that neither Bayer nor others had rightly understood the *Ptolomaeum*"—which Flamsteed claimed to have emended and restored. Uffenbach, 1935, p. 22.

[61] CUL MS Add. 3958. 2, fols. 38–40^{v}.

[62] Wing, 1669, pp. 237ff. (second pagination). This means that Newton probably produced the list before 1690, the year in which Hevelius' coordinates were printed, since he subsequently used Hevelius', or at least that it predates the year when Newton first obtained Hevelius' results. Since Newton went to the considerable trouble of computing right ascensions and declinations for every one of the six pages of stars that he compiled on paper which he carefully ruled into columns, it's unlikely that he would have used Tycho's coordinates if Hevelius' had been easily available to him, though Newton did have Wing's *Astronomia*, and not Hevelius, in his library.

the corresponding right ascensions and declinations. However, though Newton specifies the result as pertaining to "ye yeare 1671," the value for precession which he later used (in both the chronology and the *Principia*[63]) was one degree every 72, not 71, years. And since Wing clearly marks the table as "ad annum Incarnationis M.DC. completum,"[64] 72 years would yield the end of 1672 as the epoch, not 1671. However, whereas in the 1651 *Harmonicon* Wing had a value for precession equivalent to 70.6 years, in the *Astronomia* he produced a final result of 71.4[65]—and 71 is an exact mean between these two numbers. Consequently there's little doubt that Newton's table of stellar positions for 1671 was produced using Wing's numbers, perhaps even by taking a cut midway between the precession rates of the *Harmonicon* and the *Astronomia*. Moreover, the table must have been done before the mid-1680s given the value for precession in the *Principia*. Hevelius' coordinates were more accurate.

In any event, since Newton did not have versions of Flamsteed's own coordinates for the relevant stars for some time, it is not surprising that he calculated first with Hevelius. Having received in 1705 the charts sent by Flamsteed—who listed more stars than Hevelius—Newton faced the dilemma of how to choose among them. Consider the constellation *Perseus*. In the published *Chronology* (as well as in manuscript[66]), Newton mapped Bayer's star η to Hipparchus' "star on the right hand" despite Bayer's forearm placement.[67] Bayer's χ is better placed in Bayer's depiction than η, although it is at best what Ptolemy's catalog calls a "nebulous mass." Flamsteed did provide coordinates for χ,[68] but Hevelius did not, and so initially Newton would not have been able to choose it at all. But suppose he had done so—suppose that Newton had had all of Flamsteed's star coordinates available from the beginning of his work, as he certainly did years before he and Halley published their purloined edition of Flamsteed's coordinates in 1712. What would he have found?

[63] Newton, 1999, p. 887.

[64] The English *Harmonicon* titles the table "Tycho Brahe, and by him rectified to the beginning of the year of mans redemption 1601" (Wing, 1651, p. 240). The title page of Tycho's work also reads "Observationes, Canonica Determinatio, *Ad Annum completum*, 1600," with italic emphasis in the original (Brahe, 1602 (1969), p. 257).

[65] Wing, 1651, p. 117. Wing, 1669, p. 191 lists thirteen values for the annual rate of precession calculated by comparing the difference between the loci from two historical reports (e.g., between Ptolemy and Tycho), concluding somehow ("hinc conclusionem habemus") that the "exact" rate is 50″25‴5⁗ per year, or 71.4 years per degree. This number is not the result of an average, which would give instead 50″1‴6⁗, which would be just less than 72 years per degree. Wing was an excellent computer, so the large (24′) discrepancy between his strangely exact result and the average cannot be due to his having incorrectly computed a mean, which would in any case be highly unlikely given contemporary ways of handling discrepant values. Wing must have had some sort of procedure for combining the numbers, but he did not explain what it was. Riccioli, 1651, p. 479 used 72 years per degree, or 50″ per year, as did Newton from the time of the *Principia*.

[66] See, e.g., appendixC.i.d.COLURES 4.

[67] Ptolemy's *Almagest* specified it as being on the "right elbow." Newton probably never looked at the *Almagest* catalog itself, using instead Flamsteed's descriptors.

[68] In the purloined catalog (Flamsteed, 1712) χ is listed as "In Capulo Ensis, ad manum Aust.," that is, explicitly on the south (corresponding to the right) hand holding the sword with longitude ♉19°56′48″ and latitude 40°43′20″.

Precessing back, we obtain the (to Newton) unacceptable date of −323. He likely decided in the end not to use χ because η gave him the much better result of −826.[69] Here we have our first example of a common feature in Newton's work with astronomical words and images. As far as he was concerned, both Hipparchus' remarks and Bayer's drawings were just sign-posts to the appropriate stars; neither word nor image provided the kind of data that Flamsteed or Hevelius had produced. Words from the past had to be manipulated in appropriate ways in order to turn them into data, and images based on words—fanciful pictures that, Newton thought, referred to historical individuals or events—were even less inviolate. The integrity of word or picture took second place to more pressing Newtonian requirements. These were above all his theory of civilization, and, especially, the need to ensure coherence among the data items that Newton used.

To comprehend how he worked we will again begin at the end, with Newton's printed *Chronology*, which published the locations of the ancient colures for ten distinct items.[70] The first four columns of table 8.2 give respectively one of Hipparchus' Eudoxan remarks (as quoted by Newton and modified by 'E' for an equinoctial and 'S' for a solstitial), the star Newton chose (creatively identified using Bayer), its latitude, and the epoch 1690 longitude (taken from Flamsteed's coordinates as published in 1712). The fifth column gives accurately computed dates, and the final column provides the dates that follow from Newton's computed value for the longitude in 1690 of the point intercepted on the ecliptic by the ancient colure.[71] Both columns use a value for precession of 72 years per degree.[72] Note again that Newton in print utilized a total of ten distinct elements of Hipparchan data, though he had listed twenty altogether. We'll return to the omitted items below.[73]

[69] Newton did not give dates for each of his calculated stars, but we can use formula 1 in appendix D to recapture the equinoctials; the solstitial dates require only simple arithmetic. Newton did provide the longitudes at which the equinoctial colures cut the ecliptic in 1689 (the year of Flamsteed's data) from 1.

[70] See appendix D for Newton's method of computation.

[71] There's little difference between the solstitial dates using Newton's instead of more accurate computations because nothing beyond simple arithmetic is needed here.

[72] For important reasons, as we shall see, Newton kept the equinoctials and solstitials separate from one another, but the average date that results from all of the entries in the last column of table 8.2 taken together is −937. Replacing Newton's equinoctials with the accurately computed ones from the second-to-last column instead produces −933.6. The three-year difference between the two values reflects the inaccuracies inherent in Newton's use of logarithmic tables for the tangent and sine in calculating equinoctials. As we noted above (note 65), Newton's value for precession, 72 years per degree or 50″ per year, is the same as Riccioli's (Riccioli, 1651, p. 479), as well as Wing's (Wing, 1669), and is also, to the nearest second, the value Newton later computed from theory (Newton, 1999, p. 887: Book III, Prop. 39, Prob. 20).

[73] *Aries:* Newton's published value for the longitude of ν (♉ 9°38′45″) was apparently misprinted from Flamsteed (who has ♉ 9°48′35″) by inversion of the first digits between the minutes and seconds. Newton's MS (New College MS 361.2 fol. 82) has the Flamsteed value, but apparently in recomputing for the printed volume he used the inverted value. Using the correct longitude the date becomes −983.5 instead of −972.7.

Cetus: Newton's chosen stars appear on the figure's neck and mane respectively, not its head, in Bayer. Here we have a double problem in computation, because we don't know Newton's result for the coordinate of the midpoint between these two stars. Horsley computed the midpoint at ♉ 3°35′22″, lat

TABLE 8.2.
Newton's colure stars with their coordinates, magnitudes, and the dates that result from them. See Flamsteed, 1712.

Hipparchus	*Chosen Star*	*Latitude*	*1690 Longitude*	*Accurate Date*	*Newton's Date (from his computed colure longitude)*
E:Aries back	Aries ν mag 6	6°7′56″	♉ 9°38′45″	-971.7	-972.7
E:Cetus head	Cetus ν and ξ mags 4, 4	-9°12′26″ -5°53′07″	♉ 4°3′9″ ♉ 3°7′37″	-966.2	-972.6
E:Eridanus flexure	Cetus ρ mag 4	-25°15′50″	♈ 25°22′10″	-989.2	-989.2
E:Perseus head	Perseus τ mag 5	34°20′12″	♉ 23°35′30″	-924.6	-924.7
E:Perseus right hand	Perseus η mag 4	37°26′50″	♉ 24°25′27″	-828	-826
S:Cancer Middle	Cancer δ mag 4		♌ 4°23′40″		-786.4
S:Hydrus Neck	Hydrus δ mag 4		♌ 5°59′3″		-900.9
S:Argo between poop and mast	Argo ι mag 3		♌ 7°5′31″		-980.6
S:Sagitta Middle	Sagitta ϑ mag 6		♒ 6°29′53″		-937.9
S:Capricorn Middle	Capricorn η mag 5		♒ 8°25′55″		-1077.1

Let's next compare Newton's choice of stars with one made recently by a historian of ancient astronomy who sought to use the combined data in Aratus and Hipparchus to locate the latitude and date at which the original observations were made.[74] Table 8.3 lists his alternative star choices next to Newton's, as well as the dates that result from using Newton's method of computation.[75]

The star choices agree with one another for only one entry in the list—the very last, for *Capricorn*. And the results are suggestive: where Newton's dates produced an average year of −937, with a standard deviation of only 84, the alternative star choices (ignoring the date for Argo) give a mean of −985, with a huge standard deviation of 468. Had Newton chosen these stars, instead of the ones that he did, his chronology would have been much shakier, since he would have immediately seen that there was not a single date close to the average (and since, as we will shortly see, Newton always computed averages).

Newton's choices appear in stark outline if we list his stars together with Ptolemy's, and the corresponding entries in Flamsteed's 1712 and 1725 catalogs, from which we see that two of Newton's stars (*Argo* ι and *Sagitta* ϑ) do not appear at all in Ptolemy's catalog[76] (table 8.4).

−7°32′46″, which would produce the "accurate date" listed in the table (Horsley, 1779–1785, vol. 5, p. 69). There is about a 5′ difference from Newton's result, which is considerable in terms of computational accuracy.

Eridanus: Newton Cetus ρ is in the whale's chest, forming there part of a stellar square. He justifies the choice by asserting that in "the extreme flexure of Eridanus, rightly delineated, is a star of the fourth magnitude, of late referred to the breast of Cetus, and called ρ by Bayer" (Newton, 1728b, p. 88).

Perseus head: Newton's published 1690 longitude was ♉ 23°25′30″, an obvious misprint since his result is nearly correct for the (Flamsteed) longitude of ♉ 23°35′30″. Specifically, Newton computed the original colure to have cut the ecliptic at ♉ 6°18′57″; an accurate computation using Flamsteed's longitude put it at ♉ 6°18′47″. If the printed longitude were used instead, the ecliptic point would have been ♉ 6°8′46″, which is much beyond the accuracy of Newton's computations in general.

Hydrus: In both Bayer's and Flamsteed's atlases, *Hydrus* δ appears on the serpent's head or forehead, not its neck.

[74] Schaefer, 2004, pp. 171–73. Our purpose is not to judge the cogency of Schaefer's arguments, but rather to use his stars as a test bed to see just how labile the choice is. The comparison is particularly interesting because Schaefer was apparently unaware of Newton's calculations, which means that his choices are not contaminated by prior knowledge of Newton's. And so we have a perfect experimental configuration, since neither star-chooser knew about the other.

[75] "Newton's Date" column is computed from the equinoctial and solstitial coordinates on the ancient sphere that Newton calculated and printed. "Alternative Date" column numbers are not precisely the dates that Newton would have obtained had he used Flamsteed's data for these different stars because they are computed by Schaefer from the right ascension and declination for an epoch 2000 star catalog and not from Flamsteed's coordinates.

Aries θ is on the ram's nose, not its back, in both Bayer and Flamsteed, though the nose reaches over the back.

Cetus γ is on the serpent's neck in Bayer but at the top of its open mouth in Flamsteed.

Eridanus η is on the upper flexure as depicted by Bayer; see figure 8.13.

[76] The Ptolemaic descriptors are from Ptolemy, 1998.

Eridanus: Newton could as well have chosen τ in *Eridanus* proper, considering the original remarks, which is also 4th magnitude, but that would have put the equinox date at about −1477, which explains

TABLE 8.3.
An alternative set of stars

Hipparchus	*Newton's Star*	*Newton's Date*	*Alternative Star*	*Alternative Date*
E:Aries back	Aries ν	−972.7	Aries θ (Sch 158H)	−617
E:Cetus head	Cetus midpoint of ν and ξ	−972.6	Cetus γ (Sch 157H)	−1218
E:Eridanus Flexure	Cetus ρ	−989.2	Eridanus η (Sch 156H)	−1610
E:Perseus head	Perseus τ	−924.7	Perseus midpoint of τ and χ (Sch 159H)	−656
E:Perseus right hand	Perseus η	−826		
S:Cancer Middle	Cancer δ	−786.4	Messier 44 (Sch 143H)	−677
S:Hydra Neck	Hydra δ	−900.9	Hydra θ (Sch 144H)	−1620
S:Argo between poop and mast	Argo ι	−980.6	(Modern Vel) = Argo ψ (Sch 145H)	−3686[?]
S:Sagitta Middle	Sagitta ϑ	−937.9	Sagitta δ (Sch 147H)	−402
S:Capricorn middle	Capricorn η	−1077.1	Capricorn η (Sch 146H)	−1077.1

Did Newton just pull out whatever stars gave him what he wanted, regardless of Bayer's charts, Hipparchus' wording, or Ptolemy's list? To find out, examine the connections between his chosen stars, Hipparchus' words, and Bayer's charts. Consider the stars ν and ξ that he chose for *Cetus*, where, according to Hipparchus, the equinoctial passed through the constellation's "head." Newton set the colure at the midpoint between these two stars; yet Bayer's graphic (figure 8.11) unambiguously shows Newton's two stars on the upper neck of the beast, not on his head.

Two of the stars that are depicted as on the head in figure 8.11 (α and λ) produce dates, respectively, of 1589 and 1490 BCE. If we turn to Ptolemy's star catalog, we find that these two are (again respectively[77]) on "the end of the jaw" (in Bayer, on the

Newton's comment that the one he chose "is the only star in *Eridanus* through which the colure can pass." 'praeced.' is "In Occipite."

Perseus: This star is not however the *Bayer* η: see note c to table D.1.

[77] And according to Toomer's correspondences in Ptolemy, 1998, pp. 381–82.

TABLE 8.4.
Newton's star choices compared with Flamsteed's and Ptolemy's lists

Hipparchus	*Bayer Letter*	*Ptolemy [Translation]*	*1712 Catalog*	*1725 Catalog (Ptolemaic stars)*
E:Aries back	ν	Star on the rump	In lumbis Borea	In Renibus
E:Cetus head	ν and ξ			
E:Eridanus extreme flexure	Cetus ρ	Quadrilateral in chest: northernmost star on the advance side	In [Quadrato] Pectoris Borea praeced.	In [Quadrato] Pectoris praecdentis Lateris Borea
E:Perseus head	τ	Star on the head	In capite Persei	In capite
E:Perseus right hand	η	Star on the right elbow	In superiori brachio	In dextro brachio
S:Cancer Middle	δ	Quadrilateral containing nebula—southernmost of these two	Austrina in [quadrato]	Duarum sequentium in [quadrato] Borea, Asselus Austrinus
S:Hydrus Neck	δ	The northernmost of the 5 stars in the head, above the eye	In capite praecdentium Borea	5 in Capite Borea supra Oculum
S:Argo between poop and mast	ι	Not listed	In suprema Puppi sequens	Not listed
S:Sagitta Middle	ϑ	Not listed	Cuspidem sequentium trium Media	Not listed
S:Capricorn Middle	η	Northernmost of 3 stars in middle of body	In medio corpore 3um Borea	Trium in medio corpore sequences - reliquarum duarum Borea

snout), and on "the tip of the nostrils" (in Bayer, just above the eye). Either one, given Bayer's drawing, seems to fit Hipparchus' words. But Ptolemy's catalog has more to say. According to it, Newton's two chosen stars (ν and ξ) are respectively "on the eyebrow and the eye" and "to the north of this, about on the hair."[78]

Given Ptolemy's wording, Newton's choice seems to fit Hipparchus' words even better than the alternatives, despite Bayer's drawing. Newton may have directly consulted Ptolemy's catalog when, as in this case, he found discrepant results using Bayer's images (which suggest α and λ) and thereby discovered that he could reasonably

[78] The alternative (γ) appears in Ptolemy as "in the middle of the mouth."

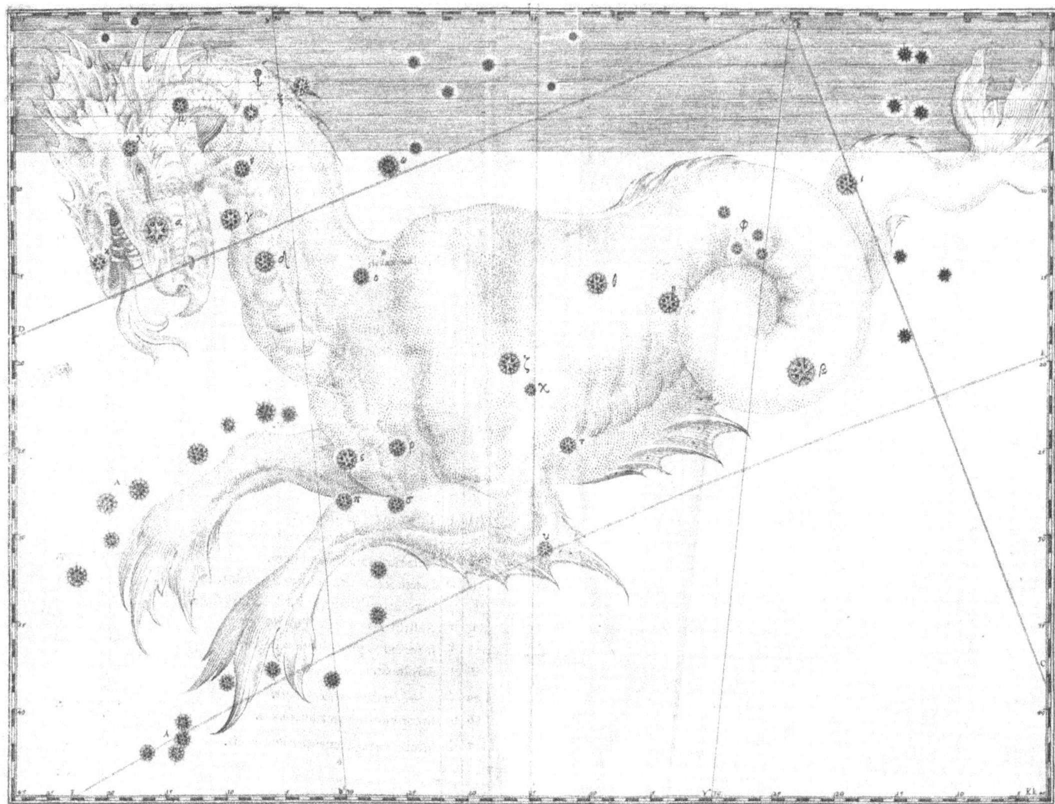

Figure 8.11. Bayer's star map and image for *Cetus.*

change his choices. Newton did not even need Ptolemy's catalog proper, because the descriptors that were included in the 1712 catalog based on Flamsteed's manuscript work just as well: ν and ξ are there respectively described as above the eye and on the mane, essentially as Ptolemy had described them. Since Newton gained access to both Hevelius and Flamsteed as soon as he started serious computation, he must have decided early on that the Bayer drawings themselves were sufficiently malleable that, for example, no reliable distinction could be dawn between stars very near and stars actually on the head of *Cetus*, especially given what he called the "coarse" observations of the ancients.[79]

More information about Newton's method for choosing stars emerges from one of the two most peculiar entries in his list—the solstitial colure through the "middle" of Sagitta. As we saw in table 8.4, the star that he chose here doesn't exist at all in Ptolemy's catalog, but it does in Flamsteed's, who however designates it as on the arrow's

[79] Newton, 1728b, p. 89. We have already seen that his choice of star η for Perseus could not be ruled out, and his star τ is in fact on Perseus' head in Bayer, and is so described by Ptolemy. Aries ν is squarely on the back in Bayer (and is described as "on the rump" in Ptolemy, who gives no star on the back per se).

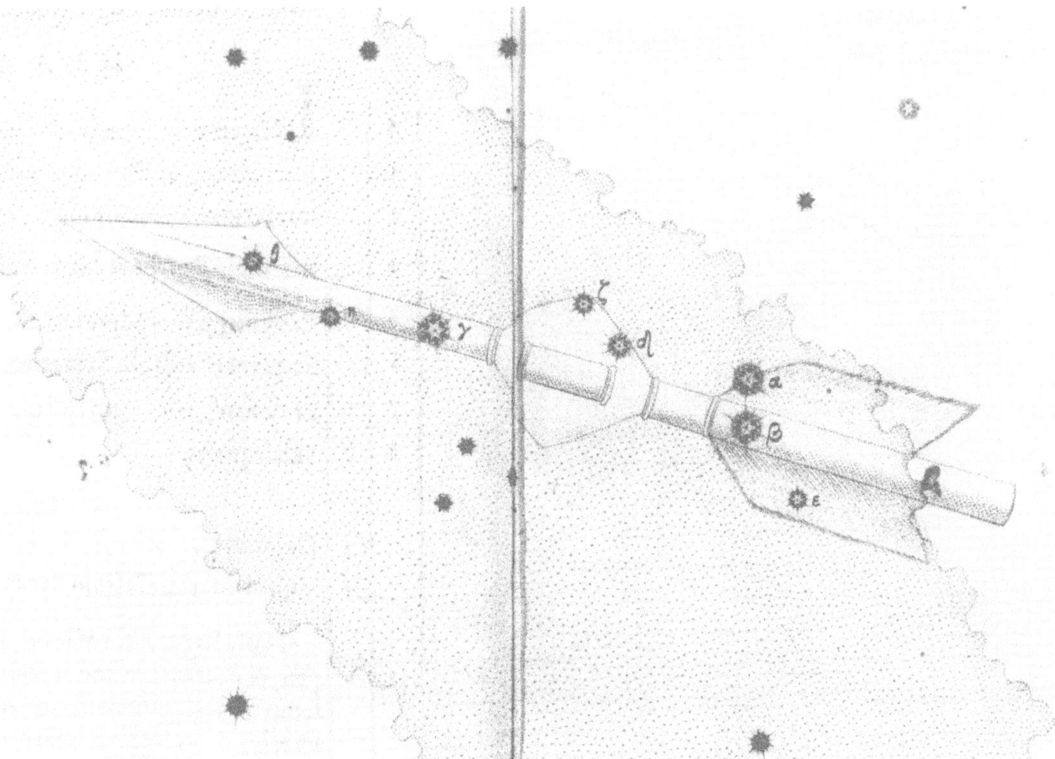

Figure 8.12. Section of Bayer's star map and image for *Sagitta.*

tip ("cuspidem"). It is worthwhile looking again to the Bayer image to understand what Newton was doing (especially since Whiston took him to task on it). Newton's chosen star ϑ appears in Bayer's figure 8.12 on the arrowhead. Hipparchus quoted Eudoxus as having placed the colure through the middle of the arrow, which, in the drawing, corresponds to star δ (and which Flamsteed explicitly describes as in mid-shaft). But that star, as we have already seen in table 8.3, yields a date of 402 BCE. This date might approximately fit a conjecture that Eudoxus had himself done at least some of the observations, but at the cost of making it a bad numerical outlier. Newton might instead have chosen star ζ, which is also near mid-arrow, but that produces the even worse date of 140 BCE. Stars γ and η would set the original globe at 202 BCE and 504 BCE, respectively. Those at the arrow's tail put the date well into the Common Era, and so the only possible choice for Newton was the one that he made, despite the mismatch of Hipparchus' wording to Bayer and Flamsteed (to say nothing of Ptolemy[80]). Among the equinoctials one star raises a serious question: Newton's choice of *Cetus* ρ for the colure though *Eridanus*' "flexure." Newton claimed that

[80] If, as seems likely, Newton turned directly to the copy of Ptolemy's catalog borrowed from Flamsteed, then he had to know that the catalog does not include ϑ Sagitta. Bayer listed the stars depicted in his maps on their backs, but he nowhere indicated which ones were also in Ptolemy's catalog.

Figure 8.13. Bayer's (left) and Hevelius' renditions of *Eridanus* (Hevelius, 1687 (1690)).

Cetus ρ is the correct choice because only ("of late") had it been moved from *Eridanus* to *Cetus*. He cited no evidence to that effect, but it seems likely that he followed Hevelius—whose 1690 star atlas shows the *Eridanus* flexure completely overlapping *Cetus*' chest. It's instructive to compare Bayer's rendition with Hevelius' (figure 8.13). Bayer includes his designators but omits the relation to neighboring asterisms, as he does in all of his star maps. Hevelius on the other hand omits the Bayer designators but shows nearby constellations. Moreover, Hevelius' *Eridanus* is, in Newton's terminology, "rightly delineated" (i.e., flipped left to right), whereas Bayer's is not. From Hevelius' map Newton could tell that the "extreme flexure" of *Eridanus* overlapped *Cetus*' chest. He could then turn directly to Bayer's map of *Cetus* to find the appropriate designator for the star and thence its coordinates.[81]

Newton commented that ρ (actually in *Cetus*) was "the only star in *Eridanus* through which this [equinoctial] *Colure* can pass." What did he mean by this? Was it just a statement that any other star in *Eridanus* gave dates that conflicted with his chronological scheme? Or is it possible that Newton chose the star because of a reasoned position that might be called *proto-statistical*: namely, that any other star in *Eridanus* would not cohere at all well with the other members of his data set *itself*, which is certainly true. He had after all supposed from the outset that Hipparchus' Eudoxan remarks all referred to observations that were performed at nearly the same time. If (as he also assumed) the observations were "coarse" but accurate to within a degree at the very worst, then all, or at least most, of the dates had to lie within some comparatively small interval, say three-fourths of a century. Since Hipparchus' words and stellar maps usually offered more than one possible star for any given remark, Newton may have felt on firm argumentative grounds in choosing a star that fell closest in line with the other members of the set. Whiston would later challenge the selection.

[81] Newton's particular choice is certainly problematic nevertheless. Ptolemy lists a star in *Eridanus* that is "the first star in the bend of the river, which [star] touches the chest of Cetus" (Ptolemy, 1998, p. 385). Flamsteed gives epoch 1690 coordinates of this star (one of Bayer—there are two nearby labeled) as longitude 27,43,50 and south latitude 32,46,03 (Flamsteed, 1712, p. 52). That would give a date of about −1495, far outside the other members of Newton's set.

Though Newton did not comment on his star choices, either in print or in manuscript, we do find him engaging in careful separation of data with different reliabilities. For he explicitly took *separate averages* of the data sets for the solstices and the equinoctials. Each set represented observations done under different circumstances, and moreover, as Newton certainly knew, the solstitials were liable to be much less accurate than the equinoctials. To produce his final result from the five equinoctials, Newton accordingly separately formed "the fifth part of the sum of the places in which these five *Colures* cut the Ecliptic."[82] And for the solstitials he formed "the fifth part of the sum of the three first Longitudes, and of the complements of the two last to 180 Degrees."

Newton did not put the two results together into a single date as yet (table 8.5). Instead, he first listed the equinoctial and solstitial loci that he had found, effectively isolating each set from the other. Proceeding in this way emphasized what would have been obvious to anyone who worked with Newton's data: although both sets averaged separately to nearly the same date, the solstitial dates differed much more among one another than did the equinoctial dates, just as one would expect given the comparative difficulty of pinpointing the solstice.[83] If Newton had not carefully separated the two data sets, then his conclusion could have been challenged on the grounds that he had improperly mixed observations of hugely different reliability. Keeping them separate, and then pointing out that each yielded as a set the same date—namely 937 BCE—shielded his conclusion from criticism:

> The two *Colures* therefore, which in the time of the *Argonautic* Expedition cut the Ecliptic in the Cardinal Points, did in the end of the year 1689 cut it in ♉.6°.29′; ♌.6°.29′; ♏. 6°.29′; and ♒.6°.29′;[84] that is, at the distance of 1 Sign, 6 Degrees and 29 Minutes from the Cardinal Points of *Chiron*; as nearly as we have been able to determine from the coarse observations of the Ancients: and therefore the Cardinal Points have gone back from those *Colures* one Sign, 6 Degrees and 29 Minutes; which, after the rate of 72 years to a Degree, answers to 2627 years. Count those years backwards from the end of the year 1689, or beginning of the year 1690, and the reckoning will place the *Argonautic* Expedition, about 43 years after the death of *Solomon*.[85]

We can spy Newton in an even more intricate engagement with star data. In the printed *Chronology*, Newton first mapped the point in the "back" of Aries (which

[82] Newton, 1728b, p. 89.

[83] Specifically, the standard deviation among Newton's equinoctial dates is 67 years, while among the solstitial it rises greatly to 107 years. Newton would not of course have calculated standard deviations, but he would certainly have immediately recognized that the scatter among the solstitial dates signaled their comparative unreliability.

[84] The average for the solstitial colure was ♒ 6°28′46″, and the equinoctial average yielded ♉ 6°29′15″.

[85] Newton, 1728b, p. 89. Which places the death of Solomon within a year of 980 BCE. Newton never actually furnished a date for Solomon's death, only that the reign of his son Rehoboam began in 979 BCE. Hence, the ambiguity in the phrase "the end of the year 1689, or beginning of the year 1690," signals his uncertainty as to when Solomon's reign ended and Rehoboam's commenced.

TABLE 8.5.
Newton's locations for the original colures

1690 Loci of Original Colures	*Ecliptic point 1*	*Ecliptic point 2*
Equinoctial longitudes	♉ 6°29′	♏ 6°29′
Season	spring	autumn
Near the center of constellation	Aries	Libra
Solstitial longitudes	♌ 6°29′	♒ 6°29′
Season	summer	winter
Near the center of constellation	Cancer	Capricorn

Hipparchus referred to as the midpoint) in both longitude and latitude between the two stars that seem to bracket the constellation in Bayer's drawing. This gave him a date of 25 years after Solomon.[86] But, he continued, "It is not necessary that the middles of the Constellation of *Aries* should be exactly in the middle between the two Stars called *prima Arietis* and *ultima Caudae*: and it may be better to fix the Cardinal points by the Stars, through which the *Colures* passed in the primitive Sphere, according to the description of Eudoxus above recited." Newton's results for colures follow in the printed text. These words, which reflect his early thoughts concerning the original locations of the equinoxes and solstices in respect to the constellations, hint at Newton's grappling with issues concerning star placement, and they suggest as well a concern with errors. A page from his manuscripts takes us further. For there we find Newton discarding a solstitial colure based on finding the midpoint between two stars at the extremes of a constellation for one based on two stars, each of which should pass through the colure because "it may be better to depend upon the Colures drawn through several stars so that the errors of the starrs [*sic*] on either side of each Colure may correct & balance one another."[87]

Here we see Newton telling himself to calculate in such a way as to cancel out the errors he assumed to exist in the original reports. He never discussed this in print, probably for the very good reason that many of his contemporaries would have regarded his procedure as casting doubt on the integrity of the data that he was using. Yet by the early 1700s, Newton had long developed complex methods for generating what he considered to be trustworthy knowledge from otherwise doubtful data. Exhibited strikingly in the *Principia*, as well (in an early form) as in his optical manuscripts, Newton's carefully worked methods carried over into his astronomical chronology, his alchemy, and even into his investigations of prophecy.

In Newton's opinion all data—wherever it originated, and whoever produced it—was inherently unreliable. No single item or data bit could be used to generate trust-

[86] Newton, 1728b, p. 87. In general, Newton considered the Eudoxan colures to run approximately through the "middles" of the asterisms: "Eudoxus ... placed the solstices and equinoxes in the middles of the constellations of Aries, Cancer, Chelae, and Capricorn, as is affirmed by Hipparchus Bithynus" (pp. 82–83). The specific remarks from the Hipparchan commentary follow immediately. The general placement in the "middles" appears in section 3 of Petavius' edition of Hipparchus' commentary on Aratus.

[87] New College MS 361.2 fol. 131^{v}.

worthy knowledge because each one was subject to doubt. But doubt could be effaced by means of a mutual assistance pact. Taken by itself, a piece of data shook with uncertainty, but put pieces together and a steady, dependable product might emerge. Sure data, for Newton, was inescapably built out of unstable elements.

What is more, a royal hierarchy, as it were, ranked kinds of data among one another. Ranking supreme would be the securely produced number—a number fabricated out of measurements done with instruments that were nicely adapted to the task and that had been skillfully worked by those who knew how properly to use them. Of this sort were Newton's own optical observations, or, later, his experiments with the motion of bodies in air, water, and mercury. Of equal rank were astronomical coordinates generated by observers such as Hevelius or, better still, Flamsteed, who, whatever faults he might have had in Newton's eyes (and in Flamsteed's case there were many), nevertheless understood how to produce trusty results—as Newton knew because he had seen Flamsteed's instruments and working methods for himself.

Next in rank would be numbers whose reliability was uncertain for a variety of reasons, ranging from instrumental inadequacy through the undependability of their producer to complete absence of knowledge concerning their provenance. The astronomical bits and pieces that Newton managed to extract from Hipparchus' Eudoxan remarks were of just this sort. Even more inferior were items that could be properly trusted only when they came in sufficiently well-populated groups, and preferably if the groups were themselves of different kinds. Even then, adjustments might have to be made in order to effect coherence among them all. Hipparchus had given Newton just what he needed here, because the data that Newton forged from his remarks fell into two unequally reliable sets (equinoctial and solstitial), and each set was reasonably well-populated. Yet lower in rank would be a singleton—a data bit that, though it could be assimilated to a number, had no support from other numbers.

Least reliable of all, and decidedly lowest in the Newtonian data hierarchy, were purely textual remarks that required interpretation to turn into proper data. Newton's jaundiced views of individual testimony, forged at the mint during his interrogation of coin-clippers, underscored the suspicion of words that he shared with many of his English contemporaries. Hierarchy prevailed even here, for some words demanded more respect than others. Biblical reportage ranked highest in this otherwise uncertain group, though Newton spied everywhere the polluting effects of human intervention—of the unfortunate but unavoidable fact that events could not present themselves in their pristine form, unsullied by passage through the distorting lens of language. Nevertheless, words from the Hebrew past could be transformed into useable evidence by proper collation with other words, which would themselves have to undergo a suitable conversion.

This "Newtonian philology" presumed that all texts are suspect, but not primarily, or rather solely, because of corruption by copying and misrepresentation over the centuries, as Humanist scholars had it. Corruption of this sort certainly occurred, but it was an inevitable by-product of something deeper and fundamentally ineradicable. Language, Newton thought, must distort by virtue of its nature from the very moment of utterance, because words cannot reproduce events in time and space. For language represents not events but something out of time and place—an idea, perhaps an image, but always a memory. And singular memories must be forever unreliable.

Observations or experiments generate numbers that are also distorted by human fallibility, but here one distortion can compensate another by means of the numerical average, whereby an automatic procedure—an algorithm—was thought by Newton to efface error. Multiple linguistic sources can, in superficial similarity, mitigate the problem of human unreliability because here too errors might be considered mutually compensating. Yet language doubles unreliability, because words are not instrumentally generated numbers. In the ideal case they could be changed into numbers, as, we shall see, Newton felt that they could with Hesiod and Thales. In other cases, in fact the majority, numerical transformation was at best difficult and perhaps impossible, in which case words could only be corroborated by other words, and that could require elaborate and tendentious argument. Which is precisely what Newton provided in the wordy bulk of his *Chronology*, with its elaborate maze constructed out of mutually reinforcing bits and pieces of ancient verbiage extracted *ad libitum* from Biblical sources cross-referenced to other texts.

The numbers in table 8.5 were the ones that Newton printed, and so they represent the public face of his conclusions. Nowhere did Newton point out that he had been forced to make hard choices, and that if Hipparchus' words were mapped as closely as possible to the Bayer drawings then dates inimical to his new chronology would have emerged. Although Newton's chronological calculations followed his earliest publications in optics by at least three decades, and his published chronology followed them by half a century, he continued here much the same pattern that he had established early on—and probably for essentially the same reasons. His first printed work in optics left behind a trail of perplexity, not least because Newton neither argued for, nor (at first) even explained, the subtleties of his experiments with prisms. Decades later, he failed to explain why he chose certain stars rather than others for calculation. In fact, he left the impression that the stars rather chose themselves, writing for example that in "the right hand of *Perseus*, rightly delineated, is a star of the fourth magnitude, called η by *Bayer*."[88] No trace of doubt, no hint of ambiguity, troubles these words.

But Newton's calculated numbers were not the only observations that he *cited* in print. He mentioned others from Hipparchus' remarks as well, ones that, he claimed, confirmed his computed results. "The same [solstitial] *Colurus* [just computed to pass through ♌.6°.29′ in 1689]," he asserted, "passes also in the middle between the Stars η and χ, of the fourth and fifth Magnitudes, in the neck of the *Swan*; being distant from each about a Degree: it passeth also by the star κ, of the fourth Magnitude, in the right wing of the *Swan*; and by the Star ο, of the fifth Magnitude, in the left hand of *Cepheus*, rightly delineated; and by the Stars in the tail of the *South-Fish*; and is at right angles with the *Colurus Aequinoctiorum* found above: and so it hath all the characters of the *Colurus Solstitiorum* right drawn."[89] The claims here correspond to four of the Hipparchan remarks that Newton had initially mentioned. Why didn't he use these for computation, instead of just mentioning that they confirm his results? And what of the other remarks that Newton had mentioned but that he used for neither calculation nor confirmation?

[88] Newton, 1728b, p. 89.

[89] Ibid., p. 90.

The answer to our first question is simple: Newton didn't publish the dates for these numbers because they are not particularly good, producing an average (excluding *Cepheus* for a moment) of 1062 BCE. Newton didn't hide the fact, though he didn't emphasize it either, noting that the colure (for example) passes "about a Degree" from the stars he chose in the Swan's (Cygnus) neck,[90] and stating only that it "passeth also" by his other chosen stars—but not how far. In fact the colure misses them by an average of about 1⅔ degrees, corresponding to the 125-year difference between his calculated mean date and the one specified by this confirming data that give Newton's colure "all the characters" that it ought to have.

Still, a century and a quarter difference for *solstitial* data need not be overly disturbing, particularly since five of the solstitial data points work well with the equinoctial data. It only means that these particular observations were not as well done as the others, which, for solstices, might not be surprising. The same could hardly be said for the star on the "left hand of Cepheus," because it cannot even remotely pass by either a solstitial or an equinoctial colure.[91] But here we probably have a simple lapse on Newton's part: it's likely that he conflated the solstice point of 90° with an equinoctial at 0°, since precessing the star to the equinoctial (and ignoring its latitude, as one does for a solstitial) produces a reasonable date (for Newton) of 881 BCE. Had Newton noticed the problem he would likely have left Cepheus out of consideration altogether as an impossible outlier.[92]

We're left finally with six omitted Hipparchan remarks: one for Ursa Major, two for Arctophylax (modern Bootes), one for Chelae (Libra), and two for Centaurus. The omission of Centaurus is simply explained given Newton's computational procedure: there are no useable coordinates for them in Flamsteed, though Hevelius mentions one star that might be possible for Centaurus.[93] One point to bear in mind is that Newton was apparently prepared to consider Hipparchan remarks that did not have suitable nearby stars to be subsidiary to ones with which he could associate stars, provided that the remark very broadly fit the locus of the corresponding colure as determined by his chosen stars. Consider for example the (omitted) case of the remark that the solstitial colure passes through the middle of Ursa Major. Newton's final result requires that the colure should intersect the ecliptic at an epoch 1690 longitude of ♌ 6°30′. The resulting colure passes near the right shoulder of Bayer's depiction of the Great Bear (the closest neighboring star being Bayer ϕ of the fourth magnitude). In this case, Newton seems to have decided that the preponderance of the five remarks concerning the locus of the solstitial colure that he did use, remarks to which he associated stars, outweighed the apparent pictorial displacement from Ursa's center to his shoulder.

[90] Newton did explicitly compute for this colure.

[91] Flamsteed gives the longitude of o in Cepheus as 35°42′36″. The solstitial colures are either 90° or 270°. Choosing the closest of the two (the other being utterly impossible) yields a date of 2219 BCE.

[92] Which is how Schaefer dealt with it: "The left hand of Cepheus (and all parts of Cepheus) is far from the solstitial colure for all plausible epochs. This item is an outlier (by more than seven sigma), and so we are easily justified in not using it in the fit" (Schaefer, 2004, p. 180).

[93] Hevelius marks a star "in genu sinistro; sive potius ad ungulam" of Centaurus, which might fit the colure description (Hevelius, 1690, p. 315). It has epoch 1660 coordinates that would, however, give it a date of about 1470 BCE.

It's hard to explain Newton's omission of the Bootes remarks. The brightest star in Bootes, namely Arcturus, lies right between the constellation's legs. If we use Arcturus (Bayer α), then the resulting date would be 848 BCE. This was certainly a bit low for an equinoctial, but he did have at least one other datum that he did use and that produced an even lower date, namely the right hand of Perseus, which yields 826 BCE. That would have decreased Newton's overall result, but only to 930 BCE, so it is difficult to understand the omission. In any case, Newton's colure passes through Bootes right between the legs of the asterism, and near as well to his scythe-bearing right hand, all of which is consistent with the Hipparchan remarks.

As for Libra, it looks as though Newton avoided it because the dates produced by any star chosen in it vary radically from the ones generated by his other data. In particular, the two brightest stars are α and β. The former, which is depicted in Bayer as right at the constellation's central fulcrum, produces a date of 1258 BCE, the latter yields the impossible date (for Newton) of 1826 BCE. The less bright Libra γ, through one of the pans, produces 1405 BCE. Whiston would later catch the problem in his extended critique of the *Chronology*. Had he lived to hear Whiston's critique, Newton might have argued that the colure was originally set through Aries and then its position through Libra was only mentioned very approximately, namely to within several degrees or so.

Throughout the more than two decades after 1703 or 1704, Newton worked and reworked his colure calculations. Overall the tendency in the manuscripts is to put the date of the Argonautic expedition somewhat more than a generation after Solomon's death. Table 8.6 shows the results (among others) in the manuscripts.

These extracts are a subset of the ones that Newton produced over the years, but they illustrate the flexibility that he found as he calculated and recalculated dates. Which raises the question of why Newton recalculated with Flamsteed's data since he initially used Hevelius' coordinates. The answer might seem obvious: Newton decided to use Flamsteed's data because he knew that it was more accurate, and more precise, than Hevelius', which would thereby better enable him to pinpoint the date of the original colure. But the obvious answer cannot be correct, because there was nothing to choose between them: their differences were vastly smaller than the accuracy that he was prepared to grant the original, "coarse," data. One might argue that he could not have known this before seeing Flamsteed's data, but this will hardly do because Newton certainly knew from Halley that Hevelius' longitudes were good to better than a half minute of arc.

So Newton would not have shifted from Hevelius' to Flamsteed's data just to achieve a negligible increase in accuracy. The most likely reasons for the change are, first, that he wanted to use the best possible modern values in order to eliminate potential arguments based on their inaccuracies. Notoriously averse to public controversy, Newton knew that his chronology was sure to generate a great deal of it, as it most certainly did. Eliminating one possible source of criticism, however ill advised such a critique would have been given Hevelius' quite obvious accuracy, meant one less issue to contend with. Second, Flamsteed provided many more star coordinates than Hevelius, which gave Newton a much wider choice of candidate stars. Finally, in view of the growing rift between them, Newton expected Flamsteed to subject his reasoning from astronomy to harsh criticism, especially as Flamsteed made public

TABLE 8.6.
Dates of the original globe computed from Newton's MS

MS Locus	*Epoch*	*Displacement*	*Date*
Appendix C.i.a. COLURES 1	1660	35.52.36	−923
Appendix C.i a. COLURES 1	1660	36.41.34	−982
Appendix C.i.a. COLURES 1	1690	37.7.55	−984
Appendix C.i.c. COLURES 3.	1689	36.44	−956
Appendix C.i.c. COLURES 3.	1689	36.45.32	−958
Appendix C.i.d. COLURES 4.	1660	35.50.16	−920
Appendix C.i.e. COLURES 5.	1660	35.53.30	−924
Appendix C.i.f. COLURES 6.	1660	35.54	−925
Appendix C.i.g. COLURES 7.	1690	36.44	−955
Appendix C.i.h. COLURES 8.	1690	36.44	−955

his views regarding ancient astronomy.[94] Hence, Newton's determination to gain access to Flamsteed's observations was motivated, in no small part, by the desire to stave off possible charges by his antagonist that he used inferior (and foreign) results.[95]

■

Newton backed his colure calculations with four confirmations that were also based on astronomical arguments, but each of which, he knew, required subsidiary assumptions and so could only be used to support the conclusions that had been reached on the firmer grounds provided by the colure data. Of these three, only one required extensive computation, as is evident from the manuscripts, and yet it takes the least space of all in the printed *Chronology*. By examining these four, together with manuscript remarks, we can gain even clearer insight into Newton's methods and style of argument.

The first of Newton's four confirming pieces of astronomical evidence concerns the ancient astronomer Thales and nicely reveals the complexity of Newton's work with data, because at first sight the confirmation seems to be circular. The argument requires three bits of evidence: first, the locus of the colure at the time of the original star globe; second, an observation by Thales of the morning setting of the *Pleiades* "upon the 25th day after the autumnal equinox"; and third, that Thales was a young

[94] As it happened, Newton outlived Flamsteed, who died in 1719.

[95] Newton surely recalled Flamsteed's complaining to him in 1700 of Gregory's intention to use Cassini's observations when expounding his theory of comets. "[I]t was not onely an Injury to me," fumed Flamsteed, "but the Nation to Rob our Observatory of what was due to it and further to bestow it on the French." Flamsteed, 1995–2001, vol. 2, p. 817.

man (and, in Newton's words, "fit to apply himself to Astronomical studies") during the 41st Olympiad.[96] From the first and second bits and the rate of precession he obtains the time interval between the original globe and Thales, and then from the third he reasons back to the globe's date. This has the appearance of circularity because, after all, the locus of the original colure itself provided the original globe's date *via* precession, and Newton had to choose a particular star in the *Pleiades* and then find its locus at the time of the original colure. But if we pay close attention to the argument's form, then we can see just how careful he was trying to be, or perhaps it would be better to say how clever Newton was in handling data.[97]

What Newton did was to separate the pure data concerning the colure loci at the time of the original globe from the globe's date; that is, he took the Eudoxan remarks as establishing the colure loci without presuming anything at this point about precession rates (and so about dates). He then followed Petavius in associating a particular star in the *Pleiades* (namely, the *Lucida Pleiadum*) with the Thales observation of the asterism's morning setting to find its longitude at Thales' time. Taking *Lucida*, Newton then asked what its longitude would have been when the colure had its original locus. To find that, he used his computational criterion, namely that "the place of any Star in the Primitive Sphere may readily be found, counting backwards one Sign, 6°.29′. from the Longitude which it had in the end of the year of our Lord 1689."[98] This last procedure did not depend on the rate of precession per se, but solely on the longitudinal shift between 1689 and the time, whatever it may have been, of the colure's original position.

Only then did Newton turn to chronology: he used the rate of precession to translate the displacement into a time interval, and to this he added Thales' date, itself obtained from the date of the 41st Olympiad (see appendix E.ii for details and analysis). The result should cohere with the date obtained previously for the Argonautic expedition, and indeed it does. Newton himself never carried out the required computation; he relied instead on Petavius, who had. The procedure is essentially the same as the one required for Hesiod's remark, to which we will turn below, and Newton probably chose not to repeat the tedious calculations when someone else had already done it for him, with the added benefit of being able to cite an independent authority.

This line of reasoning could even be run in a different direction: Newton's date for the expedition, combined with Thales' observation and the place of the original colure, could be used to compute the date of the 41st Olympiad. Accordingly his claim

[96] Newton, 1728b, p. 92. Newton extracted the datum concerning the *Pleaides* from Pliny's *Natural History*, book 18 (Pliny 1949–1971, vol. 5, p. 325). Newton, we shall see, shared Scaliger's opinion that Thales wrote about equinoxes and solstices precisely because he had noticed precession, though not Scaliger's view that Eudoxus had himself discovered it (Grafton, 1993, pp. 481–84). Thales has been a pregnant source for chronologers—or, rather, a remark by Herodotus that Thales predicted a solar eclipse has been (Herodotus, 1992, p. 27: book I.74). Argument about Thales' eclipse is not confined to the sixteenth and seventeenth centuries: see, e.g., Mosshammer, 1981, who avers that the "reliability of literary eclipses is poor in general" (p. 151).

[97] The MS versions of the argument (see appendix Cii), though organized differently, are not fundamentally at odds with their published form, though the printed version contains more specific data concerning Thales' observation.

[98] Newton, 1728b, p. 91.

for coherence is not at all circular, because Thales' observation, together with the date of the Olympiad, might very well have provided an utterly different date for the original globe.[99] If that had occurred, then the implication would have been that either the Eudoxan remarks (or Newton's translations of them into numbers) were themselves faulty in some way, or else one or both of the two pieces of Thales data would have to have been.

Newton's second "confirmation" concerns the observation of a summer solstice by Meton and Euctemon in Nabonassar 316 (that is, 432 BCE) "in order to publish the Lunar Cycle of nineteen years."[100] Further, Columella's extensive first-century CE work on farming, which provided astronomical advice, "tells us that they [viz. Meton and Euctemon] placed it in the eighth Degree of [the constellation] Cancer."[101] Since Newton was certain that the equinoxes and solstices were originally located in the middles of the constellations, this meant that they had regressed "at least seven Degrees" by 432 BCE. At 72 years per degree, this put the original globe's formation at 936 BCE, or in "the 44th year after the death of *Solomon*, or thereabout."[102]

Newton wrote nothing more about this, and we shall see later that it caused a considerable amount of argument among his French critics. There is, however, little doubt as to what he had in mind. Chiron had created the first sign system for locating the positions of stars, but Chiron knew nothing of precession. Moreover, the system of reference that prevailed at the time of Hipparchus—or so Newton and others thought—was tied to an abstraction: namely, to the great circles of the ecliptic and the equator. Such a system could prevail even were there no stars at all, because it is specified by the diurnal and annual motion of the sun, which can be determined by zenith and horizon observations. Chiron, the first to create a sign system, would naturally have linked it directly to the stars themselves. On the day of an equinox he might have looked to the setting sun to see what constellation emerges after its light grows dim. At Newton's date for Chiron, the stars in the Ram's horns would have preceded the sun in setting, so that a good portion of the constellation already lay

[99] The observation, combined with the star's original locus and the rate of precession, provides a time interval only. The crux of the argument is that the known date of the Olympiad moved back through the time interval agrees with the previously calculated date.

[100] Petau 1703. The lunar, or *Metonic*, cycle, in which 19 tropical years are equal to 235 synodic months. Scaliger argued that Meton's solstice observation indicated that, like Scaliger's Thales, he knew about precession (Grafton, 1993, p. 481).

[101] Columella, 1745: in Book IX, chap. XIV Columella mentions that the winter solstice occurs in "the eighth part of Capricorn" according to "the calendars of Eudoxus and Meton." Euctemon was a contemporary of and collaborator with Meton.

[102] The placing of the equinoxes and solstices at various points in a sign system in fact dates back to the Babylonians and persisted among some commentators into the first century CE: see Neugebauer, 1975, vol. 1, pp. 593–600. Newton's construal of Columella's remark did, however, overlook the latter's assertion that the difference between the Eudoxan-Metonic system and that of Hipparchus might have to do with the formers' calendar "and those of the ancient astronomers [being] adapted to the public sacrifices; because husbandmen are both better acquainted with that old opinion which has been commonly entertained: nor yet, is the niceness and exactness of Hipparchus necessary, to the grosser apprehensions, and scanty learning, of husbandmen" (Columella, 1745, p. 410). Fréret would later use this in a letter to Halley to argue against Newton's use of the datum.

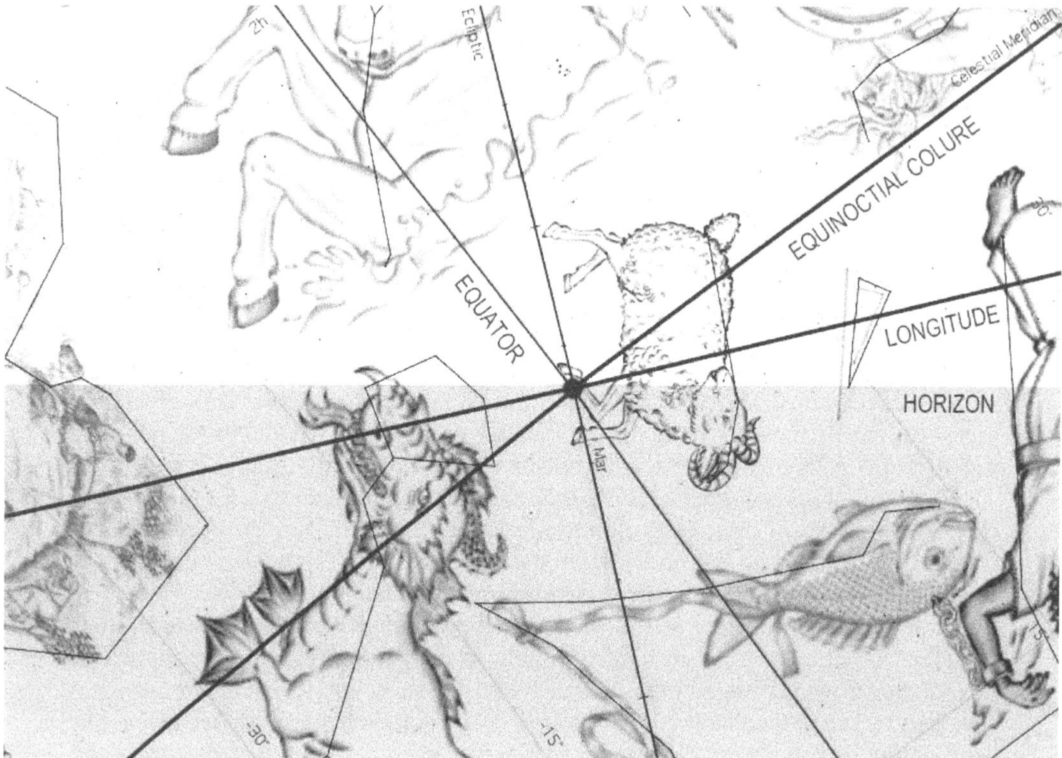

Figure 8.14. The sun (black circle) setting in southern Grecian latitudes on the day of the vernal equinox in 939 BCE, the year when Newton thought Chiron delineated the first celestial sphere. Note that the equinoctial colure runs through the back of *Aries*, while the zero longitude axis runs near its nose.

below the horizon. Chiron, Newton apparently reasoned, had decided to create a scheme consisting of twelve equal divisions along the zodiac that were each connected to a specific constellation, a system whose first sign included the position of the sun among the stars on the day of the vernal equinox. The question was where to set the beginning of the first sign. Instead of locating it at the equinoctial sun, Chiron chose to make the sun the *center* of his first sign, not only because a good part of the constellation to which the sign was connected (the Ram) had already preceded it in setting, but also because the *equinoctial colure* ran through the back of *Aries* (figure 8.14).

Because Chiron linked the signs to the stars proper, over time the entire reference system would move forward past the equinox in the same direction as the annual motion of the sun. Newton thought that for centuries the Chironic scheme continued to prevail, up to and perhaps past the time of Meton himself. Consequently, Meton's assertion that the cardinal points had decreased their positions by eight degrees could be translated directly via precession into a date.

Newton's use here of precession in respect to the Chironic system especially troubled one of his French critics because he had not also used remarks by Hipparchus in the *Commentary* that seemed to entail a very different result for Chiron's date. For example, Hipparchus had remarked that "the final and brightest star of [Ursa Minor] lies at about 18d of the Fishes, or—as Eudoxus divides the zodiacal circle—at about 3d of the Ram [Aries]."[103] This and similar remarks in the *Commentary* could be taken to imply that between the time of the sphere that Eudoxus referred to (viz. Chiron according to Newton and many of his critics) and the time of Hipparchus (or perhaps Eudoxus himself), the star had shifted fifteen degrees. In which case the original sphere would have been created only 1,080 years earlier (assuming the shift represented longitude and not right ascension, otherwise a correction would be needed), upsetting Newton's chronology.[104] We shall see that Newton's French critic Fréret made precisely that point. Newton, however, interpreted Hipparchus in a different way.

In Hipparchus' time the first star of Aries (*prima Arietis*) was located on the equinoctial colure.[105] And there he placed the origin of his sign system, fixing it to the equinoctial point proper and not—as Newton apparently thought Chiron had—to the stars themselves. Hipparchus knew that Eudoxus instead used a system that referred locations to the middle of asterisms and not to points near their beginnings. Newton assumed that in giving positions both in Hipparchus' own, equinox-originating system, and "as Eudoxus divides the zodiacal circle," Hipparchus simply shifted the origin of coordinates fifteen degrees without any implication that the original system had been constructed in like fashion to Hipparchus' equinoctially fixed own.

Thus far Newton believed that he possessed two independent bits of evidence, which apparently cohered perfectly with one another and with the date he had obtained through his colure statistics. In the case of Thales, he could even refer to an unbiased (because dead) source for the critical element in the confirmation (the longitude of the chosen star in the *Pleiades*). The Thales argument did itself depend on the loci of the original colures obtained from the Eudoxan remarks, together with the date of the 41st Olympiad. Meton and Euctemon also required the Eudoxan remarks, this time to set the cardinal points in the "middles" of the constellations Aries, Cancer, Libra, and Capricorn.

The coherence was certainly remarkable since it seems to be entirely unforced. And from Newton's point of view it was. But his form of presentation rather hid the method of argument. He prefaced the confirmations with a paragraph that was based on his colure statistics but that reads as though it stands alone as a procedure for computing the star positions on the original globe. "By the same *method*," he wrote, "the place of any Star in the Primitive Sphere may readily be found, counting backwards one Sign, 6°.29′. from the Longitude which it had in the end of the year of our

[103] Petavius, 1630, p. 188, translated in sec. 1.6.4 of Mcfarlane and Mills, 2010.

[104] One might think the date to be taken on this construal would be that of Hipparchus and not Eudoxus, given the *Commentary*'s wording. We will see below that Fréret instead took Eudoxus' date as the moment to move back from. This was because the first star of *Aries* (namely *prima Arietis*) had longitude 0° in −388, making that the year in which the signs would, he supposed, have been fixed in relation to the similarly named constellations.

[105] Specifically, *prima* lay on the colure in −159. However, its *longitude* reached 0° in −388, which coincides with the period when Eudoxus flourished. We will see that this has implications for Fréret's construal of Hipparchus' *Commentary*.

Lord 1689."[106] Yet the confirming bits lack meaning in the absence of the very set of evidence (the colure locations) that was itself used to calculate a date on the basis of the same factor (the rate of precession) that must also be used with the independent bits.

Newton had framed his confirmations in a particularly crafty way, for he might have put it differently. He might have written that he could use his computed date for the original globe to *predict* the dates of Thales' *Pleiades* observation and Meton and Euctemon's solstitial one, which he could then have "confirmed" from the dates given in the very same texts that provided the astronomical evidence. Yet such an argument would certainly not have appealed to Newton because it would have made altogether too apparent that the evidentiary bits each provided two separable pieces of evidence (an astronomical observation and a date). What he wanted was to bind these two pieces tightly together, with each piece apparently supporting the other, and with both together supporting his original date.

But there is more, because it seems from his manuscripts that Newton did not at first intend to use the Thales observation as a confirmation of his computed date for the original globe, but to confirm something else instead—namely, his belief that Thales, whom Newton admired as the first true astronomer, was also the first to displace the equinoxes and solstices from what Newton presumed to be their position on the original globe at the centers of the constellations. To do this he took Hevelius' longitude for *Lucida* in 1660 (♉ 25°15′51″) and precessed back at 72 years per degree to the "42th year of Thales," viz. to the 41st Olympiad. This puts *Lucida* in ♈23.57′ (only four minutes more than Petavius' computation). That in turn meant Thales had indeed carried out the observation reported by Pliny. Since Thales also wrote a book about "Tropics & Equinoxes," Newton concluded that he must have known that the cardinal points could not be in "ye 15th degree of the signs" but, instead, must be in the 12th degree.[107]

Here we see Newton the astronomical philologist at work combining numerical and textual arguments, and along the way forgetting to distinguish terminologically between the fixed, 30-degree *signs* of the zodiac and the constellations with which they are associated, and which move through the signs as a result of precession. The argument amounted to this: Thales had to know that pre-Trojan war "Astronomers" had put the cardinal points in the middle of the constellations.[108] But Newton's Thales was unwilling to take numerical data on faith, and so he decided to do his own ob-

[106] Newton, 1728b, p. 91 (emphasis added).

[107] Newton's "12th degree" is not quite correct (whether Hevelius' or Flamsteed's data are used), though he apparently insisted on it and even provided a supporting argument (see immediately below).

[108] Newton's precise locations for the original colures placed them imprecisely in the "centers" of their respective constellations—imprecisely because the "center" of an asterism is an inherently vague locus. Nevertheless, we shall see below that these vague claims were first spread about and generated the initial reactions and critiques, some of which orbited about the meaning of "middle" and might accordingly have been different had his computed loci been known. Newton's argument for placing the original colures more or less in the constellation centers amounted to this, that the "first Astronomers, who formed the Asterisms," having observed that "the first month of the Luni-solar year, by reason of the Intercalary month, began sometimes a week or a fortnight before the Equinox or Solstice, and sometimes as much after it" accordingly decided "to place the Equinoxes and the Solstices in the middles of the Constellations of Aries, Cancer, Chelae [Scorpio], and Capricorn" (Newton, 1728b, p. 82).

servations, of which we unfortunately know only the one concerning the morning setting of *Lucida via* Pliny. Thales discovered as a result that the cardinal points had moved about 3 degrees away from the constellation centers, and he accordingly wrote a lost book about them.

This is hardly a firmly grounded argument—and neither was Scaliger's before Newton. Nothing remains of the book or books that Thales supposedly wrote, and the only observation that Newton could refer to does not involve the cardinal points at all. Peering into the distant past, he had found in Thales a man like himself, a critical observer and perhaps even a calculator. Such a man would certainly have undertaken a coherent program of observation and computation.

Newton did not altogether abandon this first argument when he drew up the manuscript for publication. There, however, he evolved a slightly different conclusion. In print he took the date of Thales (the 41st Olympiad, or 616 BCE) and used it to find the loci of the cardinal points in his time, to conclude that "in the days of *Thales*, the Solstices and Equinoxes, by this reckoning, will have been in the middle of the eleventh Degrees of the Signs."[109] Here again we find that he conflated sign with asterism (albeit only terminologically), since the cardinal points are forever fixed at the beginnings of the signs for Aries, Cancer, Libra, and Capricorn. However, he knew from his final colure computations using Flamsteed's 1689 data that the midpoint between the stars *Prima Arietis* and *Ultima Caude* had longitude ♉ 6°44′. Taking this as the midpoint of the constellation *Aries*—not the sign of course—he would have precessed it backwards to 616 BCE. Doing so puts the midpoint of *Aries* at ♈ 4°43′8″. If the constellation proper extends over 30 degrees then the equinox at the time of Thales would have been displaced about 10°17′ from the constellation's beginning, putting it near the middle of the asterism's "eleventh Degree" (whereas according to Newton at the time of the Argonautic expedition the midpoint would have been 15° away). But, Newton continued, "*Thales*, in publishing his book about the Tropics and Equinoxes, might lean a little to the opinion of former Astronomers, so as to place them [the cardinal points] in the twelfth Degrees of the Signs."

Newton apparently wished to mitigate the discomfort that someone like (Newton's) Thales might have felt in altering others' values. But he also had a different reason for choosing the twelfth instead of the mid-eleventh degree, and this provided him with yet another confirming bit for his overall structure. Petavius' *De Doctrina*, like his *Uranologion*, included a Latin translation of a fragmentary item on Aratus' *Phaenomena* entitled the *Isagoge* by the obscure Achilles Tatius. It is, to say the least, a rather uncertain source since its date remains unknown, though it must come after, perhaps shortly after, the Aratus *Phaenomena* of the mid-third century BCE.

Nevertheless, in section 23 of this *Isagoge*, Newton read that "some anciently placed the Solstice in the beginning of Cancer, others in the eighth degree of Cancer, others about the twelfth degree, and others about the fifteenth degree thereof."[110]

[109] Ibid., p. 92.

[110] Ibid., p. 82. Scaliger had assigned Cleostratus (ca. −520) as the originator of the 8th degree position (Grafton, 1993, p. 463). The original remark reads "Hanc porto conversionem, sive solstitium alii initio cancri fieri volunt; alii in octave parte; alii circa duodecimam; alii circa decimamquintam" (Petavius, 1630, p. 146).

These differences were due, he assumed, to precession "then not known to the Greeks"—rather than, say, to convention or to highly inaccurate observation. Thales, Newton evidently thought, knew of others near his time who had located the cardinal points in the twelfth degrees of the constellations. So, he continued, in writing his "book" Thales would likely have moved his data from the mid-eleventh degree to the twelfth to accord with them. He had no evidence whatsoever for this claim, beyond the statement from the *Isagoge*, and this evidently put him in a difficult position. Thales, Newton's first true astronomer, would not have betrayed his numbers without cause; mid-eleventh degree is just not twelfth to a good observer. But the Achilles Tatius mentioned unspecified "others" who placed the points in the twelfth degree, and he had not mentioned anything about the mid-eleventh degree. Since he would have known or heard about Thales' book (Newton presumed), it follows that Thales must also have written of the twelfth degree. And the only possible reason for someone as acute as Thales to have manipulated his own data would have been to satisfy the claims of "others" at Thales' time who had specified the twelfth degree.

What an extraordinary tissue of assumptions went into this largely unvoiced argument. Its very beginning rests on Newton's image of a heroic Thales observing and calculating as Newton himself would have done. Nothing beyond Thales' long-lost book or books and the singular report from Pliny supports the claim that he had ever found a new locus for the cardinal points. And even if he had, there was no evidence at all to indicate that he would have bothered about a half-degree's difference in specifying them. In fact, nothing supported the presumption that in Thales' time the zodiacal signs had even been divided into degrees by the Greeks.

Yet Newton was hardly alone in his vision of Thales. A century before the Master of the Mint canonized Thales, Scaliger—who more than anyone else joined chronology to astronomy—developed a similar notion.[111] Scaliger, notes Grafton, became infatuated with the notion that "the independent astronomers of Greece, from Thales on, had arrived at their new theories about the solstices, the equinoxes, and other celestial phenomena" at a very early date indeed.[112] In fact, according to Scaliger, Thales was as extraordinary an astronomer as Newton later made him. "Even before any Chaldean observation of eclipses had reached the Greeks," Scaliger remarked, "Thales of Miletus had come to understand the eclipses of the sun and moon without the help of a barbarian teacher, and had made such progress that he predicted eclipses long in advance."[113] Kepler, with whom Scaliger corresponded, demurred, noting that Thales must have learned "the art of computing eclipses from the Babylonians."[114] Despite the fact that there was no evidence at all that early Greek

[111] Thales had reputedly predicted a solar eclipse, whose date the Protestant theologian Heinrich Bünting computed as May 28 585 BCE: Grafton, 1993, p. 138.

[112] Ibid., p. 450.

[113] Ibid., p. 472.

[114] Ibid., citing the translation in Jardine, 1988b, p. 158. Kepler's remark presumes that Thales actually did predict an eclipse, a point that Kepler was apparently prepared to cede to Scaliger. The two sources for the claim were Herodotus and Pliny. In addition, Scaliger cites Diogenes Laertius (Diogenes Laertius, 1925 to the effect that Thales had written books on solstices and equinoxes). Newton cited the same remarks and to the same effect (Newton, 1728a, p. 92). Diogenes wrote that Thales produced two books "and no more" on the solstices and equinoxes. Scaliger had early on troubled over the remark because he thought

astronomers had divided the zodiacal signs into degrees, Scaliger decided that they must have done so in order to have observed as precisely as, he was convinced, they had. Indeed, these early Greek calculators would certainly have noticed the motion of the equinoxes long before Hipparchus. Newton never referred to Scaliger, relying instead on the handy 1703 edition of Petavius. He did have a dog-eared copy of Scaliger's *De emendatione temporum* (1583) but not of the 1629 edition. In the latter he would surely have found much of interest, though they would have parted company over Scaliger's odd notion that the entire machinery of precession was utterly mistaken.[115]

Newton's attempt to combine astronomical computation with textual exegesis went even further when he brought forward Hesiod as a witness to the date of the Argonautic expedition, who "tells us that sixty days after the winter Solstice the Star *Arcturus* rose just at Sunset: and thence it follows that *Hesiod* flourished about an hundred years after the death of *Solomon*, or in the Generation or Age next after the *Trojan* war, as *Hesiod* himself declares."[116] We will return to the question of whether Hesiod had in fact "declared" himself to be living a "generation or age" after the Trojan War. Newton presumed that he had, and that this meant a "generation" of human years. Establishing Hesiod's date independently, using the astronomical claim, would accordingly constitute another potential confirmation. In fact, Hesiod's date seems to be the very earliest computation that Newton did for astronomical chronology, since it first appears in the manuscripts in 1701 (see appendix c.iii.f.HESIOD 6). The published chronology mentions the claim only in this single paragraph. Yet when we examine Newton's manuscripts we find that this computation, which is essentially the same one that Newton would have to have done for Thales' *Pleiades*, occupied him nearly as much as the colures over the years and involved vastly more complicated calculations.

Despite the brevity of the printed remarks, in manuscript Newton returned to Hesiod again and again, trying each time to tie down the elements of the calculation as though he were forever uncertain. And well he might have been, because horizon-related phenomena require intricacies of spherical trigonometry that led Newton to turn to a lengthy and clear handbook by Riccioli that laid out methods of computation through examples, including the very one that Newton found in Hesiod.[117] Repeatedly Newton worked and reworked the calculation, realizing in the end that it was so intricate, and might even seem to be circular, that he simply could not find a short and persuasive way to present anything more than the bare claim in print. What were the problems?

---

that the passage continued with the phrase "considering everything else to be soluble," which, Scaliger wrote, "I fail to understand" (Grafton, 1993, p. 363). Isaac Casaubon suggested in 1584 changing "soluble" to "insoluble," and eventually Scaliger changed his version to end "incapable of being apprehended" (ibid., p. 463).

[115] Grafton, 1993, chap. 3.4.

[116] Newton, 1728b, p. 95. Hesiod's words concerning *Arcturus* are: "When Zeus has completed sixty wintry days after the solstice, the star Arcturus is first seen rising, shining brightly just at dusk, leaving behind the holy stream of Oceanus. After this, Pandion's daughter, the dawn-lamenting swallow, rises into the light for human beings, and the spring begins anew." Hesiod 2006, p. 133.

[117] Riccioli, 1651, pp. 463–64.

In order to attach a date to Hesiod's claim, Newton had to see whether *Arcturus* at Hesiod's horizon could rise as the sun set "sixty days after the winter solstice." The first step required calculating the longitude of the sun at the appropriate moment, which then yields the rising point of the ecliptic. From that, given Hesiod's latitude, Newton had to compute the point of the horizon that intersects the (celestial) latitude of *Arcturus*, which in turn required a value for the angle between the ecliptic and the horizon, necessitating a number of steps, each involving the use of log tables.[118]

Newton and Petavius were scarcely the first to ponder the implications of Hesiod's remark, for Scaliger himself had come to it in 1594. Taking the date of the winter solstice at Hesiod's time to have been "around 7 or 8 January," thereby setting Hesiod's sixty days to "around 10 March," and asserting as well that in the Olympic era Arcturus did rise around March 10, Scaliger (who at the time still believed in precession, an opinion he was to alter within four years) could date Hesiod to about 776 BCE (that being the date of the first Olympiad).[119] Note that Scaliger's argument depended on the assertion that Arcturus rose on March 10 both in Hesiod's time and at the time of the first Olympiad. Newton would not have put much weight on the assertion even had he known about it[120] because it was inferential—though he might have used it as supporting evidence.

Little wonder that Newton left all the details out of the printed *Chronology*. Had he put them in, someone might have noticed the slight circularity arising from the need to use Hesiod's date to establish the total proper motion of the sun's apsidal line. In the end it makes almost no difference at all to the final result whether or not the line is moved, but to argue the point Newton would have been pushed too far into elaborate technicalities. He chose instead to pack everything into a single "thence." Someone might also have noticed (as we see from the manuscripts) that Newton's calculation actually places Hesiod indistinguishably close to the very time that he had calculated for the Argonautic expedition proper, and not a "generation" after the Trojan War.[121] But Newton's "thence" hid more than astronomical complexity and approximation; it also hid an utterly doubtful textual interpretation.

[118] Newton did not explicitly include precession in computing the position of the solar apogee, which would seem to be a significant oversight since the solar apsides move a considerable distance between 880 BCE and AD 1700. However, the final effect of the precession and proper motion of the apsides scarcely affects Newton's final result given the level of accuracy that he was prepared to accept. See appendix E.i.e for discussion.

[119] Grafton, 1993, p. 474.

[120] Which he might not have since it appeared in a 1594 letter to Tycho; ibid., n.44.

[121] Specifically, Newton used the presumptive date of 880 BCE only to compute the sun's apsidal motion, which, as we remarked, makes little difference in the end. After that he computed what Arcturus' longitude had to be for it to rise as the sun sets sixty days past the winter solstice. The result he obtained in MS was within a minute and a half of where it would have been at Newton's date for the Argonautic expedition itself—which would have placed Hesiod essentially at the time of the Argonauts. Horsley redid the computation, though he gave no details (and may not have included a correction for refraction), but he obtained essentially this same result, remarking that "the conclusion from this passage of Hesiod should rather been that he flourished in that very age when the Greeks first formed their sphere; that is, according to Sir Isaac Newton, in the age of the Argonautic expedition" (Horsley, 1779–1785, vol. 5, p. 75).

In the *Works and Days* Hesiod had written, "When Zeus has completed sixty wintry days after the solstice, the star Arcturus is first seen rising, shining brightly just at dusk, leaving behind the holy stream of Oceanus," and here Newton felt that he was on sure ground.[122] Concerning Hesiod's sense of his own position in time, Newton noted that Hesiod claimed to have lived in "the fifth generation an iron one full of trouble," following "four generations or ages of the Gods." This generation would die off "when they grew grey bearded."[123] The "fourth generation of semigods," Newton continued, "were destroyed partly at Thebes in fighting for the wealth of Oedipus & partly in sailing beyond sea to war against Troy." Since Hesiod lived "one generation later," Newton concluded, he followed a generation after Troy.

There was more. According to Newton's insistent naturalism, we noted earlier, even Hesiod's four ages of the gods had to refer to matters of historic fact—to "the four first kings of the Monarchy of Egypt founded by the expulsion of the shepherds." But what then of this fourth age, just before Hesiod? How do we know that it ends with Troy, besides what Hesiod himself may have said? Newton had a second data source to mine: Ovid. In the *Metamorphoses* Ovid wrote "Piety/goodness lies vanquished, and the virgin Astraea leaves, the last of the gods, the earth dripping with slaughter/gore."[124] Astraea, or Justice, was identified with Isis, and Newton decided that this passage linked Hesiod's fourth age to the fourth Egyptian shepherd king. None of this last reached print.[125]

But even what Newton did publish betrays the text, because Hesiod had not written what Newton claimed. Hesiod's fifth age did not end with grizzled beards but with gray-haired infants, monsters that signal a breach in the natural order and therefore an appropriate turning point. "But Zeus will destroy this race of speech-endowed human beings too," wrote Hesiod, "when at their birth the hair on their temples will be quite gray." But Newton would have nothing to do with unnatural events of this sort, and so he transformed monstrous birth into natural aging.[126]

[122] Referring to Wenskus, 1990, p. 46, Grafton remarks that the passage is not altogether transparent. Wenskus notes that Hesiod might have meant that sixty days will pass after the solstice before Arcturus rises in complete darkness, which would alter a computation based instead on the assumption of its first evening visibility. In further discussions with David Pingree, Wenskus, however, agreed that the passage does refer to first visibility. The question of how, or even whether, to use Hesiod's remark has been much debated. West, 1978, pp. 299–300 notes five potential problems, concluding that the "uncertainties are such that it is futile to hope for any closer dating from this passage than can be reached on other grounds." An important point for Newton's calculation would be the question of what to use for the distances the sun must be below the horizon, and the star above it, for the star to become visible. This depends on several factors in addition to refraction, though Newton considered only the latter. See appendix E.i.e for discussion.

[123] Appendix C.iii.i. HESIOD 9

[124] Ovid, *Metamorphoses*, 1:149–50. We thank our colleague Mac Pigman for the reference and translation, who also notes that Aratus at the beginning of the *Phaenomena* told of Astraea/Dike.

[125] "And they called her Justice," Aratus, 1997, p. 81. She is the Maiden constellation "beneath the two feet of Bootes." Newton would have read these lines in the *Phaenomena* as well as in the *Metamorphoses*.

[126] Hesiod, 2006, p. 103. Newton was hardly alone in corrupting Hesiod. William Whiston, among his bitterest critics, also thought Hesiod to have written of gray hair and not monstrous births—only Whiston used this, *contra* Newton, to argue for the usual chronology since Whiston thought Hesiod to have supposed men before his own era to have lived extraordinarily long lives (Whiston, 1727–1728, pp. 1027–28).

Newton had picked and chosen his way among the ancient remarks, because the very passage that he cited from Pliny to the effect that Thales had the "occasus matutinus" of the *Pleiades* twenty-five days after the spring equinox also remarks that Hesiod had placed it at the equinox itself. In which case, if no further argument is made to downgrade the Hesiod remark in comparison to the one about Thales, then, since Hesiod did write that he lived after the Trojan War, the Argonautic expedition would have taken place some 1,500 years before Newton's date for it.[127]

Newton's way with the Hesiod text was hardly the only point that betrays his attitudes toward ancient words, for the very foundation of the claims that Chiron had delineated the asterisms and that Musaeus had "made a sphere" are themselves questionable. Moreover, Newton's contemporaries certainly thought so for the second claim and likely for the first as well. For Chiron, we saw above, Newton cited the *Stromata* of Clement of Alexandria, but, half a century earlier Thomas Stanley, the nephew of John Marsham, seemingly rejected the very existence of Orpheus (traditionally the friend or teacher of Musaeus) as fabulous, figuring that philosophy began with Thales. According to Stanley one should "omit the dark traditions of the Athenians concerning Musaeus, of the Thebans concerning Linus, and of the Thracians about Orpheus," so that "it is manifest that the original of Greek philosophy is to be derived from Thales, who traveling into the east, first brought natural learning, geometry, and astrology thence into Greece." His 1655 *History of Philosophy* was reprinted in 1701, and this version was in Newton's library. Moreover, Stanley considered Thales to be the first true philosopher, and the originator of astronomy and geometry, as did Newton himself.[128]

Newton's anti-poetic, literally prosaic turn of mind shows itself as well in his remark that "Chiron delineated σχήματα ὀλύμπου Asterisms." He immediately transformed the Greek phrase into a pictorial representation of the heavens of the sort that an astronomer mapping the skies might produce. The phrase likely means nothing of the kind. Two centuries later the historical geographer Jean-Antoine Letronne, an expert Grecian, citing Herodotus and Xenophon, with backup from Euripedes and Varro, remarked that in antiquity the word χήματα referred to the dance-like motion of the heavens, not to a static graphical image. But Newton rejected anything that reeked of poetic license, and so he re-fashioned the text to suit his own view of what it had to be.[129]

And finally (though penultimate in print order), Newton managed to turn Hipparchus' underestimate of the magnitude of precession into a proof for his own chronology. "Great astronomer" that he was, how could Hipparchus possibly have found a rate of one degree per century instead of the considerably different 72 years? He "first of any man," Newton wrote, "comparing his own Observations with those of

[127] Horsley noted the problem and tried to rescue Newton by asserting that to "this specious objection, our author would probably have replied; that Pliny reports the reason of the morning-setting of this star [the *Lucida Pleiadum*], from a book of astronomy, of which he says only, that it was extant *under the name* of Hesiod: and that this book could not be Hesiod's" (Horsley, 1779–1785, vol. 5, p. 73). Perhaps Newton would have so argued—but he didn't, preferring to obscure the issue beneath the same bush that hid his reading of Hesiod's own remark concerning *Arcturus*.

[128] Stanley, 1701, Preface.

[129] Letronne, 1846, p. 104, note.

former Astronomers," concluded "that the Equinoxes had a motion backwards in respects of the fixt Stars."[130] Hipparchus did observe the stars, and very accurately, thought Newton, but he only had an "opinion" that they "went backwards one Degree in about an hundred years." Since he was much too good to have relied uncritically on previous observations, how did Hipparchus obtain such a result? He had after all certainly observed the equinoxes himself some 286 years after Meton and Euctemon had observed the summer solstice.[131] To reach his result, Newton argued rather elliptically, Hipparchus first found the equinox to lie in about "the fourth degree of Aries." And he also knew (having after all been the one to comment on Aratus and Eudoxus) that the original colure went through the middles of the constellation *Aries*. That, Newton evidently assumed (though he himself, of course, thought differently), was translated by Hipparchus as referring to the fifteenth degree of the sign for *Aries*, that sign originating at the equinoctial point. Further, "according to the Chronology of the ancient *Greeks* then in use," the Argonauts sailed 1090 years before Hipparchus. Taking eleven degrees in 1090 years gives a rate of "about 99 years, or in the next round number an hundred years to a Degree." And so the very inaccuracy of the great astronomer's rate of precession reflects his reliance on the inadequacies of Greek chronology in his time. Even in antiquity, it seems, those whom Newton saw as his intellectual forebears were misled by the inadequate methods and beliefs of historians and chronologers.

Although we shall see that Newton's astronomical claims generated considerable reaction both in England and France, they were hardly his only forays into proto-statistical argumentation. For Newton aimed not merely to upset ancient chronology on the basis of novel evidence, but also to explain how it was that the Greeks and others had gone so badly astray in their own dating. His remarks on Hipparchus proved, Newton thought, that the Greeks took their chronology so seriously that Hipparchus had used it to compute precession itself. But if they were so certain, how could they have gone so very wrong? Was it just another instance of "poetical fancy" trumping natural order? Indeed, Newton thought, Greek chronologers did trump natural order, but not in this case because words, poetic or prose, deceived them.

Newton averred in the very first sentence of the first chapter of his printed *Chronology* that "All Nations, before they began to keep exact accounts of Time, have been prone to raise their Antiquities; and this humour has been promoted, by the Contentions between Nations about their Originals."[132] That universal tendency explained the impulse to overstate antiquity, but a more specific defect led the Greeks in particular astray. The error occurred when first Thucydides,[133] "followed" by Eratosthenes

[130] Newton, 1728b, p. 93.

[131] Here again Newton takes an average: the observations, he writes, were made between Nabonassar 586 and 618, and then "the middle year is 602."

[132] Newton, 1728b, p. 43.

[133] Newton provides no reference but in the chapter on the early history of the Hellenes in Thuycidides' *History of the Peloponneisan War* we read that "Sixty years after the capture of Ilium, the modern Boeotians were driven out of Arne by the Thessalians, and settled in the present Boeotia, the former Cadmeis; though there was a division of them there before, some of whom joined the expedition to Ilium. Twenty years later, the Dorians and the Heraclids became masters of Peloponnese," which makes (in Newton's words) "eighty years to the return of the Heraclides."

and then Apollodorus incorrectly computed the time from "the return of the Heraclides into Peloponnesus" after the Trojan War until the first Olympiad. "*Diodorus* tells us," wrote Newton, "that the times were computed from the Kings of the *Lacedaemonians*; and *Plutarch* tells us, that *Apollodorus*, *Eratosthenes* and others followed that computation: and since this reckoning is still received by Chronologers, and was gathered by computing the times from the Kings of the *Lacedaemonians*, that is from their number, let us re-examin that Computation."[134]

Newton began his reexamining with the Egyptians, who, he claimed, equated regnal length to a generation, setting three generations to a century. So too "did the Greeks and Latins," who "accordingly" took reigns to be "one with another thirty and three years a-piece." And that, Newton asserted, was the source of the problem: the "35 or 40 years a-piece, one with another" that Newton deduced from Greek accounts for the regnal lengths of Greek kings "before the times of the Persian empire … is a length so much beyond the course of nature, as is not to be credited."[135] The Greeks had gone wrong when they translated generations into years.[136]

To arrive at the unnatural "35 or 40 years a-piece" for pre-Persian war Greek regnal lengths, Newton added up reigns and then took the lowest and highest among the eight averages that he computed. He did not, however, list the specific reigns of each of his Greek kings, but provided only the total length for groups of nine or ten. Newton printed the same computation for the kings of Israel and Judah, and yet here he generated a considerably different result—one more in tune with the true "course of nature" as he saw it.[137]

Newton was not one to rely on nature's course without having ascertained it—in this case, without having other evidence for his claim that regnal lengths did not (except in unusual cases) run to the length of a generation. And indeed he did have evidence, drawn from the chronicles of the Hebrews. The "eighteen kings of Judah who succeeded Solomon," Newton asserted, "reigned 390 years, which is one with another 22 years a-piece." And the "fifteen kings of Israel after Solomon, reigned 259 years, which is 17¼ years a-piece." This, together with similar results for the kings of Baby-

[134] Newton, 1728b, pp. 50–51. See Manuel, 1963, pp. 54–56, who cites New College MS 361.1 fol. 9; the MS wording also appears in the printed version in the introduction to the *Short Chronicle* (p. 3). We will see that the MS opens a window on another point that the printed version keeps firmly closed.

[135] Newton, 1728, p. 52.

[136] Compare Newton's view to that of a like-minded contemporary. In an essay on the family of Esau, Richard Cumberland noted that the Horite kingdom existed for 420 years, with ten successions averaging forty-two years a reign. Not that he supposed that each reign lasted that long, Cumberland explained, "but, because we have no foundation to go upon, whereby an accurate and real division of this interval may be made; and because such an unaccurate and conjectural division may afford a little light into the times, in which the life and death of Esau, and of his Sons, *Eliphaz* and *Reuel*, might happen in the land of *Seir*." Cumberland, 1724, pp. 13–14.

[137] Contrast Newton's procedure here with that of Scaliger's a century before, who with Herodotus took three generations to a century as "the mean period of human life within which a man can generate and produce children, not for the whole duration of a life" (cited in Grafton, 1993, p. 610). Note Scaliger's use of "mean." The sense would have been the common one of the time, namely not an average, properly speaking, but the usual span according to natural norms, so that $3n$ generations would in the normal course of events occupy $100n$ years.

lon, Persia, the successors of Alexander in Syria, Egypt, and Macedonia, and finally for the kings of England and France all supported the claim that "by the ordinary course of nature kings reign, one with another, about eighteen or twenty years a-piece: and if in some instance they reign, one with another, five or six years longer, in others they reign as much shorter; eighteen or twenty years is a medium."[138]

As with the pre-Persian kings of Greece, for Israel and Judah Newton printed only combined regnal lengths, not the names or durations for each king separately. However, in his manuscripts Newton twice listed all post-Solomonic kings along with the durations of their reigns. The first list[139] combines the two kingdoms into a single chronological sequence, the second (figure 8.15) keeps them separate. Both lists are unusually neat and carefully done, which probably reflects the special importance of these reigns for him. Those familiar with Biblical chronology will quickly recognize from the MS list (table 8.7) that Newton omitted the last four post-Solomonic kings of Judah.[140] This, however, simply reflects Josiah's reign as the last before Judah fell under the control of Egypt and then Babylon following Josiah's death in his ill-conceived attempt to foil Egyptian aid to the Assyrians at Megiddo. But the manuscript does hold a true surprise.

Newton listed nineteen kings of Israel reigning for a combined total of 343 years, and sixteen kings of Judah reigning altogether for 372 years—producing thereby thirteen and twenty-three years respectively as mean regnal lengths, though he did not (in manuscript) compute the averages. Yet in the printed *Chronology* Newton had, for Judah and Israel respectively, eighteen and fifteen kings at 390 and 259 years, producing regnal means of 22 and 17¼ (which he did print). Four kings of Israel have vanished, and two kings of Judah have suddenly appeared, unnamed. The effect was to lower the Judah mean by a year to 22, and to raise the Israel mean by over 4 years, thereby bringing both much closer to Newton's "ordinary course of nature" of "about eighteen or twenty years a-piece." What had Newton changed?

The regnal lengths that Newton provided in manuscript were generally in use at his time (and today) with one exception—Pekah, whom Newton has reigning for twenty years, whereas the usual length is thirty-nine. That changes the total duration in Newton's list for Israel to 267 from 243, which is close to his printed value of 259. But the printed list refers to only fifteen kings, not nineteen, and there is simply no way to remove four kings from the manuscript version to obtain the printed number, even though there is clear evidence from the ten kings ascribed to Persia that Newton dropped reigns of a year or less.[141] If we instead retain Newton's incorrect reign for

[138] Newton, 1728, p. 52.

[139] New College MS 362.2 fol. 127^{v}.

[140] Specifically Jehoahaz (608 BCE, 1 year), Jehoiakim (608–597 BCE), Jehoiachin (597 BCE, 1 year), and Zedekiah (597–587 BCE).

[141] Including Cyrus, there were in fact thirteen (or fourteen, counting Artaxerxes V, the assassin of Darius III) Persian kings. Of these, three reigned a year or less, and Newton clearly dropped them. There are three kings who reigned a year or less (Zimri, Zechariah, and Shallum) in Newton's MS list for Israel, which lowers the count to 16 and (to be generous) 264 years. If we remove in addition one of the four kings whose reigns lasted only two years, we obtain 15 kings and 262 years, which is still not Newton's 259 and produces a mean reign of 17½. To come closer to 259 we should remove one of the other 2-year reigns, obtaining now 14 kings and 260 years, which is nearly correct for the printed total length but now

480 years from the coming out of {Mesopotamia / Egypt} to the building of the Temple 505

A works David

	Solomon, reigns 40 years 1 King. 11. 42			
Solomons 0	Solomon, reigns 40 years 1 King. 11. 42	0	Jeroboam 22 years 1 King. 14. 20	69
40	Rehoboam 17. 1 King 14. 21. 2 Chron. 13. 1	22	Nadab. 2 years 1 King. 15. 25	
57	Abijah 3. 1 King. 15. 2. 2 Chron. 13. 2.	24	Baasha 24. 1 King 15. 33.	
20	Asa 41. 1 King. 15. 10 2 Chron. 16. 13	48	Elah 2. 1 King. 16. 8.	
61	Jehosaphat 25. 1 King 22. 42. 2 Chron 20. 31	50	Zimri 7 days. 1 King. 16. 15. Tibni 5 Omri 6	
86	Jehoram 8. 2 King. 8. 17. 2 Chron. 21. 5, 20	50	Omri 12. 1 King 16. 23	
94	Ahaziah 1. 2 King. 8. 26. 2 Chron. 22. 2.	62	Ahab. 22. 1 King. 16. 29	
95	Athaliah	84	Ahaziah 2. 1 King. 22. 51.	
		86	Jehoram 12. 2 King. 3. 1	
		98	Jehu	
0	Athalia 6½ 2 King. 11. 3, 4. 2 Chr. 22. 12	0	Jehu 28 2 King. 10. 26.	
6½	Joash 40 2 King. 12. 1. 2 Chr. 24. 1	28	Jehoahaz 17. 2 King. 13. 1.	
46½	Amaziah 29. 2 King. 14. 2. 2 Chron. 25. 2	45	Jehoash 16. 2 King 13. 10	
75½	Azariah 52. 2 King 15. 2. 2 Chr. 26. 3	61	Jeroboam 41 2 King 14. 23	
127½	Jotham. 16. 2 King. 15. 33 2 Chr. 27. 1	102	Zechariah 6 months. 2 King. 15. 8	
143½	Ahaz 16. 2 King. 16. 2. 2 Chr. 28. 1.	102½	Shallum 1 month 2 King. 15. 13	
159½	Hezekiah 29. 2 King. 18. 2. 2 Chr. 29. 1	102½	Menahem. 10 2 King. 15. 17	
188½	Manasseh 55. 2 King 21. 1. 2 Chr. 33. 1	112½	Pekahiah 2 2 King. 15. 23	
243½	Amon. 2. 2 King. 21. 19. 2 Chr. 33. 21	114½	Pekah 20. 2 King. 15. 27	
245½	Josiah 31. 2 King. 22. 1 2 Chr. 34. 1.	134½	Hoshea 9	
276		143½		

Jeroboam began in the 15th year of Amaziah 2 King. 14. 23
Azariah began in ye 27th of Jeroboam. 2 King. 15. 1.
Zechariah began in ye 38th of Azariah. 2 King. 15. 8
Shallum began in ye 39th of Uzziah or Azariah 2 King 15. 13
Menahem began in ye 39th of Azariah 2 King 15. 17
Pekaiah began in the 50 of Azariah 2 King 15. 23
Pekah began in the 52d of Azariah. 2 King. 15. 27
Jotham began in ye 2d year of Pekah 2 King. 15. 32
Ahaz began in ye 17 y. of Pekah. 2 King. 16. 1
Hoshea began in ye 12th of Ahaz. 2 King 17. 1
Hezekiah began in ye 3d of Hoshea 2 King. 18. 1

Abijah began in ye 18th of Jeroboam 1 King 15. 1
Asa began in ye 20th of Jeroboam. 1 King. 15. 9
Nadab began in ye 2d year of Asa 1 King. 15. 25
Baasha began in ye 3 year of Asa 1 King 15. 33
Elah began in ye 26 of Asa 1 King. 16. 8.
Zimri began in ye 27 of Asa 1 King. 16. 10, 15
Omri reigned after Tibni in 31th of Asa 1 King. 16. 23
Ahab reigned in ye 38th of Asa 1 King. 16. 29
Jehosaphat reigned in ye 4th year of Ahab 1 King 22. 41.
Ahaziah began in ye 17th of Jehosaphat 1 King. 22. 51.
Jehoram began in ye 2d year of Jehoram son of Jehos. 2 King 1. 17
Jehoram son of Ahab in 18th of Jehosaphat. 2 King. 3. 1.
... Jehosaphat began in 5t of Joram son of Ahab. 2 King 8. 16.

Rehoboam 17¼
Abija 2½
Nadab 1⅔
Baasha 23¾
Elah 1⅓
Omri 5⅓ + 6⅓.

Figure 8.15. MS list of regnal lengths for the kings of Judah and Israel. New College MS 361.3 fol. 69. By permission of the Warden and Scholars of New College, Oxford.

TABLE 8.7.
Results of Newton's MS list for Israel and Judah

Regnal Lengths for the Kings of Israel and Judah				
King	*begin*	*end*	*Newton's regnal length*	*kingdom*
Jeroboam	−933	−911	22	ISR
Nadab	−911	−910	2	ISR
Baasha	−910	−887	24	ISR
Elah	−887	−886	2	ISR
Zimri	−886	−886	0	ISR
Omri	−886	−875	12	ISR
Ahab	−875	−854	22	ISR
Ahaziah	−855	−854	2	ISR
Joram	−854	−843	12	ISR
Jehu	−843	−816	28	ISR
Jehoahaz	−820	−804	17	ISR
Joash	−806	−790	17	ISR
Jeroboam II	−790	−749	41	ISR
Zechariah	−748	−748	1	ISR
Shallum	−748	−748	0.08	ISR
Menahem	−748	−738	10	ISR
Pekah	−748	−710	20	ISR
Pekahiah	−738	−736	2	ISR
Hoshea	−730	−721	9	ISR
Rehoboam	−933	−916	17	JUD
Abijah	−915	−913	3	JUD
Asa	−912	−872	41	JUD
Jehoshapha	−874	−850	25	JUD
Jehoram	−850	−843	8	JUD
Ahaziah	−843	−843	1	JUD
Athaliah	−843	−837	7	JUD
Joash	−843	−803	40	JUD

(*continued*)

TABLE 8.7.
Continued

Regnal Lengths for the Kings of Israel and Judah				
King	*begin*	*end*	*Newton's regnal length*	*kingdom*
Amaziah	−803	−775	29	JUD
Uzziah	−787	−735	52	JUD
Jotham	−749	−734	16	JUD
Ahaz	−741	−726	16	JUD
Hezekiah	−726	−697	29	JUD
Manasseh	−697	−642	55	JUD
Amon	−641	−640	2	JUD
Josiah	−639	−608	31	JUD
19 ISR kings reign a total of				243 years
16 JUD kings reign a total of				372 years
mean ISR reign				13 years
mean JUD reign				23 years

Pekah (20) then the problem is even worse. As for the kings of Judah, Newton might perhaps have moved past Josiah to include the previously excluded four and then have dropped the two among them with reigns of only a year (although this would leave Ahaziah, who also reigned for a single year). That would have given him eighteen kings with a combined reign of 396 years, using Newton's twenty years for Pekah, or 415 years with Pekah's corrected length. Neither matches the printed value of 390, which cannot be produced in any reasonable way from Newton's manuscript list. It is moreover striking that only one entry in the manuscript—the incorrect one for Pekah—corresponds to an actual length of "eighteen or twenty years."

How did Newton obtain his "eighteen or twenty years"? It is not after all an average, but a spread; in fact, the mean among his several regnal lengths (Israel, Judah, Persia, etc.) is nineteen years. Steven Stigler pointed out that dividing the standard deviation among these lengths by the square root of the number of kingdoms yields one.[142] If Newton's list represented, as it were, a procedure such that each kingdom constituted one randomly chosen sample, then 19 ± 1 would correctly represent the range of expected regnal lengths. Of course, Newton had no such notion, though he clearly

too low by one king, producing a mean reign of 18½. And what determines which among the 2-year reigns should be dropped?

[142] Stigler, 1977.

thought that taking an average compensates in some way for experimental or observational error. But in the *Chronology*, Newton did not average his regnal lengths—he thrice cited the "eighteen or twenty years" spread, as Stigler noted, and when he did provide a specific length he used twenty years, not nineteen.[143] Nevertheless, it seems entirely in keeping with Newton's general procedure for him to have computed the average and then to have assumed a year's general discrepancy.

Still, we are left with the peculiar mismatch between Newton's manuscript list for Israel and Judah and his printed results, or for that matter between generally accepted biblical chronology of the time and Newton's specific choices. None of his critics noted the discrepancy, though most did attack, on several grounds, his reasoning concerning generational and regnal lengths, as we shall see. Certainly, none of the critics saw Newton's manuscript lists. We must conclude, therefore, that Newton adjusted his numbers for Israel and Judah to produce the desired result, for he needed his "eighteen or twenty years" to shrink Greek chronology in particular by the right amount.

■

In his introductory dedication to the posthumous publication of the *Chronology*, John Conduitt, the husband of Newton's niece, gushed, "Your Majesty will see Astronomy, and a just Observation on the course of Nature, assisting other parts of Learning to illustrate Antiquity; and a Penetration and Sagacity peculiar to the great Author, dispelling that Mist, with which Fable and Error had darkened it; and will with pleasure contemplate the first dawnings of Your favorite Arts and Sciences, the noblest and most beneficial of which He alone carried farther in a few years, than all the most Learned who went before him, had been able to do in many Ages. Here too, Madam, You will observe, that an Abhorrence of idolatry and Persecution (the very essence and foundation of that Religion, which makes so bright a part of Your Majesty's character) was one of the earliest Laws of the Divine Legislator, the Morality of the first Ages, and the primitive Religion of both Jews and Christians; and, as the Author ads, ought to be the standing Religion of all Nations; it being for the honour of God, and good of Mankind."[144]

Conduitt has put it all together. Newton, who had accomplished in science in "a few years" more than all before him, had also swept away "Fable and Error" from the past, thereby "assisting" "other parts of Learning to illustrate Antiquity," and in the process revealing the true, idolatry-free primitive religion. How had he done all this? By means of his penetrating insight into the "course of Nature." The implication was clear: the "other parts of Learning," ignorant of nature's course, had best alter their methods lest they remain mired in the obscuring "mists" of error. In France and England, we shall see, the warning was heard early on (albeit in different ways), and a vigorous defense was mounted.

For Newton's *Chronology* was not just a radical redating of the past, startling enough though that alone would have been. It was far more than that, since, as Con-

[143] Newton, 1728b, p. 60.

[144] Newton, 1728b, pp. viii–ix.

duitt's words suggest, Newton had proposed to consign the methods of humanistic inquiry to the same dust bin that held Cartesian mechanism, and for precisely the same reason. Structures that are not tightly integrated with data, and that do not entail specific consequences (as, e.g., did Newton's astronomy in respect to modifications of Kepler's laws) simply do not qualify as knowledge. And because data are always subject to doubt, no proper art or science can be constructed on a foundation riddled with "fable and error" unless the data can be purified. The best, purest data had to be intrinsically numerical and available in more than one instance so that Newton's error-expunging algorithm—the average—could be applied. Failing that, error reduction required multiple, independent items, and each of these had to be stripped of the obscuring patina inevitably imparted by passage through the cloaca of the human mind. As Frank Manuel evocatively remarked, Newton's "passion for factual detail shriveled the past to a chronological table and a list of place names."[145] In what appears to be an early exercise in social science history, Newton had incinerated the fiber and substance of human life in a numerical furnace. But in so doing, and despite his reliance on numbers, Newton trod much the same path that several of his erudition-despising English predecessors had beaten throughout the seventeenth century. Moreover, Newton's astronomical chronology, despite its seeming concordance with the spirit of an exact and experimental science, was designed to enslave antique pagan words in the service of a theory for the very origins of civilization.

[145] Manuel, 1963, p. 10.

Publication and Reaction

The July 1754 issue of the *Philosophical Transactions of the Royal Society* included an article by the Oxford orientalist and astronomer George Costard, devoted to the eclipse Thales is said to have predicted. *Inter alia* Costard controverted Newton's dating of the eclipse but promptly "absolved" him of blame on the grounds that *The Chronology of Ancient Kingdoms* "never had the finishing hand of its great author, and it is well known now in what manner it came abroad."[1] Upon reading the article, Zachary Pearce, then Bishop of Bangor, approached Costard via an intermediary to inquire about the basis for such an assertion. He had been informed, Costard responded, that fifteen drafts of the *Chronology* were found in Newton's study, "of no one of which it could be affirmed that it was perfect, as not to have received further corrections and improvements, if Sir Isaac himself had lived to have published it." Costard had also learned that Princess Caroline had solicited a copy from Newton shortly before his death, which she then lent to the French ambassador. The latter stealthily arranged for a copy to be made, which was dispatched to Paris "where it was immediately translated into French, and animadverted on by Souciet." Alarmed, Newton's executors chose to immediately publish an "authentic" edition of a book that Newton himself might have withheld from publication.

Costard's account prompted Pearce to compose his own recollections of the events relating to the publication of the *Chronology*. Five months before his death in March 1727, Pearce wrote, Newton had paid him a visit in which their conversation revolved mostly around chronological issues. Newton related that Princess Caroline wished to see his chronology and he had prepared an abstract of it, "thinking it in that shape the properest for Her Perusal." Caroline, in turn, lent the manuscript to the Abbé Conti, who made a copy and had it published in France without Newton's knowledge or consent. Newton had gone on to reminisce that he'd been working on ancient chronology for some three decades and recently had begun to arrange the material. He had rewritten the manuscript several times over, albeit introducing only minor alterations "for the sake of shortening it . . . and leaving out in every later Copy some of the authorities and references, upon which he had grounded his opinions." Pearce now lamented such an editorial policy, for he feared that in so doing Newton had cast

[1] *Philosophical Transactions*, vol. 48, part 1 (1754), p. 19. According to Costard—who assigned the eclipse to 603 BCE—in dating the event to 585 BCE, Newton "rather followed others, than adopted it after any examination of his own." He was unaware of Newton's change of mind on the matter.

the *Chronology* in the "same method" he'd approved for the *Principia*, namely, "concealing his proofs, and leaving it to the sagacity of others to discover them." As a result, Pearce claimed, not only were certain claims set down as assertions, but the absence of supporting evidence prompted the editors of the *Chronology*—Martin Folkes and Thomas Pellet—to insert in the margins of the published book references of their own making, which quite possibly did not always refer "to the very same places, upon which he founded his assertions." The matter was not insignificant, Pearce insisted,

> because two or three of the places referred to in the Margin of his work have been thought by good Judges not to speak fully to the point for which they are brought, and therefore Sir *Isaac*'s Credit in this particular has suffered with some persons: But proofs he may have had, which he chose to conceal, though what now stands in the Margin in those few places may have come from another hand, and may not account to a full proof, as it pretends to do.

Pearce proceeded to claim credit for persuading Newton to offset the printing of the faulty *Abrégé* and its appended critique by allowing publication of the complete *Chronology*. He'd persisted despite Newton's protestations that he was too old to publish a book destined to generate heated controversies—and ultimately prevailed. Indeed, John Conduit intimated to him that Newton himself avowed that it was to Pearce and his persistence that he owed his resolve to publish. Pearce visited Newton a few days before the great man died, encountering there the publisher John Innys who was just leaving. Newton confirmed that "he was preparing his chronology for the Press, and that he had written the greatest part of it over again for that purpose."[2]

Pearce's account is illustrative of the confusion surrounding the precise events leading to the composition and publication of both the "Short Chronicle" and the *Chronology*. Newton's indictment of Conti as the false friend, who had betrayed his trust by first allowing his copy of the "Short Chronology" to circulate freely and then by becoming party to its publication, has been generally accepted as accurate. In fact, Newton was not averse to publishing his chronology, nor was he concerned to keep the nature and content of his researches private. Long before Conti's arrival in England, Newton routinely communicated to friends and acquaintances his evolving opinions regarding ancient history and theology, and no evidence exists to suggest that he imposed the condition of secrecy. We noted earlier his open exchange with John Locke on *An Historical Account of Two Notable Corruptions of Scripture* that nearly turned into publication, as well as his uninhibited exchange of ideas with David Gregory.

Gregory undoubtedly shared some of the information he had received with his close friend Archibald Pitcairne, who benefitted independently from Newton's generosity. Pitcairne visited Newton in Cambridge in March 1692 when he was not only given a copy of "De natura acidorum," but regaled with several of Newton's notions pertaining to religion and ancient history. Pitcairne was so fired up by what he learned that he decided to emulate Newton. In September 1694, he divulged to a friend his

[2] Pearce, 1777, vol. 1, pp. xl–xliv.

"vast propensities" to compose a "Relligio mathematici," toward which end he entreated Gregory, a month later, to procure for him "a scheme of Mr Neuton's divine thoughts"—which included "papers about the mythologies, & Christian religion"—so that he might compose "a demonstration for our religion." Pitcairne wished to reciprocate with a promise to translate into Latin Newton's projected treatise on optics.[3]

Newton's openness with ideas was a distinguishing characteristic of his relations with disciples and acquaintances alike. William Whiston, who succeeded Newton as Lucasian professor of mathematics at Cambridge, was beneficiary of Newton's largesse from the late 1690s until the two became estranged a decade and a half later. Whiston recalled in his autobiography that he and Roger Cotes—the editor of the second edition of the *Principia*, who died in 1716—discussed ancient chronology with Newton, though, he claimed, he found Newton's notions "very weak."[4] In fact, Whiston attributed the rift with Newton to the latter's inability to be contradicted—on matters of chronology in particular—asserting that he'd purposely refrained from publishing his refutation of the chronology fearing it might kill Newton.

William Burnet and Alexander Cunningham were other acquaintances of Newton made privy to his religious and historical ideas. Burnet, member of the Royal Society committee charged with determining the matter of the calculus dispute, and future governor of New York, published in 1724 *An Essay on Scripture Prophecy* that was grounded—as he informed Cadwallader Colden—upon "principles which ... he had received from Sir Isaac Newton."[5] Less known is the historian and diplomat Alexander Cunningham, a former colleague of Gregory at Edinburgh University. The two probably left Scotland at the same time, and in London Cunningham befriended Fatio de Duillier and Locke, and met Newton in the early 1690s.[6] By the time he left for Venice in October 1715 as British ambassador, Cunningham must have been well informed of Newton's researches into ancient history, for on May 1716, he related to Newton the freedom he had taken in broadcasting Newton's ideas: "I have been talking to some of the scatches, you have been pleased to communicat to me in conversation, of chronologie, Every body says tis a pity you doe not put them in writing and I am Sr Isac, of their mind. I wish it wer possible for you sr to get the better of your inflexible Modestie."[7]

In light of this habitual openness, it was inevitable that Conti, too, once in Newton's confidence, would be made privy to Newton's ideas. The studious scion of a

[3] Pitcairne, 1979, pp. 18–20.

[4] Whiston, 1753, p. 35.

[5] Colden, 1868, pp. 214–15. Newton's *Observations upon the Prophecies of Daniel, and the Apocalypse of St. John* was published only in 1733.

[6] Grant, 1884, vol. 2, p. 280; Locke, 1976–1989, vol. 4, pp. 339, 778, 792, 796–97. For a detailed attempt to differentiate Cunningham from his namesake, the legal scholar, see Cairns, 2001.

[7] Newton, 1959–1977, vol. 6 p. 331 In an earlier letter Cunningham articulated his own admiration of the *Principia*—"such profundity and judgement as far surpassed both the genius and discoveries of antiquity, and the capacity of his own contemporaries"—before informing on the deep impression that the book had made on numerous acquaintance of Poleni, who followed Newton in studying nature. He continued: "I find all that speak of you, which are many, have a true sentiment not only of your sublime learning, but alsoe of your solid judgement and candure, and these they say they draw from your way of writing." Ibid., pp. 279–80.

noble Venetian family, Conti joined the Oratorian Order in 1699. Nine years later he left the Order, having determined his unsuitability for the clerical state in the face of a consuming passion for philosophy and metaphysics. Michelangelo Fardella had fired his interest in Cartesian and Malebranchian philosophy, while the lectures of Jacob Hermann in Padua had introduced Conti to the new differential calculus. Leibniz heard about the talented and ambitious Venetian from Hermann and other correspondents, and instinctively formed an apt estimation of him as one desirous of fame: "We must allow him this spur of glory of wishing to be original," he wrote Louis Bourguet on March 22, 1714, "provided he gives us one day something beautiful of his own."[8]

Eager for Leibniz's approbation, Conti communicated to Hermann and Christian Wolff a critique of Leibniz's philosophy even before leaving for Paris in 1714, where he befriended Nicholas Rémond, who considered Conti's remarks profound. On the eve of his departure for England, in the company of the French delegation brought together to observe the total eclipse of April 15, 1715, Conti wrote a long letter to Leibniz, in which he presented his side in the philosophical dispute with Francesco Maria Nigrisoli over the latter's rejection of a mechanistic interpretation of life in preference for "seminal light" and plastic nature. Conti regarded plasticity as indistinguishable from an occult quality, and he proceeded to liken Newton's "universal spirit" to a plastic force. Conti sought to endear himself to Leibniz by boasting that he was the first to publicly support pre-established harmony in Italy, further pledging to uphold Leibniz's cause in London, just as he had in Paris.[9]

Within a month of his arrival, Conti had sufficiently endeared himself to Newton to receive a visit from the great man. Their conversation on May 21, 1715 revolved mainly around physics and metaphysics, but Conti also recorded certain reflections made by Newton on the slow development of civilizations. The Phoenicians "were the first to navigate towards the equinoctial point" where they found "a forested island." Italy remained covered in forests well into Roman times, and so had Germany, "as far as the Rhine and the Danube. At first there were small princes or Kings in Italy and afterwards great Kings; one sees a progression." Conti's correspondence with Rémond—the details of which were promptly communicated to Leibniz—corroborates his growing intimacy with Newton. On August 15, 1715, Conti related that he visited Newton thrice weekly, waxing eloquent on Newton's erudition: "You have no idea how learned he is in ancient history and how reasonable and accurate are the reflections he makes on the facts. He has read much and meditated a great deal on the Holy Scriptures, and he speaks about them with great wisdom and good sense, stripping the words of their allegorical meaning and reducing them to history." Newton also knew a great deal about the ancient history of the Egyptians and thee Phoenicians, Conti continued, "or rather he meditated deeply about what Herodotus, Diodorus, Eusebius, Manetho and others have said about these people. He makes use of the character and genius of these people in order to better explicate the sense of Scriptures that was set down by these Orientals, that is to say by men who spoke the same way."[10]

[8] Leibniz, 1875–1890, vol. 3, p. 568.

[9] Ibid., vol. 3, pp. 618, 626; Leibniz, 1899, vol. 1, pp. 258–62; Ferrone, 1995, pp. 99–101.

[10] Iliffe et al., 2006, vol. 1, p. 241.

Conti communicated a similar mixture of admiration and information to other correspondents as well. Writing to Muratori on June 22, 1716, he exclaimed that Newton's mind was undoubtedly one of the vastest and soundest that ever existed, eminently deserving the kind of eulogies that Avveroes conferred on Aristotle, or that the French shower on Descartes. In contrast to Aristotle and Descartes, however, who copied and imitated the philosophers that preceded them—and when philosophizing concocted suppositions and not demonstrations—there was nothing unoriginal in Newton; everything that he expounded was rigorously demonstrated. Furthermore, he stressed, Newton "has an astounding grasp of the ancient history of the Egyptians and Greeks, [and he] has corrected many things" in Sir John Marsham's *Chronology*. Conti went so far as to add that "everything that is most sound and original" in Locke's *Human Understanding* derived from Newton![11]

Initially, Rémond appeared amused by Conti's infatuated embrace of England. Reporting to Leibniz in September 1715 about the abbé's frequent letters, as well as with his boasting that Newton kept "nothing hidden from him," Rémond observed that such openness would enable him to be better informed "about the private views of the English philosophers." Before long, however, he became fearful lest such infatuation blight Conti's fidelity to the Leibnizian cause. In fact, in one letter Rémond went so far as to liken Conti to a chameleon. Outwardly at least, Leibniz appeared unconcerned. "I am not angry that Conti turned a bit English for a time," he responded in December, "provided he remains the same to his friends." Surely the Venetian would revert to his former self once "the charm of the present impressions passes." Conti's disposition might even serve him to better benefit from what the English could offer. Still, to play it safe, Leibniz dispatched Conti with "a small preservative, so that the contagion does not affect him too strongly." That "preservative" took the form of the delayed response to Conti's eight-month-old letter to Leibniz, to which was added the fateful appendix in which Leibniz forcefully staked his position vis-à-vis the calculus priority dispute. The letter arrived too late to shield Conti from Newton's charm, albeit it emboldened him to assume the role of peacemaker. By early 1716, Conti dined with George I and, as Rémond informed Leibniz, the King expressed the wish to be informed of the "history of the quarrel" between his two illustrious subjects. Such a royal request provided Conti with the cue to assemble several ambassadors in London to review Newton's manuscripts, as well as to compel Newton—for the first time—to make his own case public.[12]

The details of the final salvos of the calculus dispute are beyond the scope of this book, but the ramifications of Conti's efforts to effect reconciliation between Newton and Leibniz proved fateful. Newton grew increasingly resentful of his direct embroilment in the dispute—especially as Leibniz opened a second front in the form of his exchange with Samuel Clarke—and he soon came to blame Conti. For his part, Conti left for the Continent in late summer 1716, planning to join the Royal Court in Hanover and to meet Leibniz. Since he tarried, Conti arrived at Hanover only to discover that Leibniz had just died. On November 10, (NS) he informed Newton of Leibniz's

[11] Conti, 1739–1756, vol. 2, p. 27, trans. in Iliffe et al., 2006 vol. 1 p. 241; Leibniz, 1875–1890, vol. 3, p. 655, partly translated in Manuel, 1968, p. 325; Badaloni, 1968, p. 239, n.57.

[12] Leibniz, 1875–1890, vol. 3, pp. 650, 662, 671. For the details, see Hall, 1980.

death and proceeded to declare that the dispute was over. He was wrong. When Conti returned to England several months later, nothing of his former intimacy with Newton remained. Indeed, it is quite likely that Conti informed Caroline of Newton's historical researches in an attempt to ingratiate himself again with Newton by garnering for him royal attention. In any case, sometime during the second half of 1717,[13] Conti wrote Newton to apprise him of Caroline's wish to meet and to learn more about his chronology. During the ensuing meeting the princess requested to see the work and Newton promised to prepare an abstract of it, which he delivered within a few days.[14] Not daring to approach Newton directly, Conti had entreated Caroline to prevail on Newton for a copy, and Newton had consented—on condition, he later claimed, that it be kept secret.

Once he released the text for public view, Newton appears to have acquiesced in its distribution. True, he would later maintain that Conti was alone in possessing a copy, but the Abbé was quick to rebut the claim. He pointed out that not only did Pierre Coste prepare one for a "lady of quality"—presumably the baroness von Kilmansegge—before Conti even received his,[15] but that three other copies existed in England prior to his departure for Paris in early 1718: Caroline's copy, Coste's copy, and one made for a young man who left for the Continent, where he died. There may have been a sixth, Conti noted, for Coste was not very careful to guard his.[16] That Coste indeed continued to be liberal with his copy is evident from the experience of the Italian physician Antonio Cocchi, who resided in England from 1723 to 1726. Cocchi befriended Coste and was granted free access to the latter's library where, among other things, Cocchi helped himself in late May 1725 to a copy of the "Short Chronicle."[17] Zachary Pearce, too, was aware of several copies of the work, including his own, which was copied from the earl of Macclesfield's copy.[18] Several copies of the abstract chronology still survive today but their provenance is uncertain. One, possibly belonging to an unidentified E. Jekyll, is preserved in King's College, Cambridge; two are preserved in the British Library; and one existed at St. Omer's College in the early nineteenth century, allegedly "with corrections by [Newton's] own hand."[19]

Conti did not deny Newton's other charge, namely, that he had revealed the content of the "Short Chronicle" in Paris, even allowing copies to be made. Evidence for his rapid diffusion of Newton's historical and theological ideas may be gleaned from a fragment manuscript diary by the Parisian magistrate Antoine-Robert Pérelle. Pérelle

[13] Newton stated in 1725 that the request had reached him "about seven years" earlier.

[14] Newton's original copy appears to be Yahuda MS 28.

[15] Giuseppe Toaldo, Conti's biographer, claimed that Conti's exemplar was copied from that of the Baroness. Conti, 1739–1756, vol. 2, p. 58.

[16] Conti, 1726, pp. 15–16.

[17] Guerrini, 2002, pp. 59–61. Cocchi's copy, "transcribed from a copy of Mr. Coste's hand," is in the Biblioteca Nazionale Generale, Florence, Fondo Magliabechiano, Ms. XXIII.89.

[18] Pearce, 1777, p. xlii. Thomas Parker, first earl of Macclesfield, was a close friend of Newton, and one of the pallbearers at the latter's funeral. He appears not only to have received a copy of the "Short Chronicle" from Newton but the "Prophecies of Daniel" as well. He appears to have been collating "two distinct and very different MSS" of the treatise at the time of his death on April 28, 1732. Edwards, 1864, p. 341.

[19] Keynes MS 136.1; BL MS Lansd. 788 fols. 44–56—with a critical commentary in Latin and BL MS Sloane 3208 fols. 49–67^{v}; Phillips, 1834, p. 264.

befriended the abbé during Conti's first stay in Paris, and it was partly owing to such familiarity that Pérelle became converted to Newtonian science. No sooner had Conti returned to Paris than the two resumed their meetings. In early April 1718, Pérelle was regaled with numerous particulars regarding Newton's life and ideas. In addition to scientific information, Pérelle was told that Newton was esteemed a great theologian in England, and that it was believed that he was the inspiration behind Benjamin Hoadley's famous 1717 sermon *The Nature of the Kingdom, or Church, of Christ*. It was further believed that those who had taken Hoadley's side in the ensuing Bangorian controversy were agents of Newton, though the great man himself was loath to become personally involved in scholarly disputes. Conti also elaborated on Newton's conviction that facts alone must serve as the foundation of religion; that he established the epoch of the Argonauts in a manuscript given the Princess of Wales (which he did not wish to be printed during his lifetime); that Troy fell much later than commonly held. The antiquity of Egyptian dynasties was also recalibrated by Newton, so that the Egyptian kings during Moses' time were deemed to be minor rulers, akin to those battled by Abraham. The Bible itself, in Newton's view, though the most ancient of books to survive, was nevertheless poorly written, with limited use for history, and, to make matters worse, was considerably revised by Samuel and Ezra. The *pièce de résistance*—which Pérelle jotted down in English—was the juicy gossip that Newton "believe that J.-C. was a man, not God's son, who hath given us a very good Morale."[20]

Pérelle does not seem to have been a recipient of the "Short Chronicle"—at least at this early stage. But soon Conti began sharing the manuscript with other friends. In part, his largesse masked a self-aggrandizing impulse to parade intimacy with the social and intellectual elite—in kind not unlike his eagerness to circulate among such friends as Pérelle and Rémond letters he had received from Lady Montagu, the wife of the English ambassador to Constantinople.[21] Yet Conti was a true admirer of Newton who often made a nuisance of himself by extolling the virtues and ideas of the Englishman everywhere in Paris. As he argued years later, his inability to procure from Newton clarifications regarding the grounds upon which the Chronology was based drove him to seek enlightenment elsewhere. He first sought to obtain the desired information by pressing Coste to approach Newton. When Coste failed to respond, Conti turned to the Jesuit Etienne Souciet, who in 1719 was given a copy of the "Short Chronicle." After studying the manuscript, the Jesuit formulated several queries that were communicated to Newton through Brook Taylor, who was then in Paris. Far from becoming infuriated, Newton instructed John Keill to inform Souciet that the treatise was "but a short abstract of a much longer work, and he had not set down the proofs" before volunteering a brief précis of its rationale:

> According to his best remembrance, he found that the Ancients had recorded that at the time of the Argonauts Chiron had found the equinoctial points to be in the middle or 15th degree of the constellation Aries. In Meton's time it was found to be in the 8. and in Hipparchus's in the 4. degree of that constellation. Hipparchus counted the precession to be a degree in a hundred

[20] Briggs, 1934, pp. 499–504 (on Newtonian science), pp. 510–11 (diary).

[21] Montagu, 1967, vol. 1 pp. xv–xvi, 395–96.

years, and they generally founded their Chronology upon that computation. But the equinoxes move a degree in 72 years and by that means, if we compute, we shall find the time of the Argonautical expedition to have fallen out at the time Sir Is. Newton puts it.

Six years later Souciet claimed that though he'd been unsatisfied with Newton's response, he adhered to "la religion du secret" and desisted from pursuing the matter further.[22]

By the time the exchange had occurred, Conti had struck up a close friendship with Henry St. John, Lord Bolingbroke, who fled to France in 1715 as impeachment proceedings were brought against him. In exile, Bolingbroke immersed himself in the study of ancient history and chronology, with Pierre-Joseph Alary serving as his mentor.[23] By fall 1719, Bolingbroke heard from Conti about the "Short Chronicle" and on November 8, he entreated the abbé to send him a copy. "You can be certain," Bolingbroke assured Conti, "that it will not be seen by anyone."[24] Conti evidently complied and the perusal of Newton's manuscript was destined to make a profound, albeit negative, impression on the earl, who henceforth became increasingly disillusioned with ancient chronology. As he informed Brook Taylor on November 23, 1721,

> I have very near done, for my whole life, with all inquiries into remote antiquity. My intention was to see the foundations of those historical and chronological systems, which have been erected with so much learned pains in our western world. I have seen them, these corner stones, and I think I have examined them enough to be sure, that he who cannot content himself to employ his time about consequences, drawn from principles evidently begged, ought not to employ it in this kind of erudition. If ever we meet, I'll trust to your candour what I have observed, and what has fixed my thoughts, and put an end to all my curiosity on this subject.[25]

Meanwhile, Conti once more drew down Newton's wrath, following the publication in 1720 of the *Recueil de Diverses Pièces sur la Philosophie, la Religion Naturelle, l'Histoire … par Mrs. Leibniz, Clarke, Newton, & autres Auteurs Célèbres*. In addition to reprinting the Leibniz-Clarke correspondence, Pierre Des Maizeaux, the editor, published other documents pertaining to the calculus dispute, many of which were supplied by Conti. Newton, who initially approved the project, changed his mind and soon convinced himself that Conti had engineered it. Newton made his displeasure

[22] Souciet, 1727, pp. 55–56, 49, Manuel, 1963, p. 23.

[23] See Bolingbroke, 1808, vol. 3, pp. 3–6, 21–35. An account of Alary's life can be found in Clément, 2002.

[24] The letter makes clear that Conti had shown Bolingbroke a copy of the letter that Newton wrote Leibniz in 1716, detailing his side in the priority dispute, for Bolingbroke also requested Conti to send him Leibniz's response to Newton's letter: Morrison, 1882–1893, vol. 1, p. 318.

[25] Taylor, 1793, pp. 126–27. Bolingbroke had articulated his burgeoning historical Pyrrhonism already a year earlier in "The substance of some Letters, Written Originally in French, about the Year MDCCXX, to Mr. De Pouilly," Bolingbroke, 1977 [1757], vol. 5, pp. 221–311, esp. 232–82.

known, and Conti's correspondence with Brook Taylor attests to the abbé's desperate efforts to exculpate himself using both threats and flattery. On May 22, 1721, Conti sent Taylor an account of his involvement in the affair, expressing bewilderment at Newton's continued animosity toward him. He honored Newton and acted on his behalf, to the detriment of his relations with the Leibnizians, Conti wrote. Yet, if Newton persisted with his accusations, he would be forced to vindicate himself by making the truth of the matter public, proving that he was not eager to make a name for himself as Newton charged.[26]

A subsequent letter to Taylor, dated July 14, 1721, suggests that Conti's attempt to appease Newton was partly motivated by his continuing desire to learn more about the chronology. In a decidedly more obsequious tone, Conti beseeched Taylor to pay his respects to Newton and convey the "profound veneration" with which Conti had always held him. Conti promised to send some stanzas he composed, part of a design to render the Newtonian philosophy into an epic poem in four books.[27] Conti continued: "If Mr. Newton wished to send me the entire manuscript of the history that he communicated to the Princess of Wales, it would give me the greatest pleasure in the world. I'ld communicate it only to Lord Bolingbroke who wishes avidly to see it, and Mr. Newton can be certain on my word of honor that the manuscript would not leave our hands."[28] It is hard to imagine that Newton could have gratified the request; no manuscript of the complete chronology yet existed. Nor was he mollified. As late as January 15, 1724 (NS) Conti was still trying to regain Newton's favor by sending a short poem which, he hoped, would please Newton by its invocation of attraction, denoted by love, as governing the system assigned by Descartes to Phaeton. Conti prefaced his offering by informing Taylor of a rumor that Newton was about to publish his "Chronologie Raisonnée," eagerly awaited by everyone.[29]

Writing to Taylor three weeks earlier, Bolingbroke had referred favorably to Conti's undertaking a "Philosophical Poem," in which Newton's system would "make the principal beauty of it." Bolingbroke thereupon bid a second farewell to chronology, worth citing in full for the evidence it offers of an early reaction to Newton's re-ordering of ancient chronology—based not only on the "Short Chronicle" but on conversations that Bolingbroke had had with Conti, Taylor, and Lévesque de Pouilly, all privy to additional information from Newton:

> I have gone through all that I proposed to myself, in the way of studying, wherein I was when you gave us your good company. I never intended to do more than to examine, as well as I was able, the foundations on which those Systems of Chronology and Ancient History, which obtain in our western world, are built, a fin de savoir a quoi m'en tenir. I have done this; and I have no more desire to pursue this study any further than to be a proficient in

[26] RS MS 82, fols. 47–48^{v}, in Taylor, 1793, pp. 121–25.

[27] Conti managed to compose only 200 lines of the projected epic, which was conceived in emulation of Charles-Claude Genest's *Principes de philosophie, ou, Preuves naturelles de l'existence de Dieu et de l'immortalité de l'ame* (Paris, 1716).

[28] RS MS 82, fols. 49–49^{v}.

[29] Ibid., fols. 53–54, in Taylor, 1793, pp. 140–41.

> Judicial Astrology. Who can resolve to build, with great cost and pains, when he finds how deep soever he digs, nothing but loose sand? Some have been so pleased with an high and lofty situation, that they have ventured upon this project; for my part, I incline not to imitate them: and to carry the similitude a little further, when such buildings are raised, I may be tempted to take a cursory view of them, but I can by no means resolve to dwell in them, *a limine salutanda sunt.*[30]

Before continuing to evaluate the impact of Newton's ideas on French and English savants, we should clarify the circumstances surrounding the publication of the "Short Chronicle." The catalyst for the publication may have been commercial, at least in part. In 1722 there appeared in Amsterdam a French translation of Humphrey Prideaux's *The Old and New Testament Connected in the History of the Jews and Neighbouring Nations*. It proved an instant bestseller and the Parisian publisher Guillaume Cavelier the younger wished to capitalize on its success by issuing a rival edition. To differentiate his version of the *Histoire des Juifs*—apart from alleging a purer French—Cavelier thought to commission several supplementary essays, including a chronicle of the eras that preceded and followed those narrated by Prideaux. It quickly became clear, however, that such a project would require greater effort than might be expected. And since a dry summary of the facts seemed inappropriate—and time constraints precluded commissioning an account that could approximate Prideaux's precision—it was hoped that Newton's chronology might well serve the purpose: a treatise written by a celebrated author, which offered a very curious overview of profane history leading to the destruction of the kingdoms of Judea and Israel. Accordingly, on April 30, 1724 (NS) Cavelier informed Newton that a copy of the "Short Chronicle" had fallen into his hands and that he wished to publish it. However, as he was also informed that the manuscript was defective, Cavelier asked Newton to furnish him with a corrected version. Several other individuals who also owned defective copies, the Frenchman added, expressed a similar wish.[31] It does not appear that at this stage there was a plan to publish a critique of the "Short Chronicle" as well; at least in his first letter Cavelier did not mention it.[32]

Curiously, two weeks later, on May 13, Philip Henry Zollman—then secretary to Horace Walpole, British ambassador to France—apprised James Jurin, Secretary of the Royal Society, that the current issue of the *Journal des Savants* announced the imminent publication of Newton's *Chronology*, and that it was "eagerly awaited." The source of the literary news was not given, except to indicate a London origin. It may have been a regurgitation of the rumor that Conti had mentioned three months earlier or, more likely, it was based on more recent information that had reached Paris. Either way, what is significant here is that Newton's response to Zollman's letter was a calm affirmation that, indeed, he "had some thoughts of printing that Piece, as soon as he [had] finish'd his Principia, which [would] be done before ye end of ye Year." Zoll-

30 Taylor, 1793, pp. 135–36.

31 Newton, 1725, pp. 48–49; Newton, 1959–1977, vol. 7, p. 279.

32 Newton, 1959–1977, vol. 7, p. 279.

man undoubtedly made the response public and the news reverberated in learned Parisian circles.[33]

We can only speculate on the precise sequence of events in Paris. Quite possibly, the rumors regarding Newton's intention to publish his chronology at long last helped alleviate considerations of confidentiality that bound those who possessed copies of the "Short Chronicle." Assuming that Cavelier wrote Newton as soon as he obtained a copy, we may further conjecture that whoever had furnished him with the text expected Newton's acquiescence to Cavelier's request—or at the very least confirmation that Newton would publish the work himself—thereby enabling a public debate about the chronology. (Certainly, according to Giuseppe Toaldo, Conti's biographer, French savants had conspired to "smoke" Newton out by publishing the abstracted chronology.)[34] When Newton appeared unwilling to take the bait, the schemers turned in late 1724 or early 1725 to Nicholas Fréret to provide a critique of the "Short Chronicle" in order to exert additional pressure on Newton to publicly defend his chronology.[35] Yet, it was probably not Fréret's copy that fell into Cavelier's hands.[36] Rather, we believe, Cavelier was indebted to the Parisian Jesuits for the text—provided either by Souciet himself or, indirectly, by his co-religionist René-Joseph de Tournemine. Cavelier co-published in 1719 a revised edition of Giovanni Stefano Menochio's popular commentary on the Bible edited by Tournemine, to which the Jesuit appended twelve of his own dissertations, including his celebrated system of chronology. Tournemine was actually the editor of the *Histoire des Juifs*, to which he contributed an introduction that exposed Prideaux's doctrinal errors, as well as two dissertations: 1. on the destruction of Nineveh and the duration of the Assyrian Empire, and 2. on the Protestant rejection of the canonicity of certain Old Testament Books.

Given that Newton's chronological system was diametrically opposed to the one upheld by Tournemine and Souciet, their determination to battle the Englishman is hardly surprising. In view of Souciet's known connections to Conti and his previous exchange with Newton, however, it was probably considered prudent to enlist a non-Jesuit opponent of Newton's chronology to confute the "Short Chronicle." Fréret, a protégé of Tournemine, was the obvious choice for the task. Nevertheless, the Jesuits offered assistance. Fréret was given a copy of Souciet's initial criticism of the work, while Tournemine availed himself of the *Mémoires de Trévoux* (which he'd edited) in order to taunt the Englishman. Thus, the January 1725 issue of the journal included the following caustic remark: "Newton's famous chronology is forever promised. It must bring marvelous changes to the science of time. Why else would such a celebrated mathematician have given his principal application to it for so many years?"[37]

[33] Jurin, 1996, pp. 246, 257–58. *Journal des Sçavans* (June, 1724), p. 413: "Le célébre Mr Newton va bien-tôt faire imprimer sa Chronologie. Tous les Savans attendent cet Ouvrage avec impatience."

[34] Conti, 1739–1756, vol. 2, p. 58.

[35] That the subsidiary ploy was concocted at a relatively late stage may be surmised from the hurried nature of Fréret's observations—to be analyzed below.

[36] Fréret was allowed to copy L'Evesque de Pouilly's copy.

[37] *Mémoires de Trévoux* (January 1725), p. 175: "On promet toûjours la fameuse Chronologie de Monsieur le Chevalier Newton: elle doit, dit-on, apporter des changemens merveilleux dans la Science des tems. Sans cela, pourquoi un si grand Mathematicien y auroit-il donné sa principale application depuis tant d'années?"

Newton, meanwhile, ignored Cavelier's letter. Not to be rebuffed, the publisher repeated his request in October 1724, and when Newton again failed to respond, Cavelier dispatched a third letter on March 20, 1725 (NS), notifying Newton of his intention to interpret silence as consent for publication of the manuscript in its present form:

> Sir, six months ago I had the honor of informing you that a copy of your chronology had fallen into my hands. I asked you to inform me whether you had any additions or corrections to make in it because of errors on the part of the translator. Since the savants await anything which comes from a man as talented as you with great eagerness, Sir, I have the honor of writing you this second letter to ask you to inform me immediately if you have something to change in it. If I do not hear from you I shall take your silence for consent and let it appear as it is and I shall give it to the public with remarks.

Newton waited for more than two months before firing back. On May 27, (OS) he penned a curt rebuke of Cavelier for intending to print a text that had been composed "for a particular friend, on condition that it should not be communicated." Since it was not intended for publication, Newton neither granted permission nor wished to "meddle with that which hath been given you under my name." By the time Newton's letter arrived in Paris, the book was already in press. On May 20, (NS) Cavelier received official permission to publish both Newton's manuscript and Fréret's reflections on it, and on June 1, he obtained a Royal privilege for three years to print the *Abrégé* separately or together with other works, as often as he wished.[38] The seventh and final volume of the *Histoire des Juifs*, which included the abstracted chronology and Fréret's censure of it, was in press in August 1725,[39] and both Prideaux's history and the free-floating edition of the *Abrégé de la Chronologie* went on sale in early fall. Newton received a complimentary copy on November 22.

As we shall see below, the brunt of Newton's fury at the sight of the *Abrégé* was directed at Conti, for he regarded the French critique as trifling. He was right. Not only was the *Observations* a hurried work, but Fréret's condescending attitude toward the "Short Chronicle" prevented him from fully comprehending key elements of Newton's argument. For the same reason, Fréret's attempt at refutation neither did justice to his great erudition nor did it succeed, in any sense of the word.

The son of a *procureur* for the Parlement of Paris, Fréret had managed to contravene his father's wish that he pursue a legal career, and followed instead his passion for history. His budding talent was recognized early, securing his election in March 1714 as *élève* of the Académie des Inscriptions at the young age of twenty-six. Within eight months of his admittance, Fréret had infuriated certain of his colleagues by reading a memoir confuting the mythical origins of the Franks as civilized descendents of the Greeks (or Trojans), and suggesting instead their German (and barbaric) origination. Such audacity earned him half a year incarceration in the Bastille—despite a

[38] Newton, 1959–1977, vol. 7, pp. 311, 322; Newton, 1725.

[39] *Journal des Sçavans*, August 1725.

valiant effort by Tournemine to rescue him.[40] Outwardly reformed, Fréret devoted himself thenceforth to retired studiousness and the presentation before the Académie of an ever-increasing number of historical and chronological communications, which culminated in his spirited rejoinder to Pouilly's dissertations on the uncertainty of history, delivered in December 1722 and December 1724. It was against the backdrop of this critique—which partly revolved around Fréret's interpretation of the chronology of the ancient near-Eastern empires—that we must read his reaction to Newton's "Short Chronicle." The Pyrrhonist position that Pouilly defended originated, in part, as a reaction to Newton's intervention in ancient history—analogous to the response of Pouilly's skeptical friend, Bolingbroke. Fréret, who received a copy of the "Short Chronicle" from Pouilly, fully appreciated the need to undermine Newton's authority as part of his endeavor to preserve the solid foundations of chronology, especially as his understanding of ancient history, too, was the exact opposite of Newton's.[41]

Fréret defended the unauthorized publication of Newton's text by claiming that the wide interest in the illustrious Englishman's work, combined with the singularity of the ideas contained in the "Short Chronicle," merited its rescue from the common fate of "pièces fugitives." Fréret conceded that his objections might not be original. Any among the handful competent to illuminate such matters—provided they read the text with care—was capable of arriving at similar conclusions. Hence, he was hardly surprised to discover that Souciet had anticipated him on certain issues. Nevertheless, Fréret insisted, while they pursued a similar objective, they took different routes, and he had not borrowed anything from Souciet. Everyone is endowed with a particular way, Fréret reasoned, and such a way differentiates those who make their path from those who follow the common track. Having thus distanced himself from Souciet's critique, thereby also staving off charges of plagiarism, Fréret turned to excuse his critique of Newton. While he recognized that it was necessary to wait for the publication of the complete chronology before passing judgment on Newton's claims, he (Fréret) should not be faulted for raising objections to the proofs that emboldened Newton to abbreviate ancient Greek history by four centuries. The years he had devoted to studies of that nature had earned him the right to controvert the Englishman by furnishing the authorities Newton had ignored.[42]

After summarily dismissing the credibility of Plutarch's remarks concerning the uncertainty of chronology, cited approvingly by Newton, Fréret turned to refute Newton's reckoning of 18 to 20 years per generation in lieu of the ancient value of three generations per century. As proof that Newton's revision could not be correct, Fréret offered several stretches of French history. His method was simple. Fréret divided the time between the birth of one king and the death of another by the total number of "generations" that separated them in order to obtain a figure for a generation. The

[40] For Tournemine's letter, see Grell and Volpilhac-Auger, 1994, pp. 26–27 n.6.

[41] Ostensibly, Pouilly's dissertation was devoted to the first four centuries of Roman history. The broader implications of his argument, however, were crystal clear. For the debate, see Grell and Volpilhac-Auger, 1994, 99–161; Grell, 1995, vol. 2, pp. 826–51. According to Grell, Pouilly's first dissertation plagiarized a treatise by the abbé Lacourt, p. 829.

[42] Newton, 1725, pp. 48–50.

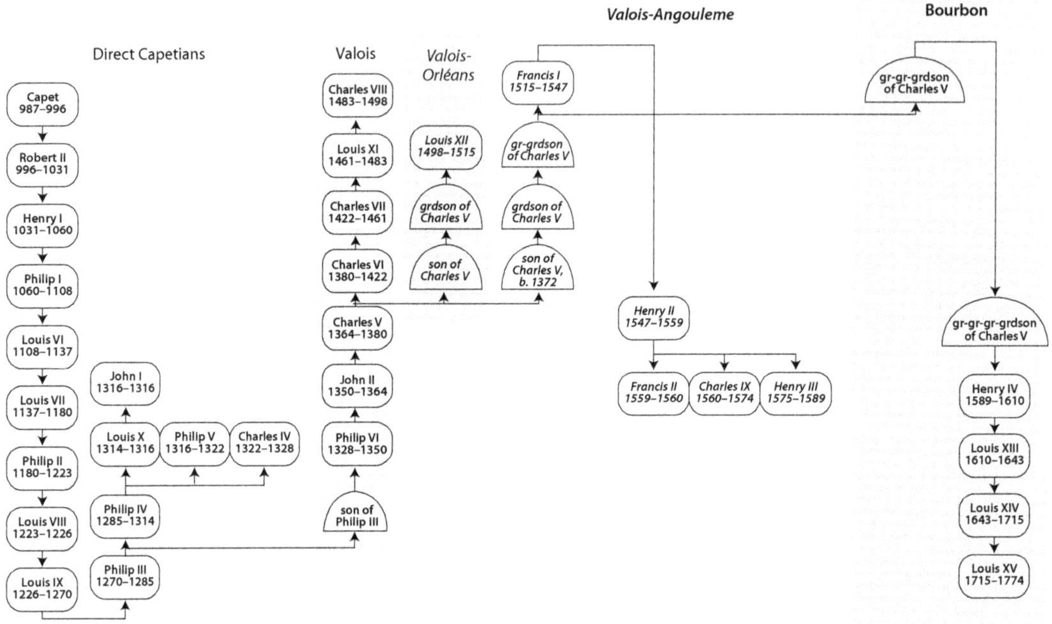

Figure 9.1. Genealogy of French kings from Hugh Capet through Louis XV.

kings list, and their dates, from Hugh Capet through Louis XV were of course not in doubt. Capet was born in 940, and Louis XV in 1710, an interval of 770 years. Their reigns are separated by a combined total of thirty other reigns, including Capet but dropping the five-day reign of the infant son, John I, of Louis X in 1316.[43] Figure 9.1 lists all of the kings, with their mutual affiliations, from Capet through Louis XV. The rectangular outlines bracket the several dynasties, from direct Capetians then Valois, Valois-Orléans (one king only), Valois-Angouleme, and finally the Bourbons. Note that a trio of brothers were successive kings in both the direct Capetian line and in the Valois-Angouleme.

Fréret began by rather confusingly remarking that there were "24 generations by the Bourbon branch, from Robert de Clermont, son of St. Louis (Louis IX)." He meant by this that the Bourbon line, which derived originally from Louis IX's son, reached back through 24 "generations" to Capet. Fréret then divided 770 by 24 to obtain a "generation" length of 32 years. Yet there are thirty distinct reigns separating Capet from Louis XV. His rationale for reducing 30 reigns to 24 "generations" becomes clear from his next paragraph. There he indeed noted that there were "30 reigns" in "this interval." This was so, he continued—implicitly explaining the apparent contradiction with 24 "generations"—because "of the three sons of Philip IV, and the three sons of Henry II, who, having died without children, were succeeded by

[43] Namely Louis VII (1137–1180), Louis VIII (1223–1226), Louis X (1314–1316), John I (1316), Philip V (1316–1322), and Francis II (1559–1560).

princes of their same age." This produces 24 reigns by combining brothers to form a single generation (which must have meant, in effect, the time interval from a king to the birth of his first son), with the other sons coming along soon thereafter (and so justifying Fréret's combination of brothers to form a single "generation").

And, indeed, his next datum is similar to this one, as Fréret proceeded by cutting up the reigns into different segments to see whether the result remained consistent—indicating his awareness that the choice of a range could markedly affect the result. He counted 12 generations from the birth of Capet to the death of Charles IV in 1328, again compacting the three sons of Philip IV into one generation. And so Fréret divided the interval (this time 388 years) by 12 to obtain once again his 32 years. Fréret next tabulates 20 generations from the birth of Capet through the death of Henry III, again yielding 32 years per generation. Here we see directly that he literally counted the direct father-son sequence: Figure 9.1 shows that there are 20, compacting Henry III with his three brothers, of course, into a single generation. The succeeding example works equally well, following the same procedure from the son of Charles V through the Valois-Angouleme line to Henry III, this time producing, correctly, 36 years per generation. Fréret next mistakenly calculates 673 years from Capet's birth to the death of Louis XII in 1515, instead of 575 years, and arrives thereby at a generational length of 40.[44] No doubt a hurried mistake, but perhaps one that indicates just how eager Fréret was to trump Newton. And his final example, from Capet through Charles VIII, correctly produced a 31-year generation. He wrote in conclusion that "from the several different calculations we find that the generations contained sometimes 40 Years, and never less than 30."[45]

Having demonstrated to his own satisfaction that Newton's estimate of generations was flawed, Fréret turned to expose his errors regarding the ancient reckoning of time. The "Short Chronicle" offered few details illustrative of Newton's method and little evidence to support his re-dating of Greek history. Fréret decided, in effect, to lecture the Englishman on the nature of Greek calendars, feeling it reasonable to do so because Newton had explicitly invoked astronomy only once. He had postulated that it was Chiron who placed the solstices and equinoxes at the 15th degree, or toward the middle, of the constellations Cancer, Capricorn, Aries, and Scorpio. And since in 433 BCE Meton, Newton had written, observed that the summer solstice had receded seven degrees since the Centaur's time, he concluded that the Argonautic expeditions must have commenced on 937 BCE.

In Fréret's view, ancient sources warranted no such assumption. The Greeks produced only "country calendars" (*calendriers rustiques*), whose purpose was simply to accommodate the understanding and prejudices of rural readers by furnishing them with the modicum of information necessary for their daily lives. Hence, authors of such calendars—Eudoxus and Meton included—felt no need to fix the cardinal points in their "true Places." When Meton "plac'd the Solstices and Equinoxes in the 8th Degree of the Constellations, it was not because he had observed they were gone 7 Degrees back since *Chiron*'s Time; for then he must at the same Time have observ'd, that those Points were in Days in the first, and not in the 8th Degree of the Constel-

[44] The text misprints 27 for what must clearly be 17 as the generation count.

[45] Fréret, 1728, pp. 52–55; 55–59.

lations. All he meant, was to follow the calendars then in use, which he has not meddled with in this particular." In fact, Fréret swiped, Meton was a pitiable astronomer and it was thus incumbent on Newton to furnish the grounds for his assertion that this Athenian actually possessed the skills necessary for making a precise determination for the position of the summer solstice in 432 BCE. "Till he does that, we shall believe we have good Reason to hold that *Meton* did not regulate his Calendar in this Particular, by Truth, as *Euctemon* had done, but by the Opinion receiv'd in his Days, amongst the Country People, for whose use it was made."[46]

Fréret accepted Newton's premise that Chiron was the author of the most ancient Greek Calendar, though he fixed its composition half a millennium earlier on the basis of Herodotus's claim that his own birth, in 482 BCE, postdated by some 900 years the birth of Hercules—who, in turn, was a generation younger than Chiron. More invidious was Fréret's insinuation of incompetence on Newton's part. If the first star of Aries was located fifteen degrees from the equinoctial colure—as Fréret at this time incorrectly took to represent Newton's argument—then precession would have mandated the passage of 1080 years before that star lay on the colure. This, he (again incorrectly) asserted, occurred in 388 BCE, with the inescapable conclusion that Chiron flourished ca. 1468 BCE.[47] Likewise, since Newton maintained (again according to Fréret) that the first star of Aries was seven degrees closer to the colure in Meton's time, the Athenian should have lived 576 years before that realignment, i.e., in 964 BCE—an anachronism of 532 years since Newton dated Meton's observation to 432 BCE. Clearly, then, Fréret alleged, Newton's dating of Chiron's epoch to 937 BCE—fully cognizant that the equinoctial colure was positioned fifteen degrees from the first star of Aries—was intended for no other reason than to destroy ancient chronology.[48]

Fréret also challenged Newton on several issues pertaining to Egyptian history. How could Newton maintain that the Egyptians formed their religion only a century after Solomon had built his Temple, when Scripture plainly mentions the Egyptian kings in the time of Abraham, 1100 years earlier? And what was Newton's justification for rejecting the authority of Egyptians' chroniclers relating their own history—notably Manetho—while accepting the fables that the Greek poets related about Egyptian and Greek deities? Much more could be said about Newton's unwarranted conjectures regarding Greek history, Fréret contended, but he believed that his objections to Newton's manipulation of generations and to Newton's dating of Chiron's epoch sufficed to ring a cautionary note; readers of the *Abrégé* would be wise to suspend judgment until such time as Newton published his proofs. Not yet done with his critique, Fréret floated a veiled charge of plagiarism. Even as he commended Newton for offering "several very ingenious things" relating to Sesostris's great deeds, he stressed that Henri, comte de Boulainvilliers, had already expounded these very notions a decade and a half earlier, and his manuscripts circulated freely. Nor was New-

[46] Ibid., 59–74.

[47] *Prima Arietis* lay on the equinoctial colure in 163 BCE. To get his date, Fréret has ignored the effect of *prima*'s latitude, which retrogresses the star to place it on the colure. His date would hold only if *prima* were on the ecliptic itself.

[48] Fréret, 1728, pp. 81–87. A final proof to substantiate Chiron having lived ca. 1500 BCE was provided by citing Seneca to the effect that he recognized the ancient Greek calendars "were more antient than himself by 1500 Years." Ibid., pp. 87–89.

ton's identification of Sesostris as Sesac original, for it had already been made by Marsham. It was instead simply wrong, for both Tournemine and Boulainvilliers had demonstrated that the great Egyptian ruler flourished 560 years before Solomon's Temple.[49]

Fréret concluded the *Considerations* by reiterating his respect for Newton and vowing that he would consider it an honor to convert to the Englishman's sentiments if and when Newton's proofs were found to be as solid as his friends maintained. Until such time, however, he would retain the "Right to believe the antient Writers of *Greece*" who were better informed of their own history than Fréret's own contemporaries, two millennia later, who "have no other Materials but what they have left us."[50]

Newton embarked on a counter-refutation immediately upon receiving a copy of the *Abrégé*, as the several surviving drafts attest. He had been perfectly justified to ignore Cavelier's letters, Newton claimed. He had written the "Short Chronicle" to gratify "a particular Friend [Princess Caroline], on condition that it should not be communicated." In view of the fact that he had not set eyes on the manuscript Cavelier wished to publish—and that the translator had conspired to refute him—what man "could be so foolish as to consent to the publishing of an unseen Translation of his Papers, made by an unknown Person, with a Confutation annexed, and unanswered at their first Appearance in Publick"? Newton's fury was merely a prelude to his assault on the person whom he regarded as the true culprit in the affair, Conti—whose manuscript, Newton inferred from the prefatory material of the *Abrégé*, was the one that had been used. Newton's fury might well have been fanned by the visit he had received five months earlier, on July 1, 1725, from Pierre-Joseph Alary. The Abbé spent nine "very agreeable" hours with Newton, much of the time devoted to ancient history.[51] Less agreeable, perhaps, certainly in retrospect, were the details that Alary undoubtedly imparted regarding the circulation of Newton's chronological ideas among the Bolingbroke circle as far back as the late 1710s. In short, Newton had every reason to believe that Conti was back to his old tricks.

In the published version of his response, Newton recalled that "about seven years ago"—i.e., following Conti's return to England from Hannover in mid-1717—the abbé had written that Caroline wished to meet him. The formality of epistolary communication indicated a coolness had crept into their relationship, and increased the likelihood, as suggested above, that Conti intended the Royal attention to restore him to Newton's good graces. During the arranged meeting Caroline expressed the wish to see what Newton had written about chronology. Though his papers were yet in an "imperfect and confus'd" state, Newton replied, within days he would draw up an abstract, on condition that it remain a secret. He presented the "Short Chronicle" some time later, and Caroline desired that Conti might also be given a copy. Conti, Newton insisted, was the only person to possess a copy, and despite his awareness of its private nature he'd dispersed copies freely in France, eventually engaging an antiquary to translate and confute it. After animadverting on Fréret's errors—to be discussed below—Newton accused Conti of duplicity, since, during his sojourn in

[49] Ibid., pp. 89–102.

[50] Ibid., pp. 102–3.

[51] Brewster, 1965, vol. 2, p. 389.

England, he had portrayed himself as Newton's friend, all the while aiding Leibniz in entangling Newton in disputes. More egregious still, Conti acted in a similar manner in Italy, encouraging his friends to attack Newton's optical experiments.[52]

In his draft manuscripts Newton had belabored Conti's crimes even further. Key among these was the repeated charge that the abbé incessantly conspired to embroil Newton in dispute, which he (Conti) then volunteered to mediate. Explicit in at least two drafts was Newton's firm belief that Giovanni Rizzetti's attack on the *Opticks* was engineered by Conti, who then proposed to dispatch the young Oxford mathematician James Stirling to Italy to defend Newton, provided Newton "contributed to [Stirling's] maintenance" there. As Newton articulated Conti's modus operandi: "His first step was to insinuate himself into my acquaintance, & in a while he began to be at work in engaging me in disputes & sometimes finding persons to defend me. And because I do not trust him he has now found out one to oppose me, & to soften the business had lately writ a Poem upon me under the colour of a friend. But I hope that this & the perpetual motion will be the two last efforts of the friends of Mr Leibniz."[53]

In comparison to the abuse showered on Conti, the response to Fréret's critique was measured, albeit dismissive. In his eagerness to wreak havoc on the two poles upon which the "Short Chronicle" rested, the Frenchman had obviously misunderstood Newton's positions. As regards the expedition of the Argonauts, Fréret was simply wrong to claim that Newton had placed the colure "fifteen Degrees from the first Star of *Aries*," which would have necessitated predating the expedition by half a millennium. In fact, Newton retorted, he placed the colure in the middle of the constellation, which is not 15 degrees but about seven and a half from that star. Consequently, since Fréret had nevertheless granted Newton's key suppositions—including the attribution of the invention of the constellations to Chiron as well as Newton's interpretation of Eudoxus's and Hipparchus's concurrence with the delineation of the colures in the middle of the constellations—surely Fréret had no grounds to object to Newton's dating of the expedition.

Fréret's claim that the first star of Aries lay on the *equinoctial colure* in −388 is incorrect. It is certainly true that *prima Arietis* does coincide in longitude with the vernal equinox at that date. But it does not also then lie on the equinoctial colure because, unlike the solstitial colures, the equinoctials are orthogonals only to the equator, and not also to the ecliptic, while *prima Arietis* hardly lies on the ecliptic, having nearly eight degrees north latitude.[54] Newton had not written a word in the abstracted chronology about the colures; he had mentioned only that Chiron had placed the "solstitial and equinoctial points" in the middles of the constellations.[55]

[52] Newton, 1726, pp. 315–17, 320–21.

[53] Yahuda MS 27, fol. 8^v.

[54] Not that the error made much of a difference to Fréret's critique. Had he done it correctly, he would have placed *prima Arietis* on the equinoctial colure in about −150 instead of −388, thereby altering Fréret's calculation for Chiron from 1480 to 1230 BCE, still far from Newton's date.

[55] Newton, 1726, p. 318. Still, even Newton could be terminologically obscure. In his 1726 reply he avowed to follow "Eudoxus, and, by doing so, place the equinoctial colure about 7gr. 36′ from the first star of Aries." However, in 939 BCE (Newton's date for the Chironic sphere), *prima Arietis* was about that distance in longitude from the equinoctial point proper, with somewhat more than three degrees' difference in distance to the point of the colure at *prima*'s latitude. Specifically, in Flamsteed's catalog, *prima* has

Fréret's comprehension of Newton's claim regarding generations was equally misguided. Newton had been admittedly cryptic in the abstracted chronology, but he had there explicitly written "... of these generations, or rather of these reigns." In his reply to Fréret's claims, Newton noted that he did "agree with the Antients in reckoning three generations, at about an hundred years. But the reigns of kings I do not equal to generations."[56] Moreover, though Newton did not take the trouble to explain, there are in fact a total of 36 reigns from the beginning of Capet's in 987 until the end of Louis XIV's in 1715. That produces an average regnal length of 20 years. That one among the reigns was abbreviated, and that others were of succeeding brothers, would to Newton have been entirely appropriate—since presumably *on average* the same kind of distribution would have taken place in antiquity. Of course, Fréret was not thinking in terms of an average with its concomitant notion of deviations that more or less cancel out over time. That was still comparatively uncommon even among natural philosophers. So instead of elaborating the point, Newton decided to take a longer series of French kings, dating far back before Hugh Capet, to produce further regnal periods. He pointed out that the 64 kings that ruled France—from the legendary Pharamond in AD 428 to Louis XIV (d. 1715)—reigned for 1287 years, an average of 20 years per reign, adding that "they, that examine the matter, will find it so in other kingdoms." It's worth recalling that manuscript evidence shows that Newton had been selective in choosing among the reigns in calculating a mean regnal length for the kings of Israel and Judah.

Fréret had not only mistaken his meaning on the "two main Arguments on which the Whole is founded," Newton complained, but he had translated, and sought to confute, a treatise that he barely grasped. Newton concluded by dismissing the *Observations* with yet another instance of Fréret's misconstruals—this one to do with Egyptian religion and calendar—before denying that he had ever intended his chronology for publication: "When I lived at *Cambridge* I us'd sometimes to refresh myself with History and Chronology for a While, when I was weary with other Studies: but I never told him, that I was preparing a Work of this Kind for the Press."[57] Clearly, Newton was less than truthful about that, but from this point on his alleged unwillingness to publish the chronology prior to the unauthorized publication of the "Short Chronicle" became the official line in England.[58]

By early 1726, Newton had printed a French version of his response and dispatched it to Paris. Conti, still in town, quickly penned a reply. Noel Pissot undertook publication of both letters and obtained the censor's approbation on March 13. The pamphlet

---

longitude 7°8′58″ at epoch 1689. Precessing back to 939 BCE puts it in *Pisces* at about ♓ 22°10′, which is therefore 7°40′ from the equinoctial point, not from the point at the same latitude as *prima* on the corresponding colure. Though we have of course seen that in finding his dates from equatorial colure loci Newton had carefully carried out the appropriate correction. And in 1726, Newton was after all very old and would die the next year. His computations had for the most part been completed years before. More on this when we turn to Souciet's subsequent critique.

[56] Newton, 1726, p. 319.

[57] Ibid., pp. 317–20.

[58] See Conduitt's claim that Newton was of two minds about publishing his *Chronology*, but following the publication of the *Abrégé* he resolved to do so, though he resolved making it "as privately as possible & keep the copies in his own possession." Iliffe et al., 2006, vol. 1, pp. 93, 101.

appeared a few weeks later, and was subsequently inserted into the July-August issue of *La France Savante*.[59] The English version of Newton's letter was inserted into the thirty-third volume of the *Philosophical Transactions*, which was presented to the Royal Society on May 12, 1726.[60] It was also excerpted as a separate publication for wider distribution, as may be gathered from a letter by the physician Pierre Noguez who in October 1725 was to be appointed keeper of the Natural History Cabinet at the Jardin du Roi. Noguez may well also have been the actual translator of the "Short Chronicle" into French, for he was a prolific translator of English scientific works during the 1720s, mostly published by Cavelier. Noguez befriended Conti during the latter's sojourn in Paris, and the two remained in contact following Conti's return to Italy. On August 19, 1726, Noguez informed his friend that he had delivered one copy of Conti's response to Souciet and another to Pierre-Jean Burette, editor of the *Journal des Savants*, for inclusion in that periodical. Noguez also intimated that an Englishman had recently arrived from London, carrying sixty copies of Newton's critique of the *Abrégé* for distribution in Paris, as well as the issue of the *Philosophical Transactions* wherein Newton's censure was inserted. Noguez continued: "It seems that the English take Mr. Newton's part with a great deal of harsh words. This celebrated geometer has entirely taken to religion, he took a death's head for his coat of arms which he bears on his carriage."[61]

Newton may even have explored the possibility of inflicting an official French censure on Conti. Evidence to suggest such a scheme may be gleaned from a letter sent by the oculist John Thomas Woolhouse—who resided in Paris during the first three decades of the eighteenth century—to Jurin on May 4, 1726, obviously in response to a letter by Jurin. Woolhouse reported that he had met the Abbé Bignon, who was "highly displeas'd that Sr Isaac's goodness has been so ill treated, as appears by ye philosophical transactions & by a printed account, brought hither from London, & printed here before at Pissot." Nevertheless, Bignon had made clear that as he was president of neither of the two royal academies to which he belonged—nor was Conti a member—no official course of action could be initiated against the Italian.[62]

Conti's response to Newton took the form of measured indignance. After correcting Newton on the issue of whether he alone was in possession of the "Short Chronicle," Conti queried the very basis for Newton's claim to have imposed a condition of secrecy. What would have been the point in allowing him a copy of the manuscript if he was barred from seeking clarifications? Nevertheless, Conti came to understand before even leaving England how useless it would be to approach Newton, who had made clear his unwillingness to respond to anyone, loath as he was to enter into scholarly controversies. Hence Conti turned first to Coste, then to Father Souciet. But once Newton responded to the Jesuit, they desisted from pursuing the matter further; only the publication of the *Abrégé* by a third party revived the dispute. As far as Conti was concerned, Newton himself was to blame for the publication. Had he responded

59 Conti, 1812, pp. 29–30; *La France Savante, ou Histoire Litteraire de la France* (Amsterdam, 1726), 173–93.

60 RS Journal Book vol. 12, p. 654. A summary of Newton's letter appeared in the *Bibliothèque Anglois* 14 (1726), 66–70.

61 Albertan and Chouillet, 2003, p. 222. The *Journal des Savants* summarized both letters in its August issue (pp. 459–64).

62 RS Early Letters W.3.116.

to the publisher's letters—as Conti had suggested in a letter to Taylor—it might well have been averted.

Conti was unapologetic about sharing the "Short Chronicle" with others. As several copies were already in circulation in England, he saw no reason not to make the manuscript known in France as well. Was this a crime, Conti asked wryly. Anyone but Newton would have been pleased to learn of the high opinion he had formed of the Englishman's work and his desire to celebrate him as enlightened an historian as he was profound mathematician and natural philosopher. Conti feigned disinclination to indict Newton of perpetrating the same practices for which he now stood accused—namely, publishing without permission papers by Leibniz that the abbé had communicated to Newton. Nevertheless, Newton's allegation that Conti had engaged someone to translate and refute the "Short Chronicle" bordered on calumny. On a more personal note, Conti expressed astonishment at his incrimination as a false friend. Had he not been as true a friend to Newton in Paris as he was in England? Had he not defended Newton's mathematics and physics with vehemence, as Fontenelle, Fraguier, Saurin, Nicolle, and many others could attest? Surely, Bolingbroke and his wife could confirm the verity of his sentiments. Once more Conti recounted his role in the calculus priority dispute; rejected the accusation that he had invited Italian savants to oppose Newton's optics; denied that he had asked Newton to nominate him Fellow of the Royal Society. If anything, it was Newton who had conferred the honor on Conti, "perhaps as a reward of my so-called mediation."

Finally, Conti fashioned himself as one who sought out study not for enrichment or fame, but for pleasure: after all, he esteemed mathematics and natural philosophy no more than the quadrille or the hunt. Anticipating an argument that Diderot would make famous three decades later, Conti envisaged that with the exception of fifteen or twenty useful problems, one day these studies would become as despised as the very issues (void, atoms, time, and universal harmony) that Newton had scorned—a clear swipe at Newton's repeated complaints that Conti had embroiled him in disputing these topics with Leibniz.[63]

It is not easy to reconstruct the more general reaction to the publication of the *Abrégé* and Fréret's critique. Reviewing the *Histoire des Juifs* in February 1726, the *Journal des Sçavans* praised its editor for bestowing on the public Newton's treatise—which he proceeded to summarize along with Fréret's response—before endorsing the latter's hardly forthright statement regarding the necessity to suspend judgment until the publication of the complete chronology. Were the work to demonstrate what the *Abrégé* proposed, the reviewer pronounced, Newton would be honored for revolutionizing chronology as he had previously revolutionized mathematics. For the most part, however, contemporaries remained non-committal, if not downright skeptical. When it came to choosing between Conti and Newton, French public opinion tended to side with Conti. Members of the Society of Jesus were certainly expected to do so, judging by the protestation of Father Louis-Bertrand Castel—then vying for election to the Royal Society—that he refused to pronounce against Newton even though he resolutely sided with Conti.[64]

[63] Conti, 1726.

[64] *Journal des Sçavans*, Feb 1726; RS EL C2 #77.

Fontenelle's carefully worded mention of chronology in his éloge of Newton is indicative of its lukewarm reception in France. After summarizing the argument regarding the epoch of Chiron, the Secretary of the Académie des Sciences exculpated Fréret and Souciet of the charge that they had launched too hasty a critique of the "Short Chronicle" instead of awaiting the complete chronology. Didn't such haste "do honour to Newton"? Fontenelle asked. Indeed, in his opinion the two Frenchmen moved swiftly against Newton purposely in order to reap the honor of engaging such an eminent adversary. Aware that Edmond Halley had defended the astronomical foundation of the chronology (to which we shall return), Fontenelle still considered the dispute unfinished: "the rather restricted public capable of making judgments on it has not yet pronounced; and should it happen that the strongest arguments were on one side, and Newton's name on the other, perhaps this public would remain for some time undecided, and that perhaps would be understandable."[65]

Reaction in England was more mixed. Zachary Pearce, then vicar of St. Martin-in-the-Field, enthusiastically availed himself of the "Short Chronicle" in an appendix to a sermon he delivered on October 20, 1726 upon the consecration of a new parish church. In it he found inspiration (or support) for an argument against the existence of "true" temples—i.e., enclosed and roofed structures—prior to the building of Solomon's Temple. Pearce recognized the conjectural nature of Newton's claim that Belus' Temple in Babylon, as well as Vulcan's temple in Memphis, postdated the Temple in Jerusalem, yet he expressed confidence that when

> that Excellent Genius, the Glory of our Isle, would find Leisure to give us his Observations upon Chronology more at large than we have yet seen, and supported by all those other Proofs which, no doubt, he has long since collected, we should find a new and a strong Light breaking in upon the remotest and darkest Parts of the ancient Chronology: even now his Epitome is sufficient to guide us in many Cases, if we tread warily, and look to our Steps and to incline us to think, that his Discoveries will give him as great a Superiority in this sort of Learning, as he is by the Judicious Part of Mankind allowed to have in Natural Philosophy and the Mathematicks.[66]

Thomas Hearne read Pearce's treatise in late December 1726 and found it a mean effort. Dismissive of the vicar's shallow scholarship, the Oxford antiquarian also took exception to the lavish praise that Pearce showered on the Master of the Mint. Hearne had been informed as early as January 1724/5 that Newton, who "does not understand a bit of Classical Learning," had studied chronology as relaxation. Now, confronted with Pearce's animated endorsement of the "Short Chronicle," Hearne turned more critical. Newton might well be a "very great" mathematician, he noted in his diary, but in chronology he merely "advanceth Paradoxes & new opinions, and being no Classical Scholar (as I am well assured he is not), he must be at a loss for Reading,

[65] Hall, 1999, pp. 70–71.

[66] Pearce, 1727, pp. 33–34.

to know w^{t} the Antients delivered of such and such Affairs, and 'tis too late to begin Reading now at his great Age, tho' I cannot find that he much thinks of Death." Shortly after Newton's death, Hearne reflected again on what he considered to be an indifferent understanding of chronology on Newton's part, likening it to the performance of another great mathematician—John Wallis—whose inaptitude (or so Hearne believed) was in plain view in the contribution he made to John Fell's edition of St. Cyprian's works.[67] Indeed, prior to the publication of the *Chronology of Ancient Kingdoms*, passing judgment on Newton's ideas was often predicated on an assessment of Newton's qualifications to make such a revolutionary intervention.[68]

The publication of the *Abrégé* prompted some scholars to communicate constructive criticism to Newton. The Genevan scholar Firmin Abauzit, who had met Newton during a visit to England in 1699, is a case in point. According to Abauzit's biographer, the two maintained a correspondence—albeit no letters appear to have survived. The Swiss appears to have taken Newton's side in the calculus priority dispute, judging by Newton's alleged commendation of Abauzit as "a very fit person to judge between Leibnitz and me." In addition to his considerable scientific attainments, Abauzit was also reputed to be "a living library" who "knew ancient Greece and Egypt better than the ancient Greeks and Egyptians had known it." The biographer further notes that although Abauzit was eminently qualified to frame an "excellent system of chronology," he "contented himself with correcting the mistakes" of eminent chronologers, among them Newton. According to a note by Abauzit, it was his intervention that convinced Newton to correct his dating of the famous eclipse predicted by Thales from 610 to 585 BCE.[69]

As noted above, Newton confirmed in May 1724 his intention to publish the *Chronology* as soon as the third edition of the *Principia* was ready. He probably reached that decision in early 1723 following a serious illness that served to remind him of his mortality. Whether he decided at the same time to revise the *Principia*, too, is unclear. Perhaps that decision came later in the year as a reaction to the unauthorized reprinting in Amsterdam of the second edition of the *Principia*. Either way, by the second half of 1723, following the arrival on the scene of a willing and enthusiastic editor, Henry Pemberton, work on the *Principia* had taken precedence over the chronology. Printing commenced in October 1723, but progress proved slower than Newton had anticipated. It finally appeared in February 1726. Newton's inability, or unwillingness, to work simultaneously on both projects may be gathered from Antonio Cocchi's intimation to Conti on May 21, 1725 that he had recently met Newton, who did not appear to consider publication of the chronology but was preparing a new edition of the *Principia*. Ten months later, however, on March 6, 1726, Cocchi bore different news. At last, Newton's friends had managed to persuade him to publish the *Chronology*. (Cocchi also conveyed his inability to comprehend why Newton felt it necessary to keep his conjectures mysterious, especially over such a common matter as history.)[70]

[67] Hearne, 1885–1921, vol. 8, p. 326; vol. 9, pp. 244, 332. For Wallis' work, see Cyprian, 1682.

[68] Ouvry, 1874, p. 81; Atterbury, 1783–1787, vol. 1, pp. 180–81.

[69] Abauzit, 1774, pp. ix, xvi–xvii.

[70] Conti, 1812, pp. 29–30.

A month later James West, too, learned about the imminent publication of the *Chronology*, a sample sheet of which had already been printed by John Innys.[71]

As it turned out, Newton did not live long enough to see the published *Chronology*. Nor was Innys ultimately entrusted with its publication. Within days of Newton's spectacular funeral at Westminster Abbey, John Conduitt commissioned the erudite physician Thomas Pellett to inspect Newton's surviving papers. Three days sufficed for Pellet to determine that only five manuscripts, the *Chronology* included, were publishable. The rest he deemed "not fit to be printed." The grounds for his verdict were purely commercial, without a whiff of value judgment regarding the propriety of revealing Newton's theological or alchemical preoccupations. Reassured, Conduitt lost little time before handing over the *Chronology* to the publisher Jacob Tonson for the princely sum of £350.

How Tonson became known to Conduitt is unclear. Perhaps Alexander Pope—who advised Conduitt on the epitaph for Newton's tomb—introduced them.[72] Or it was owing to the intercession of John Arbuthnot, whose *Table of Ancient Coins* was published by Tonson in 1727. Certainly, the buzz surrounding the chronology convinced the enterprising Tonson that purchasing the manuscript would prove a good commercial investment, and he managed to beat Innys to it. Innys undoubtedly felt betrayed. After all, he served not only as printer to the Royal Society, but he published the third edition of the *Principia*—and was future publisher of Newton's *System of the World*. Innys attempted to fight back, imploring Pearce—who, we may recall, had met Innys at Newton's house—to corroborate his claim that Newton had promised him publication rights. Pearce refused. By April 4, 1727, Richard Mead informed William Stukeley that Conduitt had been busy preparing the *Chronology* for publication, to which he intended to append Newton's life. Conduitt confirmed his intention to write Newton's life in a letter to Fontenelle a week earlier, in which he also offered to furnish Fontenelle with material for Newton's éloge. Conduitt dispatched that material in late July. But though he reiterated his commitment to write Newton's life, Conduitt never proceeded beyond the draft memoirs he prepared for Fontenelle. At some point he also relegated the task of overseeing the publication of the *Chronology* to Martin Folkes and Pellett.[73]

[71] Ouvry, 1874, p. 81. West proceeded to comment that it was "very short," which may imply that Innys presented Newton with a portion of the "Short Chronology."

[72] Pope, 1956, vol. 2, pp. 457–59.

[73] Pearce, 1777, p. xlv; Stukeley, 1882–1887, vol. 1, pp. 424–25.

The War on Newton in England

Tonson's willingness to fork out a sizeable sum of money to secure the publication rights of the *Chronology* attests to the considerable public interest the book generated. The poet Edward Young informed a friend on June 5, 1727 that Londoners anticipated "with Impatience" the publication of Newton's theological works; five months later, the Scottish divine Robert Wodrow learned from Colin Maclaurin that the *Chronology* was in print "and there are very great expectations from it."[1] Testimony to the burgeoning excitement is also signalled by the efforts of other publishers to capitalize on public interest. In late September 1728, John Peele published an English translation of the *Abrégé de la chronologie*, together with Fréret's critique.[2] In his preface to *Sir Isaac Newton's Chronology, Abridged by Himself*, Peele felt it necessary to justify his printing of a vernacular critique of Newton:

> I hope, upon a mature Consideration of the Merits of the Argument, which ought always to weigh any Personal Affection, the World will not take it ill that an Englishman has made Sir Isaac's Antagonist speak the Language of this Country. For if the Observer has Truth on his side, every candid Enquirer will be fond of embracing it. But if our learned Mathematician's System can stand the Test of his Objections, I shall expect the Thanks of his Friends, for having thus put it in their power to defend him. And as Truth itself is my only Concern; so I publish the Chronology and Observations together with no other View than that some able Pen may undertake the Cause, and settle the weighty Points in question, to the Satisfaction, and for the Information of the Learned World.

Tonson was sufficiently alarmed by this venture to advertise at once in the London newspapers: Newton's genuine and complete chronology was in press, and what Peele had published amounted to "only a Translation of a very bad French Translation from an erroneous and imperfect Copy of the short Chronicle, which is not a tenth Part of the whole Work above mentioned." Judging by Peele's apparent failure to sell his

[1] Young, 1971, p. 56; Wodrow, 1842–1843, vol. 3, p. 461.

[2] Advertised in the *Whitehall Evening Post* (September 26, 1727) and the *London Journal* (September 30, 1727).

edition—unsold copies were handed over to another bookseller, Thomas Warner, who in February 1728, republished the *Abrégé* with a new title page and without the preface—Tonson's aggressive campaign proved effective.

More dramatic still was the visual representation of Newton's chronology as it was engraved on some of John Senex's more expensive globes. From the start of his independent career in the early eighteenth century, Senex—a leading publisher of maps, astronomical charts, and scientific books as well as maker of terrestrial and celestial globes—collaborated with Edmond Halley, John Theophilus Desaguliers, and William Whiston. By 1705 he had become Halley's preferred bookseller, publishing the latter's *Miscellanea curiosa* as well as his *Synopsis of the Astronomy of Comets*. Their partnership would remain fruitful for three decades. It was probably through Halley that Senex also became Whiston's publisher, beginning with *A Scheme of the Solar System* (1712). Thereafter Senex handled most of Whiston's scientific publications, and not a few of his theological works—either because he shared Whiston's religious views or because he expected Whiston's lucubrations to turn a profit. Or perhaps both. By the late 1710s, Senex began publishing Desaguliers' original works and translations as well.

Senex's production of globes commenced around 1707, when he advertised globes measuring 12 and 16 inches in diameter, the celestial one being "laid down from the Observations of Capt. Halley, Mr. Havelius, and other eminent Astronomers, with 19 Constellations never before printed on any Globe."[3] Six years later, in collaboration with John Maxwell, Senex set his sights on a pair of 30-inch globes. But though they were advertised as forthcoming from February to July 1713, the project apparently faltered, owing perhaps to the dissolution of the partnership the following year. In 1718, Senex again attempted to construct large globes, this time alone. His *Treatise of the Description and Use of Both Globes* announced the imminent completion of 26-inch globes, the celestial one "laid down according to Mr. *Flamsted*'s New Catalogue." This globe, Senex boasted, would include not only double the number of stars contained in other globes, "but these will likewise have their Positions fix'd, and their Magnitudes determin'd to a much greater Exactness. The whole will be finish'd to the utmost Perfection, and will both for Elegancy and Accuracy far exceed any Thing of that kind that ever yet made Publick." As Dekker suggests, this ambitious project sought to replace Willem J. Blaue's 68-cm globes, which had dominated the market for decades. Senex's aim may also be inferred both from the size of the projected globes and from his borrowing much of the nomenclature inscribed on Blaue's celestial globe—albeit with additions. Years later, Mary Senex claimed that her husband had constructed his globe "under the Direction of the great Dr. *Halley*"; in all likelihood it was at Halley's suggestion that Senex employed the Bayer notation of stars.[4]

Senex was emboldened by his close connections to leading members of the Royal Society to seek election as Fellow. To this end, he dedicated during the 1710s or early 1720s a pair of 16-inch globes: "To the great philosopher and geometrician Sir Isaac Newton, Knight, most worthy President of the Royal Society of London, for the promotion of knowledge, and to the Committee and Members of the same Society." On

[3] *Observator* (November 29, 1707).

[4] Senex, 1718, pp. [115–16]; Dekker, 1999, pp. 120–21, 493–95; Senex, 1749, pp. 290–91.

June 20, 1728, his efforts were rewarded. Halley, seconded by President Sloane and Desaguliers, proposed him for a Fellowship, to which he was elected the following week. On July 4, Senex was officially admitted Fellow.[5]

In late 1727, or early 1728, Senex decided—or was persuaded—to augment his large globes with a representation of Newton's chronological system. Prudently perhaps, he never publicized such augmentation. The advertisement he appended to the 1730 edition of Nieuwentijt's *The Religious Philosopher*, which he also published, simply publicized a pair of 28-inch globes, "fit to adorn the Libraries of the Curious. On the … Celestial are placed all the Stars in Mr. *Flamsteed's Catalogue*, as published by Dr. *Halley*, &c. being above 2000 more than ever were inserted upon any Globe. The Asterisms are designed so as to answer the Description of the Antients, and the Letters of Reference made use of by *Bayer*, in his Tables, are inserted. The Price of these handsomely fitted up is 25 Guineas." In Samuel Johnson's day, such a sum would also have bought about sixty-six bottles of champagne at New Spring (later Vauxhall) Gardens, making the globes a decidedly luxury purchase. An identical advertisement appeared in the updated edition of *A Treatise of the Description and Use of Both Globes* (1740).[6] Senex's partnership with Halley no doubt explains his continued use of Halley's unauthorized 1712 edition of Flamsteed's star catalogue, notwithstanding Flamsteed's retrieval and destruction of 300 of the 400 copies in 1716 and the posthumous publication of Flamsteed's authorized edition of the catalogue (by his former assistants) in 1725. In 1749, however, when defending the superiority of her husband's globe, Mary Senex noted that it not only facilitated the easy determination of "the ancient Observations without the Labour of a tedious *Calculus*," but "the famous Astronomical Argument likewise of Sir Isaac *Newton*, in his *Chronology* … may hereby be more particularly enquired into, and considered."[7] The celestial sphere was large, colorful, and beautifully mounted (figure 10.1).

Just above Aries, and parallel to a dotted line running through the back of the figure, appeared the following label: "Colurus Aequinoctialis, ut situs erat Anno 939 ante Christum, quo tempore secundum Newtonum, Chiron formavit Constellationes in usum Argonautarum." To Aries' lower right appeared another dotted line, perpendicular to the first and with the same label, except that "Colurus" was replaced with "Circulus."[8] On the ecliptic the two dotted lines intersected. These were, of course, respectively, Newton's colure and equinoctial circle at the time of the Argonauts (figure 10.2). This vivid visualization of Newton's most celebrated discovery in the *Chronology of Ancient Kingdoms* no doubt explains why William Whiston found Senex's globe particularly suited to illustrate the force of his sharp critique of Newton's *Chronology*.

More about the particulars of Whiston's critique in due course. For now, its particularly pointed character and the wide-ranging scope of this critique demands a close scrutiny of Whiston's vexed relationship with Newton, especially in light of the

[5] Stevenson, 1921, p. 151; RS JBC XIII, fols. 233, 241.

[6] Nieuwentyt, 1730, p. [1101]; Wright, 1740, opposite title page, [102].

[7] Senex, 1749, pp. 291–92.

[8] The "equinoctial circle" marks the position of the equator on the celestial sphere at the time of Newton's Argonauts.

Figure 10.1. The Senex celestial globe (from the collections of the National Maritime Museum at Greenwich. Photo by JZB.)

prevailing assumption of intimacy between the two until the mid-1710s. In his 1749 autobiography, Whiston relates that he first encountered Newton in 1694, boasting that he was Newton's choice to succeed him as Lucasian professor of mathematics, and that he had enjoyed "a large portion of [Newton's] favour for twenty years together." Nevertheless, the precise nature of such "favour" is left vague, and a careful reading of Whiston's voluminous writings suggests that the reality was otherwise. Consider Whiston's claim in his 1749 autobiography that the manuscript of his *New Theory of the Earth* (1696) was shown to Richard Bentley and Sir Christopher Wren, "but chiefly laid before Sir *Isaac Newton* himself, on whose Principles it depended, and who well approved it." In 1728, however, when some people who would have known better were still alive, Whiston was more precise, and less grandiose. Newton's approval had then been tethered only to his endorsement of Whiston's notion that the Deluge had been caused by a comet.[9] Nor apparently did the *New Theory* suffice to recommend Whiston to Newton as his successor in 1700. As Whiston himself acknowledges, his appointment was secured at Richard Bentley's behest. Indeed, Whiston's less than stellar qualifications for the professorship led some "Malitious people" to quip "that Augustus left a Tiberius to succeed him purposely to render his own fame the more illustrious."[10]

[9] Whiston, 1753, pp. 250, 38; Whiston, 1727–1728, vol. 2, p. 1069.

[10] Whiston, 1727–1728, vol. 2, p. 1071; Flamsteed, 1995–2001, vol. 2, p. 929.

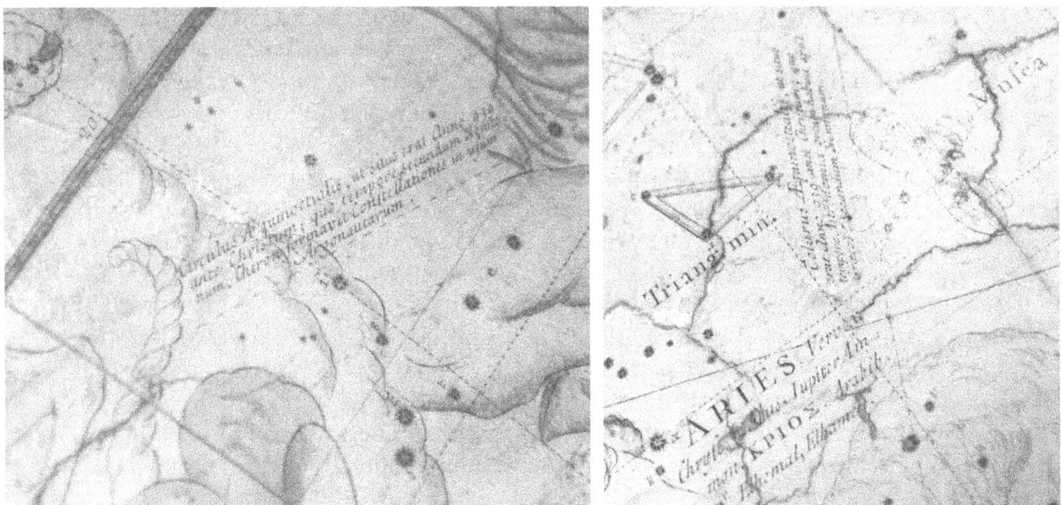

Figure 10.2. Newton's equinoctial circle (left) and colure on the Senex globe.

As much as Whiston implied friendship and unity of mind with Newton, he carefully sidestepped the depth of his familiarity during the Master's lifetime. Prior to Whiston's removal to London in 1711, geographical separation alone would have prevented the possibility of more than intermittent meetings. By then, as we shall see below, Newton had already cast Whiston aside. Certainly, in his writings Whiston recorded only a handful of meetings with Newton during which non-scientific issues may have been discussed. Once, in the company of Newton and Roger Cotes, the conversation veered toward chronology, with Whiston judging Newton's notions to be exceedingly "weak"—or so Whiston related in his autobiography. In 1706, Whiston supposedly spent four hours alone with Newton discussing the Apocalypse. He "could hardly assent to more than one of his expositions," Whiston recalled years later, by which date he deemed Newton's errors in chronology to be even more egregious. Whiston also reveals that Newton had furnished him with several "most valuable Intimations" in theology, "which tho' I have hinted at already sometimes; yet did not I think fit publickly to name Sir I. N. before now." These included Newton's fidelity to the Apocrypha; his opinion concerning "the lascivious Nature of the Book of *Canticles*"; the argument "against the Allegorical or double Interpretation" of Old Testament prophecies; and the notion "concerning the *Geographical* as well as *Chronological* Distinction of the four Monarchies in the Language of Scripture."[11] Nor did Whiston ever ascribe to Newton paternity for his own anti-Trinitarian heresy. The mentor he names is Samuel Clarke. By all appearances, then, Whiston was not a member of Newton's inner circle; much of what he absorbed of Newton's religious and chronological ideas appears to have derived from conversations with such friends in common as Clarke or Hopton Haynes.

[11] Whiston, 1753, pp. 35–36; Whiston, 1727–1728, vol. 2, pp. 1071–72.

Most importantly, Whiston's acquaintance with Newton did not endure until 1714, though he implies otherwise in his autobiography, which attributes the rupture to Newton's inability to tolerate his genius and independence of mind: "he then perceiving that I could not do as his other darling friends did, that is, learn of him, without contradicting him, when I differed in opinion from him, he could not, in his old age, bear such contradiction; and so he was afraid of me that last thirteen years of his life." In truth, Whiston had fallen out of favour almost a decade earlier, when he'd set himself up as a prophet, and, as many thought, an insufferable one at that. Already in his *Essay on the Revelation of Saint John* (1706), Whiston predicted the advent of the Millennium in 1716, "the time of the Commencing of Christ's Kingdom, upon the first fall of Antichrist." Whiston did allow the possibility of an alternate date—1736—since the coincidences he discovered could not amount to "certain demonstration." Nevertheless, Whiston pretty much settled on his original interpretation: the "exactness in the Coincidence of such large and express numbers seems to me not a little remarkable, and worthy of more than ordinary consideration." Contemporaries proved eager to interpret Whiston as he wished to be interpreted.[12]

Within days of the publication of the *Essay* in May 1706, the Oxford antiquary Thomas Hearne was informed that Whiston held that the "World [would] be at an End" in about nine years "when there will be a great Eclipse of the Sun"—an opinion shared by William Lloyd, Bishop of Worcester. The following year John Lacy, one of the most prominent backers of the "French Prophets," claimed that Whiston's book had inspired his own vision of an imminent new cosmic order—the arrival of a new "State of the Church on Earth, more resplendent than ever yet"—while Alexander Pope damned in a poem the "wicked Works of *Whiston*." Such notoriety prompted the more moderate millenarian divine Robert Fleming to complain in 1708 that Whiston's *Essay* "is extremely pleasing" to the French Prophets, who admire in particular "his Grand Notion, viz. *That Antichrist will fall, and that the Kingdom of God will eminently take place about the Year, 1716*."[13] Whiston thereby came perilously close to being marked a fellow traveler, so much so that he felt obliged to publicly repudiate the Camisards in the course of his 1707 Boyle lectures, which treated the interpretation of Scripture prophecies: "If any Persons, in this Age, who pretend to a Prophetick Spirit, do foretel Events, whether of Mercy or of Judgment, which do not come to pass accordingly, we have the Warrant of God himself for their rejection."[14] Obviously, expediency informed his reprimand. Whiston was loath to relinquish his role as a latter-day prophet—especially as he had just begun unearthing the "genuine" constitutions of Christianity as well. On April 1713, he advertised the publication of "A Scheme of the Scripture Prophecies . . . Wherein the Downfal of the French Monarchy within Four Years is foretold." When 1716 came and went peacefully, Whiston without comment fell back on the 1736 date. He remained undaunted when that year, too, passed without drama, and revised again. And again. In 1748, at age eighty-one, he "assured" Edward Young, that "the Jews [would] be converted" within eighteen years

[12] Whiston, 1706, pp. 62, 84, 222, 248, 270–72 and passim.

[13] Hearne, 1885–1921, vol. 1, p. 248; Lacy, 1708 , p. iv; Pope, 1956, vol. 1, p. 26; Fleming, 1708, vol. 2, p. xxiii.

[14] Whiston, 1708, p. 230.

and that "twenty years hence the Millennium [would] begin." Mercifully, Whiston did not live to be 101 and witness the failure of these prophecies as well.[15]

Newton could hardly have approved of Whiston's new-found enthusiasm. His own extensive research into the Books of Daniel and Revelation was grounded on the premise that Scripture prophecies could be correlated to historical events only retrospectively. Only once did he propose a date for the commencement of the Millennium—the year 2060—at the same time making clear (perhaps with Whiston in mind) that his intent was more facetious than serious: "This I mention not to assert when the time of the end shall be, but to put a stop to the rash conjectures of fansifull men who are frequently predicting the time of the end, & by doing so bring the sacred prophesies into discredit as often as their predictions fail. Christ comes as a thief in the night, & it is not for us to know the times & seasons wch God hath put into his own breast."[16] More to the point, Whiston's positioning himself as a prophet occurred when Newton was in the act of severing his intimate and longstanding friendship with Fatio de Duillier, precisely because Fatio had fallen in so completely with the French Prophets. Surely, then, the idea of Whiston turning into an enthusiast offended Newton.

Equally offensive to Newton would have been Whiston's garnering ill repute on account of his "love for *wrangling and Disputing*." The traveler Zacharias Conrad von Uffenbach witnessed Whiston perform at a coffeehouse during a visit to Cambridge in 1710: "a man of very quick and ardent spirit," who "is very fond of speaking and argues with great vehemence." Both Francis Hare and Anthony Collins lampooned the hotheaded Whiston for believing that Christ and his apostles condemned "prudence" and for learning from them that it was a "gross prevarication and hypocrisy to conceal the discoveries he conceives he has made." Collins also commented on how the "warmth" of Whiston's temper "disposes him to receive any sudden thoughts, any thing that strikes his imagination, when favourable to his preconceiv'd scheme of things, or to any new schemes of things, that serve, in his opinion, a religious purpose. And his imagination is so strong and lively on these occasions, that he sometimes even supposes facts, and builds upon those facts." Samuel Parker commented in 1709 on the futility of engaging such a person in argument: "To what purpose you dispute with Mr. *Whiston* about What is in Scripture, when he runs out such a further length of Scepticism, as to make it a Doubt What is Scripture? how many, not only Parts, but BOOKS, of the Old Canon are quite lost and extinct? and how many are yet to be Added to those still Extant and Receiv'd?"[17] This is hardly a person Newton would tolerate.

If any vestige of familiarity survived this episode, it surely vanished in the aftermath of Whiston's expulsion from Cambridge in 1710 owing to his open espousal of anti-Trinitarianism. What Newton found particularly offensive was not Whiston's religious sentiments—strands of which he shared—or even the vociferousness with which they were promoted. Rather, the very recklessness of his defense infuriated

[15] *Daily Courant*, April 1, 1713; Young, 1971, p. 302.

[16] Yahuda MS 7.3g, fol. 13^{v}.

[17] Scriblerus, 1731, p. 12; Mayor, 1911, p. 179; Hare, 1721, p. 17; Collins, 1724, pp. 274–75, 278–79; Parker, 1709, pp. 613–14.

Newton, for he found himself publicly implicated. Whether Whiston was naïve in thinking that the mention of prominent individuals who purportedly shared his views would help his case, or whether he was reacting in anger at Newton's refusal to come to his aid, is unclear. What is clear is that in his *apologia pro sua vita* of 1710 he inserts a thinly veiled allusion to Newton:

> an excellent Friend of mine whom I have not liberty to name, who discoursing with my self and another Person of great Eminence about such Matters, and particularly about the then so much disputed Doctrine of the Trinity, He began with this Declaration of his Mind, "That for his part, had it not been for the Church's farther Determination, he had been contented with the *Arian* Scheme." Which words at that time a little shock'd us both: tho now I have examined that Matter to the Bottom, I am more shock'd that the same excellent Person does not more freely declare the Reasons of such his ancient Sentiments, and more freely endeavour the Alterations of such Things in our Church, as he cannot but know or suspect to be unsupported by the Christian Revelation in these Matters.[18]

The following year Whiston rendered the oblique reference explicit when he called on Newton and other like-minded individuals to attest "the truth and fairness of [his] Quotations and Assertions" regarding the Church's Arianism during the first three centuries. Incumbent on these "great Masters of Primitive Antiquity," he thundered, was to "speak their minds fairly and fully, *as they will answer it to our common Lord another day*, when no political, prudential, or temporal Regards will be admitted against the *plain Demands of Conscience and Sincerity*."[19]

By 1712, Whiston understood that Newton had cast him completely aside. When he sought to learn Newton's opinion of his *Primitive Infant-Baptism Reviv'd*, he was reduced to turning to Hopton Haynes for information. Nor did Newton lift a finger to help establish Whiston as lecturer of experimental philosophy in London. Indeed, so alienated had Newton become that, several years later, he threatened to resign the Presidency of the Royal Society if Whiston was put up for election as Fellow. Colin Maclaurin remarked in 1729 that the two had been "at great variance" for the previous twenty years.[20] Whiston's subsequent behavior confirms that, if little else, he learned his lesson well. His rhetoric following Newton's death notwithstanding, he

[18] Whiston, 1712, vol. 1, p. ix (of the "Historical Preface"). This might be the meeting that Whiston mentions, when discussing "the ancient Manner of Election and Ordination." The language he uses concerning Newton thoroughly examined state of the Church in the fourth century, and his discovery of Athanasius' role in changing views regarding the Trinity, makes it clear that Whiston received the information second hand: "This was occasionally known to those few who were intimate with him all along; from whom, notwithstanding his prodigiously fearful, cautious, and suspicious Temper, he could not always conceal so important a Discovery." Whiston, 1727–1728, vol. 2, pp. 1075–77.

[19] Whiston, 1712, vol. 4, p. 196. In 1742, Whiston cited the passage again in an appeal to another person named, John Potter—by then Archbishop of Canterbury—reiterating that Newton, Clarke, and Whitby "were heartily on [his] Side," while most others "by their obstinate Silence imply'd their Inability to contradict that Account." Whiston, 1742, p. 4.

[20] Whiston, 1727–1728, vol. 2, p. 1075; Wodrow, 1842–1843, vol. 4, p. 59.

lived in dread of Newton and took precautions never to provoke him again. Nevertheless, resentment surely simmered, and humiliation lingered. He who had fearlessly taken on the Church establishment did not dare to stand up to Newton—which may well account for his bitter excoriation of Newton beyond the grave. A quarter of a century after Newton's death, Whiston still vented, trying to excuse away his failure to defy Newton:

> He was of the most fearful, cautious, and suspicious temper, that I ever knew: and had he been alive when I wrote against his chronology, and so throughly confuted it, that no-body has ever ventured to vindicate it, that I know of, since my confutation was published, I should not have thought proper to publish it during his life-time; because I knew his temper so well, that I would have expected it would have killed him.[21]

No sooner was Newton interred in Westminster Abbey and plans confirmed to publish his non-scientific manuscripts than Whiston launched a fierce anti-Newtonian campaign. In his autobiography he recorded how he sought to warn friends that Newton's notions regarding chronology were "so weak" that he expected little from their publication, "yet would none of them believe me at that time." Evidence of his rants abounds. On December 7, 1727, John Byrom notes in his diary seeing Whiston holding forth at Tom's Coffeehouse, discussing *inter alia* Newton's chronology, which he branded as "the merest romance that could be, and that he would write against it when it came out." A month later Martin Folkes notes that many sceptics came around once they perused the *Chronology*, except for Whiston, who "continues in the same way, and declares he shall overturn it so easily that he shall not be able to extend the whole confutation to a sheet of paper."[22] Not content with coffeehouses, Whiston publicized his intention to demolish Newton's chronology in the London papers. A news item in the *Weekly Journal* for January 27, 1728, for example, announces: "We are certainly informed that Mr. Whiston is about to answer the main Parts of Sir Isaac Newton's Chronology, and that he will particularly shew, that the Astronomical Argument about the Place of the Colures at the Argonautick Expedition, demonstrates that it happened at least 360 Years before Sir Isaac determines it." Similarly, maddened by the positive reception of the book in *The Present State of the Republic of Letters*—to be discussed below—Whiston convinced its editor to publish in February 1728 the following note: "'tis said that Mr. Whiston, who in some other cases readily yields even to the lowest degree of probability, is not yet convinced that Sir Isaac is right, in the astronomical part of this work; and that he designs very soon to demonstrate his mistake. If he does, it will be a great triumph both over Sir Isaac and Dr. Halley, who had undertaken his defence."[23]

[21] Whiston, 1753, pp. 250–51. "As Dr. *Bentley*, Bp. *Stillingfleet's* chaplain told me, that he believ'd Mr. *Lock's*, thorough confutation of the bishop's metaphysicks about the Trinity, hastened his end also." Whiston is also the source for the related story that Clarke heard Newton boast that he broke Leibniz's heart.

[22] Whiston, 1753, p. 35; Byrom, 1854–1857, vol. 1, p. 278; Edleston, 1850, p. lviii.

[23] *The Present State of the Republic of Letters*, vol. 1 (1728), p. 167.

All the while Whiston worked on his refutation, which became far longer than a "sheet of paper." He chose to append it to his current venture, *A Collection of Authentick Records Belonging to the Old and New Testament*, the first volume of which was published in early July 1727. The second volume was stayed until Whiston had had an opportunity to obtain a copy of the *Chronology of Ancient Kingdoms* and compose his response. It ended up taking him nearly five months to write the 48,000-word refutation, which he published in late May 1728, as the penultimate appendix to the second volume of *Authentick Records*. Concision and focus were never Whiston's strong points. His writing flows as though he were speaking onto the page, putting down whatever occurs to him, in whatever order, seemingly never revising—in stark contrast to Newton's celebrated precision and constant revision.

As incumbent on a man who for nearly twenty-five years had claimed credentials as a Newtonian philosopher, Whiston commenced his critique with a well-trodden epigraph—"Amicus Plato, Amicus Socrates, Amicissimus Newtonus, sed magis Amica veritas"—before getting down to business: his reasons for rebutting the great Newton:

> Since I have long used to determine my self by the real original Evidence in all ancient Facts, and to have no regard to any such modern Schemes or Hypotheses as are *unsupported*; especially not to those that are *contradicted* by the old proper Evidence; since, according to that plain, honest and satisfactory Method of Determination, I have several Years ago stated the most Ancient Chronology of the World, Sacred and Profane, in my *Essay on the Old Testament*, and my *Chronological Table*: From which general Scheme, I have hitherto seen no manner or Reason to recede; since the intire Support of the several Assertions in this *Collection of Authentick Records*, depends on the Truth and Authority of such Ancient History and Chronology.

Though "a Work of vast Learning, and very uncommon Sagacity," Newton's *Chronology* is based "not upon ancient Evidence, and the Testimonies of Historical Authors; but partly upon the Poetick Stories of Mythologists, laid together by himself; and partly, nay principally, as it seems to me, upon fond Notions, vehement Inclinations, and groundless Hypotheses of his own; and that generally in open contradiction to the most authentick Records of ancient Times." Proceeding with his indictment, Whiston spins Newton's claim that his chronology was made to conform "with the course of Nature, with Astronomy, with Sacred History, [and] with *Herodotus*." Far from it, Whiston retorts. Newton "almost lays aside the Authority of the *Samaritan* Pentateuch, of the Septuagint Version, of *Joesphus, Sanchoniatho, Herodotus, Dicearchus, Manetho, Eratosthenes, Abydenus, Berosus, Varro, Ptolemy of Mendes,* the *Egyptian Obelisk*, the *Parian* Marble, with *Hesiod, Homer, Aristotle*, and the rest of the most authentick original Historians and Antiquries." To Whiston's thinking, Newton's "Novel System" subverted "almost all the most ancient Chronology we have hitherto had; and bids fair, if it prevail, for the introduction of the utmost Scepticism into all ancient History whatsoever."[24]

[24] Whiston, 1727–1728, vol. 2, pp. 962–63.

Whiston clearly set himself up as judge of the evidence proffered by Newton: whether it "be at all like the Evidence he used to give us in Mathematicks, and Natural Philosophy? Or, whether the whole be not rather an Imaginary or Romantick Scheme . . . built upon no manner or real foundation whatsoever." His answer is unequivocal. On nearly every point the *Chronology of Ancient Kingdoms* is doggedly faulted: the treatment of philology and ancient history, the calculation of regnal lengths, and, most elaborately, the astronomical reasoning Newton furnished. In short, Whiston sets up his own Scripture-centered *Short View of the Chronology of the Old Testament* (1702) as a counterpoint to Newton's "secular" attempt to reform universal history. In particular, Whiston derided Newton's "Misfortune" in grounding his chronology on the English version of Scripture—based on the Masoretic Hebrew version—"the worst Copy of all." Sir John Marsham's reliance on that copy, noted Whiston, must have informed Newton's decision, or "at least to confirm him in that Error." Whiston regarded this obdurate attachment to the Masoretic version to be Newton's "*fundamental Mistake*," from which all else flowed. It caused Newton to embrace Marsham's "Dream"—the identification of *Sesostris* with the Biblical *Sesac*—and to follow the "natural Consequences" of this identification even further than Marsham by lowering the date of the Argonautic expedition.

Whiston conceded that he, too, had long been misled into such error. Fortunately, by the late 1710s he had corrected that "pernicious Mistake"—"which Happiness" Newton "never had"—and the more secure foundations he laid down permitted him to amend both sacred and profane chronology. And so, unlike Newton, Whiston resisted "later Supposals or Imaginations," always adhering to the "oldest and most authentick Testimonies themselves, taken in their native Simplicity."[25] Basing his chronology on the Samaritan version of the Pentateuch—which allowed for nearly 900 years between the Deluge and Abraham—Whiston accordingly dated the Trojan War to the mid-fourteenth century BCE, a century before, and not three centuries after, traditional reckonings. Undoubtedly, this accounts for his determination to undercut every element in Newton's text. To help his cause, and clarify for the reader, Whiston distilled the book's structure into eight "principal general arguments."[26] He may actually be the only one to this day who has read and argued about every single passage in that stupefyingly tedious *Chronology*. Rather than follow Whiston through each step of his wearying counterattack, we have listed these eight in table 10.1 and proceed directly to what Whiston clearly thought to be his trump card: namely, that Newton's astronomy could not be fit to celestial requirements.

Whiston devoted a particularly large amount of verbiage to countering Newton's astronomical arguments. With Newton, he did accept that the Eudoxan remarks noted by Hipparchus refer to the colures at a time "when Chiron observed them, and taught them, together with the stars in the celestial sphere, to Hercules his pupil, and to the rest of the Argonauts." Indeed, the idea of so using the colures is "highly worthy the uncommon sagacity, of the great Sir I. N. and in its own nature a chronological character truly inestimable."[27] So much for the compliments. Everything

[25] Ibid., pp. 1041, 1055.

[26] Ibid., pp. 989–91.

[27] Ibid., p. 991.

TABLE 10.1.
Whiston's list of Newton's arguments

1	That the original colures produce a date for the Argonautic expedition of about 937 BCE[a]
2	That the generations from Hercules the Argonaut to Hippocrates the Physician set the expedition 300–400 years later than usually thought
3	That ancient chronologers mistook regnal lengths by conflating kingdom with familial successions
4	That Sesac, the Biblical pillager of Jerusalem in the reign of Solomon's son, was the Egyptian Sesostris
5	That Hesiod dates to a generation after the Trojan War
6	That Homer knew a Mentor who was at the Siege of Troy
7	That poets of the Argonauts say some things that place the date later than assumed
8	That Virgil made Aeneas, just come from Troy, contemporary with Queen Dido, the Tyrian, who according to the Annals of Tyre in Josephus was three centuries after usual date for Troy

[a] *In the* Chronology *Newton explicitly specifies 936 BCE in his short list (p. 26), but his subsequent calculation (p. 91) gives "about" 937 BCE, as Whiston claims.*

Newton did thereafter was wrong-headed and misleading, perhaps deliberately so, Whiston insinuated.

The thread that winds through Whiston's critique, which often backtracks as something new occurred to him—probably while writing—was drawn from his own reconstrual of the Eudoxan remarks. The signal that he was about to qualify, elaborate, and perhaps rework a point is always advertised by his stock phrase, "we must note however." Rambling and logorrheac Whiston may have been, but he, perhaps uniquely, had read Newton carefully enough to dig out the weak points in his argument.

Newton, recall, had transformed every Eudoxan remark into a claim concerning an asterism proper. Whiston countered that many of the remarks could not have referred to the asterisms at all. If, for example, the equinoctial colure passed through the middle of Aries (which Newton took it approximately as doing, though he refined the point), then it could not run anywhere through the middle of Chelae (Libra), Whiston noted, if both are taken to be asterisms, though this was another Eudoxan datum. It would in fact miss any *star* at all in Libra.[28]

Several problems can be rectified, Whiston asserted, by construing the original wording in a different way than Newton had. Hipparchus had also written that Eu-

[28] Newton had, we have seen, disregarded the Libra datum altogether, though his colure does pass close to Libra, albeit not near to its center. Again, the only two stars in Bayer that have reasonable magnitudes and that are near mid-Libra are Bayer α and β. Newton's colure misses even the closest of these two (α) by about four degrees in right ascension. To have a workable star, Newton would have to have chosen one in the Flamsteed catalog that is not marked by Bayer and that has a magnitude of either 4.5 or 6, which would be much too dim. Still, one of the magnitude 6 stars would have yielded a useable date of –963 (its latitude being 9° south, 1690 longitude 11°). The intersection of Newton's colure with the ecliptic at the autumnal equinox has 1690 longitude 216.5°. This puts the colure just past one end of Libra.

doxus had placed the solstices and equinoxes in the "middles" of Aries, Cancer, Libra, and Capricorn, a point Newton had of course noted.[29] He had interpreted this as referring to asterisms, to relations among the stars proper, as he similarly interpreted the Eudoxan remarks concerning the colures. Whiston demurred. According to him, when the adjective "middle" alone qualified a statement, then it referred not to an asterism but to an abstractly fixed *sign* whose specific relation to the asterism that it embraced required knowing the epoch in which the system had been delineated. But when a statement contained the qualifiers "breadthways" or "lengthways," as many did for the colures, then it could refer to the actual asterism. For the qualifiers really made sense only if the corresponding object had length and breadth, whereas the signs had only lengths, being distances along the ecliptic. When this was the case, the colure remark would not have referred to a precise position but only to a vague location somewhere within the asterism or near to it.

How, though, did this help Whiston counter Newton and support the alternative that, we shall see below, he had in mind? If, as Whiston asserted, the colure remarks were deliberately intended to refer to vague positions within the asterisms, whereas only the remarks about the loci of the equinoxes and solstices referred to specific points—namely those within the abstract signs—then it was hardly surprising that Newton's colure, which passed through the middle of the asterism Aries, would miss the asterism Libra altogether. It was simply the case that a Eudoxan *colure* remark did not refer to any particular point within an asterism. Newton had over-specified the data. In which case Whiston could construct alternative loci that captured all the Eudoxan remarks, with these remarks suitably interpreted as referring to any reasonable locus within the corresponding asterism, without regard to any particular star therein. Moreover, even if one took Newton's view that particular stars were involved in the ancient specifications, even then, Whiston continued, the procedure was flawed. For Newton had chosen stars that "are all but one so small [in magnitude]" as unlikely to have been remarked, especially since the original use of the constellations was as an aid to sailors, who would choose the brightest stars, ones that would have been visible even in moonshine. Whiston went further, critiquing on various grounds Newton's several star choices, most especially the claim that it was appropriate to choose a star in *Cetus* for the specification of the colure passing through the flexure of *Eridanus* (a point we discussed above). Indeed, Whiston remarked, he might agree to this if Newton "could produce some original evidence that these stars once belonged to Eridanus." The point is interesting, because Hevelius' stellar atlas does clearly show overlap between *Cetus* and *Eridanus* (figure 8.13). Whiston suggested that if stars had to be chosen, then it would be better to select ones near the mid-head of *Cetus*. Though he did not make the point explicit, doing so would put the colure much closer to his own desired location, which we will come to momentarily.[30] And if this is done, he continued, "Sir I. N.'s position of the colures will be utterly overthrown."[31]

[29] Newton, 1728b, pp. 82–83.

[30] If this is done, placing say *Cetus* γ, which is pretty much at its mid-head, on the equinoctial colure, then the corresponding date would be −1227, which is within a year of Petavius' for the Argonautic expedition (Petavius, 1659, p. 28). We'll see below what date Whiston actually had in mind.

[31] Whiston, 1727–1728, vol. 2, p. 1007.

Whiston claims to have heard from an unspecified source (perhaps Zachary Pearce?) that Newton had actually attempted to overcome these problems by refiguring the asterisms so that they "might better suit his purpose." That rumor allowed Whiston to capitalize on his major theme, that Newton had effectively violated all his long-stated principles when he turned from natural philosophy to chronology. "This strange attempt of Sir I. N. to change the very nature of the asterisms for his own purposes," Whiston accused, "together with those *eighteen* several copies of the first chapter of his Chronology which he left behind him, are marks of the strongest and longest attachment to an *Hypothesis* that almost ever appeared among mankind."[32] Were Newton alive, he would have bridled in anger, for Whiston had, perhaps deliberately, resurrected the very thing that had so exacerbated the young Newton when Robert Hooke took his early work on optics to be grounded on hypothesis rather than proved by experiments. But, of course, Newton was dead, and Whiston safe from retribution.

Thus far Whiston had not laid out his own rationale for placing colures, but he hastened to assert that his would work where Newton's failed, and that where in one instance both his and Newton's failed it was because the modern delineations of the constellation figures are ambiguous or because the observations could not have been made accurately. Whiston had of course accused Newton of playing around with the delineations, but he, unlike Newton, could advance a rationale for any such changes. It is worth briefly examining his reasoning here, since it shows just how far Whiston would go to undermine Newton's authority.

The equinoctial colure was supposed to pass through the Centaur's right hand and fore knee. According to Whiston, Newton's colure reached the figure's wrist or elbow only, missing the knee altogether. His own colure, however, passes directly over the right hand's extremity. Still, Whiston admitted, the "main part of the hand" lies between his and Newton's colures, so that it is not possible to be precise about this. Moreover, neither his nor Newton's colures capture the fore knee. This would seem to place both on equally dubious grounds. But Whiston had an explanation that, though it did not here raise his determination over Newton's, nevertheless explained why the particular datum could be ignored.

There were, Whiston argued, no stars in the Centaur's knee noted in antiquity, and so "their position is only laid down by the fancy of delineators, and ought to be altered so as to agree with this description of Eudoxus." Which is to say that whereas Newton had no right to play around with the constellation figures (not that there is any evidence that he had, despite Whiston's reported rumor), he, Whiston, was prepared to do so here. Why? Because there were no stars listed there in antiquity.[33] This

[32] Ibid., p. 1008. Whiston must have heard something about this, if not from Pearce directly then from someone who had spoken with him. Though according to Costard's later account, there were some fifteen, not eighteen, copies floating around Newton's study.

[33] In fact Ptolemy does list one star in the knee, the 36th in his list (Ptolemy, 1998, p. 396). That star, however, would not have worked for either Newton or Whiston, and Bayer in any case marked no stars there. Hevelius did provide an apposite star, "in genu sinstro: sive potius ad ungulam," for which the corresponding date would have been −1439, which at once explains why Newton never used it. It might, however, have helped Whiston, whose dating was closer, as we shall see, but it seems that he never examined Hevelius' catalog, using only Flamsteed's from the Halley-Newton 1712 excoriated by Flamsteed.

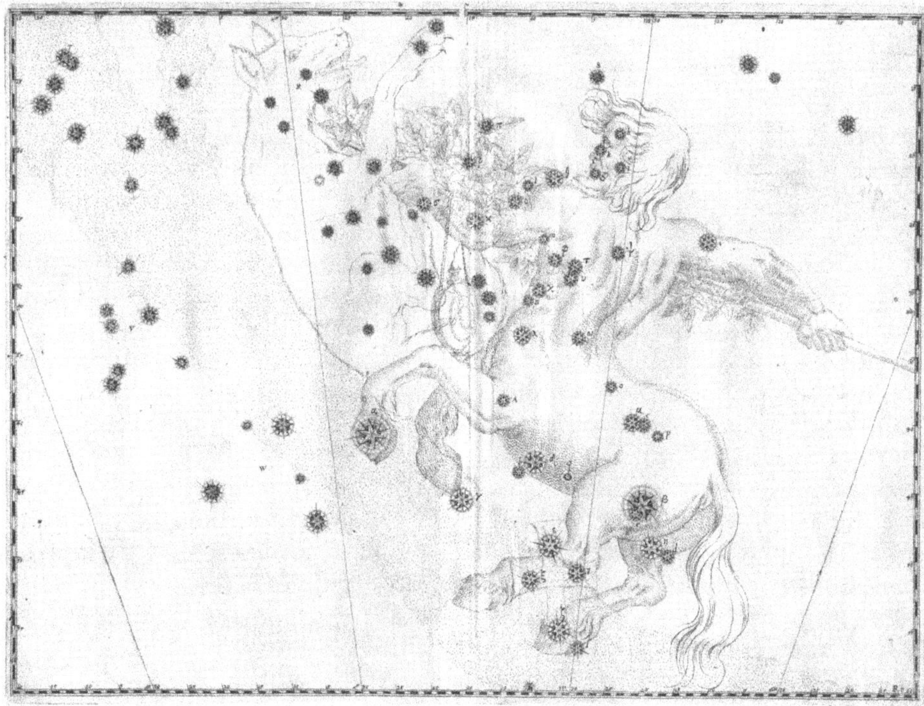

Figure 10.3. Bayer's figure for *Centaurus.*

presumably reflected Whiston's conviction that stellar loci specified the constellation figures but not the ancient colures, and that modern delineations could reflect ancient figures only where it was known that the ancients had assigned similar stars. This was a convenient argument for Whiston, because Newton's colure does in fact come rather close to the Centaur's knee as depicted in Bayer (figure 10.3)—whose delineation Whiston had just rejected as fanciful.[34] And Whiston's own colure, to which we will turn in a moment, was actually much farther away from the Bayer-depicted knee than was Newton's. Moreover, Newton, we saw above, never did use the knee remark for calculation, no doubt precisely because there were no useful stars there.

That was not all, for Whiston went further in justifying his own colure's failure to capture the datum by recurring to the ancients' presumptive observational accuracy. The constellation Centaurus is located quite a distance below the ecliptic. So far, indeed, that observation cannot place its stars with the accuracy possible near the ecliptic—"the degrees," he wrote, are there "at such a great distance from the equinoctial and the ecliptic, where degrees are comparatively small, [that they] cannot afford us the exactness which those give us which are under or near those circles, where

[34] Newton's colure seems to be only about half a degree in right ascension away from the knee as depicted by Bayer, which translates into a comparatively small discrepancy of 46 years in Newton's dating.

such degrees are much larger and more sensible: as every astronomer will readily understand."[35]

Whiston's invocation of "every astronomer" was a clue to the reader that he was not expected to have that kind of knowledge, a kind possessed by *cognoscenti* like Whiston himself. Not that Whiston was alert to the intricacies of figurative convention. Newton and Flamsteed well knew that Bayer had inconsistently depicted his figures, some as though seen in sky view, others in globe view. Whiston was unaware of this. He was accordingly altogether puzzled by Newton's phrase "rightly delineated," by which Newton meant that the Bayer depiction had to be interpreted as a mirror image. Whiston thought the phrase referred somehow to Newton's rumored alteration of the asterisms. Newton had for example chosen the Bayer star δ as the one to use for the remark that placed the colure through the "neck" of *Hydra*. This had to be wrong, Whiston argued, because "this star in all delineations is the very first *star* in the *head* of *Hydrus*, and near *seven* degrees distant from the first part of his *neck*: Nor can that be a *right delineation* of a *Snake*, where the neck comes before the head."[36] Whiston was certainly correct about the star's locus in all extant depictions (figure 10.4), but then Newton had his reasons for referring what he thought to be vague verbal descriptions to nearby stars. And as to "rightly delineated," why, Whiston averred, what "lies at the bottom of this phrase, five times here used, of asterisms rightly delineated" could only be "delineated as his hypothesis requires."[37] Once again, Whiston's chronologizing Newton had altered data to fit "hypothesis."

Having accused Newton of violating his own strictures, and claiming *pari passu* that he himself had a solution that nicely fit all that could be asked for in that way, Whiston at last explained how to find his replacement colures. He had remarked along the way that his colures were some 5¾ degrees removed from Newton's, but he had not explained more than that. How in fact had Whiston found his colures? Moreover, how did he know just what parts of constellations Newton's own colures passed through, when, after all, Newton had only provided the points on the ecliptic that they intersected? To know where they lie elsewhere on the celestial sphere is no simple matter. Further, Whiston abjured the use of stars, which meant that he had had to work directly with constellation figures, drawing his colures right through the delineations. How had he done this? To calculate without choosing stars requires instead determining the 1690 coordinates of some chosen point in an asterism that conjecturally fits a Eudoxan remark. Then one would compute the resulting date for a colure passing through it and proceed from there to try other asterisms. This is nearly impossible to do in any reasonable way without a roomful of assistant calculators. Did Whiston ever try?

There is clear evidence that Whiston never did calculate the actual loci of the ancient colures. To do so for the equinoctial colures, he would have had to make a compensatory adjustment for latitude. In the course of his critique, Whiston felt called upon to answer Halley's defense of Newton against Souciet. We shall see below that Whiston's reply provides clear evidence that he never did undertake to *calculate* the colures. How then did he determine what to do?

[35] Whiston, 1727–1728, vol. 2, p. 1009.

[36] Ibid., pp. 1007–8.

[37] Ibid., p. 1005.

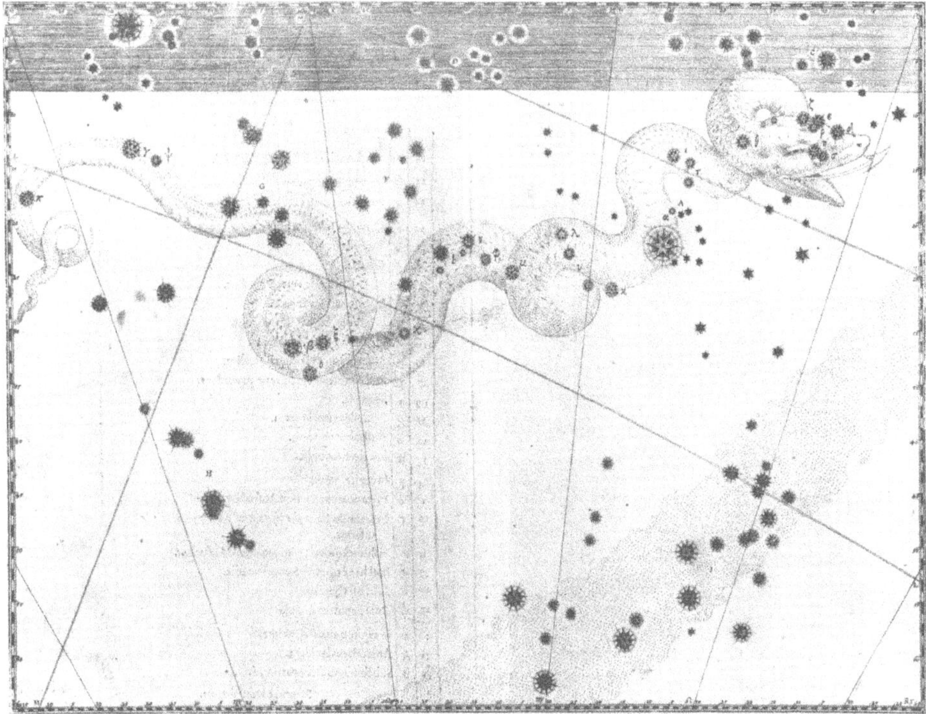

Figure 10.4. Bayer's figure for *Hydra.*

Avid to spread his confutation of Newton as widely and as convincingly as possible, and wishing to avoid intricate computations that would hardly be persuasive to such an audience, Whiston looked for an easy way out, one that would satisfy all of these conditions. He found it in Senex's globe, in particular the one recently made upon which Newton's colures were inscribed. "All who have astronomy enough to examine this matter farther," Whiston advised, "may yet very easily satisfy themselves in [the colures'] position, by examining it by that very large and very exact celestial globe which Mr. Senex has lately made, and upon which Sir I. N.'s colures are accurately depicted ... And to that globe I appeal in this matter, I mean to all that are not properly skillful in astronomy themselves. For astronomers can satisfy themselves when they please more authentically, without any farther assistance from me."[38]

Whiston undoubtedly used the globe to do his own colure construction, since that was the only way to work without choosing, as Newton had, specific stars. Whiston's critique provides instructions that could even be used by those who did not have a Senex sphere, with its printed Newtonian colures. To draw these colures on any other sphere, Whiston explained, begin by placing two pins into it, one each at the points where his colures intersect one another, for these will mark the poles of the ancient equator on the modern sphere—since the modern sphere is peppered with stars, the positions among these drawn stars of the equatorial axis' poles shift with time. Now

[38] Ibid., pp. 1013–14.

wrap a thread or a silk all the way around both pins, producing two strands. Separate the strands so that one is over Newton's ecliptic locations for the equinoctial colures in Taurus and Scorpio, while the other is over his positions for the solstitials in Leo and Aquarius. Then, Whiston claimed, it will be "impossible" also for these colures to satisfy the other Eudoxan requirements. Whereas setting the pins instead for his, Whiston's colures, and repeating the operation would show that the Whiston colures, but not Newton's, nicely satisfy the requirements.

This was all well and good, but to do something like this requires knowing just where the equatorial poles are on the modern sphere for either Whiston's or Newton's periods. And to do that first requires marking the position on the modern globe of its own ecliptic pole. Then the moveable positions of the equatorial pole could be drawn in relation to the fixed ecliptic pole. Unfortunately there is no star near the ecliptic pole, so its location on the sphere has to be constructed. Whiston did not say how to do so, but it can be done comparatively simply in the following way.

First procure a compass whose legs are equal in length to the radius of the Senex globe (14 inches), or any other. Then, using a protractor, separate the legs of the compass by 23.5 degrees and lock the compass at that angle. Next place the point of one of its legs on the globe at its north *equatorial* pole, which is fixed by the physical pivot points. Then place the tip of the other leg so that it touches the globe somewhere. Keeping the point of the first leg carefully fixed, and the point of the other leg ever on the globe, rotate the compass around, drawing a circle on the sphere. The pole of the globe's ecliptic will then lie somewhere on this circle. To find where, locate the position of the summer solstice. This is easy because most globes, include Senex's, have equatorial marks every ten degrees, and also have the ecliptic drawn on them. So just move 90 degrees away in the appropriate direction along the equator from its vernal intersection with the ecliptic. Now take a thread and pin one end of it to the globe's north pivot. Stick a second pin through the south pivot and tie the string's other end there, stretching it tightly along the globe. Slide the string along until it covers the summer solstice. Mark the point where the thread intersects the circle previously drawn with a compass. That gives the location of the ecliptic's pole for the epoch that the particular celestial sphere represents.

We can now proceed to find the positions of the colures on our present sphere in the past. To do that, we again use our compass, with its legs still fixed at a 23.5 degree angle, just as we did before, only now we draw a circle on the globe about the newfound pole of its ecliptic. The position of the equatorial pole in the past will always lie on that circle. To find out where, we proceed as follows. First quickly calculate how many degrees the pole has moved from the given globe's epoch to the desired date by dividing 72 into the year difference; then mark the appropriate point on the drawn circle. This can also be done quite simply in good approximation using a ruler.[39] Using

[39] For example, from 1690 to Newton's date for the Argonautic expedition requires about 36.5 degrees of precession. On Senex's 14-inch radius globe, a 23.5 degree circle about its pole has a circumference of about 35.1 inches, in which case the pole for Newton's date would be located about 3.6 inches around the circle from its position in 1690. A simple way to mark the point would be to take a ruler divided in tenths of an inch and then a division at a time along the circle from the 1690 pole. Since the difference between a linear tenth and a circular arc tenth is small, the accuracy is effectively perfect (for the example here, the

Figure 10.5. The Senex globe with Newton's colures. The black ring about the *ecliptic* pole represents what Whiston might have drawn with a compass. The axis through it depicts the pole of the *equator* for Newton's date, while the unlabeled gray rings are his colures, with the solstitial running through both poles.

a procedure like this, or something similar to it, Whiston could easily mark the pole for his chosen date and for Newton's on any celestial sphere. Do this for both the southern and northern poles. The colures can then at once be drawn using pins through the newly found poles, as Whiston suggested. Figure 10.5 illustrates the result.

According to Whiston, the best fit to the Eudoxan specifications occurs for a pole position located about 5¾ degrees away from Newton's, giving a date for Whiston's colures, and so for the Argonautic expedition, of 1353 BCE, or about a hundred years *earlier* than the date usually assigned in his day, and of course over four centuries earlier than Newton's proposed revision.[40] He urged the reader to work with a globe

total deficit in moving successively a linear tenth of an inch from the 1690 position to Newton's would amount to about four millionths of an inch).

[40] Whiston, 1727–1728, vol. 2, p. 1011. The vernal intersection of Newton's equinoctial with the ecliptic is at longitude 36.5°, while Whiston's shifts to longitude 42.25°.

to see how much better his own colures were than Newton's. It is not actually necessary to do so in order to get a reasonable sense of at least some of the differences between Whiston's and Newton's constructions.

Overall, Whiston's colures do not seem to match the Eudoxan data very much better than Newton's. In one case at least it was much worse. Newton's colure passes very close to the middle of Aries' back as depicted by Bayer, whereas Whiston's runs nearly through the constellation's rump. Whiston certainly knew this, which was no doubt why he insisted that the colure could go through any part of the back at all, including apparently an altogether extreme end. There is in the end comparatively little physical displacement between Whiston's and Newton's colures, amounting to a maximum of about one and half inches at the equator even on the large Senex sphere, and less the farther off the ecliptic the asterism lies.

Whiston was nevertheless sufficiently confident to press forward with a further critique, this time against Newton's calculation of generations. As always, Whiston's prolix argument reached the inevitable conclusion—that not only was Newton wrong to postdate the Argonauts, but everyone else was wrong in the other direction, for the proper date is a century before the one generally accepted. To reach this conclusion on the basis of Biblical "generations," Whiston decided that human life spans must have decreased after the time of Solomon. At the time of Hercules, Whiston averred, humans lived 124 years, but only 93 somewhat later, for a "mean" of 108. Sons were born at the halfway point, or 54 years of age. After Solomon humans lived only about 72 years, so sons were then born at age 36. Taking a new mean between 36 and 54, Whiston obtained 44 years to a "generation." He agreed with Newton that there were about 18 generations from Hercules to Hippocrates, and so he had 836 years for the interval, putting the Argonauts "somewhat beyond the highest chronology of the death of Hercules," and in any case altogether removed from Newton's redating.[41] As for Newton's further argument that ancient chronologers conflated generations with reigns, thereby greatly exaggerating the latter, this is "an unjust and criminal accusation." Why, Whiston went on, how could Newton, "so great a searcher into, and so firm a believer of the sacred books," possibly have forgotten that "mankind generally lived between 900 and 1000 years before the Flood."[42] And human life span only gradually decreased through the time of Solomon.

There was more. Like Marsham, whom Whiston blamed for the conflation, Newton confused the Biblical Sesac with the Egyptian Sesostris. There was no support for this beyond the similarity of names. Whiston himself claimed to have shown that Sesostris was none other than the Pharaoh who drowned in the Red Sea. Newton had also recurred to Hesiod's remark that, in Whiston's own words, Hesiod's own fifth age in the mythical sequence would end "when his contemporaries should grow old and grey headed." Whiston felt it to be "evident" that Hesiod assigned 500 years to an age, which would inevitably place Troy 400 years before him. Note that Whiston had, like Newton, naturalized Hesiod's remark about an age ending when children are born with gray beards.

[41] Ibid., pp. 1019–20.

[42] Ibid., 1021–22.

On two occasions Whiston all but charged Newton with blasphemy. Newton appears to suggest that the Ninivites heeded Jonah's preaching on account of the small size of the city—120,000 inhabitants—which rendered it vulnerable to invasion. To Whiston, such an explanation was tantamount to "profane contempt of Divine Power and Providence," as if God could not procure Nineveh's destruction if it were at its height. In fact, Whiston claimed, Nineveh was at its height—which Newton refused to accept owing to his devotion to the short Hebrew chronology. Newton further misinterpreted Scripture, for the 120,000 in question refers only to very small children yet unable to "discern between their Right-hand and their Left hand," thus implying for Whiston that overall population must have been much greater. Likewise, Newton's attribution of naturalistic means to Senacherib's loss of 185,000 people (either by plague, lightning, or fiery wind) was for Whiston a "direct Contradiction to the solemn Predictions of the Prophets concerning the Author of that terrible Destruction." "I am very much ashamed," Whiston concluded sanctimoniously, "that the great Sir I. N. so perpetual a Student in, and so firm a Believer of the sacred Prophecies and sacred History, should suffer himself to mention such a *weak* and *profane* Hypothesis."[43]

Whiston's other ramblings through Greek history, Homeric myth, and the byways of antiquity yield little but tedium. Surely only the most dedicated and heavily caffeinated reader of the day, if even such a one, could possibly have followed Whiston (or Newton) down these twisting lanes. We can perhaps breathe a sigh of relief that Newton was not available to counter Whiston, for even a brief page or two from him in reply would surely have generated an immense flood of verbiage from his critic—if, that is, Whiston would ever have had the temerity to answer while the author of the *Principia* lived, which is doubtful. Nevertheless, Whiston's long rambles are witness to the fact that once Whiston had demolished the *Chronology* to his own satisfaction, he could then set about reconstructing Newton as a proper Christian scientist after his own fashion.

In his autobiography Whiston waxes lyrical about "the wonderful *Newtonian* philosophy" which he always regarded "in an higher light than others, and as an eminent prelude and preparation to those happy *times of the restitution of all things*." As for his own aspirations, he wished no more than his own "most important discoveries concerning true religion, and primitive Christianity, may succeed in the *second* place to [Newton's] surprizing discoveries" and help bring about the Second Coming. And, indeed, from the beginning of his career, Whiston had hailed Newtonian philosophy as an "eminent prelude and Preparation to those happy times of the Restitution of all things," corollary of his own attempt to make Genesis, as well as Newton, the foundation of his "scientific" works.

Now that Newton was dead, Whiston claimed to possess better insight than his contemporaries into Newton's "true" scientific beliefs—"the impossibility of solving *Gravity* mechanically," in particular. According to Whiston, Newton was "always firmly persuaded, that this *Gravity* was deriv'd from the immaterial Presence and Power of the Deity." And though Whiston acknowledged that the queries of later editions of the *Opticks* might be interpreted to suggest otherwise, he remained adamant that

[43] Newton, 1728b, pp. 270, 282; Whiston, 1727–1728, vol. 2, pp. 1062–63.

they were contrary to Newton's own position. Whiston took every occasion to present Newton's scientific motivation as informed by deep religious convictions. On one occasion, for example, he relates asking Newton why in the *Principia* he did not draw the same "Principles" of natural religion that Bentley had drawn in his Boyle lectures, and that he, Whiston, had in his writings? To which Newton allegedly replied: "He saw those Consequences; but thought it better to let his Readers draw them first of themselves." Whiston hastened to reassure his own readers that Newton did, in fact, draw such consequences explicitly in subsequent editions of the *Principia*. In recounting the similarity of Newton's religious outlook to his own, Whiston continued for several pages. The moral and religious depravity of modern courtiers; the nature of primitive infant baptism; the ordination of ancient bishops; and, most importantly, anti-Trinitarianism. Such a happy congruence permitted Whiston one last salvo to remind those who had prosecuted him for his Arianism that he had been banished and persecuted "for the very same Christian Doctrines which the great Sir I. N. had discovered and embraced many Years before [him]."

Perhaps the best measure of Whiston's ego was his inability to perceive any contradiction between his savaging of the *Chronology* and his claiming for himself the role of Newton's one true disciple. With no apparent qualms, Whiston prepared for publication a collection of quotes culled from the *Principia*, the *Opticks*, and the *Chronology* intended to demonstrate the sort of religiosity that he claimed for Newton. The pamphlet, *Sir Isaac Newton's corollaries from his philosophy and chronology, in his own words*, was published shortly before Whiston's *Authentick Records*. In an effort to confer on the *Corollaries* greater credit, and wider distribution, Whiston sought to persuade Henry Pemberton to append the pamphlet to Pemberton's *View of Sir Isaac Newton's Philosophy*—his rationale being that they were "of the greatest value for the support of natural and revealed religion"—but to no avail. Undaunted, Whiston published the *Corollaries* himself, using the same paper and size as Pemberton, and proceeded to advertise it as a work suited to be bound along with it. The passage Whiston excerpted from the *Chronology* was devoted to Newton's discussion of the original religion bestowed by God on Abraham and his progeny. In his gloss, Whiston expressed his dissent from Newton's view that "the Earth was thinly peopled" when language and the arts and sciences were introduced into Europe—an unfounded assumption if one accepts Whiston's far longer chronology—yet he was happy to join Newton's conclusion that "Mankind could not be much older than is represented in Scripture."[44]

[44] Whiston, 1728, p. 16.

The War on Newton in France

Whiston's English critique was at least founded directly on Newton's printed *Chronology*. The same cannot be said of the second French response to reach print, namely that of the Jesuit Etienne Souciet, for it appeared the year of Newton's death. Souciet had been implicated in the affair of the *Abrégé*, though it was Fréret who had been inveigled into writing a critique perhaps with Souciet's assistance. Once Newton published his dismissive response to Fréret's critique, Souciet himself joined the fray, responding at extraordinary length in his 1727 *Receuil*. Newton was still alive while Souciet was writing his reply, and Souciet undoubtedly expected Newton to see it. Which makes not only his wild excursions into astronomy all the more astonishing, but Newton's mild response to them as well. Conduitt reports that when he received a copy "being apprehensive [of] the manner in wch he might affect" the old Newton, he (Conduitt), "had an extract made of all the real objections & shewed them to Sr Isaac, & had the pleasure of finding the only effect they had upon him was to convince him of the ignorance of the authour."[1] Ultimately, we shall see, Edmond Halley, now Astronomer Royal, took it upon himself to expose the Frenchman's ignorance.

Since Fréret had already written along similar lines, albeit anonymously, Souciet felt obliged to explain why he had entered the lists. In fact, he explained, it was the (unnamed) Fréret who might be suspected of having copied from Souciet, and not the other way around. His language was polite but barbed. The "able Academician" who had added "observations" to the published *Abrégé*, had thereby apprised "the public in the most obliging and polished way possible of the remarks that I made about this new system of chronology almost as soon as it crossed the sea, as you will remember, M[onsieur, viz. Conti], and before whomever it may be translated or saw it." He admits that he had "seen my writing, and he even tries to forestall the suspicion that might arise that he has used [it]." Souciet never contemplated making such an accusation. Had the "able Academician" indeed done so, then no harm done, for "who would complain?" What Father Sirmond said of a similar subject holds true here: "The Muses are sisters; they live on good terms, and all their goods are common property."[2] Having nicely disposed of Fréret, insinuating that he had more than likely just picked up whatever Souciet had told him—while obligingly allowing that as a scholar he could take no offense at such a thing—Souciet turned to his own critique.

[1] Iliffe et al., 2006, vol. 1, pp. 93, 101–2.

[2] Souciet, 1727, pp. 48–49.

Souciet divided his remarks into five "dissertations." The first tries to counter "astronomical proofs"; the second provides "historical" arguments; the third uses medallions to counter Newton's dating; the fourth critiques his use of regnal lengths; and the fifth, which is by far the longest, addresses Newton's response to the *Abrégé*, while advancing Souciet's own theory.[3] The inaugural preface expends a good deal of verbiage in an effort to shield Souciet from Newton's potential wrath: the dissertations were not intended "to attack Newton" but rather "to defend and justify [his] own views on the duration of the world."[4] Souciet feigned astonishment at finding this "first geometer of Europe" engaged in chronology, for he thought that Newton had limited his studies to "mathematics and philosophy." His "singular esteem" for the Englishman was "infinitely" increased thereby, despite being compelled to disagree with him.[5] Perhaps Souciet was sincere, perhaps he was trying to fend off an enraged counter-attack, most likely both. It is impossible to miss the defensive tone in Souciet's remarks. Nor was he simply defending his own and traditional chronology against an admittedly iconoclastic view. Souciet aimed to defend the practice of *historical* chronology from an increasingly powerful interloper—geometry in alliance with natural philosophy. If Newton were correct, then the analysts of texts and the evidence derived from coins, inscriptions, and medallions would have to cede primary place to calculators of times past: to those who could deploy the tools of astronomy. Words would have to take second place to numbers, which alone would set chronology's essential framework. This could not be allowed to stand, and so Souciet decided to rebut Newton by using words to counter numbers, by undercutting astronomy using ancient astronomical remarks themselves. Astronomy would refute itself. Certainly a clever ploy, but one that, we shall see, required Souciet to have mastered the tools of his antagonist.

The principal target of Souciet's attack concerned the locations of the cardinal points. The first dissertation was apparently written before Souciet had seen Newton's *Philosophical Transactions* reply, for at this point Souciet worked with Newton's remark, as translated in the *Abrégé*, that Chiron "placed the solstices and equinoxes at the fifteenth degree of these constellations."[6] Without comment Souciet asserted that the beginning of the Chironic sign for *Aries* must have been taken as the "star in the ear of the Ram," i.e., *prima Arietis* (Bayer γ). He then interpreted Newton to have meant that at Chiron's time the equinox was distant from *prima Arietis* by an amount equal to the longitude of that star increased by fifteen degrees—just as Fréret claimed. Souciet could then use the rate of precession to find out when this would have taken place. To do that he turned to the tables of Philippe de la Hire for the star's longitude in 1700.[7] Adding fifteen degrees to this and then calculating how long it would take

[3] The first and fifth dissertations together comprise 91 of the 123 pages devoted to the five "dissertations." They are preceded by a preface and Souciet's own chronology from Creation up to Christ.

[4] Souciet, 1727, page 7 of the (unpaginated) preface.

[5] Ibid., pp. 50–51.

[6] Newton, 1725, p. 29.

[7] La Hire, 1727. He gives the ecliptic coordinates of *prima Arietis* in 1700 as 29°1′0″ in longitude and 7°9′17″ north in latitude. Flamsteed's 1690 had it at 28°51′0″ longitude and 7°8′58″ latitude. Precession over ten years accounts for all but about a minute and two-thirds of the difference, the rest presumably being a result of different observations.

for that much longitude to be generated from an initial value of zero at 72 years per degree, Souciet arrived at a date of 1470 BCE—more than half a millennium before Newton's.[8]

That raised the question of how the great geometer could possibly have decided that Chiron had located the cardinal points in the middles of constellations. Certainly he could not have concluded this from the locus of *prima Arietis*. And indeed, Souciet thought, he had not. Newton must instead have used Columella's assertion that Meton had placed the equinox in 433 BCE (316 Nabonassar) in the eighth degree of Aries. Precessing for seven degrees, to Aries fifteen, then yields the Newtonian date of 939 BCE. Souciet may have also learned from Keill's letter to Taylor, as we saw above, that Newton had apparently used the Meton placement in the abbreviated chronology, and that whereas Meton placed the point in Aries eight, in Hipparchus' time it was in Aries four. Furthermore, according to Newton the entire chronology of the Greeks was defective precisely because Hipparchus took precession at a degree every hundred, and not the correct seventy-two, years.

Souciet objected. Meton had not set the point at Aries eight, neither had Hipparchus at Aries four, and Greek chronology was not founded by them on the erroneous Hipparchan rate of precession. If their chronology had been so founded, then there ought to have been four centuries between Meton and Hipparchus, whereas there were at most three. Further, in Meton's time, and then in the time of Hipparchus, the cardinal points were certainly not respectively in the eighth and fourth degrees.[9]

About Hipparchus Newton was therefore simply mistaken. But what of the original remark from Columella about Meton? What could that have meant if not the eighth degree? Souciet had an answer for that, one that allowed him to deploy the kind of expertise that a geometer like Newton might not have thought to use. The Greeks, Souciet noted, wrote numbers using letters, capitals in fact. Their letter for the numeral one, "A," is after all very similar to the letter for eight, "H." Just open the top of "A" and you have it. Columella's remark, Souciet implied, was a mistranscription of no consequence. "Astronomy pointed the error out to us," he remarked, but "*la Critique*, coming to the rescue, will lead us to discover the causes."[10] That nicely reversed the Newtonian order of precedence, in which textual criticism took a decidedly second place to astronomy.

In all of this we can spy the effects of third-hand transmission from Newton to Souciet *via* Keill and Taylor. As we have already seen, Newton thought the Metonic remark referred to the ancient system, which had originated at the time of Chiron and was fixed with respect to the stars. The same applied to the Hipparchan claim: the points were, respectively, in the eighth and fourth degrees at the times of Meton and Hipparchus—*in the Chironic system*. It was that system to which Newton applied the correct value for precession, yielding only a workable 288 years, not 400, between their two dates. And Columella's remark referred to that system, undoubtedly recalling some ancient memory. Or so Newton would have argued had he been asked. Souciet could not have known this, though he might have suspected that Newton had

[8] Souciet, 1727, pp. 52–53.
[9] Ibid., pp. 57–65.
[10] Ibid., p. 68.

something other than what Souciet thought had Souciet applied his own methods of textual criticism to the letter from Keill.

As noted above, Souciet wrote his fifth dissertation, by far the longest of all, in direct response to Newton's critique of the *Abrégé* commentary (which, recall, had been written by Fréret), and, again, he felt it necessary to defend himself from Newton's anger. Addressing himself to Conti, Souciet claimed that he had always maintained secrecy and that he had no hand in the publication of the *Abrégé* and the commentary. But this didn't really matter. These are useless quarrels. Grateful for the additional information Newton had provided, Souciet felt empowered to complete his own critique by landing the "final blows" on the geometer's chronology."[11]

Four points were agreed to by both Newton and Souciet, he opined in the later, fifth, dissertation that Chiron had formed the constellations, or had at least taught them to the Greeks; that he had placed the cardinal points at the centers of constellations; that, consequently, to find Chiron's date required calculating precession; and finally, that to do this required knowing where the center of Aries lay, and therefore where Chiron had placed the beginning of this "first constellation of the zodiac." Newton, Souciet continued, placed it much to the east of *prima Arietis*—given Newton's remark in the *Philosophical Transactions* that in Chiron's time the "equinoctial colure" was "about" 7°36′ from *prima*—a remark that Souciet, correctly as it turns out, interpreted as referring to the equinoctial point proper and not to the colure.[12] He, Souciet, would instead choose that star itself as the beginning.

This much differed little from Souciet's first critique, but now he had a specific number from Newton to hand, and he thought he could work with that to undercut the entire system by playing Newton's own game with astronomy. He admitted that he had previously been wrong about Newton's placing of the equinox (i.e., at mid-Aries), but after all had not Newton himself been misleading? Souciet proposed to examine—astronomically—all of the possible places in which Chiron could have placed the cardinal points, confident in his ability to prove that every one of them contradicted Newton's choice, and he would do so using astronomy. This was, to say the least, a reckless course to sail. As we shall presently see, Souciet's ship foundered on the shoals of his inadequate knowledge of mathematics and astronomy.

The first thing to do, remarked the intrepid explorer, would be to find out just how Newton had come to fix the equinoctial point in relation to the constellations. The geometer had not let that be known, but Souciet decided to find out. Why the 7°36′ from *prima Arietis*? There is after all no star there. However, Souciet remarked, there may once have been a star at about that distance from *prima*, and that star may have been considered the true first star of Aries: "I found that in those times there was a star of the first magnitude in the forward foot of the Ram, which was the first of that constellation, and the commencement of the entire Zodiac."[13] For Hipparchus had

[11] Ibid., p. 115.

[12] This was a seemingly loose remark on Newton's part because at his date (−939) 7°36′ is just about the longitudinal distance of *prima* from the equinoctial point proper—it is not the distance from the star to the point of intersection of the colure with the star's latitudinal circle. There is however no vagueness here, but rather an issue of terminology. See note 55, chapter 9 above and, below, for Halley's similar meaning.

[13] Souciet, 1727, p. 119.

written that "the first star [of Aries] to rise is the star in the front foot." Moreover, Hyginus had also mentioned a star in that region.[14] According to Souciet, that star would have been west about forty-five minutes in right ascension from the Ram's ear.[15] Still, this hardly differed from Newton's value for the equinoctial point, and so Souciet let the matter drop—having shown that he was sufficiently versed to beat the famous geometer at the astronomical game. Halley, we shall see, would later turn this around.

Next, Souciet took Newton's date for forging of the ancient sphere by Chiron (−939), and began his critique by computing the precession for *prima Arietis* (Bayer *gamma*) down to 1700. That put *prima*'s longitude, according to Souciet, 7°38′10″ from the equinoctial point in −939, and into the sign of Pisces—which is reasonably correct given the coordinates that Souciet used. This demonstrated that he knew how to multiply, divide, and subtract. The calculation gave him a second chance to challenge Newton's competence. After all, Souciet wrote, Newton had produced a different number, namely 7°36′, and that cannot be correct for −939. How unfortunate to find so great a geometer erring in so simple a matter, though Souciet would not quibble over such a trifling issue.[16] Here we spy Souciet, the textual expert, not having carefully read Newton's words in the *Philosophical Transactions*. The year in question, −939, was indeed the year of Chiron for Newton, but it was not the year of the Argonautic expedition, which was −937 in his reckoning. Newton was explicit on the matter, and the difference of two years nicely accounts for much of the apparent discrepancy.[17]

But where did that 7°36′ come from in the first place, Souciet wondered, since the geometer had certainly not used the now-invisible star in the Ram's front leg (though he should have, Souciet insinuated)? Herein lay the crux of his critique. For Souciet was quite certain that Newton derived his number from Hipparchus' remark in the *Commentary*, that the "knot" of the constellation Pisces had been located at 3°15′ "of Aries [sign]" (the star in the "knot" of the Fishes is Bayer *alpha* in Pisces).[18] That same knot, Souciet continued, is 4°21′ "west" of *prima*.[19] Which together make 7°36′. "I see nothing in all of Hipparchus' work," Souciet asserted, "that is stronger or more plausible in favor of M. Newton."[20]

[14] Mcfarlane and Mills, 2010, p. sec. 3.3.10 and Hyginus, 1960, pp. book 3, sec. 19 of the *Poeticon Astronomicum*.

[15] Hipparchus remarked that the "leading star of the three bright stars in the Ram's head" is 1/20 an hour "behind the colure" (Mcfarlane and Mills, 2010, p. sec. 3.5.18). However, a star in the Ram's foot is south of one in the head, and a southerly star cannot rise at the same time as a northerly one unless it is very far to the latter's west, in which case its right ascension would differ from *prima*'s by a good deal more than Souciet's forty-five minutes. We shall see below that de la Nauze, who critiqued Souciet, understood this quite well, and we shall also see that Halley criticized Souciet on the point, using a different argument.

[16] Souciet, 1727, p. 133.

[17] "I follow Eudoxus, and, by doing so, place the equinoctial colure about 7gr.36′ from the first star of Aries. But the observatory represents, that I place it [recall that Newton here intends the equinoctial point and not the colure's distance from *prima* at the latter's latitude] fifteen degrees from the first star of Aries, and thence deduce that I should have made the Argonautic Expedition 532 years earlier than I do. Let him rectify his mistake, and the Argonautic expedition will be where I place it." Newton, 1726, p. 318.

[18] The passage in question occurs in Mcfarlane and Mills, 2010, p. sec. 1.11.18.

[19] Souciet, 1727, p. 133.

[20] Ibid., p. 132.

There is at once a problem here, because Souciet's purported distance between *prima* and *alpha* contradicts every contemporary catalog. According to Flamsteed's, the distance in longitude should be 3°48′27″, not 4°21′. A difference of more than half a degree is simply not possible on any contemporary stellar catalog. Neither could this correspond to a difference in right ascensions, for that would be about 2°22′ in 1689, and not much different a decade later. Souciet himself furnished the longitudes for *prima* and *alpha* in his own time, and the resulting distance between them was consequently 3°49′. Moreover, he also calculated the longitudes for the year −143 (using tables from de la Hire), which produces a difference of 3°58′47″, and since both stars precess at the same rate, the longitudinal distance between them never changes—except for small proper motions.[21] It is therefore unclear from where Souciet obtained the 4°21′, though he might possibly have had it from Ptolemy's own catalog, wherein *prima* is 6⅔ and *alpha* is 2½, for a difference of 4⅙ or 4°10′ which is at least close to Souciet's 4°21′. In any case there is no *technical* reason for his having attributed such a number to Newton, who explicitly worked with longitudes. But there may be another reason.

Figure 11.1 marks the positions in longitude of *prima* and *alpha* at three dates, starting with Newton's −939. In order for *prima* to be 7°36′ from the equinox, when *alpha* is 3°15′ from that point, both stars must obviously be to the *east* of the equinox (using Souciet's designations of east and west). For were *alpha* instead to the *west*, then *prima* would be 1°7′ *east* of the equinox. In that case it's at once obvious that the corresponding date would be hundreds of years from Newton's, since his date obviously holds only if *prima* is *west* of the equinox (in the sign of Pisces). Newton could not possibly have come to his number in the way Souciet suggested.

Souciet undoubtedly knew this, because he remarked that Newton's date required the equinox to be "to the east of the ear of Aries [*prima*]," and since *alpha* is by Souciet's own reckoning west of *prima*, it could not then be 3°15′ from the equinox. In fact, the locus that Souciet attributed to Newton (with *alpha* at 3°15′ of Aries and *prima* then 7°36′ from the equinox) could hold only in −121, not −939. Souciet cleverly decided to gloss over the point, though he certainly realized this. Had he done so, it would have been quite obvious to any careful reader that he ascribed an utterly elementary blunder to the English geometer. That would hardly have been plausible, and so Souciet simply let stand the implication that this was how Newton had obtained his date.

Behind all this posturing lay the real reason for Souciet's ambiguity. Undoubtedly intent to demonstrate that he could outshine the vaunted geometer, Souciet figured a way to turn the Hipparchan remark in his own favor. He might have done as much simply by using the remark directly and leaving Newton out of it, arguing instead that Newton had overlooked a nice bit of evidence. But then Souciet would have left hanging the source of Newton's 7°36′, leaving the uneasy implication that Souciet might himself have missed some evidence. Attributing the 3°15′ and 4°21′ to Newton as the sources of his 7°36′—even though Souciet had to know that Newton could not possibly have used them—and then turning the result to his own advantage would look like a *coup de grace*. Here, ironically, the core of his argument concerning the mean-

[21] Ibid., pp. 134–35.

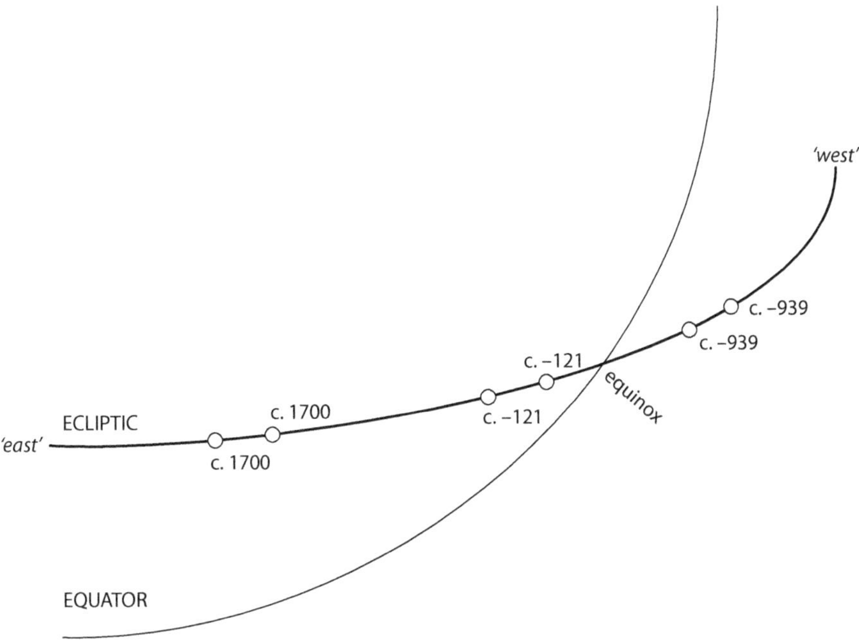

Figure 11.1. Longitudes of *prima* and *alpha* at three dates: the circles that represent *prima* are "east" of the ones that represent *alpha*: *alpha* trails *prima* as time progresses. Here the sign of *Pisces* is to the west of the equinoctial point, and the sign of *Aries* to the east.

ing of the Hipparchan datum was probably correct (though not of the longitudinal distance between *prima* and *alpha* that Souciet attributed to Newton), but his attempt to prove it mired him in errors that were obvious to anyone versed in astronomical calculation, or even to anyone who knew how to work an armillary sphere.

Newton, Souciet argued, erred in assuming that Hipparchus had used longitude. Suppose instead that he had used right ascension. To see whether this might work, Souciet had to calculate the stars' equatorial coordinates in antiquity. Unfortunately, he clearly had no idea how to transform from one system of coordinates to the other—that after all required trigonometry, not just arithmetic—and so he took a shortcut. He first looked up the stars' right ascensions from contemporary tables. Then, in sixty years, he asserted, right ascension changes by 49′43″, a number he presumably obtained by taking de la Hire's approximate value for the decanal change of *prima*'s right ascension in 1700, and using it for both *prima* and *alpha*.[22] Then, he concluded, in the year −143 (we shall see in a moment why he chose this year) the two stars would have had the longitudes and right ascensions listed in table 11.1.

Therefore, in −143 *alpha* was *east* of *prima* in right ascension, but it was *west* of it in longitude. To illustrate the point he produced a diagram (figure 11.2).

[22] De La Hire gives 8′18″ for *prima*'s change in right ascension over ten years which yields 49′48″ in sixty, nearly Souciet's value.

TABLE 11.1.
Souciet's coordinates in −143 for *prima* and *alpha*

In −143 according to Souciet	*Longitude*	*Right Ascension*
Prima	3°25′30″ of Aries	28°51′19″20‴ "of Pisces"
Alpha	29°36′30″ of Pisces	1°12′17″3‴ "of Aries"

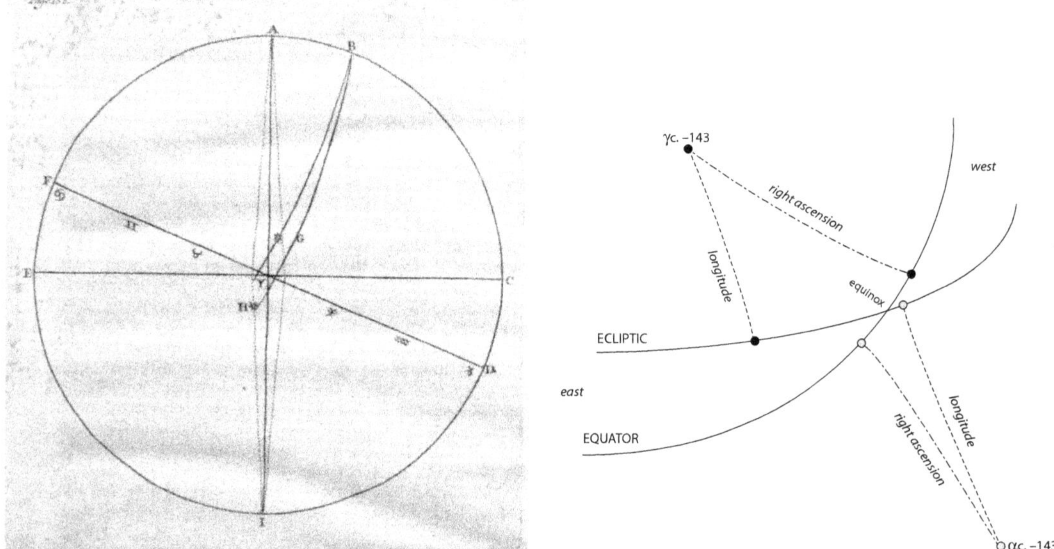

Figure 11.2. Souciet's illustration (left) and a rendition of what he had in mind, showing that, though *prima* is always east of *alpha* in longitude, in −143 it was west of it in right ascension.

The true aim of the exercise was to capture what Souciet had (illogically but craftily) asserted to be Newton's use of the Hipparchan 3°15′ for his own purposes. According to Souciet's calculation, the distance between *prima* and *alpha* in right ascension was 2°15′57″43‴ in the year −143. This alone was incorrect arithmetically, since the proper difference using Souciet's numbers is, in fact, 2°20′57″43‴. Further, Souciet decided that Hipparchus in his time had found *prima* at 45′—in right ascension—"from the colure," this being (he thought) the distance of *prima* from the now-invisible star that marked the equinox in Hipparchus' time. Consequently, Souciet concluded, *alpha* would be a satisfying 3°0′57″ from that point. This was not Hipparchus' 3°15′, but he thought it to be close enough—after all, Souciet wasn't completely certain of the epoch that Hipparchus was using, having just selected the midpoint of the time during which he flourished.[23] Little else was needed. Newton had gone wrong in assuming that the Hipparchan datum referred to longitude, when it actually meant

[23] Souciet, 1727, pp. 141–42.

right ascension. Making the correction moved the datum in question right into the time of Hipparchus himself, taking it altogether away as a proof of Newton's impossible dating scheme. In which case, Souciet's claim that *prima* was about fifteen degrees from the equinoctial point in Chiron's time remained untouched, putting that era back more or less where it used to be. According to Souciet, this was really the only astronomical bit of evidence that Newton could have used to obtain his value of 7°36′.

We noted already that Souciet must have known that Newton could have done no such thing. He had set up a false trail, apparently with the intention of showing that its dead end could be avoided by properly interpreting Hipparchus' meaning. Ironically, Souciet was almost certainly correct in assuming that Hipparchus had meant right ascension and not longitude. But his own calculation was flawed owing to ignorance of astronomy and carelessness. Because precession takes place around the ecliptic, and not the equator, the corresponding rate of change in right ascension varies over time. It is not constant. Souciet apparently thought the difference always to be small, which it is not in general, and so his calculation using a constant rate of shift along the equator failed. Using Souciet's values for their longitudes in −143 (and contemporary values for their latitudes) *prima* and *alpha* respectively have right ascensions of 0°16′26″, 3°17′28″. These values differ considerably from the ones that Souciet provided (table 11.1). In fact, Souciet's equatorial coordinates are inconsistent with the ones that he himself provided for the stars' longitudinal loci, as we can see by transforming Souciet's equinoctial values. Souciet himself could do nothing of the kind, which is why he resorted to a specious calculation of motion in right ascension.

Worse still, Souciet had Hipparchus placing *prima* 45′ "from the colure." Souciet had then added to this the distance in right ascension between *prima* and *alpha* to place the latter over three degrees from the colure. Now, according to his own set of right ascension coordinates for −143, *prima* was then *west* of the equinox by slightly more than a degree. At that date, again on Souciet's own account, *alpha* was *east* of the equinox by about the same amount. Consequently, in order for Souciet to place *prima* at 45′ from the equinox with *alpha* then being more than three degrees from that point, both stars would have to be moved nearly two degrees from west to east from their loci in −143, placing *prima east* of the equinox, with *alpha* even farther to the *east*. And so, according to Souciet's way of calculating, Hipparchus must have been very accurate in measuring the distance in right ascension between *prima* and *alpha*, but stupendously inaccurate in determining the distance of *prima* from the equinox.

Almost nothing in Souciet's astronomical arguments makes much sense, and these make up the bulk of his critique. In attempting to trump Newton, he had knowingly, it seems, attributed to him something that the English geometer could not possibly have used, and he had then appropriated the impossible datum for his own purposes. Souciet thus appeared to have beaten Newton at his own game. Along the way he had even suggested that Newton should have used a different star altogether than the one that Souciet thought to lie at the basis of his calculation—though he "magnanimously" allowed that it made little difference. All of this was preceded by effusive praise of the English geometer. Given Souciet's subsequent arguments, these too must have been deliberately crafted to mislead the unwary reader. One wonders how he could have thought that the ruse would go undetected, when even slight acquaintance with precession would have shown that Newton could never have used the Hipparchan datum. Perhaps Souciet hoped that readers would be so impressed by his own use of the datum that they would fail to notice the impossibility of Newton himself having done so.

Still, we have seen that at least a part of Souciet's difficulties in trying to understand Newton was due to apparently confusing terminology on Newton's part—even though the wording would have been clear to anyone who, like Halley, was versed in astronomical usage. In his reply to the *Abrégé*, Newton had asserted that at Chiron's time the equinox was in "the middle of the constellation, and the middle is not fifteen degrees from the first star of Aries"; that Eudoxus "drew" a colure through "the back of Aries"; and that this colure "passes through the middle of Aries, and is but eight degrees from the first star of Aries." Now, the third statement cannot hold if the first two do, unless by the colure's being about eight degrees "from" *prima* Newton meant that the star was that far from the equinoctial point proper—and not from a point on the colure at *prima*'s own latitude (see above, note 12). As for mid-constellation, that is not fifteen degrees from *prima*, just because the *asterism* is not in fact thirty degrees in length, taking with Newton *prima* and *ultima caude* as bracketing it.[24]

The terminology can be confusing, but it reflects the astronomical significance of the two issues at stake: first, to determine the year on which a given star actually lay *on* the colure (which provides the star's longitude at the year in question), and, second, to specify the longitude of some other star by its distance from the first one. The distance of this second star from the colure proper at the star's own latitude was not important, and so, where Newton referred to a distance "from" a colure he meant its distance along the ecliptic to the equinox itself. Souciet missed this, even after reading Newton's reply to the *Abrégé*, and in Souciet's own analysis the colures proper play no role, though he mentions them. Hardly surprising that he missed altogether that the *colure remarks* themselves contained the answer as to why Newton had placed the equinoctial point at Chiron's time nearly eight degrees from *prima*. Even if Souciet had suspected as much, he surely could not have calculated the effect, since it required knowledge of spherical trigonometry.

However much Souciet may have been genuinely puzzled over the source of Newton's 7°36′, his astronomical farrago amounted to a devious attack on the Englishman's competence. This was not a wise move. Edmond Halley fully understood what Souciet had done, and shortly after Newton's death he composed a potent response. The first paper was read at a meeting of the Royal Society on June 29, 1727.[25] Too polite to openly expose the Jesuit's duplicity, Halley instead destroyed his argument along with any semblance of astronomical competence on Souciet's part. Angry that Souciet had not waited until the publication of the *Chronology*, Halley was further incensed by his mistaken belief that the Jesuit had also written the commentary to the *Abrégé*. Newton had answered, but that had only given the opportunity for publishing a further critique in the form of "five dissertations." Since the first and last of these

[24] Since the longitudinal distance between these two stars in Flamsteed's catalog is just 20°12′42″.

[25] RS JB XIII (1726–1731), pp. 107–9.

were "chiefly astronomical," Halley as Astronomer Royal and friend of the deceased Newton felt called upon to rise in defense.

At first, Halley recalled, he had himself not been averse to Souciet's claims since he too thought on traditional grounds that the Argonautic expedition took place hundreds of years before Newton's new date. But that was only while he had assumed the Jesuit's calculations to be correct—and probably before he read deeply enough to see how Souciet had tried to set Newton up for a fall. Thereupon he "resolved to examine the matter with due attention, especially since the good Father seems to triumph over his adversary, and to treat a man of his figure in the Commonwealth of Learning in a very ludicrous manner, notwithstanding the several fine things he says of him to palliate it."[26]

Halley went right to the core of the matter. The only real issue, he wrote, concerned what part of the back of Aries the equinoctial colure passed through. The principal point, accordingly, was when the colure cut through the Ram's back.[27] Newton, Halley asserted, must have worked in some manner with a locus on the back of Aries, with some star in fact. To do so requires understanding that whatever star it may have been cannot simply be precessed back from its contemporary longitude to the equinoctial point since the colure is inclined to the ecliptic. Account must be taken of that fact. Not knowing which star Newton had used, Halley decided to illustrate the point by doing a calculation for *prima* in the following way. He asked, first, at what date in the past *prima* itself would have been on the colure. Taking careful account of the correction due to the star's latitude (see figure D.2 and relation D.1 in appendix D), Halley determined that *prima* would have been on the colure in 162 BCE. Starting with Newton's 7°36′ as *prima*'s distance from the equinox proper at Chiron's date, Halley precessed it back from −162 to obtain a date of −934 by *adding* the latitude correction to the star's longitude in −162.[28] To illustrate the point another way, Halley decided to choose a specific star, namely Bayer ν, which is in the Ram's mid-back. Again taking account of the corresponding latitude correction, he obtained 984 BCE. Either way produces a result close to Newton's, and there was no mystery about Newton's 7°36′ for *prima*, since he had obviously worked with some locus on the back of the constellation that must have been close to Bayer ν.[29]

Souciet had done nothing like this, and Halley suspected that he could not have owing to his being ill-grounded in astronomical methods. The English Astronomer Royal made the point quite explicit. In countering Newton, Souciet had moved the Chironic equinox to a point near *prima*, where the now-invisible star in the Ram's

[26] Halley, 1726/7, p. 206.

[27] Although Halley had clearly been in contact with Conduit, it seems probable that he had not seen the manuscript of the *Chronology*, which reached print the following year. Consequently he remained at this point unaware that Newton had produced his date (and so the locus for *prima*) by taking averages among the set of remarks about the Eudoxan colures in Hipparchus' *Commentary*.

[28] Halley used an ecliptic obliquity of 23°29′ and Flamsteed's epoch 1690 coordinates for *prima*. The correction EK due to latitude (figure D.2) was 3°7′30″, which is the star's distance *into Aries* from the equinox at that date. Consequently, to find Newton's date Halley could simply precess directly by an amount equal to the sum of 7°36′ and 3°7′30″.

[29] Newton's 937 BCE for the Argonautic expedition differs from Halley's 934 BCE (calculated using the 7°36′ distance of *prima* from the equinox) because Newton's number was avowedly approximate ("about").

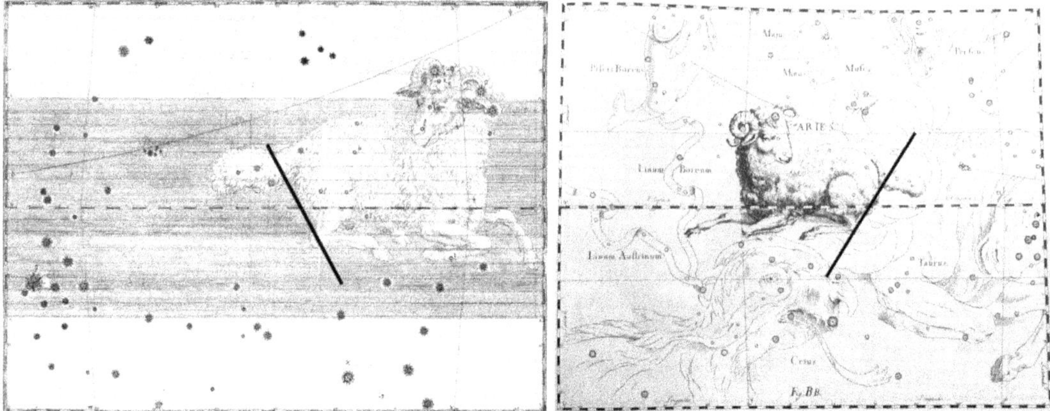

Figure 11.3. Souciet's equinoctial colure (heavy black line) added to Bayer's (left) and Hevelius' maps for Aries.

foreleg, which presumably began the sign, was located. Such a position can't possibly be correct, Halley pointed out, since then the colure proper would cut right between the rump and tail of the Ram, contradicting Hipparchus' remark in the *Commentary*. Souciet, he implied, had apparently not looked at a stellar atlas, since even a straightedge applied to Bayer's or any of the others (including Hevelius') would at once have shown the problem (figure 11.3).

Halley had more. Souciet's lengthy discussions in the fifth dissertation concerning Newton's 7°36′ further illustrated his astronomical deficiencies, for in his calculation Souciet "ought to have deducted 3°7½′ [the latitude correction for *prima*] out of the 15 degrees he assumes for the distance of his colure from the first star of Aries." Further, apropos Souciet's apparent ignorance over how to convert between ecliptic and equatorial coordinates, Halley curtly entreated him "in the next edition of his Dissertations to be a little more careful of his numbers than he has been ... and inform himself in the spherics [viz. spherical trigonometry], so as to give us the right ascensions of the stars truly, from their longitudes and latitudes."[30] He soon decided to sink the knife in deeper by doing to Souciet what the Jesuit had tried to do to Newton. Recall that Souciet had sought to undermine confidence in Newton's argument by claiming that the appropriate star in Aries to set at the beginning of the constellation was not *prima* but another, no longer visible, that preceded it and that Hipparchus

[30] Halley, 1726/7, p. 209. Though on the point concerning the latitude correction Halley perhaps went a bit too far, because Souciet was, it seems, arguing in terms of the equinoctial point—though in so doing Souciet had altogether transgressed the Hipparchan remark concerning the locus of the colure through the back of the Ram. Halley was aware that Newton had himself been somewhat vague. Halley used terminology to obscure the point by distinguishing between the position of the colure "in consequence of the first star of Aries"—meaning the distance of *prima* from the equinoctial point—and a star being literally "under the colure." Newton, Halley opined, "no doubt, had reason to place this colure 7°36′ in consequence of..."

had set in the forefoot of the Ram. He had thereby captured Newton's 7°36′ for his own purposes. Halley used it against Souciet.

In his second rebuttal of Souciet's dissertations (and still before having seen the printed *Chronology*), Halley remarked that when writing his previous one he did not have access to Hipparchus' *Commentary*. Since then "by the favour of a good friend" he had seen the 1567 Florence edition. In it he found something that the Jesuit "must confess will bring the Argonautic Expedition full as low as Sir Isaac Newton makes it."[31] From the *Commentary*, Halley concluded that Souciet's supposedly vanishing star is still visible and is Bayer η in the constellation Pisces. He determined this by considering carefully Hipparchus' further remarks concerning what degrees of the ecliptic were on the meridian when Aries rose and set, given that Hipparchus had placed his star in the Ram's forefoot as the one that first did so. A thorough calculation yielded Bayer η in Pisces as the star in question, which is the twenty-second in Ptolemy's catalog for that constellation. It did not matter why Hipparchus had placed it instead in Aries, though Halley suggested that perhaps his *Commentary* preceded his own stellar catalog, when "he might change his opinion." In any case, picking up Souciet's supposition that such a star was in Chiron's time fifteen degrees from the vernal equinox, and then precessing Bayer η, yields 1010 BCE. However, the proper procedure (given that the *Commentary* remarks refer to colures) would be to ask when a locus fifteen degrees away from the star *but at its own latitude* would have been on the equinoctial colure, which requires taking account of the latitude correction. Including it, Halley showed, changes the date in question to 842 BCE, close enough to Newton's. What more needed to be said? Souciet had been trumped by his own suggestion.

Halley's defense against Souciet, and especially his turning the claim for an alternative star to Newton's advantage, was, Whiston soon realized, dangerous. He accordingly decided to rebut it, not in defense of Souciet, but in opposition to Halley. And here we find that Whiston, who was vastly more competent than the mathematically and astronomically ignorant Souciet, himself never calculated colures either.

If anyone thinks that Halley had vindicated Newton, he averred, he would "take the freedom to undeceive him."[32] Let's agree with Halley, Whiston began, that the first star of Aries should be Bayer η. If it is chosen, however, then the colure date should be 229 years later than Newton claimed. How so, when Halley had obtained a thoroughly different result? Whiston wrote nothing about his calculation, but there's little question as to how the result emerged—Whiston followed Newton's procedure in the *Chronology*, or at least tried to do so. Like Newton, he took a midpoint between the first star of Aries (now taken *per* Halley to be Bayer η) and the last star, *ultima caude*. The former has (1690) longitude of 22°29′20″, while the latter's longitude is 49°3′42″. Consequently the longitude of a point midway between the two would be 35°46′16″. Newton's own first star for Aries (Bayer γ—*prima*—which has longitude 28°51′0″) produces a midpoint with longitude 38°57′21″, putting the new midpoint about 3⅙ degrees in longitude from Newton's (more precisely, 3°11′4″ distance). This, Whiston concluded, lowers the colure date by 229 years, "to the utter subversion of

[31] Ibid., p. 296.

[32] Whiston, 1727–1728, vol. 2, p. 1016.

Sir I. N.'s entire chronology"[33] How so? At 72 years per degree, the decrease would seem at once to follow.

Whiston certainly knew that Halley would not have made an elementary mistake, and he needed therefore to shield himself from any counterclaims. Consequently he grudgingly excused the error in an oblique way: "as for the scheme itself, as given us by Sir I.N. this hypothesis of Dr. Halley's, though I suppose without his intention, overturns one of its main foundations."[34] Halley, Whiston maintained, was undoubtedly unaware that Newton had gone midway between a first and last chosen star, and not just 15 degrees from the first star. He, Whiston, corrected the error, which obviated altogether any help to Newton on this score.

The claim was both misleading and incorrectly calculated. Misleading because once Newton's procedure of cutting between two stars is used, then his own chosen first star (Bayer γ) continues to work perfectly well. The alternative used by Halley, Bayer η, fails only if the cut method is used. It fails, but not with the displacement in years that Whiston computed. To obtain Whiston's number requires ignoring the latitude difference between the first and last stars, which in this case is $2°46'17''$. When that is properly taken into account by choosing a locus midway between the two in both latitude and longitude (as Newton's procedure requires), the displacement in years changes from Whiston's 229 to 200. This would still invalidate the use of Bayer η, but that would be a moot point in any case, given Newton's actual choice of stars, as Halley would by the date of Whiston's critique certainly have known. The fact that Whiston simply ignored the latitude effect certainly suggests that he was reluctant to undertake the kinds of computations that precise results demanded. Whereas Souciet did not know what to do, Whiston opted not to.

At about the same time that Whiston was working on his critique, Fréret wrote a letter to Halley, in some dismay concerning Newton's published reaction to the *Abrégé*.[35] Fréret corrected Halley's misattribution of the *Abrégé* remarks to Souciet, insisting that he alone was responsible for both the translation and the observations upon it. Although it might seem that Souciet and he had worked on similar lines, given Souciet's remarks in the *Receuil*, they, in fact, "follow very different routes." Moreover, Fréret had never intended to write words that "merit the odious name of refutation." He merely wished to propose some "doubts and difficulties," offered "with all the regard due to a man as respectable as M. Newton."[36]

Clearly concerned to deflect the anger that he sensed in Halley's double critique of Souciet, Fréret trod carefully at first. "Here then, Monsieur," Fréret offered, "are the basis of the clarifications I asked of M. Newton [in the comments to the *Abrégé*] with all the care necessary to avoid wounding him. Because I knew the sensitivity of this great man to contradictions." Unfortunately, "despite these efforts he could not refrain from demonstrating his bitterness against me in the response that he published;

[33] Ibid., p. 1017.

[34] Ibid.

[35] The letter, signed by Fréret, is undated and unaddressed. However, it immediately refers to the addressee's publication in the *Philosophical Transactions* "against Souciet" and goes on to make clear in further remarks that this is Halley.

[36] CUL MS Add. 9597/7/12, pp. 1–3.

when fundamentally he pretty much contented himself with saying that he hadn't been listened to and he didn't explain himself more clearly." Having obtained a copy of the English *Chronology* from Fontenelle, Fréret "read it and reread it with attention, and I was extremely surprised to find that this reading, far from clarifying my former difficulties, instead gave me new ones. I saw that M. Newton was obliged at each step to add new suppositions to his first ones. Hypotheses to square the facts with his system. It even seemed to me that he gave no solid proof of his fundamental hypotheses. In such a situation I believed, Monsieur, that I could not avoid proposing my difficulties to you for you to judge before publishing them."[37] He had not as yet published these "difficulties" because he felt it important to await the publication of the *Chronology*'s French translation in order to give the (French) public an opportunity to judge their respective claims in a matter so important for all "men of letters."

Fréret implored Halley to consider his critique, sweetening the request with flattery. "Decreasing the antiquity of the Greeks by five hundred years, and of Egyptian antiquity by two thousand years?" Fréret cries. "You will not be indifferent to that question. To you, Monsieur, who is no less versed in ancient literature than profound in the most sublime geometry and in that ingenious astronomy which owes its birth to new methods which have scarcely appeared beyond your isle."[38] After all, Halley had himself indicated (in his first article) that he had initially doubted Newton's redating.

Yet Fréret would not challenge Newton's astronomical arguments. Quite the contrary—he accepted the great man's calculation of a date for an ancient star globe, and he agreed that Chiron invented Greek astronomy and was the author of the first such sphere. He accepted as well Newton's fundamental premise: namely, that the most ancient date for such a globe that can be deduced from Hipparchus' *Commentary* must be attributed to Chiron. All this was perfectly valid, Fréret admitted.[39] There was but one problem with Newton's argument: where Newton had calculated the date of one stellar sphere from the *Commentary*, other remarks in it demonstrate that there must have been *two* such spheres, and one of them was older than Newton's by about enough to salvage standard chronology.

Fréret's entire point depends upon the occurrence in the *Commentary* of two sorts of remarks. The one sort, which was what Newton had used, referred qualitatively to colures and risings and settings. But there are other remarks in which Hipparchus provided two numerical locations for a star. One of them provided the star's locus as Hipparchus had it, but the other had reference to Eudoxus. Take, for example, the constellation *Ursa Minor*, of which Hipparchus writes that "the final and brightest star of that constellation lies at about 18° of the Fishes, or—as Eudoxus divides the zodiacal circle—at about 3° of the Ram."[40] That fifteen-degree difference appears in other Hipparchan remarks as well.

We have already seen what Newton assumed about these constant fifteen degrees—namely, that the Eudoxan value merely represented Hipparchus' expression of the

[37] Ibid., pp. 4–5.
[38] Ibid., pp. 6–7.
[39] Ibid., pp. 8–10.
[40] Mcfarlane and Mills, 2010, p. 1.6.4.

star's position in his (Hipparchus') time but referred to a positional system whose origin had been set at a different point, namely fifteen degrees from Hipparchus' own system.[41] And we also mentioned there Fréret's alternative interpretation: namely, that the fifteen degrees reflected a difference in precession, not just in where to set the sign system with respect to the equinoctial point, that Eudoxus must according to Hipparchus have been working with positions from a much earlier time than Eudoxus' own. In which case, Fréret figured, the fifteen degrees of precession must be taken from Eudoxus' period and not from Hipparchus. That would put the date of the star globe corresponding to these remarks at −1468 or thereabouts, almost precisely at the time of the Argonautic expedition according to standard chronology (which Souciet put at −1467). And so the Hipparchan remarks must refer to two different star globes. There was the one that Newton calculated, but there was also a much older one, and this last must therefore have been the original stellar sphere delineated by Chiron. Hipparchus himself had saved chronology.

Fréret knew quite well that there were several possible objections to his claim, the most significant in his view being that his argument required the fixed signs *as used by Hipparchus*—those which begin at the equinoctial point proper—to have been created by Chiron. Newton obviously thought otherwise, since he had taken it for granted that in that early day the cardinal points were placed within the centers of the asterisms. "In M. Newton's hypothesis," Fréret argued, "the beginning of the signs or dodecatemoria is not marked by anything sensible. It could only be determined by means of a retrograde calculation. That of Aries for example is fixed at the 15th degree before the spring equinox and the determination of the sun's entry into this sign depends on observation of the equinox, which can only be done when the sun is at the 15th degree of the sign."

At Newton's date of −939 for the creation of the sphere by Chiron, the equinoctial *colure* passed not far from the mid-back of *Aries* (figure 8.14), and a good deal of the asterism had moved below the horizon by sunset on the day of the equinox. This was why, Newton may have reasoned, Chiron had set his origin point at the asterism's center, near where the colure passes, and not at the (invisible) locus of the sun in *Aries* on that day, or even at the first star of the constellation. Further, in Hipparchus' time the colure ran through *prima Arietis*, whereas (*pace* Souciet) no star in *Aries* then lay at the equinoctial point. Newton accordingly thought that Hipparchus had produced the signs with regard to the colure in his own day, as had Chiron in his nearly a millennium before. Precession had produced the change. Where Newton consistently reasoned with colures, Fréret was convinced that his own date was much better than Newton's despite the fact that, a millennium before Eudoxus, *Aries* was scarcely near the equinox, whether colure or equinoctial point (figure 11.4), thereby excluding the Hipparchan remark that in the original sphere the colure passed through the Ram's back. He went on at some length, developing his point and arguing that Newton's assumption concerning the locus of the signs' origin was "purely gratuitous," that there was "not the least vestige" of evidence for it, indeed that "this hypothesis is absolutely contrary to the text and to the idea of Hipparchus."

[41] See chapter 8.

Figure 11.4. Fréret's configuration for Chiron ca. −1470 (top), with no star in *Aries* either on the colure or at zero longitude, and (bottom) the configuration at the time of Hipparchus, with *prima Arietis* on the equinoctial colure. Note that *prima Arietis* has zero longitude at the time of Eudoxus, ca. −380385.

Halley seems never to have answered Fréret, but that mattered little because Fréret altered his argument after reading Whiston's critique. He accepted completely Whiston's challenge to Newton's colure argument, for it meant that even on that basis the date for the Argonauts reached back to the fourteenth century BCE, making it consistent with Fréret's fifteen-degree calculation. Fréret translated into French large sections of Whiston's critique, remarking afterward that he "could have stopped there in respect to the astronomical proof of M. Newton." He pleaded for indulgence on the part of the reader, asking for "pardon" to introduce his own considerations.[42]

Fréret's final words on the subject were printed posthumously in 1758 and ran to more than 500 pages. Somewhat more than a fifth of these pages present astronomical and calendrical arguments. The bulk of his lengthy "defense" consisted of a wide-ranging canvass of ancient texts for evidence designed to demonstrate that Newton's own associations cannot be reconciled with them. Starting with the "chronologies" of Herodotus and Thucydides, Fréret ranged over the "families" of Cadmus, Euphemus, Ajax, the Heraclides and others; he considered the history of Athens, and of Egypt. He tore down Newton's attempted identification of Sesostris with the Biblical Sesac; he attacked as absurd "the system of Euhemerus," on which Newton had so heavily relied to naturalize mythologies. Along the way he repeated his arguments against Newton's count of generations, rejecting Newton's replacement of a count by generations with one by reigns.[43] "I think to have destroyed the fundamental principles" of Newton's system, he concluded, and he left it to "the public" (of letters) to decide whether he or Newton was in error.[44]

Fréret had distanced himself from Souciet, and he implored Halley to consider his own objections to Newton. Halley had himself attacked Souciet's astronomical arguments in considerable detail, but he was not the only one to do so. At about the same time that Halley's critique reached print, Louis Jouard de la Nauze defended Newton in Paris at great length. Born in 1696, and so two decades younger than Souciet, de la Nauze came from a noble family distinguished "less by opulence than by virtue," according to his éloge.[45] Educated by the Jesuits at their college in Agen, de la Nauze distinguished himself both by his dedication to study and by his desire to outshine his fellow students, which must hardly have endeared him to them. The Jesuits saw in him a likely candidate, to which de la Nauze eagerly responded, and against the wishes of his family he joined the Society at Bordeaux. There he soon became disillusioned, most likely because the exigencies of religious matters did not square well with his desire to immerse himself in ancient history. Evidently as persistent in his efforts to leave the Jesuits as he had been to join them, de la Nauze was released. The Jesuits tried to entice him back, and de la Nauze almost acceded, no doubt because he had trouble finding a suitable position, until he was liberated by an offer to become tutor to the family of Louis Antoine de Pardaillon de Gondrin, the first Duc d'Antin (so created by Louis XIV in 1711). De la Nauze remained with the family for the rest of his life, finding sufficient leisure to produce works on Greek antiquity, on the Egyptian calendar, and on the Hebrew Kabbala. He was particularly concerned with chronology. His work on the Egyptian calendar would require a reasonable knowledge of astronomy, which he had clearly acquired by the mid-1720s.

When Souciet's critique of Newton's abrégé appeared, de la Nauze reacted at once and at length. He did so through a series of five "letters" addressed to Souciet, but not

[42] Fréret, 1758, pp. 442–43.

[43] He had written a separate piece on the issue of generations that was published years before (Fréret, 1743).

[44] Fréret, 1758, p. 502.

[45] Anonymous, 1780.

sent, the first of which he lent an (unnamed) Oratorian to read. At the time de la Nauze was, or would shortly be, presented by friends for election to *Inscriptions et Belles Lettres*, of which Fréret was a prominent member, so it is perhaps not surprising that he may have been reluctant to go into print. De la Nauze had available all five of Souciet's dissertations, and he proceeded to address them in order in five letters. The Oratorian to whom he had given the first letter died shortly thereafter. De la Nauze then asked one "Abbé G**," a mutual friend of his and of Pierre Nicolas Desmolets, to request that Desmolets retrieve the letter, since Desmolets had also been a friend of the deceased Oratorian.[46]

The Abbé in question thought the letter to be worth publishing and asked de la Nauze's permission to have Desmolets do so. This took place sometime after October 1726, which indicates that the Souciet *Recueil* had been available by late fall of 1726, despite its title page date of 1727 (the various permissions having been obtained between June and September 1726). Given the situation, de la Nauze was reluctant but eventually agreed. Given as well the rather vehement tone of the critique, Desmolets decided first to ask Souciet, whom he knew, for his consent. Souciet gave it, having no other desire than "to establish the truth," wrote Desmolets, whereupon de la Nauze sent along his second letter, followed by a further three.

De la Nauze's critique savaged Souciet's arguments, and it did so on grounds altogether similar to those which Halley provided, independently, at very nearly the same time. Although de la Nauze certainly lacked Halley's expertise, nevertheless he had studied enough astronomy to work an armillary sphere in order to examine stellar risings and settings.[47] He made good use of the device—which nicely avoids the necessity of elaborate calculations—to undercut every one of Souciet's astronomical arguments. He then continued by attacking Souciet's historical claims, followed by an even more pointed critique of Souciet's effort to undercut Newton on regnal lengths.

De la Nauze's astronomical critique was simple but pointed. The only issue was whether Newton was correct in locating the equinoctial point where he had it, namely at 7°36′ from *prima Arietis*, or whether Souciet's reasons for arguing otherwise were correct. About this de la Nauze had no doubts at all. He rejected every one of Souciet's various claims, and especially his efforts to find a star that could replace *prima* and set chronology back to what it had been. The main thrust of his critique, to which de la Nauze returned time and again in various ways, concerned Souciet's effort to set the equinoctial point at some point near *prima*, eventually at the now-invisible star in the leg of the Ram. He then buttressed his counter to Souciet with a defense of Newton, all still based just on the *Abrégé* since the *Chronology* had not yet been printed.

What then of Souciet's claims concerning the locus of the equinoctial point and the vanished star? That depended on precisely where the star in question had been. De la

[46] Desmolets was secretary to Montesquieu and known as an editor of others' works, including those of the Flemish writer, lawyer and financial counselor Albert Henri de Sallengre, who had died of smallpox at the age of thirty in 1723, having by then edited four volumes of a literary journal that he had founded in the Hague in 1713. Sallengre had met Desmolets during his sojourn in Paris from 1717 to 1719, and Desmolets took up the journal's continuation following Sallengre's death. The fifth volume appeared in 1728, with the sixth following the next year. Desmolets provided an account of the letters' route to publication in an introduction to them: Nauze, 1728, pp. 332–34. On Sallengre see Saint-Philippe, 1722.

[47] Nauze, 1728, p. 352, where he suggested using either a "globe or a planisphere of stars."

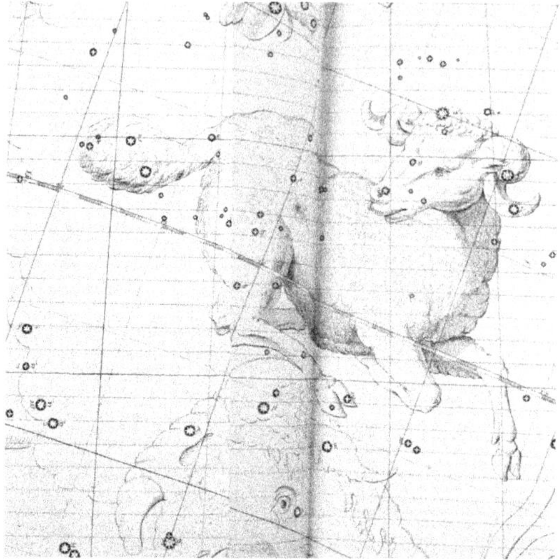

Figure 11.5. From Flamsteed's 1729 atlas, showing the head of Cetus near the center of Aries.

Nauze went right to the heart of the matter. Hipparchus, he of course agreed, told that the first-rising star of the Ram occupied its front foot; Hyginus had also remarked that the front feet rise first. But Hyginus had further noted that the Ram's feet nearly touch the head of Cetus (the whale or sea monster). "And the whole world knows," remarked de la Nauze, that the whale's head is "at the center of the Ram." Flamsteed's stellar atlas shows this particularly well (figure 11.5).

Obviously any star in the Ram's foot is much to the south of *prima*. In order for such a star to rise before *prima*, in the Northern Hemisphere it must also be considerably to its west (i.e., to the right in figure 11.5). For, de la Nauze noted, in those latitudes the diurnal arc of the sun is the smaller the farther south a star lies, and the less the diurnal arc the later the rise. Consequently, position to the west must compensate a southerly location. This much would be obvious even to a relatively unsophisticated commentator provided that he had an armillary sphere available, which de la Nauze indeed did. In fact, he was able to put the device to particularly good use in order to demonstrate that Souciet's location for his putative star marking the Chironic equinox had to be utterly wrong. Souciet had placed his star at 45′ in right ascension west of *prima*. That, de la Nauze wrote, would make it rise ahead of *prima* by three minutes of time provided both stars had the same declination. But they do not, and Souciet's star would have to be several degrees even further west to rise first. In this first letter de la Nauze was not more precise than this, but in the fifth he tightened the argument with greater numerical and graphical detail.[48]

Souciet's putative star, on which his argument for replacing Newton's use of the Hipparchan remark concerning the rising times of the Ram, failed on this single point alone. But that was not all, for de la Nauze had already noted another critical

[48] Nauze, 1728, pp. 343–45 and Nauze, 1729, pp. 377–87 for greater detail.

point against Souciet. The Eudoxan remarks cited by Hipparchus asserted that the equinoctial colure at the time of the original sphere passed through the back of the Ram. Halley, we have already seen, pointed out that Souciet's original location for the equinoctial point would instead pass the colure through the Ram's rear instead. Granted, Souciet later changed his mind, but only to find a star with nearly the same right ascension that did not avoid the original problem. De la Nauze made the same point as Halley, noting that the Hipparchan remark fits only Newton's system.[49] Then there was Souciet's criticism that Newton had falsely assumed that Greeks from the time of Hipparchus had used the incorrect value of a hundred years per degree for precession to calculate their chronology. That, de la Nauze continued, was ancillary to Newton's claims, just a side remark as it were, but Souciet had inadvertently supported the very same point in citing Seneca on the constellations having been set in Greece a millennium and a half before his time. Given Seneca's own attested date, the millennium and a half corresponds to a hundred years per degree, and there was no other way, de la Nauze was certain, for Seneca to have determined the period in question.[50]

Considerably more on astronomy, Hipparchus, and the Eudoxan remarks followed, with each of Souciet's claims taken apart one by one. De la Nauze turned as well to Souciet's attempted use of medallions—a characteristic feature of the new ways with evidence of *Inscriptions et Belles Lettres*—but these he dismissed as irrelevant since no such object reached back before the foundation of the Persian Empire by Cyrus, and Newton's *Abrégé* "ends with the overturning of that empire by Alexander the Great." Consequently, Souciet's critique here recurred only to the question of the foundation of Rome itself, which Newton had moved forward 134 years. Against this Souciet had a medallion from the period of Hadrian and several others, "none of which is very convincing." Hadrian's medallion, for example, referred to the foundation of Rome as having taken place 874 years beforehand, which, given Hadrian's date, would place it at the customary point in time, not at Newton's. But the centuries referred to by the medal can hardly be taken at face value. The interval is "too large for a historical monument to set the epoch."[51]

Equally caustic were de la Nauze's remarks on Souciet's critique of Newton's regnal calculations. In his fourth "dissertation" Souciet had argued that, if the method must be used, then in any case Newton had it wrong, that regnal lengths were about twenty-seven or twenty-eight years, not Newton's nineteen, though we shall see in a moment that Souciet rejected the entire procedure. He had reached this number by considering the lengths of "modern monarchies." De la Nauze took this apart as well. Where Souciet had provided a few, scattered numbers, he went into precise detail, listing the total durations and reigns of thirty monarchies. Dividing each total duration by the corresponding number of reigns produced a regnal length for each monarchy. De la Nauze counted a total of fifteen different such numbers, and of them he then formed the "mitoien" by summing them all and dividing by fifteen. The word connoted an

[49] Nauze, 1728, pp. 357–58.

[50] Ibid., pp. 365–66, from Souciet, 1727, pp. 127–28. The remark occurs in book VII, chap. 25 of Seneca's *Natural Questions*.

[51] Nauze, 1729, pp. 3–8. For the medallion critiques see the third dissertation, Souciet, 1727, pp. 84–91.

"intermediate," as usual for the time, suggesting the notion of a normal rate that is perturbed by adventitious circumstances. The result was Newton's nineteen.[52]

Souciet had not it seems considered all the available regnal data, but there was more to it than that, and de la Nauze was acutely sensitive to what lay behind Souciet's dislike of such a calculation, whatever the result might be. According to Souciet the variations in lengths between reigns "suffice to destroy the new theorem and the new rule that [Newton] proposes to us." The "theorem" or "rule" that Souciet referred to was, he wrote, the following:

> It doesn't matter what the most ancient and least suspect historians, the most venerable monuments of antiquity have said, there [in Newton's numbers] lies the true course of nature; the rule of proportion [i.e., calculation] contravenes all the historians, and geometry must instead be believed.
>
> Here's a truly extraordinary, new rule of history, of chronology, and of criticism. It's astonishing what use, or rather what abuse, many people make with these sorts of proofs. Because one has a genius for geometry, and makes it the main point of one's studies, it's applied everywhere and everything is reduced to it. The most inconstant and bizarre effects of nature, history, ethics, even religion, of facts that depend only on the free will of God or of man, and often on an infinity of different wills, all are measured by rules of proportion, all are brought back to these rules, all must bend to them. Is there anything less reasonable? Each science has its region and its own subject; and each subject, that is to say each truth, is not susceptible to the same proofs as others. Historical facts are proved only through testimonies, they do not fall within the competence of either geometry or arithmetic, any more than the problems of geometry lie under the jurisdiction of history. What we have just said, and the natural variations that we have just pointed out from M. Newton himself, suffice to destroy the new theorem.[53]

Here we have it: Souciet tried to use astronomy to counter Newton's own calculations, but that was at least a matter of the heavens' natural regularity. Nothing like that could hold for human affairs because, he continued, "if the nature [of such affairs] is so changeable and so contrary to herself, what foundation can we make of it? If in the same times, in the same countries, there are men so similar in habits and ways of living, and yet this nature was so different among them, can we reasonably suppose that after centuries, in such different climes, among men so opposite in habits and ways of living, that such natures would be more alike?"

Souciet made no attempt to gauge the variability in question; he simply rejected the possibility of applying calculation to human affairs, having gone into the effort solely to counter Newton's use of it. He seemed even to imply that doing such a thing

[52] See the fourth letter: Nauze, 1729, pp. 17ff.

[53] Souciet, 1727, p. 109.

constituted an attack on human liberty, morality and even religion itself.[54] Numbers could not, *should not*, be used to constrain human moral freedom, freedom granted by the deity. The implication was clear: historical matters must be left to judgments grounded in text and material evidence, all of which testified to the inherent liberty of human affairs from the tyranny of numbers—and what numbers represent, namely the hijacking of history by the calculating denizens of the new philosophy. De la Nauze saw this well enough, and he spurned it. Why, "the theorem must in effect be truly new, to interfere with morality and liberty," he averred. What has this to do with religion, he asked? After all, Souciet had himself done such a calculation, defective though it was (missing the point, or deliberately ignoring it, that Souciet objected to the entire method where human affairs are concerned).[55] There was more, but the opposition between the two men—the one a prominent Jesuit, the other trained by them—is strikingly apparent. De la Nauze not only admired Newton as a geometer and natural philosopher, he willingly set calculation ahead of ancient testimony, whether transmitted through texts or inscribed on medallions. Souciet cast calculation to the side; he defended the methods and evidentiary base deployed by himself and other members of *Inscriptions et Belles Lettres*. Admitted to their ranks in 1729, de la Nauze nevertheless remained an outlier. Unlike Souciet, he was altogether taken by the Newtonian phrase, "the course of nature," to the extent that his interest in something that might seem altogether out of nature's course, namely the mysticism of Kabbala, caught his interest precisely because he thought its aims fit the course of nature, though not its veracity. The methods employed by the Kabbalists in their efforts to secure the secrets of religion and of nature, were, he wrote,

> not usual ones. The human sciences all turn on experience or on reason; [the Kabbala] only employs a combination of letters, of numbers, or some other symbols.
>
> It seems at first that such a method is arbitrary, and that it has no connection with the purpose proposed by the Kabbalists. But they judged otherwise; *they held that it is founded on the nature of things* [emphasis added], and on divine revelation, because God, they say, established different degrees of analogy and of subordination between himself and the angels, between the angels and the stars, and between the stars and sublunary bodies, that he impressed the character of this relation on letters, numbers, and symbols, and that he revealed the manner of consulting these symbols to find the connection between all real beings.[56]

De la Nauze had much to say about the Kabbala's origins and contemporary form, and he was certainly not alone in his interest, since the Protestant Jacques Basnage

[54] Souciet also remarked that Newton had not provided enough to judge the conformity of his regnal claims with sacred history, in particular with the kings of Israel and Judah. The latter, as we have seen, appeared in considerable detail in the published *Chronology*.

[55] Nauze, 1729, pp. 63–72.

[56] Nauze, 1736, pp. 37–38.

had written a good deal about it, and both Newton and Leibniz, for different reasons, were interested in the subject.[57]

Souciet may have rejected Newton's methods and certainly his conclusions, whereas de la Nauze accepted both, but the two agreed that the investigation's foundation was entirely correct: namely, that Chiron had formed the original sphere. Whiston, too, never questioned the claim, and neither did Fréret or Halley. They disagreed only over when the event had taken place. Let's recall what the evidence for this was. (see Chapter 8, note 7). The Hipparchan *Commentary* on the Aratean *Phaenomena* provided remarks concerning the colures which he attributed to Eudoxus. These descriptions, it was commonly agreed, could not pertain to the time of Eudoxus himself. This much had nothing to say about Chiron proper, and so could not be used in themselves to reconstruct chronology, though they might be used to determine just when the ancient sphere was first figured. The link to Chiron depended almost entirely on a single piece of evidence: namely, the following remark by Clement of Alexandria in his *Stromata*:

> Hermippus of Berytus calls Chiron the Centaur wise; about whom, he that wrote The Battle of the Titans says, "that he first led the race of mortals to righteousness, by teaching them the solemnity of the oath, and propitiatory sacrifices and the figures of Olympus." By him Achilles, who fought at Troy, was taught. And Hippo, the daughter of the Centaur, who dwelt with Aeolus, taught him her father's science, the knowledge of physics.
>
> Euripides also testifies of Hippo as follows:—
>
> "Who first, by oracles, presaged, And by the rising stars, events divine."[58]

What of Chiron the Centaur? The most extensive remarks about him were in the *Argonautica* of Apollonius, which Newton frequently referred to. There were other references to him as well, though Newton did not cite them. Apparently he had good reason not to, since the one French critic who ignored all the tortured arguments over chronology argued on the basis of these other sources that Chiron was no astronomer at all.

Jean Hardouin was certainly unusual among his fellow Jesuits. Born in 1646, and so only four years younger than Newton, Hardouin was appointed librarian of Lycée Louis le Grand in 1683, where he also served as professor of theology. He remained there to the end of his life in 1729. He was in several respects similar to his contemporaries at *Inscriptions et Belles Lettres*: he edited texts and later worked in numismatics, whose probative character he placed above that of texts. Grafton wrote of him that, like other antiquaries devoted to material evidence, he was nevertheless "committed to traditional ways of studying texts and stimulated by new and radical theologies," that he "still lived in a world of encyclopedic humanism."[59] Yet Hardouin's commonality with his contemporaries only went so far, for he convinced himself that

[57] On Newton and Kabbala, see Goldish, 1998; Manuel, 1974.

[58] Clement of Alexandria, 1869, Book I, Chap. XV.

[59] Grafton, 2001a, pp. 200–201, which provides the most insightful account of Hardouin in his time.

almost all known texts of Greco-Roman antiquity were forgeries produced by a sinister, and still active, cabal. The head malefactor of this impious deception was apparently the pseudonymously named Severus Archontius, a late fourteenth-century Italian. Coins that were missing or in conflict with the forged texts provided some of the clues to what they had done in their clever efforts to undercut the power of specifically Christian revelation by implying through their interwoven forgeries that the coming of Christ had long been known among Jews and pagans; standard literary techniques of the time, suitably if over-rigorously worked, provided others. With perhaps a touch of exaggeration, given the broader scope of Hardouin's work, Momigliano wrote that he "carried the contemporary bias for non-literary evidence and the contemporary suspicion of literary evidence well beyond the verge of madness."[60]

Hardouin thought only five sources from antiquity to be genuine: Homer, Herodotus, Cicero, the *Georgics* by Virgil, Horace's *Satires* and *Epistles*, and Pliny the Elder's *Natural History*, which Hardouin had himself edited. All else had been the work of the "impious coterie [who] had in their service mathematicians who computed eclipses, lawyers who framed codices and laws, medical men who wrote on medicine, poets who put forth their poems, linguists and interpreters of great skill, who turned their Latin writings chiefly into Greek."[61]

Hardouin died on September 3, two years after Newton, at the age of eighty-three. That same month the *Journal de Trévoux* carried what must have been among his last works, a heated critique of Newton's *Chronology* that he signed the previous March 18. Age had certainly not dampened Hardouin's critical vigor. In nineteen pages he dismantled stone by stone the *Chronology*'s foundation. "Will disputes over the age of the world never cease?" he lamented. "In our day, some have wanted to make it excessively old: and recently an Englishman named M. Newton, on the contrary, by means of astronomical calculations wishes to remove about 534 years from Greek antiquity: which no annalist before him had dared to think. He nevertheless has his defenders, it is said, who undertake to uphold the foundation and the entire structure of his system." The author of the *Principia* and the *Opticks*, recently deceased (though Hardouin seems unaware of the fact), was some "Englishman named M. Newton."[62] Hardouin knew perfectly well who Newton was, as we shall see, but he chose to begin by putting in his place someone whom he saw as unqualified to pronounce on history, and he set out not only to show that Newton was wrong, but that he, and others who chose like him to calculate with history, were flagrantly mistaken.

Hardouin came right to the point. Both Newton and his critics accepted that Chiron had designed the constellations, and the cardinal points, for the use of Argonauts as they navigated to Colchis. "I maintain," Hardouin objected, "that the astronomer Chiron is a phantom: that these two [contesting] parties both build their systems on the false idea that they have of Chiron, and that they employ all of their astronomical erudition on a chimerical supposition." Chiron himself was no chimera, but he was unquestionably not an astronomer. He was in fact a doctor, knowledgeable principally in botany and the medicinal properties of herbs. To prove his point Hardouin

[60] Momigliano, 1950, p. 303.

[61] Translated in Grafton, 2001a, p. 193.

[62] Hardouin, 1729 (September), p. 1567.

cited three of the ancient sources that he thought genuine, namely the *Iliad*, the *Georgics*, and the *Natural History*.

The *Iliad* speaks of the doctor Machaon who, to ameliorate Menelaus' wound, "wiped away the blood and applied some soothing drugs which Chiron had given to Aesculapius out of the good will he bore him." Further, Euryplus said to Patrochlus, "cut out the arrow from my thigh; wash the black blood from off it with warm water, and lay upon it those gracious herbs which, so they say, have been shown you by Achilles, who was himself shown them by Chiron, most righteous of all the centaurs." The *Georgics* writes "Nor change of pasture might any more avail: The flock, for misery came of every one: Of the precious healing arts; their heads did fail,—Amythaonian Melampus, and the son of Philyra, Chiron." Pliny remarked that Chiron taught Achilles medicinal herbs. Hardouin adduced "two or three other witnesses, which one ought not impugn, though I put them much lower than the first." Diodorus Siculus wrote that Hercules had the misfortune to kill Chiron, who was admirable for his medicine. Plutarch in the *Symposiacs* noted that the people of Magnesia venerated Chiron as the first to give medicine to humans. Finally, Suidas wrote of Chiron the Centaur who first found medicine through botany and taught the precepts thereof to Achilles.[63]

Newton had relied entirely on Clement of Alexandria, and that source, Hardouin asserted, merely "reported fables." Clement gave no author for the *Titanomachia*, which provided a nice opportunity for sarcasm. "Where," Hardouin wondered, "did he obtain what he here attributes to a writer, whose name he dares not pronounce, that Chiron exposed the constellations of heaven to the eyes of men? All antiquity knew nothing of that. The unknown author he cites, or himself, did he think, when he wrote, that Chiron made known to men three equally important points to live justly by: oaths, sacrifices, constellations? Are these three not nicely matched to form a good man? Is this a maxim to attract the title of a sage of Greece for its author?" Constellations have nothing to do with just behavior, and in any case even if Chiron (or whoever may have done it) did first design figures for asterisms, "is that a truly new discovery, if that's all there is to it? Does it suffice to make Chiron a *practical astronomer* [*sic*], as he is called by M. Newton?"[64]

In a particularly acute remark, Hardouin anticipated Letronne in questioning the very sense that Newton had attributed to the phrase σχήματα ὀλύμπου.[65] Letronne demonstrated that the phrase refers to the dance-like motion of the heavens. Hardouin did not make that connection, but he queried whether it had anything at all to do with equinoctial points. They served only to inspire both Newton and his critics with an unwarranted zeal for calculation: "It is no less surprising [that these two words] … seized [enlightened people] with a precipitate joy to find a good occasion for displaying their erudition in astronomy and calculation: the ones, to push Greek

[63] Hardouin, 1729 (September), pp. 1570–73. *Iliad*: on Machaon, Book 4, verse 218; on Euryplus, Book 11, verse 830. The *Georgics*: Book 3, verse 549. The *Natural History*: Book 26, section 19. Diodorus: Book 4. Hardouin further remarked that he had "somewhere read" that "Chiron taught Jason himself, leader of the Argonauts: what! Astronomy? No: medicine." He thought he might have read as much in Moreri, 1711, but neither the entry there on Chiron nor the one on Jason makes the connection.

[64] Hardouin, 1729 (September), pp. 1574–76.

[65] On Letronne see chapter 8.

antiquity far back; the others, to bring it closer to us." Moreover, what possible navigational purpose could the cardinal points have served when the Argonauts, who surely did not invent the astrolabe for finding latitude from stellar height, in any case always hugged the coast in traveling from Thessaly to Colchis?

Why, Hardouin went on, did Newton in referring to Clement, change the latter's mention of the *Titanomachia* to the *Gigantomachia*? These were two different things since the Giants were brought forward by Gaia to challenge the Gods in their prior victory over the Titans. Was it, Hardouin insinuated, perhaps "because the Titans suggest a fable?" Was "Newton afraid that this idea might enfeeble the poem's evidence concerning Chiron. He would thereby have challenged his own cause." The reasoning here, though unexpressed, was likely Hardouin's understanding that Newton was an inveterate Euhemerizer, and Hercules (or Newton's multiple Hercules candidates), unlike the potentially more fabulous denizens of the *Titanomachia*, was for Newton based on a real person. Clement himself was in any case uninterested in the cardinal points, having written of constellation risings just because of their significance for their signing the seasons. In fact, Hardouin claimed, Clement's main point was to show that the philosophy of the most ancient sages of paganism and of Greece deployed utterly common knowledge of their time, and that only the philosophy of Moses was truly profound. In the end, "Clement of Alexandria is not one of those Sainted Fathers incapable of or exempt from pious fictions, in order to establish what they put forward."[66]

And so the entire scheme was based on a fallacy, one amply attested to by the ancient sources that Hardouin deemed reliable. Chiron was not an astronomer but a talented purveyor of herbal remedies. "Newton's system," he concluded, "is imaginary and chimerical: it's a phantom either embraced or heatedly countered. So many calculations, disputes, questions, useless reasoning, all destroyed by this single principle historically demonstrated!" How unfortunate, he went on, that "this *great man, this first geometer and mathematician of Europe* [*sic*] has in the end built nothing but a frivolous system. How much we value those who have acquired a reputation."[67]

Hardouin's sarcastic dismissal of Newton's revised chronology suggests that he disapproved altogether of geometers and mathematicians invading the precincts of history, which resonated with controversy and claimed expertise. Hardouin had good reason for his sarcasm, since by the time of his death, historical erudition, whether based on texts or on coins and medallions, had faded in comparison to the growing reputation of natural philosophy, and especially of the "sciences exactes." This exercised members of *Inscriptions et Belles Lettres*, who fully realized that their discipline was increasingly marginalized. An anonymous 1743 article in the society's journal referred to a recent defense of letters by the Jesuit Jean François Du Bellay Du Resnel in mounting, not an attack on the new sciences, but a lament for the downgrading of

[66] Hardouin, 1729 (September), pp. 1578–81. Newton had several Euhmerized Hercules, among them "the general of the forces of *Thebais* and *Ethiopia* whom the Gods or great men of *Egypt* called to their assistance, against the Giants or great men of *Libya*, who had slain *Osiris* and invaded *Egypt*: for *Diodorus* saith that *when* Osiris *made his expedition over the world, he left his kinsman* Hercules *general of his forces over all his dominions, and* Antaeus *governor of* Libya *and* Ethiopia." Newton, 1728b, p. 233.

[67] Hardouin, 1729 (September), pp. 1578–81: "Que ne s'en tenoit-il à ce qu'il avoit acquis de réputation?"

letters, whose importance the author sought to rescue by presenting a list of the subject's particular virtues.[68] Fréret himself defended the study of ancient history against the potential inroads of calculation in March 1724, the same year in which Newton had rebuked the publication of his abbreviated chronology in France, which had of course born Fréret's own, unsigned critique. Fréret wrote:

> Ancient history has its merit, its study has its advantages; and I have difficulty understanding how smart people do not see that all the efforts they make to disparage the subject serve only to arm ignorance with new weapons. The taste for antiquity has been much weakened by the latest disputes in which those who attack the ancients are perhaps not those who have struck them the most dangerous blows: to destroy that taste requires those who unite erudition with the exact knowledge of geometry and of [natural] philosophy to declare against the study of antiquity.[69]

Newton had certainly not declared against the study of ancient history, but he had submitted it to the unyielding, and in Fréret's view unwarranted, demands of calculation.

[68] Anonymous, 1743.
[69] Fréret, 1729 [1724], p. 189.

The Demise of Chronology

A few weeks before the publication of Whiston's attack, Arthur Bedford's far longer *Animadversions upon ... the Chronology of Ancient Kingdoms* appeared. Consumed with religious zeal and passionate for music, Bedford believed that "the restoration of church music had a providential purpose and would play a role in the conversion of the Jews to Christianity." Equally fervent was his crusade against stage-plays and on behalf of the reformation of manners and morals—so much so, that he was seen by some "as a crazed man."[1] In his introduction to *The Scripture Chronology Demonstrated by Astronomical Calculation* (1730), Bedford recounted the background to the composition of his *Animadversions*. For several years, following Archbishop Ussher's recommendation, he had been engaged on "a more exact Method of a *Chronological* System of the *Sacred Scriptures*," and in 1727 he printed a prospectus advertising his project. Upon learning of the impending publication of Newton's *Chronology*, however, Bedford desisted, "expecting great Assistance in [his] Design from that most able Hand." As it turned out, he found himself greatly disappointed. Not only did Bedford discover few astronomical observations in the *Chronology* (and even these unsatisfactory); Newton's book proved so "contrary to all Mankind, and utterly destructive of the *Scripture* History" that Bedford—presumably consumed by the same ardor that had previously prompted him, as a Christian and a clergyman, "to renounce the Devil and all his Works"—found himself obliged "to confute it."[2]

Bedford's *Animadversions*—advertised as forthcoming throughout February 1728, and published on May 1[3]—proved hasty and intemperate. In retrospect, his vehemence appears quite possibly shaded as much by resentment of a formidable "competitor," to whose "astronomical" chronology he might be presumed to have been beholden, as by piety. Certainly, at least one reader of Bedford's prospectus commented on his "publishing Chronologicall Tables calculated from astronomicall observations," which "some way bottomed on Sir Isaac Neuton's hints given this way."[4] Be this as it may, having summarily dispensed with the obligatory encomium to the "greatest Man in the World, not only in this Age, but even in any Age," and to New-

[1] Hearne, 1885–1921, vol. 10, p. 7.

[2] Bedford, 1706, sig. A2; Bedford, 1730, pp. i–iii. A quarter of a century later Bedford still preached against the disease. See his sermon with almost identical title, *The Evil and Mischief of Stage-Plays* (London, 1730, second edition 1735).

[3] *Daily Post*, Wednesday, May 1, 1728.

[4] Wodrow, 1842–1843, vol. 3, p. 428.

ton's contribution to natural philosophy, Bedford launched his attack. Basically, he turned against Newton the very disclaimer that the great man had made regarding his chronological studies when denouncing the unauthorized publication of his *Abrégé*. Might not Newton have been mistaken in history and chronology, studies in which he had conceded he had been but "accidentally engaged, and usually when he was tired with other studies"? Might not Newton's "absolute refusal" to publish the *Chronology* during his lifetime have indicated his realization that the chronology was "uncapable of that Demonstration, upon which his other Writings are founded"—as Conduitt noted when dedicating the book to the Queen—as well as his unwillingness to engage in controversies that he surely knew his singularity would engender? To such spins on Newton and Conduitt's statements, Bedford added his own: Since the *Chronology* was Newton's last work, he may well have commenced it "when his Judgment and Memory failed him, being probably weaken'd not only by his great Age, but also by his former indefatigable Labours." Certainly, Newton did not "begin to fit it for the Press, until he was above fourscore Years of Age."[5]

What confounded Bedford most was that whereas heretofore disagreements between chronologers amounted to a decade or two—sufficiently small differences that might be reconciled by computational adjustment—Newton's reckonings differed so drastically that no reconciliation was possible. Newton's estimation of Ogyges' Flood varied from Ussher's account by 680 years; of the Fall of Troy by 280 years; of the foundation of Rome by 121 years. As for the emergence of the Egyptian and Assyrian monarchies, Newton diverged from Bishop Cumberland's reckoning by 633 and 1300 years, respectively. For all these reasons, Bedford balked: "either all Mankind… have been grosly mistaken in all these Particulars, or he must be mistaken." More confounding still, were Newton's "new *Hypothesis*" to be admitted, its irreconcilableness with other accounts would "shake the Authority of the holy *Scriptures*." In particular, Newton's rewriting, and contracting, of the history of several nations everywhere contradicted the Mosaic account delineating the orderly (and rapid) manner by which the posterity of Noah peopled the world in the aftermath of the dispersion from Babylon—a century and a half after the Deluge. Case in point for Bedford was Newton's claim that as late as Samson's time (twelve hundred years after the Deluge), Greece and Europe were barely inhabited, thereby putting "an End of all" that Moses related of the division of the "isles of the Gentiles" among Japheth's sons (Genesis 10:5).[6]

With much the same result, Newton's contention that the Edomites, fleeing King David, changed their name to Phoenicians and disseminated their arts and sciences throughout the Mediterranean and Europe, became tantamount to saying that all learning originated with them and that no arts and letters existed previously, surely an implication pregnant with "a Train of ill Consequences"—not least, that their predecessors were but "brute Beasts." If true, how could one account for the Building of Noah's Ark or the Tower of Babel? With this challenge in mind, Bedford devoted many pages to confuting the claim that the Edomites were the inventors of letters, astronomy, and of navigation—a claim Newton never made. Bedford further demurred at Newton's privileging, as the benefactors of mankind, the posterity of the

[5] Bedford, 1728, pp. 2, 4.
[6] Ibid., pp. 5–7, 8.

"cursed" race of Ham over the blessed sons of Japhet. *Pace* Newton, he insisted that since Moses was the sole authority for the ancient peopling of the world, a literal reading of Scripture-based genealogy must be accepted—including the antiquity of the colonization of Greece by Japhet's sons, and their early mastery of navigation—lest "all the Labours, which have been hitherto used to clear up, and vindicate this Part of the *Mosaical* History" be demolished.[7]

More noxious to Bedford even than violating the Scriptural account of the peopling of the earth was Newton's pinpointing the beginnings of Greek idolatry in King David's time, with Ceres allegedly the first to be worshipped as a deity. In so doing, Bedford charged, Newton ignored the extensive evidence furnished by Scripture regarding the pervasiveness of idolatry ever since Nimrod's day. Here Bedford resorted to distortion, for Newton explicitly stated that "Idolatry began in *Chaldaea* and *Egypt*, and spread thence into *Phoenicia* and the neighbouring countries, long before it came into *Europe*." Unfazed, Bedford enumerated at great length the rise and progress of idolatry according to Scripture, before damning Newton who, "*without the least* Shadow of Proof, contradicts all at once, and brings both the *Scripture* History and *Chronology* into the outmost Confusion."[8]

Newton, in Bedford's eyes, blundered on other matters as well. Astronomy arose in Babylon prior to the dispersion of Noah's progeny, not in Egypt—as Newton wrote—in the days of Ammon. Similarly, Newton's interpretation of Hesiod's "four ages" to mean that four generations (or 140 years) had elapsed between Deucalion's Flood and the Trojan War, was unfounded; he obviously misunderstood the text, taking γένος ("kind of men") for γενεά (generation or age). Furthermore, Chiron was not a practical astronomer but a physician who surely could not be expected to delineate correctly the positions of the constellations on a sphere—for which there exists no other evidence in any case—especially when Newton himself admitted that astronomy at the time was in its infancy and observations "but coarse." To "expect an Exactness of Astronomical Observations in those Days, is to expect Impossibilities." As for Hesiod's age, Bedford embarked on some obscure calculations in order to prove that 1014 BCE was the closest date for Arcturus to rise 60 degrees after the winter solstice, and that, accordingly, Hesiod could have been born "at any Time within 245 Years either before or after."[9]

Bedford's garrulous and dismissive critique of the *Chronology* was everywhere in evidence. Were he to examine Newton's quotations, he scoffed, "it would be the Means of Protracting a Book to be sixteen shillings Price, which would not be worth sixteen Pence." He repeatedly misinterpreted Newton's chronology, as when he scorned Newton's reliance on the scholiast on Apollonius, who represented Isis and Osiris to be two generations older than Sesostris. Bedford retorted that since Newton placed Sesostris ca. 1002 BCE, and since Osiris "is allowed by all the *Chronologers*" to have been Ham's son Mizraim—who died in 2080 BCE—there elapsed "1080 Years between the Beginnings of the Reigns of the Father and Son, which is hardly consistent" with Newton's provision of only "eighteen or twenty Years to a Generation." Had he "designed

[7] Ibid., pp. 55–67.

[8] Ibid., pp. 87, 105; Newton, 1728b, p. 160.

[9] Bedford, 1728, pp. 80–81, 71–72, 160–61, 171–72.

to prove, that *too much Learning had made [Newton] mad*," gloated Bedford after some additional manipulations of the duration of Egyptian monarchy, he "could not have pitched upon a more unlucky Argument than this Quotation."[10] Clearly, Bedford's intention was to denigrate the *Chronology*'s evidentiary basis by any means possible. More menacingly, he set his sight on the *Chronology*'s perniciousness. Difficult to know, he ponders, whether Newton was in earnest or in jest when he boasted that he had made his chronology conform to sacred history. If in jest, "he should have wrote like other Romantick Writers"; if in earnest, "it is a sad Sign, that he had never studied the *Scriptures*, or that he wrote this Treatise, when he had forgot them." Newton's deviation from Scripture chronology by nearly a thousand years "brings the utmost Confusion" into it, so much so, that were Newton's chronology to be "put into the Hands of Youth," they would undoubtedly become prejudiced against Scripture "to the Rooting out of Religion, and Bringing in of Infidelity.[11]

Judging by a report from Paris, Bedford's flippant account of the *Chronology* did not go unnoticed. The same savants who judged Whiston's critique of Newton "extremement fort," deemed it far superior to Bedford's who, though knowledgeable, had neither penetrated the material nor given due regard to Newton's merit. Bedford, however, persisted in his anti-Newtonian campaign, both in his mammoth *The Scripture Chronology Demonstrated* and in his review of that book, which he probably communicated to the editor of the *Historia litteraria*:

> how great soever the Name of Sir *Isaac Newton* may be, yet, as he made *Chronology* not his peculiar Study, but an Handmaid only to a nobler Science, there is no reason to suppose, but that he might be mistaken in Disquisitions of this nature, as well as another Man and since, (as our Author in his Preface shews) there is more than Suspicion that he was thus mistaken, 'tis his Glory and Commendation, I think, to be reckon'd a Person *nullius addictus jurare in Verba Magistri*.[12]

Bedford's *Scripture Chronology* appeared almost simultaneously with the publication of the second volume of Samuel Shuckford's *The Sacred and Prophane History of the World Connected*. The first volume was published in 1728 and gave no indication that its author was aware of, or concerned with, Newton's *Short Chronicle*. Here Shuckford proposed to offer "a View of the History of the World" from the Creation to the dissolution of the Assyrian empire, concluding, like Bedford, where Prideaux's *The Old and New Testament Connected in the History of the Jews and Neighbouring Nations* commenced. Shuckford's expressed aim was to form "a Judgment of the Truth and Exactness of the ancient Scripture-History, by showing how far the old Fragments of the Heathen Writers agree with it."[13] Nevertheless, two years later, in a lengthy preface to the second volume, he specifically targeted Newton. He promised to cor-

[10] Ibid., pp. 183, 222–23.

[11] Ibid., pp. 142–43.

[12] *Present State of the Republick of Letters*, vol. 3 (1729), p. 314; *Historia Litteraria*, vol. 1 (1731), p. 133.

[13] Shuckford, 1728–1730, vol. 1, pp. i–ii.

rect Newton's specific errors in the body of the text. Nevertheless, since several of Newton's arguments were "of a more extensive Influence," impinging on the entire body of ancient chronology, Shuckford deemed it requisite to consider Newton's general arguments first, so that the reader might judge his correctness in ignoring Newton's overall scheme. Shuckford mentioned neither Bedford nor Whiston, even as the criticism he hurled followed along a similar track, albeit with less depth.

Shuckford focused on three main issues: First, the spurious nature of Newton's attempt to determine the epoch of the Argonautic expedition on the basis of Chiron's sphere. The Greeks were simply too ignorant of astronomy prior to Thales' sojourn to Egypt, Shuckford insisted. Even if Chiron had constructed a sphere, his "draughts of constellations" (Σχήματα ὀλύμπου) were sure to have been inaccurate; he would have been unable to place the solstices on the sphere without erring by several degrees. Had such a scheme survived, Shuckford was confident, "we should be able to demonstrate nothing from it, but the great Imperfection of the ancient Astronomy."[14] Second, he adjudged equally erroneous Newton's rule regarding the lengths of reigns and generations. Not only did Newton fail to realize that the Greek term γενεα signifies both generation and age—the latter of which is of indeterminate length—but, more crucially, Newton remained oblivious to the fact that the duration of reigns and generations must be measured according to "the common Standard of the length of Men's Lives in the Age they belong to." Given that man's life span was shortened to its current length of 70–80 years only during King David's reign, it follows that reckoning ought to be more capacious for earlier periods. Accepting Newton's rule regarding an average of 35 years per generation, and 20 years per reign, for the post-Davidic period, Shuckford argued that the ratios for the earlier period should have been doubled. Thus, he pointed out, 978 years spanned from Abraham's birth to David's death, an average of seventy years per generation; hence, the length of reigns during that period ought to have been computed as 40 years. As "proof," he noted that the average reign of the first twelve kings of Sicyon, and the first eight kings of Argos, was 44 years and 46 years, respectively.[15] Third, Shuckford rebutted Newton's adaptation of Sir John Marsham's theory regarding the short duration of the Assyrian monarchy by denying that Herodotus licensed such a reading. Nor had Newton been correct to dismiss Ctesias' enumeration of Assyrian reigns, as the latter derived his information from authentic records kept by the Persians. Indeed, Newton's supposition that ancient chronologers invariably computed lengths of reign artificially—according to what they imagined they might be—could not be conceded; most "took their Account either from Monuments, Stone-Pillars, or ancient Inscriptions, or from other Antiquaries of unsuspected Fidelity."[16]

Curiously, Shuckford embraced Marsham's identification of Sesac as Sesostris, omitting to mention that Newton had done so as well. By way of "demonstration" Shuckford argued that from the death of Menes (=Mizraim)—287 years after the Deluge—to Sesak (=Sesostris), there elapsed 1090 years and fifty-five successions. In his eagerness to make such a claim, Shuckford violated his own principle regarding

[14] Ibid., vol. 2, pp. iii–vi.

[15] Ibid., vol. 2, pp. xiv–xx.

[16] Ibid., vol. 2, pp. xxv–xxix, lxi.

the more capacious length of reigns prior to King David; his computation unwittingly produces an average reign of less than twenty years—commensurate with Newton's rule![17] We shall return to Egyptian chronology below.

The critical attention directed by Whiston, Bedford, and Shuckford to the *Chronology* indicates the significance contemporaries attributed to the volume. Viewed as the most sweeping attempt since Scaliger to revolutionize the discipline, the *Chronology* set down a radical interpretation of antiquity that made its underpinnings germane to any investigation of ancient history—not least because Newton's growing stature conferred considerable authority on the new system, or, at the very least, prevented its marginalization.

Telling of the buzz surrounding the *Chronology* is the extensive attention it received in periodicals. Consider *The Present State of the Republic of Letters* launched by the aspiring Andrew Reid in January 1728. Newton figured prominently in the new monthly, the very first issue of which included an account of Newton's *Optical Lectures* as well as a translation of Fontenelle's éloge, to which Reid added corrections of factual errors or perceived blemishes on Newton's reputation. The next month's issue announced the publication of the *Chronology*, "wherein the great Author supports his opinions with such solid and numerous arguments, that his system seems now to need no further defence against the objections which a partial view of it had raised." Also announced, as noted above, was Whiston's forthcoming critique of Newton. In March followed Fatio de Duillier's poetic eulogy of Newton, while the entire April issue (100 pages) was devoted to an annotated summary of the *Chronology* dedicated to John Conduitt.[18]

In all likelihood, the summary was written by the rising Independent minister Philip Doddridge, who read the *Chronology* shortly after its publication. Though declining to serve as Reid's assistant editor, on February 14, 1728, Doddridge gave advice on how best to run the new journal. He also volunteered to regularly furnish reviews, starting with a review of the *Chronology*, which he proposed to submit by the end of the following week. In it he intended to include "a general character of the book" together with an enumeration of "some of the principal maxims" laid down by Newton "as the foundation of his calculations," complete with explanatory notes. Out of modesty—or perhaps owing to qualms regarding the propriety of such literary activity for a dissenting minister—Doddridge requested that Reid keep anonymous all his contributions.[19]

Doddridge—assuming he is the author—enthusiastically endorsed Newton's innovative distinction between the length of reigns and of generations, proclaiming that the rule of averages was demonstrated "beyond contradiction from the *certain* histories of all nations." Equal enthusiasm was expressed for Newton's astronomical arguments. Indeed, in his opinion, Newton should have emphasized even more forcefully how the rectification of Hipparchus' error concerning the rate of precession established the correct date of the Argonautic expedition. To Doddridge, this argument appeared an "absolute demonstration, and alone sufficient to establish all that Sir Isaac

[17] Ibid., vol. 2, pp. xxxiv–xxxv.

[18] *Present State of the Republick of Letters*, vol. 1 (1728), pp. 52–84, 166, 237–42, 251–352.

[19] Doddridge, 1829, vol. 2, pp. 399–400, 413–22.

had advanced before."[20] Doddridge's sympathetic summary and useful annotations of Newton's historical arguments undoubtedly helped render the *Chronology* more accessible to a broad audience, and in 1732 Reid actually issued it as a separate publication.

The June issue included a clarifying note—presumably by Doddridge again—on Newton's precise meaning of average length of generations, no doubt in response to Whiston's critique of the rule.[21] Two months later, an anonymous member of Wadham College, Oxford—quite possibly the young George Costard—also sprang to Newton's defense. Calling Newton perhaps the "greatest genius ... of all ages since the creation of the world," he considered him as well the inaugurator of "a new science in Chronology." Newton, to be sure, had been "attacked by some persons in their writings, and by many more in their private conversations," especially by the admirers of Sanchoniathon's Phoenician history. Without considering the authenticity of Sanchoniathon's history, Costard endeavored to demonstrate the indefensibility of the three foundation blocks upon which it rested: 1. the identification of Noah with Ouranus (or Jupiter Uranius), and of his son Ham with Jupiter Belus (or Ammon)—which contradicted Newton's identification of Sesostris/Sesac with Ammon's son. 2. the commencement of idolatry as early as Noah who, following his death, became the first object of idolatrous worship. 3. the denotation of Ham, Ammon, and Zeus as the same person on accoount of their being "words of a near signification."

Costard considered Newton's identification of Sesostris and Sesac beyond dispute, thus deeming any additional defense "entirely superfluous." He attributed Sanchoniathon's deep appeal to the complete profane history he provided from Adam to Toth the second, when Eratostenes picked up the narrative and carried it to the first Olympiad—thus supplying a combined narrative that conformed to Scripture. To undermine the credibility of the Phoenician history, Costard noted that it had been preserved by the heathen Porphyry, who found Sanchoniathon's record of monarchs and events that long predated Moses serviceable, offering "proofs" that "Idolatry was the first and oldest religion," and thereby debunking the alleged antiquity of the Jews. "And yet," Costard enthused, Sanchoniathon actually served "as the chief evidence for the opposite side": by acknowledging that mankind originated from a single pair and that a mere thirteen or fourteen generations elapsed from Misor to Toth—"a great attestation to the truth of the Mosaick history" and to Newton's Chronology.[22]

Such sustained and favorable exposure to Newton's *Chronology* rankled its opponents, who demanded that their objections also be aired. Reid acquiesced. In February 1729 he announced: "it is now equally incumbent upon us, in order to do impartial justice to the Authors on both sides ... that we impart to the Publick a like faithful and exact Account of what objections and difficulties have been urged against it." Bedford's *Animadversions* had appeared first and hence it was noticed first. The précis recited the key points of disagreement before concluding: the entire "foundation of the Controversy" rests on Bedford's conviction that Newton's chronology "entirely contradicts" Scripture, and "that if the one is true, the other must be false." Additionally,

[20] *Present State of the Republick of Letters*, vol. 1 (1728), pp. 260, 282.

[21] Ibid., vol. 1, pp. 499–500.

[22] *Present State of the Republick of Letters*, vol. 2 (1728), pp. 210–29, 362–71.

whereas such luminaries as Bochart, Prideaux, Ussher, Lloyd, Cumberland, and "all the best *Chronologers* and Commentators" backed Bedford, Newton's scheme was "wholly new," ignoring the clear and contrary evidence produced by others. Lastly, to quote Bedford, if one were to rely on Newton's mode of argumentation, he might prove as well that Hannibal, Belisarius, and the Duke of Marlborough "were the same person."[23]

Whiston's turn came two months later. Reid undoubtedly incorporated into his prefatory note "promotional" material provided by Whiston, who bitterly complained about the slight attention his critique of Newton had received in England, as it had been "buried" among the appendices to his *Authentick Records*. Nonetheless, he boasted, it had been "read and examined ... with particular attention and approbation" abroad. The reference was clearly to Fréret, who had availed himself of Whiston's critique when planning a further attack on Newton's *Chronology*. In all likelihood, the letter from Paris published by Reid in the same issue—which praised Whiston's critique while slighting Bedford's, was communicated to him by its proud recipient. Reid also permitted Whiston to furnish his own précis of his attack on Newton—an opportunity Whiston made full use of. The novel system, he thundered, "lays aside the authority" of the Samaritan and Septuagint versions of Scripture as well as "most authentick original Historians and Antiquaries"; were it to prevail, he cautioned, "all ancient history whatsoever" would permeate with "the utmost scepticism." He followed this admonition with nearly a complete transcription of the text published two years earlier.

Costard again sprang to Newton's defense. In the June issue of the periodical, he wrote that the more he inquired into the *Chronology*, "the more reasonable and just" it appeared to him, and the more convinced he became that all the objections to it were "no more than mere Cavils." Thus, to Bedford's repeated complaints that Newton ignored, or differed from, the approved authorities, Costard responded: "'tis not a mere opposing any received opinion, or contradicting many great men, that makes any new System erroneous, provided it be supported by superior reason and argument." Newton's system "may have some faults," but on the whole it was "true."[24] As to specific points, Costard noted that when Newton argued for the existence of many kings in Egypt in Joshua's time, he meant that the Egyptian empire—which included not only lower Egypt, but Thebes, Libya, and Ethiopia—did not come into being until Solomon's day. Nor had Newton contradicted Scripture in regard to the "Isles of the Gentiles." Even more frivolous was the objection that Newton had suggested that the arts and sciences necessary for human life were unknown before the time of the Edomites. Newton had intended the progress in astronomy and the liberal sciences of the sort enjoyed by the Chinese following the arrival of the Jesuits.[25] By way of conclusion, Costard dismissed Whiston's attempt to erect his own chronological system on the indefensible basis of the Septuagint and Samaritan Pentateuch. "But this is not at all to be wonder'd at, when we consider that strange Notions have for many years been advanc'd by this Gentleman, in Chronology, Critick and Divinity; which, what-

[23] *Present State of the Republick of Letters*, vol. 3 (1729), pp. 98–99, 110.

[24] Ibid., pp. 402–3.

[25] Ibid., pp. 406, 411–12.

ever Specimens they may be of his warm Imagination and extensive Reading, are (in the opinion of the generality of learned Men) none of his Judgment."[26]

Whiston, though, had the last word. In December 1730, the *Present State of the Republic of Letters* published a review of the astronomical part of Bedford's *The Scripture Chronology Demonstrated*, almost certainly written by Whiston. Nearly half of the review was devoted to a critique of Newton, culminating with the claim that in the *Authentick Records* Whiston had challenged Halley—who had twice defended Newton against inferior adversaries—to examine the astronomy of Newton's *Chronology* yet again, this time in light of Whiston's critique. Halley never responded. From his silence, the reviewer contended, "we may rest pretty well satisfied ... that he sees it incapable of being defended; and if Dr. *Halley* has given it up, I think we may venture to say, there is scarce any other left hardy enough to undertake it." Hence, he continued, we may now "look upon this controversy as in effect ended" for those few qualified to judge have given their decision—in his favor, of course, is the insinuation. Grudgingly, however, the reviewer conceded: "though it should happen that the strongest arguments were on one side, and only the name of Sir *Isaac Newton* on the other; we are not to wonder if the others, who are always the great majority, should remain some time in suspense, and undetermined; nor indeed can they be blamed, but are very excusable in so doing."[27]

As we remarked, Bedford and Shuckford's contrasting views on the duration of the Egyptian monarchy are indicative of the incertitude surrounding the historical record as well as of the ideological uses to which the ancient history of the Middle East was subjected within the contemporary religious context. This context spurred an acrimonious pamphlet war that inadvertently embroiled Newton's *Chronology*. The fracas originated with the publication in 1730 of Daniel Waterland's critique of Matthew Tindal's *Christianity as Old as the Creation*: Tindal's deist manifesto that posited the sufficiency of natural religion for the regulation of moral life, thereby challenging the orthodox position of revealed religion as the prerequisite grounding for morality. Waterland, one time Master of Magdalene College, Cambridge, had distinguished himself in the previous fifteen years as a champion of Trinitarianism, a crusade he directed primarily against Samuel Clarke. Increasingly, he came to regard Newton, too, as sharing Clarke's views, and this prejudice would inform his behind-the-scene campaign against Newtonian science and religion. In 1730, however, he set out to defend revealed religion against Tindal's attempt to "vilify" Scripture and "magnify the Law of Nature."[28] Waterland's plodding defense of the historicity and authority of the Old Testament, *Scripture Vindicated ... Part I*, focused primarily on the Genesis account of the Fall of Man, the history of Abraham, and the institution of circumcision—the latter as an instance of the Divine origin of Jewish religion and rites and directed against Tindal's claim that circumcision was borrowed from the Egyptians. Waterland promoted the position that circumcision was unknown in Egypt prior to Abraham's arrival there.[29]

[26] Ibid., p. 415.

[27] *Present State of the Republick of Letters*, vol. 6 (1730), pp. 412–13.

[28] Waterland, 1730, p. 3.

[29] For a summary, see Holtby, 1966, pp. 81–89.

The pamphlet occasioned a savage critique from Waterland's old Cambridge foe, Conyers Middleton, who lashed out at Waterland for exposing Scripture "to fresh ridicule and contempt" in his inept response to Tindal, raising "new scruples" rather than quieting "old ones." Far better than to attempt a futile defense of the historical accuracy of Scripture, Middleton argued, would have been to point out how all "*great Reasoners* of the *Heathen World*" had already attested to the insufficiency of reason for the regulation of religious, moral, and civic life, in one blow undercutting Tindal's aggrandizement of the "Religion of Nature." Controverting Waterland's defense of the Genesis account of the serpent in the Garden of Eden, Middleton invoked St. Stephen's statement (Acts 7:22) that Moses was "learned in all the wisdom of the Aegyptians." This suggests that since the Egyptians had invented hieroglyphics before Moses' time and since the symbol of the serpent "was of more common use with them, than any other animal," it stood to reason that "Moses's account of the fall might [have been] drawn from principles and notions imbibed in his youth in the Schools of the Aegyptians." Such a conjecture also led him to take issue with Waterland's defense of the divine institution of circumcision by claiming the rite did not originate with the Egyptians. By way of rebuttal, Middleton discussed Egypt at some length, noting the power and greatness it consolidated as well as the reputation it garnered as a center of wisdom and learning. The Jews, in contrast, "were an obscure contemptible people, famed for no kind of literature; scarce known to the polite world, till the *Roman* Empire dispersed them." For this if for no other reason Middleton deemed it highly improbable that a nation that valued itself for wisdom "should borrow so remarkable a custom from a Nation they always hated and despised." Undoubtedly, then, the Jews borrowed this practice, as well as others, from the Egyptians.[30]

Zachary Pearce, whom we've encountered earlier as Newton's champion, leaped to vindicate Waterland's honor, exposing the "many Falshoods" "both in quotations and Historical Facts" that abounded in Middleton's pamphlet. Chief among such factual falsehoods—defended with "more Zeal than Judgment"—was Middleton's defense of the cultural superiority of the Egyptians over the Israelites. By way of rebuttal, Pearce marshaled Newton's claim in the *Chronology*: Egypt was thinly populated and the Jews were nearly as strong as their masters, hence Pharaoh's determination to decimate them. Pearce further embraced Newton's claim that the posterity of Abraham alone made use of letters until the time of King David. Civilization originated with the Jews.[31]

Middleton would have done well to heed to Newton, "Great Author, this true Freethinker," Pearce pronounced, who had spent thirty years in studying antiquity and who had demonstrated the lateness of both Greeks and Egyptians. "Till you have confuted Sir *Isaac Newton*'s Account, you should never have argu'd from the Vulgar Hypothesis with as much Assurance as if it had never been called in Question, much less as if it had never been Demonstrated to be False." Until Middleton "confuted this new System of Chronology," Pearce repeated, he could not maintain that Egypt was "a great and flourishing Kingdom." If Newton's Discoveries carry "any weight with You, he has shewn that not only *Learning and Politeness* did not flourish, but even

[30] Middleton, 1752, vol. 2, pp. 137, 151–52, 154–55.

[31] Pearce, 1732a, pp. 6, 41–43.

Letters were not in use, in *Egypt*, 'till many Ages after the Period where [Middleton had] fixt it." Pearce felt compelled to add that Newton himself had intimated to him his belief that "the Kingdom of David was the most considerable Kingdom, that Then was or had been erected in the World."[32]

In his rejoinder, Middleton denied the insinuation of infidelity. He insisted on his intent to defend Christianity upon "more rational and consistent principles" than Pearce seemed to profess. Nor was his scholarship wanting. And since the only matter of fact that Pearce hurled against him involved the *time* when Egypt was a great kingdom—an issue that he had not addressed in his original pamphlet—Middleton was happy to accept the challenge. He'd tackle what Pearce "absurdly call[ed] the *Demonstrations and Discoveries of Sir Isaac Newton*." His opponent, Middleton noted, misunderstood the meaning of demonstration:

> You had heard much talk of Sir *Isaac*'s *Demonstrations* in Mathematicks and his great *Discoveries* in Natural Knowledge; and imagined perhaps that those Words signified nothing more that *Conjecture* or *Opinion*. For had you reflected what a *Demonstration* meant, you could not have applied it to a *System of Chronology*, however probable, or preferable to all others, which from the Nature of Things can never reach Certainty or admit of *Demonstration*.

Newton discovered no new texts. Nor did he invent "a *new Telescope*, to pry into remote and dark Antiquity with more Accuracy than had been practicable before." Hence, "*all his Discoveries* can amount onely to *Conjecture*; which, like that of all other Men, will always be fallible and uncertain, as we shall find more especially in the very Instance before us."[33]

Of course, Middleton professed no intention of "weakening the Authority or lessening the Character of Sir *Isaac*"; he honored Newton as much as did Pearce. He considered him "as one of the greatest Philosophers the World has ever produced." Nonetheless, Newton's authority in mathematics and philosophy, as well as in other parts of learning, could not be "advanced above that of all others." If Newton was superior "within his proper Character," others surpassed him in "a thorough Knowledge of Antiquity, and the whole Compass of *Greek and Aegyptian* Learning." In fact, Pearce himself had weakened Newton's authority—just as he had weakened Moses'—"by forcing and stretching it to a Point whither it can't be extended, to a length it never can reach." As for the issue of fact, Middleton dismissed as unfounded Newton's assertions that Egypt was thinly populated in Moses' time and that Thoth, Sesostris' secretary, had invented letters. Likewise, Middleton considered the identification of Sesostris as Sesac as mere conjecture. More egregious in his mind was Newton's suggestion that the great kingdom of Egypt was deprived of letters during the millennium that stretched between Abraham and Sesac, while the "petty state of the Jews" possessed such letters. He also chided Pearce for slighting the testimony of St. Stephen—"a Person *inspired by God, full of the Holy Ghost*"—that Moses was "learned in all the

[32] Ibid., pp. 50–51, 40–51.

[33] Middleton, 1752, vol. 2, p. 225.

Wisdom of the Aegyptians." In short, he derided Pearce for demolishing the very Foundation of all Revealed Religion "in order to get rid of a slight Difficulty."[34]

Pearce responded with a more muted defense. Though Middleton's opinion regarding the antiquity of the Egyptians "had long prevailed in the World," Newton's *Chronology* offered a different account "with such Appearance of Probability against the common Sentiment" that it was hardly "fair and equitable" for Middleton to flaunt the "old Principles in Chronology, as undoubted Truths, and as if nothing had been said by that Great Author to confute or weaken them." Pearce also justified his use of the term "demonstration" to account for Newton's accomplishment. Surely, the term was not restricted to the domain of mathematics. One need only to recall Eusebius' *Demonstratio Evangelica*, or the statement of Bishop Patrick that Ussher "has demonstrated his Annals." In terms of rebuttal, however, he did little more than refer Middleton to an unpublished treatise by John Woodward, who showed that the Egyptians had "borrowed all their Learning from the *Hebrews*." After complaining about the flippant manner in which Middleton had dismissed Newton's identification of Sesostris and Sesac, Pearce concluded (rather lamely) that surely someone would "attempt to do Justice" to Newton's proofs. And with such certitude he left Middleton to his opinion.[35]

Sniffing victory, Middleton pounced. He excoriated Pearce for "such rashness" shown by his treatment of St. Stephen's inspired testimony regarding Moses' debt to Egyptian learning, accepted even by Waterland. Middleton further claimed for himself the right to judge "and of thinking differently even from Sir *Isaac*, or any man else, as oft as the force of reasons oblige me." As for Pearce's leaving him to his own opinion, he called it "a poor childish evasion." Since Egyptian learning was the "single historical fact" contested, Pearce's failure to "do justice" to Newton's "proofs" spoke volumes. Whereas Middleton had produced arguments that would "be thought by every body else sufficient and effectual to confute" Pearce, the latter "plainly disclosed" that he was equally convinced, yet he "stood condemned in [his] own mind, and punished by a *silly pride*" for failing to admit his error.[36]

Upon receipt of Pearce's first pamphlet, Waterland expressed gratitude. He'd been delivered from the need to respond to such an "Amalekite"—deeming Middleton an adherent of the "high Leader" Tindal—and congratulated Pearce for having "demolished him at once." He'd effectively disabled the scholarly foundation upon which Middleton's argument rested. "And though I do not myself follow Sir Isaac Newton's Chronology," added Waterland, "yet I am very well pleased to see it so strongly pressed upon one who perhaps does."[37] To such a remarkable expression of opportunism—Waterland openly stated that very year, in his second installment of *Scripture Defended*, that Egypt was "the most great and flourishing Kingdom of the ancient times"—were added later suggestions of how best to respond to Middleton's rebuttal. In February 1732, following Pearce's second response, he furnished his defender with a suggestion for an equivocating response to St. Stephen's problematic statement:

[34] Ibid., pp. 225–56, 207.

[35] Pearce, 1732b, pp. 40–42.

[36] Middleton, 1752, pp. 246–48.

[37] Churton, 1868, pp. 5, 7.

Though St. Stephen "mentions Moses' Egyptian learning, he says nothing of the extent or degree of it. He had the best education that Pharaoh's court could give, and as much wisdom as the Egyptians could teach, be it little or more." Such an interpretation of St. Stephen "interferes not with Sir Isaac Newton's hypothesis." Notwithstanding his qualms, Waterland availed himself of Newton's *Chronology* a couple of months earlier, when claiming that the Egyptians had embraced circumcision from the Ethiopians ca. 946 BC.[38]

Side-by-side ran another theological controversy that ultimately embroiled Newton: whether the mention by the Greek chronicler Phlegon of Tralles in the second century (as recorded by Eusebius) of an extraordinary solar eclipse and an earthquake in Bithynia—which occurred in the fourth year of the 202nd Olympiad—attested to the verity of the great darkness that accompanied Christ's crucifixion. Samuel Clarke had invoked the widely accepted testimony in the course of his 1706 Boyle lectures, *A Discourse Concerning the Unchangeable Obligations of Natural Religion*. Shortly before his death in 1729, Clarke prepared a new edition of the work, in the course of which he heeded the suggestion of his friend Arthur Ashley Sykes to delete the reference. Sykes believed he had sufficient evidence to disprove that the eclipses were identical. The book, without the passage, appeared only in 1732, but Whiston berated the "groundless Suggestion" made by Sykes, and his introducing a "*Supposal* in a thing that was capable of *Certainty*," already in his self-serving *Historical Memoirs of... Samuel Clarke* (1730).[39]

Sykes, however, remained convinced that Phlegon's account bore no relation to the Scriptural one. By removing a false argument he had made an important contribution to Christianity, for he'd removed as well from its adversaries' arsenal "an opportunity of confuting what is brought as a principal confirmation of the truth." More to the point, Sykes reasoned that since Christ was crucified during Passover, when the moon is nearly full, there could have occurred no solar eclipse. So, too, the total darkness mentioned in Scripture lasted for nearly three hours, whereas the longest solar eclipse cannot produce complete darkness for more than six minutes. After minute scrutiny of the sources, Sykes concluded that corruption must have crept into the relation of Phlegon's lost text. The chronicler must have assigned the eclipse to the *first* year of the 202nd Olympiad—an event that could be corroborated by calculation as Kepler had shown—rather than to the *fourth* year of that Olympiad, the year in which Christ died.[40]

Whiston retorted swiftly with a pamphlet in which he enumerated, as was his wont, all the "original Testimonies" that seemed, to him if not to others, germane for the confirmation of Phlegon's ccount. Also in character was Whiston's manner of dealing with the fact that no total solar eclipse could be dated to Olympiad 202.4: a comet must have obscured the face of the sun "without the interposition of the moon," giving an illusion of an eclipse and accounting for the extraordinary duration of the event. Sykes undoubtedly derived considerable pleasure from refuting such an outlandish suggestion, which even Whiston only insinuated. He could only recommend

[38] Waterland, 1731, p. 143; Churton, 1868, pp. 23–24, 13.

[39] Clarke, 1706, p. 325; Clarke, 1732, p. 357; Whiston, 1730, pp. 116–17.

[40] Sykes, 1733, pp. v–vi; Sykes, 1732.

to Whiston's consideration "how big the comet must be that cou'd cause a three hours darkness," and how near to earth without causing the kind of havoc that Whiston regularly associated with comets.[41]

Whiston opted not to respond, but Sykes met with other opponents who objected to his singularity on more traditional grounds. Responding to them, Sykes availed himself of the recently published *Observations upon the Prophecies* in order to draw support from Newton's determination there that Christ died the year after Olympiad 202.4—in AD 34. And "if it be a crime in [him] to oppose" Phlegon's testimony—a charge hurled by one of his opponents—it was a crime that Newton, Vossius, and Kepler "(Men of great Learning, and true believers of the Christian Religion) may all be equally charged with, as much as [he could] be."[42]

Such embroilment in heated theological debate affected adversely the subsequent reception of the *Chronology* among certain orthodox divines, especially as Middleton appeared to gain the upper hand on the matter of Egyptian chronology and the publication of the *Observations upon the Prophecies* confirmed in the minds of certain divines at least a propensity for heterodoxy in the *Chronology* as well. Daniel Waterland's attempt to mastermind a public campaign against Newton's *Prophecies*—without appearing himself in print—is telling of the theological temper of the mid-1730s. A new chapter on that front commenced with the publication of William Warburton's *Divine Legation of Moses*.

Like Newton, Warburton belonged to that group of scholars who considered their ingenuity necessary and sufficient for establishing more solid grounds for "orthodoxy" in the war against the enemies of Christianity—analogous in temper to Isaac Vossius who sought to substitute the authority of the Septuagint for the Masoretic text in order to diffuse Isaac de la Peyrère's challenge to Biblical chronology. Warburton's ingenuity manifested itself in his willingness to acquiesce to the deists' claim that Judaism lacked a doctrine of a future life—which they paraded as proof against the divine origination of the Mosaic law—but by turning the argument on its head. The very fact that Judaism lacked a "Doctrine of a Future State of Reward and Punishment," he insisted, attested to God's extraordinary providence and his direct involvement in the affairs of the Hebrews, an involvement that rendered unnecessary the revelation of the doctrine of future state at such an early stage. Thus, operating "on the Principles of a Religious Deist," as the subtitle of the *Divine Legation of Moses* proclaimed, Warburton boasted of his ability to refute the argument for the human origins of Mosaic religion—and, by extension, the verity of Christianity that depended on the Jewish "dispensation." Indeed, he believed that his demonstration of the theory fell "very little short of mathematical Certainty."[43]

Central to Warburton's project was the proposition that "the most wise and learned Nations of antiquity" promoted the doctrine of the future state, Egypt in particular, whose religion Warburton regarded as "prototype of all pagan religions." Indeed, "the high Antiquity of the Arts and Empire of Egypt . . . illustrates and confirms the Truth

[41] Whiston, 1732, pp. 38–39; Sykes, 1733, pp. 48–49.

[42] Sykes, 1734 , pp. 4–5. Newton did not mention Phlegon when revising the date of Christ's death: Newton, 1733, p. 168.

[43] Warburton, 1738–1741, vol. 1 p. 7.

of the Mosaic Religion."[44] Understandably, Newton's advocacy of the lateness of Egyptian civilization stood diametrically opposed to Warburton's design. He had articulated his disapproval as early as 1733, shortly after embarking on his "*Moses vindicatus*" which, he promised William Stukeley, would ensure "that Moses' Divine Legation shall never be called in question again by impartial men." Warburton fully concurred with Stukeley's negative view on the *Prophecies of Daniel*: Newton may well have been "a prodigy in his way, yet [he] never expected great things in this kind (which requires a perfect knowledge of antient Literature, History, and Mankind), from a man who spent all his days in looking through a telescope."[45] Warburton elaborated on such a conviction in the second volume of the *Divine Legation*, a long section of which he devoted to a refutation of that part of the *Chronology* that concerned Egypt.

Like his friend Middleton, Warburton began by extolling the "incomparable" Newton, "whom Science and Virtue seemed to be at strife about which should render him most illustrious, while they equally concurred to make him the Ornament of human kind." Nevertheless, Newton ought not be regarded as a Midas who turned everything he handled into a demonstration. When Newton turned "to correct Old Time"[46] of Egyptian chronology, he "suffered himself to be drawn away from the *Goshen* of Moses into the thickest of the *Egyptian* Darkness, by little lying *Greek* Mythologists and Story-tellers." Newton was led astray because his Egyptian chronology "was fashioned only to support his *Grecian*; which he erected on one of those sublime Conceptions peculiar to his amazing Genius"—an example of the pernicious effect of devotion to a "favourite Hypothesis." And since Newton's discourse "contradicts every thing Holy Scripture delivers concerning that ancient People," Warburton felt it necessary—"for the sake of Scripture, and for no Fondness to any private Opinion"—to refute an hypothesis that could gain prominence owing to Newton's reputation.[47]

Warburton presented Newton's reasoning for the low antiquity of Egypt in the form of a syllogism:

> *Osiris* advanced *Egypt* from a State of Barbarity to a Civil Policy. But *Osiris* and Sesostris were the same. Therefore *Egypt* was advanced from a State of Barbarity to Civil Policy in the time of *Sesostris*.

Further to fix Sesostris' time, Newton identified him with Sesac, an identification that Warburton claimed not to be concerned with, determined as he was to disentangle that of Osiris and Sesostris. At best, he continued, Newton raised several "Difficulties" against "the best established Fact of History," by stringing together "a number of Circumstances, from ancient *History*, to prove the Actions of *Osiris* and *Sesostris* to

[44] Ibid., vol. 1, pp. 6–7; vol. 2, "contents"; Assmann, 1997, p. 101.

[45] Nichols, 1817–1858, vol. 2, pp. 19, 21.

[46] Warburton borrowed part of Alexander Pope's homage to Newton in Essay on Man: "Go, wondrous Creature! mount where Science guides, / Go, measure Earth, weigh Air, and state the Tides,/ Instruct the Planets in what Orbs to run,/ Correct old Time, and regulate the Sun." Pope, 1733, p. 6.

[47] Warburton, 1738–1741, vol. 2, pp. 206–7.

be *greatly alike*; and a number of Circumstances from ancient *Fables*, to prove the Gods, whom he supposes to be *Osiris*, were *about the Age* of *Sesostris*." Using Newton's method, Warburton quipped, one may argue from the similarity of circumstances and the similitude of names and actions, the sameness of virtually all great heroes of remote antiquity—King Arthur and William the Conqueror, for example. Newton should have taken cognizance of the ancients' predilection to honor great heroes by bestowing them with the names of earlier gods; identity could not be inferred from likeness. Thus, to make Osiris a great conqueror simply because Sesostris was one necessitated believing that his great deeds were carried out while Egypt was barely emerging from barbarism. Conversely, to make Sesostris "the Inventor of Arts, and the Civilizer of a rude and barbarous People," simply because Osiris was, ignored centuries during which they "had been a flourishing and powerful Empire."[48] In other words, once the identification was made, Newton interchangeably credited one king with the actions of the other.

Confusion among ancient historians regarding the histories of Osiris and Sesostris, Warburton wrote, facilitated Newton's enterprise, while the mingling of Egyptian history with Greek mythology further strengthened his argument. Case in point—the identification of Bacchus and Osiris. The ancient Greeks, Warburton posited, worshipped only natural divinities. Once they came in contact with the Egyptians, however, the Greeks also took to deifying dead men. But though they embraced Egyptian customs, initially the Greeks did not borrow Egyptian gods; they continued to worship their own. In time, they adopted certain Egyptian gods as well and, naturally, followed this adoption with the construction of a mythology that made their gods "conformable to the *Egyptian* History of them." Hence, the Greeks made their Bacchus two generations older than the Trojan War—which he was—and Osiris the conqueror of India, which he was not.[49] Such a prolegomena enabled Warburton to detail at some length Newton's mistakes in deducing the time of Sesostris from the genealogies of the gods and heroes in Greek mythology.[50]

Warburton considered these instances as false premises that engendered erroneous conclusion. He next turned to erroneous conclusion that generated false premises. Thus, he noted that Newton accepted the claims of Diodorus Siculus that Osiris abolished human sacrifices, built the city of Thebes, and conquered many nations. Yet Diodorus also stated that Osiris invented the cultivation of wine, outlawed cannibalism, and gave Egypt its first legal system, all of which Newton rejected, thereby exhibiting a penchant for selectiveness. More serious still, Warburton wondered, how could Newton assert that Vulcan lived through the Trojan War and beyond, furnishing the Egyptians with arms—for which they deified him—when Homer as well as Scripture attest to the far greater antiquity of arms? How could Newton claim that Egypt was furnished with horses only after Sesostris conquered Ethiopia, when the Bible is explicit on Pharaoh's pursuit of the Israelites with chariots centuries earlier?[51] Warburton proceeded to pound Newton on matters such as the division of land in

[48] Warburton, 1738–1741, vol. 2, pp. 208, 214–15, 221–22.
[49] Ibid., pp. 229–43.
[50] Ibid., pp. 243–55.
[51] Ibid., pp. 255–68.

Egypt, the late invention of letters there, and the alleged thinness of the Egyptian population, before concluding: "by the casual Confounding the distinct Actions of *Osiris* and *Sesostris* with one another, each came to be, at the same Time, the *Inventor* and *Perfecter* of the Arts of Life." Newton should have known better, for such alleged simultaneity contradicts "the very *Nature of Things*." More intolerable still, in so doing Newton made Sesostris' extraordinary age "distinguished from all others by an *inseparable Mixture* of *Savage* and *Polished* Manners. Which is so unnatural, so incredible, so impossible a Circumstance, that were there only *this* to oppose against his System, it would be a sufficient Demonstration of its Falsehood."[52]

The *Divine Legation* met with a chorus of detractors, chief among them Arthur Ashley Sykes' 1744 critique of several key tenets, which also included a lengthy and valiant effort to defend *The Chronology of Ancient Kingdoms*. Newton "has shewn so very great Sagacity in adjusting and reconciling the fabulous Stories of the Antients," Sykes began, "that if he has not hit upon the *Truth*, he has made the whole so very *probable*." The *Chronology* conformed so well to nature and to Scripture, "that it is not easy to shake so compact and well united a Building." Certainly Warburton "has not done it, even supposing Sir *Isaac* to have been mistaken in his Chronology of *Egypt*."[53] This lackluster introduction was followed by highly rhetorical quibbles: Warburton often disparaged the same ancient annals that he had accused Newton of ignoring. Hence, if the Egyptian priests had deceitfully extended their annals tens of thousand of years, how could they be of any value for setting "the Time when any Man lived"? In fact, the Egyptians were no less crafty mythologers and storytellers than "the little Lying Greeks" that Newton relied on in order to interpret Egyptian history. Nor could Herodotus avail himself of any authentic ancient records, as Warburton claimed, for Cambyses carried away all of them when he conquered Egypt. Herodotus only recorded what he had learned from the priests, while subsequent historians relied on forged records.[54]

Not satisfied with his effort to "weaken" Newton's account of Egyptian chronology, Sykes charged, Warburton sought to undermine the very "Quality of the Evidence"—the testimony of poets and mythology—that Newton utilized for the elucidation of Greek chronology. Sykes demurred at Warburton's refusal to accept that a sagacious man might extract useful information from such records—by separating truth from myth—and dismissed such evidence as inadmissible. All that Newton sought to accomplish by so doing was to extract the number of generations and the time between events; on what grounds did Warburton base his claim that fables cannot yield such information? Was a reliance on fables wrong for Newton but legitimate for Warburton? As for the lack of identity between Osiris and Sesostris, which annalists on the matter should be believed? And what constituted the strong "collateral evidence" that, Warburton claimed, contradicted Newton's account?[55]

Disputes over Scripture chronology and dogma, however, were hardly the exclusive domain of the public debate over Newton's *Chronology* during the two decades

[52] Ibid., pp. 277–78.

[53] Sykes, 1744, p. 223.

[54] Ibid., pp. 238, 241.

[55] Ibid., pp. 251–53.

that followed its publication. Scholars of all stripes viewed Newton's intervention with a mixture of excitement and concern, envy and admiration, many deeming him an interloper who meddled in domains beyond his expertise. Typical in this respect was Thomas Hearne. In a hostile comment in April 1726, he suggested that in addition to being beholden to Robert Hooke and Christopher Wren for his science, Newton understood "not one bit of Classical Learning, nor can he, as [he heard], write Latin, but [was] beholden to others to do that for him"—even though his books "be only Mathematical Latin."

Hearne repeated such disparagement on other occasions. Reacting to Zachary Pearce's praise for Newton's chronology in his "Essay upon the Origin and Progress of Temples," Hearne objected: though Newton was a "very great Mathematician," in Chronology he merely "advanceth Paradoxes & new opinions." Indeed, not raised a classical scholar, Newton "must be at a loss for Reading, to know w^{t} the Antients delivered of such and such Affairs, and 'tis too late to begin Reading now at his great Age." Hearne likened Newton to another great mathematician, John Wallis, who, the antiquarian maintained, also proved himself a poor chronologer—evidence for which Hearne found in Wallis' inadequate contribution to the Oxford edition of Cyprian. The final salvo came in April 1730, after Hearne read a description of Newton's monument at Westminster Abbey. Newton "was certainly a very great mathematician, but had little skill in other Things," Hearne reiterated. "He had no classical Learning," and "little Religion" besides, and "therefore could not be a good Interpreter of Scripture. He was little versed in Antiquity. His Chronology is exploded by the most orthodox and judicious Readers."[56]

"'Tis but seldom that men are really universal Scholars," Hearne opined on another occasion, articulating a widespread consensus. Typical was Francis Atterbury's reaction to learning from reading Fontenelle's éloge of Newton's historical and theological *nachlas*. Atterbury wished "for the honour of our country that they may be as excellent in their kind as those he published," the exiled Bishop of Rochester noted wearily. Yet he feared that the case would turn out otherwise; Newton would "be found to have been a great Master only in that one way to which he was by nature inclined. It is enough for us poor limited creatures," he continued, "if we remarkably excel in any one branch of knowledge. We may have a smattering of more; but it is beyond the lot of our nature, to attain any perfection in them."[57]

Such a perception set the grounds upon which it became possible, and necessary, to criticize Newton. Having strayed beyond mathematical learning, lesser mortals forced to take cognizance of the *Chronology* could nevertheless challenge Newton as equals—correct his errors or outstrip him. Of course, technical chronology and astronomy lurked in the background. Yet scholars managed to get around the thorny issue of astronomy and focus on philology and history. Whiston, Bedford, and Shuckford had already singled out certain literary and historical fault lines in the Newtonian edifice. Now it was up to the "professional" scholars to address more fully the evidence and the conclusions raised by Newton. Appropriately, Hesiod led the way.

In 1728, Thomas Cooke published the first English translation of Hesiod, to which he prefaced a short historical introduction. At the time, Cooke was unaware of New-

[56] Hearne, 1885–1921, vol. 9, pp. 111, 243, 290; vol. 10, p. 412.

[57] Ibid., vol. 11, p. 263; Atterbury, 1783–1787, vol. 1, pp. 180–81.

ton's *Short Chronicle*, or chose to ignore it. In a brief section devoted to Hesiod's age, he understood the issue to be unproblematic. He cited Longomontanus' interpretation of certain lines of the *Works and Days* in conjunction with the Arundelian marble to place Hesiod thirty years before Homer. He further cited Herodotus' statement that Hesiod and Homer had flourished four hundred years before his time as sufficient to establish the age in which Hesiod lived. This, of course, was Newton's position, too. In July 1729, however, Cooke communicated to the *London Journal*—under the pseudonym Atticus—a brief essay that challenged several details of the *Chronology*. Cooke sought not "to strike at the Foundation of that Work," but only to elucidate "such Passages as appear like Spots on a refulgent Shield." One such spot, he held, was the assumption that Homer flourished a generation after the Fall of Troy, misguided on the grounds that 1. a generation hardly sufficed to account for the degeneration of body and strength mentioned by Homer; 2. Homer and Hesiod wrote "before the Greeks and Egyptians corrupted their Antiquities"; 3. each of Hesiod's celebrated four ages were meant to denote a generation.[58]

Cooke did not incorporate this material into the 1740 edition of his translation, nor did he revise the paragraph based on the testimonies of Longomontanus and Herodotus regarding Hesiod's age. Instead, he inserted a couple of pages that sought to mobilize against Newton a discovery made by Samuel Clarke in his edition of Homer. Clarke, wrote Cooke, came across a conclusive argument for the greater antiquity of Homer on the basis of the quantity of the word χαλοζ (fine).[59] Whereas Homer used the word more that 270 times, "and every place made the first syllable long," Hesiod "frequently makes it long, and often short." Hence, the two could not have been contemporaries—unless speaking different languages or separated by great geographical distance—for Hesiod took it for granted that "the liberty of making the first syllable of χαλοζ short" had become practice long after Homer. "This argument of Dr. *Clarke's*, founded on a single quantity of a word," Cooke marveled, "is entirely destructive of Sir *Isaac Newton's* system of chronology." Newton claimed that Hesiod flourished a generation after the fall of Troy, and was Homer's contemporary—an argument made "to support his favourite scheme of reducing all to Scripture chronology."[60] Noteworthy is that the argument, as we shall see below, had been rejected three years earlier by Robinson in his edition of Hesiod. Before discussing Robinson, however, we should analyze the earlier critique of Jean Masson.

Masson was a French emigré who, in the early years of the eighteenth century arrived in England, where his reputation for erudition and chronology procured him the patronage of Bishop Lloyd. Not everyone shared Lloyd's high opinion of Masson, however. The censorious Thomas Hearne thought him in 1710 "a dull, pretending, heavy, muddy-headed man." A year later he further reflected on that "vain Gentleman" who "pretends to confute the Errors of a great Number of Able Chronologers, & braggs of strange Discoveries."[61] This very penchant manifested itself in Masson's critique of the *Chronology of Ancient Kingdoms*. Newton, he sneered, made "a great shew" when setting up a new chronology on the basis of astronomy, preying on unskilled readers.

[58] Cooke, 1731, pp. 56–62.

[59] Homer, 1729–1732, vol. 1, p. 42.

[60] Cooke, 1731, pp. 56–62; Hesiod, 1740, pp. xxix–xxxii.

[61] Hearne, 1885–1921, vol. 3, pp. 21, 37, 133–34.

Yet any discerning reader would recognize the weakness of his argument and its reliance on certain "*suppositions* which are certainly very precarious, if not absolutely false." With this distinction in mind, Masson promised to expose "with all the impartiality imaginable" the chief among these suppositions, starting with the claim that Chiron and Musaeus constructed a sphere for the Argonauts' use.[62]

Like Hardouin before him, Masson charged Newton with grounding his case on an arbitrary reading of the phrase σχήματα ὀλύμπου in Clement's *Stromata*. He had gone astray by holding Chiron to be a practical astronomer—errors to which Fréret and Souciet also subscribed. The phrase, Masson argued, signified neither stars nor constellation, and should have been understood instead as a moral signifier: "Chiron was the first who taught mankind righteousness, *having shewed them how to be faithful in their promises, or oaths, the manner of religious sacrifices, and the beings that are in heaven.*" So, too, Clement expressly ascribed to Chiron the practice of medicine, not astronomy. Had Newton been more proficient in Greek, Masson chastized, and had he consulted accepted authorities, he would have avoided making such errors. A perusal of John Potter's edition of Clement's *Opera* (Oxford 1715) would have informed Newton that Orpheus was not Musaeus' disciple, but his master (Newton, of course, was aware of this). Likewise, Newton mistook the meaning of the verb ποιῆσαι; it denoted not "making a sphere" but "writing of it," or describing it. Even if Musaeus made a sphere, it did not follow that he constructed it jointly with Chiron, or that the latter had delineated on it the position of the constellations, or that they were delineated at the same time. Moreover, the names of the constellations indicate renown garnered by the Argonauts during and after the voyage; consequently, Chiron could not have been in a position to know it before they set sail. Equally misguided was Newton's determination that Eudoxus described Chiron's sphere. Might it not be more reasonable to conclude that Eudoxus, having studied astronomy in Egypt, described what he had acquired there?[63]

Nor was Masson done. He came close to charging Newton with falsifying his sources when citing Eudoxus and Hipparchus as stating that the colures passed through a star of the sixth magnitude in the back of *Aries*—designated ν by Bayer—which was simply not there: "We see here the force of prejudice, since such a man as Sir I. could represent things so strangely wide from what he found in the Authors he took for his guides." Similarly, Newton distorted Hesiod's phrase regarding Arcturus' rise "just at sun-set." These words are not in the text, Masson insisted—"neither expressly, nor indirectly"—and Hesiod was not even an astronomer. All that he intended was to impart to farmers practical information relating to the coming of spring: namely, that about sixty days after the winter solstice the rising of Arcturus in the evening signifies the arrival of the new season. But no astronomical exactitude was expected, as the rising of the star varied from year to year.[64]

More revealing than his scrutiny of Newton—which benefited from Whiston and Bedford's critiques—was Masson's admission that he composed it "at the desire of a

[62] Jortin, 1731–1732, vol. 2, p. 150.

[63] Ibid., pp. 151–58.

[64] Ibid., pp. 163–68. The remarks concerning the star ν reflect Masson's either having misunderstood or simply rejected Newton's methodological assumption that the colures must have been fixed with respect to stars and not simply at vague points within asterisms.

person, to whom [he owed] all the respect and duty imaginable."[65] The identity of that instigator cannot be determined. Nevertheless, enlisting experts to assist in a critique of Newton appears to have been common. Among them we find Thomas Robinson, who enlisted the scientific expertise of Joseph Atwell—FRS and Rector of Exeter College, Oxford—in order to dismantle the astronomical grounding of Newton's account of Hesiod's age.

Atwell covered the ground plowed by earlier critics but with greater thoroughness. Astronomical reasoning regarding Hesiod's age, he argued, would be expected to be accurate. The ambiguity of Hesiod's phraseology, combined with the primitive state of astronomy during his time, made it impossible to believe that the acronychal rising of Arcturus could be conventionally defined, or the true dates of the solstices determined. Nor was it clear that the event Hesiod described was actually observed in the poet's own time, or in his hometown, Ascra. Atwell proceeded to enumerate the difficulties involved in translating Hesiod's words into numbers. Should the solstice itself be included in the count of sixty days? Did the poet mean "true" or "apparent" acronychal rising of Arcturus? What exactly is meant by "shining" or "rise"? And did Hesiod have in mind the rise of Arcturus or of the constellation Bootes?

Having weighed words, Atwell turned to calculations. In so doing he concluded that in 870 BCE—the date, Atwell incorrectly assumed, Newton assigned for Hesiod's flourishing—the acronychal rising of Arcturus took place at Ascra when the sun was entering the 4th degree of Pisces, at evening or the beginning of the third day of March. The previous year, the winter solstice fell in the afternoon of December 29. Hence, the acronychal rising of Arcturus in the region of Ascra took place at evening on the 64th day after the completion of the solstice, or the 65th day if we count both extremes. So how could Newton claim that, according to Hesiod, the acronychal and true rising of Arcturus occurred that year? "Could that eminent man, whose name should always be honored, have been mistaken (let no one attack him unfairly)? Or did he undertake his calculation by a different method, for a region farther to the north, computing sixty days from the middle of the season of winter, and leaving out the extremes from the total? Since the author himself is silent on this subject, [he could] say nothing with certainty." Atwell discussed several additional phenomena mentioned by Hesiod, concluding that from the phenomena reported to us by the poet, nothing could be inferred which conflicts with his age according to the Parian Marble—ca. 942 BCE.[66]

Robinson took over from Atwell in order to examine the relation of Hesiod's era to the era of the Trojan War and to Homer's age. Regarding the former, Robinson saw no reason to revise the widespread and established opinion that pinned the fall of Troy to 1200 BCE, about 300 years before Hesiod. Such dating is supported by Homer's attribution of superior strength to the men of the Trojan War—compared with his contemporaries—and one could not accept that such a remarkable diminution of human capacity might occur in less than two centuries. Nor did Homer know the number of the ships, or the name of their captains, information he would have surely known had he lived within a generation of the sack of Troy—a clever borrowing of Newton's argument that oral tradition could survive no longer than a century.

[65] Ibid., pp. 170–71.

[66] Hesiod, 1737, pp. xxvii–xxx. In manuscript Newton dated the event to 880 BCE.

From these arguments, Robinson inferred that no small space of time intervened between the Trojan War and the century of Homer, and this applied with equal weight to Hesiod who, Robinson argued later, was Homer's contemporary, or even his elder.[67]

Robinson accounted for the divergence of his opinion from the hypothesis held by Newton—who, lopping a full three hundred years from ancient chronology, did place the flourishing of Hesiod at 870 years before Christ, a mere generation from the Trojan War—by noting four principal points. First, Newton erred in supposing γενεᾶς (*generations*) to equal about thirty years. True, Herodotus and others confirmed that three generations corresponded to a hundred years, but Hesiod's "generations" must be considered as an indeterminate period. For example, Homer stated that eight generations spanned between Deucalion and the Trojan War. If we compute, with Newton, three generations for a hundred years, then the heroic age lasted 266 years. Nor did Deucalion live before the Golden Age—as Newton would have it—not only because Apollodorus and Proclus attest that Zeus sent that flood to destroy the men of the Brass Age, but because Newton himself related that iron was discovered "ten years after Deucalion"—and iron was only discovered toward the end of the Brass Age.[68]

Equally faulty was Newton's translation of γεινόμενοι πολιοκρόταφοι τελέθωσιν to mean that the age of men who fought in Troy would end when they "grew hoary and dropt into the grave." To "become gray" was, Robinson averred, actually a metaphor for the shortness of life, and the correct translation should have been "when they turn gray being scarcely born." Hesiod, therefore, did not live in the generation after Troy—just one example of the difficulties Newton encountered when attempting to establish mythic periodization. In particular, Robinson made merry with Newton's predilection to establish the age of a "god" (Jupiter) from the known age of his "sons." Newton, he pointed out, failed to take into account that flatterers of kings often dubbed them "sons" of the ancient hero, even when living centuries apart. Using Newton's line of reasoning, one might lengthen and shorten history at will. Suppose Robinson wished to lengthen Roman history. He might argue that the hero who went by the name of Mars was Sesostris, who flourished ca. 1000 BCE; but Mars was Romulus' father; so Rome must have been founded earlier than 900 BCE. Conversely, he could just as easily argue that Romulus flourished ca. 600 BCE and, therefore, his father (Mars = Sesostris) was only a generation older.[69]

Hesiod himself helped determine his distance from the Trojan War, Robinson pointed out, when he reported that his father migrated to Ascra from Cuma. That town was founded only during the third Aeolian migration—or 150 years after the Trojan War—which would make Hesiod flourish about fifty years later. How could Newton, then, claim that Hesiod flourished so many years before the migration? Finally, Robinson laid down the (literary) grounds that convinced him that Hesiod and Homer were virtually contemporaries. *Inter alia* he rejected Clarke's contrary thesis as inconclusive, owing to the absence of any other contemporary poets whose writings might have shed light on the usual practices of that century.[70]

[67] Ibid., pp. xxx–xxxi.

[68] Ibid., pp. xxxii–xxxiii; Newton, 1728b, p. 14.

[69] Hesiod, 1737, pp. xxxiv–xxxvii; Newton, 1728b, p. 32.

[70] Hesiod, 1737, pp. xxxviii–xlii.

The intention to synchronize sacred and profane history that informed Newton's *Chronology* also irked other classical scholars. In 1741, while still an undergraduate at Trinity College, Cambridge, the twenty-year-old Thomas Francklin published a translation of Cicero's *On the Nature of the Gods*, to which he appended a brief "Enquiry into the Astronomy and Anatomy of the Antients." Francklin was keen to prove the antiquity and proficiency of ancient astronomy. He believed that the Greeks had turned astronomy into a science more than a millennium before Christ, and he was equally confident in the verity of the account regarding Calisthenes importing from Babylon observations made two thousand years before Alexander the Great. Such ancient observations were doubted, he believed, only because they "may break into prevailing System of Religion." On such grounds, he objected to Newton's *Chronology*, the design of which, he wrote, "was to reduce all Chronology to Scripture Chronology, and thereby to establish, in that Point, the Authority of those Books. On this weak Hypothesis of his own he proceeds to knock down all former Chronology; and by this single instance of his wild Inference from the Colures moving over the back of *Aries* we may see how the most considerable Men in particular sciences may be blinded by their Favour to particular Systems." To substantiate his dismissal of the *Chronology*, Francklin availed himself of Whiston's refutation of Newton, the first to argue the impossibility of determining through which part of Aries' back the colure passed in Chiron's time.[71]

The focus on specific aspects of Newton's system for evaluation, both positive and negative, informed much of the eighteenth-century discussion of the *Chronology*. In many instances scholars simply noted the variance between their own and the great man's narratives and proceeded about their business. Consider Temple Stanyan's *Grecian History*. When in 1739 he embarked on a revised edition of the work that had appeared three decades earlier, he felt compelled to take cognizance of *Chronology*. Stanyan expressed admiration for the "infinite Pains" Newton had taken in setting out the *time* of kings and events, while noting that Newton did not undertake "to determine them." So, too, he drew comfort from Newton's acknowledgment of the difficulties involved in correctly fixing the chronology of the fabulous ages—and his choice to leave such matters "to be further examin'd"—as it allowed Stanyan to follow suit. Thus, at various junctures Stanyan simply noted Newton's opinion, before proceeding with traditional history. Case in point is his discussion of the length of the reigns of the early kings of Sicyon. Stanyan interrupted the narrative to present Newton's rule regarding regnal years, only to resume the narrative "according to the common receiv'd Accounts of these Kings."[72]

Partly through such debates, partly through the general dissemination of his scientific ideas, Newton's chronological ideas became part of the Enlightenment cultural landscape. Newton's most important early scientific popularizers, Francesco Algarotti and Voltaire, also proved staunch supporters of the *Chronology*. In a digression prefacing the fourth dialogue of *Sir Isaac Newton's Philosophy Explain'd for the Use of the Ladies*, for example, Algarotti praised the Northern regions "where the Liberty of the People is made compatible with the Superiority of the Nobles, and the Authority

[71] Cicero, 1741, pp. 273–77.

[72] Stanyan, 1739 , vol. 1 "Advertisement,"; pp. 23–4.

of the Sovereign," before offering a paean to Newton's reordering of past time: "The Chaos of Chronology and History has from these Observations received its Light and Order. Sir *Isaac Newton*, that divine Philosopher, who may be regarded as the Founder of human Knowledge, has from Observations drawn chiefly from the ordinary Course of Nature, ranged historical Facts in a certain Series, by joining Epochas, which the Ignorance or Pride of Mankind had set at a great Distance from each other, in the same Manner as a judicious Observation had united the Boundaries of the Earth in Geography."[73]

For his part, Voltaire gave an admiring account of the *Chronology* in the *Letters on England*. Newton may have "toyed" with chronology "for relaxation," and the book may have proved to be "more adapted to the Capacity of the human Mind," yet it exhibited the same "creative Genius" that informed all his researches—not surprising, as Newton was destined "to change the Ideas and Opinions receiv'd by the rest of Men" in whatever domain he investigated. Moreover, as one accustomed "to unravel and disentangle Chaos, Newton was resolv'd to convey at least some Light into that of the Fables of Antiquity which are blended and confounded with History, and fix an uncertain Chronology." Voltaire summarized Newton's contributions as belonging to two kinds: the ordinary course of nature—the rule regarding regnal years—and the argument from astronomy. The former helped determine the correct lengths of eras lacking fixed dates, while the latter furnished scientific evidence for shortening Greek history by about 500 years. "I don't know whether this ingenious system will be favorably receiv'd," Voltaire concluded with thinly veiled irony, "and whether these Notions will prevail so far with the learned, as to prompt them to reform the Chronology of the World." Perhaps they would consider it "too great a Condescension" to permit one man the glory of reforming the sciences as well as history. "This would be a kind of universal Monarchy, which the Principle of Self-Love that is in Man, will scarce suffer him to indulge his Fellow-Creature."[74]

Voltaire incorporated Newton's time line into his contemporaneous *Essay on Epic Poetry*, where he accepted Newton's claim that Hesiod and Homer were contemporaries—living in the generation that followed the fall of Troy—so that the "historic Homer had spoken as a boy with men who had seen Ulysses, Menelaus, and Achilles face to face." Voltaire introduced a more critical tone to a chapter devoted to the precession of the equinoxes which he added to the revised edition of the *Elements of Newtonian Philosophy*. The fact that Hipparchus—who flourished ca. 60 BCE—was the first to make the discovery, Voltaire reasoned, attested to the inferiority of earlier Greek astronomy. A single author of the second century credited Chiron with fixing the beginning of spring when determining that the earth's ecliptic intersects with the equator in the fifteenth degree of Aries. Five hundred years later, he continued, Meton and Euctemon discovered that at the beginning of summer the sun entered into the eighth degree of Cancer, and Newton established part of his chronological system on this observation. Nevertheless, objected Voltaire, had Meton and Euctemon found such a "palpable" discrepancy, surely they would have been able to discover the precession of the equinox—for that, simple mathematics would have sufficed. Since the

[73] Algarotti, 1739, vol. 2, pp. 17–18.

[74] Voltaire, 1733, pp. 155–65, quotes at pp. 155–56, 164.

phenomenon was not discovered until Hipparchus, Voltaire doubted whether Chiron knew as much as was reported of him. Only in the aftermath of the discovery was it believed that he had fixed the equinox; it was imagined that he did so because he *should* have done so. This observation appeared to Voltaire to "shake," to a certain degree, Newton's chronology. (Newton could have argued that Meton might not have thought that Chiron set the equinox at mid Aries.)[75]

Notwithstanding his doubts, Voltaire continued to promote the *Chronology* for years by turning from stars to men. In particular, he repeatedly expressed utter confidence in Newton's regnal rule. He regarded it as "indubitable" in elective kingdoms, such as ancient Rome, and applicable to Chinese rulers before Hiao. Likewise, he found confirmation for the rule in the annals of the Holy Roman Emperors, and believed it would be found equally true in states subjected to frequent revolution.[76] Following the publication of Fréret's posthumous *Défense de la Chronologie* in 1758, Voltaire dismissed rumors imparted to him by a correspondent that "Fréret has pulverized Newton." These rumors, he scoffed, could surely be traced to the ill-informed. He himself had read Fréret's initial response to Newton, and found that the author "did not know what he was talking about." He doubted "whether he has learned more since."[77]

The controversies over Newton's new system took shape as polite society gradually lost interest in the minutiae of erudition, particularly of the sort exemplified by chronology. Charles de St. Evremond helped set the tone as early as 1671 when he sneered at the "abundance of Criticks" in his day, few of whom revealed themselves to be "good Judges" as well. Learned men, he complained, "rack their brains to restore a Reading, which is not mended by the restitution," all the while blissfully ignoring "what's really worth knowing." They are as incapable of grasping the "Delicacy of a Sentiment," as they are unaffected by the "Fineness of a Thought." He continued: "They may succeed well enough in expounding Grammarians, who applied themselves to the same study, and whose genius was the same: But they can never hit that of a polite, well-bred man among the Antients, because theirs is diametrically opposite to it. In History, they neither mind Men, nor Affairs: They lay the whole stress on Chronology; and for the Date of a Consul's Death, will neglect the knowledge of his Character, and of the transactions during his Consulship." This blistering attack by an advocate of the "man of taste" gained support a generation later with the erudite Jean Le Clerc. In his review of St. Evremond's *Oeuvres Mélées*, Le Clerc wrote: "If our Grammarians knew how to reason and write after this manner on the Antients, they would make everybody envious to study them. But their learning consist of only the knowledge of Words, Customs, and, at most, of Chronology, joined by a blind admiration of all they read." Hardly surprising then that such learning "gives Gentlemen a disgust for Literature."[78]

The obsessiveness with which the learned busied themselves with trivia also informed Sir William Temple's "Essay on Ancient and Modern Learning." Noting learned disagreements over the antiquity of the Assyrian monarchy, Temple dis-

[75] Voltaire, 1992, pp. 491–93.

[76] Voltaire, 1877–1885, vol. 18, p. 343; vol. 11, p. 166; vol. 13, p. 613.

[77] Manuel, 1963, p. 179.

[78] St. Evremond, 1728, vol. 2, p. 71; *Bibliothéque choisie*, vol. 9 (1706), p. 328.

missed chronology as irrelevant. Since numerous scholars had already "spent their whole time and pains" in a futile attempt to reconcile sacred and profane chronology, Temple scarcely expected "to see it done to any purpose" in the future.[79] The essay famously launched the "ancients and moderns" controversy, in which chronology became directly involved in determining the genuineness of the *Epistles of Phalaris*. Repeatedly Bentley repudiated errors of chronology which, he was certain, sufficed to damn the book. Thus, for example, he summarily rejected the supposition that Lucian authored the book simply because the latter would not have "forfeited all learning and wit," or committed those "gross blunders in Chronology, and that wretched pedantry in the Matter," which pervaded *Phalaris*. More generally, in discussing the authorship of the epistles of Euripides, Bentley insisted that "in a Disquisition of this nature, an inconsistency in Time and Place is an argument that reaches every body. All will cry out, that *Phalaris*, &c. are spurious, when they see such breaches upon Chronology."[80]

Francis Atterbury and his Christ Church friends who rose to Phalaris' defense disagreed. As far as they were concerned, the age in which the epistles were written proved "a very inconsiderable Point; and which a wise man would grudge the throwing away a weeks thought upon." Bentley's objections to particular letters, which were "grounded upon *Chronology*," they dismissed, resorting variously to the obscurity of ancient times and to canny misrepresentations of Bentley's argument. They also invoked the above-mentioned disparagement of chronology by St. Evremond and, for good measure, insinuated heterodoxy against Bentley: to argue that Phalaris could not have been the author of the letters because of anachronisms is the same kind of evidence used by Spinoza and others to "ruine the Authority of *Moses*'s Writings." Bentley should have been careful not to promote such arguments "in a Doubting Age."[81]

In response, Bentley confronted head-on the cutting dismissal of his enterprise as "a very inconsiderable point, which a wise man should grudge the throwing away a week's thought upon." Those whose interests laid elsewhere, he acknowledged, would find such argument unappealing. Nevertheless, the genuineness of *Phalaris* was of singular importance, and to devalue its significance simply "because it does not suit to one's own studies, is to quarrel with a circle because it is not a square." Later he sneered at his critics who, though many of the letters were "refuted from chronology, would still *comfort* himself with the cold hopes, *that the rest may be genuine*." More to the point, he reiterated that "an inconsistency in time and place is an argument that reaches every body. All will cry out, that Phalaris, &c. are spurious, when they see such breaches upon chronology."[82]

The changing standards of erudition and taste help us to understand the contrasting reactions to chronology by Viscount Bolingbroke and Edward Gibbon. Bolingbroke embarked on the study of history and chronology while in exile, as he was approaching forty. His guides were Pierre Joseph Alary and Lévesque de Pouilly, who

[79] Temple, [1814] 1968, vol. 3, p. 499.

[80] Bentley, 1697, pp. 64–65, 122–23; Bentley, [1836] 1971, vol. 2, pp. 172, 214.

[81] Boyle, 1698, pp. 24, 118–22, 227–28.

[82] Bentley, [1836] 1971, vol. 1, pp. lxix, 78; vol. 2, p. 214.

furnished him with books and engaged him in heated discussions. Bolingbroke probably received a copy of Newton's "Short Chronicle" from the abbé Conti. The progress he made was sufficiently impressive for Voltaire to praise him in 1722 as joining perfectly English erudition with French politeness, proving himself as knowledgeable in ancient Egyptian history as in English history. Nevertheless, by this date the intricacies and tedium involved in such study dampened his enthusiasm. "I have very near done with all inquiries into remote antiquity," he informed Brook Taylor in November 1721. His only wish had been to view "the foundations of those historical and chronological systems, which have been erected with so much learned pains in our western world." By now he realized that he "who cannot content himself to employ his time about consequences, drawn from principles evidently begged, ought not to employ it in this kind of erudition." Two years later, Bolingbroke reiterated his disinclination to proceed beyond examining the foundations of chronology and ancient history "a fin de sçavoir à quoi m'entenir." Having done so he felt "no more desire to pursue this study any further than to be a proficient in Judicial Astrology. Who can resolve to build with great cost and pain, when he finds, how deep soever he digs, nothing but loose sand?"[83]

His position with respect to chronology hardened with age. In 1731, he confided in Jonathan Swift that the precarious nature of Marsham's authority derived from the precarious nature of the sources Marsham relied on. Nor did Bolingbroke believe it was possible to erect other chronological systems without falling into the same pitfalls; "the defect is in the Subject not in the writer." The building blocks of chronology, he explained, "are disjointed & broken," an inevitable consequence of the ravages of time and "the unfaithfulness of those who have transmitted them down to us"—Eusebius in particular. Indeed, so fragmentary and confused were such records, that "a man may with tolerable ingenuity ... prove almost any thing by them." He himself had once composed a dissertation wherein he proved, "by the same set of authoritys, that there had been four Assyrian monarchys, that there had been but three, that there had been but two, that there had been but one, & that there never had been any."[84]

Noteworthy is that Bolingbroke all but exempted Newton from censure. He did lament on one occasion Newton's losing "himself in the vague probabilities of chronology, after having pursued with so much success the certainty of mathematical demonstration." But when he expressed "a thorough contempt" for "all the researches into antiquity, for all the systems of chronology and history," that are owed to the "immense labors" of Scaliger, Bochart, Petavius, Ussher, and Marsham, he avoided including Newton among the chronologers—an omission that was not lost on one of his critics. Bolingbroke again justified his position by pointing out that given the "moral impossibility" of chronologers coming up with new materials, they are left to combine and recombine the fragments at their disposal, in the most arbitrary manner, leaving "no liberty untaken." Ultimately all their systems amounted to "many enchanted castles; they appear to be something, they are nothing but appearances: like them too, dissolve the charm, and they vanish from the sight." Hence, he would

[83] Bolingbroke, 1808, vol. 2, p. 456; vol. 3, pp. 21–31; Voltaire, 1968–, vol. 85, p. 143; Taylor, 1793, pp. 126–27, 135–36.

[84] Swift, 1963–1965, vol. 3, pp. 486–87.

rather "make as many anachronisms as a Jewish chronologer, than sacrifice half [his] life to collect all the learned lumber that fills the head of an antiquary."[85]

The passion that originally inflamed Bolingbroke flared again a generation later in Gibbon. In his autobiography Gibbon recalled how a passion for history had taken hold by the time he turned sixteen in 1753. And while his "vague and multifarious reading" could not teach him how to properly think or write, the one principle "that darted a ray of light into the indigested chaos, was an early and rational application to the order of time and place." He continued:

> from Strauchius I imbibed the elements of chronology: the Tables of Helvicus and Anderson, the Annals of Usher and Prideaux distinguished the connection of events, and I engraved the multitude of names and dates in a clear and indelible series. But in the discussion of the first ages I overleaped the bounds of modesty and use. In my childish balance I presumed to weigh the systems of Scaliger and Petavius, of Marsham and Newton, which I could seldom study in the originals; and my sleep has been disturbed by the difficulty of reconciling the Septuagint with the Hebrew computation. I arrived at Oxford with a stock of erudition, that might have puzzled a doctor, and a degree of ignorance, of which a schoolboy would have been ashamed.

Evidence for his early preoccupation can be found in a youthful essay written in French—"Critical remarks on the new chronological system of Newton"—that Gibbon composed in Lausanne in January 1758. Newton's name "stirs up the idea of a profound, luminous, and original genius," Gibbon exclaimed. "His system of chronology alone suffices to ensure him immortality." Experience and astronomy were Newton's threads and, combined with a reliance on the "purest monuments of antiquity," occasioned a "host of most singular consequences." Newton's results varied considerably from his predecessors', and were hotly contested both in England and France. Gibbon proposed to comment on several key points of the system, offering "such remarks as may show a simple love of truth: I may contradict myself: here I might appear as the ardent defender of the system, there armed for its ruination." The more favorable parts of the system included Newton's synchronization of one of the most contested issues in ancient literature: Virgil's embracing the gross anachronism of making Dido and Aeneas lovers, when traditional chronology set them 300 years apart. Newton's setting the fall of Troy to 904 BCE, however, and Dido's building of Carthage twenty years later, ensures that even if the two never met—let alone loved—they were nevertheless contemporaries. Equally propitious, Newton's chronology resolved the tradition that Pythagoras served as Numa's tutor—whereas traditional chronology kept them a century apart. If Numa ascended to the throne only in 610 BCE, as Newton claimed, then the anachronism disappears. (Gibbon discounts Fréret's dating of Pythagoras' birth to 608 BCE, preferring to accept the tradition that in 612 BCE the philosopher had discovered that the evening and morning stars are one and the same. He also preferred to attribute the great deeds of Numa to the influ-

[85] Bolingbroke, 1967 [1844], vol. 3, p. 334; vol. 2, pp. 175–76; Leland, 1753, p. 21.

ence of Pythagoras.) Other significant discoveries by Newton, albeit more problematic, included the identification of Sesostris and Sesac, and the determination that four generations spanned between Cadmus and the fall of Troy.

Turning to criticism, Gibbon made explicit that when he praised "that original spirit which shines in the works of Newton," he had in mind only the first two chapters. The others were different. Not that they lacked many curious remarks, but "the soul of Newton was made to found and to destroy empires and not to pick at minute details." Insofar as Greek history was concerned, Gibbon raised a curious objection to Newton's regnal rule, which he probably learned from Shuckford. Might it not be possible that the rule applied only to "historical" times, not to remote antiquity when, according to the most respectable monuments, the lives and reigns of men were far longer than at present? Thus, according to traditional chronology, Phoroneus was a contemporary of Jacob, Cecrops of Moses, and Cadmus and Danaus of Joshua. The Hebrews during that time lived for a hundred years and more, and the same should hold for their Greek contemporaries. Perhaps only during Solon's time was the human life span set to seventy. Other problematic issues raised by Gibbon included Homer's age and a variety of topics relating to Roman history. Gibbon concluded by expressing the hope that in Fréret's forthcoming posthumous chronology, "the questions in dispute will receive every degree of illumination to which they are susceptible through the combat of these two great men."[86]

In the event, the publication of the *Défense de la chronologie fondée sur les monumens de l'histoire ancienne, contre le système chronologique de M. Newton* disappointed him. "Prejudiced greatly in his favour," Gibbon wrote, he "eagerly ran over" Fréret's reply, only to find that "it by no means answered my expectations. I see nothing new in that piece, if we except the principles of a new theology and chronology, which, however, we already possessed; some defective and inconclusive genealogy; a few minute researches into the chronology of Sparta, an ancient system of astronomy, which I do not well understand, and the elegant preface of M. de Bougainville, which indeed I peruse every time with additional pleasure."[87] Small wonder that, like Bolingbroke, Gibbon opted to steer clear of deep antiquity—albeit without disparaging chronology and ancient history—and proceeded to channel his talents to more recent, and better documented, history. Given that *The Decline and Fall of the Roman Empire* treats the history of a period to which Newton devoted little attention, it is hardly surprising that Gibbon found it unnecessary to discuss Newton's system. Nevertheless, a few references attest to the continued hold that Newton's ideas exerted over the historian. Thus, in the first volume Gibbon notes in a footnote that while Varro's received view fixes the foundation of Rome to 754 BCE, "so little is the chronology of Rome to be depended on, in the more early ages, that Sir Isaac Newton has brought the same event as low as the year 627." Elsewhere, he noted with some surprise that no fewer that sixty emperors ruled over Byzantium in the course of the six hundred years that spanned from Heraclius to Isaac Angelus—an average of ten years per reign—"far below the chronological rule of Sir Isaac Newton, who, from

[86] Gibbon, 1814, pp. 152–69.

[87] Gibbon, 1764, pp. 122–23n.

the experience of more recent and regular monarchies, has defined about eighteen or twenty years as the term of an ordinary reign."[88]

Gibbon's dismissal of Fréret's novelty reflects a generation of criticism of Newton's *Chronology* on the Continent. Over thirty years had elapsed since the publication of Fréret's original critique of Newton, and the intervening years had witnessed a torrent of publications that took issue with, or defended, the *Chronology*— testimony to continued widespread interest in Newton's work, not to mention deep concern over its conclusions. It would hardly be an exaggeration to assert that several members of the Académie had targeted Newton in their specialized researches into ancient history, albeit rarely by name. Fréret himself had held Newton constantly in sight, again avoiding direct reference except for the occasional jibe. Case in point: Fréret wove into his analysis of the Bellerophon myth—which Newton had not discussed—his ever present mantra: that acceptance of Newton's abbreviated chronology necessitates utter rejection of all ancient Greek genealogies. Antoine Banier, too, avoided engaging Newton directly in papers that specifically concern the Argonautic expedition. He reserved his direct critique of Newton's dating of the expedition to the *History of Fables* (1738), wherein he cited Hardouin as having proved "unanswerably, that *Chiron* was never considered by any of the antient Authors . . . but as a physician." Banier further argued that the asterisms mentioned by Clement of Alexandria attest, at best, to a rudimentary knowledge of astronomy, which would have not sufficed for Chiron to construct a sphere.[89] Such criticism of Newton did not go unchallenged within the Académie; de la Nauze continued his rear-guard defence of the Englishman throughout the 1730s, with Fréret rising instantly to rebut him.[90]

One example of the veiled stance against Newton's chronology within the Académie must suffice. In 1751 the magistrate Charles de Brosses read a paper on the monarchy of Nineveh, the foundation of which (by Nimrod-Belus) he dated to 2263 BCE. Newton went unmentioned, except for the concluding paragraph. There de Brosses reflected on the seductiveness of errors committed by great men, attributing it to the high degree of probability they manage to confer on their paradoxes. For example, though Newton's system did not conform to historical truth, it could have been produced only by a powerful genius who, nonetheless, approached history as a "*physicien*" eager to subject all ancient records to his calculations. Absorbed as he was in his mathematical studies, Newton had failed to acquire detailed knowledge of ancient history. Neither did he take contradictory results into account. For just these reasons, de Brosses concluded, "the most ingenious chronological system" completely lacked foundation, for Newton "pretended to contravene the entire narrative of antiquity, and to restrain the duration of empires and of our historical knowledge, more than any before had ever done."[91]

[88] Gibbon, 1994, vol. 1, p. 212; vol. 3, p. 83.

[89] Fréret, 1733 , pp. 88–89; Banier, 1736 , p. 95; Banier, 1739–40 , vol. 3, pp. 541–44.

[90] See, for example, the heated debate between the two over the chronology of Pythagoras, during August and September 1738. *Histoire et Mémoires de l'Académie Royale des Inscriptions et Belles Lettres*, vol. 14 (1743), pp. 375–504 (second pagination).

[91] Brosses, 1761, p. 82n.

An agenda to dismiss Newton's *Chronology*, and his competence as a historian, characterized many Continental critiques. Consider the *Réflexions critiques sur les histoires des anciens peuples* (1735) by Etienne Fourmont. This academician, and professor of Arabic at the Collége de France, had elevated Euhemerism to new heights in order to systematically force all ancient myths to conform with Scriptural narrative. Convinced that the surviving fragments of Sanchoniathon offered a key to deciphering the origins of idolatry, Fourmont developed an imaginative theory about the process through which the deification of Old Testament patriarchs had been initiated by the Phoenicians, from whence it migrated to the Egyptians, to the Greeks and to the Romans. Thus, for example, he equated Noah with Sydak, Terah with Ouranus, Abraham with Kronus and Saturn, Sarah with Rhea, Isaac with Sadid or Jupiter, Esau with Osiris and Bacchus, and Jacob with Typhon. Furthermore, in an effort to make Chinese dynasties conform to Scripture, he also embraced the chronology of the Septuagint. Understandably, Fourmont proved hostile to Newton's *Chronology*. Great philosopher though he was, Fourmont wrote in the introduction to the *Réflexions*, Newton was a geometer rather than a historian. He clung to generalizations that failed to lead him toward his goal, while every one of his novelties suffered either from too much reasoning, or too little.[92] In the body of the work Fourmont made only a handful of references to Newton by name, further signaling his refusal to engage with Newton's system. Three years later, Alphonse Des Vignoles proved even more dismissive. Ignoring Newton entirely in his massive *Chronologie de l'histoire sainte*, des Vignoles articulated his utter contempt for the Englishman's system in a manuscript critique. Here is a sampling of his rhetoric:

> It is no less true, and I am not afraid to declare it, that the geometrical genius does no way appear in Sir Isaac's new *Chronology*. The principles upon which he has founded it, are neither demonstrated, nor incontestable, nor admitted to be true. The materials have there no connection, nor have I been able to discover any method. It is nothing but a confused system, loaded with an infinity of conjectures, heaped upon one another, without order, and repeated even to tiresomeness.... In one word, it is an edifice built of stones, ill proportioned, and ill bound together, without any determined plan; and this upon an ancient and solid building, which people imagine they can overthrow with the wind of their mouth, as if it were a castle built with cards.[93]

Clearly, for these as for earlier critics—Bedford, Whiston, and Shuckford in England, Souciet, Hardouin, and Fréret in France—opposition to the *Chronology* stemmed from the determination to defend their own schemes. Souciet freely admitted the role of self-interest in 1726 when protesting the innocence of his intentions: he sought not to attack Newton, only to defend his own notions regarding the age of the world,

[92] Fourmont, 1735, vol. 1, pp. xvii–xviii.
[93] *Nouvelle Bibliotheque Germanique*, vol. 18, part 1 p. 421, in Steuart, 1805, pp. 324–25.

while vindicating the proofs for the epochs underlying his system.[94] Equally important, as Manuel pointed out, was the rare opportunity presented "to chronologists and antiquarians to engage the greatest man on earth in intellectual combat. If defeated, there was no disgrace; if victorious—and the weakness of the Newtonian ramparts was soon revealed—the triumph was the more for the little men who had challenged the giant of the age." Fontenelle had already noted a motivation in his éloge, when he suggested that Souciet and Fréret had "moved the more rapidly [against Newton] only because they might thus have the fame of [engaging] such an adversary. They will find others ready to take his place."[95] Such a backhanded compliment became a favorite trope among Newton's detractors. Having dismissed the *Chronology* as contrary to reason and historical evidence, Fourmont deigned to credit its author with a single "positive" contribution: Newton's errors furnished savants of the first order, such as Souciet and Fréret, with the occasion to embark on their more exact researches. Half a century later, Juan Andrés wrote along similar lines. Irrespective of the matter of right or wrong, he asserted, the ensuing disputes helped elucidate aspects of astronomy and chronology. Had Newton pursued erudition more seriously, the ex-Jesuit contended, Newton would have undoubtedly excelled in chronology as well. As it happened, however, it remained to Fréret to become the Newton of ancient chronology.[96]

In addition to partiality in the cause of self-interest, many critics found it difficult to seriously engage with the *Chronology* because they had publicly committed themselves to an anti-Newtonian stance on the basis of the "Short Chronicle." Previous commitment made it awkward for them, if not downright impossible without losing face, to modify their criticism following the publication of the complete *Chronology*. It could have been otherwise, as the example of James Logan of Philadelphia proves. At first he differed little from the French. The Colonial mathematician did avow great veneration for Newton, even though he had once entertained "little resentments" on account of Newton's ill treatment of Flamsteed and Leibniz. Upon reading the "Short Chronicle," Fréret's critique, and Newton's rebuttal in the *Philosophical Transactions*, however, Logan found himself "deeply affected": Newton, "had already acquired Immortal fame in subjects of which he was an absolute master, wherein he had proceeded with so much caution that he would admitt of no Hypothesis whatever." Hence, Logan found it "astonishing" that Newton "should hazard his reputation" in a study "wherein he had no advantages ... above other men." Indeed, one would have expected that in shortening ancient history by some 400 years, Newton would only proceed on a secure foundation that "Could not justly or rationally be controverted." Yet, as far as Logan could determine, "nothing [could] be more imaginary or groundless." The idea that Chiron "ever fixed the Aequinox" in the manner Newton described was "perfectly Chimaerical," while Newton's reliance on Geminus and Meton in order to establish the era of the Argonauts appeared to Logan unwarranted. Even "more astonishing" was Newton's shortening the era of Rome's foundation by 130 years, in

[94] Souciet, 1726, preface. Souciet dated the Creation to 4020 BCE; Ninus' reign in Assyria 2162 BCE; Chiron's formation of the Zodaical signs for the Argonauts at 1470 BCE; and the fall of Troy at 1388 BCE.

[95] Manuel, 1963, p. 168; Hall, 1999, p. 71.

[96] Fourmont, 1735, vol. 1, pp. xvii–xviii; Andrés, 1785–1822, vol. 3, pp. 533–34.

view of the infinite labors taken by Varro and Cato to fix it. "What then must be thought of one Starting up above 1800 years after, when all their old monuments are utterly destroyed?" he asked. Like Newton's French critics, Logan thought it prudent to wait for the complete chronology before passing final judgment. Yet, like the French again, he felt confident to conclude on the basis of what he had read that there existed "more than enough to convince us that it is never to be Defended." Newton's system "may shew the sharpness of his thought, yet it must necessarily tarnish that Glory which, had he gone off the Stage at the age of 70, would ever have attended his name."[97]

However, in June 1728 Logan began reading the *Chronology* and found himself transported. He "seldom mett wth any thing of the Kind that gave [him] a more sensible pleasure or more solid Entertainment." Quite clearly, Newton's abilities "were not confined to Mathematical and Physical subjects only but the vastness of his Genius extended to every part of Literature." Indeed, what for Newton amounted to "Diversions & Parerga" equaled, if not surpassed, the choicest labors of those illustrious men who had "Applied themselves professedly to ye like Inquiries." Having seen how Newton handled ancient astronomy, Logan concluded that Newton had furnished "very good Reasons" for reducing the age of the Argonauts and the Fall of Troy. Nevertheless, doubts lingered. Though he "admired" Newton's reasoning vis-à-vis the ancient sphere, Logan wondered whether "the bare Authority of the writer of the Gigantomachia" suffices to establish Chiron's formation of the Constellations. Surely, he added, the Argonauts could not have been immortalized before they ever set sail.[98]

In addition to suggesting a somewhat earlier date for the expedition, Logan continued to entertain other doubts. He still could not accept the late dating for Rome's foundation, still less for the late invention of writing. He also found it difficult to accept, or properly follow the reasoning behind, Newton's rules regarding regnal reigns and generations. Newton may have been correct to abridge the reigns of the Latin kings at Alba, he protested, but not the reigns of the seven Roman kings. When their successions are considered individually, "tis hard to assign a reason why their Years should suffer such a vast reduction only from that one Rule that the successions of Kings of wch we have any Certainty seldom exceed 18 or 20 years one with another." While Newton did not satisfy him "in all particulars," Logan reiterated his admiration for the *Chronology*: "a wonderful performance & speaks the Genius of the Man much greater than I could ever have conceived before. His penetration in finding out his materials, his exactness in laying them together, and his Art in the whole Structure, in which he leaves the Scaligers, Petavius's &c almost out of sight behind him."[99]

Had he been aware of the manner in which Newton employed Hipparchus in order to establish his "valuable Accot. of stars by wch. the Colures pass'd in the first spheres," Logan intimated at one point, he would have been spared from writing "a good deal of what [he had] said in a former Letter." Logan, it appears, found it relatively unproblematic to modify his position, in part because his earlier critique had never appeared

[97] Wolf, 1974, pp. 338–40.
[98] Ibid., p. 342.
[99] Ibid., pp. 342–44.

in print and in part because of his commitment to Newtonian science. In contrast, Newton's Continental critics not only failed to share Logan's willingness to re-evaluate a system that they had already excoriated in print; they entertained considerable misgivings regarding Newton's natural philosophy as well. In fact, as the *Chronology* appeared at the height of the animated debate over Newtonian science, its reception became irrevocably entangled with the fortunes of the *Principia* and the *Opticks*.[100] The Jesuits, in particular, were as united in their opposition to Newton's system of the world as they were in rejecting his reckoning of time. According to Tournemine, for example, in the *Principia* "Newton has substituted conjecture for certainty, embarrassment for clarity, and ignorance for evidence"—a verdict he would have certainly extended to Newton's *Chronology*. Tournemine also considered it unpatriotic for a Frenchman to side with an Englishman, an opinion widely shared by his countrymen.[101] For his part, when Fourmont sarcastically asked whether Newton sought to establish a sect of chronologists, he clearly drew an analogy with the prevailing French perception of the Newtonians as a sect.[102] More directly, Des Vignoles opened his scathing critique of the *Chronology* by asserting Leibniz's priority in the calculus dispute. Several pages later he inserted disparaging remarks on both universal gravitation and the theory of colours, offering them as damning parallels to Newton's inability to distinguish between cause and effect in the *Chronology*.[103]

Small wonder, then, that the seemingly successful assault on the *Chronology* by Frenchmen emboldened some to challenge Newtonian science as well. Consider the Oratorian Joseph-Etienne Bertier, who published in 1760 a Cartesian theory of comets. He drew his inspiration for contesting the prevailing Newtonian theory from the example of Fréret, who, Bertier averred with some exaggeration, had demonstrated that while Newton the geometrician was invincible, Newton the astronomer and chronologer was not.[104] Others availed themselves of Newton's forays into chronology as means to rehabilitate Descartes. The astronomer Pierre Estève, for example, rebuked those belittling the value of Descartes' *Principles of Philosophy*—simply because it contained errors in physics—with a crushing rhetorical question: what would the English think of those who judged Newton's worth solely on the basis of the *chronology* and his commentary on the apocalypse? More pointedly, the eccentric Delisle de Sales, who had located Atlantis in Sardinia, argued that whereas Descartes remained the same great man in all his endeavors, Newton ceased to be a Newton when he turned to comment on the Apocalypse.[105] By the turn of the eighteenth century, some thought the coupling of the two domains had outlasted its usefulness. Once Newton's system of the world had become a "science," rather than "Newtonian science," stated

[100] For the debates, see Feingold, 2004; Feingold, 2013.

[101] Wade, 1969, pp. 415–16. Tournemine's coreligionist, Antoine Laval, combined a dismissive critique of the *Principia* and *Opticks* with a rejection of the *Chronology* —the latter of which, he admitted, he had not seen. Laval, 1728, pp. 155–91.

[102] See, for example, Castle 1743, p. 13. In 1729, Fontenelle praised Molières' valiant attempt to defend Cartesian vortices which were challenged "by the formidable objections" of Newton "and his numerous sect." *Histoire de l'Académie Royale des Sciences* (1731 [1729]), p. 87.

[103] *Nouvelle Bibliotheque Germanique*, vol. 18, part 1, pp. 409–11, 425–27.

[104] Bertier, 1760, pp. vii–viii.

[105] Estève, 1755, vol. 1, p. 347; Delisle de Sales 1766, 15–16.

the protagonist of *Idéologie*, Destutt De Tracy, one could safely subscribe to it without the obligation to also believe in Newton's chronology or his explication of prophecies.[106]

Particularly unsettling for critics was the way in which both the *Principia* and the *Chronology* had a seemingly unbreakable structure. Readers of the latter would have recognized the near impossibility of picking and choosing select elements of the new chronology; Newton's "chain of evidences" compelled assent to the system as a whole—a requirement that did not exist in any other chronological system. For contemporaries, such a compulsion marked the distinctive mind-set of a mathematician, a mind-set Fontenelle berated in 1686:

> Grant ever so little to a lover, and presently you are forced to grant him a great deal more, and so on till you don't know how to stop. In like manner admit any principle a mathematician proposes, he then draws a consequence which you are obliged to admit, and from that consequence another, and thus before you are aware he carries you so far that on a sudden you wonder where you have got to.[107]

A generation later, a contributor to the *Journal de Trévoux* reiterated the point: "if one admits Newton's [physical] premises, his system is so well linked together that there remains no further difficulty."[108] This constraint prompted Father Castel to caution Cartesians against attempting to refute Newtonian science on mathematical grounds. "One must refuse all his principles or accept all their consequences," Castel thundered. Newton "is a geometer and a consequential one." Nevertheless, "he is not a *physicien*. Thus, one must stop him at the first step and show him that he makes suppositions, and false suppositions."[109] Along similar lines, the abbé Conti recalled how he had impressed on Souciet that if one were to successfully undermine Newton's chronology, the focus must be on the literary sources upon which the system rested, not on any attempt to rival the philosopher with astronomical calculations.[110]

Notwithstanding such advice, Newton's earliest critics, as we have seen, paid particular attention to the astronomy-based arguments, perceiving these to be foundational to the revised chronology. By 1730, however, the criticisms hurled by Souciet, Fréret, and Whiston—and the attempted rebuttal by Newton, Halley, and de la Nauze—had effectively exhausted the technical side of the debate. Instead, commentators turned their attention to questions such as whether Chiron was an astronomer—or, more critically, whether he existed at all—and then, more attentively, to the primitive state of Greek learning as insufficient in itself to nullify Newton's claims. By 1758, Antoine-Yves Goguet felt confident to take Newton to task for presuming that the Greeks were adequately proficient in astronomy at the time of the expedition of

[106] Destutt de Tracy, 1992 , pp. 245–56.

[107] Fontenelle, 1803, pp. 114–15.

[108] *Journal de Trévoux* (1714), pp. 550–51.

[109] Castel, 1743, p. 302; Gingras, 2001, p. 401.

[110] Conti, 1739–1756, vol. 2, p. 91.

the Argonauts to form the constellations or fix the points of the equinoxes and solstices. Goguet deemed such an opinion to be "contrary to all that ancient history teaches us of the little knowledge the Greeks had of astronomy in the heroic times." Prior to Thales, he contended, "the Greeks had no idea of astronomy considered as a science." In a similar vein, Robert Wood argued that had "the relaxations of that extraordinary Genius from more severe and important studies permitted him to have considered Homer in the light we have attempted to place him," Newton would never have ascribed astronomical profundity to the Greeks, who possessed neither "instruments fit to take an observation, [n]or knowledge to use them." In contrast to the censorious Goguet, however, Wood believed that greater immersion in Homer would have actually bolstered Newton's opinion regarding the duration of Greek antiquity: Homer "entirely coincide[s] with him in that part of his system, which contracts the distance of time between the Argonautic Expedition and the Siege of Troy."[111]

Only toward the end of the eighteenth century did a few astronomers return to consider the technical side of Newton's chronological arguments. Jean-Sylvain Bailly deemed Newton's assumptions and ensuing calculations to be groundless, and proceeded to date the formation of the ancient sphere to the late fourteenth century BCE instead. Still, Bailly conceded that however misguided he might have been, Newton's idea to reform chronology by determining the ancient positions of equinoxes and solstices "was beautiful, grand, and worthy of a man of genius." Jean-Baptiste Joseph Delambre, Bailly's younger contemporary, proved less sympathetic. Insofar as the controversy between Newton and Fréret was concerned, victory belonged to the Frenchman. Nonetheless, Delambre thundered, the participants of the protracted and futile debate over Newton's chronology overlooked the very foundation of the matter at hand when they accepted as unproblematic the alleged positions of the solstices and equinoxes furnished by Aratus and Eudoxus. His own computations of such positions, Delambre asserted, had "proved invincibly" that the ancient sphere indicates nearly as many epochs as it contains stars, thereby rendering it useless for both astronomy and chronology. Before Hipparchus, Delambre insisted, the Greeks simply lacked the means and the instruments necessary for carrying out careful observations. And if even the great Hipparchus could scarcely achieve an accuracy of half a degree, how can one pretend to have fixed the epoch of Chiron and the Argonauts?[112]

As most eighteenth-century commentators steered clear of calculations, the linguistic foundations of Newton's astronomical arguments came to the fore. As noted previously, nearly all critics took issue with Newton's invocation of σχήματα ὀλύμπου to mean asterisms. Jean Masson, for example, insisted on comprehending the phrase as "figurative"—and "consequently ambiguous"—"requiring therefore something else to shew that the signification of *Asterisms*, assign'd here by [Newton] is the only necessary one they can bear in that place." Nevertheless, Newton could not be faulted for mistranslating the testimony of Clement of Alexandria, for even the latter's eighteenth-century editor, John Potter, translated the phrase as "coeli figuris." Equally controversial was Newton's translation of the verb ποιῆσαι when crediting Musaeus with constructing the sphere. As noted above, Masson had already disputed Newton's

[111] Goguet, 1761, vol. 2, pp. 278–84; Wood, 1769, p. liv.

[112] Bailly, 1781, pp. 507–15; Delambre, 1817, vol. 1, pp. x–xii, xii–xiii (second pagination).

translation, and Larcher took the matter up half a century later, arguing that the phrase "made a sphere" is a mistranslation of Diogenes Laertius. Again, however, the Greek text Newton used, as well as the facing Latin translation of ποιῆσαι as "fecisse," certainly warranted his rendering.[113] More, therefore, was necessary in order to impugn Newton's scholarly competence.

The opportunity presented itself in 1738 when Angelo Maria Quirini published an essay on the ancient history of the island of Corfu, where he exposed a notable blunder by Newton. An admirer of the Englishman—"undoubtedly the prince of the mathematicians of this age"[114]—Quirini took issue with Newton's reliance on the entry in Suidas' *Lexicon*, in which the grammarian Anagallis is cited as crediting Nausicaa, daughter of king Alcinous, with the invention of a "Σφαιρα." Newton interpreted the term to mean a celestial sphere, further adding: "it's most probable that she [Nausicaa] had it from the *Argonauts*, who in their return home sailed to that Island, and made some stay there with her father."[115] Quirini devoted a whole chapter to Nausicaa, surveying all available ancient sources—from Homer and his scholiasts Spondanus and Eusthatius to Athenaeus and later authorities—in order to demonstrate that in this particular context the term denotes a ball (pila).[116] Newton thus stood convicted of ignorance in both Greek language and literature.

Quirini's book was widely reviewed, with Newton's "manifest error" invariably given prominence. Writing in the *Giornale de' Letterati d'Italia*, for example, Apostolo Zeno marvelled at the sight of the "preeminent mathematician of our age" not only showing himself ignorant of ancient testimonies, but misreading Suidas' explicit statement regarding Nausicaa's playing a game of ball. And yet, as Zeno saw it, this mistranslation served Newton as the foundation for establishing the epoch of the Argonautic expedition.[117] Critics of Newton eagerly availed themselves of this error. Angelo Maria Ricci incorporated Quirini's discussion into his *Dissertationes Homericae*, displaying his glee at Newton's ignorance by substituting Quirini's "insignis hallucinatio" (manifest delusion) with the more pointed "foede hallucinatus" (disgraceful delusion). Five years later, Gian Rinaldo Carli made equal use of Quirini's discovery in his scathing critique of Newton's account of the Argonautic expedition.[118] Not content with proving to his own satisfaction how Newton blundered in ascribing to Chiron the construction of a celestial sphere, Carli believed he had exposed a notable

[113] Diogenes, 1664, p. 1; Herodotus, 1802, vol. 2, p. 287. See our discussion of the issue above, pp. 247–48, 298. For Larcher, see Buchwald and Greco Josefowicz, 2010, pp. 122–25.

[114] Quirini met Newton at Richard Bentley's residence in the Royal Library, while he visited England in early 1711. Quirini, 1749–1750, p. 53.

[115] Newton, 1728, p. 86.

[116] Quirini, 1738, pp. 122–30. Newton should have known better. He owned the 1705 edition of Suidas' *Lexicon*, where the correct definition is given: "Pila. Philae inventionem Anagallis Corcyrae, Grammatica, Nausiccae Alcioni filiae attribuit." Suidas, 1705, vol. 3, p. 415. Newton also owned Casaubon's edition of Athenaeus' *Deipnosophistarum* where Nausicaa was credited with the invention of a play-ball, not a sphere. Athenaeus, 1657, p. 14; Harrison, 1978, pp. 246, 92.

[117] *Giornale de' Letterati d'Italia*, vol. 30 (1739), pp. 40–41: *Journal des Sçavans* (March 1739), p. 166; *Mémoires des Trévoux* (July 1739), pp. 1407–8.

[118] Ricci, 1740–1741 , vol. 1, pp. 205–6; Carli, 1745, pp. 39–60. Carli continued his anti-Newtonian campaign throughout the second volume of his *Lettere Americane*. Carli, 1780.

contradiction in Newton's reckoning of the fall of Troy. On the one hand, Carli argued, Newton dated the event to seventy-five years after Solomon's death. Yet Newton also cited the historian Justin as stating that Troy fell a year after the foundation of Tyre—which, Newton stated elsewhere, occurred 140 years before Troy fell! "Here is Newton against Newton," Ricci enthused, and the reformed chronology destroyed by its very author. In his eagerness to claim victory, Ricci conveniently ignored the fact that Newton had cited Justin simply in order to establish that Sidonian refugees had built Tyre. He certainly did not interpret Justin's "ante annum" to mean "one year" before the fall of Troy. Rather, he took the phrase to mean "before the year" in which Troy fell, further indicating the indeterminate date by inserting into the quotation two asterisks after "ante annum."[119]

Quirini's discovery remained central throughout the eighteenth century, as it evinced Newton's fallibility as well as encouraged challenge of his command of erudition more generally. Thus, when Gilbert Charles Le Gendre, marquis de S. Aubin-sur-Loire—who believed that not a single truth could be found either in Newton's system of the world, his optics, his mathematics, or his chronology—issued in 1741 a new edition of his *Traité historique et critique de l'opinion*, he made sure to insert a note publicizing Newton's error.[120] In 1760, Johann August Ernesti jeered more pointedly at Newton's failure to consult Homer, who had made it clear that Nausicaa played a game of ball with her maids. Newton's "absurd fantasy," he added, nevertheless served him as the basis "for fixing the epoch of the Argonauts from the position of the stars at the time when the Argonauts landed in Corcyra. Such are the demonstrations of Philosophers and Mathematicians who are ignorant of letters when they apply themselves to history."[121]

We noted earlier how the seemingly closed structure of the *Chronology* made it difficult to accept parts without assenting to the system as a whole. Nevertheless, some contemporaries were tempted to extract useful elements without necessarily subscribing to the system as a whole. Consider again the celebrated anachronism that Virgil introduced into Book Four of the *Aenead*, where he made Dido and Aeneas contemporaries, notwithstanding that according to traditional reckoning three centuries spanned the Fall of Troy and the founding of Carthage. Admirers of the poet have endeavored for centuries to remove such a blemish on his reputation. In 1697, for example, on the eve of Newton's embarking on his chronological studies, John Dryden condoned Virgil's poetic license in the dedication to his translation of the *Aeneid*. While Virgil's "great judgment made the laws of poetry," the poet "never made himself a slave to them." In fact, Dryden claimed, chronology is

> at best … but a Cobweb-Law, and [Virgil] broke through it with his weight. They who will imitate him wisely, must chuse, as he did, an obscure and a

[119] Carli, 1745, pp. 59–60; Newton, 1728, p. 103. Fréret, too, recognized the lacuna in Justin's text, though he claimed that Trogus, who consulted the records of the Phoenicians, elsewhere dated the fall of Troy to 1184 BCE. Fréret, 1758, p. 277.

[120] Le Gendre, 1741, vol. 5, p. 561; vol. 1, p. 643.

[121] Homer, 1759–1764, vol. 3, p. 222.

> remote Aera, where they may invent at pleasure, and not be easily contradicted. Neither he, nor the *Romans* had ever read the Bible, by which only his false computation of times can be made out against him.[122]

Small wonder, then, that Newton's own reckoning of a mere twenty-one years between the two events offered solace to those who found it difficult to accept, on literary or aesthetic grounds, that the great poet could have blundered so spectacularly. Edward Gibbon, for one, wished to see Newton's "plausible system" render the episode less "irreconcileable to chronological truth." "Whoever condemns the episode of Dido," Gibbon determined, "must have more philosophy or less taste than I have."[123] Denis Diderot concurred. He announced his preference for Newton's chronology precisely because, according to the latter's calculations, Dido and Aeneas were contemporaries.[124]

In 1788, Juan Andrés devoted an entire pamphlet to absolving Virgil from the charge of appropriating poetic license. Just as Homer is generally acknowledged reliable on matters of historical fact and chronology, the ex-Jesuit argued, so Virgil's fidelity to historical evidence cannot be doubted. As proof, Andrés invoked the authority of the "great Newton," who had established that Troy fell in 904 BCE, twenty-one years before Dido had founded Carthage. If that "rigorous calculator" credibly narrowed the span separating the two events, Andrés asked, why blame Virgil for making Dido and Aeneas contemporaries?[125] Six years earlier, an anonymous Frenchman had published a similar defense, except for reckoning thirty-eight years separating the fall of Troy and the foundation of Carthage.[126]

Naturally, many demurred. Charles Lamotte—for whom Virgil's anachronism constituted "a Stain that sullies the other Beauties, and palls the Pleasure of the Reader"—admitted that were Newton's solution to rest "upon a solid Foundation," then the problem is resolved. "But as his Opinion has been attack'd by some very considerable Writers, & *adhuc sub judice lis est*," Lamotte concluded, "I will not here pass my Judgment, nor interpose in the Dispute, but leave it wholly to Time to clear up the Difficulty.[127] Six decades later, François Xavier de Feller's entry on "Dido" in the *Dictionnaire historique* presented Newton's theory as a mere curiosity; Newton's *Chronology* was "peu estimée." Besides, the dispute concerning the epoch of Dido's reign is futile:

[122] Dryden, 1956–, vol. 5, pp. 299–300.

[123] Gibbon, 1764, p. 71 and note. Two decades later the Scottish historian John Gillies argued in a similar fashion: "every man of taste will be desirous of extending the duration of this dark and unknown period, to have the pleasure of believing that Æneas and Dido were contemporaries: an opinion not altogether improbable, since it is adopted by Sir Isaac Newton in his *Chronology*." Gillies, 1786 , vol. 1, p. 457n.

[124] Diderot, 1875–77, vol. 12, p. 91.

[125] Andrés, 1788, pp. 7–9, 15. As late as 1813 Lorenzo Pignotti, in *The History of Tuscany* felt it necessary to argue that Newton's authority alone lends credit to the claim, and such authority had been "confuted by the astronomical observations of Whiston, and the reasoning of Fréret and others." Pignotti, 1813–1814, vol. 1, pp. 22–23n.

[126] *Journal des Sçavans* (1782), pp. 31–32. The author does not mention Newton in this context, though he refers to him when discussing the length of generations (pp. 35–6).

[127] Lamotte, 1730, p. 4 and note.

Aeneas never existed, nor did the city of Troy, nor did the Greeks wage war against that city.[128]

More enticing still was Newton's rule regarding the length of regnal years. Francesco Algarotti placed the issue front and center in his *Saggio sopra la durata de' Regni de' re di Roma* (1745)—a youthful production, which he originally presented to the Bologna Academy in 1729. Algarotti lamented the failure of Newton's chronology to receive the accolades it so eminently deserved: learned men resented seeing their researches challenged by a mathematician, while common men refused to accept that a single man could be right about every subject. Algarotti viewed the decoupling of the duration of regnal reigns from the length of generations as one of Newton's greatest achievements. Such decoupling enabled the great Englishman to demonstrate, "by an unanswerable calculation," that while a century was comprised of three generations, reigns, both ancient and modern, averaged 18–20 years, in conformity with the laws of nature. Algarotti sought to substantiate the rule's verity by providing the scholarly evidence for Newton's shortening the duration of the first seven Roman kings from 244 to 133 years. The new reckoning would remove all historical difficulties, exculpate Virgil from committing a gross anachronism, and lend credence to the tradition that Numa had been the disciple of Pythagoras.[129]

Algarotti's provocation did not go unchallenged, especially in Italy, where, years later, Gian Francesco Galleani Napione published a forceful refutation—albeit respectfully praising Newton's scientific eminence and acknowledging his ingenuity in forging his chronological system. Subsequent historians of Rome followed Napione's lead.[130] Several French authors, on the other hand, were cautiously complementary. Élie Fréron dubbed the essay "interesting," while according to the editor of the *Mercure de France*, Algarotti successfully employed "Newton's principles for the reformation of chronology."[131] Nor should such a favorable view be surprising. After all, despite Fréret's confusions concerning Newton's rule, the rule still worked. Indeed, even Banier appeared willing to come closer to Newton's reckoning when he accepted the average reign of kings to be between twenty-two and twenty-five years.[132] Tacit acceptance of the rule is even apparent in playful references to it. Thus, Frisi pointed out how "in Italy there is no happier exception" to the rule than the eight Medici dukes and the twelve dukes of Este, whose average reigns spanned twenty-six and twenty-eight years, respectively. The perpetual Secretary of the Berlin Academy of Sciences made a similar point in an address commemorating the thirtieth anniversary of the reign of Frederick the Great in 1770; he congratulated the monarch on having already exceeded the law established by Newton and Fréret (*sic*!) The reigns of George II and

[128] Feller, 1790–1794 , vol. 3, pp. 534–35. Feller's indictment was newly inserted. Missing in the first edition a decade earlier: Feller, 1781, vol. 2, p. 515.

[129] Algarotti, 1778–1784, vol. 4, pp. 105–38.

[130] Napione Galeani, 1773 .

[131] *L'année littéraire*, vol. 8 (1768), p. 177; *Mercure de France* (January 1769), p. 68. For additional notices, see *Journal Étranger* (November 1758), pp. 118–24; *Journal des Sçavans* (January 1771), pp. 392–94. Noteworthy is that as early as 1737 an anonymous author published in Paris "Remarques sur la durée de l'Etat Monarchique de Rome, pour justifier la Chronologie de M. Newton." *Mercure de France* (July 1737), pp. 1537–57.

[132] Banier, 1739–40, vol. 3, pp. 341–42.

Louis XIV, added the Secretary, also exceeded the bounds of the rule, but without diminishing thereby anything from its verity.[133]

Still, few, if any, of those who embraced Newton's rule proceeded to embrace the chronology as a whole. With the passage of time, and with the mounting historical and literary evidence that cast doubt on all chronological systems, it became increasingly convenient to take literally Newton's disclaimer: he had pursued chronology merely as relaxation from more serious studies. Such an admission could be used as readily by those eager to exculpate Newton from engaging in an embarrassing course of study as by those who pounced on the admission to red-flag Newton's less than adequate mastery of erudition. Thus, having praised Newton for introducing "illumination into the obscurity of dates and of the ancient epochs," Paolo Frisi stressed that in contrast to the success of the *Principia* and *Opticks*, on the *Chronology* "opinions are still divided." The reason for the incongruity was not hard to find: Newton "never thought that his chronological and historical ideas would ever pass into the recollection of posterity." Properly speaking, then, the *Chronology* should be viewed merely as "a playful piece, or rather a learned whimsy written by an octogenarian, who having been accustomed all his life to spend so many hours of the day in reading and writing," sought to find in history and chronology "a relaxation from his exaltations in physics and mathematics."[134] Commenting on this very same matter, the Leipzig mathematician Abraham Gotthelf Kästner remarked wryly that only a Newton could have chosen to study chronology for relaxation. More mediocre minds would have settled on chess.[135] Others waxed more gloomily in their lament of Newton's lapse of judgment. The young Johann Gottfried Herder, for example, quipped that when Newton, in old age, had turned to chronology, the great genius thereby became "childish." Years later Herder remained bemused at the thought that so many great men—"among whom even a Newton himself is to be reckoned"—had permitted "the Jewish chronology and the Apocalypse rob [them] of time, that might have been employed in more useful inquiries.[136] The same reflection had already been articulated by Hardouin in 1729, when he castigated Newton for wasting his final years: "This Great Man, this first Geometer and Mathematician of Europe, has in his last years constructed nothing but a frivolous system. Why did he not stick by the one that had earned him a reputation?"[137]

By the early nineteenth century the tenor of such reflections hardened. Delambre perpetuated the tradition that Newton composed most of his writings on theology and chronology during the final thirty years of his life, during which period he had become a mere shadow of his former self. As for the philosopher Johann Gottlieb Bühle, he believed it best to consign the *Chronology* to oblivion in order to preserve Newton's reputation.[138] Such remarks, however, fail to do justice to the substantial impact that Newton's *Chronology* exerted during the Enlightenment—only a sample

[133] Hall, 1999, p. 167; *Nouveaux Memoires de l'académie Royale des sciences et belles-lettres Année 1770* (Berlin, 1772), p. 37.

[134] Hall, 1999, pp. 166–70.

[135] Kästner, 1751, p. 488.

[136] Baasner, 1986, p. 77; Herder, 1800, 334.

[137] Translated in Manuel, 1963 , p. 177 from Hardouin, 1709, p. 1586.

[138] Delambre, 1827, p. 60; Buhle, 1816, vol. 4, p. 99.

of which we've been able to present in this chapter. Perhaps Louis Elisabeth de la Vergne de Tressan best captures Newton's impact in success as well as in failure. An admirer of Newton's "sublime philosophy" and one for whom the *Principia* signified the "chef-d'oeuvre de l'esprit humain," de Tressan wrote in 1776: "The obstinate and minute writer who wishes to probe more deeply into the darkness" of antiquity, is destined to exhaust himself in a vain pursuit. Many have tried to establish systems of chronology and failed. "If anyone was worthy to succeed, it would have been the sublime Newton. But his chronology, the weakest work of this great man, appears to be the price to be paid for the feebleness of humanity. If even Newton was wrong, what could be hoped by those who labored after him, than to obscure more and more the order, or rather the appearance of order, of events and eras, which, it must be admitted, are almost entirely lost to us?"[139]

[139] Tressan, 1822–1823 , vol. 9, pp. 316, 358–59; vol. 10, p. 117.

Evidence and History

Throughout these pages we have traced Newton's ways with evidence, whether in his laboratory, in the mint, or within the ancient texts that he pored over and copied again and again. We followed the young Newton at Cambridge, where he developed his early thoughts and techniques for working with the discrepant data that observation and laboratory work produced. There he developed a related way of thinking about the inevitable errors introduced by the senses, one that led him to a radically unusual procedure for producing trustworthy numbers out of unreliable ones. That method eventually inflected his novel ways with ancient history, as he sought to turn a morass of conflicting words into probative evidence. This evolved over time as Newton's focus changed, and as he experienced in later years the conflicting testimonies of the counterfeiting miscreants brought before him as Master of the Mint, for this solidified his skepticism concerning words from the past that could not be balanced against one another or, better yet, turned into numbers. Newton's radical revision of chronology—radical both for its methods and for its results—generated much controversy at the time, controversy that extended even into the nineteenth century.

In 1855, an anonymous reviewer of George Grote's *History of Greece* contrasted that monumental work with Newton's *Chronology*, better to underscore the great "advance which historical science" had made in the previous hundred years. Newton—"that miracle of misdirected ingenuity"—had morphed his gods into historical personalities, thereby grounding his entire Greek mythology on a mistranslation of Egyptian hieroglyphics. The reviewer enumerated Newton's extravagances before concluding: "This theory is the *reductio ad absurdum* of all efforts to convert myth and legend into history. Had it been manufactured by Swift and Pope for the purpose of throwing ridicule on all such attempts, it would have formed no unworthy pendent to their satire on the scholastic philosophy in the 'Memoirs of Scriblerus.' By Newton, however, it is promulgated with as much parade of demonstration as the best established theorem in the 'Principia.'"[1]

The attack—Grote himself had never so much as mentioned Newton—illustrates that Newton's ideas remained current, if highly controversial, in historical debates into the second half of the nineteenth century. There is a certain irony to the reviewer's dismissal of Newton's approach to the past as contrary—even pernicious—to historical norms, for Newton's purpose in bringing calculation to bear on history was

[1] *The Dublin University Magazine* 45 (1855), pp. 478–79.

meant to ground it upon a more secure foundation. Moreover, for a time the *Chronology* mattered a great deal. A modern reader may find the text disagreeable and unworthy of serious attention. To Enlightenment readers, however, the very dryness of Newton's book and its matter-of-fact format of presentation only heightened its revolutionary content.[2] Versed in the sources upon which Newton drew, they recognized his inventiveness and ability effectively to handle the relevant literature. As Augustus De Morgan—an appreciative critic of Newton—put it in 1846: "his ideas on chronology, founded on the assumption of an accuracy in the older Greek astronomers which nobody now allows them, are rejected and obsolete. But the work does honour to his ingenuity and his scholarship, showing him to be not meanly versed in ancient learning."[3] And given Newton's stature, the new solution he had offered to the vexing problem of ancient chronology inevitably polarized the republic of letters between admirers and critics—apathy was never an option.

Just as the centrality of Newton's contribution to eighteenth- and early nineteenth-century scholarly debates has gone largely unnoticed by historians, so did the long, drawn-out, and complex process of composition that ultimately resulted in the *Chronology of Ancient Kingdoms Amended*. Newton's initial forays into remote antiquity around 1685 intended to establish the Ancients' comprehension of the principles of heliocentric cosmology and atomistic natural philosophy as well as to argue that their grasp of the "true frame of nature" derived from its being an integral part of the "true" primitive religion. Newton embarked on this investigation as part of an abortive effort to include in the *Principia* a defense of the orthodoxy of the new science. By 1686, he thought better of the idea and, incensed by Robert Hooke's priority claims, determined to render Book III ("On the System of the World") technically difficult, purposely to avoid "being baited by little Smatterers in Mathematicks." Concurrently, the focus of his historical interests began to shift. Whereas initially Newton sought to chronicle the progress of idolatry—the process through which the primordial monotheistic religion had been perverted, with the concomitant corruption of the correct understanding of the natural world—he increasingly immersed himself in the attempt to forge a coherent postdiluvian history of mankind: hence his creative correlation between the Genesis narrative and ancient myth.

Beginning with the assumption that Hesiod's Four Ages of Man encapsulates, in the guise of parable, the political history of the first four generations after the Deluge—or, more precisely, the annals of Noah's immediate progeny—Newton labored for over a decade to reconstruct a coherent narrative that harmonized conjectural sacred and profane histories. He embarked on this enterprise armed with the habit of purposeful reading conjoined with an ironclad system of synchronization, which had already enabled him, or so he believed, to successfully correlate prophetic images and symbols with historical events. The seeming success of his method in lifting the veil from the visions of Daniel and Revelation informed Newton's anticipation of a similar feat

[2] Jean-Pierre de Bougainville articulated the sense of alarm that informed Fréret's attack on Newton: "At the prospect of a revolution about to change the face or at least the perspective of the historical world it was natural, shall we say it was correct, that M. Fréret should become alarmed and that he should move to the frontier to reconnoiter the terrain." Fréret 1758, p. x, cited in Manuel 1963, p. 24.

[3] De Morgan, 1914, p. 36.

with respect to historical reality, obfuscated as it was by pagan fables. He began with Noah's division of Asia, Europe, and Africa among his three sons, before proceeding to correlate the mythical *Titanomachia* with a conjectured internecine conflict among the posterity of Ham, which subsequently pit Chus and his son Nimrod against the posterity of Shem. These conflicts gave rise to the world's first great empire, stretching from North Africa to India—and with Noah yet alive. Chronology proper was irrelevant to this reconstruction of postdiluvian history, except for the presupposition that the alleged events had occurred within a century and a half after the Deluge. For evidence, Newton drew exclusively on hermeneutics and the synchronization of a few verses from Genesis with pagan myths.

Before long, however, Newton began to entertain serious doubts regarding the cogency of the tight web he had woven. In particular, considerations of time—as well as a steadfast belief in the regularity of nature—impressed on him the need to entertain a protracted process for the propagation of mankind, the emergence of kingdoms, and the flowering of civilization. It was at this point, when he relocated the focus of his research to 1200 years after the Flood, that Newton became a technical chronologer. His ambition to provide a universal history of the ancient world necessitated furnishing not only a solid evidentiary basis for the radical postdating of the origination of kingdoms, but a strict and comprehensive correlation of reigns and events—not to mention the updating of his synchronization of myths. Newton thereby became the first since Scaliger (as revised by Petavius) to propose a new system of technical chronology. However, in contrast to his predecessors, who focused primarily on calendric systems and eclipses,[4] Newton began by imposing quantification on chronology. First, he deployed his novel methods for handling discrepant data in order to determine the proper lengths of reigns and generations. Later, he deployed the same methods in conjunction with astronomical precession to establish the epoch of the Argonautic Expedition—and thereby the dating of the Fall of Troy—on the basis of an ambiguous sentence in a literary text. In so doing, Newton resembled Scaliger, who boasted openly of having been "the first to reveal" the true epoch of the Olympiad "on the sole basis" of a verse in Pindar.[5] Not surprisingly, both chronologers were criticized for their presumption in erecting monumental edifices on such meager, and debatable, grounds.

The originality and radical methods that characterized Newton's approach to profane and sacred history could scarcely be gleaned from the published *Chronology* proper, though not a few contemporaries perceived the dangers to traditional historical methods that Newton's work posed. Consider for example the centrality of synchronisms to his system. He outdid contemporaries in his reliance on them, primarily because he was unique in pursuing, simultaneously, technical chronology and Biblical prophecies. As he saw it, the seeming equivalence of prophetic and mythical symbolism warranted the rigorous application of the same interpretative methodology to both domains. Only Newton's manuscripts, and not the *Chronology* itself, reveals the extent to which synchronization and the quest for coherence informed his historical studies for more than four decades. He was not content merely to harmonize

[4] Eclipses rarely concerned Newton; there existed no reliable records of eclipses for the period he studied.

[5] Grafton 1993, p. 394.

conflicting versions of myths. Or to argue that multiple deities denote a single historical figure. Or to correlate pagan myths to Biblical events. For Newton, an effective use of his methodology mandated that the establishment of the historical time line of one nation necessarily established the correct time line of others. Hence the centrality of Greece to his chronological system. Since Greece had been colonized (and civilized) by the Egyptians and by the Shepherds (i.e., Phoenicians) who were driven out of Egypt, forging an exact chronology for Greece would help determine the chronology of Egypt. As Newton put it: "let us now try to rectify the Chronology of *Egypt*, by comparing the affairs of *Egypt* with the synchronizing affairs of the *Greeks* and *Hebrews*."[6] It did not suffice to synchronize dates, events, and dynasties without paying careful attention to whether the events in question fit the periods in which they had purportedly occurred.

The great amount of time and effort that Newton devoted to synchronizing myths attests to his assent to Scaliger's belief—more recently endorsed by Richard Simon[7]—that the Old Testament in itself was insufficient to establish an exact chronology of the ancient world. Like Scaliger, Newton too concerned himself primarily with secular chronology and with the annals of pagan nations—albeit embedded within the confines of Masoretic time reckoning. This crucial aspect became obscured in the published *Chronology*, the structure and reasoning of which gave the impression that Newton intended to prioritize and aggrandize the Jewish contribution to world civilization. The antiquity and affairs of the Jews in themselves were, however, secondary to Newton's interests, which had a much broader aim into which the ancient Hebrews fit as significant exemplars. From the outset, he concerned himself with the origin of kingdoms, and of the Egyptian empire in particular, for the latter was a prerequisite for establishing a new universal history of the ancient world. His new time line, linked as it was to a theory for the origin of civilization, postulated a near simultaneity for the rise of the Kingdom of Israel and of the Egyptian Empire, thereby requiring synchronization of the new dating system with events drawn from Jewish history. Readers of the *Chronology* accordingly have misconstrued his intentions, which are nevertheless amply apparent in his working papers.

Well into the early years of the eighteenth century Newton considered St. Stephen's statement regarding Moses as one "learned in all the wisdom of the Egyptians" sufficiently authoritative to conclude that letters and civilization flourished in Lower Egypt before the birth of Moses. Early on he even believed that all Eastern tongues—Hebrew, Chaldaic, and Syriac included—were merely dialects of the Egyptian language.[8] This helps explain Newton's initial embrace of John Spencer's controversial theory regarding the grounding of early Jewish belief in the monotheistically correct parts of Egyptian culture.[9] The revised chronology could hardly accommodate such a

[6] Newton, 1728, p. 191.

[7] Simon, 1682, sig. pp. 5–6.

[8] Yahuda MS 16.2 fol. 59.

[9] As Newton explained in an early version of the *Original of Kingdoms*, "the religion w[ch] Moses taught y[e] Jews was no other then y[e] religion of Noah purged from the corruptions of y[e] nations. For D[r] Spencer has shewn y[t] Moses retained all y[e] religion of y[e] Egyptians concerning y[e] worship of y[e] true God, & rejected only what belonged to y[e] worship of their fals Gods the Sun Planets & Elements, Jupiter Hammon, Osyris, Isis, Orus & y[e] rest, & that y[e] Mosaical religion concerning y[e] true God conteins little else besides what

theory given its novel time line. Without repudiating it, therefore, Newton never again mentioned Spencer—just as he passed over in silence Marsham's parallel claims regarding the derivation of all Jewish laws and customs from the Egyptians. Overall, Newton, unusually for the period, avoided attributing to the ancient Hebrews any significant contribution to arts and letters proper, excepting high architecture, embodied as it was in Solomon's Temple.[10] He arrived at this conclusion in part as a result of his researches into the size, shape, and symbolism of the Jerusalem Temple, and in part because the temporal priority of Solomon's Temple enabled him to post-date all similar structures elsewhere in the ancient world—a crucial element of Newton's chronology. It was never part of his aim to argue that the cultural effects of monotheism per se had birthed arts, letters, and science, for he became focused on the coordination of historical evidence in the light of his evolved theory for the origin of civilization.

Once Newton came to the conclusion that populations and kingdoms had evolved gradually, the origination of learning required a more careful parsing. For even if Noah and his sons were in possession of arts and letters, that knowledge must have faded away during the centuries of savagery and conflict that followed the Deluge. If so, how to account for, and whom to credit with, the rediscovery of learning? In his manuscripts Newton credited the Egyptians with a variety of practical discoveries, ranging from the cultivation of corn and manufacturing metals, to masonry, painting, and navigation. As late as the early years of the eighteenth century, he further thought that navigation and certain arts and sciences—albeit not writing—had originated in Egypt, whence they propagated eastward to the Arabians and Chaldeans. He changed his mind by the time he composed the final draft of the *Chronology*; there the Egyptians were credited solely with the invention of corn, and with advancing the arts and sciences invented by others.

While Egypt remained central to Newton's revised chronology, the late dating he assigned to its unification precluded crediting its inhabitants with an early attainment of advanced civilization—for this was a level of development he associated only

---

was then in use amongst the Egyptians. And if so, then it's certain that ye old religion of the Egyptians was ye true religion tho corrupted before the age of Moses by the mixture of the worship of fals Gods wth that of ye true one: & by consequence ye religion of ye Jews was no other then that of Noah propagated down in Egypt till ye age of Moses" (Yahuda MS 41 fol. 5). Newton further informed David Gregory in 1694 that "Religion is the same at all times, but religion which they received pure from Noah and the first men, the nations debased by their own inventions. Moses began a reformation but retained the indifferent elements of the Egyptians (it was the Egyptians who most of all debased religion with superstition and from them it spread to the other peoples). Christ reformed the religion of Moses." Newton 1959–1977, vol. 3, p. 338.

[10] When, for example, attempting to locate the origin of atomism with Moschus, Newton declined to follow most contemporaries in further identifying the Phoenician with Moses. Neither did he credit either Moses or Abraham with the invention of letters. And in the *Chronology* he took care to attribute the invention of writing and astronomy to the non-Jewish posterity of Abraham, never to the Patriarch himself. Newton accepted as unproblematic the early presence of arts and sciences, perhaps as part of that "common tradition" that had passed down from Noah to Abraham, a tradition known to the gentiles as well. Since such putative knowledge had existed before God's Covenant with Abraham, the Jews as such were irrelevant to the origins of learning.

with a flourishing kingdom. Precisely for that reason no cultural superiority could have been ascribed to the Jews either, for they hardly existed as a nation prior to the expulsion of the Shepherds from Egypt. Hence the elevation of the posterity of Ishmael, Midian, and Esau to a position of cultural prominence. While not constituting a major kingdom, their unique position as merchants could explain their invention of writing, along with the rudiments of sciences necessary for navigation and account keeping. Newton's focusing on these people is better understood when bearing in mind that he considered them to be the ancestors of the Phoenicians who, as cultural mediators, played such a crucial role in the *Chronology*. Equally important, Moses' sojourn of forty years to the land of Midian explains how he came to possess the broad learning ascribed to him by St. Stephen.

To a considerable degree, Newton's mature vision of the early history of mankind heralded the stadial model of societal progress, which would become the hallmark of such Enlightenment figures as Turgot and Adam Smith. Without explicitly delineating four distinct developmental stages, Newton nevertheless considered the postdiluvian evolution of humanity to follow successive phases of growth, based on modes of subsistence, everywhere in the world. The savage age—when mankind lived in the open or in caves, and subsisted from hunting or from the fruit of the land—preceded the pastoral period. The agricultural stage followed, along with new forms of cohabitation and governance, which gave rise to towns and kingdoms. Following these three sociopolitical and economic stages came the commercial activity of the "Phoenicians," which engendered the rise of early civilization.

Newton assembled the building blocks for his system from several sources, both ancient and modern. Ancient authors such as Plato and Lucretius furnished him with important clues regarding the savage condition of mankind in its infancy, as well as alerted him to the considerable span of time needed for the transition to higher stages of subsistence, and then for the development of cities, laws, and the arts.[11] More recent authors, notably Selden, Vossius, Grotius and Pufendorf, added both materials and theoretical insights to the framework. Consider, for example, the attempt of Gerardus Vossius to resolve the problem of Manetho's dynasties by resorting to a type of conjectural history to account for the proliferation of concurrent kingdoms in Egypt:

> If anyone finds this hard to accept, let him consider that all great empires have come together as wholes from several smaller units. Not to mention ancient Assyria, everyone knows that before the Persian empire there were first the different Medean, Babylonian, Assyrian, and others. How many empires did the Macedonian swallow? How many did Rome? Into how many kingdoms was Spain originally divided? And only think of Britain and Gaul.[12]

Newton appropriated the type of conjectural history he found in Vossius (and in Bochart), expanding on it to create a powerful explanatory structure that underpinned his chronological system. One may reasonably apply to Newton's scheme for

[11] Plato, 1961, pp. 167–79; Lucretius Carus, 1969, pp. 161–68.
[12] Vossius, 1668, p. 211, cited in Grafton, 2012, p. 68.

ancient history what Dugald Stewart wrote of Adam Smith's theory for the growth of language:

> In this want of direct evidence, we are under a necessity of supplying the place of fact by conjecture; and when we are unable to ascertain how men have actually conducted themselves upon particular occasions, of considering in what manner they are likely to have proceeded, from the principles of their nature, and the circumstances of their external situation.... In examining the history of mankind, as well as in examining the phenomena of the material world, when we cannot trace the process by which an event *has been* produced, it is often of importance to be able to shew how it *may have been* produced by natural causes.... To this species of philosophical investigation, which has no appropriated name in our language, I shall take the liberty of giving the title of *Theoretical* or *Conjectural History*.[13]

In Newton's chronological studies, as in his mechanics and optics, we find a careful demarcation between what appears to approach mathematical demonstrativeness and what may be regarded as plausible conjecture. He repeatedly denounced the accepted chronology of the Greeks as grounded on mere "reasoning & conjecture."[14] But once the time line of that chronology had been rectified, its history could be conjecturally reconstructed on the basis of less robust, but still probative, evidence, which was how Newton developed the correlation between dominion and idolatry. And once he had established by calculation *when* the Egyptians had embarked on astronomical observations, Newton felt secure in explaining *why* such interest arose among them.[15]

These examples illustrate Newton's deployment of a variety of mutually supportive types of data. Astronomical evidence furnished the highest degree of certitude, and Newton went to great lengths to turn words into numbers. We've discussed at some length his ingenuity in revising Greek chronology through just such a methodology. We might add that he attempted, more hesitantly, to accomplish a similar feat for Egyptian chronology. Taking his cue from Herodotus, who ascribed the regulation of the year into 365 days to the Egyptians, Newton determined that this regulation postdated the Exodus—for the Israelites carried with them the luni-solar year—and even the reigns of Ammon and Sesostris, since in the latter's Temple near Philae the priests

[13] Stewart, 1854–1860, vol. 10 pp. 33–34.

[14] Keynes MS 146 fol. 14; MS 361.1 fols. 102, 103^{v}; MS 361.3 fol. 63^{v}.

[15] "The kings of cities were worshipped by the cities & the kings of Nations by the nations. Whence the largeness of their dominion may be conjectured by the extent of their worship. So from ye extent of the worship of Ammon it seems probable to me that he subdued not only Libya & all Ammonia but also Ethiopia & the Indians above Egypt & Arabia foelix." And: "How ye court of Egypt came to be so much addicted to this study I do not find in history, but if room may be allowed for conjecture I suspect that the Merchants trading upon the red sea were the first that found out letters numbers & Arithmetic & observed the stars ... & that when the Court of Edom fled from David into Egypt they carried these things with them to ye Court of Pharaoh, & on that account were there enterteined wth extraordinary favour." New College MS 361.3 fols. 63^{v}, 129^{v}.

were enjoined to fill a bowl of milk every day, for 360 days, in order "to count the number of days in the Calendar year, & thereby to find the difference between this & the true solar year." The Temple of Amenophis, in contrast, exhibited a new calendar.[16]

Here we have a good illustration of Newton's way with confirmations. Lest the evidence for dating the introduction of the solar year seem insufficient, based solely as it was on Newton's dating of Amenophis' reign, he offered a more potent astronomical argument. In 747 BCE, Newton argued, Egyptian priests who fled the invading army of King Sabacon of Ethiopia, carried with them to Babylon the 365-day Egyptian year, whose beginning had originally been set at the vernal equinox. They founded there the era of Nabonassar—"beginning the years thereof on the very same day with the years of Egypt." The era of Nabonassar began on February 26, which was 33 days and 5 hours before the vernal equinox. Since the Egyptian year fell back annually by 5 hours and 49 minutes, an easy calculation demonstrated that 137 years had elapsed since the original Egyptian calendar had been fixed—namely, in 884 BCE, three years before the death of Amenophis.[17] The seemingly sound astronomical argument was based, like the identification of Chiron as an astronomer, on a literary source: Hecataeus' description of the inscription on Amenophis' tomb, as cited by Diodorus Siculus. Newton deemed Diodorus' account unproblematic, as he also deemed his own supposition regarding the formation of the era of Nabonassar by Egyptian priests. Nor was he troubled by the fact that Diodorus named the Egyptian king Osymandyas, for Marsham had determined that Osymandyas was one of Amenophis' royal names.[18]

Whereas astronomically driven evidence—albeit not evidence based on the precession of the equinoxes—had been an integral part of chronology since the sixteenth century, the new mathematical tools introduced by Newton in order to estimate more accurately the lengths of reigns and generations in remote antiquity were wholly novel. The young Newton had absorbed a good deal at Cambridge, where he had developed new ways of thinking about, and working with, observational and laboratory evidence. Unlike such contemporaries as Huygens, who always sought to perfect their measurements until only the best, most reliable of them all would be presented, Newton took the radical step of generating a good number by combining all of the bad ones. The chronology's astronomical basis was grounded on this method, but so, in a broader sense, was the realization of a theory for the origin of civilization in the synchronizations and Euhemerizations that Newton eventually developed. He was

[16] Herodotus, 1975, p. 279. In the Temple of Amenophis "the golden border or circle" above the tomb had been "divided into 365 days wth the rising & settings of the stars in every day shews that they added 5 days to the 360 either in his reign or upon his death or consecration. For the Egyptians dedicated those five days to Osiris, Isis, Orus, Typhon & Nephte the wife of Typhon & therefore these days were added to y[e] year after the reign of those five Kings & Queens." New College MS 361.1 fol. 100[v]; MS 361.3 fols. 61[v], 129[v]–130, 136, 170, 184[v]. In the *Chronology* Newton partially corrected himself, now contending that the five days were actually added to the Egyptian year during the reign of Ammon, but the new calendar "was scarce brought into common use before the Reign of *Amenophis*." Newton, 1728b, p. 30.

[17] New College MS fols. 184[v]–185; Newton, 1728b, p. 31, where Chaldea replaced Babylon as the destination of the fleeing Egyptian priests.

[18] Diodorus, 1939, vol. 1 p. 175; Marsham, 1676, pp. 423–33. Osymandyas is one of the royal names of Ramesses II.

confirmed in his skepticism concerning words that derived especially from a time before "poetry" had been displaced by "prose" by the utterances of coiners and witnesses who testified before him, utterances that were quasi-poetic, as Newton would have seen it, in the sense that they related conflicting fables. Numbers and testimonies had to be set off against one another in order to generate truth out of conflicting memories.

Newton certainly did not hide his methods for combining the data that he had produced from the Eudoxan remarks provided by Hipparchus in the *Commentary*, but he did not emphasize them either. Nowhere did he explain his rationale for dividing the colural data into two separate sets, one for the solstices, the other for the equinoxes and then taking a cut between the two numbers that resulted from each set. Neither did he explain why he had combined the values in each set and then divided each set through by their respective total numbers, forming the average among them. As late as the 1720s, this way to ameliorate observational and measuring error remained extremely rare, and in fact controversial in astronomy into the 1750s. And none of this appeared in the abbreviated version that had been printed in France—though if it had, the difficulties his readers there already faced would only have been compounded. Newton's unusual ways with evidence, and his views concerning the trustworthiness of the antique records that he sought to synchronize and, where appropriate, to turn into numbers, are reflected as well in his attitude toward memory, itself a product of the senses whose unreliability he sought to mitigate.

The printed *Chronology* speaks twenty-three times of memory, invoking it in such phrases as "there could be no memory of things done above three or four ages before the first use of letters" or "in memory of *Saturn*'s coming into *Italy* by sea, the *Latines* coined their first money with his head on one side, and a ship on the other." In memory Newton sought the concealed records of forgotten events. This might seem to be just a long-standing Euhemerist way to transform myth into experience, yet the deceptive, revelatory power of memory was for Newton irrevocably bound to the very nature of human cognition. And it was not individual recollection that both deceived and revealed, but something shared, the memory of an entire culture: "In memory of this Emendation of the year they dedicated the five additional days"; "In memory of *Saturn*'s coming into Italy by sea, the *Latines* coined..."; "the *Egyptians* in memory of this action erected a statue to *Sethon*."[19] Newton invoked and used the idea of *cultural memory*, a conception that for him embodied the vagaries of evidence that could not be avoided, but that could perhaps be tamed.

The novel, and from a modern viewpoint quasi-statistical, methods introduced by Newton were designed to establish a proper conformity with the regularity of nature. During the 1690s he had become convinced that the "course of nature" governs human affairs, just as the "laws of nature" govern the motions of the heavenly bodies. And while the demonstrative (and predictive) power of the latter undoubtedly exceeded that of the former, the "course of nature" is indicative not only of how certain aspects of human life conform to law-like regularities, but how these regularities might be quantified. Small wonder then that, faced by constraints imposed by demographic and societal regularities, Newton abandoned his original schema of postdilu-

[19] Newton 1728b, pp. 79, 154, 282.

vian history as he increasingly recognized the need to allow for a far lengthier span of time for the evolution of populations and empires. In doing so, Newton, ever willing to take liberties with the written word, exhibited his determination to reject facts that appeared to violate the normal course of nature—even if Scripture appeared to condone these very facts. Just as he ignored all passages in Scripture that insinuated rapid population growth—including the benediction to Noah—so, in contrast to many contemporaries, he disregarded Biblical evidence concerning God's gradual shortening of the human life span.

These three pillars served as the frame of the new chronology established by Newton, into which subsidiary types of evidence were woven. Consider analogy for example. Repeatedly, Newton inferred from the known conditions of one nation the unrecorded affairs of another. As he stated outright on one occasion: "Had we good accounts of the earliest ages of the more eastern nations I doubt not but we might find the same constitution of things there [the Near East]," as existed in Greece and Italy. What held true for the body politic also held true for the organization into cities and the formation of empires. From observing how "Kings upon founding or much enlarging their kingdoms usually build their royal cities more large & sumptuous," Newton inferred that the aggrandizement of Tyre by King Hiram attested the establishment of "a new dominion of the Tyrians." Likewise, the correlation Newton established between the grandeur of a nation and the veneration of great men, led him to conclude that from "the extent of [Ammon's] worship [one] may gather the extent of his dominion."[20]

Synchronization ranked even higher in Newton's evidentiary structure. Though commonly used by seventeenth-century chronologers and historians, few went to the lengths that Newton did in ascertaining a seamless synchronization within and between myths—as well as between myth and Scripture—in order to determine that his meticulous synchronizations could only apply to a specific time and place. Richard Cumberland came close to describing Newton's methodology when he introduced his own efforts to synchronize Egyptian and Canaanite history:

> In general I premise, that we do not pretend here to deliver exact or perfect synchronisms, but such as have the nearest approach thereunto, which we can attain in the history of such antient times, whereof we have but imperfect records. And I am encourag'd to use such approaches to truth by the example of the mathematicians themselves, because, tho' they be the best accountants, yet, when the *Data* which they have will not suffice to bring them to a precise determination of the quantity sought, they content themselves with some methods of approximation, which they find exceeding useful in the affairs of human life.[21]

[20] Thus, he utilized Artapanus' testimony (as cited by Eusebius) on the Egyptian multitudes before the time of Moses—who "being under no order, now expelled and now set up kings"—in order to argue "that ye kingdoms of Egypt were at first like those of Greece, many in number & free fom Tyranny, the people placing & displacing their kings that is the captains of their armies till the government was new modelled & reduced into ye 36 Nomi." New College MS 361.1 fols. 94–94^v^, 120; 361.3 fols. 97, 165.

[21] Cumberland, 1724, p. 84.

A strict evidentiary structure, we argue, permeates the *Chronology of Ancient Kingdoms* and a quasi-mathematical form envelops it. Newton's methodology and manner of presentation infuriated many readers of the *Chronology*, and their reaction brings to mind Joseph Scaliger's own lashing out at mathematicians who presumed to appropriate the discipline: "Astronomy is the foundation of chronology; and the modern mathematicians are all asses, and ought not to be ranked among men of letters, but among mechanics. They make many observations, indeed, but are not conversant with good books . . . The mathematicians are offended with me for treating the subject of chronology, and think it was their province exclusively. They are mistaken. A good scholar, with little knowledge of the mathematics, will write better on chronology than a deep mathematician, who is merely such." Mathematical chronologers, he reiterated elsewhere, "collect many passages, but mindlessly (sans esprit) and without judgment."[22]

Fidelity to evidence often prompted Newton to depart from the word of Scripture—or at least from common perception of what Biblical phrasing meant. He followed Scaliger and Selden in considering pagan sources to have near parity with Scripture when the history of gentile nations is recounted and, consequently, he felt free to supplement the spare Biblical narrative with details drawn from myth or his imagination. In the eyes of many, however, his privileging of Egyptian history—and his crediting it as the birthplace of the first great empire as well as of idolatry—might be perceived as contradicting Scripture, as it appeared to suggest that Mesopotamia was the cradle of both. Along similar lines, Newton's belief in the gradual evolution of civilizations, as well as of nations, caused him to reject such widely held opinions as the divine origins of language, and the early origins of the arts and sciences.

Departures of this sort from conventional interpretations placed Newton alongside other early-modern scholarly freethinkers, such as Scaliger, Selden, Marsham, and Spencer. It would be a mistake, however, to interpret Newton's embrace of new canons of criticism, and his preoccupation with the secular history of the ancient world, as testimony to his religious heterodoxy.[23] To begin with, neither the *Theologiae gentilis origines philosophicae*, nor the "Original of Kingdoms," or for that matter the *Chronology* itself can be characterized as theological, notwithstanding their partial address of the origin of religions and idolatry. Neither was Newton a committed anti-Trinitarian for more than a decade before he embarked on the *Origines*—which might otherwise account for its subversive quality. We locate the origins of Newton's serious undertaking of the study of theology in the late 1670s, starting with Biblical prophecies, and we contend that his seeking a dispensation from taking holy orders was not motivated by religious considerations, let alone heterodoxy. Newton's descent into anti-Trinitarian heresy had been gradual, and became fully formed only in the 1690s. Heretical views did not therefore inform Newton's study of prophecies (or postdiluvian history); anti-Trinitarianism only began to take shape after Newton moved beyond the sole hermeneutical analysis of Scriptural prophecy to begin a careful study of the early history of the Christian Church.

[22] Scaliger, 1695, pp. 36–37 [translation from Garnier, 1805, vol. 2 pp. 115–16], 92.

[23] Though the contrary view has been advanced, in particular by Westfall, 1982.

A full history of Newton's evolving theological views exceeds the bounds, and the purposes, of the present book. However, one aspect merits attention. It might be assumed that Newton's heretical views not only forced him into equivocation—allegedly for fear of losing his public employments or worse—but prevented him from publishing his theological and historical writings. This, however, exaggerates the extent of the danger Newton perceived he would encounter, as well as misinterprets the public face of his heresy. Newton refused to engage in the sort of inflammatory public quarrels on which William Whiston thrived—but not out of cowardice. While not making a secret of his religious views, Newton believed that "enmity & discord in things not necessary to communion tends to schism & is contrary to the rule of charity imposed upon all men in the second of the two great commandments."[24] But if loathing of "schism" might explain Newton's reluctance to enter the lists in the debates over the Trinity, fear played no role in his failure to appear in print in other domains. His decision to withhold publication of *An Historical Account of Two Notable Corruptions of Scripture*, for example, was based on personal reasons, not panic over public reaction.[25] Insofar as his historical writings are concerned, Newton never sought to publish his *Origines* simply because he abandoned the project once he found it to be seriously flawed.

The *Chronology* proper should be viewed, fundamentally, as a contribution to secular historical scholarship: an attempt to establish a new chronological system, within the confines of the Masoretic time line, upon which a universal history might be erected. Newton never did explicitly pronounce his position on the reality of Creation, the universality of the Deluge, and the common origination of mankind; they were simply taken for granted, whereas in Query 31 to the *Opticks* he emphasized his belief that Creation could not have emerged out of chaos. Perhaps he would have changed his mind had he lived a Patriarch's length to witness the development of Big Bang and multiple universe cosmology.[26] In the *Chronology* his convictions remain largely invisible, as the book commences in media res. In contrast to earlier chronologers, whose points of departure had invariably been the Creation or the Deluge, Newton began with the expulsion of the Shepherds from Egypt, ca. 1100 BCE. His readers would have recognized that he followed the (Masoretic) time line established by Archbishop James Ussher, which set Creation at 4004 BCE and the Flood 1656 years later (2340 BCE). But in the *Chronology* proper—as opposed to the *Short Chronicle*—Newton deliberately avoided assigning dates to the events he described, orienting the readers instead by a correlation with Biblical events. Underlying Newton's unconventional approach was undoubtedly the desire to present his radical solution to ancient chronology as perfectly conformable to Scripture, and without publicly pronouncing against the chronology of the Septuagint.[27]

[24] Keynes MS 3 fols. 1, 11.

[25] To be demonstrated in a forthcoming article by Feingold.

[26] Query 31 of the *Opticks* reads: "it's unphilosophical to seek for any other Origin of the World, or to pretend that it might arise out of a Chaos by the mere Laws of Nature; though being once form'd, it may continue by those Laws for many Ages. For while Comets move in very excentrick Orbs in all manner of Positions, blind Fate could never make all the Planets move one and the same way in Orbs concentrick." Newton 1721, p. 378.

[27] Already in the 1680s Newton had come to believe that the translators of the Bible into Greek had purposely lengthened the chronology of Scripture, "in order to reach as closely as possible the Egyptian

Ironically the one person in the late seventeenth- and early eighteenth centuries to fully grapple with the implications of time for the evolution of human history, simply would not follow the conclusions of his theoretical understanding and extricate himself from the constraints imposed by Biblical chronology. Newton shared this constraint with virtually all contemporaries. Yet the cacophonous reaction that greeted his system made one thing abundantly clear: if the great Newton could not restore order to chronology, surely the task lay beyond human reach. Witness Viscount Bolingbroke's frustration: all chronological systems, he complained, amount to "many enchanted castles; they appear to be something, they are nothing but appearances: like them too, dissolve the charm, and they vanish from the sight."[28] Chronology did not vanish from sight, but its stature lessened, a shrinking exacerbated by the growing knowledge of antiquity conferred by the new sciences of archaeology, numismatics, epigraphy, comparative religion, and anthropology. Newton was the last great technical chronologer to attempt a system capable of determining with precision the temporal boundaries of ancient history in its entirety. Thereafter chronology receded into the subsidiary, if necessary, historical discipline that it maintains today, and Newton's version of it, so radical for its time, is largely forgotten—though anyone who visits his tomb in Westminster Abbey can see the Chironic colure inscribed in the globe that surmounts it (figure 13.1). Yet his heroic failure to preserve Biblical chronology as he sought to resolve Egyptian origins contributed to the eventual rejection of the hold of Scripture over chronology, a process initiated long before Newton by Joseph Scaliger.

---

numbers." At the time, however, he had concerned himself with the deeds of Noah's immediate posterity, without regard to the precise time in which these events occurred; in theory, both systems of reckoning would have served his purpose. Once the focus of his investigation shifted, however, the Septuagint chronology became problematic. Recourse to it was primarily by those who sought a longer time frame, capable of accommodating sequentially the long lists of Egyptian and Chinese rulers. Newton, following Marsham, believed that he could fit the Egyptian dynasties into the Masoretic time frame—by deeming many of them to be collateral—and therefore found no need for the eight postdiluvian centuries detailed by the Septuagint chronology. In principle, Newton would have found it easy, even beneficial, to admit the additional centuries into his system, as they could provide more time for population growth and the growth of nations. However, if a reordering of Egyptian dynasties followed the admission of the additional centuries, Newton's entire evolutionary schema would have been jeopardized—especially his late dating of the unification of Egypt and the momentous consequences of such a unification for the entire Mediterranean world. Yahuda MS 16.2 fols. 48, 50.

[28] Bolingbroke 1967 [1844], vol. 2, p. 175.

Figure 13.1. Newton's tomb at Westminster inscribed with his colure (photo by JZB).

Appendix A

Signs, Conventions, Dating, and Definitions

i. Signs of the zodiac, sun, and moon

♈ Aries

♉ Taurus

♊ Gemini

♋ Cancer

♌ Leo

♍ Virgo

♎ Libra

♏ Scorpio

♐ Sagittarius

♑ Capricorn

♒ Aquarius

♓ Pisces

☉ Sun (Sol)

☽ Moon (Luna)

Each of the twelve zodiacal signs designates (sequentially) a 30° segment of the ecliptic. Newton usually followed convention in giving longitudes as position within a sign, e.g., ♏25°15′23″ is equivalent to a longitude of 235°15′23″ with respect to the vernal point at the beginning of Aries.

ii. Definitions

Acronychal — pertaining to sunset.

Anomalistic year — the number of solar days that it takes the sun, starting from aphelion (or perihelion), to return to aphelion (or perihelion).

Aphelion (solar) — the greatest distance between the sun and the earth.
Asterism — a group of stars that together form a notable configuration, such as a constellation.
Cardinal points — the equinoxes and solstices.
Colures — the two great circles which respectively include the poles and equinoxes, or the poles and solstices.
Dates — are given as unsigned numbers for CE (Common Era) and as negatively signed numbers for BCE (Before the Common Era).
Declination — a star or planet's distance from the equator along a great circle orthogonal to it.
Ecliptic — the great circle on the stellar sphere along which the sun travels.
Epoch — the year for which the coordinates of a star or planet are observed or calculated.
Equator — the great circle on the stellar sphere which is perpendicular to the sphere's axis of diurnal rotation.
Equinoxes — the points in which the ecliptic and the equator intersect. When the sun reaches the equinoxes, the lengths of day and night are as close to equal as they ever become.
Globe view — a depiction of the celestial sphere seen from outside looking inward. Constellations should be depicted from the rear. *See sky view.*
Heliacal — pertaining to sunrise.
Julian date — the number of solar days since noon, January 1, 4713 BCE, a convention introduced by J. J. Scaliger in 1583.
Latitude — see *tropical coordinates.*
Longitude — see *tropical coordinates.*
Lunar month — see *synodic month.*
Metonic cycle[1] — a calendar suggested by Meton in −432 to be based on the relation 6940 days = 19 tropical years = 235 mean synodic months.
Nabonassar, Era of[2] — 1 Thoth, Year 1 of Nabonassar = 26 February, −747.
Perihelion (solar) — the least distance between the sun and the earth.
Precession — the millennia-long rotation of the stellar sphere about an axis perpendicular to and through the center of the ecliptic in the same direction as the sun's yearly motion through the zodiac, i.e., forward through the signs as time increases.
Right ascension — a star or planet's distance along the equator from the vernal point. Because of precession the object's right ascension changes with time, so that the epoch of the observation or calculation must also be specified.
Sidereal coordinates — the coordinates of a star or planet's distance from another star chosen as the point of reference. Since (ignoring small proper motions) the stars do not change their distance with respect to one another, these coordinates are not affected by precession.
Sidereal year — the number of solar days that it takes the sun, starting from a given star, to return to it. Precession produces a difference between the sidereal and tropical years.
Sky view — a depiction of the celestial sphere as seen from the center. Constellations should be depicted from the front. See *globe view.*
Solar day — the interval between two successive returns of the sun to the local meridian.
Solstices — the northernmost and southernmost points of the ecliptic. In the Northern Hemisphere when the sun reaches the northernmost point, the length of the day is longest, and the night shortest, and vice versa when the sun reaches the southernmost point.
Stellar sphere — a sphere centered on the earth that appears to carry the stars and planets.

[1] On the Metonic cycle see Evans, 1998, p. 185 and Pedersen, 1974, p. 128.
[2] See Evans, 1998, pp. 163–204 for calendars and sample computations.

Synodic — the time between two events of the same kind in respect to the appearance of a celestial object in relation to the sun (e.g., a maximum elongation).
Synodic Month — the time from one new moon to the next, also called the *lunar month.*
Tropic — either of two circles on the stellar sphere which are parallel to the equator and which respectively touch the solstices.
Tropical coordinates — a star or planet's distance along the ecliptic from the vernal point (*longitude*), and its distance from the ecliptic and orthogonal to it (*latitude*). Because of precession the object's longitude changes with time, so that the epoch of the observation or calculation must also be specified.
Tropical year — the number of solar days that it takes the sun, starting from a tropic, to return to it.
Vernal point — the spring intersection of the ecliptic with the equator.
Zodiac — the band of stars that form the twelve constellations arrayed along the ecliptic.

iii. Rising and Setting Terminology

TABLE A.III

Latin phrase	*Latin Equivalent*	*English*	*Meaning*
Ortus Heliacus	Ortus matutinis	heliacal rising	star rises at sunrise
Occasus Heliacus	Occasus matutinis	heliacal setting	star sets at sunrise
Ortus Acronychus	Ortus vespertinus	acronychal rising	star rises at sunset
Occasus Acronychus	Occasus vespertinus	acronychal setting	star sets at sunset

iv. Conventions and Dating

BCE, CE, and (-) dating

Except when quoting, we adopt *BCE*, *CE*, respectively, to signify times *Before the Common Era* (conventionally *BC*) and the *Common Era* (conventionally *AD*). For textual simplicity, dates *BCE* will usually be designated here with a minus (-) sign, e.g., -4004 denotes 4004 *BCE* (and not 4005 BCE).

Egyptian Calendar

Ancient Egyptians divided the year into twelve months of exactly thirty days, plus five extra or *epagomenal* days. The Egyptian year was closely linked to, and probably first began at, the *heliacal rise of the dog-star Sirius*, or "the going forth of Sopdet (*pr.t spd.t*)." Throughout ancient Egyptian history this occurred nearly the same number of days before the summer solstice, which in turn marked the period of the Nile's annual flood. Since the Egyptian year is short about a quarter day, it moves out of synchrony with astronomical and seasonal events by one day every four years. This has occasioned much scholarly discussion.[3] Never-

[3] On which see, among others, Clagett, 1995; Lowdermilk, 2000; Neugebauer and Parker, 1960–69; Parker, 1981.

theless, a year of exactly 365 days makes it simple to keep track of time and was accordingly used by Ptolemy in the *Almagest* as an absolute counter.

English Old-Style Calendar

The old-style English calendar differed by eleven days from the Gregorian by the eighteenth century, and it also set the New Year on March 25 instead of January 1. This remained the case in England, Ireland, and the British colonies until 1752. So, e.g., March 25, 1727 old-style corresponds to April 5, 1727 Gregorian, while March 24, 1726 old-style corresponds to April 4, 1727 Gregorian.

Julian Period

A system of dating devised by the influential chronologer Joseph Justus Scaliger and published in 1583. The Julian period begins at noon, January 1, −4713. The date was old enough to include known history and was based on the synchronous occurrence of three calendrical cycles.[4]

v. Months of the Egyptian Year

Although Newton does not use Egyptian months, Ptolemy did in the *Almagest*, and they are accordingly worth listing for reference:

Month 1: *Thoth*
Month 2: *Phaophi*
Month 3: *Athyr*
Month 4: *Choiak*
Month 5: *Tybi*
Month 6: *Mechir*
Month 7: *Phamenoth*
Month 8: *Pharmouthi*
Month 9: *Pachon*
Month 10: *Payni*
Month 11: *Epiphi*
Month 12: *Mesore*

The *Almagest* uses the *era of Nabonassar* (−747) as its epoch since its Babylonian observations date from the reign of King Nabonassar.[5]

[4] Grafton, 1993. For calendar history and methods of computation see Richards, 1998.

[5] Ptolemy, 1998, p. 9.

Appendix B

Newton's Computational Methods

B.i. Use of logarithms

There are many calculations scattered throughout Newton's manuscripts on chronology. Some of these are simple multiplications and divisions that hardly require the use of logarithms even when the numbers are large. Others require the use of trigonometric tables, and here logarithms were essential.

Henry Briggs originally computed base 10 logarithms to 14 decimal places for the integers from 1 through 20,000.[1] These were extended to include the integers from 90,000 to 100,000 by Adrian Vlacq, whose results were printed in 1624.[2] In 1628, Vlacq published a new version that expanded the set to all integers from 1 through 100,000, though the logarithms were given only to ten decimal places.[3] This edition also included a table (again to ten decimal places) for the logarithms+10 of the sine, cosine (listed as the sine of the angle's complement), tangent, cotangent (listed as the tangent of the angle's complement), and secant (the reciprocal of the cosine) for every minute from 0° through 90°. An English language version appeared in 1631,[4] and a new version of the 1628 Latin was printed in 1633 with a different title that emphasized the trigonometric tables.[5] Newton owned this latter edition. There were a number of other printed tables that would have been reasonably available, including one in Streete's *Astronomia Carolina* that provided the logarithms for the sine, cosine, tangent, and cotangent, as did Vincent Wing's English *Harmonicon* of 1651 (but not his Latin *Britannica* of 1669, which provided log tables for astronomical motions[6]). Newton possessed Wing's texts, both of which he annotated, the *Harmonicon* extensively, and he took notes on Streete's *Astronomia*, which he must have borrowed, near the time of composition of the *Questiones quaedam philosophicae*.[7] Newton also owned, and was clearly

[1] Briggs certainly did not think in terms of bases but just required the logarithm of 10 to be one.

[2] Briggs, 1624.

[3] Vlacq, 1628.

[4] Briggs, 1631.

[5] Vlacq, 1633.

[6] Wing's *Britannica* contains for example a table of logistical logarithms. These tables, which were in common astrological use, provide the logarithms of the angles through which the mean sun has moved (viz. at the rate of 1° per 4 minutes of time): each of Wing's columns represents a step of 24 minutes of time (6° angular), and each of his rows represents a step of 24 seconds of time (6′ angular) (Wing, 1669).

[7] Streete's *Astronomia* was first printed in 1661. It was reprinted in 1663, 1664 (by a different printer, as well as a separate appendix), and then after Streete's death in 1689 in 1705, 1710, and 1716. The young Newton

quite familiar with from at least 1665, Oughtred's 1657 *Trigonometria*, which contained logs for the sine and tangent.

These trigonometric tables made astronomical computations particularly simple. The tables are constructed for the lengths of the sides of right triangles drawn on a hypotenuse which is itself the radius of a circle, with the radius set to be 10^x units. One of the legs is, in modern notation, $10^x\sin(\theta)$. And then the sine tables in effect provide $10^{x-y}(10 + \log(\sin(\vartheta))$ and similarly for the cosine and tangent, where the value of y depends on the maximum logged sine desired.[8] The logs in Newton's chronology MS are to eight or nine digits, which points to his having used the Vlacq tables. For an example of a division that yields an angle *EK* through its sine see figure D.3, where we see that Newton even wrote separately the required addition of 10.

B.ii. Computation of Stellar Coordinates

In order to calculate the coordinates of the comet of 1664/early 1665, the twenty-two-year-old Newton required the angular distances between the comet and each of two stars, which he measured, and he required as well the coordinates (either longitude/latitude or right ascension/declination) of the stars, which he obtained from Tycho's catalog (updated, we shall see, for precession). The expression for the cosine of a side in a spherical triangle in terms of the other two sides and the angle between these latter then solves the problem. We do not have direct evidence for Newton's knowledge at this time of spherical relations. Wing's *Harmonicon*, with which he was almost certainly familiar by then, does contain a table of relations for spherical triangles, which Wing applies to a numerical example in which an angle is sought given the triangle's three sides.[9] His procedure is, however, extraordinarily awkward computationally since it requires introducing a third arc drawn from the vertex of the angle sought and perpendicular to the latter's opposite side, thereby forming an inner pair of right spherical triangles to which the simpler relations governing the latter can then be applied. It is certainly possible that Newton followed Wing's procedure, but he also might have known other sources at this time. Whiteside remarks that Newton's library also contained "well-thumbed" copies of Oughtred's 1657 *Trigonometria* and Norwood's 1645 *Trig-*

---

used Streete's *Astronomia* in late 1664/early 1665 (cf. ULC 3996, ff26v–30v, transcribed and discussed in McGuire and Tamny, 1985, pp. 354–65). Streete's table sets precession at 48″ per year, or 80′ per century, which is only 58′ in 72 years (Streete, 1663, p. 22)—compared to the value that Newton later used of one degree in 72 years (50″ per year), or indeed the one he used in his comet calculations of late 1664/early 1665, which he probably obtained from Wing's *Harmonicon*.

[8] The 1631 Briggs provides integer logs with a maximum value for the logged sine of 10^{11} on a radius of the same magnitude, and so he tabulated $10^{10}(10+\log(\text{function}))$, i.e., y is 1. Oughtred's and Streete's tables have a maximum value of 10 on unspecified radii, which means that they computed 10+log(function). Since the largest number of digits (excluding the first value) in their logged sine tables is 7, the corresponding radius must have been 10^7 (though Oughtred intended to include 8 digits, the last one was omitted; cf. Oughtred, 1657b, p. 37). Wing's tables reach 10^7 on an equivalent radius, and so he tabulated $10^6(10+\log(\text{function}))$, yielding a y of 1 as with Briggs but with no more than 7 digits for the logged sines and cosines. Gunter wanted his logs to reach 10,000 on the same radius, so he tabulated $10^3(10+\log(\text{function}))$. Gunter's table runs to four decimal places with a four-place integer, so that his radius was effectively 10^8 (x 8, y 1).

[9] Wing, 1651, p. 16.

onometrie. Oughtred provides an example for finding an angle given the three sides, but like Wing he introduces the extra arc with the resulting inner triangles.[10]

We do know that at some point Newton took notes on Gunter's "Description and use of the sector …"; Whiteside suggests about 1665–66.[11] Gunter also introduced the auxiliary arc along the way, and he too solved the problem, but unlike Wing or Oughtred he achieved a formulation that eliminates the arc in the end, which greatly simplifies computation.[12] Newton reproduced this result by working with the auxiliary arc, explicitly noting the solution as Gunter's.[13] With reference to figure B.1, it has the form

$$\frac{r^2(\text{versin}(\widehat{ED}-\widehat{EP}) - \text{versin}(\widehat{PD}))}{\sin(\widehat{ED}) - \sin(\widehat{EP})} = \text{versin}(\angle \text{DEP})$$

Since the versed sine is just 1-cosine, and r is the standard introduction of the radius and may be set to one to find the modern expression,[14] this is equivalent to

$$\cos(\widehat{DP}) = \cos(\widehat{ED})\cos(\widehat{EP}) + \sin(\widehat{ED})\sin(\widehat{EP})\cos(\angle DEP)$$

which is just what Newton needed to efficiently find the comet's coordinates. It is therefore possible, and perhaps probable (particularly in view of the likelihood that he used a cross-staff for measurement, also described at some length by Gunter), that Newton had worked through this material by December 1664.

In figure B.1, S_1 and S_2 are stars of known coordinates λ and β, C is a comet of unknown coordinates, and the distances $\widehat{CS_1}$, $\widehat{CS_2}$ are known from observation. $\widehat{EFD}$ is an arc of the ecliptic, and P is the ecliptic's pole. Newton observed the comet in several different configurations with respect to reference stars. The following pertains to his observation of Dec. 27 (=Jan. 6), in which S_1 was Aldebaran, and S_2 was Rigel. We are given: λ_1, λ_2, β_1, β_2, $\widehat{CS_1}$, $\widehat{CS_2}$ and we are to find λ_c and β_c. To do so we apply the cosine formula five times:

$$\begin{aligned}
\angle \mathbf{S_1PS_2} &= \lambda_1 - \lambda_2 \\
\widehat{\mathbf{PS_1}} &= 90° - \beta_1 \\
\widehat{\mathbf{PS_2}} &= 90° - \beta_2 \\
\cos(\widehat{\mathbf{S_1S_2}}) &= \cos(\widehat{PS_2})\cos(\widehat{PS_1}) + \sin(\widehat{PS_2})\sin(\widehat{PS_1})\cos(\angle S_1PS_2) \\
\cos(\widehat{PS_1}) &= \cos(\widehat{PS_2})\cos(\widehat{S_1S_2}) + \sin(\widehat{PS_2})\sin(\widehat{S_1S_2})\cos(\angle \mathbf{S_1S_2P}) \\
\cos(\widehat{CS_1}) &= \cos(\widehat{CS_2})\cos(\widehat{S_1S_2}) + \sin(\widehat{CS_2})\sin(\widehat{S_1S_2})\cos(\angle \mathbf{CS_2S_1}) \\
\angle \mathbf{CS_2P} &= \angle CS_2S_1 - \angle S_1S_2P \\
\cos(90° - \beta_c) = \cos(\widehat{\mathbf{CP}}) &= \cos(\widehat{CS_2})\cos(\widehat{PS_2}) + \sin(\widehat{CS_2})\sin(\widehat{PS_2})\cos(\angle CS_2P)
\end{aligned}$$

[10] Oughtred, 1657b, pp. 32–34.

[11] Whiteside, 1967–, pp. 466–73, vol. 461.

[12] Gunter, 1655, pp. 95–96.

[13] "ait Gunter" = "Or so says Gunter": Whiteside, 1967–, pp. 469, n.421, vol. 461.

[14] Since the "right sine," in contemporary terminology, is geometrically half the chord of twice the angle, then the radius of the defining circle is just the sine of 90°.

$$\cos(\widehat{CS_2}) = \cos(\widehat{CP})\cos(\widehat{PS_2}) + \sin(\widehat{CP})\sin(\widehat{PS_2})\cos(\angle\mathbf{CPS_2})$$
$$\lambda_c = \lambda_2 - \angle CPS_2$$

These, or related (depending on the configuration), formulas were also deployed by Tycho and Hevelius to produce their stellar catalogs. They yield right ascension and declination, or longitude and latitude, depending on which set of coordinates is known for the reference stars. Once accurate clocks became available this procedure was infrequently used. Instead, astronomers measured the altitude of an object and the time of its passage across the local meridian to determine right ascension and declination. Longitude and latitude would then be computed. This replaces two distance measurements with an altitude measurement and a clock reading and avoids the proliferating computational inaccuracies that inevitably arise in using spherical trigonometric formulae multiple times.

On December 27, "before midnight Sirius being 16^d high ye distance of ye comet from Aldeboran was 28^{d}11′. From Rigel 38^{d}36½. Its longitude was 37^d.4′.13″. its latitude south 10^d,20′,47″. At 9^{h}8′ at night." The comet was in fact at longitude 36°50′42″, latitude 10°3′30″ south, and its respective distances from Aldebaran and Rigel were 28°21′23″, 36°55′52″. Newton was accordingly extremely good in measuring the distance to Aldebaran (10′), but over a degree and a half off in the distance to Rigel. His computed longitude differs from the comet's by about 50′, and his computed latitude is off by less than 20′.

We can use these results to estimate not only Newton's observational accuracy, but his computational one as well, and further, whether or not he was using values for stellar longitude that were reasonably well updated for precession. To do so we determine the coordinates of Aldebaran by using Newton's cometary distance and his computed locus for the comet. We need only assume that he took the stellar latitudes from Tycho's catalog, which was conveniently available in Wing's *Harmonicon*. This gives $\widehat{PS_1}$ from the catalog, $\widehat{CS_1}$ from Newton's measurement, and $\widehat{CP}$ from his own computation. We find (assuming a correct computation on his part) that the longitude for Aldebaran would accordingly have been ♊5,7,30, for a difference from Tycho's longitude (♊4,12,30) of 55′. From the end of 1600—the epoch of Tycho's tables—through the end of 1664 marks 63 years. Clearly the difference between Newton's and Tycho's longitudes for Aldebaran would then be due to precession since it is close to a degree. The precession rate that follows from an accurate calculation of the longitude that would yield Newton's result for the comet's coordinates is about 52.4″ per year.

Newton, however, had to compute using log tables for the trigonometric functions. Our computation cannot therefore by itself specify the precession rate that he did use, and so what his level of computational accuracy was, without further information. Wing's *Harmonicon* yields it. The *Harmonicon* provided an example which entails a rate of about 51.5″, while his reprint of Tycho's catalog specifies that the catalog applies to "the beginning of the year of man's redemption 1601." Let's assume that Newton did use Wing's rate, which would produce a longitude for Aldebaran of ♊5,6,35 and therefore a value for $\angle CPS_1$ of 28°2′22″ on the basis of a highly accurate computation. Since Newton's longitude for the comet was ♉6,50,42, under this assumption he would himself have obtained it from a value of 28°3′3″ for $\angle CPS_1$, for a computational difference of only 41″. This is utterly negligible, especially since there's no evidence that Newton went to elaborate efforts in calculation, and since he measured distances only to the nearest minute in any case.

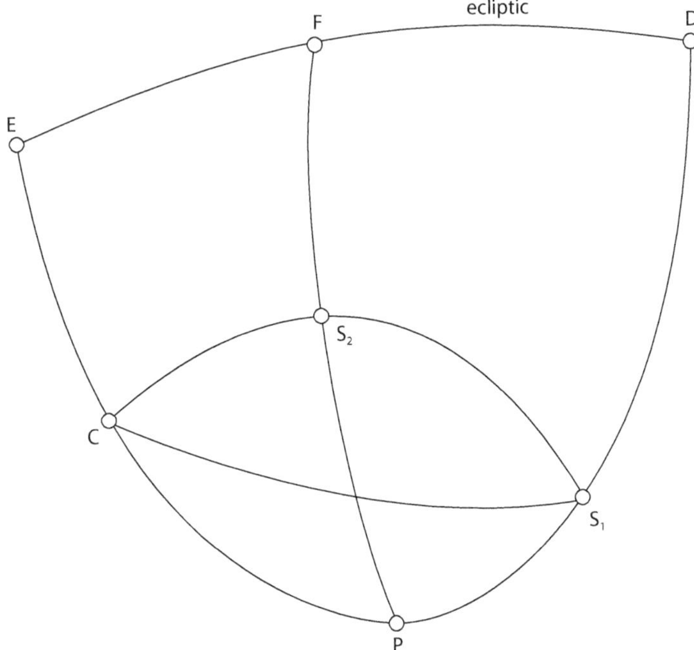

Figure B.1. Locus of an unknown object (C) given two known ones (S_1,S_2) and their distances to the unknown.

Newton would have used a similar method in his chronological computations to find the midpoint between two stars, which arises for two of his published results, though he used only one of them: the one he used specifies that the colure would have passed through the midpoint between Bayer ν and ξ in the head of *Cetus* with end of 1689 longitude and latitude respectively of ♉ 4°3′9″, −9°12′26″ and ♉ 3°7′37″, −5°53′7″. The pair that he decided not to use were *Prima Arietis* and *Ultima Caude Arietis*, with end of 1689 longitude, and latitude, again respectively, of ♈ 28°51′0″, 7°8′58″ and ♉ 19°3′42″, 2°34′5″.

We can find the midpoint between a pair of stars with given coordinates as follows (figure B.2). The longitudes and latitudes of our stars S_1, S_s are, respectively, ♈E_1, S_1E_1 and ♈E_2, S_2E_2. From the figure we find the following relationships, which determine the longitude ♈E_M and latitude ME_M of the midpoint M between them:

$$\cos(\overset{\frown}{\mathbf{S_1S_2}}) = \sin(\overset{\frown}{S_1E_1})\sin(\overset{\frown}{S_2E_2}) + \cos(\overset{\frown}{S_1E_1})\cos(\overset{\frown}{S_2E_2})\cos(\overset{\frown}{\text{♈}E_2}-\overset{\frown}{\text{♈}E_2})$$

$$\overset{\frown}{\mathbf{S_1M}} = \overset{\frown}{S_1S_2}/2$$

$$\cos(\angle\mathbf{PS_1S_2}) = \frac{\sin(\overset{\frown}{S_1E_1})\cos(\overset{\frown}{S_1S_2}) - \sin(\overset{\frown}{S_2E_2})}{\cos(\overset{\frown}{S_1E_1})\sin(\overset{\frown}{S_1S_2})}$$

$$\cos(\overset{\frown}{\mathbf{PM}}) = \sin(\overset{\frown}{S_1E_1})\cos(\overset{\frown}{S_1M}) - \cos(\overset{\frown}{S_1E_1})\sin(\overset{\frown}{S_1M})\cos(\angle PS_1S_2)$$

$$\textit{latitude} = 90^\circ - \overset{\frown}{PM}$$

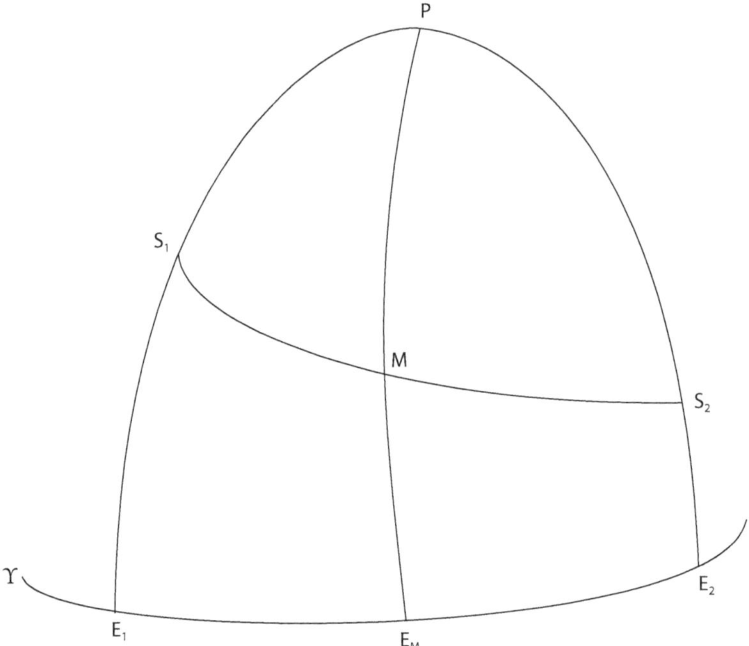

Figure B.2. To find the midpoint M between stars S_1 and S_2

$$\cos(\angle \mathbf{S_1PM}) = \frac{\cos(\widehat{S_1M}) - \sin(\widehat{S_1E_1})\cos(\widehat{PM})}{\cos(\widehat{S_1E_1})\sin(\widehat{PM})}$$

$$longitude = \widehat{\Upsilon E_1} + \angle S_1PM$$

The longitude, latitude of the midpoints between the stars in, respectively, *Cetus* and *Aries* would accordingly be ♉ 3°35′17″, −7°32′46″ and ♉ 8°59′28″, 4°56′6″.[15]

[15] For comparison, Horsley computed the *Aries* coordinates to be, respectively, ♉8°59′27″, 4°56′5″ (Horsley, 1779–1785b, pp. 67, vol. 65).

Appendix C

Commented Extracts from Newton's MS Calculations

GENERAL NOTE

Of the two sets of extensive notes and drafts for the *Chronology* (New College 361, vol. 2 and vol. 3), vol. 2 contains many folios with definite astronomical computations, while there are few if any in vol. 3. Both volumes are difficult to date because the folios in each are clearly not in anything like chronological order, nor could they be, since Newton frequently went back to a page and added or altered remarks over time. The extracts that follow contain most, if not all, of the significant computations that can be extracted from the MS. It's important to note, however, that, with a few exceptions, the remaining MS do not contain the scrap pages on which Newton must have done most of his arithmetic; neither are there any notes concerning the equations to be used, besides one reference to Riccioli for Hesiod, and so we have had to reconstruct the procedure from Newton's results. Our reconstruction is, however, firm, first because we can obtain Newton's values (or close enough to them given his use of log tables), second because we know the text that Newton had to hand (Riccioli), and third, because the problems of colures and rising/setting are not extraordinarily difficult to solve using techniques of the day.

C.i. The Colures

Throughout the second bound volume of New College 361 various numbers appear, sometimes connected to text, other times without any clear connection to nearby words. We cannot therefore guarantee to have uncovered every calculation in the MS that concerns colures, or for that matter always to date precisely the ones that we have identified. Nevertheless, familiarity with Flamsteed's and Hevelius' stellar coordinates, as well as with Newton's methods of calculation does allow us to pinpoint, and sometimes to date, his changing computations for both the colures and, we shall see, for Hesiod. New College 361, vol. 2 contains folios that either compute colures or refer to them. Conduitt referred to the folios in vol. 2 as "Loose papers relating to the Chronology," which indicates that he was himself unsure of their original order. In fact there probably is no original order for the folios proper, because Newton would keep pieces of paper for years and add calculations here and there on unused parts of a page. The proportion of folios that contain colure calculations to the entire number is not large, because most of the remaining pages follow Newton's lifelong habit of securing things in his memory by writing and then rewriting, with changes, paragraphs and whole sections of whatever he was working on. We list relevant information from all of these folios below and additionally transcribe several that are exemplary of Newton's writing and computational technique.

C.i.a. COLURES 1. New College MS 361(2)f7r

This folio immediately follows Conduitt's "Loose papers" designation and is a version of what eventually appeared in print. The folio includes coordinates for both 1660 (Hevelius) and 1690 (Flamsteed), making its time of first composition ambiguous. However, the paragraph containing the Hevelius data has no overwriting, whereas the following paragraph, which contains the Flamsteed data, overwrites a date originally written as 1660 with 1690 and alters as well the originally written longitude. The overwritten longitude is not a replication of the previous paragraph's 1660 data, and the latitude is not overwritten and is from Flamsteed. Either the first two paragraphs were composed pre-Flamsteed data, and the third added post-Flamsteed, or else this folio dates from very close to the time when Newton first acquired the relevant coordinates from Flamsteed and tried out both. This last possibility seems the most likely because the handwriting and the visible ink density are the same throughout the folio.

MS extract:

Through the middle of those Constellations For Hipparchus tells us that Eudoxus drew the Colure of the solstices through the middle of the great bear & the middle of Cancer & the neck of Hydrus, & the star between the Poop & mast of Argo & the tail of the south fish & through the middle of Capricorn & of Sagitta & through the neck & right wing of the Swan & left hand of Cepheus — & within 9′.50″ of the third. To make it pass as near as can be in the middle of these Starrs, it should cut the Ecliptic in ♌ 5.52.36, & ♒5.54.37. And if it be drawn in the middle between the two places ♌ 5.50.36 & ♌ 5.54.37 it will cut the Ecliptic in ♌ 5.52.36 & ♒5.52.36. This Colure passes also through the middle of the great Bear—described by Eudoxus.

The back of Aries through which the Equinocial Colure should pass is a star of the sixt magnitude marked η[1] by Bayer. Its longitude in the end of the year 1660 was ♉ 9°.22′.57″ & north latitude 6°.7′.20″. And the colure drawn through this star to the Ecliptick in an Angle of 66^{gr}30′ the complement of the angle in wch the Ecliptick cuts the Equator did then cut the Ecliptic in ♉ 6.41.34.

The back of Aries through wch the Equinoctial Colure should pass is a star of the sixt magnitude marked η by Bayer. Its longitude in the beginning of the year 16~~6~~90 was ♉ 9°.48′.35″ <[2]> [original minutes and seconds overwritten and illegible] ~~Lat. Bor. 6.7.96~~ & north Latitude 6°.7′.56″. And the Colure drawn through this star to the Ecliptic in an Angle of 66^{gr}30′ (the complement of the angle in wch the Ecliptick cuts the Equator) did then cut the Ecliptic in ~~♉ 6.41.34 ♉ 7^{gr}.7′.~~[illegible and crossed out] ♉ 7^{gr}.7′.55″. So then the Aequinox between the Argonautic expedition & the beginning of the year 1690 moved backwards 37^{gr}.7′.55″ which after the rate of 72 years to a degree [illegible and crossed out]

[1] Newton here copied Bayer's designator incorrectly: the star whose coordinates he gives, and which is on the back of Bayer's Aries, is ν, not η.

[2] Here Newton has correctly transcribed Flamsteed's ♉ 9°48′35″. However, in the published *Chronology* the coordinate was misprinted, or mistranscribed, as ♉ 9°38′45″.

produces an interval of 2673 years wch counted backwards from ye end of the year 1689 places ye Argonautick expedition in ye 36th year of Solomons reign.

NOTE

The epoch 1660 data put the colure on the ecliptic at ♉ 6°41′34″; the epoch 1689 data put it at ♉ 7°7′55″. The corresponding dates are −952 and −984. The epoch 1660 date worked well, but the other did not because it put the expedition "in ye 36th year of Solomon's reign," four years before his death, whereas Newton was convinced that it had to have occurred after Solomon's demise, though not more than a generation or so later. In the published *Chronology* (using the same latitude and a mistranscribed longitude for the star, now correctly designated ν), Newton obtained an ecliptic locus of ♉ 6°58′47″, which yields the somewhat more acceptable (if still a bit early) date of −973. The improvement is, however, entirely the result of the mistranscribed longitude.

C.i.b. COLURES 2. New College MS 361(2)f82r

The remarks here are on the verso of an envelope addressed "For S^{ir} Isaak Newton in German street." The envelope must have been received before 1709 since that was the year he left his house in Jermyn ("German") Street (having moved there in 1697). The recto and verso of the envelope refer to Palamedes as the originator of the sphere. The verso remarks concern much the same material as C.i.a. COLURES 1, but refer only to the year 1660.

C.i.c. COLURES 3. New College MS 361(2)f131r–131v

The recto contains dating using precession from Thales, while the verso contains colure calculations. The coordinate dates are from 1689, so these calculations (and probably the Thales remarks as well) postdate Newton's definitive adoption of Flamsteed's data. Chiron has not as yet replaced Palamedes as author of the sphere.

MS extract, recto:

<text upright>
Thales revived Astronomy among the Greeks & wrote a book of the Tropicks & Equinoxes. He began to flourish about 324 years after the time when we have placed the Argonautic expedition & in that ~~Aequinoxes~~ time the card. points would go backward 4^{gr}30′ & so be ~~in the~~ removed from the middles of the signes into the ~~middle of~~ the 11th degree thereof. Pliny tells us that Thales—11th degree of Cancer. But it being placed in the middles of the signes by those who formed the sphere, he might ~~favor~~ lean to their opinion so far as to place it in the beginning of the twelfth degree.

<text turned 180°>

& so might be found by Thales in ♈&♎11^{gr}36′ [illegible crossed out] or thereabouts. [For Thales wrote his book of the Tropicks & Equinoxes& therefore observed them & published his Observations [illegible] in the days of Thales be] Pliny tells us that Thales determined—11 degree of Cancer.] After the Argonautic expedition & ~~death of Palamedes~~ Trojan war we hear nothing of Astronomy till the days of Thales. He revived Astronomy & wrote a book of the Tropics & Equinoxes, & Pliny tells us that he determined the occasus matudinus[3] of the Pleiades –11 degree of Cancer. It had therefore moved backwards between 4 & 5 degrees, or about 4½ since the Argonautic Expedition & this motion after ye rate of 72 years to a degree answers to 324 years. Count these years back from the 41th Olympiad in wch Thales was a yong man for Astronomical studies: & the reckoning will place the Argonautic Expedition 42 years after ye death of Solomon as above. But Thales in publishing his book about the Equinoxes might lean a little to the opinion of former Astronomers who placed the Cardinal points in the middles of the signes

MS extract, verso:

<text upright>
And the Equinox in ye end of the year 1689 was gone back 36^{gr}44′ since the Argonautic Expedition.

<text turned 180°. The transcription omits simple arithmetic with the coordinates>

In the end of the year 1689 the star called Prima Arietis was in ♈28.51.00 with north Latitude 7.8.58 And the star called ultima Caudae Arietis was ♉ [illegible]42 with north latitude 2.34.5 And the Coluris Aequinoctis [pass]ing through the middle point between these two starrs did then cut the [Ecliptic] in ♉ 6.45.32

<on left of page below the arithmetic>

[illegible] arm of Cepheus rightly delineated is <a star of the first> magnitude called ο by Bayer. Its <longitude> in the end of the said year was ♉ 5.42.36 [illegible] Colurus passes in the middle between the starrs η and χ of y^{e} [illegible] magnitudes in the neck of the swan [illegible] about a degree from each, & the star of 4th magn. in the north wing of the Swan [illegible] from ο a star of y^{e} 5th magn. in y^{e} left arm of Cepheus. <[4]> And so it has all the characters of the Colurus solstitiorum rightly drawn $29.^{2592}_{39}.10'' = 2627.^{1689}_{938}42$ years after y^{e} d of Solomon.

<on right of page>

But perhaps it may be better to depend upon the Colure drawn through several stars so that the errors of the starrs on either side of each Colure may correct & balance one another.

In the neck of the swan are stars η and χ of the 4th & 5th mag. Their longitudes in the end of that year ♒8.37.28 & ♒4.36.37 & the longitude of the middle point between the ♒6.37.2 colurus between them a degree from each & a degree from κ in the north wing of the 4th magn. & 23′ minutes from ο a star of the 5th magn. in the left arm of Cepheus.

C.i.d. COLURES 4. New College MS 361(2)f146r

The remarks on this page (recto) follow a centered section at the top that writes out the title of the Latin *Optice*: "Optice ~~Libri tres~~ Tractatus de Reflexionibus Inflexionibus Lucis" and consequently must post-date 1706. But it also predates the use of Flamsteed's data since the longitudes are for 1660.

[3] = morning setting: see appendix A.

[4] Here we have the MS version of the only astronomical claim in Newton's *Chronology* that must be entirely rejected, since no colure passes close to this star in Cepheus at all.

MS extract

The extreme flexure or elbow of Eridanus is a star of y^{e} 4th magn. of late referred to ye breast of Cetus but anciently not. Tis the only star in Eridanus through wch this Colure can pass. Its Long anno 1660 fini [illegible] was ♈24.59.45 & south Lat 25.18.19 & the Colure Aequinox through it [illegible] cuts ye Ecliptic in ♉6.51.34. The right hand of Perseus rightly delineated is a star of ye the 4th magn. Its long anno 1660 fin ♉ [illegible] 37.26.50 & the Colure drawn through it cuts ye Ecl in ♉4.33 [illegible]. And the head of Perseus is a star of 5th m. in ♉23.18.1 north lat. 34.19.16; & the Colure drawn through it cuts ~~ye Ecl~~ in ♉5.55.56. And the Colure drawn as neare as may be through these six stars did then cut ye Ecliptic in ♈5.50.16 & ♎5.50.16[5] as [illegible] find by taking the sixth part of ye the sum of the longitudes [of] the six colures drawn severally [illegible deletion] the ~~afore~~said six stars did cut the Ecliptic. And this Colure is 90 degrees from the solstitial colure found above: wch forms [illegible] the truth of the determination. This Equin[ox] Colure thus found passes [illegible] through the first star in the left hand of Arctophylax [illegible] along the middle of his body.

C.i.e. COLURES 5. New College MS 361(2)f147r–147v

The folio uses Hevelius' 1660 longitudes, attributes the sphere to Palamedes, and goes through the list leading to the colures.

MS extract, recto

... Whence it's evident that Eudoxus did not observe the Equinox himself but placed it where [Astronomers had found it two hundred years before the days of Euctemon & Meton or above that is] in y^{e} ~~12th degree~~ 11th or 12th degree of the old astral signs where Thales & Anaximander found it ...

The Equinoctial therefore in the Sphere of the ancients described by Eudoxus & Aratus did cut the Ecliptick in the middle of the Asterisms of Aries & Libra. And this is the more manifest by what Hipparchus cites out of Eudoxus ...

Hipparchus not understanding the precession of the Equinoxes disputed against these Descriptions <by Eudoxus of the colures> because the Coluris in his days did not pass through these places. But we are enquiring where they were when the sphere was first formed And tho these descriptions are coarse yet by their help we may come partly near the truth.

[5] Newton mistakenly wrote Aries (♈) and Libra (♎) for Taurus (♉) and Scorpio (♏), which are what his "six colures" yield for the original equinoctial colure. His signs cannot refer to positions within the actual constellations, because he was convinced (and his final date fairly well implies) that the equinoxes and solstices on the original globe were set in the middles of the constellations as they then appeared. Thus in the mid-10th century BCE the vernal equinox is indeed reasonably close to the center of the constellation Aries, though just how to determine the asterism's "center" can easily become a matter of debate.

MS extract, verso

<after finding the solstitial colures from two pairs of stars, taking the mean position between the stars in each pair to find the corresponding colure> The distance between the position of the Colurus solstitiorum found by these <latter> two stars & that found by the two former stars is inconsiderable being only 4′51″. Let the Colurus have a middle position between these two positions & the solstices of Palamedes will be in ♌ 5°.53′.30″ & ♒5°.53′.30″ that is $35^{gr}53'.30''$ distant from the places where they were in the end of the year 1660.

C.i.f. COLURES 6. New College MS 361(2)f148r

Continues C.i.e. COLURES 5., now for the equinoctials.

MS extract So then this <equinoctial> Colurus by the first determination ~~passed through~~ cut the Ecliptic in ♉ 5.59.13 by the second in ♉ 6.2.[illegible] by the third in ♉ 5.41.45. The difference is inconsiderable & taking the mean (wch will be found by taking a third part of the sum of three longitudes) the Vernal Equinox will be found in ♉ 5°.14′.35″. ~~[wch conclusion agrees very well with the place of the Colurus solstitiorum [illegible] summer solstice found above in ♌ 5°.53′.30″]~~ & by consequence the summer solstice in ♉ 5°.54′.35″. ~~wch conclusion~~ This solstice we found above & determined the position of the Colurus solstitiorum to be in ♌ 5°.53′.30″. The difference is inconsiderable, & by taking a medium the Equinoxes & Solstices of Palamedes will be found in ♈ 5°.54′, ♋5°.54′, ♎ 5°.54′, ♑ 5°.54′[6] [illegible] $35^{gr}54'$ forwarder than in the end of the year 1660.

... or one degree in 72 years, the Equinox will go backwards $35^{gr}54'$ in 2585 years being counted backwards from the end of the year 1660 place the formation of the sphere by Palamedes about 57 years after the death of Solomon.

C.i.g. COLURES 7. New College MS 361(2)f148v

The page contains tables of star locations, as well as a computation for an equinoctial colure. It also displays a list of data that pertain to the colure through the "back" of Aries. In the published *Chronology* Newton had calculated its locus by first finding the midpoint between what he took to be the extreme two stars in the constellation itself, but he immediately threw the result aside for a choice based on a specific stars. Here we spy Newton in the midst of deciding which stars to choose. In print he ran the colure directly through the star Bayer ν. The MS lists eight stars through which the colure might pass. The signs on the far left in Newton's MS indicate the constellation to which the star belongs, followed by the star designator, its magnitude, and finally its longitude. The following table gives Hevelius' and Flamsteed's longitudes as well as the star designator. Two of the stars are named (*Aldeberan* and *Antares*) in Hevelius.

[6] Newton again chooses the wrong zodiacal signs here: his coordinates put the "Equinoxes and Solstices" in Taurus (♉), Leo (♌), Scorpio (♏), and Aquarius (♒). Curiously, Newton uses the correct signs in giving the separate values but, as above, miswrites in giving the final, statistical result.

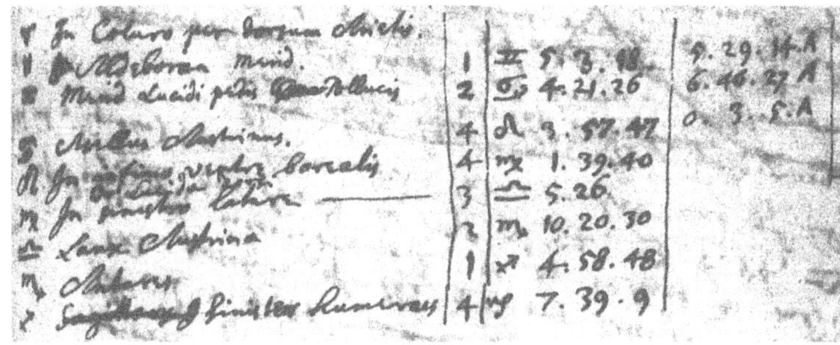

Figure C.1. Newton's MS star locations and coordinates for the colure through the "back" of *Aries*. By permission of the Warden and Scholars of New College, Oxford.

TABLE C.1.
Hevelius' and Flamsteed's coordinates for stars on the colure passing through the "back" of Aries

Constellation	*Designator*	*Location*	*Hevelius' Value Epoch 1660*	*Flamsteed's Value Epoch 1689*	*Flamsteed-Hevelius Adjusted for Precession from 1660 to 1689*[a]
Taurus	*Aldeberan*	south eye	♊5.3.18	5.27.00	−40″
Gemini	Bayer γ	bright star in foot of Pollux	♋4.21.26	4.46.18	30″
Cancer	Bayer δ	south star	♌3.57.41	4.23.40	1′37″
Leo	Bayer ρ	[Hev]furthest part of belly north [Flam]mid belly	♍1.39.40	2.03.40	−22″
Virgo	Bayer ε	left side upper arm	♎5.26.0	5.37.4	−13′18″
Libra	Bayer α	south scale	♏10.20.30	10.46.40	1′48″
Scorpio	*Antares*	Heart	♐4.58.48	5.26.04	2′54″
Sagittarius	Bayer σ	left shoulder	♑7.39.9	8.03.12	−19″

[a] *Using the more accurate precessional period of 25,800 years instead of Newton's 25,920 precession yields a total of 24′22″.*

Table C.1 provides a convenient (though still approximate[7]) list for comparing the differences between Hevelius' and Flamsteed's coordinates, particularly since Newton chose these stars and would therefore have been fully aware of the changes. Hevelius' values (adjusted for precession, ignoring the sign of the difference, and excluding the obvious outlier in *Virgo*) differ from Flamsteed's by an average of 1′10″, with a standard deviation of 58″. For a star on the ecliptic proper, the difference means that Hevelius' date for the original colure would be a bit more than one year different from Flamsteed's, which is utterly insignificant. Newton apparently thought the original observations good to between a half degree and a degree, so that Flamsteed's data clearly offered nothing at all by way of improvement in accuracy. Though Newton could not have known the actual differences until he had seen the data, he would likely have been aware that Hevelius' observations had to be very good, because he had Riccioli's text to hand. There he would have found that in 1644 Riccioli had observed *Aldeberan* (the first star in Newton's list above) at ♊4.44.8. Adjusting for precession between 1644 and 1660 (using Newton's and Riccioli's value of 72 years per degree) would put Riccioli's *Aldeberan* at ♊4.57.32 compared with Hevelius' ♊5.3.18, for a difference of only about 6 minutes. Both Riccioli and Hevelius could hardly have been vastly worse than Flamsteed by nearly the same amounts, and so any improvement that Flamsteed's data might provide would not be more than the difference between the two of them, which translates into just a few years change in the date of the original globe. Moreover, Newton was well aware of the Hevelius affair involving Hooke, and he knew that Halley had attested to the accuracy of Hevelius' observations.

On comparing all three of their longitudes for *Aldeberan* (adjusting both Riccioli and Flamsteed to 1660 using the modern value for precession) we see that Flamsteed's value lies between the other two:

Aldeberan in 1660:

Riccioli:	♊4.57.32
Flamsteed:	♊5.2.38
Hevelius:	♊5.3.18

Taking Flamsteed's as the best observation, we see that Hevelius was a bit better of an observer than Riccioli. This is an anecdotal judgment because we have not systematically compared all of the star coordinates common to the three of them. It does seem likely that Hevelius' values fluctuate around Flamsteed's, which might indicate that random variations in Hevelius' observations swamp whatever systematic difference might exist between his results and Flamsteed's. Riccioli's much smaller set of stellar coordinates probably do so as well, and perhaps with a greater spread than do Hevelius'. However, many stars have proper

[7] Since we have not taken account of a star's proper motion between 1660 and 1690. Nevertheless, the proper motions over 30 years are not large. Aldeberan, for example, changes position at the rate of .2″ per year, so over 30 years it will have moved only 6″ in addition to precession.

motions which, though not large, also need to be taken account of in estimating the comparative accuracies of these three observers.[8]

C.i.h. COLURES 8. New College MS 361(2)f156r

Uses Flamsteed data for Aries midpoint computation, suggests better to use stars through which the colures actually pass.

MS extract

In the beginning of the year 1690 the star called Prima Arietis was in ♈28^{gr}.51′, & the star called Ultima Caudae was in ♉19^{gr}.3′.42″. ~~And the middle between them was in ♉8.5′.22″.~~ And the Colurus Aequinoctiorum passing through the ~~mid~~ point in the middle between these two stars did then cut the Ecliptic in ♉6^{gr}44′ as may be collected by Trigonometry And therefore ~~the middle of~~ the middle of the Constellation of Aries wch in the time of the Argonautic Expedition was in ♈00^{gr}.00′.00″ was in ye beginning of the year 1690 in ♉6^{gr}.44′; & so had then gone back 36^{gr}.44′. It goes back one degree in 72 years & 36^{gr}.44′ in 2645 years. Count these years back from the beginning of the year 1690 & the reconing will end 955 years before the vulgar Era of X, which is about 25 years after the death of Solomon, & thence place the Arg. Exp.

But its better to deduce the positions of the Coluri from the places of the first stars through wch the Coluri passed in the primitive sphere & this method places the Argonatuic Expedition about 42 years after the death of Solomon.

C.i.i. COLURES 9. New College MS 361(2)f157r

Unambiguous date of 1720.

MS extract

Now drawing the Colures of the Equinoxes & Solstices in the positions here assembled, they will cut the Ecliptick in ♉6^{gr}, ♏6^{gr}, ♌6^{gr} & ♒6^{gr} ~~the present year~~ reckoning the Equinoctial where it is this present year 1720. And therefore the Equinoctial has moved 37^{gr} since the Argonautick Expedition But it goes back so many degrees in 37x72 years that is in 2664 years wch counted back from the year 1720 places the Argonautic Expedition 944 before Christ or about 34 years after the death of Solomon.

Now drawing the Colures of the Equinoxes & Solstices in the positions here assembled, they will cut the Ecliptick in points wch in the beginning of the year 1690 were in ♉6½^{gr}, ♏6½^{gr}, ♌6½^{gr} & ♒6^{gr}30′. And therefore in that year the Equinoctial points had moved backwards 36½^{gr} ~~in the~~ from the places in wch they were in the time of the Argonautic Expedition. Now 36½^{gr} at the rate of 72 years to a degrees answers to ~~produce give~~ 2628 years, wch counted back from the year 1690 places the Argonautic Expedition ~~in the begin~~ 938

[8] See Volkoff et al., 1971 for a discussion of Hevelius' work, including several remarks on the accuracy of his measurements.

years before the vulgar Era, that is about 42 years after the death of Solomon. The description of the Colures set down by Hipparchus is but coarse & the places of the fixt stars were but coarsely observed by the ancients; but yet they suffice for determining ~~the Longitude without~~ [illegible crossed out] Colures without erring above a degree in their Longitude: & therefore the Argonautic Expedition was not earlier than ~~the~~ [illegible crossed out] ~~year of~~ Solomons reign.

C.ii.a. Thales 1. New College MS 341(2)f43r

MS extract

~~Among~~ Of the Astronomers who followed after the times of the Trojan war Thales is reconned the oldest. He ~~observed the~~ revived Astronomy, & observed the stars & was the first who could predict eclipses & wrote a book of the Tropics & Equinoxes. Pliny tells us that he determined the Occasus matudinus of the Pleiades to be upon the 25th day after the Autumnal Equinox, & thence Petavius computes the longitude of the Pleiades in ♈ 23^{deg}53′. Now Lucida Pleiadum in the end of the year 1660 was in ♉ 25.15.51 & thence reconning backwards a degree for every 72 years (wch is the motion of the Equinox according to the opinion of Astronomers of this age) the Lucida Pleiadum will be found in ♈ 23.59′ in the [illegible number] year Before Christ that is in the 42th year of Thales. And therefore Thales did not retain the place of the Equinox determined by Astronomers who lived before the Trojan war but placed it where he found it by his own Observations. ~~that is~~ For his publishing a book about y^{e} Tropics & Equinoxes shows that he receded from the opinions of ~~former~~ Astronomers [& by consequence he was the first who removed the Equinoxes from y^{e} 15th degree of the signs & placed them in the 12$^{th.}$ ~~For to do this his [illegible] was greater than any man's~~] & if he was the first who removed the Equinoxes & Solstices from y^{e} 15th degree of the signs ~~[it must be he that placed them in the 12th degree~~ (For his authority to do this was greater than any man's) ~~who placed them in the 15th degree or middle of the signs. And if he was the first~~ we have reason to reccon him ~~the author~~ of the opinion that they were in the 12th degrees.

C.ii.b. Thales 2. New College MS 361(2)f131r

MS extract:

<text upright>
Thales revived Astronomy among the Greeks & wrote a book of the Tropicks & Equinoxes. He began to flourish about 324 years after the time when we have placed the Argonautic expedition & in that time the card. points ~~Aequinoxes~~ would go backward 4^{gr}30′ & so be ~~in the~~ removed from the middles of the signes into the ~~middle of~~ the 11th degree thereof. Pliny tells us that Thales—11th degree of Cancer. But it being placed in the middles of the signes by those who formed the sphere, he might ~~favour~~ lean to their opinion so far as to place it in the beginning of the twelfth degree.

<text turned 180°>

& so might be found by Thales in ♈ & ♎ 11^{gr}36′ or thereabouts ~~or taking [illegible]~~ [For Thales wrote a book of the Tropicks & Equinoxes & therefore observed them & published his Observations [illegible] in the days of Thales be] Pliny tells us that Thales determined—11 degree of Cancer.] After the Argonautic expedition & ~~death of Palamedes~~ Trojan war we hear nothing more of Astronomy till the days of Thales. He revived Astronomy & wrote a book of the Tropics & Equinoxes, & Pliny tells us that he determined the occasus matutinus of the Pleiades—11 degree of Cancer. It had therefore moved backwards between 4 & 5 degrees, or about 4½ since the Argonautic Expedition & this motion after ye rate of 72 years to a degree answers to 324 years Count these years back from the 41th Olympiad in wch Thales was a yong man fit for Astronomical studies: & the reconning will place the Argonautic Expedition 42 years after ye death of Solomon as above. But Thales in publishing his book about the Equinoxes might lean a little to the opinion of former Astronomers who placed the Cardinal points in the middles of the signes

C.iii.a. HESIOD 1. New College MS 361(2)f10r

DATING

<recto>
Addressed to Sir Isaac Newton in 'German' Street, so post-1709
<verso>

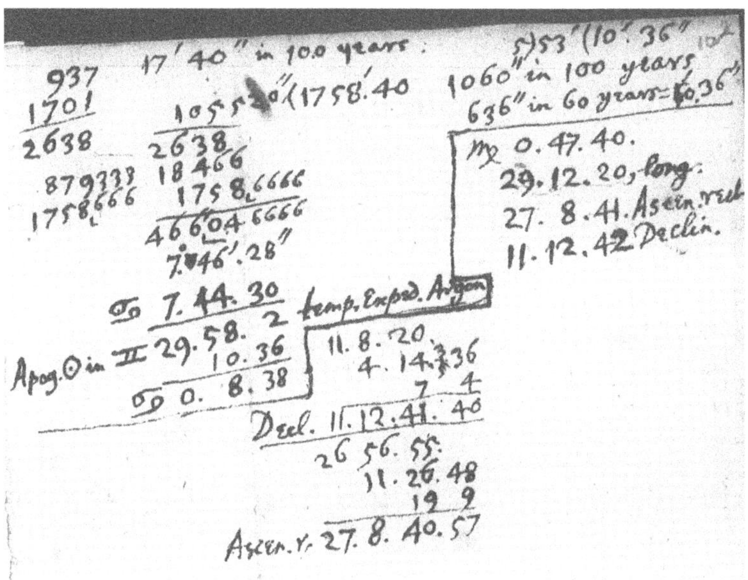

Figure C.2. Newton MS calculation for Hesiod, post-1709. By permission of the Warden and Scholars of New College, Oxford.

C.iii.b. HESIOD 2. New College MS 361(2)f43r

MS

Hesiod flourished in [illegible] mountain Helicon near Atlas in a latitude of about 37°44′ [illegible] He tells us that when sixty days after the winter solstice are past Arcturus has his ortus vespertinus, that is, as authors interpret he rises at sunset sixty days after the day of the winter solstice. How many hours the solstice happens before y^e sun sets is uncertain. If at a middle reconning we take 12, then will be 60½ days from y^e solstice to y^e ortus vespertinus of Arcturus. The Apogee of the sun was the<n> in ♉ 25° or thereabouts & in 60 days the sun moved 62°.

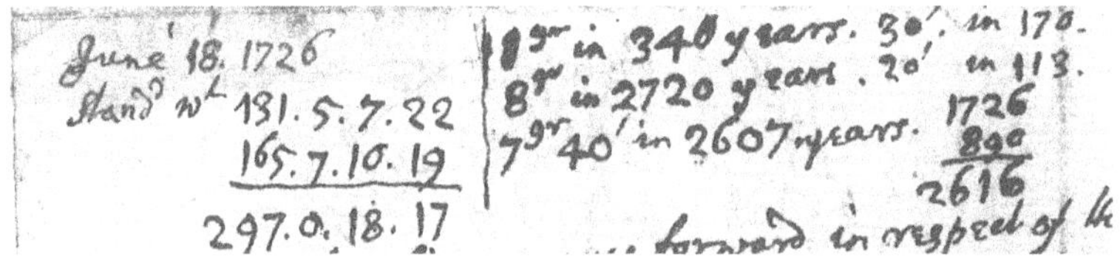

Figure C.3. Newton MS calculation for Hesiod, 1726. By permission of the Warden and Scholars of New College, Oxford.

C.iii.c. HESIOD 3. New College MS 361(2)f60v

NOTE

The calculations transcribed below were done in 1726: first, we have the date in Newton's by-then shaky hand on the left; more significantly, he thought the Trojan war to occur around the time of Hesiod, which he puts circa –880, and this is to be 2,607 years before the time of writing.

The earth's Aphelion moves forward in respect of the fixt starrs 1^{gr} in 340 years, 8^{gr} in 2720 years , $7^{gr}40'$ in 2607 years, i.e. since y^e ~~Argona~~ Trojan war.

The earth's Aphelion moves forward 17′40″ in 100 years in respect of the fixt starrs & 53 in 300 years & 477′ in 2700 years, that is $7^{gr}57'$ in 2700 years & $7^{gr}.39'.20''$ in 2600 years & $7^{gr}.40'.34''$ in 2607 years.

<on the same page but rotated 90°>

Apoge ☉ is in ♋ $0^{gr}.4'.17''$.<[9]>

Aph. ☉ is in ♑ $0^{gr}.4'.17''$.

C.iii.d. HESIOD 4. New College MS 361(2)f80r

MS

Perige Apoge was then in ♑ ♋ $0^{gr}10'$. In those sixty days the suns ~~would move~~ mean motion was $1^s29^{gr}.8'.9''.48''$, & his Aequation $1^{gr}.39'.41''$ [& ~~his~~ by consequence his whole mo-

[9] These are the coordinates in 1700 of the apogee's locus in –880.

tion $2^{s}0^{gr}.48'$, ~~And the opposite point of the ecliptic w^ch^~~ & his ~~place in ♓.$0^{gr}48$~~] & therefore in those sixty days the sun would move from the winter solstice into ♓ $0^{gr}48'$: & the opposite point of the Ecliptick wch rose at the same time with ~~the Ecliptic~~ Arcturus would be in ♍ $0^{gr}48'$. The north Latitude of Arcturus is $30^{gr}57'$, & the elevation—[~~illegible~~] excess is $10^{g}.36'$. wch being added to ♍0.48 gives the longitude of Arcturus ♍ $11^{gr}.24$—~~And there be 62 &~~ latitude is is 45′¼ And these 62′ & 41¼ minutes amount unto 103¼ minutes— gives its correct Longitude in ♍13.~~48~~7¼. The longitude of this starr at the time of the Argonautic expedition was ♍13°.24′.52

C.iii.e. HESIOD 5. New College MS 361(2)f91r

MS

Hesiod tells us that he lived in the fifth age & that age should end when the men then living should grow old & drop into the grave & therefore he & Homer fluorished within thirty or forty years after the taking of Troy. Some ancient authors placed Homer in the days of Orestes the son of Agamemnon & Cornelius Nepos placed him an hundred years before the first Olympiad, where also we place him a Apud Euseb. in Chron.

C.iii.f. HESIOD 6. New College MS 361(2)f107r

NOTE

This folio was written in late 1700 or early 1701, the date for which the solar apogee is given, since Newton had the apogee before May, 1700. Moreover, he had it from Flamsteed.[10] Newton might have obtained Arcturus' position from either Hevelius' 1660, Flamsteed's 1681, or updates of Flamsteed's 1681 coordinates. If Newton used Hevelius' coordinates he might have obtained them directly from Flamsteed (there being no copy of Hevelius' catalog in the register for his library).

Newton gives the longitude of Arcturus in 1700 as ♎20°.18′.52″, which in the last paragraph on this folio he has changed to ♎20°.3′.2″. The very next folio (appendix C.iii.g. HESIOD 7) has " [~~♎20.18′.52″+12″~~] = ♎20°.02′.14″," indicating a corrected calculation. Flamsteed sent Molyneux the longitude of several stars for epoch 1681 in 1683,[11] and we also have his results for 1689/90, as well as Hevelius' for epoch 1660. If we use the value that Newton always employed for precession (50″ per year), then in 1700 and 1701 we obtain the following longitudes, computing for the end of 1700, early 1701:

[10] Flamsteed had a copy of pages Newton had written for the Royal Society in May of 1700 on calendrical issues (Forbes, et al., 1995–, p. 822)—namely some version of MS Yahuda 24. There Newton wrote "In the Royal Observatory at Greenwich in the coming year 1701, at midday on the 1st of January, Old Style, the mean motion of the sun will 9s.21°.42′.38″, of his apogee 3s.07°.44′.30″ …." Flamsteed remarked, towards the end of the copy, "J F's Solar Tables which the most celebrated Newton used."

[11] Forbes et al., 1995–, pp. 116–17.

TABLE C.2.
Hevelius', Flamsteed's, and Newton's coordinates for Arcturus

	Epoch	*Longitude*	*Precession to 1700*	*Longitude in 1700*
Hevelius	1660	♎19°29′6″	34′10″	♎20°3′16″
Flamsteed	1681	♎19°46′30″	16′40″	♎20°3′10″
Flamsteed	1689	♎19°53′52″	10′0″	♎20°3′52″
Newton	(1701)			♎20°3′2″ *or* ♎20°2′14″

MS

The suns Apoge was in ♋7.44.30 in the year of Christ 1700 & an hundred years goes forward 17′.40″ in respect of the fixt starrs & therefore about 880 years before Christ (when Hesiod flourished) was

In the last day of December at noon ~~the~~ A.C. 1700 the suns Apoge was in ♋7.44.30 [~~It moves forward 17′.40″ in respect of the fixt starrs in an hundred years~~] And the star Arcturus was in ♎20°.18′.52″ wth north Latitude 30.57′.00″. And the suns Apoge moves forward 17′.40″ in an hundred years in respect of the fixt starrs & therefore about 880 years before Christ when Hesiod flourished, that is 2580 years before the year of Christ 1700~~, the suns Apoge was Apoge~~ Arcturus was distant from ~~Arcturus~~ the suns Apoge in longitude [~~3^{s}12°.34′.22″ + 7°.35.48~~] 3^{s}20°.10′.10″

In the last day of December at noon A.C. 1700 the suns Perige was in ♑7.44.30 And the star Arcturus was in ♎20°.18′.52″ with north Latitude 30°.57′.00″. And the suns Perigee moves forward 17′.40″ in an hundred years in respect of the fixt starrs, & therefore about 880 years before Christ when Hesiod flourished, that is about 2580 years before the year of Christ 1700 ~~Arcturus was distant from the suns Perige in longitude~~ [~~3^{s}20^{grad}.10′.10″~~] ~~2^{s}9^{gr}.49′.50″~~ the Suns Perige was distant from Arcturus in longitude 2sign.9^{gr}.49′.50″.

In the last day of December at noon A.C. 1700 the star Arcturus was in ♎20°.18′.52″ with north Latitude 30^{gr}.57′.00″, & the suns ~~Apoge~~ Perige was in ♑7^{gr}.44′.30″ being 2^{sign}.17°.26′.52″ distant from Arcturus <in> longitude. ~~The suns Perige moves forward 17′.40″ in an hundred years~~ Hesiod flourished about 880 years before Christ or 2580 years before the year of o^r Lord 1700 & the ~~Apo~~Perige of the Sun moves forward 17′.40″ in an hundred years in respect of ye fixt starrs & 7^{gr}.35.48″ in 2580 years , & therefore in Hesiods days the Perigee of the sun was 2^{sign}.9^{gr}.49′.50″ distant from Arcturus in longitude.

In the last day of December at noon A.C. 1700 the winter solstice was in ♑12^{gr}.20′.35″.13.‴ & the first star of ♈ in ♈28.^{gr}59′.20″

In the last day of December at noon A.C. 1700 the star Arcturus was in ♎20^{gr}.3′.2″ with north Latitude 30^{gr}.57′.00″, & the suns Perige was in ♑7^{gr}.44′.30″ being 2^{sign}.17°.41′.28″ distant from Arcturus in longitude, & the ~~Sun's~~ winter solstice was in ♑12^{gr}.20′.35″. ~~the same being~~ & by consequence 2^{sign}22°.17′.33″. distant from Arcturus in longitude. Hesiod flourished about 880 years before Christ & 2580 before the year of o^r Lord 1701. And the Perigee of the Sun moves forward 17′40″ in an hundred years in respect of the fixt starrs,

& $7^{gr}.35'.48''$ in 2580 years ; & therefore in Hesiods days was distant from Arcturus ~~$2^{sign}.25^{\circ}.17'.16''$~~ $2^{s}.10^{\circ}.5'.40''$. And the winter solstice of the sun ~~was~~ moves backwards 50″ in a year & [~~illegible~~] $35^{gr}50'$ in 2580 years & therefore ~~was in ♒18°.10′.35″ in the days of Hesiod, being then $3^{sign}23^{\circ}.31'.28''$ distant from Arcturus. $3^{sign}23^{\circ}.7'.33''$ distant from Arcturus in longitude~~ in the days of Hesiod was distant from Arcturus ~~$3^{sig}15^{\circ}.55'.40''$ in longitude~~ $3^{sign}.28^{gr}.7'.33''$ in longitude.

C.iii.g. HESIOD 7. New College MS 361(2)f107v

These words are scribbled on an envelope addressed to "The Right worshipful Sr Isack Newton m^{r} of y^{e} mint"

MS

Apoge was then in ♊29. In these sixty days the sun would move from the winter solstice into ♓ $0^{gr}.48'$ counting the signes ♑ ♒ ♓ ♈ ♉ ♊ & c from the middle of the constellation of Capricorn. And the opposite point of the Ecliptic wch rose at the same time with Arcturus would be in ♍ $0^{gr}.48'$—that this excess is Which being added to ♍ $0^{gr}.48'$ gives the longitude of Arcturus ♍. When the sun sets visibly his upper limb is 33′ below the horizon—gives it correct longitude in ♍ 13^{gr}—would be in ♍ $0^{gr}.48'$. In the end of the year 1700 Decemb. 31 at sunset Arcturus was in [~~♎20.18′.52″+12″~~] = ♎20°.02′.14″. And in the 2638 years ~~before when Hesiod flourished the Argonauts made their expedition Arcturus was the solstice [illegible] Aequinoxes were~~ since the Argonautic expedition the equinoxes were gone back 2638x50″ or $36^{gr}.38'.20''$. And therefore in the time of that expedition Arcturus was then in ♍13.23.54.

C.iii.h. HESIOD 8. New College MS 361(2)f151r–151v

<recto>

Hesiod tells us that sixty days after the winter solstice the star Arcturus rose just at Sunset. Till his days & long after the solstices were placed in the middles of the Constellations of Cancer & Capricorn, their motions not being known. And the suns Apoge was then in ♋0°.8′.38″ or thereabouts. In those days the Sun would move from the winter solstice ~~into ♓ $0^{gr}48'$ placing [illegible] beginning of the reconning in~~ $60^{gr}47'40''$ & so be in ♓ $0^{gr}48'$ according to the reconning of the ancients. And the opposite point of the Ecliptic wch rose at the same time with Arcturus would be in ♍ $0^{gr}48'$. The north latitude of Arcturus is $30^{gr}57'$, & the elevation of the Pole at mount Helicon neare Athens where Hesiod lived was $37^{gr}45'$ according to Ptolemy. And thence Ricciolus (lib vi Almagest. Cap. Xx. Prob. Viii) teaches how to compute the excess of the longitude of Arcturus above the longitude of the said opposite point of the Ecliptic. And by the computation I find that this ~~longitude~~ excess is Which being added to ♍ $0^{gr}48'$ gives the longitude of Arcturus ♍ 13^{gr} When the sun sets visibly his upper limb is 33′ below the Horizon being so much elevated by the refraction of the Atmosphere & his center is still 16 minutes lower, in all 49 ′ below the Horizon. And the part of the Ecliptick between the Horizon & the center of the sun is an arch of 62′. And when the star rises visibly it is 33′ below the Horizon being so much elevated by the refraction. And the arch between the Horizon & the star in the parallel of the starrs Latitude is 41 minutes. And these 62 & 41 minutes amount unto 103 minutes. Which being added to the longitude of the starr found above, give its correct longitude in ♍13^{gr} The longitude of the starr at the time of the Argonautic expedition was ♍13.^{gr}24.′52″, as above. And the difference is so small as scarce to be sensible in the coarse observations of the ancients.

* In those sixty days the suns mean motion ~~from the winter solstice~~ was 59^{gr}.8′.10″, & his Equation ~~amounted unto~~ about 1^{gr}.39′.30″ & by consequence his whole motion ~~from the winter solstice was~~ 60^{gr}.47′.40″, & this motion was from the winter solstice according to the reconning of Hesiod & the ancients that is, from the middle of the Constellation of Capricorn to ♓ 0^{gr}.47′.40″. reconning the solstice to begin in the middles of the constellations of Cancer & Capricorn And the opposite point of the Ecliptic wch rose at the same time with Arcturus was ~~in~~ 60^{gr}.47′.40″ from the summer solstice & this solstice was then placed in the middle of the Constellation of Cancer. The north Latitude of Arcturus is 30^{gr}.57′, & the elevation of the Pole at mount Helicon near Athens where Hesiod lived, was 37^{gr}45′ according to Ptolomy. And thence Ricciolus—that the excess is . Which being added to ♍0.^{gr}47.′40″, gives the longitude of Arcturus ♍13.^{gr} .

When the sun sets visibly his upper limb is 33′ below the Horizon—that is from the middle of the Constellation of Capricorn. There the Ancients of those days placed the winter solstice; & from thence reconning the signs ♑, ♒, ♓, & c; the 60^{gr}.47′.40″ end in ♓ 0^{gr}.47′.40″. And the opposite point of the Ecliptic wch rose at the same time with Arcturus was in ♍0^{gr}.47′.40″. The north Latitude of Arcturus is 30^{gr}.57′

Hesiod tells us that sixty days after the winter solstice the star Arcturus rose just at sunset. And thence ~~it may be gathered~~ follows that Hesiod flourished about an hundred years after the death of Solomon.

<verso>

Hesiod tells us that sixty days after the winter solstice, the star Arcturus rose just at sunset. Till his days & long after the solstices were placed in the middles of the Constellations of Cancer & Capricorn their motions not being then known. And the sun's Apoge was then in ♋00^{gr}.8′.38″ or thereabout. In those sixty days the Sun's mean motion was 59^{gr}08′.10″ & his equation about 1^{gr}39′.30″ & by consequence his whole motion was 60^{gr}.47′.40″. And this motion was from the winter solstice according to the reconning of Hesiod & the Ancients, that is from the middle of the Constellation of Capricorn. There the ancients of those days placed the winter solstice, & from thence reconning the signs ♑ ♒ ♓ ♈ &c the 60^{gr}.47′.40″ end in ♓ 00^{gr}.47′.40″. And the opposite point of the Ecliptic wch rose at the same time with Arcturus was in ♍00^{gr}.47′.40″. The north latitude of Arcturus is 30^{gr}.57′ & the elevation of the Pole at mount Helicon neare Athens where Hesiod lived was 37^{gr}45′ according to Ptolomy. And thence Ricciolus <the remainder of the page is the same as on recto after "thence Ricciolus" through "the coarse observations of the ancients.">

C.iii.i. HESIOD 9. New College MS 361(2)f162r

MS

Hesiod describes four generations or ages of the Gods ~~of Greece~~ meaning the Gods of Greece & saith that ~~the fourth [illegible] wth the~~ first was a golden ~~[illegible] in gen~~ age in wch Saturn reigns & man lived happily wthout care & trouble. When that generation was dead they were made Gods & there arose a second generation a silver one inferior to ye former & when they died & were laid in the grave they were also honoured & there arose a third generation of brass vibrant strong warlike & fierce having armour & weapons of brass yet they died & there arose a fourth generation of semigods who were destroyed by war partly at Thebes in fighting for the wealth of Oedipus & partly in sailing beyond sea to the war against Troy ~~in fighting~~ for [illegible] of Helena. And <there> arose a fifth generation an iron one full of trouble & [illegible] in wch Hesiod himself lived & which when they grew grey bearded should also dye [illegible] in wch Hesiod himself lived. Hesiod was therefore one generation later than the war against Troy & ~~that of the seven captains~~ Thebes.

The Greeks honoured many stories of the Gods from the Egyptians & applied them to ~~their~~ men of their own nation whom they deified of [illegible] kind in this story of the four ages. ~~of gold silver brass & iron~~ In The fourth age Mercury fram^{d} ~~& Hesiod here borrows the story of the four ages from~~ the figures & [illegible] of the Gods of Egypt & this is usually called the iron age. [several few lines of extreme deletions]

Hesiod lived one ~~age~~ generation later & therefore ~~calls the fifth age again here~~ mentions a fifth age & calls that the iron age. The four ages of the Egyptian Gods seem to be the origins of the four first kings of the Monarchy of Egypt founded by the expulsion of the shepherds viz Saturn, Jupiter Ammon [illegible] ~~Sesotris~~ Osiris & Orus. ~~In the fourth age Mercury framed the figures & was king of the Gods of Egypt Thoth pursued <?> Saturn with his [illegible]~~ The fourth ~~age [illegible]~~ age may comprehend not only the short reign of Orus but all the time from the death of Osiris to the ~~expulsion~~ end of ye civil wars in Egypt, wch lasted about 30 years. For in the fourth age as Ovid tells us ~~[illegible] say~~ [illegible] he was Astraea reliquit, wch is to be understood of Isis after the death of Orus. And thus the fourth age of the Egyptian Gods will end about the same time with the fourth age of the Grecian, ~~and comparing the time that we have alloted to the taking of Troy~~ at wch time Troy was taken.

Appendix D

Placing Colures on the Original Star Globe

To find the original time at which a colure passed through some point of a constellation, Newton first located what he thought to be an appropriate star within the asterism, or (once) the midpoint between a pair of stars. He then looked up its tropical coordinates, originally in Hevelius' *Prodromus*,[1] which gives coordinates for epoch 1660, and eventually in Flamsteed's manuscript catalog, published by Halley and Newton in a purloined edition,[2] and whose coordinates are for epoch 1690. The locations are given in both places precisely as Ptolemy had—namely, as the position of a star within one of the 30-degree-long signs of the zodiac.

At the time of the original globe, the star lay on a colure. If the colure was solstitial, then it is a simple matter to find the date: take the contemporary longitude of the star and precess it back to either 90° for a summer solstice, or to 270° for a winter solstice. Since the solstitial colures are perpendicular to both ecliptic and equator, precessing the star back to a solstice simply moves it parallel to the ecliptic (figure D.1). We need only the difference $S^{cont}S^{orig}$ (where S^{orig} is either 90° or 270°) between its contemporary and original longitudes, for then the precession rate determines the original date.

Equinoctial colures can be more complex.[3] If the star is on the ecliptic, then it's still a simple matter to precess backwards to the original colure. If, however, the star is off the ecliptic, then Newton had to take account of its latitude ($S^{orig}K$ in figure D.2). For a star north of the ecliptic, simply precessing it back until it has the same longitude as the equinoctial point E (whose longitude is 0° for a vernal, and 180° for an autumnal equinox) would place it incorrectly at S′ rather than on the colure at S^{orig}. To reach the colure therefore requires subtracting the distance EK from $S'S^{cont}$, that is from the contemporary longitude of the star in question. For a star south of the ecliptic EK must instead be added. Newton gave neither the expression for calculating the shift, nor did he discuss the point at all, but the numbers he obtained, as well as the steps in his computation, where given, are

[1] Hevelius, 1690.

[2] Flamsteed, 1712.

[3] Horsley, 1779–1785b, pp. 68, vol. 65 provides a method that also results in (D.1). His technique was to draw a great circle through the star's position in 1690 that is orthogonal to the equator (and so that forms an angle of 90°-ε with the ecliptic, ε being the angle between equator and ecliptic). Having marked the point *A* where this great circle intersects the ecliptic, Horsley then subtracted the distance between *A* and the ecliptic projection of the star from its 1690 longitude, which yields the longitude when the colure passed through it.

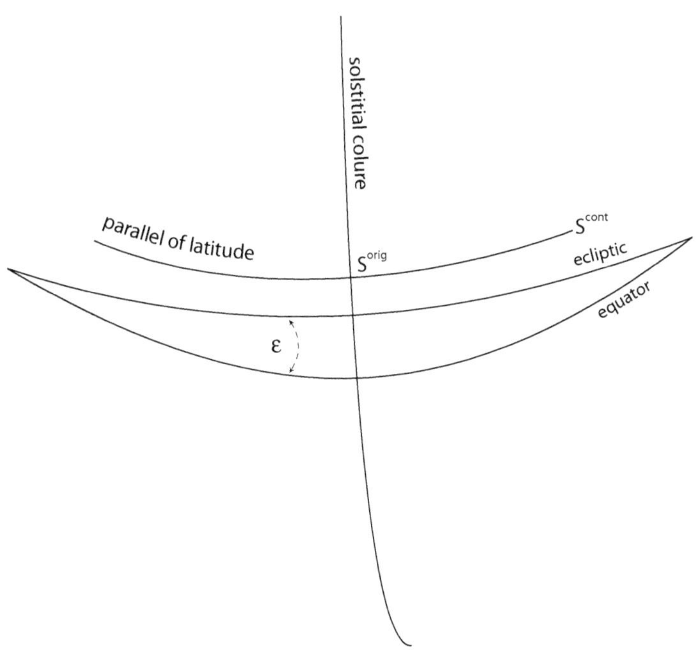

Figure D.1. Solstitial colures.

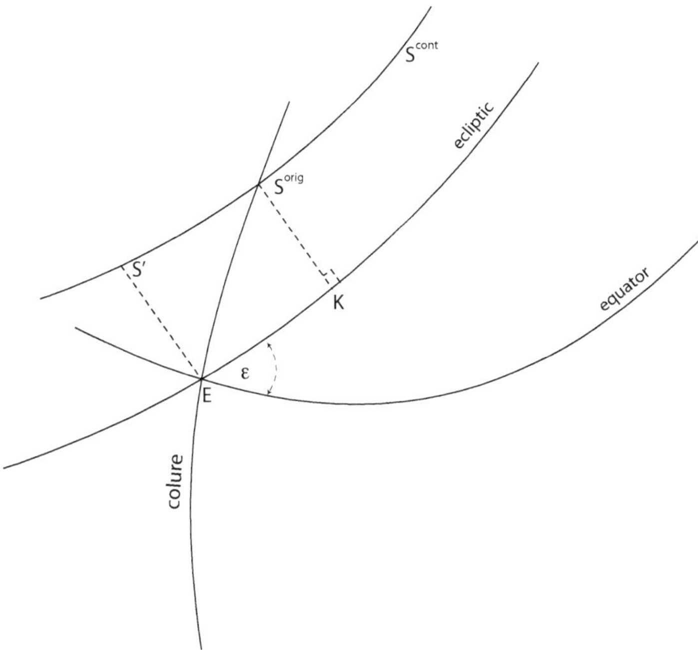

Figure D.2. Equinoctial colures

consistent with the following simple relationship. Since $\Delta S^{orig}KE$ is a right spherical triangle, and since the colure is perpendicular to the equator, EK is given by (D.1):

$$EK = \sin^{-1}\left(\frac{\tan(S^{orig}\,K)}{\tan(90^\circ - \varepsilon)}\right) \qquad \text{D.1}$$

Newton's manuscripts, as well as his published *Chronology*, provide solid evidence that he used (D.). Consider for example the following excerpt:[4]

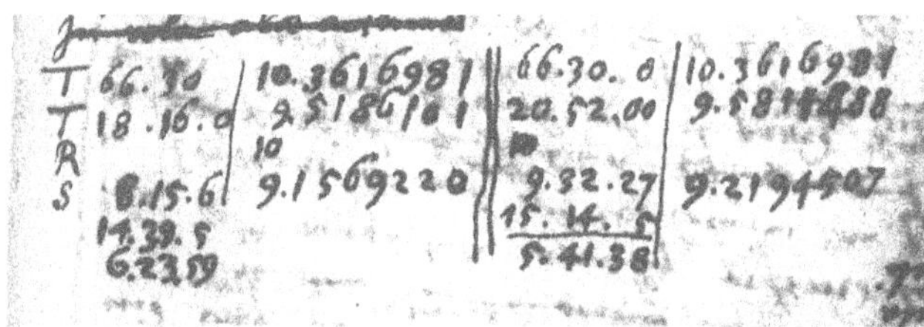

Figure D.3. An MS colure computation. By permission of the Warden and Scholars of New College, Oxford.

Here the number next to the first 'T' is 66.5, which is the complement of Newton's value for the obliquity of the ecliptic when he needed to compute quickly;[5] the number next to that is the logarithm of its tangent (plus 10). The number beside the second 'T' is the latitude of a star ($S^{orig}\,K$), with its log (again plus 10) next to it. 'R' represents the radius used in the log tables Newton had, and then the '8.15.6' next to the 's' is precisely what results from our formula, i.e., it is the angle that must be used to compensate the star's longitude when it sat on an equinoctial colure.

[4] New College MS361(2)f148v.

[5] From 1694, Newton had a more precise value of 23°29′12″ (Kollerstrom and Yallop, 1995 citing Forbes et al., 1995–, p. 75), but a minute's difference in the obliquity has no important effect on chronology. In epoch 2000 the ecliptic's obliquity was 23°26′21″. The value changes over time at a variable rate, currently running close to half a second per year decrease, and returning to its original value after about 40,000 years. It changes as well due to the earth's nutation with about an 18-year period; nutation alters the obliquity by about 9″ above and below its mean value. The phenomenon was discovered by James Bradley, who made the finding public in 1748.

TABLE D.1.
Newton's, Horsley's, and accurately computed longitudes for the ancient equinox

Hipparchus	*Newton's Star*	*Newton's Longitude for the Ancient equinox*	*Accurate Longitude (Newton's Coordinates)*	*Horsley's Longitude*
E:*Aries* back	*Bayer* ν	♉ 6°58′57″	♉ 6°58′5″	♉ 7°7′56″[a]
E:*Cetus* head	*Cetus* midpoint of ν and ξ	♉ 6°58′51″	♉ 6°53′24″	♉ 6°53′29″
E:*Eridanus* Flexure	*Bayer* ρ	♉ 7°12′40″	♉ 7°12′40″	♉ 7°12′38″
E:*Perseus* head	*Bayer* τ	♉ 6°18′57″	♉ 6°8′46″	♉ 6°18′48″[b]
E:*Perseus* right hand	*Bayer* η[c]	♉ 4°56′40″	♉ 4°58′23″	♉ 4°57′33″

[a] *The difference from Newton here is due to Horsley's use of the correct Flamsteed longitude, namely B9°48′35″, instead of Newton's mistranscribed ♉9°38′45″.*

[b] *Horsley here likely mistranscribed his own result, knowing that Newton had 18′, whereas a good computation gives 8′.*

[c] *Horsley here uses different coordinates from Newton, who called the star* Bayer η, *describing it as on the "right-hand of Perseus, rightly delineated." However, in Bayer's depiction of Perseus the star designated η is on the right* forearm, *whereas a star designated χ is on the right* hand. *In Flamsteed's plates, we also have χ on the right hand, but no star depicted at all on the right forearm. Still, in Flamsteed's catalog we do have a star in* Perseus *designated η, whose 1690 coordinates are ♉24°23′27″, 37°26′50″ and which is described as "In superiori Brachio." Newton's coordinates for the star are ♉24°25′27″, 37°26′50″. Consequently it seems likely that he used* Perseus *η—meaning by "right-hand" just righthand side—and mistranscribed the minutes from Flamsteed's 23′ to his 25′. Horsley, however, decided that "this star does not appear in the Britannic Catalogue," which it doesn't if he went looking for a star on the right hand. He turned instead to Hevelius, where he found a star listed with longitude ♉24°0′29″, epoch 1660, in Perseus, which Hevelius described as "In dextro Brachio, vel inter dextrum Brachium & Caput," and Hevelius did not use Bayer designators. This certainly is the same star as Flamsteed's* Perseus *η, but Horsley, having failed to find a* Bayer *η in Flamsteed's catalog, and being free to chose an appropriate star from Hevelius, decided to update Hevelius' longitude to 1690, thereby obtaining ♉24°24′29″ instead of Flamsteed's ♉24°23′27″. This star in any case gives by far the worst date for his purposes among Newton's equinoctials (see table 8.2).*

Table D.1 lists the stars as Newton printed them in the *Chronology*, together with his (1690) coordinates for them, the resulting longitude of the vernal point that he found, and the longitude that results from a highly accurate computation using the printed values for the coordinates, as well as the values that he always used for precession and ecliptic obliquity. The last column lists the longitudes computed by Horsley from Newton's coordinates; he, however, did change the mistranscribed longitude for *Aries* ν back to Flamsteed's value, which (though Horsley could not have known) Newton had used in his notes.

Appendix E

Hesiod, Thales, and Stellar Risings and Settings

E.i. Hesiod

There are two ways to determine whether the *acronychal rise* of *Arcturus* (the day when it first appears on the eastern horizon at sunset) will occur 60 days following the winter solstice at a given date. One method calculates the longitude of the sun at the date in question to see whether *Arcturus* will then lie on the eastern horizon at Hesiod's location (which Newton took to be Mt. Helicon). This amounts to seeking the point of the eastern horizon that intersects the latitude of *Arcturus* just as the sun sets. The star should be there in about −880 (Newton's date for Hesiod). Alternatively, one could first precess *Arcturus* back to −880, put the precessed star on the horizon, and then calculate the longitude of the ecliptic point that rises with it. One would then compare the result with the ecliptic rising point 60 days after the winter solstice. In this alternative method, the longitude of precessed *Arcturus* is a given, the locus that the sun must have follows from it, and the result is tallied with a presumptive date for Hesiod's observation. Newton used only the first method, but we'll consider both in what follows since the second also applies (with a trivial modification) to the heliacal rise of Sirius.

Although it might seem that only the second method uses precession before the last step, precession has to be taken into account in both methods during the main computation and afterwards. Just where precession is used does, however, differ between the two procedures. In the first—the one that Newton (albeit, we shall see, incompletely) used—the sun's longitude sixty days after the solstice at the appropriate date in the past must be calculated in order to find the ecliptic rising point at sunset. That in itself requires precession, at least in principle, because the solar *apogee* precesses along with the stars. The apogee has in addition a *proper motion* relative to the stars that Newton took into account. Precession once again raises its head in comparing the final result with the longitude of *Arcturus* at the presumed time of Hesiod (−880). In the second method, we precess *Arcturus* at the very start of the computation, and at the end we again use precession to compare our result with the sun's longitude at the time of Hesiod's observation. Each method must precess both the sun's apogee and *Arcturus*, but the first uses an initially precessed apogee to compare with a finally precessed *Arcturus*, whereas the second does the reverse.

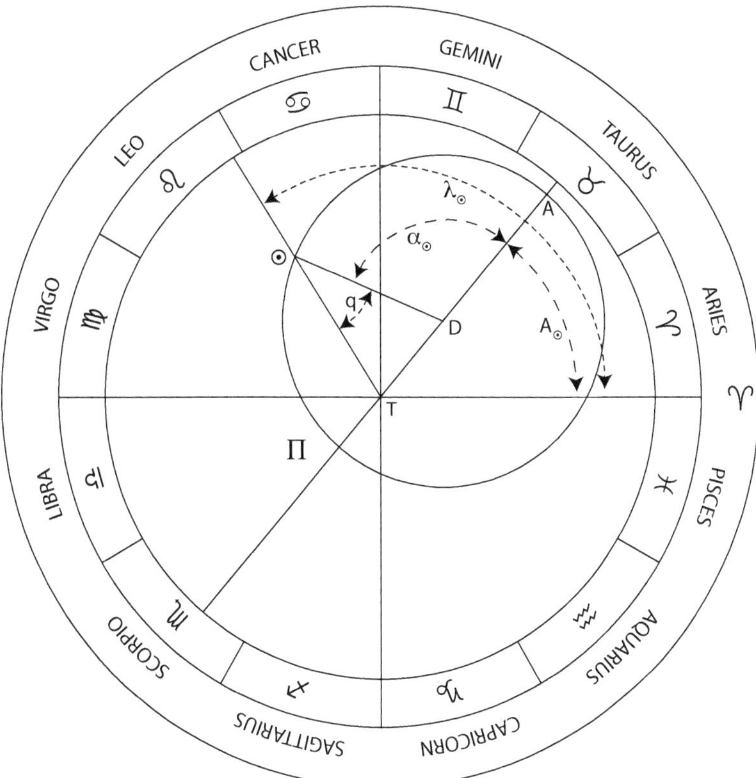

Figure E.1. Solar motion through the zodiac.

E.i.a. Sixty days after the winter solstice

In figure E.1, the ring marked with the signs of the zodiac divides the ecliptic into the twelve 30-degree segments that, Newton believed, were first constructed about the time of the Argonautic expedition, that is (in Newton's chronology) about the middle of the tenth century BCE. The outer ring names the signs, and it is certainly true that circa −950 the asterisms themselves were more or less located in the signs with corresponding names.

Ptolemaic (or, more precisely, Hipparchan) methods for computing the sun's motion remained common throughout the seventeenth century since they provide more than sufficient accuracy for almost any purpose. Newton seems to have followed the usual methods, which are nicely set out in Riccioli's text.[1] The sun (☉) rides a circle A☉Π whose center

[1] Riccioli, 1651. Kepler had altered the Ptolemaic model for the sun before introducing ellipses in two ways. First, he bisected Ptolemy's solar eccentricity (TD in figure E.1), placing the sun at the midpoint between the earth at T and Ptolemy's D. Second, he introduced a point (not marked in the figure) on the opposite side of T and also at a distance TD/2 from the earth. This "equant" point was the locus from which the sun's motion was to be uniform, thereby making the solar motion on its eccentric variable. Many objected to the device, not least because it reintroduced the complexity of the equant. In any case Newton would certainly not have used this model for his calculations since the non-uniformity of motion on the eccentric

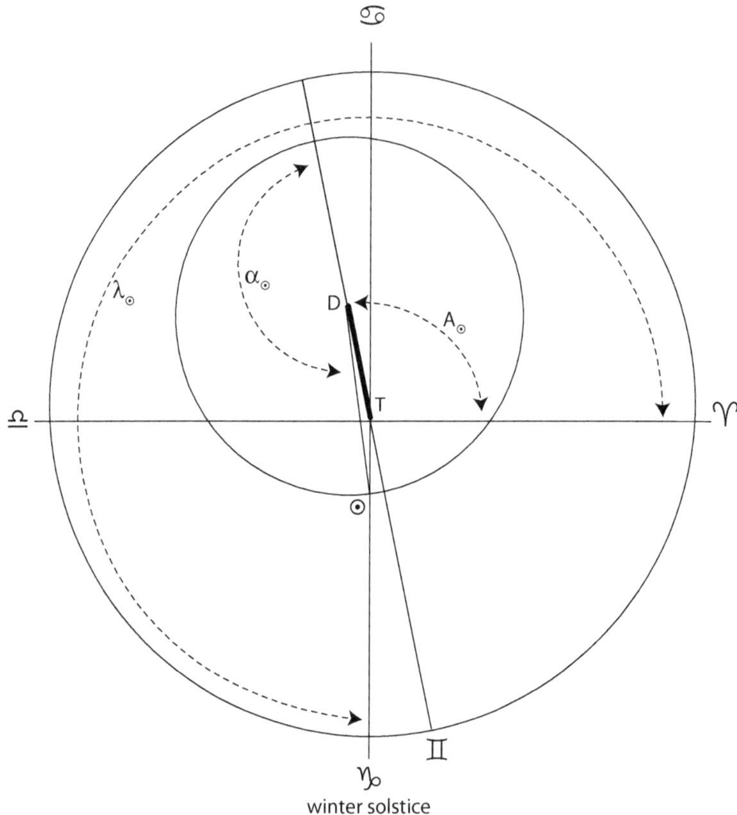

Figure E.2. Winter solstice ca. 1700 ($A_\odot$ = ♋7°44′33″)

D is displaced from the earth at T by the *eccentricity* TD. For simplicity we set the radius D⊙ of the solar *eccentric* to one. The zero-point of longitude lies at the *vernal equinox* (the intersection of the ecliptic with the equator)—the beginning of Aries, or ♈ along line T♈. At any given epoch the eccentricity TD lies at some angle $A_\odot$ with respect to T♈. During the year the sun moves uniformly around the circle A⊙∏, and as seen from the earth at T its position among the zodiacal signs lies along T⊙. This line makes an angle ∠♈T⊙ (the sun's longitude, $\lambda_\odot$) with T♈. The solar radius D⊙ forms angle ∠AD⊙ (or $\alpha_\odot$, the sun's *mean anomaly*) with the *line of apsides* A∏, along which lie the sun's maximum (AT) and minimum (∏T) distances from T. Because the center of the eccentric is displaced from the

makes computation more difficult, and unnecessary for his purposes here. During the eighteenth century Cassini measured the sun's diameter, projected on the floor of the church at San Petronio, under conditions that emphasize the difference between Ptolemy's and Kepler's models in respect to changes in the solar distance, finding that the bisected-eccentricity fit his measurements. On the derivation of Kepler's model see Wilson, 1968, pp. 5–10, and for Cassini see Heilbron, 1999. The third book of Riccioli's *Almagestum* provides a comprehensive account of the several solar models available at mid-century.

earth, the sun's longitude differs from the sum $A_\odot + \alpha_\odot$ of its apogee with its mean anomaly by the angle $\angle D\odot T$, or q, which Newton follows convention in calling the solar *equation*.

The winter solstice is located at the boundary between Sagittarius and Capricorn, the summer solstice at the Cancer-Gemini boundary, and the autumnal equinox between Virgo and Libra. As a result of precession the line of apsides moves forward through the zodiacal signs. Precession, that is, *increases* the longitude through which the sun moves in a given time because it is in the same direction as the sun's annual motion.

To compute the solar longitude sixty days later, we first find the mean anomaly $\alpha_\odot$ at the winter solstice itself (figure E.2), which can be calculated from the sun's equation ($\angle T\odot D$ or q^{ws}) and apogee according to (E.1):

$$\left.\begin{aligned} \lambda_\odot^{ws} &= \text{♑}\,0^\circ \\ q^{ws} &= -\sin^{-1}(TD\cos A_\odot) \\ \alpha_\odot^{ws} &= \lambda_\odot^{ws} - A_\odot + q^{ws} \end{aligned}\right\}\text{ at the winter solstice} \qquad \text{(E.1)}$$

Sixty days later the sun will have moved much of the way to the vernal equinox at ♈:

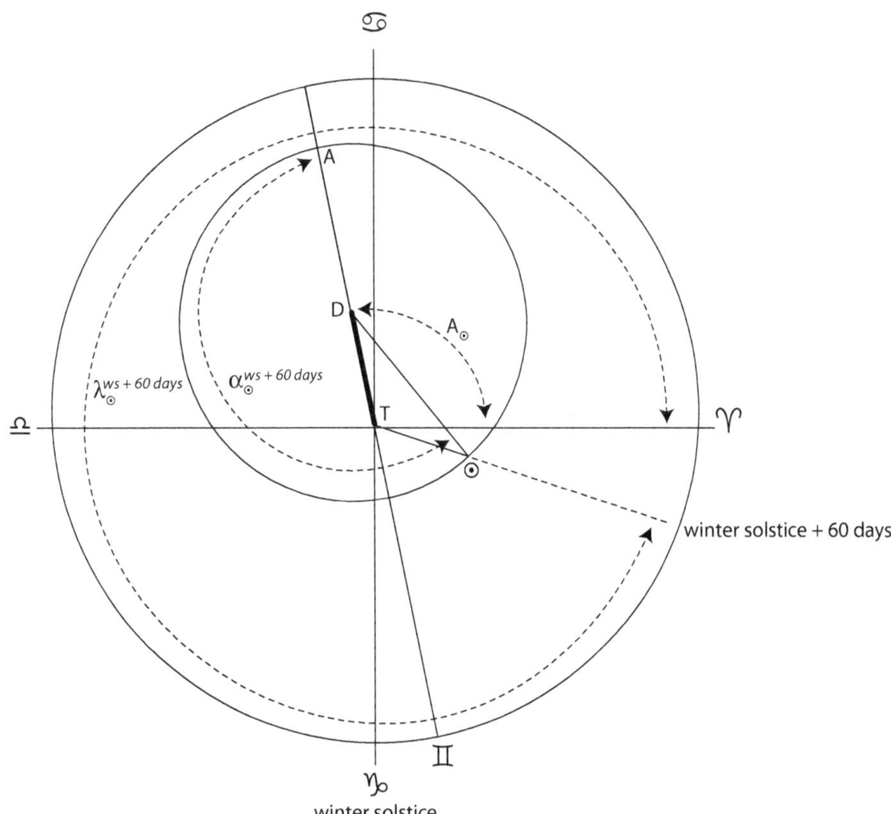

Figure E.3. 60 days past winter solstice ca. 1700.

If we fix the line of apsides and so consider only the mean solar velocity $\omega_\odot$ (since the apsidal motion is tiny and has an even smaller effect on the mean motion), then over sixty days the sun would move $60d\omega_\odot$ degrees around its eccentric, its mean anomaly will be $\alpha_\odot^{ws} + 60d\omega_\odot$, and its equation and longitude become (with $q^{ws+60days} = \angle D\odot T$):

$$\left.\begin{aligned} \alpha_\odot^{ws+60days} &= \alpha_\odot^{ws} + 60d\omega_\odot \\ q^{ws+60days} &= -\sin^{-1}\left(\frac{TD\sin(\alpha_\odot^{ws+60days})}{\sqrt{1 + TD^2 + 2TD\cos(\alpha_\odot^{ws+60days})}}\right) \\ \lambda_\odot^{ws+60days} &= \alpha_\odot^{ws+60days} + A_\odot + q^{ws+60days} \end{aligned}\right\}\text{ at winter solstice + 60 days} \quad (E.2)$$

Since the apogee $A_\odot$ both precesses and has its own proper motion among the stars, a complete calculation of the longitude at a given epoch in terms of the fixed signs will take account of both effects. Note that the corresponding velocities are very different from one another. For precession Newton in his chronology work always used a value of 72 years per degree (which is 50″ per year), and for the proper motion he gave 17′40″ per century. Although these motions have the same directions (forward through the signs as time progresses), they obviously have considerably different magnitudes, since the ratio of precession to proper motion is about 4.7 to one. Whether or not even precession has a substantial effect on the comparison with *Arcturus*' position can be determined only by direct calculation for a given solar eccentricity. The smaller the eccentricity, the less the effects of precession and proper motion on the solar locus.

E.i.b. Newton's Method

According to one reading of Hesiod,[2] *Arcturus* first appears on the eastern horizon just as the sun sets 60 days after the winter solstice. Since the horizon and the ecliptic are both great circles, the point of the ecliptic that should rise with *Arcturus* on that day must have longitude $\lambda_\odot + 180^\circ$, where $\lambda_\odot$ refers to the sun. The question that Newton had to answer was this: what is the longitude of a point that has the same latitude as *Arcturus* and that lies on the horizon at this moment? Newton left few numerical hints of his computation, which is not entirely simple, but he did refer to Riccioli,[3] and he explicitly sought the difference in longitude between the rising points of the ecliptic and of *Arcturus*' latitude. Furthermore the procedure is not only laid out in Riccioli but it is there applied specifically to Hesiod's remark, albeit with slightly different numbers from the ones that Newton used.[4] Riccioli obtained a date of −953. Since he put Hesiod at −775 on other grounds, this meant that the Greek poet was writing about a preceding period and not about his own. Newton certainly read this, and Riccioli's result probably stimulated him to include corrections for refraction,

[2] But see chapter 8, n.122.

[3] New College MS 361(2)f151r: "And thence Ricciolus (lib. VI Almagest. Cap. XX. Prob. VIII) teaches how to compute the excess of the longitude of *Arcturus* above the longitude of the said opposite point of the Ecliptic." The value that Newton used for the rate of precession in calculation agrees with Riccioli's (1 degree in 72 years). See below for discussion of the solar eccentricity that he used.

[4] Riccioli, 1651, p. 463.

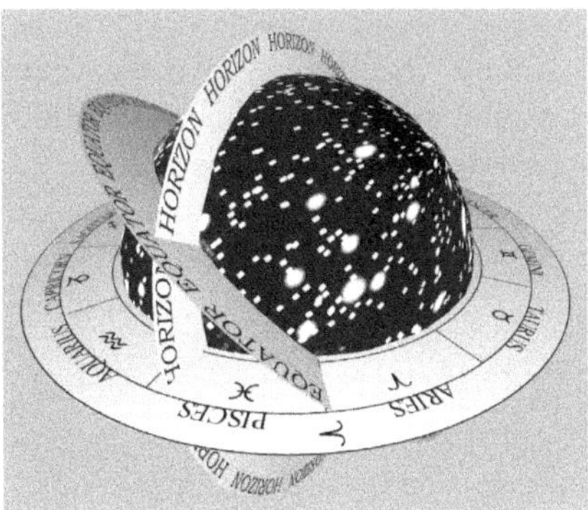

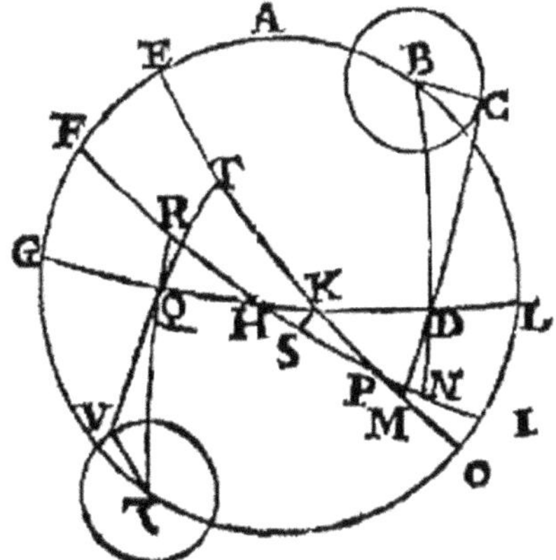

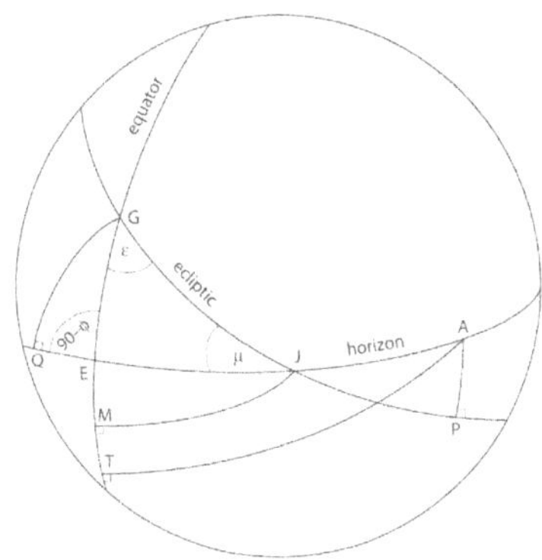

Figure E.4. Illustration of the configuration (top). Riccioli's diagram (bottom left), and Newton's method.

because that pushes the date forward just enough for Newton's purposes (given, we shall see, that he ignored the precession of the sun's apogee). For clarity figure E.4 (top) illustrates a moment of the day at which the equator has carried the vernal point below the horizon. Riccioli's figure for the problem (bottom left) does not well represent the configuration that Newton assumed (though in the end the computation is indifferent to the graphical representation as long as it is done consistently), and so for historical clarity we use the figure at bottom right instead, which does tally with Newton's remarks.

Both figures represent the configuration of great circles at sunset sixty days past the winter solstice, when the intersection G of the equator and the ecliptic (that is, the vernal point and origin ♈ of longitudes) lies above the eastern horizon $\widehat{QJA}$. Point J marks the intersection of the ecliptic with the horizon, while point A marks the latter's intersection of the small circle of *Arcturus*'s latitude. Newton had already computed the longitude GJ of the ecliptic's rising point as the sun sets (albeit, we shall see, incorrectly), and now he must find the longitudinal distance JP between J and A. If his conjecture concerning Hesiod's date is correct, then *Arcturus* itself will have the same longitude (GP) as A near the year −880.

Though it requires several steps, the computation of JP can be obtained from the diagram that represents Newton's method. There we see that the ecliptic intersects the horizon at some angle μ, which we must find in order to compute GP. The two right spherical triangles ΔGQE, ΔGJQ share the same side GQ as well as the angle μ, from which we find μ in terms of the terrestrial latitude φ, the rising time GE, and the longitude GJ of the ecliptic point J:

$$\mu = \sin^{-1}\left(\frac{\sin GE}{\sin GJ}\sin(90^\circ - \varphi)\right) \tag{E.3}$$

We find GE in terms of the equatorial coordinates of J—its declination JM and right ascension GM. JM yields EM through right triangle ΔJEM, in which ∠JEM is 90° – φ, and then GE is just the difference GM-EM:

$$\begin{aligned} EM &= \sin^{-1}(\tan JM \tan \varphi) \\ GE &= GM - EM \end{aligned} \tag{E.4}$$

The equatorial coordinates of J follow from well-known transformations (or from tables) in terms of GJ and the obliquity ε of the ecliptic:

$$\begin{aligned} JM &= \sin^{-1}(\sin GJ \sin \varepsilon) \\ GM &= \sin^{-1}\left(\frac{\tan JM}{\tan \varepsilon}\right) \end{aligned} \tag{E.5}$$

Finally, to find the distance JP consider right triangle ΔAPJ, whose side AP represents the latitude of *Arcturus*:

$$JP = \sin^{-1}\left(\frac{\tan AP}{\tan \mu}\right) \tag{E.6}$$

And, at last, we have the longitude of point A in terms of JP and Newton's previously computed value for GJ:

$$GP = GJ + JP \tag{E.7}$$

We now precess *Arcturus* back to the conjectured date for Hesiod and see whether it has longitude close to GP (though just what "close to" might mean was a matter for argument, as we see from William Whiston's critique of Newton's chronology). Alternatively, we can use the value for GP explicitly to find the date by computing how many years it would take to move *Arcturus* from GP to its longitude in 1690.

Although Newton's method involves several steps, each of which introduces potential computational errors and approximations due to the use of logarithmic tables for trigonometric functions, it is both simpler and less subject to error than the alternative (which has the same potential inaccuracies)—namely, to compute the locus of J from A, and which we will now examine because it was later used by a commentator.

E.i.c. An alternative calculation

We begin with the tropical coordinates of *Arcturus* at a given epoch (1690 for Newton, using Flamsteed's data) and precess back to –880. The question to be answered is whether this precessed *Arcturus* will lie on the horizon at Mt. Helicon at sunset 60 days after the winter solstice. To find out, we compute the locus of the ecliptic point that rises simultaneously with precessed *Arcturus* and then see whether the sun will be 180° around the ecliptic from there.

In figure E.4 bottom right, as before J represents the rising point of the ecliptic, which we seek given *Arcturus*' (A) longitude GP and latitude AP. We first find the equatorial coordinates GT, AT. This alone is more complicated than any step in Newton's method because, since *Arcturus* does not lie on the ecliptic, we need to include the effect of its latitude:

$$\sin AT = \sin AP \cos \varepsilon + \cos AP \sin GP \sin \varepsilon$$
$$\tan GT = \frac{-\sin AP \sin \varepsilon + \cos AP \sin GP \cos \varepsilon}{\cos AP \cos GP} \tag{E.8}$$

Next we find the rising time GE, which follows from the right triangle ΔAET (where, as in Newton's method, ∠AET equals ∠GEQ and so is 90°-φ) and the difference GT-ET:

$$\sin ET = \tan AT \tan \varphi$$
$$GE = GT - ET \tag{E.9}$$

To find the equatorial coordinates GM, JM of J requires EM, since then GM is just GE + EM, and JM will follow from GM and the obliquity ε. Again the procedure is more involved than Newton's method, precisely because we must now work to J from A. EM can be found in terms of GE, φ and ε by equating the two expressions for the right ascension JM that follow from right triangles ΔJGM, ΔJEM respectively:

$$\tan JM = \sin(GM)\tan(\varepsilon)$$
$$\tan JM = \sin EM \tan(90^\circ - \varphi)$$
$$GM = GE + EM \tag{E.10}$$

and so

$$\tan EM = \frac{\sin GE \tan \varphi \tan \varepsilon}{1 - \cos GE \tan \varphi \tan \varepsilon}$$

and then GM follows as GE + EM. Finally, we find the longitude GJ as:

$$GJ = \cos^{-1}(\cos(GM)\cos(JM)) \tag{E.11}$$

We now test whether the sun reaches a point 180° round the ecliptic from J 60 days after the winter solstice in about –880.

This alternative method is more tedious than Newton's (or, more accurately, than Riccioli's). Moreover, the procedure still requires the angle μ between ecliptic and horizon in order to calculate corrections for refraction (as Newton did), even though μ is not required for GJ. The most intricate (and so error-prone) calculation in Newton's method is for μ, and even that one involves only two multiplications and a division in addition to table look-ups. In contrast, the alternative method additionally requires three lengthy computations (for AT, GT, and EM), each of which is prone to computational and table-lookup errors. Both methods must compute the locus of the sun 60 days past winter solstice.

Horsley undertook this alternative, though he provided no details beyond the results, remarking that "taking the longitude and latitude of Arcturus, each such as it was on the primitive sphere; namely the longitude ♍13°26′52″ the latitude 30°52′18″; I find, that when the star rose at sun set, the sun's true place must have been ♓00°59′. And according to the situation of the aphelion of the earth's orbit, which obtained in that age of the world, when the equinoxes were in ♉ 6′27″[5] and ♏6′27″[6] of the sphere of 1690; the sun was in this place sixty days after the winter solstice. So that the conclusion from this passage of Hesiod should rather be, that he flourished in that very age when the Greeks first formed their sphere; that is, according to Sir Isaac Newton, in the age of the Argonautic expedition."[7] Given Horsley's printed values for the ecliptic coordinates of Arcturus, an accurate computation puts the sun at ♓2°34′25″ when Arcturus rises. Assuming that he carried out the computation correctly, then his result (♓00°59′) may have included a total decrease of about 1.5° in the calculated solar longitude as a result of refraction, which is about the same as Newton's manuscript value.

E.i.d. Correcting for refraction

Given Riccioli's result, which placed Hesiod's observation almost exactly at the time of Chiron, and therefore too early, Newton decided to include corrections for refraction. For sunset Newton took the visible position of the sun's upper limb as it just touches the horizon. Because the limb is raised by refraction through some amount (see remarks below) the sun will actually be located below the horizon at setting. Newton added in the sun's radius, which he set at 16′, since the computation for longitude presumably holds for the sun's center.

In figure E.5, *Arcturus* appears on the horizon at A^v but is actually located at A. Newton's calculation absent refraction placed *Arcturus* on the horizon at A^c, where its latitude circle $\widehat{A^cA}$ intersects the ecliptic. Consequently the star's true longitude at rising is not given by JP^c but by JP, and so the previously computed value must be increased by an amount P^cP. Newton does not specify how he computed the additional longitude, but the procedure is simple.

Since $\angle A^vA^cA$ is just μ (the angle between the ecliptic and the horizon at sunset) from the right triangle ΔAA^vA^c we obtain:

$$P^cP = A^cA = \sin^{-1}\left(\frac{\sin A^vA}{\sin \mu}\right) \tag{E.12}$$

[5] A misprint for ♉ 6°27′.

[6] Again a misprint: this should be ♏6°27′.

[7] Horsley, 1779–1785a, pp. 74–75.

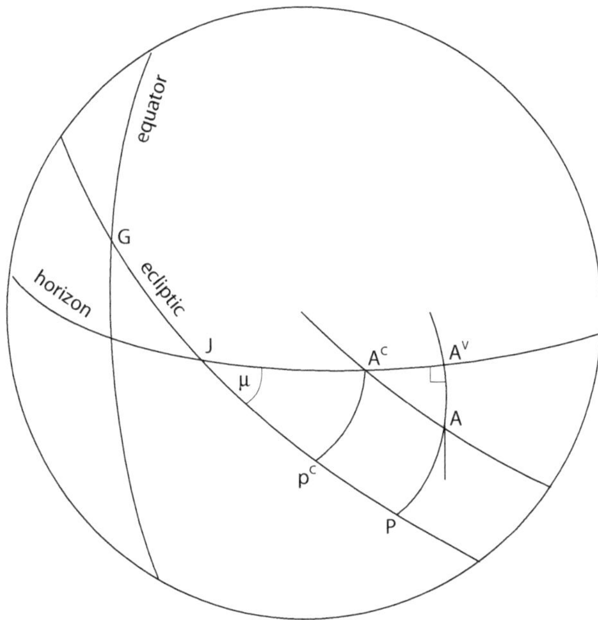

Figure E.5. The effect of refraction on *Arcturus.*

This then is the amount by which the calculated longitude of *Arcturus* must be changed to account for the effect of refraction. A similar correction must be made for the sun, though since Arcturus is rising its longitude must be increased, while the setting sun's must be correspondingly decreased.

In figure E.6 (upper) the sun is seen at S^v when it is actually located at S^T, whereas our calculation had assumed the sun to be at S^c. We must therefore alter the solar longitude by the amount S^TS^c, which we can find from the right triangle $\Delta S^vS^TS^c$:

$$S^TS^c = \sin^{-1}\left(\frac{\sin S^TS^v}{\sin \mu}\right) \tag{E.12}$$

For the distance through which refraction will raise a point below the horizon onto it Newton always used a value of 33′.[8] To this arc he added the radius of the sun, since the center is located 16′ below the upper limb *U*.

[8] Riccioli, 1651 , Book 1, chap. 9 ("De refractionibus solis") p. 115 cites Tycho as giving 34′ for the "Refractio Horizontalis" of the sun. Newton had developed a theory for atmospheric refraction based on his particle model for light, and he had computed a table of refraction from it (Whiteside, 1980). The force acting on a particle was in effect set equal to the atmosphere's density gradient, and was therefore always directed to the center of the earth since the atmosphere forms a concentric shell. A table that Newton computed and sent to Flamsteed on November 17, 1694 (Newton, 1959–1977, pp. 49, vol. 44) has four columns. The first lists the apparent altitude of a star, the second the amount of its refraction in summer, the third the refraction in spring and fall, and the fourth in winter. Hesiod's observation occurred 60 days after the winter solstice, which would put it about 2/3 of the way to the spring equinox. Newton's values

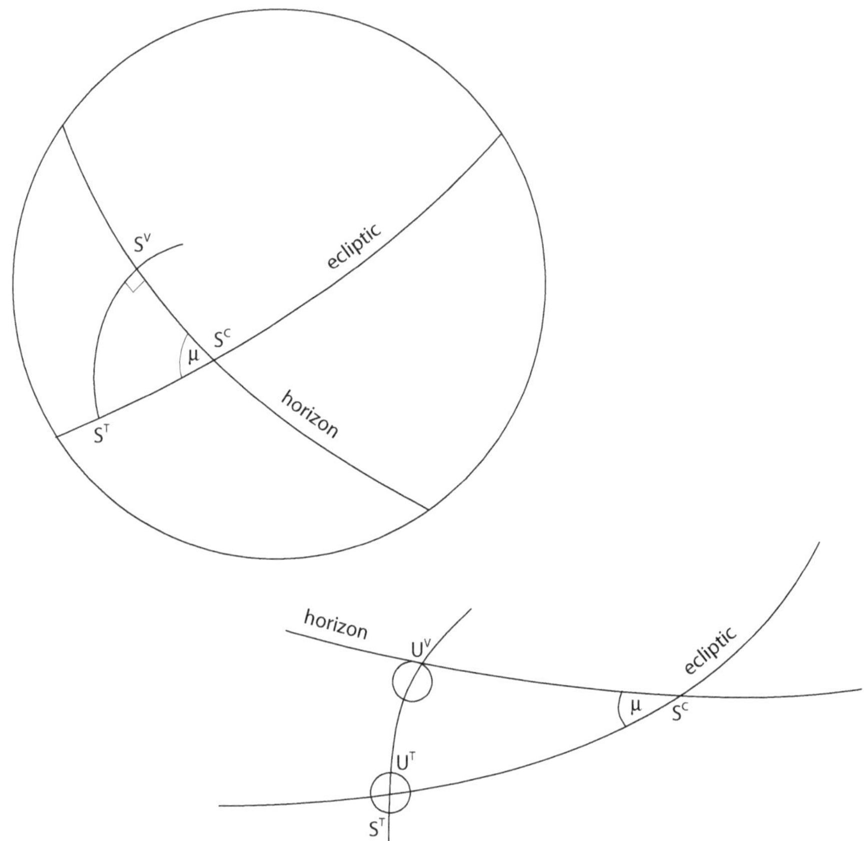

Figure E.6. The effect of refraction on the sun.

E.i.e. Newton's computations

Although Newton gave no details in the printed *Chronology* of how he arrived at his confirming date for Hesiod, the manuscripts provide considerable evidence. The details vary, and Newton clearly struggled time and again with the numbers and how best to frame the results. Nevertheless the most elaborate account (appendix C.iii.h. HESIOD 8.), when combined with appendix C.iii.f. HESIOD 6., provides enough partially to recover his computations. From the latter we see that Newton decided not to compute absolute longitudes, that is longitudes relative to the fixed signs, for that would have required introducing precession right at the beginning of the calculation in order to find the absolute location in Hesiod's day of the solar apogee.

Newton apparently asked what the position of the solar apogee would be *relative to* the fixed stars in Hesiod's time. To do that he assumed a date of −880 for Hesiod, computed the

for the spring and winter refractions are, respectively, 33′20″ and 35′10″. Newton, however, used 33′, which is the nearest minute to his "spring" value. From this we may conclude that Newton chose the value for the date that is closest in time to the observational report.

amount of proper motion between 1700 and that time, and then moved the locus of the apogee to the resulting longitude, thereby placing the line of apsides at ♋0°8′38″ (and apparently ignoring the apsidal line's precession, which is much larger).[9] He then calculated where the sun would be 60 days after the winter solstice. From that, taken as the sun's locus at sunset, Newton moved 180° around the ecliptic to find the rising point of the ecliptic. Next, he found the point of *Arcturus*' latitude circle which rises simultaneously with the ecliptic point. Newton then corrected for refraction. This gave him the longitude of the point at *Arcturus*' latitude that rose at Mt. Helicon with the setting sun 60 days after the winter solstice in Hesiod's time. And finally, in the MS Newton compared this longitude with the longitude of *Arcturus* in −880 to see if they agreed reasonably well. Newton's method, one might think, was flawed, because, though he certainly did precess Arcturus, he seems not to have done so for the sun's line of apsides. In fact it makes little difference whether one or both of the apsidal line's proper motion or precession is included in the computation.

In order to carry out his calculations Newton also needed a good value for the solar eccentricity, though he did not specify one in the MS. In Riccioli we find a table that lists a series of values for the eccentricity, including Kepler's (.036) as well as .0346 (strictly, .03465), which latter according to Riccioli appears in Copernicus, Rheticus, and Peurbach.[10] However, two pages before the table he wrote, next to a marginal notation which reads "Determinatio Eccentricitatis & Apogei," that the observations he has just gone through in previous examples, as well as 'others,' when taken together give a best eccentricity of .0346.[11] It seems moreover that in 1700 Newton had the eccentricity at .03385.[12] A correct calculation using all of Newton's specified values requires an eccentricity of .0332 in order to produce a value of 1°39′30″ for the equation 60 days past solstice.[13] Newton additionally

[9] Newton gives the solar apogee in 1700 at ♋7°44′30″. At his rate for proper motion of 17′40″ per century the apogee would have shifted 7°35′48″, which Newton correctly computes, and so would have been at ♋0°8′42″ in −880 if only proper motion were to be considered; Newton somehow subtracted to find ♋°8′38″ "or thereabouts."

[10] Riccioli, 1651, pp. 156, book 156, chap. 124.

[11] This is not, however, an average but reflects a judgment of some sort. The original reads "Ob has & alias observations a quibus nunc tempero, statui medio electo," which translates unidiomatically as "on account of these and other observations from which I now divide equally, the middle having been chosen for the position." We thank our colleague Mac Pigman for the translation. Tycho's value in 1588 was .0345 about, while Thomas Harriot (ca. 1560–1621) a few years later had about .0361 (Roche, 1981, p. 255).

[12] In a letter written to Hans Sloane on April 25, 1700, and sent after revision by Flamsteed to Leibniz on July 4, 1700 (Newton, 1959–1977, pp. 328–30,vol. 324) Newton wrote that the maximum value of the solar equation (called by Newton here the "prosthaphaeresis," following traditional terminology, viz. for an earth-centered computation the maximum absolute value of q in Equation E.1, which occurs when the mean anomaly is 90°) "should be" ("debet esse") 1°56′20″, which is the value determined by Flamsteed in 1692 and was used by Newton in his 1702 *Theory of the Moon's Motion* as well as by Whiston and Halley. The corresponding value for the solar eccentricity, TD, is then .03385.

[13] This is approximately the eccentricity that would produce Newton's value for the 'equation' 60 days past the winter solstice using his position for the apogee (♋8°38′) and his value for the mean motion in 60 days (59°8′10″). If Newton had used the eccentricity that follows from his 1700 letter to Sloane (cf. the previous note, .03385), then he would have obtained 1°41′29″ for the equation instead of 1°39′30″. It's certainly possible that, in computing either the mean anomaly at the winter solstice, or in computing the equation 60 days later, or in both, Newton's errors accumulated to about 2′.

needed the angle between the ecliptic and the equator, which he always took to be 23°30′.[14] Table E.1 provides Newton's results.

According to Newton's calculation the position of the sun 60 days past the winter solstice should be ♓0°47′40″, which he approximated as ♓0°48′ (showing that he was indifferent to as much as 20′). Going 180° around the ecliptic gives the rising point of the ecliptic as ♍0°48′. From this we can now find the point on the horizon (terrestrial latitude 37°45′) at *Arcturus*' latitude (30°57′) that rises simultaneously. We find that this point's longitudinal distance from the rising point of the ecliptic (*JP*) is 11°11′23″; in appendix C.iii.d. HESIOD 4 Newton obtained 10°36′, for a difference of about 35′.[15] He computed refraction corrections to the longitude of 62′ and 41′, whereas (using his values) we obtain 52′ and 35′, all of which added together gave Newton ♍13°7′ for the longitude of the rising point at the latitude of *Arcturus* (we obtain ♍13°26′, to the nearest minute, again using his values). He approximated to the nearest degree, and then compared this result with his previously calculated value for the longitude of *Arcturus* at the time of the Argonautic expedition, which was ♍13°24.′52″. He remarked in appendix C.iii.h. HESIOD 8 that "the difference is so small as scarce to be sensible in the coarse observations of the ancients."

Newton's result is in principle vitiated by his having neglected to precess the line of apsides, despite including its proper motion, but the degree of accuracy that Newton was prepared to accept utterly swamps the combined effects of precession and proper motion on the line of apsides. He need not have bothered with either effect. Any calculation of Hesiod's date based on the *Arcturus* remark moreover involves other effects than refraction. Newton had assumed that the *arcus visionis* could be accounted for entirely by the 33′ he allowed for solar refraction, which meant among other things that the star at first evening visibility lay precisely on the horizon. In fact the star will become visible when it lies a certain distance above the horizon, and the sun a certain distance below it, as a result of refraction and conditions due to the transparency of the air. The precise number to use was (and remains) somewhat controversial,[16] but all agree that it is vastly larger than Newton's apparent 33′, amounting perhaps to as much as 9.1° for the sum of the distances of the star above and the sun below the horizon. This would greatly complicate the computation because we must now raise the star as well as lower the sun.[17]

[14] Towards the end of the eighteenth century the angle was computed to have been 23°48′30″ around the time of Newton's date for the Argonauts (Horsley, 1779–1785a, pp. 72, vol. 74). Horsley found this value "by [Tobias] Mayer's tables" (on Mayer see Wepster, 2010). The rate of change was not clear in Newton's time. The difference is not consequential, producing in the end about a decade's alteration.

[15] He termed *JP* the "elevation excess," reflecting its origin in point *A*'s distance from the ecliptic.

[16] In his 1592 *Tractatus*, which became a standard on the subject of globes in the seventeenth century, Robert Hues explained how to use a celestial globe to find a heliacal rising or setting, noting that to do so required choosing an elevation "above the horizon 12, 13, 14, or any other number of degrees that the magnitude of your Starre shall require for distance from the sunne" (Markham (1889), p. 111).

[17] We can deploy a computer program ("Planetary, Lunar and Stellar Visibility" by Rainer Lange and Noel M. Swerdlow) that generates rising and setting dates and times to estimate the effects of Newton's assumptions concerning refraction. Choosing for convenience the latitude of Babylon, we find that Newton's value of 33′ would put the first evening rising of Arcturus in −880 at 18:01 on March 5, with the sun visibly setting at 17:58. Using instead an *arcus visionis* of 9.1° changes the moment and date of acronychal rise to 18:35 on February 25, with the sun visibly setting at 17:52. In −880 the winter solstice occurred on December 29. Using the March 5 date which results from Newton's refraction correction produces a

TABLE E.1.
Calculation results for Hesiod

Solar longitude at winter solstice:	♑ 0°
Locus of the solar apogee in 1700:	*Newton*: ♋7°44′30″ Appendix C.iii.f. HESIOD 6.
Presumptive date for Hesiod:	*Newton*: −880 Appendix C.iii.f. HESIOD 6.
Rate of proper motion of apogee:	*Newton*: 17′40″ per century Appendix C.iii.f. HESIOD 6.
Proper motion of apogee from −880 to 1700:	*Newton*: 7°35′48′ Appendix C.iii.f. HESIOD 6.
1700 longitude of *Arcturus*	*Newton*: ♎20°.18′.52″ Appendix C.iii.f. HESIOD 6.
−880 distance between *Arcturus* and solar apogee:	*Newton*: 110°10′10″ Appendix C.iii.f. HESIOD 6.
Resultant locus of solar apogee in −880 ignoring precession but including proper motion:	*Newton*: ♋8°38′ "thereabouts" Appendix C.iii.h. HESIOD 8.
Presumptive solar eccentricity using Newton's apsidal line:	.0332
Value of solar equation at winter solstice with Newton's apsidal line:	17″
Value of mean anomaly at winter solstice with Newton's apsidal line:	179°51′35″
Mean solar motion in 60 days:	*Newton*: 59°8′10″ Appendix C.iii.h. HESIOD 8.
Value of mean anomaly at winter solstice+60 days with Newton's apsidal line:	238°59′49″
Value of solar equation at winter solstice+60:	*Newton*: 1°39′30″ Appendix C.iii.h. HESIOD 8.
Total motion in longitude in 60 days from winter solstice with Newton's apsidal line:	*Newton*: 60°47′40″ Appendix C.iii.h. HESIOD 8.
Angle between ecliptic and horizon	72°4′
Longitudinal distance between intersections of horizon with ecliptic and with latitude of Arcturus (JP in figure E.4)	*Newton*: 10°36′ Appendix C.iii.d. HESIOD 4

All of this depends on the assumption that Hesiod knew the date of the solstice to within a day or two, and that he observed true evening rising. To see just how difficult it is to pinpoint a date from Hesiod's remark, let's arbitrarily and absurdly assume that Hesiod made the observation in −1100. In that year the winter solstice occurred on December 30, and the acronychal rising (using Babylon and a proper *arcus visionis*) on March 3, for an interval of 63 days—which is better than the 66 days for −880 that would result using only Newton's value for refraction. These differences are much too small to conclude anything reasonable about Hesiod's date from the Arcturus remark. If he were off by several days in the solstice (which is hardly unlikely given the slow motion of the sun at the solstices), then the computed date would shift by more than a century. And that assumes a perfect observation of the acronychal rising—whereas it's commonly remarked that heliacal (dawn) phenomena (such as Sirius' first morning rise) were generally more important than dusk phenomena.

In print no trace of Newton's computation remains, and we read only that given Hesiod's remark "it follows that *Hesiod* flourished about an hundred years after the death of *Solomon*." Newton's ♍13° puts the date about half a century after the expedition, and so a century after Solomon.

E.ii. Thales

Newton used the following items concerning Thales to "confirm" his computation of the date of the original colure:

1. a report from Pliny that Thales 'determined' the morning setting of the *Pleiades* "to be upon the 25th day after the Autumnal Equinox"[18]
2. that Thales was a "young man fit to apply himself to Astronomical studies" in "about the 41st Olympiad" (which would be −616 since the first was −776).
3. that the longitude of the star *Lucida Pleiadum* in the *Pleiades* at the time of the original colure (and so of the Argonautic expedition) was ♈19°26′8″.

The third claim comes directly from Newton's final result, based on the Eudoxan remarks, that the colures at the time of the remarks were 36°29′ in advance of their position in 1689. So that, he wrote, "by the same method the place of any Star in the Primitive Sphere may readily be found, counting backwards one Sign, 6°.29′. from the Longitude which it had in the end of the year of our Lord 1689."[19]

Newton assumed (with Petavius) that the Thales remark must refer to the brightest star in the *Pleiades*, which was then called the *Lucida Pleiadum*. Its modern name is *Alcyone* (magnitude 2.87); denoted *n* by Bayer and Flamsteed, it was located in their time and today in *Taurus*. Flamsteed gives its (epoch 1689) tropical coordinates as ♉25°40′8″, *lat* 4°0′37″, and Hevelius (epoch 1660) gives them as ♉25°15′51″, *lat* 4°1′18″, most of the change being

---

66 day interval between solstice and acronychal rise. Using the more accurate date of February 25 produces instead an interval of 58 days, which is in even closer agreement with Hesiod's 60 days.

[18] Newton, 1728b, p. 92.

[19] Newton, 1728b, p. 91.

TABLE E.2.
Apsidal longitudes for Thales

Unprecessed line of apsides	*Precessed line of apsides*
$\lambda_{\odot}$ = ♎25°28′32″	$\lambda_{\odot}$ = ♎25°23′0″

due to precession over 29 years (the rest reflecting their observational differences).[20] Its epoch 2000 longitude is ♉29°59′24″. Precessing Flamsteed's longitude back to −616 would put the star at about ♈23°39′, or, using Hevelius' longitude, at about ♈23°15′. These are reasonably close to Petavius' result, cited by Newton, of ♈23°53′. Petavius, however, did not use Thales' date, which would have begged the question, but instead obtained the longitude by calculation from Pliny's report concerning the star's morning setting.

We can use the procedure that Newton found in Riccioli for Hesiod, applied to Thales' situation. In fact, the very next problem after the Hesiod example in Riccioli's text (*Problema 9*) is entitled "Ortum & Ocassum Heliacum Stella Investigare In Data Altitudine Poli,"[21] which Newton certainly saw. The only difference between the two situations is that *Arcturus* rises as the sun sets for the Hesiod remark (*Arcturus' Ortus Acronychus* in Riccioli's traditional terminology), whereas *Lucida* sets as the sun rises for the Thales remark (its *Occasus Heliacus* or heliacal setting).

We need first to determine the solar longitude 25 days past the autumnal equinox. To do this properly, we should precess the sun's line of apsides from, say, 1700 back to Thales' epoch of −616, add in the apogee's proper motion, set the sun at autumnal equinox and then move it along by 25 days. However, as in the case of Hesiod the apsidal position does not strongly affect the sun's longitude: taking the apsidal line at its 1700 locus of ♋7°44′33″, and using the procedure described above for Hesiod (but excluding proper motion) and adapted to 25 days past the autumnal equinox, we find the results given in table E.2.[22]

The difference of a bit more than five minutes is insignificant for Newton's dating purposes, as we can see by using the computational procedure for *Arcturus* adapted to the case of the *Lucida Pleiadum* as follows.

Figure E.7, top is a schematic of the situation. The setting point *J* of the ecliptic that lies on the horizon is 180° around from the rising sun. The vernal point *G* sits below the horizon, the setting *Lucida Pleiadum* is at *A*, and *PG* is *Lucida's* longitude. Here, then, we must subtract the distance *JP* from the longitude *GJ* of *J* in order to find *PG*, whereas in the case of the acronychal rise of *Arcturus* we had to add it (see figure E.4).[23] Figure E.7, bottom shows the configuration when the *Pleiades* set at the time of Thales' remark. The light, dotted

[20] In manuscript Newton referred specifically to the Hevelius longitude.

[21] Riccioli, 1651, pp. 463–64.

[22] Using epoch 2000 parameters and procedures the solar longitude of *Alcyone* 25 days past the autumnal equinox in −616 would be ♎23°30′43″.To adapt the procedure we used above for Hesiod to this case of the autumnal equinox, note first that the mean anomaly at autumn equinox will be $\sin^{-1}(e\sin A_0) + 180°$ $- A_0$ since $\lambda_{\odot}^{ae}$ is 0°. Add 25 days of mean motion to this and proceed as before. We again use Newton's presumptive value of .0332 for the eccentricity, and his value for the solar year of 365.25–13min2sec. See table E.1.

[23] Since, in addition, point M now lies between points G and E, to compute GE we must now add EM to GM instead of subtracting it.

TABLE E.3.
Longitudes of *Lucida Pleiadum* (*Alcyone*)

Longitude of Lucida *using unprecessed apsidal line*	*Longitude of* Lucida *using apsidal line precessed to -616*	*Longitude of* Lucida *using epoch 2000 procedures and parameters*
♈24°13′43″	♈24°8′16″	♈24°30′43″

lines form the horizon grid (altitude and azimuth), with the local horizon passing through the *Pleiades*. The equatorial grid is solid, and the dot-dash black line represents the ecliptic. Carrying out the appropriately modified computations yields the results (uncorrected for refraction), given in table E.3, choosing Athens as our latitude.[24]

Clearly it makes little difference whether or not we precess the apsidal line. Moreover, the value that Petavius provided, ♈23°53′, is very close to the results obtained using the methods described by Riccioli. Any differences between these results and those of Petavius are no doubt due to slight differences in parameters and computational inaccuracies.

Although Newton did extensive calculations for the similar problem posed by Hesiod's remark concerning *Arcturus*, his manuscripts contain no evidence that he ever carried out a computation on *Lucida* for himself. Instead, he referred solely to Petavius for the (quite good) result of ♈23°53′. Newton may already have known how complicated the similar calculation was for Hesiod, and so instead of going through the process again he decided just to rely on Petavius. Although the MS is certainly not organized, overall, in the chronological order of the pages' production (which would in any case be impossible since some pages were used more than once over the years), nevertheless we find major remarks on Thales on the very same sheet, but following, a paragraph on Hesiod whereas in the printed text Hesiod follows Thales.

Subtracting the original longitude (♈19°26′8″) from Petavius' ♈23°53′ implies that *Lucida* had precessed through 4°26′52″, which requires slightly more than 320 years. Taking the date of the 41st Olympiad at −616 and subtracting 320 gives the date of the original colure as −936, or, as Newton wrote, "44 years after the death of *Solomon*" in −980.

E.iii. The heliacal rise of Sirius

The problem posed by the *heliacal rise* of Sirius (its first appearance on the eastern horizon shortly before sunrise) plagued many otherwise competent people, including a number well-versed in mathematics and even in astronomy, for nearly 250 years after 1600. Although Newton himself did not deal with it, the techniques for calculating the rise are effectively the same as the ones that he deployed for Hesiod and *Arcturus* and so are worth setting out since this was an issue that repeatedly arose both before and after Newton in respect to Egyptian chronology.

Since Sirius first rose with the sun after its long absence from the night sky near the time of the Nile's annual inundation, its reappearance marked an important point of the year for

[24] We could use the latitude of Miletus on the coast of Asia Minor, where Thales was located, but the latitude difference from Athens is insignificant, producing in the end at most a few minutes' change in *Lucida's* computed longitude.

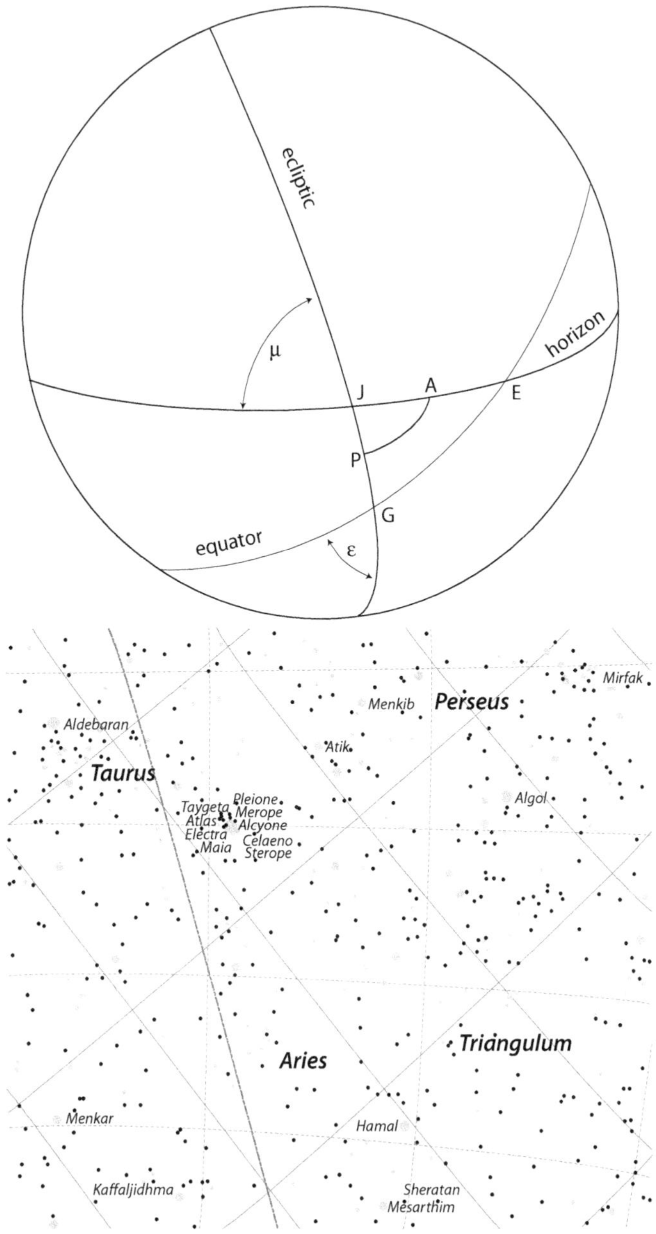

Figure E.7. Top: schematic for the morning setting of the *Lucida Pleiadum*. Bottom: star chart of the *Pleiades* setting at sunrise on October 25, –616.

the ancient Egyptians. Precession will move any star along in longitude, so it seemed natural to many (and it still does) that one could probe Egyptian chronology by calculating how long precession takes to decouple the heliacal rising from its original link to the inundation. Sirius is, however, not a zodiacal star, and so its latitude complicates the problem. Moreover, heliacal and acronychal risings and settings are horizon phenomena and so may depend strongly on terrestrial latitude, as we saw above for *Arcturus*. The question, then, reduces to this: if at some date the sun sits at a point that rises simultaneously with Sirius, will this reoccur yearly with reasonable stability over many centuries or not?

To answer the question we can use the alternative method for the acronychal rise of *Arcturus*.[25] We are given Sirius' longitude and latitude, and that it is on the horizon. We then seek the longitude of the simultaneously-rising ecliptic point, and we imagine the sun there for some heliacal rise. Finally, we determine whether, assuming this to have occurred at some specified date in the Egyptian past, the point remains stable in succeeding years. If it does, then we cannot use precession and this phenomenon for purposes of Egyptian chronology.

From the tropical coordinates (λ_s,β_s) of Sirius at our date in the past we obtain its equatorial coordinates (α_s,δ_s). We then compute the longitude $\lambda_\odot$ of the ecliptic point that rises with it at latitude φ as follows, with reference to figure E.4 (wherein A now represents the locus of Sirius, and J that of the sun):

$$\begin{aligned}
\delta_s &= \sin^{-1}(\sin\beta_s\cos\varepsilon + \cos\beta_s\sin\lambda_s\sin\varepsilon)\\
\alpha_s &= \tan^{-1}\left(\frac{-\sin\beta_s\sin\varepsilon + \cos\beta_s\sin\lambda_s\cos\varepsilon}{\cos\beta_s\cos\lambda_s}\right)\\
ET &= \sin^{-1}(\tan\delta_s\tan\varphi)\\
GE &= \alpha_s - ET\\
EM &= \tan^{-1}\left(\frac{\sin GE\tan\varphi\tan\varepsilon}{1-\cos GE\tan\varphi\tan\varepsilon}\right)\\
GE &= GE + EM\\
\lambda_\odot &= \tan^{-1}\left(\frac{\tan GM}{\sin\varepsilon}\right)
\end{aligned} \tag{E.14}$$

Equations (E.14) can give the date on which a heliacal rising will occur in a given year: we need only determine when, in that year, the sun will reach longitude $\lambda_\odot$. Alternatively, if we are given any day in a specific year we can determine whether or not it coincides with a heliacal rising by calculating the day's solar longitude.

Newton's computations for Hesiod and Thales, as well as the one here for Sirius, could have been done much more quickly, if with considerably less accuracy, using a precession globe. Ptolemy described the design of such an object in the *Almagest*.[26] A standard armillary

[25] Wing, 1669, pp. 91–92 provides a related method given the angle between the ecliptic and the horizon and the "depression" of Sirius below the horizon.

[26] Ptolemy, 1998, pp. 404–7, Book 408, sec. 403. Neugebauer discusses the globe's construction in Neugebauer, 1975, pp. 890–92, vol. 892. Toomer corrects one element of the reconstruction in his Ptolemy, 1998, pp. 405, n.181.

Figure E.8. Armillary sphere, Flemish ca. 1550 (Museum of the History of Science, Oxford). The second ring in is for precession; the third ring is for trepidation[27] and is pinned to the two circular plates visible in the figure on the right. Photos by JZB.

sphere can itself be used to answer stellar rising and setting questions for a particular epoch. Suppose for example that we want to know the day on which a given star rises heliacally. Since we know the tropical coordinates of the star, we can mark a point on the armillary (or attach somehow a marker) at the star's location. Then we turn the device until the marked point lies on the eastern side of the armillary's latitude ring. Look next at the zodiacal band and note the constellation which is rising, as well as the approximate position of the rising point within the constellation. We can then look up the day on which the sun will be at that point (or compute it), which accordingly gives the date of the star's heliacal rising. Adjustments can be made for refraction and other effects if greater accuracy is desired.

A precession globe would enable the same procedure for any epoch at all, as long as we have the star's coordinates for one: simply rotate the device around the globe's extra axis (which will be perpendicular to the plane of the zodiacal ring) to the desired epoch and operate as before. The result will always be highly approximate, since it does not take account of the proper motion of the solar apse, and since positions can only be estimated (with the latter depending on the size and scale markings of the globe). Many armillary spheres by the early 17th century do contain rings whose axes are orthogonal to the plane of the zodiac, but the purpose was to represent the motions of the sun and moon. The zodiacal band itself was usually fixed to the colures. Figure E.8 illustrates two of these: one dates to 1602 and was constructed by Giovanni Paolo Ferreri (fl. 1600–1625); the other, by

[27] Trepidation was a theory originally developed by Theon of Alexandria (or, possibly, by Hipparchus or even by "Chaldeans" or "Babylonians"—see chapter 9, n.16) according to which the stellar sphere oscillates over an arc of some eight degrees: see Neugebauer, 1975, pp. 631–34, vol. 632. Thabit Ibn'Qurra (836–901) is traditionally said to have combined trepidation with a complete rotation of the sphere, which this armillary illustrates, in *On the Motion of the Eighth Sphere*. However, Ragep, 1993, pp. 398–408 presents evidence that the author of *On the Motion* was not Ibn'Qurra but his grandson, Ibrahim ibn Sinan (908–46). For a discussion of trepidation in Islam see Ragep, 1996.

Manfredo Settala[28] (1600–1680), dates to 1646. Spheres that were capable of representing precession were, to judge by their scarcity today, rare, but they did exist. One, for example, was made ca. 1550 by a Flemish craftsman (figure E.8).

Scaliger had decided that precession could not exist because his studies of chronology indicated that the star Sirius must have risen on the same day of the year throughout more than a millennium and a half of Egyptian history.[29] In 1598, Tycho, in friendly and admirably clear admonishment, tried to disabuse him of the belief that the stability of Sirius' heliacal rise was inconsistent with precession:

> ... I too, having fitted my calculations geometrically to the motions made by this star in the meantime [since, that is, Scaliger had written], find that in the horizon of the Greeks [which is not sufficiently different from that of Egypt to make much of a difference], where the pole is elevated more or less 40 degrees, its rising with the sun moved forward by only one day of the Julian year in 1,500 years computed backwards from the time of Ptolemy. This change would have been almost imperceptible to them, since they did not have a very accurate grasp of these things. But that conformity comes about accidentally in the case of this star, for its declination has varied in that interval of centuries by 2⅔ degrees, just the right amount to correct and, as it were, destroy the change that results from the disagreement of the Julian and sidereal years. But this does not happen in the same way to other stars, because the alteration of their declinations takes a wide variety of forms.[30]

We saw above that it's certainly not simple to compute the effect of precession on heliacal and acronychal risings and settings, though Tycho probably did so since there is no indication that he owned a precession globe. A half century later, the English astronomer John Bainbridge showed how to do the computation,[31] as did Riccioli.[32]

[28] Little is known about Ferreri, though he probably worked in Rome. Dekker and Lippincott, 1999, p. 154; Dekker, 1999, p. 154 suggest that he may be "the Spanish architect and mathematician Johannes Ferrerius." Settala in Milan maintained one of the most famous of the period's cabinets of curiosities. Renowned for his abilities with a lathe, he was expert in using it to fabricate instruments and curiosities such as miniature scenes and intricately-nested geometric designs turned from ivory. The lathe was a liminal device—often a plaything of nobility during the Renaissance as well as an artisanal instrument: see Connors, 1990.

[29] For a thorough and amusing account of Scaliger's adventures through the Egyptian year and precession, see Grafton, 1993, pp. 197–209 and 459–88.

[30] Given in the original Latin and translated in Grafton, 1993, p. 204.

[31] On this see Grafton, 1993, p. 207. The procedure in Bainbridge, 1648, pp. 59–74, is lengthier but less direct and clear than Riccioli's three years later.

[32] Riccioli, 1651, p. 464.

BIBLIOGRAPHY

Abauzit, Firmin (1774). *Miscellanies of the Late Ingenious and Celebrated M. Abauzit.* Trans. E. Harwood. London: T. Becket.

Adam, Charles, and Paul Tannery, eds. (1897–1910). *Oeuvres de Descartes.* 12 vols. Paris, Léopold Cerf.

Addison, Joseph (1989). *The Freeholder (1715–1716)*, Oxford, Clarendon Press.

Addison, Joseph, and Richard Steele (1965). *The Spectator*, ed. Donald C. Bond, 5 vols. Oxford, Clarendon Press.

Albertan, Christian, and Ann-Marie Chouillet (2003). "Autographes et Documents," *Recherches sur Diderot et sur l'Encyclopédie* 34: 213–30.

Algarotti, Francesco (1739). *Sir Isaac Newton's Philosophy Explain'd for the Use of the Ladies*, 2 vols. London: C. Caves.

———. (1778–1784). *Opere*, 10 vols. Cremona, Lorenzo Manini.

Allen, Thomas (1659). *A Chain of Scripture Chronology; from the Creation of the World to the Death of Jesus Christ. In VII Periods*, London.

Allix, Peter (1711). *Remarks upon Some Places of Mr. Whiston's Books, either Printed or in Manuscript*, London, John Wyat.

Almond, Philip C. (1993). "Henry More and the Apocalypse," *Journal of the History of Ideas* 54: 189–200.

Alpers, Svetlana (1983). *The Art of Describing*, Chicago, University of Chicago Press.

Andrés, Juan (1785–1822). *Dell'Origine, progressi e stato attuale d'ogni letteratura*, 8 vols. Parma, Stamperia Reale.

———. (1788). *Dissertazione sull' episodio degli amori d'Enea e Didone*, Cesena, Eredi Biasini.

Anonymous (1743). "Reflexions générales sur l'utilité des belles-lettres; et sur les inconvéniens du gout exclusif, quit paroit s'établire en faveur des Mathématiques & de la Physique," *Histoire de l'Académier Royale des Inscriptions et Belles-Lettres* 16: 11–37.

———. (1780). "Éloge de M. de la Nauze," *Histoire de l'Académie des Inscriptions et Belles-Lettres* 4: 232–42.

Aratus (1997). *Phaenomena*. Tr. Douglas Kidd, Cambridge, Cambridge University Press.

Arbuthnot, John (1701). *An essay on the usefulness of mathematical learning*, Oxford, Oxford University Press.

Archibald, Thomas, Jed Z. Buchwald, and Kurt Moller Pedersen (1991). "Erasmus Bartholin. Experiments on Birefringent Icelandic Crystal. With a Facsimile of the Original Publication 1669," *Acta Historica Scientarum Naturalium et Medicinalium* 40.

Argelander, F. W. (1842). *De fide Uranometriae Bayeri dissertatio academica*, Bonn, Caroli Georgii.

Ariew, Roger (1999). *Descartes and the Last Scholastics*, Ithaca, Cornell University Press.

Ariew, Roger, and Marjorie Grene (1999). "The Cartesian destiny of form and matter," in *Descartes and the Last Scholastics*, Ithaca: Cornell University Press, pp. 77–96.

Aristotle (1968). "On dreams (De Somniis)," in *The Basic Works of Aristotle*, edited by Richard McKeon, New York: Random House, pp. 618–25.

Aristotle (1654). *Opera omnia Graece et Latine*, ed. Guillaume Duval. Paris, Jean Billaine, Siméon Piget, and Fredéric Léonard.

Armogathe, Jean-Robert (1987). "L'Arc-en-ciel dans les Météores," in *Le Discours et Sa Méthode*, edited by Nicolas Grimaldi and Jean-Luc Marion Paris: Presses Universitaires de France, pp. 145–62.

———. (2000). "The rainbow: a privileged epistemological model," in *Descartes' Natural Philosophy*, edited by Stephen Gaukroger, John Schuster, and John Sutton, London and New York: Routledge, pp. 249–57.

Asmis, Elizabeth (2005). "Epicurean epistemology," in *The Cambridge History of Hellenistic Philosophy*, edited by Keimpe Algra et al., Cambridge: Cambridge University Press, pp. 260–94.

Assmann, Jan (1997). *Moses the Egyptian: The Memory of Egypt in Western Monotheism*, Cambridge, MA, Harvard University Press.

Athenaeus (1657). *Deipnosophistarum libri quindecim*, Leiden, Ioannes Antonius Hugvetan & Marcus *Antonius* Ravaud.

Atterbury, Francis (1783–1787). *The Epistolary Correspondence, Visitation Charges, Speeches, and Miscellanies*, 5. vols. London, John Nichols.

Axtell, James L., ed. (1968). *The Educational Writings of John Locke*, Cambridge: Cambridge University Press.

Baasner, Rainer (1986). "Johann Gottfried Herder: Anfangsgründe der Astronomie," *Lias* 13: 69–97.

Badaloni, Nicola (1968). *Antonio Conti: Un Abate libero pensatore tra Newton e Voltaire*, Milan, Feltrinelli.

Baily, Francis (1835). *An account of the Revd. John Flamsteed, the first astronomer-royal*, London, Printed by order of the Lords Commissioners of the Admiralty.

———. (1845a). "The catalogues of Ptolemy, Ulugh Beigh, Tycho Brahe, Halley, Hevelius, deduced from the best authorities.," in *The Catalogue of Stars of the British Association for the Advancement of Science*, edited by Francis Baily, London: J. E. Taylor.

———. (1845b). *The Catalogue of Stars of the British Association for the Advancement of Science*, London, J. E. Taylor.

Bailly, Jean Sylvain (1781). *Histoire de l'Astronomie Ancienne, depuis son origine jusqu'a l'établissement de l'école d'Alexandrie*, Paris, De Bure.

Bainbridge, John (1620). *Procli Sphaera. Ptolemaei De hypothesibus planetarum liber singularis, nunc primum in lucem editus. Cui accessit ejusdem Ptolemaei Canon regnorum*, London, G. Jones.

———. (1648). *Una cum demonstratione ortus Sirii heliaci, pro parallelo inferioris Aegypti*, Oxford, Thomas Robinson.

Baker, Thomas (1700). *Reflections upon Learning, wherein is shewn the Insufficiency thereof, in its Several Particulars*, London, A. Bosvile.

Banier, Antoine (1736). "Les Argonautes, ou Dissertation sur la Conqueste de la Toison d'Or," *Histoire et Mémoires de l'Académie Royale des Inscriptions et Belles Lettres* 9: 54–96.

———. (1739–1740). *The Mythology and Fables of the Ancients, Explained from History*, 4 vols. London, A. Millar.

Barlow, Thomas (1693). *The Genuine Remains*, London, John Dunton.

Barrington, John Shute (1828). *The theological works of the first Viscount Barrington*, 3 vols. London, C & J. Rivington.

Barrow, Isaac (1683–1687). *The Works of the Learned Isaac Barrow*, 4 vols. London.

———. (1859). *Theological Works*, 9 vols. Cambridge, Cambridge University Press.

Bartholin, Erasmus (1669). *Experimenta Crystalli Islandici Disdiaclastici Quibus mira & insolita Refraction detegitur*, Hafniae, Danielis Paulli.

Bayer, Theophilus Siegfried (1730). *Museum Sinicum* St. Petersburg, ex typographia Academiae imperatoriae.

Bedford, Arthur (1706). *The Evil and Danger of Stage-Plays*, Bristol, W. Bonny.

———. (1728). *Animadversions upon Sir Isaac Newton's Book, Intitled The Chronology of Ancient Kingdoms Amended*, London, Charles Ackers.

———. (1730). *The Scripture Chronology Demonstrated by Astronomical Calculations*, London, James & John Knapton et al.
Bedini, Silvio (2005). "Along came a spider—spinning silk for cross-hairs," *The American Surveyor*, March/April.
Beeley, Philip, and Siegmund Probst (2005). "John Wallis (1616–1703): Mathematician and Divine," in *Mathematics and the Divine: A Historical Study*, edited by T. Koetsier and L. Bergmans, Amsterdam: Elsevier, pp. 441–57.
Belenkiy, Ari, and Eduardo Vila Echagüe (2005). "History of one defeat: reform of the Julian calendar as envisaged by Isaac Newton," *Notes & Records of the Royal Society* 59: 223–54.
Bennett, Jim (2003). "Hooke's instruments," in *London's Leonardo. The Life and Work of Robert Hooke*, edited by Jim Bennett, Michael Cooper, Michael Hunter, and Lisa Jardine, Oxford: Oxford University Press, pp. 63–104.
Bentley, Richard (1697). *A Dissertation upon the Epistles of Phalaris, Themistocles, Socrates, Euripides, and others; and the Fables of Aesop*, London, J. Leake.
———. ([1836] 1971). *Works*, 3 vols. Hildesheim, Georg Olms.
Béroalde, Matthieu (1575). *Chronicum, Scripturae Sacrae Autoritate Constitutum*, Geneva, Antoine Chuppin.
———. (1590). *A short vievv of the Persian monarchie, and of Daniels weekes*, London, Thomas Orwin.
Berosus (1612). *Antiquitatum Libri V*, Wittenberg, Martin Henckely.
Bertier, Joseph-Etienne (1760). *Physique des comètes: dans le sentiment de l'impulsion & du plein*, Paris, Imprimerie Royale.
Beveridge, William (1669). *Institutionum Chronologicarum Libri II*, London.
Biagioli, Mario (2006). "From prints to patents: living on instruments in early modern Europe," *History of Science* 44: 139–86.
Birch, Thomas (1968). *The History of the Royal Society of London*, 4 vols. New York, Johnson Reprint Corp.
Bochart, Samuel (1681). *Geographia sacra, cujus pars prior Phaleg... pars posterior Chanaan*, Frankfurt am Main, Bathasar Christoph Wustius for Johann David Zunner.
Bodin, Jean (1969). *Method for the Easy Comprehension of Histor*. Tr. B. Reynolds, New York.
Bogan, Zachary (1658). *Homerus Hebraizaon: sive, Comparatio Homeri cum Scriptoribus Sacris*, Oxford, H. Hall for T. Robinson.
Bolingbroke, Henry St , John, Viscount (1808). *Lettres historiques, Politiques, Philosophiques et Particuliéres*, 3 vols. Paris, Dentu.
———. (1967 [1844]). *Works*, 4 vols. New York, Augustus M. Kelley.
———. (1977 [1757]). *The Philosophical Works*, 5 vols. New York, Garland.
Boswell, John (1738). *A Method of study*, London.
Boyer, Carl B. (1959). *The Rainbow. From Myth to Mathematics*, New York, Thomas Yoseloff.
Boyle, Charles, i.e. Francis Atterbury et al. (1698). *Dr. Bentley's Dissertations on the Epistles of Phalaris and the Fables of Aesop, Examin'd*, London.
Boyle, Robert (1662). *A Defence of the Doctrine Touching the Spring and Weight of the Air*, London, Thomas Robinson.
———. (1664). *Experiments and Considerations Touching Colours*, London, Henry Herringman.
———. (2006). "The Boyle Papers Online." The Boyle Project.
Brack-Bernsen, L. and H. Hunger (1999). The Babylonian zodiac: speculations on its invention and significance," Centaurus 41: 280–92.
Brahe, Tycho (1602 (1969)). *Astronomiae Instauratae Progymnasmata*, Uraniburgi Daniae, Bragae Boehmiae.
———. (1913–1929). *Tychonis Brahe Dani Opera Omnia*, Hauniae, Libraria Gyldendaliana.

Brandt, Frithiof (1928). *Thomas Hobbes' Mechanical Conception of nature*, Copenhagen, Levein & Munksgaard.
Brewster, David (1965). *Memoirs of the life, writings, and discoveries of Sir Isaac Newton*, 2 vols. New York, Johnson Reprint Corp.
Briggs, E. R. (1934). "L'incrédulité et la pensée anglaise en France au début du dix-huitième siècle," *Revue d'histoire littéraire de la France* 41: 497–538.
Briggs, Henry (1624). *Arithmetica Logarithmica sive Logarithmorum Chiliades Triginta*, London, William Jones.
———. (1631). *Logarithmicall Arithmetike.*, London, George Miller.
Brosses, Charles de (1761). "Second Mémoire sur la monarchie de Ninive," *Histoire et Mémoires de l'Académie Royale des Inscriptions* 27: 1–83 (of the *Mémoires*).
Broughton, Hugh (1588). *A Concent of Scripture*, London.
———. (1592). *An Apologie in Briefe Assertions Defending that Our Lord Died in the Time Properly Foretold to Daniel*, London.
Brown, Harold I. (1985). "Galileo on the telescope and the eye," *Journal of the History of Ideas* 46: 487–501.
Browne, Thomas (1964). *The Works of Sir Thomas Brown*, 4 vols. Chicago, University of Chicago Press.
Buchwald, Jed Z. (2006). "Discrepant measurements and experimental knowledge in the early modern era," *Archive for History of Exact Sciences*, 61: 565–649.
———. (2007a). "Huygens' methods for determining optical parameters in birefringence," *Archive for History of Exact Sciences*, 61: 67–81.
———. (2007b). "Descartes and the rainbow," *Annals of Science*, 64:1–46.
Buchwald, Jed Z., and I. Bernard Cohen, eds. (2001). *Isaac Newton's Natural Philosophy*, edited by Jed Z. Buchwald: Dibner Institute Studies in the History of Science and Technology (Cambridge, MA: MIT).
Buchwald, Jed Z., and Greco Josefowicz, Diane (2010). *The Zodiac of Paris*, Princeton, Princeton University Press.
Buhle, Johann Gottlieb (1816). *Histoire de la philosophie moderne, depuis la Renaissance des Lettres jusqu'à Kant*. Tr. A.J.L. Jourdan, 6 vols. Paris, F. J. Fournier.
Bunyan, John (1989). *Solomon's Temple Spiritualized*, Oxford, Clarendon Press.
Burnet, Thomas (1684). *The Theory of the Earth*, London, R. Norton for Walter Kettilby.
Burnett, D. Graham (2005). *Descartes and the Hyperbolic Quest. Lens making machines and their significance in the seventeenth century*, Philadelphia, American Philosophical Society.
Burton, Robert (1652). *The anatomy of melancholy*, Oxford, Henry Cripps.
Buxtorf, Johannes (1689). *Lexicon Hebraicum et Chaldaicum*, Basel, Johann Ludwig König.
Byrom, John (1854–1857). *The Private Journal and Literary Remains of John Byrom*, Manchester, Charles Simms.
Cairns, John E. (2001). "Alexander Cunningham's Proposed Edition of the Digest An Episode in the History of the Dutch Elegant School," *Tijdschrift voor rechtsgeschiedenis* 69: 81–117, 307–59.
Calvin, John (1578). *A Commentarie of John Calvine, upon the First Booke of Moses Called Genesis* London, For John Harrison & George Bishop.
Carli, Gian Rinaldo (1745). *Della Spedizione degli Argonauti in Colco*, Venice, Giambattista Recurti.
———. (1780). *Delle Lettere Americane*, 2 vols. Cosmopoli [Florence].
Casini, Paolo (1984). "Newton: The Classical Scholia," *History of Science* 22: 1–58.
Castel, Louis Bertrand (1743). *Le vrai système de physique générale de M. Isaac Newton*, Paris, Sebastian Jorry.

Cave, William (1676). *Primitive Christianity*, London, J. G. for R. Chiswell.
Cellarius, Andreas (1661). *Harmonia macrocosmica*, Amsterdam, Joannem Janssonium.
Chapman, Allan, ed. (1982). *The preface to John Flamsteed's Historia coelestis Britannica, or British catalogue of the heavens (1725). Based Upon a Translation by Alison Dione Johnson*, London: Trustees of the National Maritime Museum.
Charleton, Walter (1654). *Physiologia Epicuro-Gassendo-Charltoniana, or, A fabrick of science natural, upon the hypothesis of atoms founded by Epicurus*, London, Thomas Newcomb.
Churton, Edward (1868). *Fourteen Letters from Daniel Waterland to Zachary Pearce*, London, James Park & Co.
Cicero (1741). *Of the Nature of the Gods*, tr. Thomas Francklin, London, R. Francklin.
Cicero (1999). *On the Commonwealth and on the Laws*, Tr. James E. G. Zetzel, Cambridge, Cambridge University Press.
Clagett, Marshall (1995). *Ancient Egyptian Science: A Source Book. Volume 2: Calendars, Clocks, and Astronomy*, Philadelphia, American Philosophical Society.
Clark, John W. (1904). *Endowments of the University of Cambridge*, Cambridge, Cambridge University Press.
Clarke, Desmond M. (1989). *Occult Powers and Hypotheses*, Oxford, Clarendon Press.
Clarke, John (1731). *An Essay Upon Study*, London.
Clarke, Samuel (1706). *A Demonstration of the Being and Attributes of God*, London, William Botham for James Knapton.
———. (1732). *A Demonstration of the Being and Attributes of God*, London.
Clement of Alexandria, (1869). *The Stromata, or Miscellanies*. Tr. Alexander Roberts and James Donaldson, 2 vols. Edinburgh, T. & T. Clark.
———. (1919). *The Exortation to the Greeks*. Tr. G. W. Butterworth, London, William Heineman.
Clément, Nicolas (2002). *L'Abbé Alary (1690–1770). Un homme d'influence au XVIIIe siècle*, Paris, Honoré Champion.
Cohen, I. Bernard (1980). *The Newtonian Revolution*, Cambridge, Cambridge University Press.
———, ed. (1958). *Isaac Newton's Papers & Letters on Natural Philosophy*, Cambridge, MA: Harvard University Press.
Cohen, I. Bernard, and George E. Smith, eds. (2002). *The Cambridge Companion to Newton*, Cambridge: Cambridge University Press.
Colden, Cadwallader (1868). "Letters on Smith's History of New York," *Collections of the Historical Society of New York* 1: 179–235.
Collins, Anthony (1724). *A Discourse of the Grounds and Reasons of the Christian Religion*, London.
Columella, L. Junius Moderatus (1745). *Of Husbandry. In Twelve Books: and his Book Concerning Trees*, London, A. Millar.
Connors, Joseph (1990). "Ars Tornandi: baroque architecture and the lathe," *Journal of the Warburg and Courtauld Institutes* 53: 217–36.
Conti, Antonio (1726). *Réponse aux Observations sur la Chronologie de Mr. Newton*, Paris, Noel Pissot.
———. (1739–1756). *Prose e Poesie*, 2 vols. Venice, Giambatista Pasquali.
———. (1812). *Lettere Scelte*, Venice, Domenico Fracasso.
Conti, Natale (2006). *Mythologiae*. Tr. John Mulryan and Steven Brown, 2 vols. Tempe, Arizona Center for Medieval and Renaissance Studies.
Cook, Alan (1997). "Edmond Halley and John Flamsteed," in *Flamsteed's Stars*, edited by Frances Willmoth, Suffolk, UK: Boydell Press, pp. 167–87.
———. (1998). *Edmond Halley: Charting the Heavens and the Seas*, Oxford, The Clarendon Press.
Cooke, Thomas (1731). *The Letters of Atticus*, London, J. Chrichley.

Copenhaver, Brian P. (1988). "Astrology and magic," in *The Cambridge History of Renaissance Philosophy*, edited by Eckard Kessler and Quentin Skinner, Cambridge: Cambridge University Press, pp. 264–300.

Craig, Sir John (1946). *Newton at the Mint*, Cambridge, Cambridge University Press.

———. (1963). "Isaac Newton and the Counterfeiters," *Notes and Records of the Royal Society of London* 18: 136–45.

Cudworth, Ralph (1678). *The True Intellectual System of the Universe*, London, Richard Royston.

Cumberland, Richard (1724). *Origines Gentium Antiquissimae; or, Attempts for discovering the Times of the first Plantations of Nations in Several Tracts*, London, W. B. for R. Wilkin.

Cuvigny, Hélene de. 2005. Silver celestial globe from antiquity. In http://kugel.artsolution.net/c/c1_en.htm. (accessed 2005).

Cyprian (1682). *Sancti Caecilii Cypriani Opera recognita et illustrata*, Oxford, Oxford University Press.

D'Orléans, Chérubin (1671). *La dioptriqve oculaire*, Paris, Thomas Jolly & Simon Benard.

Daston, Lorraine (1991). "Baconian facts, academic civility, and the prehistory of objectivity," *Annals of Scholarship* 8: 337–63.

Daston, Lorraine, and Katherine Park (1998). *Wonders and the Order of Nature 1150–1750*, New York, Zone Books.

De Jong, H.M.E. (1969). *Michael Maier's Atalanta Fugiens: Sources of an Alchemical Book of Emblems*, Leiden, E. J. Brill.

De Morgan, Augustus (1914). *Essays on the life and work of Newton*, Chicago & London, Open Court.

Dear, Peter (1992). "From truth to disinterestedness in the seventeenth century," *Social Studies of Science* 22: 619–31.

———. (1998). "Method and the study of nature," in *The Cambridge History of Seventeenth Century Philosophy*, edited by Daniel Garber and Michael Ayers, Cambridge: Cambridge University Press, pp. 147–77.

Dekker, Elly (1992). "Der Himmelsglobus— eine Welt für sich," in *Focus Behaim Globus*, edited by Gerhard Bott (Nuremberg), pp. 89–100.

———. (1999). *Globes at Greenwich. A Catalogue of the Globes and Armillary Spheres in the National Maritime Museum, Greenwich*, Oxford, Oxford University Press and the National Maritime Museum.

———. (2004). *Catalogue of Orbs, Spheres and Globes*, Firenze, Giunti for the Istituto e Museo di Storia della Scienza.

———. (2008). "A 'watermark' of Eudoxan astronomy," *Journal for the History of Astronomy* 39: 213–28.

Dekker, Elly, and Kristen Lippincott (1999). "The scientific instruments in Holbein's Ambassadors: A re-examination," *Journal of the Warburg and Courtauld Institutes*, 62.

Delambre, Jean Baptiste Joseph (1817). *Histoire de l'Astronomie Ancienne*, Paris, Courcier.

———. (1827). *Histoire de l'Astronomie au Dix-Huitième Siècle*, Paris, Bachelier.

Delgado-Moreira, Raquel (2006a). "Newton's treatise on Revelation: the use of a mathematical discourse," *Historical Research* 79: 224–46.

———. (2006b). "Epistemology and Rhetorical Strategies in Newton's Theological Writings." PhD, Imperial College.

Derham, William (1713). *Physico-theology, or, A demonstration of the being and attributes of God from his works of creation*, London, W. Innys.

Descartes, René (1656). *Opera philosophica*, Amsterdam, Ludovic and Daniel Elsevir.

———. (1965). *Discourse on Method, Optics, Geometry, and Meteorology*, Tr. Paul J. Olscamp, New York, Bobbs-Merrill.

Destutt de Tracy, Antoine-Louis-Claude (1992). *Mémoire sur la faculté de penser, de la métaphysique de Kant, et autres textes*, Paris, Fayard.
Dickinson, Edmund (1655). *Delphi Phoenicizantes*, Oxford, H. Hall.
Diderot, Denis (1875–1877). *Oeuvres complètes*, Paris, Garnier frères.
Digby, Sir Kenelm (1669). *Of bodies and of mans soul*, London, John Williams.
Dijksterhuis, Fokko (2004). *Lenses and Waves. Christiaan Huygens and the Mathematical Science of Optics in the Seventeenth Century*, Dordrecht, Kluwer Academic Publishers.
Diodorus, Siculus (1939). *Library of History*, Tr. C. L. Sherman, Cambridge, MA, Harvard University Press.
Diogenes Laertius (1925). *Lives of eminent philosophers*, Tr. Robert D. Hicks. London, W. Heinemann New York G.P. Putnam's sons.
———. (1664). *De vitis dogmatis et apophthegmatis*, London, Octavian Pulleyn.
Dobbs, Betty Jo T. (2002). *The Janus Face of Genius: The Role of Alchemy in Newton's Thought*, Cambridge, Cambridge University Press.
Doddridge, Philip (1829). *The Correspondence and Diary of Philip Doddridge*, 5 vols. London, Henry Colburn & Richard Bentley.
Dodwell, Henry (1691). *A Discourse Concerning Sanchoniathon's Phoenician History*, London, Printed for Benjamin Tooke.
———. (1692). *Prælectiones academicæ in schola historices Camdeniana*, Oxford.
Donne, John (1952). *Essays in Divinity*, Oxford, Clarendon Press.
———. (1969). *Ignatius His Conclave*, Oxford, Clarendon Press.
Drayton, Michael (1961). *Poly-Olbion*, Oxford, Basil Blackwell for the Shakespeare Head Press.
Dryden, John (1956–). *Works*, Berkeley & Los Angeles, University of California Press.
Duke, Dennis (2005a). Analysis of the Farnese Globe [draft]. In http://www.csit.fsu.edu/~dduke/farnese2.pdf (accessed 2005).
———. (2005b). Dating star catalogs and globes. In http://www.csit.fsu.edu/~dduke/stats.pdf. (accessed 2005).
Duport, James (1660). *Homer Poetarum omnium seculorum facile Principis Gnomologia, Duplici Parallelismo illustrata*, Cambridge, John Field.
Eastwood, Bruce (1983). "Origins and contents of the Leiden planetary configuration (Ms. Voss. lat. Q.79. fol 93v): an artistic schema of the early Middle Ages," *Viator* 14: 1–40.
Edleston, James, ed. (1850). *Correspondence of Sir Isaac Newton and Professor Cotes,* London and Cambridge: John W. Parker & John Deighton.
Edwards, Edward (1864). *Libraries and the Founders of Libraries,* London, Trübner & Co.
Edwards, John (1693). *A discourse concerning the authority, stile, and perfection of the books of the Old and New-Testament,* London.
Elzinga, Aant (1972). *On a Resarch Program in Early Modern Physics*, Stockholm, Humanities Press.
Estève, Pierre (1755). *Histoire générale et particulière de l'astronomie*, Paris, Antoine Jombert.
Eusebius (1903). *Evangelicae Praeparationis*, 3 vols. in 5. Oxford, Oxford University Press.
Evans, James (1998). *The History and Practice of Ancient Astronomy*, Oxford, Oxford University Press.
Evelyn, John (1955). *Diary*, 5 vols. Oxford, Clarendon Press.
Feingold, Mordechai (1990). "Isaac Barrow: Divine, Scholar, Mathematician," in *Before Newton: The Life and Times of Isaac Barrow,* edited by Mordechai Feingold (New York: Cambridge University Press), pp. 1–104.
———. (1995). "Reversal of Fortunes: The Displacement of Cultural Hegemony from the Netherlands to England in the Seventeenth and Early Eighteenth Centuries," in *The World of William and Mary*, edited by Dale Hoak and Mordechai Feingold, Stanford: Stanford University Press).

———. (1996). "When facts matter," *Isis* 87: 131–39.

———. (1997). "The mathematical sciences and new philosophies," in *The History of the University of Oxford*, edited by Nicholas Tyacke, Oxford: Clarendon Press, pp. 359–448.

———. (2002). "Science as a Calling? The Early Modern Dilemma," *Science in Context* 15: 79–119.

———. (2004). *The Newtonian moment: Isaac Newton and the making of modern culture*, New York

———. (2013). "Newton's Impact on the Enlightenment," in *The Reception of Isaac Newton in Europe*, edited by Scott Mandelbrote and Helmut Pulte (London: Continuum)

Feller, François Xavier de (1781). *Dictionnaire historique*, 6 vols. Augsbourg, Matthieu Rieger.

———. (1790–1794). *Dictionnaire historique*, 8 vols. Liège, François Lémarie.

Ferrone, Vincenzo (1995). *The Intellectual Roots of the Italian Enlightenment. Newtonian Science, Religion, and Politics in the Early Eighteenth Century*, Tr. Sue Brotherton, Atlantic Highlands, Humanities Press.

Figala, Karin (2002). "Newton's Alchemy," in *The Cambridge Companion to Newton*, edited by I. Bernard Cohen and George E. Smith (Cambridge: Cambridge University Press), pp. 370–86.

Filmer, Robert (1991). *Patriarcha and Other Writings*, Cambridge, Cambridge University Press.

Fisher, Edward (1643). *The Scriptures Harmony*, London, Richard Cotes for William Hope.

Flamsteed, John (1712). *Historiae coelestis libri duo*, London, J. Matthews.

———. Flamsteed, John (1995–2001). *The Correspondence of John Flamsteed, The First Astronomer Royal*, edited by Eric G. Forbes, Lesley Murdinm, and Frances Willmoth, 3 vols. Bristol and Philadelphia, Institute of Physics Publishing.

Fleming, Robert (1708). *Christology. A Discourse Concerning Christ*, London, J. Humfreys for A. Bell.

Fontenelle, Bernard le Bovier de (1803). *Conversations on the Plurality of Worlds*. Tr. Elizabeth Gunning, London, J. Cundee.

Fourmont, Etienne (1735). *Réflexions critiques sur les Histoires des Anciens Peuples*, 2 vols. Paris, Musier at al.

Fréret, Nicolas (1728). *Some Observations on the Chronology of Sir Isaac Newton*, London, T. Warner.

———. (1729) [1724]. "Reflexions sur l'étude des anciennes histoires, & sur le degré de certitude de leurs preuves," *Memoires de Litterature Tirés des Registres de l'Académie Royale des Inscriptiones et Belles Lettres* 6: 146–89.

———. (1733). "Observations sur le temps auquel a vécu Bellérophon " *Histoire et Mémoires de l'Académie Royale des Inscriptions et Belles Lettres* 7:83–112.

———. (1743). "Sur la durée des générations dans les familles royales." *Histoire de l'Académie des Inscriptions et Belles Lettres* 14:15–20.

———. (1758). *Défense de la chronologie fondée sur les monumens de l'histoire ancienne, contre le système chronologique de M. Newton*, A Paris, Chez Durand.

Frisch, Ch., ed. (1858–1871). *Joannis Kepleri, Astronomi Opera Omnia*, 8 vols. Frankfort: Heyder & Zimmer.

Gabbey, Alan(1982). "Philosophia Cartesiana Triumphata: Henry More (1646–1671)," in *Problems of Cartesianism*, edited by Thomas M. Lennon, John M. Nicholas, and Jown W. Davis, Kingston and Montreal: McGill-Queen's University Press, pp. 171–250.

Gale, Theophilus (1672). *The Court of the Gentiles*, Oxford, H. Hall for Thomas Gilbert.

Galileo (1632 (1962)). *Dialogue Concerning the Two Chief World Systems*, Tr. Stillman Drake, Berkeley and Los Angeles, University of California Press.

Garber, Daniel (1993). "Descartes and experiment in the Discourse and Essays," in *Essays on the Philosophy and Science of René Descartes*, edited by Stephen Voss, Oxford: Oxford University Press, pp. 288–310.

Garnier, Charles-Georges-Thomas (1805). *The French Anas*, 2 vols. London, Richard Phillips.
Gascoigne, John (1989). *Cambridge in the Age of the Enlightenment*, Cambridge, Cambridge University Press.
Gassendi, Pierre (1649). *Animadversiones in decimum librum Diogenis Laertii*, Lugduni, Guillelmum Barbier.
Gaukroger, Stephen (1995). *Descartes. An Intellectual Biography*, Oxford, Clarendon Press.
Geisler, William S. (1984). "Physical limits of acuity and hyperacuity," *Journal of the Optical Society of America* 1: 775–82.
Gibbon, Edward (1764). *An Essay on the Study of Literature*, London, T. Becket & P. A. de Hondt.
———. (1814). *The Miscellaneous Works*, London, John Murray.
———. (1994). *The History of the Decline and Fall of the Roman Empire*, 3 vols. London, Penguin Press.
Gillies, John (1786). *The History of Ancient Greece*, 2 vols. Dublin, Burnet et al.
Gingras, Yves (2001). "What did Mathematics Do to Physics," *History of Science* 39: 383–416.
Goguet, Antoine-Yves (1761). *The Origin of Laws, Arts, and Sciences, and their Progress among the most Ancient Nations*, 3 vols. Edinburgh, Alexander Donaldson and John Reid.
Goldish, Matt (1998). *Judaism in the Theology of Sir Isaac Newton*, Dordrecht, Kluwer Academic Publishers.
Gowing, Ronald (1983). *Roger Cotes—natural philosopher*, Cambridge, Cambridge University Press.
Grafton, Anthony (1975). "Joseph Scaliger and Historical Chronology: The Rise and Fall of a Discipline," *History and Theory* 14: 156–85.
———. (1985). "From De Die Natali to De Emendatione Temporum: The Origins and Setting of Scaliger's Chronology," *Journal of the Warburg and Courtauld Institutes* 48: 100–143.
———. (1991a). *Defenders of the text: the traditions of scholarship in an age of science, 1450–1800*, Cambridge, MA, Harvard University Press.
———. (1991b). "Humanism and science in Rudolphine Prague," in *Defenders of the text: the traditions of scholarship in an age of science, 1450–1800*, Cambridge, MA: Harvard University Press, p. 203.
———. (1993). *Joseph Scaliger. A Study in the History of Classical Scholarship. II. Historical Chronology*, Oxford, Clarendon Press.
———. (1997). "Fragmenta Historicum Graecorum: Fragments of Some Lost Enterprises," in *Collecting Fragments—Fragmente Sammeln*, edited by Glenn W. Most, Göttingen: Vandenhoeck & Ruprecht, pp. 124–43.
———. (1999). *Cardano's Cosmos: the Worlds and Works of a Renaissance Astrologer*, Cambridge, MA, Harvard University Press.
———. (2001a). "Jean Hardouin: the antiquary as pariah," in *Bring Out Your Dead. The Past as Revelation*, Cambridge, MA: Harvard University Press, pp. 181–207.
———. (2001b). *Bring out your dead: the past as revelation*, Cambridge, MA, Harvard University Press.
———. (2003). "Some uses of eclipses in early modern chronology," *Journal of the History of Ideas* 64: 213–29.
———. (2012). "Isaac Vossius, Chronologer," in *Isaac Vossius (1618–1689). Between Science and Scholarship*, eds. Eric Jorink and Dirk Van Miert, Leiden, Brill, pp. 43–84.
Grant, Alexander (1884). *The Story of the University of Edinburgh during its First Three hundred Years*, 2 vols. London, Longmans, Green & Co.
Grasshoff, Gerd (1990). *The history of Ptolemy's star catalogue*, New York, Springer-Verlag.
Graunt, John (1662). *Natural and Political Observations Mentioned in a following Index and made upon the Bills of Mortality*, London, John Martin et al.
Greaves, John (1646). *Pyramidographia: Or a Description of the Pyramids in Aegypt*, London, George Badger.

Gregory, David (1715). *The Elements of Astronomy, Physical and Geometrical*, 2 vols. London, for J. Nicholson.

Gregory, John (1683). "De Aeris & Epochis. Shewing the Several Accounts of Time among all Nations, from the Creation to the Present Age," in *The Works of... John Gregorie*, Oxford.

Grell, Chantal (1995). *Le dix-huitième siècle et l'antiquité en France 1680–1789*, Oxford, Voltaire Foundation.

Grell, Chantal, and Catherine Volpilhac-Auger, eds. (1994). *Nicolas Fréret: légende et vérité*, Oxford: Voltaire Foundation.

Grotius, Hugo (1600). *Syntagma Arateorum, opus poeticae et astrononomiae studiosis utilissimum*, Leiden.

———. (1643). *Dissertatio altera de origine Gentium Americanarum, adversus objectatorem*, Paris.

———. (1829). *The Truth of the Christian Religion*. Tr. John Clarke, London, William Baynes.

Guerrini, Luigi (2002). *Antonio Cocchi: Naturalista e Filosofo*, Florence, Edizione Polistampa.

Guicciardini, Niccolò (2009). *Isaac Newton on Mathematical Certainty and Method*, Cambridge, MA, MIT Press.

Gunter, Edmund (1655). *The use of a mathematical instrument, called a quadrant*, London, Joseph Moxon.

———. (1662). *The works of Edmund Gunter* London, Francis Eglesfield.

Guthrie, W.K.C. (1993). *Orpheus and Greek Religion*, Princeton, Princeton University Press.

Haag, Eugène, and Emile Haag (1846–1859). *La France Protestante*, 10 vols. Paris, Joël Cherbuliez.

Hacket, John (1693). *Scrinia reserata: a memorial offer'd to the great deservings of John Williams, D. D.*, London, Edward Jones for Samuel Lowndes.

Hakewill, George (1635). *An Apologie or Declaration of the Power and Providence of God in the Government of the World*, Oxford.

Hakluyt, Richard (1599–1600). *The Principal Navigations, Voyages, Traffiques and Discoueries of the English Nation*, London.

Hald, Anders (2003). *History of Probability and Statistics and Their Applications before 1750*, Hoboken, NJ, John Wiley & Sons.

Hale, Matthew (1677). *The Primitive Origination of Mankind, Considered and Examined according to the Light of Nature*, London, William Godbid for William Shrowsbry.

Hales, John (1765). *The Works of the Ever Memorable Mr. John Hales*, 3 vols. Glasgow, Robert & Andrew Foulis.

Hall, A. Rupert (1980). *Philosophers at War: The Quarrel Between Newton and Leibniz*, Cambridge, Cambridge University Press.

———. (1990). *Henry More and the Scientific Revolution*, Cambridge, Cambridge University Press.

———. (1999). *Isaac Newton: Eighteenth-Century Perspectives*, Oxford, Oxford University Press.

Hall, Joseph (1863). *Works*, ed. Philip Wynter, 10 vols. Oxford, Oxford University Press.

Hall, Marie Boas (1966). *Robert Boyle on Natural Philosophy. An Essay with Selections from His Writings*, Bloomington, Indiana University Press.

Halley, Edmond (1693). "An estimate of the degrees of the mortality of mankind, drawn from curious tables of the births and funerals at the city of Breslaw; with an attempt to ascertain the price of annuities upon lives," *Philosophical Transactions* 17: 596–610.

———. (1704–1705). "Astronomiae Cometae Synopsis," *Philosophical Transactions of the Royal Society of London* 24: 1882–99.

———. (1726/7). "Remarks upon Some Dissertations lately Publish'd at Paris, by the Rev. P. Souciet, against Sir Isaac Newton's Chronology," *Philosophical Transactions* 34: 205–10.

Hardouin, Jean (1729, September). "Le fondement de la Chronologie de M. Newton," *Memoires pour l'Histoire des Sciences et des Beaux Arts*: 1566–86.

Hare, Francis (1721). *The Difficulties and Discouragements which attend the Study of the Scriptures in the way of Private Judgment*, London, W. Botham for James Knapton.
Harper, W., and G. Smith (1995). "Newton's new way of inquiry," in *The Creation of Ideas in Physics: Studies for a Methodology of Theory Construction*, edited by J. Leplin, Dordrecht: Kluwer, pp. 113–66.
Harrison, John (1978). *The Library of Isaac Newton*, Cambridge, Cambridge University Press.
Hartgill, George (1656). *Astronomicall Tables First Invented by George Hartgill Minister of Gods word; And Now Reduced to our Age by Timothy & John Gadbury*, London, Company of Stationers.
Harvey, Gideon (1663). *Archelogia Philosophica Nova, or New Principles of Philosophy*, London, Samuel Thomson.
Harvey, John (1588). *A discoursiue probleme concerning prophesies*, London.
Hatfield, Gary (1998). "The cognitive faculties," in *The Cambridge History of Seventeenth Century Philosophy*, edited by Daniel Garber and Michael Ayers, Cambridge: Cambridge University Press, pp. 953–1002.
Hearne, Thomas (1885–1921). *Remarks and Collections*, 11 vols. Oxford, Oxford Historical Society.
Heilbron, John (1999). *The Sun in the Church. Cathedrals as Solar Observatories*, Cambridge, MA, Harvard University Press.
Helden, Albert van (1974). "The telescope in the seventeenth century," *Isis* 65: 38–58.
Helden, Anne C. van, and Rob H. van Gent (1999). "The lens production by Christiaan and Constantijn Huygens," *Annals of Science* 56: 69–79.
Herder, Johann Gottfried (1800). *Outlines of a Philosophy of the History of Man*. Tr. T. Churchill, London, J. Johnson.
Herodotus (1802). *Histoire d'Hérodote*, Paris, Guillaume Debure.
———. (1992). *The Histories*, Tr. Walster Blanco, New York, W. W. Norton.
Hesiod (1737). *Hesiodi Ascræi quæ supersunt, cum notis variorum*, edited by Thomas Robinson, Oxford, Oxford University Press.
———. (1740). *The Works*. Tr. Thomas Cooke, London, John Wilson.
———. (1914). *The Homeric Hymns and Homerica*. Tr. Hugh G. Evelyn-White, London, William Heinemann.
———. *(2006). Theogony, Works and Days, Testimonia.* Trans. Glenn W. Most. Cambridge, MA, Harvard University Press.
Hevelius, Johannes (1647). *Selenographia*, Gdansk, Hünefeldianis.
———. (1673). *Machinae Coelestis Pars Prior*, Gdansk, S. Reiniger.
———. (1679). *Machinae Coelestis Pars Posterior: Rerum Uranicarum Observationes*, Gdansk, S. Reiniger.
———. (1687 (1690)). *Firmamentum Sobiescianum sive Uranographia* Gedani, typis J.-Z. Stollii.
———. (1690). *Johannes Hevelii Prodromus astronomiae*, Gedani, typis J.-Z. Stollii.
Heylin, Peter (1652). *Cosmographie in Four Bookes*, London, Henry Seile.
Heywood, James, ed. (1840). *Collection of Statutes for the University and the Colleges of Cambridge*, London: William Clowes & Sons.
Hicks, Robert Drew, ed. (1925). *Diogenes Laertius: Lives of eminent philosophers*, London: W. Heinemann; New York: G. P. Putnam's Sons.
Hobbes, Thomas (1655). *Elementorum philosophiae, sectio primar de Corpore*, London.
———. (1656). *Elements of Philosophy. The First Section concerning Body*, London.
———. (2008). *Translations of Homer*, Oxford, Clarendon Press.
Holtby, Robert T. (1966). *Daniel Waterland 1683–1740: A Study in Eighteenth Century Orthodoxy*, Carlisle, Charles Thurnam & Sons.
Homer (1729–1732). *Homeri Ilias, Græce et Latine*, London, J. & J. Knapton.

———. (1759–1764). *Opera omnia*, Leipzig, Theophilus Georgi.
———. (1928). *The Iliad*. Tr. A. T. Murray, London, William Heinemann.
Hooke, Robert (1665). *Micrographia: or some Physiological Descriptions of Minute Bodies made by Magnifying Glasses with Observations and Inquiries thereupon*, London, John Martyn.
———. (1674). *Lectiones Cutlerianae.* London, John Martyn.
———. (1674a). *Animadversions on the first part of the Machina Coelestis of the Honorable, Learned, and deservedly Famous Astronomer Johannes Hevelius, Consul of Dantzick*, London, John Martyn.
———. (1674b). *An Attempt to prove the Motion of the Earth from Observations made by Robert Hooke*, in Hooke, 1674. London, John Martyn.
Hordern, J. H. (1999). "Review of Aratus, Phaenomena, ed. Douglas Kidd," *Classics Ireland* 6.
Horne, Georg (1668). *Arca Mosis; sive, Historia mundi*, Leiden, Pierre Hackius.
Horne, John (1654). *A Consideration of Infant Baptism*, London.
Horsley, Samuel, ed. (1779–1785). *Isaaci Newtoni Opera Quae Exstant Omnia.* 5 vols. London, J. Nichols.
Howell, James (1657). *Londinopolis*, London, J. Streater for Henry Twiford et al.
Huet, Pierre Daniel (1690). *Demonstratio Evangelica*, Paris, Daniel Hortemels.
Hunter, Michael, and Edward B. Davis, eds. (1999–2000). *The Works of Robert Boyle*, 14 vols. London, Pickering & Chatto.
Hutton, Sarah (1994). "The Seven Trumpets and the Seven Vials: Apocalypticism and Christology in Newton's Theological Writings," in *Newton and religion: context, nature and influence*, edited by James Force and Richar H. Popkin, Dordrecht & Boston: Kluwer, pp. 165–78.
Huygens, Christiaan (1690). *Traité de la Lumière*, Leiden.
———. (1703). *Opuscula Posthuma*, Leiden, C. Boutesteyn.
———. (1888–1950). *Oeuvres Complètes*, 22 vols. La Haye, Martinus Nijhoff.
Hyginus (1960). *The Myths of Hyginus*. Tr. Mary Amelia Grant, Lawrence, University of Kansas Press.
Iliffe, Robert (1994). "'Making a shew': apocalyptic hermeneutics and Christian idolatry in the work of Isaac Newton and Henry More," in *The Books of Nature and Scripture: recent essays on natural philosophy, theology and biblical criticism in the Netherlands of Spinoza's time and the British Isles of Newton's time*, edited by James Force and Richard H. Popkin, Dordrecht: Kluwer, pp. 39–53.
———. (1995). "'Is He Like Other Men?' The Meaning of the *Principia Mathematica*, and the Author as Idol," in *Culture and Society in the Stuart Restoration*, edited by Gerald Maclean, Cambridge: Cambridge University Press, pp. 159–76.
———. (2007). *Newton: A Very Short Introduction*, Oxford, Oxford University Press.
Iliffe, Robert, Milo Keynes, and Rebekah Higgitt, eds. (2006). *Early Biographies of Isaac Newton.* 2 vols., London: Pickering & Chatto.
Imhausen, Annette, and Volker R. Remmert (2001). "The Oration on the Dignity and the Usefulness of the Mathematical Sciences of Martinus Hortensius (Amsterdam, 1634): Text, Translation and Commentary," *History of Universities* 21: 71–150.
Ingegno, Alfonso (1988). "The new philosophy of nature," in *The Cambridge History of Renaissance Philosophy*, edited by Eckard Kessler and Quentin Skinner, Cambridge: Cambridge University Press, pp. 236–63.
Inwood, Stephen (2002). *The Man Who Knew Too Much. The Strange and Inventive Life of Robert Hooke 1635–1703*, London, Macmillan.
Isaacson, Henry (1633). *Saturni Ephemerides, sive Tabula Historico-Chronologica*, London.
Jackson, John (1752). *Chronological Antiquities*, London.
Jacob, A., ed. 1987 (1662). *Henry More, The Immortality of the Soul,* Dordrecht: Martinus Nijhoff.

Jardine, Lisa (1988a). "Humanistic logic," in *The Cambridge History of Renaissance Philosophy*, edited by Eckard Kessler and Quentin Skinner, Cambridge: Cambridge University Press, pp. 173–98.
Jardine, Nicholas (1988b). *The Birth of History and Philosophy of Science*, Cambridge, Cambridge University Press.
Jenkin, Robert (1708). *The Reasonableness and Certainty of the Christian Religion*, 2 vols. London, W. B. for Richard Sare.
Jesseph, Douglas (1996). "Hobbes and the method of natural science," in *The Cambridge Companion to Hobbes*, edited by Tom Sorrell, Cambridge: Cambridge University Press, pp. 86–107.
Johnston, Stephen (1991). "Mathematical practitioners and intsruments in Elizabethan England," *Annals of Science* 48: 319–44.
Jolley, Nicholas (1998). "The relation between theology and philosophy," in *The Cambridge History of Seventeenth Century Philosophy*, edited by Daniel Garber and Michael Ayers, Cambridge: Cambridge University Press, pp. 363–92.
Jones, Harold Whitmore, ed. (1976). *Thomas Hobbes: Thomas White's De Mundo Examined*, London: Bradfor University Press.
Jortin, John (1731–1732). *Miscellaneous observations upon authors, ancient and modern*, 2 vols. London, Printed for Thomas Wotton.
Josephus (1841). *Works*. Tr. William Whiston, Belfast, Simms & M'Intyre.
Jurin, James (1996). *The Correspondence of James Jurin*, Amsterdam, Rodopi.
Justin Martyr (1867). "Hortatory Address to the Greeks," in *Ante-Nicene Christian Library*, edited by Alexander Roberts and James Donaldson, Edinburgh: T. & T. Clark, pp. 285–328.
Kargon, Robert (1964). "Walter Charleton, Robert Boyle, and the acceptance of Epicurean atomism in England," *Isis* 55: 184–92.
Kästner, Abraham Gotthelf (1751). "Réfléxions sur l'origine du plaisir," *Histoire de l'Académie Royale des Sciences et Belles Lettres [Berlin]* 5: 478–88.
Katz, Victor J. (1998). *A History of Mathematics. An Introduction*, Reading, MA, Addison-Wesley.
Katzenstein, Ranee, and Emilie Savage-Smith (1988). *The Leiden Aratea. Ancient Constellations in a Medieval Manuscript*, Malibu, CA, The J. Paul Getty Museum.
Kessler, Eckhard, and Katharine Park (2000). "The concept of psychology," in *The Cambridge History of Renaissance Philosophy*, edited by Charles B. Schmitt, Quentin Skinner, Eckhard Kessler, and Jill Kraye, Cambridge: Cambridge University Press, pp. 455–63.
Keynes, Geoffrey (1958). "Dr. Donne and Scaliger," *TLS*: 108.
King, Gregory 1804 [1696]. "Natural and Political Observations and Conclusions upon the State and Condition of England " in *An Estimate of the Comparative Strength of Great Britain*, edited by Thomas Chalmers, London: John Stockdale.
King, Gregory (1973). "The LCC Burns Journal," in *The Earliest Classics*, edited by Peter Laslett, Farnborough: Gregg International.
King, Henry C. (1955 (1979)). *The History of the Telescope*, New York, Dover Publications.
Kircher, Athanasius (1652–1654). *Oedipus Aegyptiacus*, 3 vols. Rome, Vitalis Mascardi.
———. (1679). *Turris Babel*, Amsterdam, Johannes Janssonius van Waesberghe.
Kollerstrom, Nick, and Bernard Yallop (1995). "Flamsteed's lunar data, 1692–95, sent to Newton," *Journal for the History of Astronomy* 26: 237–46.
Kronk, Gary (2006–). Cometography.http://cometography.com/.
Kwan, Alex (2006). "Cornell study of ancient volcano, seeds and tree rings, suggests rewriting Late Bronze Age Mediterranean history," *Chronicleonline*.
La Hire, Philippe de (1705). *Planisphere celeste septentrional*, Paris, N. de Fer.
———. *Tabulae Astronomicae Ludovici Magni*, Paris, Montalant.

La Loubère, Simon de (1693). *A New Historical Relation of the Kingdom of Siam*, London, F. L. for Thomas Horne.
Lacy, John (1708). *A Relation of the Dealings of God to his Unworthy Servant John Lacy*, London, Benjamin Bragg.
Lamotte, Charles (1730). *An Essay upon Poetry and Painting, with relation to the sacred and profane history*, London, F. Fayram.
Landes, David (1983). *Revolution in Time*, Cambridge, MA, The Belknap Press of Harvard University Press.
Laval, Antoine (1728). *Voyage de la Louisiane*, Paris, Jean Mariette.
Laymon, Ronald (1978). "Newton's advertised precision and his refutation of the received laws of refraction," in *Studies in Perception*, edited by Peter K. Machamer and Robert G. Turnbull, Columbus: Ohio State University Press, pp. 231–58.
Lee, Raymond L., and Alistair B. Fraser (2001). *The Rainbow Bridge. Rainbows in Art, Myth, and Science*, University Park, Pennsylvania State University Press.
Le Gendre, Gilbert Charles, Marquis de S. Aubin-sur-Loire (1741). *Traité historique et critique de l'opinion*, Paris, Briasson.
Leibniz, Gottfried Wilhelm (1875–1890). *Die Philosophischen Schriften von Gottfried Wilhelm Leibniz*, 7 vols. Berlin, Weidmannsche.
———. (1989). *Philosophical Essays*. Tr. Roger Ariew and Daniel Garber, Indianapolis & Cambridge, Hackett Publishing.
———. (1899). *Der Briefwechsel von Gottfried Wilhelm Leibniz mit Mathematikern*, Berlin, Mayer & Müller.
Leigh, Edward (1656). *A Treatise of Religion and Learning*, London.
Leijenhorst, Cees (2002). *The Mechanisation of Aristotelianism. The Late Aristotelian Setting of Thomas Hobbes' Natural Philosophy*, Leiden, Brill.
Leland, John (1753). *Reflection on the Late Lord Bolingbroke's Letters on the Study and Use of History; especially as far as they relate to Christianity, and the Holy Scripture*, London, Benjamin Dod.
Leopold, J. H. (1980). "Christiaan Huygens and his instrument makers," in *Studies on Christiaan Huygens*, edited by H.J.M. Bos, M.J.S. Rudwick, H.A.M. Snelders, and R.P.W. Visser, Lisse: Swets & Zeitlinger B. V., pp. 221–33.
Letronne, Antoine Jean (1846). "Analyse critique des représentations zodiacales de Dendéra et d'Esné," *Académie des inscriptions & belles-lettres (France) Mémoires de l'Institut national de France*. 16: 102–210.
Levenson, Thomas (2009). *Newton and the Counterfeiter. The Unknown Detective Career of the World's Greatest Scientist*, Boston and New York, Houghton Mifflin Harcourt.
Lightfoot, John (1684). *The Works*, 2 vols. London, W. R. for Robert Scot.
———. (1822–1825). *The Whole Works*, 13 vols. London, J. F. Dove.
Lipsius, Justus (1630). *Admiranda sive de magnitudine romana libri quattuor*, Antwerp, Balthaaar Moret.
Lively, Edward (1597). *A True Chronologie of the Times of the Persian Monarchie*, London, Felix Kingston for Thomas Man.
Lloid, Lodowik (1590). *The Consent of Time*, London.
Locke, John (1976–1989). *The correspondence of John Locke*, 8 vols. Oxford, Clarendon Press.
Lowdermilk, James (2000). "The Inner Workings of the Egyptian Civil Calendar," *The Ostracon. Journal of the Egyptian Study Society* 2: 7–10.
Lubieniecki, Stanislaw (1666–1668). *Theatrum cometicum daubus partibus constans* Amsterdam, Franciscum Cuperum.
Lucian (1913). *The Syrian Goddess*. Tr. Herbert A. Strong, London, Constable & Co.
Lucretius Carus, Titus (1969). *On the Nature of Things*. Tr. Martin F. Smith, Indianapolis & Cambridge, Hackett.

Luther, Martin (1958). *Commentary on Genesis*. Tr. J. Theodore Mueller, Grand Rapids, Zondervan.
Lydiat, Thomas (1605). *Tractatus de variis Annorum formis*, London, John Norton.
———. (1609). *Emendatio temporum*, London, Felix Kyngston.
Macaulay, Thomas B. (1866). *The History of England*, 4 vols. Longmans, Green.
MacPike, E. F. (1937). *Hevelius, Flamsteed and Halley*, London, Taylor and Francis.
———, ed. (1932). *Correspondence and Papers of Edmond Halley*, Oxford: Clarendon Press.
Macrobius (1952). *Commentary on the Dream of Scipio*. Tr. William H. Stahl, New York & London, Columbia University Press.
Maeyama, Yas (2002). "Tycho Brahe's stellar observations. An accuracy test," in *Tycho Brahe and Prague: Crossroads of European Science*, edited by John Robert Christianson, Alena Hadravova, Peter Hadrava, and Martin Solc, Frankfurt: Verlag Harri Deutsch, pp. 113–19.
Mahoney, Michael S. (1980). "Christiaan Huygens: The measurement of time and longitude at sea," in *Studies on Christiaan Huygens*, edited by H.J.M. Bos, M.J.S. Rudwick, H.A.M. Snelders, and R.P.W. Visser, Lisse: Swets & Zeitlinger B. V., pp. 234–70.
Mahoney, Michael S. (2004). "Drawing mechanics," in *Picturing Machines 1400–1700*, edited by Wolfgang Lefevre, Cambridge, MA: MIT, pp. 281–306.
Maier, Michael (1618). *Atalanta Fugiens, hoc est, Emblemata Nova de Secretis Naturae Chymica*, Oppenheim, Hieronymus Galler for Johann Theodor de Bry.
Malet, Antoni (2001). "The power of images: mathematics and metaphysics in Hobbes's optics," *Annals of Science* 32: 303–33.
Mandelbrote, Scott (2002). "Newton and eighteenth-century Christianity," in *The Cambridge Companion to Newton*, edited by I. Bernard Cohen and George Smith, Cambridge: Cambridge University Press, pp. 409–30.
———. (2006). ""Then this nothing can be plainer": Isaac Newton reads the Fathers"," in *Die Patristik in der frühen Neuzeit*, edited by T. Leinkauf and M. Wriedt G. Frank, Stuttgart: Friedrich Fromm Verlag, pp. 277–97.
———. (2007). "Isaac Newton and the Exegesis of the Book of Daniel," in *Die Geschichte der Daniel-Auslegung in Judentum, Christentum und Islam: Studien zur Kommentierung des Danielbuches in Literatur und Kunst*, edited by Katharina Bracht and David S. du Toit, Berlin & New York: De Gruyter, pp. 351–75.
Manetho (1964). *Manetho*. Tr. W. G. Waddell, Cambridge, MA, Harvard University Press.
Manuel, Frank E. (1963). *Isaac Newton, historian*, Cambridge, MA, Belknap Press of Harvard University Press.
———. (1968). *A Portrait of Isaac Newton*, Cambridge, MA, Harvard University Press.
———. (1974). *The Religion of Isaac Newton*, Oxford, Clarendon Press.
———. (1992). *The Broken Staff: Judaism through Christian Eyes*, Cambridge, MA & London, Harvard University Press.
Markham, Clements R., ed. (1889). *Tractatus de Globis et eorum usu. A Treatise Descriptive of the Globes Constructed by Emery Molyneux and Published in 1592 by Robert Hues*, New York: Burt Franklin.
Marsden, Brian (1972). *Catalogue of Cometary Orbits*, Cambridge, Smithsonian Astrophysical Observatory.
Marsham, John (1676). *Canon chronicus aegyptiacus, ebraicus, graecus, & disqvisitiones* Lipsiae, M. Bircknerum.
Martinich, A. P. (1999). *Hobbes. A Biography*, Cambridge, Cambridge University Press.
Mayor, John E. B. (1911). Cambridge Under Queen Anne, Cambridge, Deighton, Bell and Bowes & Bowes.
Mcfarlane, Roger, and Paul Mills (2010). "The First Book of Hipparchus' Commentaries on the Phaenomena of Aratus and Eudoxus." Private communication.

McGuire, J. E. (1968). "The origin of Newton's doctrine of essential qualities," *Centaurus* 12: 233–60.

———. (1978). "Existence, actuality and necessity: Newton on space and time," *Annals of Science* 35: 466–75.

McGuire, J. E., and P. M. Rattansi (1966). "Newton and the Pipes of Pan," *Notes and Records of the Royal Society of London* 21: 108–43.

McGuire, J. E., and Martin Tamny (1983). *Certain Philosophical Questions. Newton's Trinity Notebook*, Cambridge, Cambridge University Press.

———. (1985). "Newton's astronomical apprenticeship: notes of 1664/5," *Isis* 76: 349–65.

Mede, Joseph (1650). *The Key to Revelation*. Tr. Richard More, London.

———. (1672). *The Works*, London, Roger Norton for Richard Royston.

Meli, Domenico Bertoloni (1993). *Evidence and Priority: Newton versus Leibniz*, Oxford, Clarendon Press.

———. (2004). "The role of numerical tables in Galileo and Mersenne," *Perspectives on Science* 12: 164–89.

Middleton, Conyers (1752). *The Miscellaneous Works*, 5 vols., London, Richard Manby.

Miller, Peter N. (2005). "Description Terminable and Interminable: Looking at the Past, Nature and Peoples in Peiresc's Archive," in *Historia: Empiricism and Erudition in Early Modern Europe*, edited by Gianna Pomata and Nancy Siraisi, Cambridge, MA: MIT Press.

Milner, John (1694). *A Defence of Arch-bishop Usher against Dr. Cary and Dr. Isaac Vossius, together with an Introduction concerning the Uncertainty of chronology*, Cambridge.

Milton, J. R. (2006). "Locke and the Reform of the Calendar," *Locke Studies* 6: 173–77.

Molesworth, Sir William, ed. (1839–1845). *The English Works of Thomas Hobbes of Malmesbury*, 11 vols. London, John Bohn.

Momigliano, Arnaldo (1950). "Ancient history and the antiquarian," *Journal of the Warburg and Courtauld Institutes* 13: 285–315.

Montagu, Mary Wortley (1967). *The Complete Letters of Lady Mary Wortley Montagu*, 3 vols. Oxford, Clarendon Press.

More, Henry (1659). *The immortality of the soul*, London, J. Flesher, for William Morden.

———. (1680). *Apocalypsis apocalypseos*, London, J. M. for J. Martyn & W. Kettelby.

More, John (1593). *A Table from the Beginning of the world to this day*, Cambridge, John Legate.

Moreri, Louis (1711). *Le Grand Dictionnaire Historique ou le Mélange Curieux de l'Histoire Sacré et Profane . . .* Amsterdam, Leiden, the Hague, Utrecht.

Morgan, Paul (1968). "George Hartgill: an Elizabethan parson-astronomer and his library." *Annals of Science* 24: 295–311.

Morrison, Alfred (1882–1893). *The collection of autograph letters and historical documents formed by Alfred Morrison*, London, Strangeways & Sons.

Mosshammer, Alden A. (1981). "Thales' eclipse," *Transactions of the American Philological Association* 111: 145–55.

Muris, O., and G. Saarman (1961). *Der Globus im Wandel der Zeiten*, Berlin, Columbus Verlag.

Nadler, Steven (1998). "Doctrines of explanation in late scholasticism and in the mechanical philosophy," in *The Cambridge History of Seventeenth Century Philosophy*, edited by Daniel Garber and Michael Ayers, Cambridge: Cambridge University Press, pp. 513–52.

Napier, John (1593). *A plaine discouery of the whole Reuelation of Saint John*, Edinburgh, Robert Waldgrave.

Napione Galeani, Gian Francesco (1773). *Ragionamento intorno al saggio del conte Algarotti sopra la durata de'regni de'ré di Roma*, Turin, Mairesse.

Nauze, Louis Jouard de la (1728). "Lettres au R. P. Souciet," *Continuation des Mémoires de Littérature et d'Histoire de Sallengre* 5: 332–463.

———. (1729). "Lettres au R. P. Souciet," *Continuation des Mémoires de Littérature et d'Histoire de Sallengre* 6: 3–72 and 373–464.
———. (1736). "Remarques sur l'antiquité et l'origine de la Cabale," *Histoire de l'Académie des Inscriptions et Belles-Lettres* 9: 37–53.
Neugebauer, O. (1950). "The alleged Babylonian discovery of the precession of the equinoxes," *Journal of the American Oriental Society* 70: 1–8.
———. (1975). *A History of Ancient Mathematical Astronomy*, Berlin, Heidelberg, New York, Springer Verlag.
Neugebauer, Otto, and Richard Anthony Parker (1960–69). *Egyptian Astronomical Texts*, Providence, RI, Brown University Press.
Newman, William R. (1998). "The place of alchemy in the current literature on experiment," in *Experimental Essays—Versuche zum Experiment*, edited by Michael Heidelberger and Friedrich Steinle, Baden-Baden: Nomos, pp. 9–33.
———. (2006). *Atoms and Alchemy. Chymistry & the Experimental Origins of the Scientific Revolution*, Chicago, University of Chicago Press.
Newman, William R., and Lawrence M. Principe (2001). "Some problems with the historiography of alchemy," in *Secrets of nature. Astrology and Alchemy in Early Modern Europe*, Cambridge, MA: MIT Press, pp. 1–38.
———. (2002). *Alchemy Tried in the Fire. Starkey, Boyle, and the Fate of Helmontian Chymistry*, Chicago, University of Chicago Press.
———. (2003). "The chymical laboratory notebooks of George Starkey," in *Reworking the Bench: Research Notebooks in the History of Science*, Dordrecht: Kluwer Academic Publishers, pp. 25–42.
Newton, Isaac (1671–1672). "A letter of Mr. Isaac Newton ... containing his new theory about light and colors," *Philosophical Transactions* 80: 3075–87.
———. (1721). *Opticks*, London, William & John Innys.
———. (1725). *Abregé de la Chronologie de M. le Chevalier Issac Newton, Fait par lui-meme, & traduit sur le Manuscrit Anglois*, Paris, Guillaume Cavelier, fils.
———. (1726). "Remarks upon the Observations made upon a Chronological Index by Sir Isaac Newton, translated into French by the Observator, and publish'd at Paris," *Philosophical Transactions* 33: 315–21.
———. (1728a). *La chronologie des anciens royaumes corrigée*. Tr. François Granet, Paris, Gabriel Martin et al.
———. (1728b). *The chronology of ancient kingdoms amended. To which is prefix'd, a short chronicle from the first memory of things in Europe, to the conquest of Persia by Alexander the Great*, London, Printed for J. Tonson [etc.].
———. (1728c). *A Treatise of the System of the World*, London, F. Fayram.
———. (1733). *Observations upon the prophecies of Daniel, and the Apocalypse of St. John*, London, Printed by J. Darby and T. Browne and sold by J. Roberts.
———. (1959–1977). *Correspondence*, edited by H. W. Turnbull et al., 7 vols. Cambridge, Cambridge University Press.
———. (1978). "Of Educating Youth in the Universities," in *Unpublished Scientific Papers of Isaac Newton*, edited by A. Rupert Hall and Marie B. Hall, Cambridge: Cambridge University Press.
———. (1999). *The Principia. Mathematical Principles of Natural Philosophy. A New Translation. Preceded by A Guide to Newton's Principia by I. Bernard Cohen*. Tr. I. Bernard Cohen, Berkeley, University of California Press.
Nichols, John (1817–1858). *Illustrations of the Literary History of the Eighteenth Century*, 8 vols. London.

Nichols, William (1696). *A Conference with a Theist*, London, T. W. for Francis Saunders & Thomas Bennet.

Nicolson, Marjorie H., and Sarah Hutton (1992). *The Conway letters: the correspondence of Anne, Viscountess Conway, Henry More, and their friends, 1642–1684*, Oxford, Clarendon Press.

Nieuwentyt (1730). *The Religious Philosopher*. Tr. John Chamberlayne, 3 vols. London, W. Bowyer for John Senex.

Nisbet, William (1655). *A Scripture Chronology*, London, Joshua Kirton.

Norwood, Richard (1645). *Norwood's epitomie: or The application of the doctrine of triangles in certaine problemes, to the use of the plaine sea-chart, and mercators-chart*, London, T. Forcet for George Hurlock.

Offenberg, Adri (2004). "Dirk van Santen and the Keur Bible: New Insights into Jacob Judah (Arye) Leon Templo's Model Temple," *Studia Rosenthaliana* 37: 401–22.

Oldenburg, Henry (1670/1). "An account of sundry experiments made and communicated by that learn'd mathematician, Dr. Erasmus bartholin, upon a crystal-like body, sent to him out of Island," *Philosophical Transactions of the Royal Society of London* 53: 2039–48.

Olmsted, John W. (1949). "The 'application' of telescopes to astronomical instruments, 1667–1669; a study in historical method," *Isis* 40: 213–25.

Oughtred, William (1657a). *Trigonometria hoc est, Modus computandi triangulorum latera et angulos, ex canone mathematico traditus & demonstratus*, London, R. and W. Leybourn for Thomas Johnson.

Oughtred, William (1657b). *Trigonometrie, or, The manner of calculating the sides and angles of triangles by the mathematical canon demonstrated*, London, R. and W. Leybourn for Thomas Johnson.

Ouvry, Frederic (1874). *Letters Addressed to Thomas Hearne*, London, privately printed.

Ovid (1951). *Fasti*. Tr. James G. Frazer, Cambridge, Mass. & London, Harvard University Press & William Heinemann.

Owen, John (1850–1853). *The Works of John Owen*, 17 vols. New York, Robert Carter & Brothers.

Pacchi, Arrigo (1988). "Hobbes and the problem of God," in *Perspectives on Thomas Hobbes*, edited by G.A.J. Rogers and Alan Ryan, Oxford: Clarendon Press, pp. 171–87.

Pardies, Ignace Gaston (1674). *Globi coelestis in tabulas planas redacti descriptio*, Paris, Sebastianus Mabre Cramoisy.

Park, Katharine (2000). "The organic soul," in *The Cambridge History of Renaissance Philosophy*, edited by Charles B. Schmitt, Quentin Skinner, Eckhard Kessler, and Jill Kraye, Cambridge: Cambridge University Press, pp. 464–84.

Parker, Richard A. (1981). "Egyptian Astronomy, Astrology, and Calendrical Reckoning," in *Dictionary of Scientific Biography*, vol. 15, supplement 1, edited by Charles Coulston Gillispie, New York: Charles Scribner's Sons, pp. 706–27.

Parker, Samuel (1709). *Censura Temporum*, London, Henry Clements.

Parr, Richard (1686). *The life of ... James Usher*, London, Nathanael Ranew.

———. (1687). *The Life of ... James Usher ... with a Collection of Three Hundred Letters*, London.

Patrick, Symon (1695). *A commentary upon the first book of Moses, called Genesis*, London, Richard Chiswell.

Pearce, Zachary (1727). *A Sermon Preached at the New Parish Church of St. Martin in the Fields, Westminster, Oct. 20 1726 ... To which is added ... an Essay upon the Origin and Progress of Temples*, London, S. Harding.

———. (1732a). *A Reply to the Letter to Dr. Waterland, Setting forth the many Falshoods both in the Quotations and the Historical Facts*, London, J. Watts.

———. (1732b). *A Reply to the Defence of the Letter to Dr. Waterland*, London, J. Watts.

———. (1777). *A Commentary with notes, on the Four Evangelists*, London, E. Cox for T. Cadell.

Pecke, Thomas (1659). *Parnassi Puerperium*, London.
Pedersen, Olaf (1974). *A Survey of the Almagest*, Oxford, Oxford University Press.
Pererius, Benedictus (1601). *Commentariorum et disputationum in Genesim*, Cologne, Antonius Hierat.
Perkins, William (1609). *The whole treatise of the cases of conscience*, Cambridge, John Legat.
Petau, Denis (1630). *Uranologion, sive, Systema variorum authorum qui de sphaera ac sideribus eorumque motibus graecè commentati sunt*, Paris, Sebastiani Cramoisy.
———. (1703). *Opus De doctrina temporum*, ed. Jean Hardouin, 3 vols. Antwerp, G. Gallet.
———. (1659). *The History of the World or, An Account of Time*, London, Streater.
Peters, C.H.F., and E. B. Knobel (1915). *Ptolemy's Catalogue of Stars. A Revision of the Almagest*, Washington, DC, The Carnegie Institution of Washington.
Petty, William (1683). *Another Essay in Political Arithmetick, concerning the Growth of the City of London*, London, H. H. for Mark Pardoe.
———. (1899). *The Economic Writings of Sir William Petty*, Cambridge, Cambridge University Press.
———. (1967). *The Petty Southwell Correspondence*, New York, A. M. Kelly.
Peyrère, Isaac de la (1655). *A Theological Systeme upon that Presupposition, that Men were before Adam*, London.
Philippes, Henry (1676). *The Sea-mans Kalender Or, An Ephemerides of the Sun, Moon, and certain of the msot notable Fixed Stars*, London, Stephen Swart.
Phillips, Thomas (1834). "Observations on Some Monastic Libraries and Archives in French Flanders," *Transactions of the Royal Society of Literature of the United Kingdom* 2: 263–67.
Pie, Thomas (1597). *A Houre Glasse contayning I A Computation from the Beginning of Time to Christ by X. Articles. II A Confirmation of the same for the times controverted before Christ*, London, John Wolfe.
Pignotti, Lorenzo (1813–1814). *Storia della Toscana sino al principato*, Pisa, Didot.
Pitcairne, Archibald (1979). *The Best of Our Own: Letters of Archibald Pitcairne*, Edinburgh, Saorsa Books.
Plackett, R. L. (1958). "The principle of arithmetic mean," *Biometrika* 45: 130–35.
Plato (1875). *The Dialogues*. Tr. Benjamin Jowett, Oxford, Clarendon Press.
———. (1961). *Laws*. Tr. R. G. Bury, Cambridge, MA, Harvard University Press.
Pliny (1949–1971). *Natural History*. Tr. H. Rackham et al., 10 vols. Cambridge, MA, Harvard University Press.
Pocock, Edward (1740). *The Theological Works*, 2 vols. London, R. Gosling.
Poole, Matthew (1669–1675). *Synopsis Criticorum aliorumque S. Scripturae Interpretum*, 5 vols. London, J. Flesher & T. Roycroft.
———. (1683). *Annotations upon the Holy Bible*, London, John Richardson for Thomas Parkhurst [and others].
Poole, Robert (1998). *Time's Alteration: Calendar Reform in Early Modern England*, London, UCL Press.
Pope, Alexander (1733). *An Essay on Man ... Epistle II*, London, J. Wilford.
———. (1956). *The Correspondence of Alexander Pope*, 5 vols. Oxford, Clarendon Press.
Pope, Walter (1697). *The Life of... Seth, Lord Bishop of Salisbury*, London, William Keblewhite.
Popkin, Richard (2003). *The History of Skepticism*, Oxford, Oxford University Press.
———. (1990). "Newton as a Bible Scholar," in *Essays on the Context, Nature, and Influence of Isaac Newton's Theology*, edited by James E. Force and Richard H. Popkin, Dordrecht, Kluwer.
Preston, John (1634). *Life Eternall or, A Treatise of the Knowledge of the Divine Essence and Attributes*, London.
Prideaux, Humphrey (1698). *The true Nature of Imposture fully display'd in the Life of Mahomet*, London, printed for William Rogers.

Prince, C. Leeson (1882). *The Illustrated Account given by Hevelius of the method of mounting and erecting an Observatory. Reprinted from an original copy, with some remarks*, Lewes.
Principe, Lawrence M. (1998). *The Aspiring Adept. Robert Boyle and His Alchemical Quest*, Princeton, Princeton University Press.
Procopius (1916). *History of the Wars*. Tr. H. B. Dewing, London, William Heinemann.
Ptolemy (1998). *Ptolemy's Almagest*. Tr. G. J. Toomer, Princeton, Princeton University Press.
Purchas, Samuel (1613). *Purchas his Pilgrimage*, London, William Stansby for Henry Fetherstone.
———. (1905). *Hakluytus Posthumous, or, Purchas his Pilgrimes*, 20 vols. Glasgow, James MacLehose & Sons.
Quirini, Angelo Maria (1738). *Primordia Corcyrae*, Brescia, Giovanni Maria Rizzardi.
———. (1749–1750). *Commentarii de rebus pertinentibus* 2 vols. Brescia, Giovanni Maria Rizzardi.
Ragep, Jamil (1993). *Nasir Al-Din Al-Tusi's Memoir on Astronomy (al-Tadhkira fi cilm al-hay'a)*, New York, Springer-Verlag.
———. (1996). "Al-Battani, cosmology, and the early history of trepidation in Islam," in *From Baghdad to Barcelona. Studies in the Islamic Exact Sciences in Hounour of Prof. Juan Vernet*, Barcelona: Instituto "Millas Vallicrosa" de Historia de Ciencia Arabe, pp. 267–98.
Raleigh, Walter (1829). *The Works of Sir Walter Ralegh*, 8 vols. Oxford, Oxford University Press.
Rawlins, Dennis (2005). Farnese-Globe Trainwrecking [sic]. In http://www.dioi.org/ggg.htm. (accessed 2005).
Renaudot, Eusèbe (1733). *Ancient Accounts of India and China* London, Samuel Harding.
Rhodius, Apollonius (1960). *Argonautica. Jason and the Golden Fleece*, New York, Heritage Press.
Ricci, Angelo Maria (1740–1741). *Dissertationes Homericae*, 3 vols. Florence, Caietanus Albizini.
Riccioli, Giovanni Battista (1651). *Almagestum novum astronomiam veterem novamque complectens*, Bononiae, Typographia haeredis Victorii Benatii.
———. (1672). *Geographiae et hydrographiae reformatae libri duodecim*, Venice, J. La Nou.
Richards, E. G. (1998). *Mapping Time. The Calendar and Its History*, Oxford, Oxford University Press.
Robinson, Henry W., ed. (1968). *The Diary of Robert Hooke*, London, Wykeham Publications.
Rochberg, Francesca (2004). *The Heavenly Writing. Divination, Horoscopy, and Astronomy in Mesopotamian Culture*, Cambridge, Cambridge University Press.
Roche, John J. (1981). "Harriot's 'Regiment of the Sun' and its background in sixteenth-century navigation," *The British Journal for the History of Science* 14: 245–61.
Rochon, Père (1672). *Lettre d'un Philosophe à un Cartesien de ses Amis*, Paris, Thomas Jolly.
Rollin, Charles (1734). *The Method of Teaching and Studying the Belles Lettres*, London.
Rome, A. (1937). "Les observations d'équinoxes et de solstices dans le chapitre 1 du livre 3 du 'Commentaire sur l'Almageste' par Théon d'Alexandrie,'" *Ann. Soc. Sci. Bruxelles* 57: 213–36.
Ross, Alexander (1626). *An Exposition on the Fourteene First Chapters of Genesis* London, Bernard Alsop & Thomas Fawcet for Anthony Upphill.
Rossi, Paolo (1984). *The Dark Abyss of Time: The History of the Earth and the History of Nations from Hooke to Vico*. Tr. Lydia G. Cochrane, Chicago, University of Chicago Press.
Sabra, A. I. (1981). *Theories of Light from Descartes to Newton*, Cambridge, Cambridge University Press.
Saint-Philippe, Cartier de (1722). "Albert Henri de Sallengre," *Journal Literaire* 12: 222.
Sammes, Aylett (1676). *Britannia antiqua illustrata, or, The antiquities of ancient Britain derived from the Phoenicians*, London, Thomas Roycroft.
Sargent, Rose-Mary (1995). *The Diffident Naturalist. Robert Boyle and the Philosophy of Experiment*, Chicago, University of Chicago Press.
Scaliger, Joseph Juste (1695). *Scaligerana*, Cologne.

Scaliger, Joseph Juste, and Sébastien Nivelle (1583). *Opus novvm de emendatione temporum in octo libros tributum*, Paris, Sébastien Nivelle.
Schaefer, Bradley (2004). "The astronomical lore of Eudoxus," *Journal for the History of Astronomy* 35: 161–223.
———. (2005). "The epoch of the constellations on the Farnese Atlas and their origin in Hipparchus' lost catalogue," *Journal for the History of Astronomy* 36: 167–96.
Schaffer, Simon, and Steven Shapin (1985). *Leviathan and the Air-Pump*, Princeton, Princeton University Press.
Scheoner, Johann (1561). *Opera mathematica*, Nuremberg, Ioannis Montani & Vrici Neuberi.
Schiller, Julius (1627). *Coelum stellatum Chirsitanum*, Augusburg, Adndreae Apergeri.
Schliesser, Eric, and George E. Smith (1996). "Huygens' 1688 report to the Directors of the Dutch East India Company on the measurement of longitude at sea and its implications for the non-uniformity of gravity," *De zeventiende eeuw* 12: 198–214.
Schüller, Volkmar (2001). "Newton's Scholia from David Gregory's Estate on the Propositions IV through IX Book II of his Principi," in *Newton, And Kant, Philosophy and Science in the Eighteenth Century*, edited by Wolfgang Lefèvre, Dordrecht: Kluwer, pp. 213–65.
Scriblerus, Simon (pseud.) (1731). *Whistoneutes: Or, Remarks on Mr. Whiston's Historical Memoirs of the Life of Dr. Samuel Clarke, &c.*, London, T. Warner.
Sedley, David (2005). "Hellenistic physics and metaphysics," in *The Cambridge history of Hellenistic Philosophy*, edited by Keimpe Algra, Jonathan barnes, Jaap Mansfeld and Malcolm Schofield, Cambridge: Cambridge University Press, pp. 355–411.
Selden, John (1617). *De Diis Syris Syntagma II*, London, William Stansby.
———. (1628). *Marmora Arundelliana*, London, William Stansby.
———. (1665). *De Jure Naturali et Gentium*, Strasbourg.
Senex, John (1718). *A Treatise of the Description and Use of Both Globes*, London, John Senex.
Senex, Mary (1749). "A Letter from the Widow of the Late Mr. John Senex, F. R. S. to Martin Folkes, Esq; President of the Royal Society, concerning the Large Globes Prepared by Her Late Husband, and Now Sold by Herself, at Her House Over-Against St. Dunstan's Church in Fleet-Street," *Philosophical Transactions* 46: 290–92.
Sergeant, John (1697). *Solid Philosophy Asserted, Against the Fancies of the Idesists*, London, Roger Clavil.
———. (1698). *Non Ultra: or, A Letter to a Learned Cartesian; Settling the Rule of Truth, and First Principles, Upon Their Deepest Grounds*, London, A. Repor.
———. (1700). *Transnatural Philosophy, or Metaphysicks*, London, The Author.
Shane, A. L. (1977). "Rabbi Jacob Judah Leon (Templo) of Amsterdam (1603–1675) and his Connections with England," *Transactions of the Jewish Historical Society of England* 25: 120–36.
Shapin, Steven (1995). *A Social History of Truth*, Chicago, University of Chicago Press.
Shapiro, Alan E. (1973). "Kinematic optics: a study of the wave theory of light in the seventeenth century," *Archive for History of Exact Sciences* 11: 134–266.
———, ed. (1984). *The Optical Papers of Isaac Newton. The Optical Lectures. 1670–1672*. Vol. 1, Cambridge: Cambridge University Press.
———. (1989). "Huygens' 'Traité de la lumière' and Newton's 'Opticks': pursuing and eschewing hypotheses," *Notes and Records of the Royal Society of London* 43: 223–47.
———. (1990). "The Optical Lectures and the foundations of the theory of optical imagery," in *Before Newton. The Life and Times of Isaac Barrow*, edited by M. Feingold, Cambridge: Cambridge University Press.
———. (1994). "Artists' colors and Newton's colors," *Isis* 85: 600–630.
———. (1996). "The gradual acceptance of Newton's theory of light and color, 1672–1727," *Perspectives on Science* 4: 59–140.

———. (2003). "Newton's optical notebooks: public versus private data," in *Reworking the Bench*, edited by Frederic L. Holmes, Jürgen Renn, and Hans-Jörg Rheinberger, Dordrecht: Kluwer Academic Publishers, pp. 43–66.

———. (2005). "Newton's experiments on diffraction and the delayed publication of the Opticks," in *Isaac Newton's Natural Philosophy*, edited by Jed Z. Buchwald and I. Bernard Cohen. Cambridge, MA, MIT, pp. 48–76.

Shea, William R. (1991). *The Magic of Numbers and Motion. The Scientific Career of René Descartes*, Canton, MA, Science History Publications.

Sherburne, Edward (1675). *The sphere of Marcus Manilius made an English poem: with annotations and an astronomical appendix*, London, Nathaniel Brooke.

Sheynin, O. B. (1973). "Mathematical treatment of astronomical observations (a historical essay)," *Archive for History of Exact Sciences* 11: 97–126.

Shuckford, Samuel (1728–1730). *The Sacred and Prophane History of the World Connected*, 2 vols. London, R. Knaplock & J. Tonson.

Simmons, Alison (1994). "Explaining sense perception: a scholastic challenge," *Philosophical Studies* 73: 257–75.

Siraisi, Nancy G. (1997). *The Clock and the Mirror. Girolamo Cardano and Renaissance Medicine*, Princeton, Princeton University Press.

Smith, George E. (1999). "How did Newton discover universal gravity?," *The St. John's Review* 45: 32–63.

———. (2000). "The Newtonian style in Book II of the Principia," in *Isaac Newton's Natural Philosophy*, edited by Jed Z. Buchwald and I. Bernard Cohen, Cambridge, MA: MIT Press, pp. 249–313.

———. (2002). "The methodology of the Principia," in *The Cambridge Companion to Newton*, edited by I. Bernard Cohen and George E. Smith, Cambridge: Cambridge University Press, pp. 138–73.

———. (2005). "Was wrong Newton bad Newton?" in *Wrong for the Right Reasons*, edited by Jed Z. Buchwald and Allan Franklin, Dordrecht, Berlin, Heidelberg, New York: Springer, pp. 127–60.

Smith, Mark (2000). "Getting the big picture in perspectivist optics," in *The Scientific Enterprise in Antiquity and the Middle Ages*, Chicago: University of Chicago Press, pp. 315–36.

Smith, Robert, ed. (1738). *Hydrostatical and Pneumatical Lectures by Roger Cotes*, London: Printed by S. Austen for the editor.

Snobelen, Stephen D. (1999). "Isaac Newton, Heretic: the Strategies of a Nicodemite," *British Journal for the History of Science* 32: 381–419.

———. (2003a). Isaac Newton and Socinianism: Associations with a Greater Heresy. *A Journal from the Radical Reformation* 99: 3–15.

———. (2003b). "'A Time and Times and the Dividing of Time': Isaac Newton, the Apocalypse, and 2060 A.D.," *Canadian Journal of History* 38: 537–51.

———. (2004). "William Whiston, Isaac Newton and the crisis of publicity," *Studies in the History and Philosophy of Science* 35: 573–603.

Sokol, Barnett J. (1974). "Thomas Harriot—Sir Walter Ralegh's Tutor—On Population," *Annals of Science* 31: 205–12.

Souciet, Etienne (1727). *Recueil des dissertations ... contenant un abregé de chronologie, cinq dissertations contre la chronologie de Newton, une dissertation sur une médaille singuliere d'Auguste*, Paris, Rollin.

South, Robert (1844). *Sermons Preached upon Several Occasions*, Philadelphia.

Sparrow, John (1958). "Dr. Donne and Scaliger," *TLS*: 115.

St. Evremond, Charles de (1714). *Works*. Tr. Pierre Des Maizeaux, 3 vols. London, J. Churchill.

———. (1728). *Works*, 3 vols. London, Printed for J. & J. Knapton et al.

Stanley, Thomas (1701). *The History of Philosophy: containing the lives, opinions, actions and discourses of the philosophers of every sect*, London, W. Battersby.
Stanyan, Temple (1739). *The Grecian History*, 2 vols. London.
Steuart, James (1805). *The Works, Political, Metaphysical, and Chronological*, 6 vols. London, Printed for T. Cadell & W. Davies.
Stevenson, Edward L. (1921). *Terrestrial and Celestial Globes*, New Haven, Yale University Press.
Stewart, Dugald (1854–1860). *The Collected Works*, 11 vols. Edinburgh, Thomas Constable.
Stewart, Ian (2003). "The Statutes of the Lucasian Professorship: A Translation," in *From Newton to Hawking. A History of Cambridge University's Lucasian Professors of Mathematics*, edited by Kevin C. Knox and Richard Noakes, Cambridge: Cambridge University Press, pp. 461–74.
Stigler, Stephen (2003). *The History of Statistics. The Measurement of Uncertainty before 1900*, Cambridge, MA, Harvard University Press.
Stigler, Stephen M. (1977). "Eight centuries of sampling inspection: the trial of the Pyx," *Journal of the American Statistical Association* 72: 493–500.
Stillingfleet, Edward (1662). *Origines Sacrae*, London, R. W. for Henry Mortlock.
Strabo (1917). *The Geography of Strabo*. Tr. Horace L. Jones, London, Willian Heinemann.
Strauchius, Giles (1704). *Breviarum Chronologicum. Or A Treatise Describing the Terms and Most Celebrated Characters, Periods and Epocha's used in Chronology*. Tr. Richard Sault, London, A. Bosvile.
Streete, Thomas (1661). *Astronomia Carolina*, London, Lodowick Lloyd.
———. (1663). *Astronomia Carolina* London, Lodowick Lloyd.
Struik, Dirk, ed. (1969). *A Source Book in Mathematics, 1200–1800*, Cambridge, MA, Harvard University Press.
Stukeley, William (1882–1887). *The family memoirs of the Rev. William Stukeley, M.D.*, 3 vols. Durham, Andews & Co.
Suidas (1705). *Lexicon*, Cambridge, Cambridge University Press.
Süssmilch, Johann Peter (1741). *Die göttliche Ordnung in den Veränderungen des menschlichen Geschlechts aus der Geburt, Tod, und Fortpflanzung desselben*, Berlin, J. C. Spener.
Swan, John (1653). *Calamus Mensurans: The Measuring Reed*, London.
Swerdlow, Noel (1986). "A star catalogue used by Johannes Bayer," *Journal for the History of Astronomy* 17: 189–97.
Swift, Jonathan (1963–1965). *Correspondence*, 5 vols. Oxford, Clarendon Press.
Sykes, Arthur Ashley (1732). *A Dissertation on the Eclipse Mentioned by Phlegon. Or, an Enquiry whether that Eclipse had any Relation to the Darkness which Happened at our Saviour's Passion*, London.
———. (1733). *A Defence of the Dissertation on the Eclipse Mentioned by Phlegon*, London, Printed for James, John, & Paul Knapton.
———. (1734). *A Second Defence of the Dissertation upon the Eclipse Mentioned by Phlegon*, London, James, John & Paul Knapton.
———. (1744). *An Examination of Mr. Warburton's Account of the Conduct of the Antient Legislators*, London, Printed for J. & P. Knapen.
Tabarroni, G. (1971). "Jean-Dominique Cassini et la datation de la sphère Farnèse," in *Actes du XIIe Congrès International d'Histoire des Sciences* (Paris), pp. 149–53.
Tachau, Katherine H. (1988). *Vision and Certitude in the Age of Ockham. Optics, Epistemology and the Foundations of Semantics 1250–1345*, Leiden, Brill.
Targosz, Karolina (1988). "Firmamentum Sobiescianum— the magnificent Baroque atlas of the sky," *Organon* 24: 151–79.
Tatian (1867). "Address to the Greeks," in *Ante-Nicene Christian Library*, edited by Alexander Roberts and James Donaldson, Edinburgh: T. & T. Clark, pp. 5–45.

Taylor, Brooke (1793). *Contemplatio Philosophica*, London, W. Bulmer.

Temple, William ([1814] 1968). *Works*, 4 vols. New York, Greenwood Press.

Temporarius, Johanes (1596) *Chronologicarum demonstrationum libri tres*, Frankfurt, Andreas Wechel.

Tertullian (1868). *The five books of Quintus Sept. Flor. Tertullianus Against Marcion* Tr. Peter Holmes, Edinburgh, T & T Clark.

Thiele, Georg (1898). *Antike Himmelsbilder*, Berlin, Weidmannsche Buchhandlung.

Thieme, Ulrich, and Hans Vollmer (1965). *Allgemeines Lexikon der bildenden Künstler*, Leipzig, Seeman.

Tillotson, John (1820). *Works*, 10 vols. London, J. F. Dove for Richard Priestley.

Toomer, Gerald J. (2010). *John Selden: A Life in Scholarship*, 2 vols. Oxford, Oxford University Press.

Townley, Richard (1694). "A Letter from Richard Townley, of Townley in Lancashire, Esq; Containing Observations on the Quantity of Rain Falling Monthly, for Several Years Successively," *Philosophical Transactions of the Royal Society of London (1683–1775)* 18: 51–58.

Tressan, Louis Elisabeth de la Vergne de (1822–1823). *Oeuvres*, 10 vols. Paris, Nepreu.

Tuck, Richard (1988). "Hobbes and Descartes," in *Perspectives on Thomas Hobbes*, edited by G.A.J. Rogers and Alan Ryan, Oxford: Clarendon Press, pp. 11–42.

Uffenbach, Zacharias Conrad von (1935). *London in 1710*. Tr. W. H. Quarrell and Margaret Mare, London, Faber & Faber.

Ussher, James (1658). *The Annals of the World*, London, E. Tyler for J. Crook.

———. (1847–1864). *The Whole Works*, 17 vols. Dublin, Hodges & Smith.

Valerio, Vladimiro (1987). "Historiographic and numerical notes on the Atlante Farnese and its celestial sphere," *Der Globusfreund* 35/37: 97–124.

Verney, F. P. (1892). *Memoirs of the Verney Family During the Civil War*, London.

Véron, Jacques, and Jean-Marc Rohrbasser (2000). "Lodewijk et Christiaan Huygens: la distinction entre vie moyenne et vie probable," *Mathématiques et Sciences Humaines* 38: 7–21.

Vlacq, Adriaan (1628). *Arithmetica logarithmica*, Gouda, Petrus Rammasenius.

———. (1633). *Trigonometria artificialis*, Gouda, Petrus Rammasenius.

Volkoff, Ivan, Ernest Franzgrote, and A. Dean Larsen (1971). *Johannes Hevelius and his Catalog of Stars*, Provo, Utah, Brigham Young University Press.

Voltaire (1733). *Letters Concerning the English Nation*, London, Printed for C. Davis.

———. (1877–1885). *Oeuvres Complètes*, 52 vols. Paris, Garnier frères.

———. (1901). *The works of Voltaire: a contemporary version with notes*, 42 vols. Paris and New York, E. R. Du Mont.

———. (1968–). *Complete Works of Voltaire*, Geneva, Oxford, The Voltaire Foundation.

———. (1992). *Eléments de la philosophie de Newton, eds. Robert L. Walters and W. H. Barber*, Oxford, The Voltaire Foundation.

———. (1668). *De Theologia Gentili, et Physiologia Christiana; sive de Origine ac Progressu Idololatria*, Amsterdam, Johannes Blaeu.

Vossius, Isaac (1685). *Variarum observationum liber*, London, Robert Scott.

Wade, Ira O. (1969). *The Intellectual Development of Voltaire*, Princeton, Princeton University Press.

Wallace, William A. (1977). *Galileo's Early Notebooks: The Physical Questions. A Translation from the Latin, with Historical and Paleographical Commentary*, Notre Dame, University of Notre Dame Press.

———. (1978). "The philosophical setting of medieval science," in *Science in the Middle Ages*, edited by David C. Lindberg, Chicago: University of Chicago Press, pp. 91–119.

———. (1988). "Newton's early writings: beginnings of a new direction," in *Newton and the New Direction in Science*, edited by G. V. Coyne, M. Heller, and J. Zycinski, Vatican: Specola Vaticana, pp. 23–44.

Wallis, John (1691). *Three Sermons Concerning the Sacred Trinity*, London, Thomas Parkhurst.
Wallis, John (1699). "An Extract of Two Letters [to Thomas Tenison, Archbishop of Canterbury, and William Lloyd, Bishop of Worcester]... Concerning the Alteration (suggested) of the Julian Account for the Gregorian," *Philosophical Transactions* 21: 343–54.
Walton, Brian (1654–1657). *Biblia Sacra Polyglotta*, 6 vols. London, Thomas Roycroft.
Warburton, William (1738–1741). *The Divine Legation of Moses Demonstrated, on the Principles of a Religious Deist, from the Omission of the Doctrine of a Future State of Reward and Punishment in the Jewish Dispensation*, 2 vols. London, Fletcher Gyles.
Ward, G.R.M. (1845–1851). *Oxford University Statutes*, 2 vols. London, William Pickering.
Ward, John ([1740] 1967). *The Lives of the Professors of Gresham College*, New York, Johnson Reprint Corporation.
Ward, Seth (1654). *Idea trigonometriae demonstratæ (in usum juventutus Oxoniensis) item prælectio De cometis et Inquisitio in Bullialdi Astronomiæ philolaicæ fundamenta* Oxford, L. Lichfield.
Warner, Deborah Jean (1979). *The sky explored: celestial cartography, 1500–1800*, New York, A. R. Liss.
Waterland, Daniel (1730). *Scripture Vindicated; In Answer to a Book Intituled, Christianity as Old as the Creation. Part I*, London, W. Innys.
———. (1731). *Scripture Vindicated; In Answer to a Book Intituled, Christianity as Old as the Creation. Part II*, Cambridge, Cornelius Crownfield.
———. (1843). *Works*, Oxford.
Wenskus, Otta (1990). *Astronomische Zeitangaben von Homer bis Theophrast*, Stuttgart, F. Steiner.
Wepster, Steven (2010). *Between Theory and Observations. Tobias Mayer's Explorations of Lunar Motion, 1751–1755*, New York, Springer Verlag.
West, M. L., ed. (1978). *Hesiod. Works & Days*, Oxford: Clarendon Press.
Westfall, Richard S. (1965). "Isaac Newton's coulored circles twixt two contiguous glasses," *Archive for History of Exact Sciences* 2: 181–96.
———. (1980). *Never at rest: a biography of Isaac Newton*, Cambridge [Eng].; New York, Cambridge University Press.
———. (1982). "Isaac Newton's *Theologiae Gentilis Origines Philosophicae*," in *The Secular Mind: Transformations of Faith in Modern Europe*, edited by W. W. Wagar, New York: Holmes & Meier, pp. 15–34.
Weststeijn, Thijs (2007). "Spinoza Sinicus: An Asian Paragraph in the History of the Radical Enlightenment," *Journal of the History of Ideas* 68: 537–61.
Wheare, Degory (1685). *The Method and Order of Reading both Civil and Ecclesiastical Histories*, London, M. Flesher.
Whiston, William (1696). *A New Theory of the Earth*, London, R. Roberts for Benjamin Tooke.
———. (1702). *A Short View of the Chronology of the Old Testament, and of the Harmony of the Four Evangelists*, Cambridge, At the University Press for B. Tooke.
———. (1706). *An Essay on the Revelation of Saint John. so far as concerns Past and Present Times*, Cambridge, Cambridge University Press for B. Tooke.
———. (1708). *The Accomplishment of Scripture Prophecies*, Cambridge, Cambridge University Press for Benjamin Tooke.
———. (1712). *Primitive Christianity Reviv'd*, London.
———. (1725). *A New Theory of the Earth*, London, Samuel Tooke & Benjamin Motte.
———. (1727–1728). *A collection of authentick records belonging to the Old and New Testament*, 2 vols. London, Printed for the author.
———. (1728). *Sir Isaac Newton's corollaries from his philosophy and chronology, in his own words*, London.
———. (1730). *Historical memoirs of the life of Dr. Samuel Clark*, London, Sold by Fletcher Gyles & J. Roberts.

———. (1732). *The testimony of Phlegon Vindicated: Or, an Account of the Great Darkness and Earthquake at our Savior's Passion, described by Phlegon*, London.
———. (1734). *Six Dissertations*, London.
———. (1742). *Three Tracts*, London, J. Whiston.
———. (1753). *Memoirs of the Life and Writings of Mr. William Whiston*, London, J. Whiston and B. White.
White, Thomas (1642). *De mvndo dialogi tres qvibvs materia ... forma ... cavssæ ... et tandem definito, rationibus purèè naturâ de promptis aperiuntur, concluduntur*, Paris, Dionysium Moreav.
Whiteside, D. T. (1980). "Kepler, Newton and Flamsteed on Refraction Through a 'Regular Aire': the Mathematical and the Practical," *Centaurus* 24: 288–315.
———. ed. (1967–1981). *The Mathematical Papers of Isaac Newton*. 8 vols., Cambridge, Cambridge University Press.
Wilford, John Noble (2006). "New evidence spurs fresh thinking on ancient civilizations," *New York Times*, May 2.
Wilkins, John (1675). *Of the Principles and Duties of Natural Religion*, London.
Willet, Andrew (1610). *Hexapla in Danielem*, Cambridge, Cantrell Legge.
———. (1633). *Hexapla in Genesin & Exodum*, London, John Haviland.
Wilson, Catherine (2008). *Epicureanism at the Origins of Modernity*, Oxford, Clarendon Press.
Wilson, Curtis (1968). "Kepler's derivation of the elliptical path," *Isis* 59: 4–25.
Wing, Vincent (1651). *Harmonicon coeleste, or, The coelestiall harmony of the visible world conteining an absolute and entire piece of astronomie*, London, Robert Leybourn.
———. (1669). *Astronomia Britannica*, London, John Macock.
Wodrow, Robert (1842–1843). *Analecta: or, Materials for a History of Remarkable Providences; Mostly relating to Scotch Ministers and Christians*, 4 vols. Edinburgh, Maitland Club.
Wolf, Edwin (1974). *The Library of James Logan of Philadelphia, 1674–1751*, Philadelphia, Library Company of Philadelphia.
Wood, Anthony (1786). *The History and Antiquities of the Colleges and Halls in the University of Oxford*, Oxford, Clarendon Press.
Wood, Robert (1769). *An Essay on the Original Genius of Homer*, London.
Wordsworth, Christopher (1877 (1969)). *Scholae Academicae*, London.
Wotton, William (1694). *Reflections upon Ancient and Modern Learning*, London, J. Leake for Peter Buck.
Wren, Christopher (1651). "Tractatus ad periodum Iulianum spectans, chronologiae summe utilis," in *Theatrum historicum et chronologicum*, edited by Christoph Helvicus, Oxford.
Wright, Thomas (1740). *The use of the globes*, London, John Senex.
Yoder, Joella (1988). *Unrolling Time. Christiaan Huygens and the mathematization of nature.*, Cambridge, Cambridge University Press.
Young, Arthur (1734). *An Historical Dissertation on idolatrous corruptions in religion from the beginning of the world*, London, C. Rivington.
Young, Edward (1971). *The Correspondence of Edward Young 1683–1765*, edited by Henry Pettit, Oxford: Clarendon Press.
Zinner, Ernst (1979). *Deutsche und niederländische astronomische Instrumente des 11.–18. Jahrhunderts*, Munich, Beck.

INDEX

Note: Page numbers in italic type indicate illustrations, figures, or tables.

Milton Keynes UK
Ingram Content Group UK Ltd.
UKHW011655071224
451937UK00002B/3